A
Sourcebook
for the
Biological Sciences

THIRD EDITION

A Sourcebook for the Biological Sciences

THIRD EDITION

Evelyn Morholt ▪ Paul F. Brandwein

HARCOURT BRACE JOVANOVICH, PUBLISHERS

San Diego   New York   Chicago   Austin
London   Sydney   Toronto

Cover Illustration: "Legion Nassellaria..." by Ernst Haeckel. Taken from the *Report of the Scientific Results of the Voyage of H. M. S. Challenger During the Years 1873–76.* London, 1887. Reprinted with permission from General Research Division, The New York Public Library, Astor, Lenox and Tilden Foundations.

ISBN: 0-15-582852-5
Library of Congress Catalog Card Number: 84-81502

Printed in the United States of America

About the Authors

Evelyn Morholt is the author of some seven editions of laboratory manuals in biology used in junior and senior high school, and is coauthor with Paul Brandwein of *Patterns in Living Things* and *Teaching-Learning Strategies*. She has served as the chairman of a high school science department and as an acting examiner for licenses in science and mathematics for the New York City Board of Examiners, City Board of Education. Dr. Morholt's teaching career covers the range of general science, biology, physics, earth science, and medical techniques. She is currently preparing a laboratory manual and teacher's guide for Coronado Publishers to accompany *The Organism: A Biology.*

Paul Brandwein has served as co-Director of the Pinchot Institute for Conservation Studies, as Director of Education for the Conservation Foundation, and as Director of Conservation Projects for the Joint Council of Economic Education. Among his many books in education is *Memorandum: On Renewing Schooling and Education*. He is also the author and coauthor of some 40 books used in the schools including *Concepts in Science* (a series of 10 books), *The Gifted Student as Future Scientist*, *A Theory of Instruction*, and *A Book of Methods on the Teaching of Science*, which was published as a companion to the first edition of the *Sourcebook*. Dr. Brandwein was chairman of the science department at Forest Hills High School, New York City, has taught biology at New York University and methods of teaching science at Columbia's Teachers College, and has served as adjunct professor at the University of Pittsburgh. He is currently preparing *Science and Technology and Society*, a series of 12 textbooks for elementary and high school students to be published by Coronado Publishers.

Preface to the Third Edition
Caution to the Reader

This Third Edition of the *Sourcebook* incorporates techniques and procedures stemming from advances in the biological sciences. To be sure, we have also included material related to changes in curriculum and instruction in biology and junior high school science since the publication of the Second Edition. The content, which falls into a ready organization of patterns in biological thought and procedure, furnishes a flexible compendium from which instructors may order their own teaching. Over the years teachers and laboratory workers have used the *Sourcebook* as a kind of *vade mecum*.

Of necessity, we have included a revised section on science safety to accommodate the subject matter being taught. We are obligated to recall to readers—teachers, to be sure—that the *Sourcebook* has never been intended for use by the student. Without careful supervision, students cannot be expected to bring to biological procedure the conceptual, experimental, and experiential background that enables them to understand the safety precautions required for work in a laboratory, where flame and combustible materials may produce hazards, or where the unwarranted mixture of substances may produce unexpected and harmful results. Laboratory work calls for unexampled care.

Indeed, we go so far as to suggest that a procedure, a formula, a device should be *school-proofed*—that is, pretested before it is used in a particular classroom or laboratory. This book is not a laboratory manual for students; its percepts, precepts, and concepts are intended for the teacher's use.

The coming decades, embracing a so-called post-Industrial Revolution (or evolution), will see a flowering of biological science and biotechnology never before experienced. Muscle-assisting machines have been wedded to mind-assisting technology. Idea will be coupled to invention and necessity will feed use. More than 90 percent of the world's scientists are now alive and are at work using new technologies to conserve, transmit, rectify, and expand our knowledge of the organism and its environment.

The last clause surely coheres to the teacher's aims—but the teacher is obliged to add another: to advance the causes of the young in the fulfillment of their powers and their varieties of excellence.

With high respect and admiration for our colleagues,

EVELYN L. MORHOLT
PAUL F. BRANDWEIN

From the Preface to the First Edition

The teaching procedures—of all manner and description—offered in this book have been tested over more years than the authors care to remember in actual teaching of high school students, of different levels of ability, in general science, biology, health science, and field biology. The topics to which these procedures and techniques apply were selected on the basis of a study of fifty-eight courses in general science, biological science, and health given in typical communities (cities, towns, and country organizations) in the United States.

A good number of these procedures were also tested in a course titled "Laboratory Techniques in the Teaching of Biology" given as an in-service course in the New York City schools by Dr. Morris Rabinowitz, George I. Schwartz, and Paul Brandwein. A mimeographed text of the techniques developed in the course was also put together in 1940–42 under the title of the course.

Certainly there are in this book more techniques, procedures, demonstrations, projects, experiments, and suggestions than any teacher can use in any one year. These procedures are typical of the kinds of things that teachers do, or want to do, as they teach biological science—in the variety of forms this takes. Since the method of teaching is and must remain a personal invention, we leave to each teacher the selection of those techniques which are useful in a specific teaching situation.

We have included techniques and procedures not only for the "average" class, if there is such a thing, but for the variety of individual students who make up classes, no matter how they have been grouped. There are demonstrations involving visual effects, those requiring manipulation of various materials—from simple clay model to advanced histological techniques—those requiring reading, observing, thinking: all the many processes which make up science. There are suggestions for short- and long-range projects. It is our hope that there are here demonstrations and techniques for every student, no matter what his beginning interest or ability in science may be.

This volume is one of a series; following the Table of Contents are outlines of the three companion volumes, *A Book of Methods*, by P. F. Brandwein, F. G. Watson, and P. E. Blackwood, 1958; *A Sourcebook for Elementary Science*, by E. B. Hone, A. Joseph, and E. Victor, 1962; and *A Sourcebook for the Physical Sciences*, by A. Joseph, P. F. Brandwein, E. Morholt, H. Pollack, and J. F. Castka, 1961. The series is published by Harcourt, Brace & World.

We take this opportunity to express our appreciation to the many teachers from whom we have drawn inspiration; we particularly offer thanks to Herbert Drapkin and Leon Rintel for their careful and encouraging critiques of the book in galley proof.

We should like nothing better than to correspond with teachers who find certain of these techniques difficult to perform or difficult to apply, or wanting in any manner. Certainly if a better technique is available, or if one we have described has been improved, we should consider it a privilege to include the contribution in a revised edition with appropriate credit. The opportunity we seek is that of being of service to teachers.

Evelyn Morholt
Paul F. Brandwein
Alexander Joseph

Contents

1

Levels of Organization:
Structure and Physiology of
Representative Animals and Plants
1

2

Levels of Organization:
Cells and Tissue
121

3

Levels of Energy Utilization:
Producers and Consumers
195

4

Utilization of Materials:
Building of the Organism
265

5

Interpretation of the Environment:
Behavior and Coordination
330

6

Continuity of the Organism:
Development, Differentiation, and Growth
383

7

Continuity of the Organism:
Patterns of Inheritance, Adaptation,
and Evolution Within Ecosystems
503

8

The Biosphere: Ecological Patterns and the Limitations of Planet Earth
575

9

Maintaining Organisms for Laboratory and Classroom Activities
654

10

Certain Useful Additional Techniques
727

A
Sourcebook
for the
Biological Sciences

THIRD EDITION

1

Levels of Organization: Structure and Physiology of Representative Animals and Plants

One of the functions, pleasures, and privileges of the biology laboratory is that it permits students to observe the elegant diversity of structure among organisms. Nonetheless, within diversity, there is an underlying unity. Observations of the variety of organisms now available to the school laboratory lead to an understanding of the commonality of structure and function, as well as the beauty and economy in the patterning of organisms at successive levels of organization. When students study one or more organisms in depth, the organism becomes a focus for developing many broad concepts and conceptual schemes not only in reference to structure and broad function but to the macro- and micro-processes related to growth, homeostasis, to energy transfers, heredity and development.

The substance of chapter 1 may serve as a guide for the observation and dissection of many representative organisms, beginning with protists and higher invertebrates and vertebrates. A study of representative plant phyla is briefly introduced here, and developed in more detail in the context of photosynthesis, reproduction, and ecological relationships. Yet it must be remembered that these descriptions are only guides to an introductory study; there are many references, both textbooks and laboratory manuals, that give detailed guides to dissection of specific plants and animals. (Some are listed in the bibliography at the end of this chapter.) Nevertheless, a major function of dissection is to allow students to engage in comparisons of forms; comparative anatomy, to be sure.

A description of organisms at the cellular and tissue level which requires the use of a microscope is offered in chapter 2. Methods for cultivating and/or preserving many of the organisms under study here are offered in chapters 8 and 9. Where possible, both freshwater and marine forms are presented as representative types that demonstrate certain adaptations, as well as levels of complexity among organisms, from the invertebrates to the vertebrates. Thus, the phyla of organisms show a phylogenetic relationship, based in genetics, as well as the gradual development of special adaptations for life on land—mainly specialized means of reproduction, and of support systems among plants. A more complete study of mosses, ferns, and seed plants is offered within the context of reproduction (chapter 6) and in the invasion and capture of the land (chapter 7).

Finally, special techniques for handling bacteria and examining parasitic protozoa, as well as the life histories of parasitic worms are offered throughout the text, in various contexts. (Refer to the index for such specialized studies.)

Note: Before beginning dissections, we suggest you read the safety precautions concerning dissection tools used by students and the possible effects of preservatives as described in the section on Safety in the Science Laboratory in chapter 10. Also refer to "Will There Be Formaldehyde in Your Future?" in *Ward's Bulletin* (November 1982) published by Ward's Natural Science Establishment (see Appendix). The article stresses the need to rinse preserved specimens in running water before and during dissection to reduce excess formalin in the specimens.

Five kingdoms

Whittaker's classification of plants and animals into five kingdoms circumvents much of the criticism of older systematics.[1] There is merit in this classification based on increasing evidences of relationships in structure of organisms. Yet textbooks in botany and zoology may well use different classification systems. Consider these five kingdoms:

1. *Kingdom Monera* includes procaryotic cells such as the four divisions of bacteria and one division of blue-green algae. These cells lack DNA within a nuclear membrane, as well as mitochondria and chloroplasts. Flagella, when present, are not in a pattern of nine plus two fibrils as found among eucaryotic cells (see Fig. 1-1).

2. *Kingdom Protista* comprises eucaryotic unicellular organisms that have DNA within a nuclear membrane, and that contain organelles such as mitochondria and chloroplasts (among the algae). Their cilia or flagella are built on a plan of nine plus two fibrils (see Fig. 1-1).

   The *protozoa* within the Kingdom Protista usually include four phyla: Sarcodina (amoebae), Zoomastigina (animal flagellates), Sporozoa (sporozoans), and Ciliophora (ciliates). Boolootian and Stiles group the

amoebae and flagellated forms into one phylum Sarcomastigophora.[2]

Ideas concerning classification change as new evidence becomes available from electron micrographs, chemical analyses, life cycles, and fossil findings.

The *algae* within the Kingdom Protista include divisions Euglenophyta, Chrysophyta (chiefly marine, golden-brown plankton algae), Xanthophyta (yellow-green and diatoms), and Pyrrophyta (chiefly marine plankton, dinoflagellates).

All these algae have a cell wall of cellulose and pectin except the Euglenophyta. They contain chlorophyll a and carotenoids; some contain chlorophyll b and/or c.

3. *Kingdom Fungi* comprises the multinucleated higher fungi.

4. *Kingdom Plantae* comprises certain algae which closely resemble higher green plants. All these groups of algae have cell walls of cellulose and pectin, and some have additional substances. Their pigments all contain chlorophyll a and carotenoids. In addition, the red algae contain chlorophyll d and phycobilins, while the brown algae contain chlorophyll c and fucoxanthin. The largest group of the algae, the green algae, contain chlorophylls a and b.

   The three divisions of algae are Rhodophyta (mainly marine red algae), Phaeophyta (mainly marine brown algae), and Chlorophyta (mainly freshwater green algae).

   In this kingdom are the divisions Bryophyta (mosses and liverworts) and the Tracheophyta (vascular plants: horsetails, club mosses, ferns, and seed plants).

5. *Kingdom Animalia* comprises all the multicellular animals.

   A general family tree illustrating these relationships is shown.[3]

---

[1] R. Whittaker, "New Concepts of Kingdoms of Organisms," *Science* (10 January 1969).
   Also see L. Margulis and K. Schwartz, *Five Kingdoms* (San Francisco: Freeman, 1981).

[2] R. Boolootian and K. Stiles, *College Zoology*, 10th ed. (New York: Macmillan, 1981).
[3] Whittaker, "New Concepts of Kingdoms of Organisms."

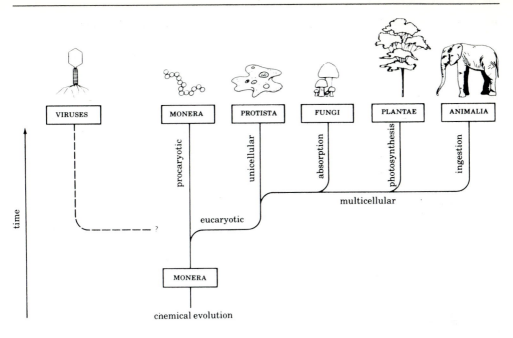

Family tree—Five Kingdoms. (Adapted with permission from Macmillan Publishing Company, from *Biology* by Clyde F. Herried II. Copyright © 1977 by Macmillan Publishing Company, Inc.)

In tracing patterns of relationships in structure and function, we shall describe protozoan protists and plant protists (here and in chapter 2) without much consideration to their specific classification. Relationships among plant protists are included in a general discussion of the evolution of sexuality and structural adaptations in plants. Here too, Kingdom Monera is compared with the eucaryotic plant protists.

These general phylogenetic patterns are used only to preserve the kind of continuity among plants and animals described in texts in botany and zoology. Of course, teachers select their own preferences.

Animal kingdoms

Protista

A view of protists under a microscope reveals their variety of form, color, and mode of locomotion. Their roles in ecosystems are a constant source of wonder. While unicellular, they are far from "simple" organisms; they are introduced as the *first level* of organization among the eucaryotic cells.

Organelles like flagella and cilia may look like thin hairs under the microscope, but electron micrographs reveal that both cilia and flagella are filaments made of two central fibrils and nine peripheral fibrils that make up the axoneme covered by an outer sheath. Refer to the excellent descriptions in many texts, and in a laboratory text, such as the one by Sherman and Sherman.[4]

Protists

This section provides a general introduction to protists; specific ideas for their use in first lessons of unicellular types are developed in chapter 2, in suggested laboratory activities and investigations. Methods for culturing and maintaining protists (as well as organisms of all the kingdoms) are given in chapter 9; behavior is described in chapter 5.

[4] I. Sherman and V. Sherman, *The Invertebrates: Function and Form.* A Laboratory Guide. 2nd ed. (New York: Macmillan, 1976).

Some flagellated protists

There is tremendous variety among the flagellated types; they seem to be an ancestral type from which other protist lines as well as higher plants and animals evolved. Ciliates are believed to have evolved from some flagellate type; suctorians, in turn, are closely related to ciliates.

Flagellates pull themselves forward by their whiplike flagella. Resembling cilia, each flagellum contains an axial fiber of eleven filaments consisting of two in the center surrounded by a circle of nine filaments (Fig. 1-1).

A widely studied flagellate, *Euglena*, swims with its one flagellum so that it rotates on its own axis a bit to one side in a spiral path since its flagellum is held at an angle to the cell (Fig. 1-2). A smaller flagellum can be observed within its reservoir. *Euglena* has an outer pellicle composed of spiral bands, not a cellulose wall.[5] Contractile fibers or microtubules enable *Euglena* to change shape so that the so-called "euglenoid motion" is visible (Fig. 1-2a).

The orange-red eye spot, or stigma, is a layer of tiny oil drops that acts to shield the photoreceptor as *Euglena* reorients its

[5] J. Wolken, *Euglena: An Experimental Organism for Biochemical and Biophysical Studies*, 2nd ed. (Appleton-Century-Crofts, 1967).

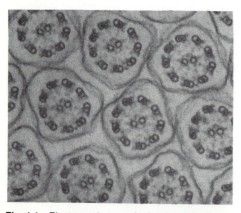

Fig. 1-1 Electron micrograph showing transverse sections of flagella (× 200,000). Each flagellum contains two inner fibers and nine outer pairs of fibers. (Dr. Peter Satir.)

Fig. 1-2 (a) *Euglena* showing changes in shape; (b) *Euglena gracilis*, an autotroph; (c) *Euglena* showing both longitudinal fission and division within a cyst. (a, from W. H. Brown, *The Plant Kingdom*, 1935; reprinted through the courtesy of Blaisdell Publishing Co., a division of Ginn and Co.; b, reprinted and adapted with permission of Macmillan Publishing Company from *The Invertebrates: Function and Form*, 2nd ed., by Irwin W. Sherman and Vilia G. Sherman. Copyright © 1976 by Irwin W. Sherman and Vilia G. Sherman; c, reprinted and adapted with permission of Macmillan Publishing Company from *College Zoology*, 10th ed., by Richard A. Boolootian and Karl A. Stiles. Copyright © 1981 by Macmillan Publishing Co., Inc.)

movement to receive light (Fig. 1-2b). The red pigment is astaxanthin. The light-sensitive photoreceptor is located at the base of the flagellum. The reservoir is not for feeding; at the base is a contractile vacuole that empties liquid wastes into the reservoir.

Reexamine Figure 1-2b to locate the gullet near the base of the flagellum. Observe scattered chloroplasts, about ten, that seem to radiate from the center. Each chloroplast contains a pyrenoid with paramylon, a polysaccharide. There are many colorless small paramylum bodies in the cytoplasm. While *Euglena* is autotrophic, with distinct chloroplasts to carry on photosynthesis, it seems to require some amino acids, thiamine, and vitamin B_{12} (cobalamin), as well as several minerals for optimum growth and reproduction. When conditions are unfavorable, many cysts of *Euglena* can be found; *Euglena* cells divide within the cysts so that two or more cells may be found within cysts. Fission is lengthwise in *Euglena* (Fig. 1-2c). (Also refer to activities on behavior, page 330; growth, page 460; culturing, page 664.)

Other fairly common flagellates, *Astasia*, *Peranema*, and *Chlamydomonas* are shown in Figure 1-3. Also refer to the wood-eating, flagellated symbionts found in the gut of termites (Fig. 1-4, and activities on p. 631).

Some ciliates

Cilia beat with an oblique stroke so that ciliated forms rotate as they swim. Neuro-

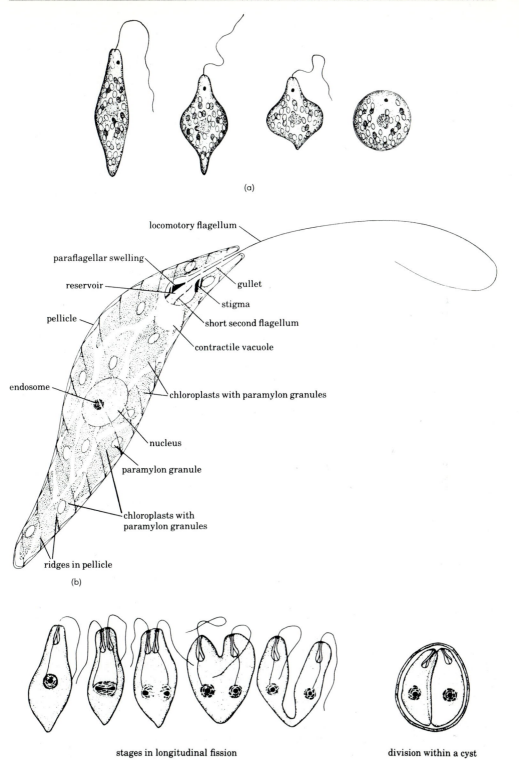

(a)

locomotory flagellum

paraflagellar swelling

reservoir

gullet

stigma

pellicle

short second flagellum

contractile vacuole

endosome

chloroplasts with paramylon granules

nucleus

paramylon granule

chloroplasts with
paramylon granules

ridges in pellicle

(b)

stages in longitudinal fission

division within a cyst

(c)

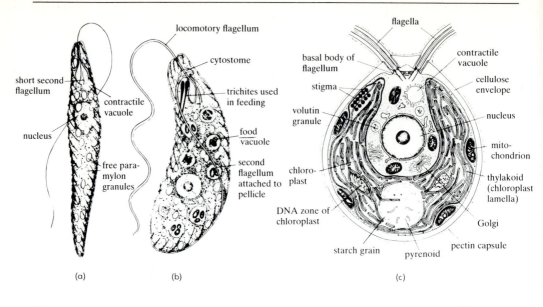

(a)   (b)   (c)

Fig. 1-3 (a) *Astasia*, similar morphologically to *Euglena* but lacking chloroplasts, is an osmotrophic heterotroph; (b) *Peranema*, a phagotrophic heterotroph, can feed on large prey organisms. Its trichites hold the prey and push it into the cytostome, an organelle found only in phagotrophic types; (c) *Chlamydomonas* (electron micrograph). (Adapted from *The Protozoa* by Keith Vickerman and Francis E. G. Cox, published by Houghton Mifflin Company. Copyright © 1967 by Keith Vickerman and Francis E. G. Cox.)

fibrils connect rows of basal bodies at the inner base of each cilium so there is excellent coordination forward and backward. The often-mentioned "neuromotor" seems to be an artifact. Action of ions affects ciliary motion (p. 139).

Ciliates such as *Paramecium* (Fig. 1-5) have a flexible covering containing chiton. Pores in the outer pellicle can be seen by staining with nigrosin (p. 137); cilia protrude through the pores. Ciliates (and Suctorians) have both a macronucleus and a micronucleus. Both nuclei divide at each asexual fission. The micronucleus functions in the several nuclear divisions involved in sexual reproduction while the macronucleus disintegrates. Well-fed paramecia reproduce by fission two to three times a day.

You may also want to explore mating types in conjugation using *P. multimicronucleatum*, or *P. bursaria* (which have zoochlorellae living symbiotically in the cytoplasm). (Refer to vital stains (p. 136) to distinguish different kinds of nuclei among protozoa, as well as their mitochondria.) Trace the changes from acid to alkaline contents of food vacuoles in *Paramecium* using Congo red-stained yeast cells (pp. 140, 236–37).

Trichocysts can be released in *Paramecium* by adding Parker ink, dilute acetic acid, iodine, 1 percent tannic acid (aqueous), or 0.5 percent methylene blue-acetic acid. Each trichocyst is located in the ectoplasm of paramecia. It consists of a small sac containing a sticky solution that forms a long thread when ejected out of a pore in the pellicle (Fig. 1-6).

Contractile vacuoles are more visible in paramecia when a drop of 10 percent nigrosin is added to a drop of culture on a slide. Apply a coverslip to examine under high power. Time the rate of contractions of the vacuoles in solutions of different salt concentration (preparation, Table 10-8, p. 747). Use shallow depression slides with different concentrations of salt to observe how the paramecia regulate the water content of the cell (pp. 139–40).

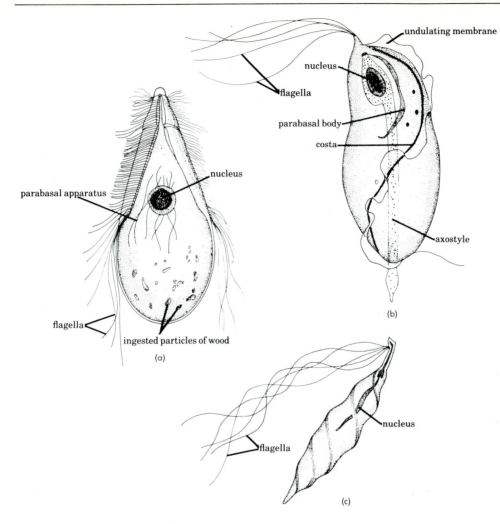

undulating membrane

nucleus

flagella

parabasal body

costa

nucleus

parabasal apparatus

axostyle

flagella

ingested particles of wood

(a)

(b)

flagella

nucleus

(c)

Fig. 1-4 Several symbiotic flagellates from the gut of termites: (a) *Trichonympha collaris;* (b) *Trichomonas termopsidis;* (c) *Streblomastix strix.* (Reprinted and adapted with permission of Macmillan Publishing Company from *The Invertebrates: Function and Form*, 2nd ed., by Irwin W. Sherman and Vilia G. Sherman. Copyright © 1976 by Irwin W. Sherman and Vilia G. Sherman.)

Colorful ciliates

Although *Paramecium* may be the classic form studied in biology, the pink *Blepharisma* is highly desirable as an organism for introductory study (Fig. 1-7). It is slow-moving, pink or rose colored due to rows of granules of the pigment blepharismin (formerly called zoopurpurin) in the pellicle. (This can be seen when *Blepharisma* is treated with special chemicals which remove the pellicle.) It has long cilia, an undulating membrane, and membranelles. Techniques for preparing glass needles and the cutting of *Blepharisma* to show regeneration are described here only briefly, and in detail in the laboratory manual by Giese.[6] For a full discussion of the physiology of this light-sensitive organism, refer to the fine text by Giese.[7]

[6] A. Giese, *Laboratory Manual in Cell Physiology*, rev. ed. (Pacific Grove, CA: Boxwood Press, 1975).

[7] A. Giese, *Blepharisma* (Stanford, CA: Stanford University Press, 1973). Also refer to R. Wichterman, *The Biology of Paramecium* (New York: McGraw-Hill, 1951), and to V. Tartar, *Biology of Stentor* (Elmsford, NY: Pergamon, 1961).

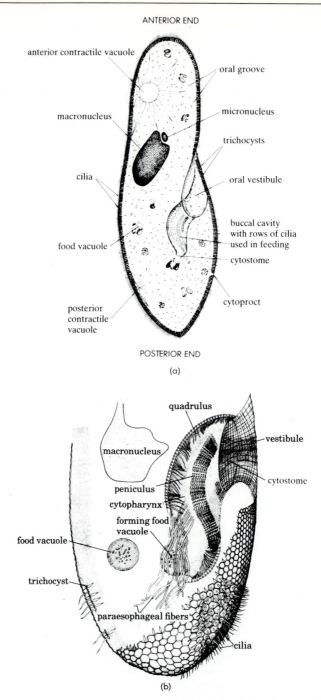

Fig. 1-5 *Paramecium*: (a) viewed from right side (the mouth is on the ventral surface); (b) detailed view of the cytopharynx of *Paramecium* showing details of ciliation in the cytopharynx (electron micrograph). (a, adapted from *The Protozoa* by Keith Vickerman and Francis E. G. Cox, published by Houghton Mifflin Company. Copyright © 1967 by Keith Vickerman and Francis E. G. Cox; b, reprinted and adapted with permission of Macmillan Publishing Company from *The Invertebrates: Function and Form*, 2nd ed., by Irwin W. Sherman and Vilia G. Sherman. Copyright © 1976 by Irwin W. Sherman and Vilia G. Sherman.)

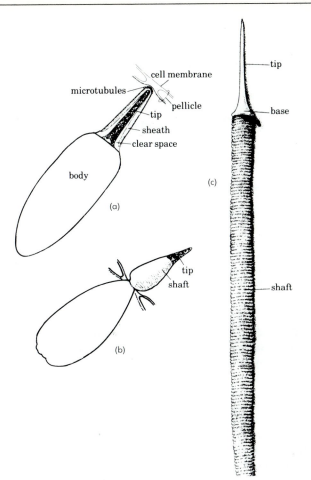

Fig. 1-6 Diagrams of a trichocyst in action (electron micrograph): (a) resting trichocyst; (b) trichocyst beginning to be extruded; (c) extruded trichocyst. (Reprinted and adapted with permission of Macmillan Publishing Company from *The Invertebrates: Function and Form*, 2nd ed., by Irwin W. Sherman and Vilia G. Sherman. Copyright © 1976 by Irwin W. Sherman and Vilia G. Sherman.)

Observe the locomotion of several protozoa. Examine the undulating membranelles of *Blepharisma* (Fig. 1-7) and of *Stentor* under the microscope (Fig. 1-8). Membranelles are composed of two or three layers of cilia that move together, giving the graceful, rhythmic motion of the membranelles. The bluish, trumpet-shaped *Stentor* contains the pigment stentorin located in lengthwise bands.[8]

[8] Tartar, *The Biology of Stentor.*

Notice the cirri in *Euplotes* (Fig. 1-9). Cirri are bundles of cilia that taper to a tip. Each cirrus consists of four to six rows of cilia with some six cilia per row. These flattened organisms use cirri as tiny, crawling limbs in locomotion.

One of the longest protozoa is *Spirostomum* (Fig. 1-10), a grayish, cigar-shaped organism with visible myonomes (bunches of microtubules); these are also found in *Vorticella* (Fig. 1-11). Use Figure 1-12 as a guide to identify several other ciliates found in pond water cultures.

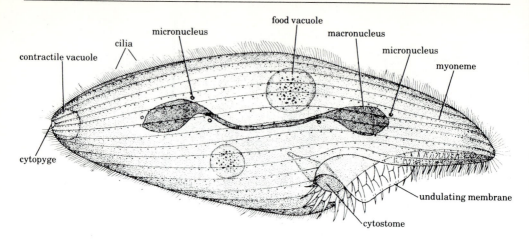

contractile vacuole

cilia

micronucleus

food vacuole

macronucleus

micronucleus

myoneme

cytopyge

undulating membrane

cytostome

Fig. 1-7 *Blepharisma undulans*, a rose-colored ciliate; note the long macronucleus with several micronuclei. (Reprinted and adapted with permission of Macmillan Publishing Company from *The Invertebrates:*

Function and Form, 2nd ed., by Irwin W. Sherman and Vilia G. Sherman. Copyright © 1976 by Irwin W. Sherman and Vilia G. Sherman.)

Some amoeboid types

Included in the Superclass Sarcodina are the amoebae that move by means of pseudopods (Fig. 1-13). Examine the locomotion of amoeba or the larger *Pelomyxa* (p. 658). Notice the clear area at the edges of the pseudopods. For an extensive description of changes from gel to sol, the hyaline cap, and the movement of granules in the cytoplasm, refer to college texts such as Sherman and Sherman,[9] Boolootian and Stiles,[10] and Villee.[11]

Many amoebae have shells such as the brownish dome shell of *Arcella* (Fig. 1-14), and the cemented sand particles of the shell of *Difflugia* (Fig. 1-15). The Foraminifera build many chambered shells with openings through which pseudopodia may extend (Fig. 1-16). Layers of shells from these organisms have formed chalk at the bottom of the oceans; the white cliffs of Dover are made of foraminifera shells. Siliceous rock at the bottom of the oceans results from the compressed skeletons of the beautiful radiolarians (Fig. 1-17).

Parasitic protozoans

While most protozoa are holozoic (consuming bacteria, organic molecules, or other protozoa), there are also parasitic ciliates such as *Balantidium* in frogs, humans, pigs, etc. (and *Opalina*,[12] a symbiont in frogs; see page 625 for methods of preparation of slides from frog tissue). A technique used to examine *Monocystis*, a sporozoan found in earthworms, is described on page 620. Refer also to the life cycle of *Plasmodium*, a sporozoan that causes malaria (p. 629).

Porifera

Sponges are masses of specialized cells that make up colonies. Sections of sponges may be examined under the microscope to study some of the specialized cells (Fig. 1-18). Note that the collar cells, choanocytes, have flagella that create a current bringing in oxygen and food particles. Wastes and carbon dioxide are eliminated in the same flow of currents. Other special cells, amoebocytes, move among the cells.

The path of water and food particles in

[9] Sherman and Sherman, *Invertebrates*.
[10] Boolootian and Stiles, *College Zoology*.
[11] C. Villee, *Biology*, 7th ed. (Philadelphia: Saunders, 1977).

[12] For the life history of *Opalina* in frogs, refer to Boolootian and Stiles, *College Zoology*. Also refer to page 625 in this *Sourcebook*.

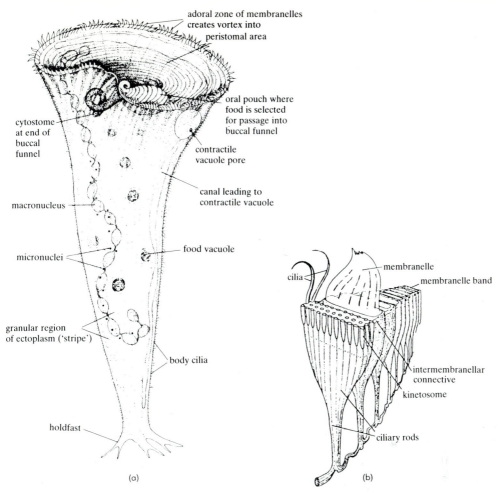

adoral zone of membranelles
creates vortex into
peristomal area

oral pouch where
food is selected
for passage into
buccal funnel

cytostome
at end of
buccal
funnel

contractile
vacuole pore

canal leading to
contractile vacuole

macronucleus

food vacuole

micronuclei

membranelle

cilia

membranelle band

granular region
of ectoplasm ('stripe')

body cilia

intermembranellar
connective

kinetosome

holdfast

ciliary rods

(a)

(b)

Fig. 1-8 *Stentor coeruleus*: (a) a large, bluish, trumpet-shaped ciliate (up to 2 mm long); note the long, beaded macronucleus and several micronuclei; (b) details of a membranelle composed of 2 to 3 rows of fused cilia (reconstructed from electron micrograph). (a, adapted from *The Protozoa* by Keith Vickerman and Francis E. G. Cox, published by Houghton Mifflin Company. Copyright © 1967 by Keith Vickerman and Francis E. G. Cox; b, reprinted and adapted with permission of Macmillan Publishing Company from *The Invertebrates: Function and Form*, 2nd ed., by Irwin W. Sherman and Vilia G. Sherman. Copyright © 1976 by Irwin W. Sherman and Vilia G. Sherman.)

living sponges can be traced by using a suspension of carmine powder in sea water. Place a living sponge in a bowl of sea water. Add a drop of carmine powder suspension; watch the currents by tracing where the carmine powder enters the sponge and where it exits.

The three basic patterns of pore animals are shown in Figure 1-19. Sponges are marine except for one family, the Spongillae, which is found in clear fresh water.

While many sponges are difficult to recognize due to their greenish, liver texture, they often are identified by the resistant gemmules that form during asexual reproduction in summer and fall (Fig. 1-20). These are balls of cells covered with a thick wall containing spicules. In the spring these dormant gemmules develop into new sponges. Sponges have a tremendous power of regeneration. For example, they can be macerated and, within

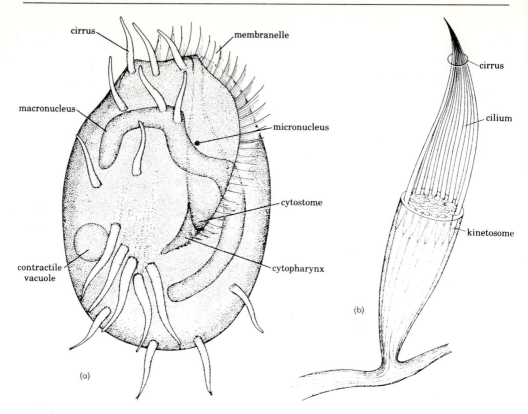

cirrus

membranelle

macronucleus

micronucleus

contractile
vacuole

cytostome

cytopharynx

cirrus

cilium

kinetosome

(b)

(a)

Fig. 1-9 *Euplotes patella*: (a) ventral view; (b) diagram of a cirrus reconstructed from an electron micrograph. (Reprinted and adapted with permission of Macmillan Publishing Company from *The Invertebrates:*

Function and Form, 2nd ed., by Irwin W. Sherman and Vilia G. Sherman. Copyright © 1976 by Irwin W. Sherman and Vilia G. Sherman.)

a month, the cells will reform in clusters and develop new canals and specialized cells.[13] Sponges also reproduce sexually, forming ciliated larvae that escape from the mouth or osculum (Fig. 1-20).

Specimens collected in the fall with gemmules may be maintained in the laboratory for a month or so. Store the gemmules in finger bowls of conditioned water in a cool, dark place. Submerge glass slides in the finger bowls for the hatching cells of the gemmules to attach themselves.

Coelenterates

This phylum[14] consists mainly of marine forms that are radially symmetrical with a two-layered body wall. The outer body wall contains stinging cells or nematocysts. A nonliving mesoglea layer is found between the two body layers that surround a central gastrovascular cavity. A nerve net coordinates the movements of these organisms.

Members of the Class Hydrozoa show alternation of generations, alternating an

[13] Described by Sherman and Sherman, *Invertebrates*.

[14] Sometimes called Phylum Cnidaria, because they contain cnidoblasts consisting of nematocysts on tentacles.

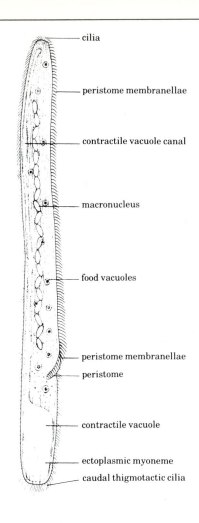

Fig. 1-10 *Spirostomum*, a large, cigar-shaped ciliate; note the long, beaded macronucleus.

asexual hydroid or polyp stage with a sexual medusa or jellyfish stage. Both stages are diploid in contrast to alternation of generation in plants. Included in this class are the small freshwater *Hydra*, and the marine colonial hydrozoans like *Obelia*, *Gonionemus*, and *Physalia* (Portuguese man-of-war). The larger, marine, true jelly-fishes of the Class Scyphozoa, are characterized by eight notches in the margin of the umbrella. The sexual stage is dominant with a reduced polyp or hydroid stage. Included in this class are *Aurelia*, and some giant forms that may reach 150 cm long. Class Anthozoa includes the

typical sea anemones, such as *Metridium*; and corals, such as sea fans, sea feathers, brain coral, and organ pipe coral. Corals are colonies of sea anemones that deposit limestone cups into which the sea anemones withdraw. There is no medusa stage.

A description of the representative members of each class follows below with illustrations that serve as guides for examination and dissection. Some interesting life histories are also shown.

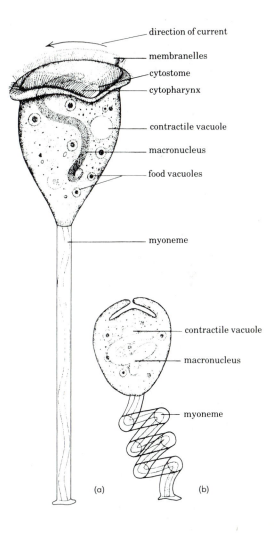

Fig. 1-11 *Vorticella*: (a) a sessile, cup-shaped ciliate; (b) with contracted stalk. Free-swimming, small forms may be found on the surface of old culture dishes of pond water.

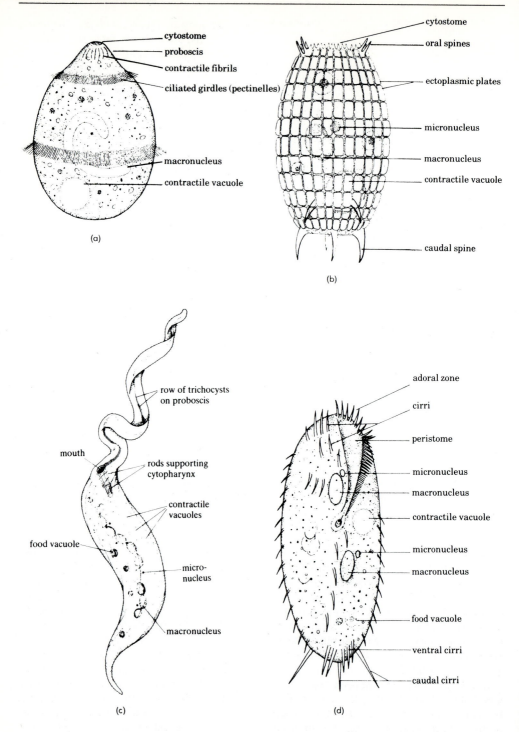

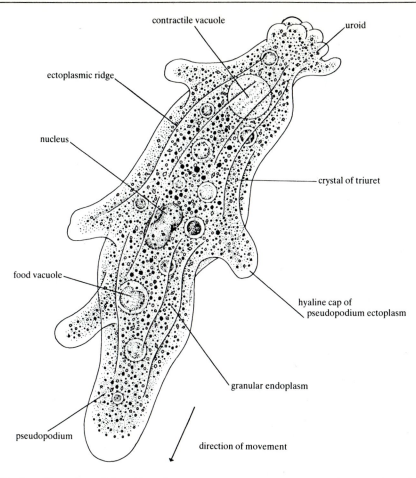

contractile vacuole

uroid

ectoplasmic ridge

nucleus

crystal of triuret

food vacuole

hyaline cap of
pseudopodium ectoplasm

granular endoplasm

pseudopodium

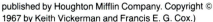

direction of movement

Fig. 1-13 *Amoeba proteus*. (Adapted from *The Protozoa* by Keith Vickerman and Francis E. G. Cox, published by Houghton Mifflin Company. Copyright © 1967 by Keith Vickerman and Francis E. G. Cox.)

Hydra

This freshwater coelenterate, which looks like a white thread about 1 cm long, has a typical hydroid body with many specialized cells. Living specimens may be found attached to submerged leaves in ponds or may be purchased from supply houses. Laboratory uses and culture methods are given in chapter 9.

Examine an undisturbed specimen in a small dish with a hand lens or under a binocular microscope (Fig. 1-21). Note the saclike hydroid body which is composed of two layers of cells, ectoderm and endoderm, separated by a thin, nonliving mesoglea. Using a toothpick, bring car-

mine particles near the peristome and watch the path of the particles. (Use prepared slides to examine specialized cells.)

Observe the six or eight long tentacles surrounding the hypostome, the only opening of the gastrovascular cavity. A simple nerve net coordinates the movements of these tentacles (Fig. 1-22). Gently squash a small part of a tentacle of a live coelenterate and mount it on a slide with a coverslip. You may have to add acetic acid (5 percent), or 5 percent NaCl solution, to cause the nematocysts, or stinging capsules, to discharge. Watch the discharge of four kinds of nematocysts (Fig. 1-23). Nematocysts which develop in cnidoblasts

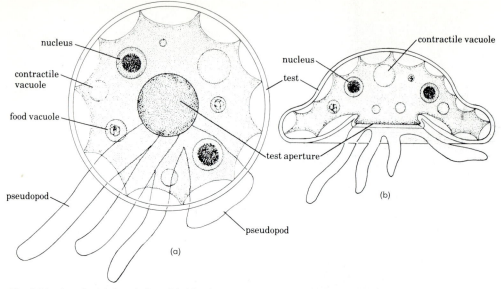

Fig. 1-14 *Arcella*: (a) dorsal view; (b) side view. (Reprinted and adapted with permission of Macmillan Publishing Company from *The Invertebrates: Function and Form*, 2nd ed., by Irwin W. Sherman and Vilia G. Sherman. Copyright © 1976 by Irwin W. Sherman and Vilia G. Sherman.)

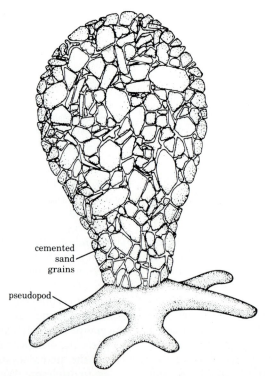

Fig. 1-15 *Difflugia*: note the shell composed of cemented sand grains. (Reprinted and adapted with permission of Macmillan Publishing Company from *The Invertebrates: Function and Form*, 2nd ed., by Irwin W. Sherman and Vilia G. Sherman. Copyright © 1976 by Irwin W. Sherman and Vilia G. Sherman.)

Fig. 1-16 Foraminifera. (Taurus Photos, New York.)

enclose a trigger, or cnidocil. Note how the coiled thread in the nematocyst is discharged.

Observe the behavior of living hydras by touching a toothpick to different parts of the animal. A nerve net made of contractile fibers found in the outer epitheliomuscular layer causes the hydra to contract, to float, or to somersault (Fig. 1-24). Students may discover a rapid response in the region where many nerves are located.

Add a few small *Tubifex* worms or some *Daphnia* to the water, or stain the food organism (pp. 238–39) before feeding living hydras. Watch how the paralyzed organisms are carried to the mouth region of the hydra during ingestion.[15]

Hydras may be gray, *Hydra vulgaris*; brown, *H. fusca*; or, if they possess symbiotic green algae (*Chlorella*, sometimes called zoochlorellae), they are green in color, *H. viridis*.

Look for asexual buds on well-fed specimens. In summer and fall, hydras develop a temporary testis and/or ovary. A testis

[15] H. M. Lenhoff and W. F. Loomis, eds. "Symposium on the Physiology and Ultrastructure of Hydra and Some Other Coelenterates" (Coral Gables: University of Miami Press, 1961).

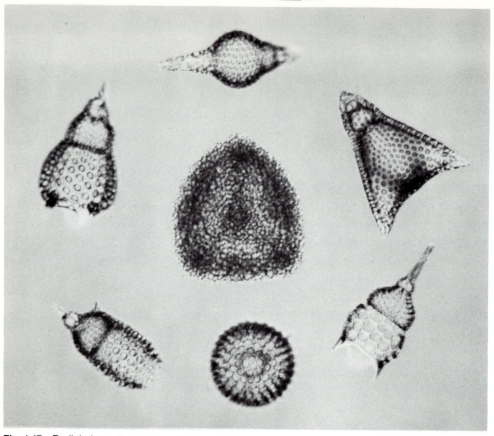

Fig. 1-17 Radiolarians. (Grant Heilman Photography.)

develops closer to the tentacles; the ovary containing one egg cell is located nearer the basal end of the stalk. However, most American species have separate sexes; some are hermaphrodites.[16]

Loomis found that an increase in the CO_2 content of pond water stimulated the development of sexual forms.[17] Living specimens can be used in experiments in regeneration (p. 496) or in studies on the effect of increased amounts of carbon dioxide on reproduction.

[16] For a description with illustrations of hatching egg stages in *Hydra*, refer to W. Pendergrass, "Hatching of *Hydra*," *Carolina Tips* (February 1970) 33:3.
[17] W. F. Loomis, "The Sex Gas of *Hydra*," *Sci. Amer.* (1959).

Marine coelenterates

Gonionemus This jellyfish, about 3 cm wide, is a marine hydrozoan available in some regions. It feeds on small crustaceans. *Gonionemus* illustrates the typical medusoid form (Fig. 1-25); the reduced hydroid or polyp reproduces by budding for a time. Then in late summer, medusae bud off and produce gametes. Fertilized eggs develop into ciliated planula larvae that settle down and grow into the hydroid stage, completing the alternation of generations of hydroid and medusa stages.

Under a dissecting microscope, or using a hand lens, examine a preserved specimen of the umbrella-shaped medusa (Fig. 1-26). Note the stinging nematocysts on the tentacles and locate the adhesive sucker on each

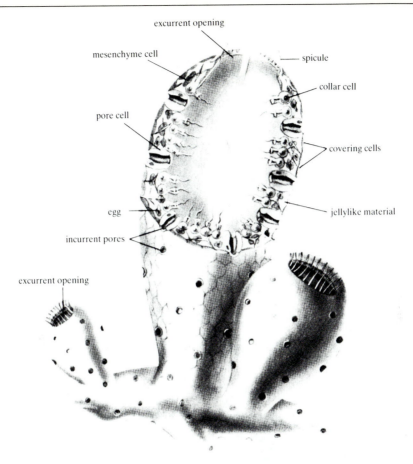

excurrent opening

mesenchyme cell

spicule

collar cell

pore cell

covering cells

egg

jellylike material

incurrent pores

excurrent opening

Fig. 1-18 Diagrammatic view of a colony of a simple sponge. (Reprinted and adapted with permission of Macmillan Publishing Company from *College Zoology*, 10th ed., by Richard A. Boolootian and Karl A. Stiles. Copyright © 1981 by Macmillan Publishing Co., Inc.)

tentacle by which the jellyfish can anchor itself. Recognize that a medusa turned upside down resembles the hydra in general pattern. In fact, *Gonionemus* swims to the surface, then turns over and sinks down with the tentacles extended and the mouth region ready to capture small crustaceans.

Obelia The colonial *Obelia*, another marine hydrozoan, is widely distributed. *Obelia* has a prominent hydroid stage with a reduced medusa stage (Fig. 1-27), the opposite of what is found in *Gonionemus*.

Examine a colony of feeding hydranths with many tentacles around a hypostome, and the reproductive polyps, gonangia, where reduced medusa buds form in spring and summer. Use Figure 1-27 to trace the life history of *Obelia*. Also examine the medusa stage to locate the four radiating canals and the four large gonads.

Physalia Another member of Hydrozoa is *Physalia*, the Portuguese man-of-war, a highly specialized colony of interdependent organisms found in warmer waters (Fig. 1-28).

Gas, composed mainly of nitrogen with some oxygen and carbon monoxide, fills the pneumatophore, but it can be expelled by the contraction of longitudinal muscles inside of the pneumatophore. When gas escapes through the apical pore, the colony submerges.

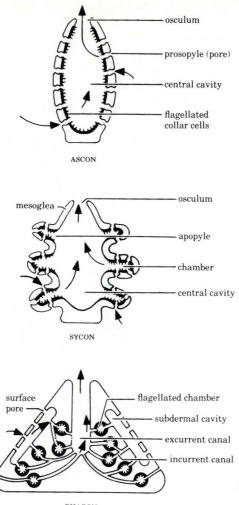

osculum

prosopyle (pore)

central cavity

flagellated
collar cells

ASCON

mesoglea

osculum

apopyle

chamber

central cavity

SYCON

surface
pore

flagellated chamber

subdermal cavity

excurrent canal

incurrent canal

RHAGON

Fig. 1-19 Patterns of systems of canals among sponges. Arrows trace the path of water through the sponges. (Reprinted and adapted with permission of Macmillan Publishing Company from *College Zoology*, 10th ed., by Richard A. Boolootian and Karl A. Stiles. Copyright © 1981 by Macmillan Publishing Co., Inc.)

Notice the enclosing structure, a coenosarc, and the cormidia at the bottom of the float which are arranged in successive series along the ventral side. Cormidia consist of dactylozooids, mouthless organisms, each of which has a long twisted tentacle armed with nematocysts for capturing food. Food is drawn up by the tentacles to the gastrozooids, organisms that have a mouth. Look for gonodendra which are treelike branched gonads.

Tentacles may be 12 meters long. Here are highly specialized members of a colony that have a different structure and function.

Physalia reproduces by asexual buds so that the colony grows larger.[18] There are also umbrellalike medusa buds that contain ovaries or testes. The zygotes grow

[18] C. E. Lane, "The Portuguese Man-of-War," *Sci. Amer.* (1960).

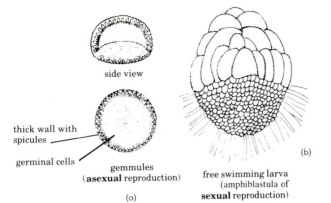

side view

thick wall with
spicules

germinal cells

gemmules
(**asexual** reproduction)

(a)

(b)

free swimming larva
(amphiblastula of
sexual reproduction)

Fig. 1-20   Freshwater sponge: (a) gemmules (asexual); (b) free-swimming larva resulting from sexual reproduction. (Reprinted and adapted with permission of Macmillan Publishing Company from

College Zoology, 10th ed., by Richard A. Boolootian and Karl A. Stiles. Copyright © 1981 by Macmillan Publishing Co., Inc.)

into ciliated planula larvae which settle down and begin new colonies.

Aurelia   One of the large surface-swimming jellyfishes (Class Scyphozoa) that feeds on plankton, *Aurelia* is widely distributed. It ranges from 7.5 to 30 cm in diameter. Examine the medusa stage (Fig. 1-29). Fertilized eggs leave through the mouth region and become attached to the mouth parts until free-swimming planula larvae develop. These planula larvae develop into the reduced hydroid stage called a scyphistoma (Fig. 1-29). Cross-fission or strobilation of the scyphistoma occurs in fall and winter whereby many buds are formed, each a small medusa.

Metridium   Shown in Figure 1-30, *Metridium* represents the widely distributed sea anemone. There is no alternation of generation, no medusa stage, in this Class Anthozoa. The sexes are separate; fertilized eggs develop into small anemones within the gastrovascular cavity.

Metridium has circular and lengthwise muscles and a well-developed nerve net so that the tentacles and the entire mouth can be retracted. Bullough describes methods of preservation with Epsom salts.[19] Also

[19] W. S. Bullough, *Practical Invertebrate Anatomy*, 2nd ed. (New York: Macmillan, 1958).

see page 142 in this text. For detailed guides to dissection, refer to Bullough, Sherman and Sherman, or Boolootian and Heyneman. (See end of chapter.)

Platyhelminthes

This phylum comprises the three-layered flatworms that have bilateral symmetry, flame cells as excretory organs, and a central nervous system. Planaria worms, about 15 mm long, belong in Class Turbellaria; they are free-living with a ciliated epidermis. Parasitic flukes and nonciliated flatworms with suckers belong to the Class Trematoda while tapeworms, complete parasites which lack a digestive system, are in Class Cestoda.

The internal anatomy of *Schistosoma* (Fig. 8-52) and of a tapeworm (Fig. 8-50) are described in a section on parasitology (p. 620). Included in this description is the lung fluke, *Hematoloechus*, found in leopard frogs and the bladder fluke, *Polystoma*, that lives in the bladder of the tree toad, *Hyla*.

Planarians

Living planarians may be collected either from ponds (p. 667) or purchased from supply houses for examination in class.

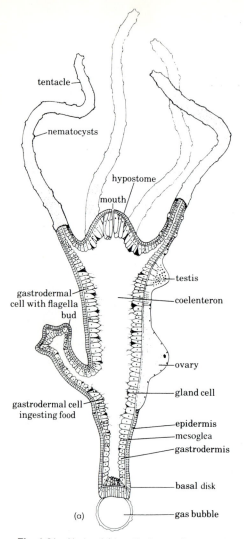

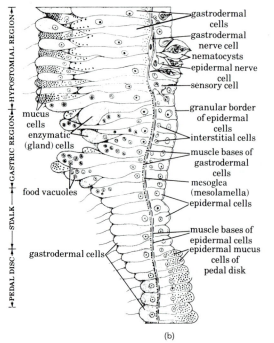

Fig. 1-21 *Hydra*: (a) lengthwise section; (b) diagrammatic view of specialized cells as seen in a cross section. (Reprinted and adapted with permission of Macmillan Publishing Company from *The*

Invertebrates: Function and Form, 2nd ed., by Irwin W. Sherman and Vilia G. Sherman. Copyright © 1976 by Irwin W. Sherman and Vilia G. Sherman.)

They are easy to maintain in the laboratory (chapter 9).

Planarians move by means of muscles and by ventral cilia. To learn the specific roles of the muscles and cilia in locomotion, refer to Sherman and Sherman, who suggest treating worms with substances that inhibit cilia and muscles.[20]

[20] Sherman and Sherman, *Invertebrates*.

For observation of behavior, place a few planarians in a little pond water in small vials or in capillary tubing sealed at both ends with plastecene or wax. Notice the pigmentation of the body, and the darkly pigmented eyespots that enable planarians to react to light stimulus. Are planarians oriented positively or negatively toward light? Also note the two earlike regions that function as tactile receptors.

To study the feeding behavior of planaria worms, introduce some brine shrimps (*Artemia* or *Enchytraeus*, Figs. 1-58 and 1-46) into the vials. Using a hand lens, or a dissecting microscope, examine the action of the hoselike pharynx as it is quickly extruded to bring the mouth near the source of food. When not actively ingesting food, the pharynx is withdrawn into a pharyngeal pouch in the middle of the body (Fig. 1-31). Also refer to feeding carmine powder and preparing a coverslip "sandwich" enclosing living planaria (p. 239).

Examine a prepared slide of a planarian, such as *Dugesia* or *Dendrocoelum* and use Figure 1-31 to trace the organ systems. Note the bilateral symmetry. The branches of the digestive tract are "blind ends" in the sense that food materials both enter and leave only through the mouth region. Enzymes pour into the branched regions where food is digested.

While planarians are representative of flatworms as a group, they are also useful in laboratory studies of behavior and in regeneration experiments.[21] For example, planaria have been used to study the role of RNA in training and memory.[22] Briefly, studies show that if planaria trained to go through a maze are bisected and allowed to regenerate, they learn the maze faster than untrained flatworms. Also, if conditioned planarians are chopped up and fed to untrained worms, the untrained worms—now possessing the RNA of the trained worms—learn the maze faster than worms fed the usual prosaic diet. (Refer to chapter 6, and to the full description of techniques for regeneration in Sherman and Sherman[23] and Boolootian and Heyneman.[24])

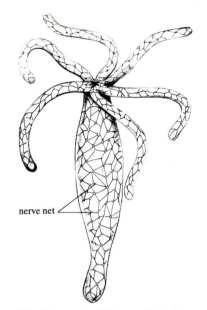

nerve net

Fig. 1-22 Nerve net of *Hydra*; note the concentration around the hypostome. (Adapted with permission from Macmillan Company, from *Biology* by Clyde F. Herried II. Copyright © 1977 by Macmillan Publishing Company, Inc.)

Some students may want to stain planarian worms and prepare permanent slides. W. S. Bullough gives instructions for fixing planaria between two slides and immersing the slides in Bouin's fluid, then into 70 percent alcohol, borax carmine, or hematoxylin.[25]

Planaria chimeras can be made to show patterns of reorganization of tissues (see p. 497).

Nemathelminthes

This is the phylum of unsegmented roundworms that shows bilateral symmetry, three germ layers, and a false coelom, a pseudocoel. Excretory organs are simple; circulatory and respiratory organs are absent. The sexes are usually separate.

Some roundworms are as small as 1 mm while some marine forms may be 5 cm long. Within this phylum are the free-

[21] H. V. Bronsted, "*Planaria* Regeneration," *Biol. Rev.* (1966) 30:65.

[22] J. McConnell, V. Jacobson, and D. Kimble, "The Effects of Regeneration upon Retention of a Conditioned Response in the Planarian," *J. Comp. Physl. Psychol.* (1959) 52:1–5.

[23] Sherman and Sherman, *Invertebrates*.

[24] Boolootian and Heyneman, *An Illustrated Laboratory Text in Zoology*, 3rd ed. (New York: Holt, Rinehart, Winston, 1975).

[25] Bullough, *Practical Invertebrate Anatomy*.

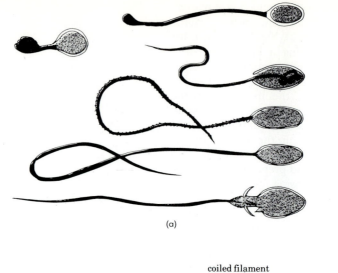

(a)

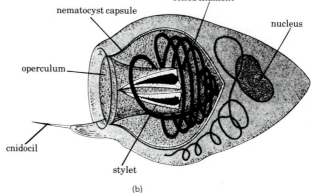

(b)

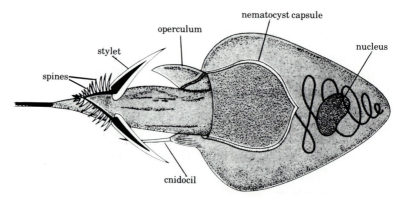

(c)

Fig. 1-23 Nematocysts: (a) several types;
(b) undischarged nematocyst; (c) discharged
nematocyst. (From I. Sherman and V. Sherman, *The
Invertebrates: Form and Function*, 2nd ed. New York:
Macmillan, 1976.)

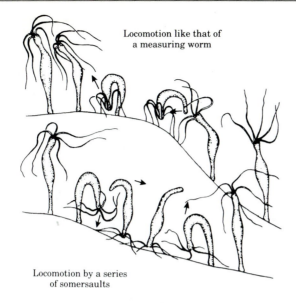

Locomotion like that of
a measuring worm

Locomotion by a series
of somersaults

Fig. 1-24   Locomotion in *Hydra*. (Reprinted and adapted with permission of Macmillan Publishing Company from *College Zoology*, 10th ed., by Richard A. Boolootian and Karl A. Stiles. Copyright © 1981 by Macmillan Publishing Co., Inc.)

living vinegar eels. Most roundworms, however, are parasitic in humans, animals, and plants. Several parasitic forms are described in the section on parasitology (p. 630). Examples include *Necator*, the hookworm (Fig. 1-32), *Trichinella*, a roundworm (Fig. 1-33), and a microfilarian worm, *Wuchereria* (Fig. 1-34).

Free-living vinegar eels are rare examples of nonparasitic roundworms. They are available in undistilled vinegar in which the vinegar eels feed upon the fungi in the medium. Or use prepared slides for detailed study. With the aid of Figure 1-35, students can identify the major organ systems of the vinegar eel, which is about 1 to 2 mm long. Both the vinegar eel, and the parasitic *Ascaris* (which is large enough to be dissected), are described as guides for comparative anatomy.

Ascaris

A larger roundworm, *Ascaris megalocephala*, is parasitic in the intestine of the horse; *A. suillae* in the intestine of the pig; and *A. lumbricoides* in the human intestine.

The female *Ascaris* is larger than the male. Females of *Ascaris lumbricoides* may be some

6 mm wide and 20 to 40 cm long. Preserved whole specimens of *Ascaris* may be pinned down in a dissection pan containing water so that the organs may float for better examination. Be sure to locate the four distinguishing longitudinal lines which serve as identifying structures in dissection; they include the dorsal and ventral lines and two brownish, thicker lateral lines (Fig. 1-36). Distinct respiratory or circulatory systems are absent in these parasitic worms. Exchanges of gases occur through the surface and open circulation of fluids throughout the body.

Rotifers

These multicellular "wheel" animalicules make up a small phylum. Some zoologists group rotifers, gastrotrichs, nematodes, and horsehair worms as a "mixed bag" in a Phylum Aschelminthes. As a phylum, they have some features in common, such as a digestive tract with a mouth and anal opening. A fluid region, a pseudocoel, lies between the digestive tract and the body wall which is covered with a cuticle. However, there are so many diverse features among

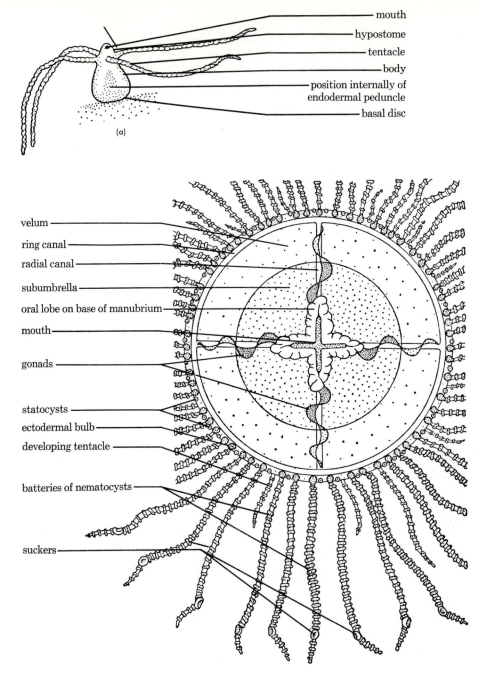

mouth
hypostome
tentacle
body
position internally of
endodermal peduncle
basal disc

(a)

velum
ring canal
radial canal
subumbrella
oral lobe on base of manubrium
mouth
gonads
statocysts
ectodermal bulb
developing tentacle
batteries of nematocysts
suckers

(b)

Fig. 1-25 *Gonionemus*: (a) diagram of hydroid stage; (b) diagram of medusoid stage as viewed from the oral underside. (W. S. Bullough, *Practical* *Invertebrate Anatomy*, 2nd ed., London, Macmillan, 1958, reprinted 1962. Copyright 1958 W. S. Bullough.)

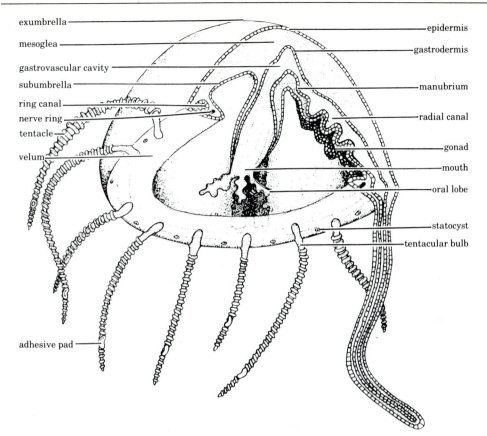

exumbrella
mesoglea
gastrovascular cavity
subumbrella
ring canal
nerve ring
tentacle
velum
adhesive pad

epidermis
gastrodermis
manubrium
radial canal
gonad
mouth
oral lobe
statocyst
tentacular bulb

Fig. 1-26 *Gonionemus:* medusa stage with part of the umbrella cut away to show the internal organs. (Reprinted and adapted with permission of Macmillan Publishing Company from *College Zoology*, 10th ed., by Richard A. Boolootian and Karl A. Stiles. Copyright © 1981 by Macmillan Publishing Co., Inc.)

them they may also be placed in separate phyla (Fig. 1-37).

There is a fine chapter on rotifers, giving illustrations of many species of genera, in addition to keys for identification, in *Fresh Water Biology*.[26] The most commonly found genera, *Philodina* and *Asplanchna* (Fig. 1-38), are often found with *Euglena* or *Chlamydomonas* in green pond water, or in old cultures of protozoa. While rotifers resemble single-celled organisms, they have a well-developed digestive tube, nervous system, gonads, and paired nephridial tubes.

Under the microscope, locate a quiet form feeding in a mass of debris. (Or add a

drop of 5 percent methyl cellulose [p. 142], to the slide to slow them down.) Notice the trochus, or ciliary disk, at the anterior end, and watch how the whirling motion of the trochus carries food material into the central mouth, which leads into a well-formed digestive tube. The mouth may be considered ventral since one side of the trunk of the rotifer is flattened (Fig. 1-37).

Vary the amount of light and try to see the flickering of the flame bulbs on each side of the mastax (pharynx). If rotifers are fed green algae, such as *Chlorella*, observe the current created by the cilia as the small green cells are whirled about. The stomach and intestine become visible as they become filled with green algae. Focus on the mastax in the anterior third

[26] H. B. Ward and G. C. Whipple, *Fresh Water Biology*, 2nd ed., W. T. Edmondson, ed. (New York: Wiley, 1959).

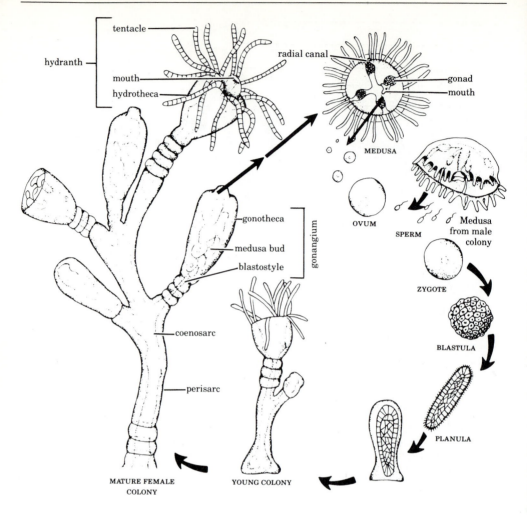

Fig. 1-27 *Obelia*: life history of a marine colonial hydroid. Note that a colony consists of feeding hydranths and reproductive gonangia. A free-swimming ciliated planula larva settles to form a new colony. (Reprinted and adapted with permission of Macmillan Publishing Company from *College Zoology*, 10th ed., by Richard A. Boolootian and Karl A. Stiles. Copyright © 1981 by Macmillan Publishing Co., Inc.)

of the body, and watch the chewing action of the jaws contained therein.

Two cement glands aid in providing attachment for the foot organ when needed. At other times, this jointed tail is "telescoped," and rotifers may be found swimming rapidly across the field of the microscope.

In the large body cavity, there is a fluid containing small granules, which probably acts as a circulating medium. There is also a large ganglion lying dorsal to the mouth and pharynx. Notice that the trunk region is enveloped in a clear lorica, which may be found as remains in very old cultures of rotifers.

The female is most often the form described; parthenogenetic females are found during the spring and summer. The large ovary of the female is connected to a vitellarium, or brood sac, which opens into the cloaca via an oviduct. The thin-shelled eggs are of two sizes. Except for a large testis, the male has reduced organ systems

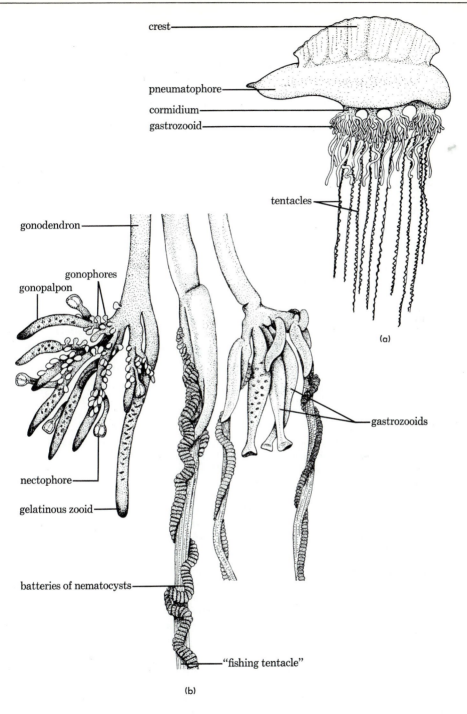

crest

pneumatophore

cormidium

gastrozooid

tentacles

(a)

gonodendron

gonophores

gonopalpon

gastrozooids

nectophore

gelatinous zooid

batteries of nematocysts

"fishing tentacle"

(b)

Fig. 1-28 *Physalia*: (a) a colony; (b) part of a colony shows specialized structure of individuals. (From D. E. Beck and L. F. Braithwaite, *Invertebrate*

Zoology: Laboratory Workbook, 2nd ed., 1962; courtesy of Burgess Publishing Co., Minneapolis.)

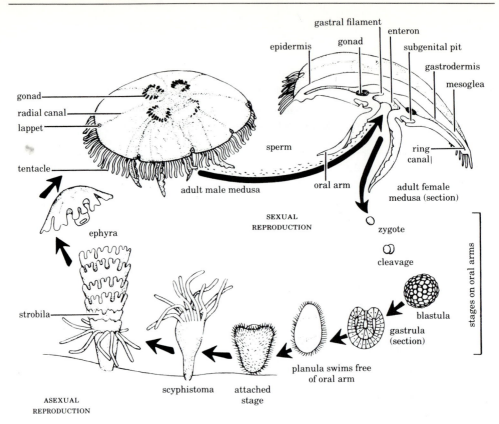

Fig. 1-29 *Aurelia*: trace the asexual and sexual stages of its life cycle. (Reprinted and adapted with permission of Macmillan Publishing Company from *College Zoology*, 10th ed., by Richard A. Boolootian and Karl A. Stiles. Copyright © 1981 by Macmillan Publishing Co., Inc.)

(no gut) and is only a quarter the size of the female.

The smaller males develop in the fall. Fertilized eggs are winter eggs; these hatch into females in the spring. Occasionally these fertilized eggs can be observed attached to the tail of the rotifer. (Also refer to page 669.)

Gastrotrichs

Gastrotrichs are grouped in the small phylum of microscopic organisms, Gastrotricha. (Or, they may be grouped with rotifers and roundworms in Phylum Aschelminthes.) They are found in quiet pools of fresh water containing *Ulothrix* and *Spirogyra*. Although gastrotrichs resemble small ciliated protozoans,

they are multicellular and have a distinguishable forked tail. They are not to be confused with rotifers because they lack trochal discs. The ventral surface is flattened and covered with two bands of cilia; these extend along the length of the body, while the dorsal region is more convex and covered with spinelike bristles.

Chaetonotus is a common gastrotrich with many species. Examine its graceful, gliding motion (due to ventral cilia) under the high power of a microscope. Look for four clusters of tactile ciliary tufts at the anterior end (Fig. 1-39). Slow down the organism with 5 percent methyl cellulose. Try to locate the thick-walled esophagus, leading from the mouth into a wide stomach; the intestine ends in a posterior anal opening. A pair of coiled nephridia termi-

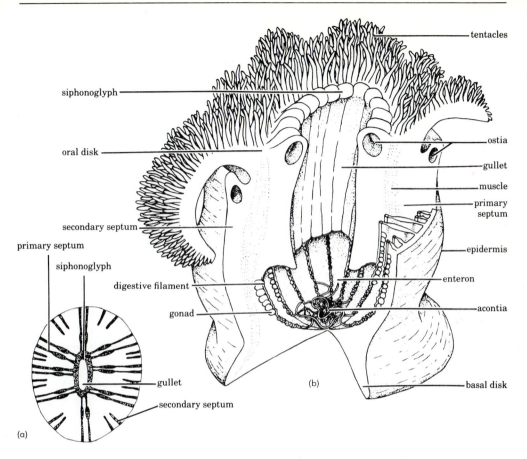

tentacles

siphonoglyph

oral disk

ostia

gullet

muscle

primary
septum

secondary septum

primary septum

siphonoglyph

epidermis

digestive filament

enteron

gonad

acontia

gullet

(b)

secondary septum

basal disk

(a)

Fig. 1-30 *Metridium*, a widely distributed sea anemone: (a) cross section through the gullet region; (b) partly dissected to show the internal organs. (Reprinted and adapted with permission of

Macmillan Publishing Company from *College Zoology*, 10th ed., by Richard A. Boolootian and Karl A. Stiles. Copyright © 1981 by Macmillan Publishing Co., Inc.)

nate on the ventral surface with flame bulbs. Vary the light to observe their flickering motion. Occasionally it is possible to distinguish the anterior ganglion, from which a pair of nerves extends ventrally along the body.

In freshwater species, only parthenogenetic females are known; locate the paired ovaries. Some marine gastrotrichs are hermaphrodites.

considered mollusks because they have a bivalve shell, they are distinguished from mollusks in having dorsal and ventral shells that are unequal in size. They are sessile, attached to rocks by a muscular peduncle (Fig. 1-40). Probably the oldest genus is *Lingula*, considered a living fossil. For an extensive description of its anatomy, refer to Hickman.[27]

Brachiopods

Members of this ancient phylum were most numerous in Cambrian times. Once

[27] C. Hickman, *Biology of the Invertebrates*, 2nd ed. (St. Louis, MO: Mosby, 1973).

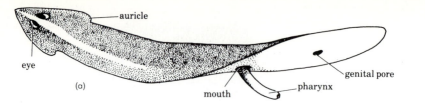

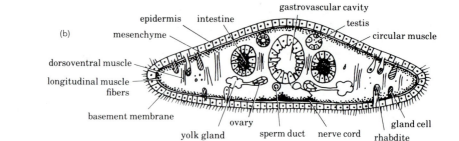

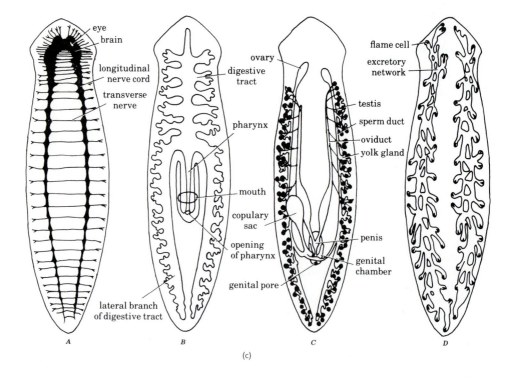

Fig. 1-31 Planarian anatomy: (a) external view; (b) cross section through pharyngeal region; (c) diagrammatic views of organ systems. (Reprinted and adapted with permission of Macmillan Publishing Company from *College Zoology*, 10th ed., by Richard A. Boolootian and Karl A. Stiles. Copyright © 1981 by Macmillan Publishing Co., Inc.)

Annelids

While most segmented worms are marine (Polychaetes), some live in fresh water and some, like the earthworm, are terrestrial (Oligochaetes). The leeches belong in Class Hirudinea. Segmented worms are characterized by a series of similar parts, known as somites or metameres. They have a large coelom, a well-developed nervous system, and a circulatory system comprising dorsal and ventral vessels joined by cross-vessels. A straight digestive system can be traced; nephridia are segmented. Examples of each of these three classes can be examined either through a microscope for the small freshwater oligochaetes, or by dissection to show similarities and specialized adaptations.

Oligochaetes

This class of segmented worms has rows of setae, or bristles, that aid in classification, especially of the smaller microscopic forms. Included here are *Aeolosoma*, which is 1 mm long, and reproduces by transverse fission, *Tubifex*, *Enchytraeus*, and many other freshwater segmented worms. These can be examined with a compound microscope or dissecting microscope. *Lumbricus*, the earthworm, can be dissected to show oligochaete anatomy.

Small freshwater oligochaetes

The microscopic, almost transparent worms that are found in pond cultures associated with protozoa and rotifers are often members of the families Aeolosomatidae and Naididae. Budding is a common asexual method of reproduction among many of them. *Aeolosoma* is characterized by yellowish oil globules in some species, becoming pale green or even blue in other species. There are 7 to 10 segments comprising the body. Notice the posterior budding (Fig. 1-41).

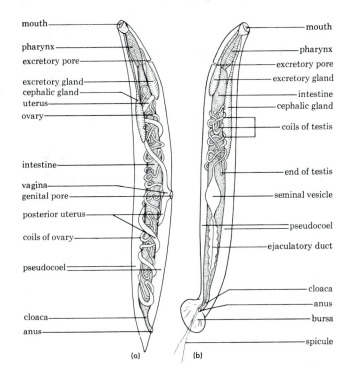

Fig. 1-32 *Necator*, the hookworm: (a) female; (b) male. (After D. E. Beck and L. F. Braithwaite, *Invertebrate Zoology: Laboratory Workbook*, 2nd ed., Burgess, 1962.)

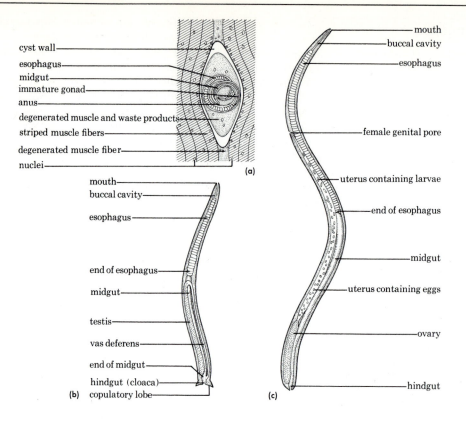

cyst wall
esophagus
midgut
immature gonad
anus
degenerated muscle and waste products
striped muscle fibers
degenerated muscle fiber
nuclei
(a)

mouth
buccal cavity
esophagus

female genital pore

uterus containing larvae

end of esophagus

midgut

uterus containing eggs

ovary

hindgut
(c)

mouth
buccal cavity
esophagus

end of esophagus
midgut

testis
vas deferens
end of midgut
hindgut (cloaca)
(b)   copulatory lobe

Fig. 1-33 *Trichinella*, a parasitic roundworm: (a) coiled larva encysted in striated muscle; (b) adult male; (c) adult female. (After W. S. Bullough, *Practical Invertebrate Anatomy*, 2nd ed., London, Macmillan, 1958.)

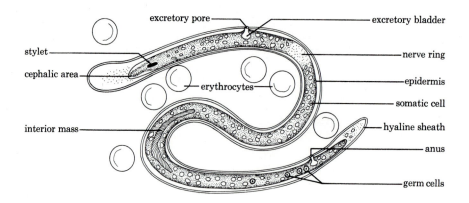

excretory pore
stylet
cephalic area
erythrocytes
interior mass

excretory bladder
nerve ring
epidermis
somatic cell
hyaline sheath
anus
germ cells

Fig. 1-34 *Wuchereria bancrofti*, a microfilarian worm. (After D. E. Beck and L. F. Braithwaite, *Invertebrate Zoology: Laboratory Workbook*, 2nd ed., Burgess, 1962.)

Commonly found members of the family Naididae are *Nais*, *Dero*, and *Aulophorus*. In *Nais*, the ventral setae of segments 2 to 5 are better differentiated than the posterior setae. This is used as an identifying characteristic. The dorsal, needlelike setae may be single or bifurcated, depending on the species. These worms are about 2 to 4 mm long and have a light brown color. *Dero*, a mud-dweller, has a posterior end modified into a gill-bearing respiratory organ (Fig. 1-42); similarly, posterior gills are found in *Aulophorus* (Fig. 1-43).

Under the microscope, without dissection, observe peristalsis, ciliary action in the intestine, circulatory system, and movement of the setae in microscopic forms like *Dero*, *Aeolosoma*, and *Tubifex*.

Tubifex   These freshwater oligochaetes, about 4 cm long, are commonly found in

lakes and ponds containing organic matter. They form tubes of mud held firm by mucus secretions (Fig. 1-44). *Tubifex* may also be purchased in aquarium shops, for they serve as live food for many aquarium fish; they are easily maintained in the laboratory (chapter 9).

Tubifex is often used in studies of regeneration (pp. 497–98). Cut off the last four segments with a scalpel; observe the rate of regeneration. *Tubifex* is also studied as a representative of the segmented worms, for it is transparent enough to show active muscular contractions of the intestine, dorsal vessel, and aortic loop. Gills or respiratory branches are located in the terminal end; when cultured in an inch of mud, *Tubifex* may be seen to have its posterior end waving from mud tubes in which the anterior part of the worm is enclosed.

Each segment, or metamere, contains a portion of the body cavity, or coelom, filled with fluid. Bundles of setae are arranged segmentally; those along the more posterior end are in pairs on each side of a segment.

Mount a specimen in a drop of pond water, and examine it under low and high power. Observe first the activity of the worm—the means of locomotion, contractions, and ingestion. Examine the length of the worm, and note the arrangement of bristle setae on each segment. Either preserve a specimen by adding a drop of magnesium sulfate solution (Epsom salts, p. 142) under the coverslip, or mount a worm in a very dilute agar medium (about

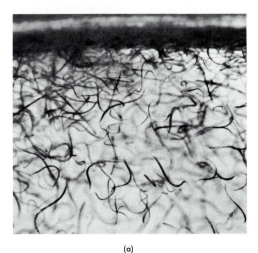

(a)

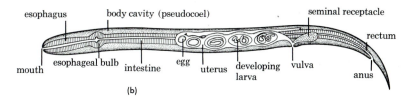

(b)

Fig. 1-35 *Turbatrix aceti*, the vinegar eel: (a) culture of these free-living transparent tiny roundworms; (b) diagrammatic view of anatomy shows digestive and reproductive organs.

(a, courtesy of General Biological Supply House, Inc., Chicago; b, after R. W. Hegner and K. A. Stiles, *College Zoology*, 7th ed., New York, Macmillan, 1959.)

1 percent). In a quiet animal trace the internal organs. Use Figure 1-45 as a guide.

As in other annelids, ganglia concentrated in the anterior end are not easily seen in whole mounts, nor are the reproductive organs readily recognized.

Enchytraeus These segmented white worms are semiaquatic, found near fresh water or on beaches at the high tide level as they feed on decaying vegetation. They are easily maintained in the laboratory

(p. 668), and can be purchased in aquarium stores. Culture them like earthworms since they are partly terrestrial.

Examine these small forms, less than 25 mm long, in a drop of water on a slide. Use Figure 1-46 as a guide to observe the segments with short setae in clusters. Trace the yellowish digestive system (covered with yellow chloragogen cells). Watch the peristaltic contraction of the anterior end of the intestine. Contractions are in the opposite direction in the pos-

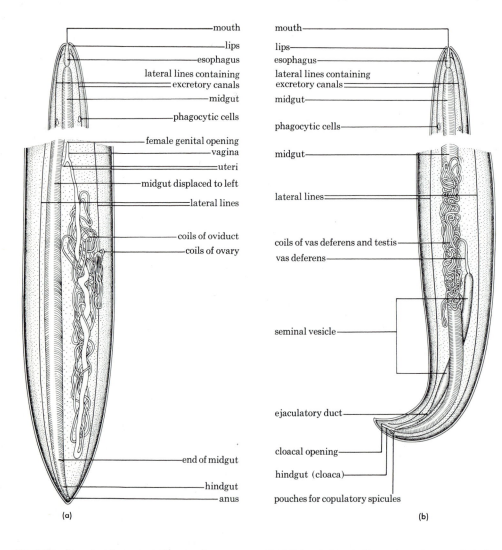

(a)

mouth
lips
esophagus
lateral lines containing excretory canals
midgut
phagocytic cells

female genital opening
vagina
uteri
midgut displaced to left
lateral lines

coils of oviduct
coils of ovary

end of midgut

hindgut
anus

(b)

mouth
lips
esophagus
lateral lines containing excretory canals
midgut
phagocytic cells

midgut

lateral lines

coils of vas deferens and testis
vas deferens

seminal vesicle

ejaculatory duct

cloacal opening
hindgut (cloaca)

pouches for copulatory spicules

Fig. 1-36 *Ascaris*, a large parasitic roundworm: (a) female with left reproductive organs omitted; (b) male showing the curved posterior end. (After W. S. Bullough, *Practical Invertebrate Anatomy*, 2nd ed., London, Macmillan, 1958.)

terior part of the intestine where water is drawn in through the anus. Watch the contraction of the dorsal blood vessel which lies above the front end of the intestine.

Find a pair of ovaries in segment 12. Eggs escape through a break in the body wall; there are no tubes to carry eggs out of the body. In segment 5 are two spermathecae that contain stored sperms from another worm. Two pairs of testes in segments 10 and 11 make up the male reproductive system. A ventral nerve cord may be visible depending on the position of the worm on the slide.

Lumbricus When living earthworms are available, their behavior in response to many stimuli, such as light, touch, sound, and moisture, may be studied (see chapter 5). Study the external anatomy of the animals. When this has been done and the animal is ready for dissection, gradually add a few crystals of Epsom salt to the water in a container holding several earthworms. Now stretch out one worm and pin it down firmly, dorsal surface up, through the first segment and the posterior end, in a waxed dissecting pan.

Cut through the skin of the back, begin-

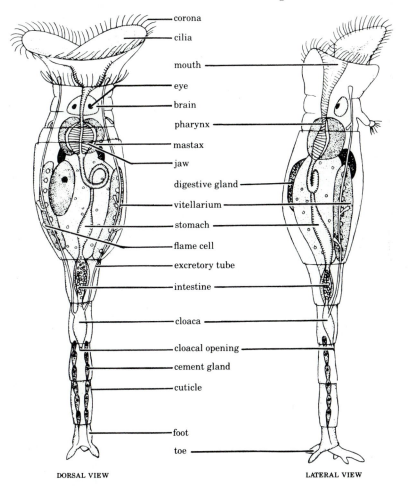

corona
cilia
mouth
eye
brain
pharynx
mastax
jaw
digestive gland
vitellarium
stomach
flame cell
excretory tube
intestine
cloaca
cloacal opening
cement gland
cuticle
foot
toe

DORSAL VIEW   LATERAL VIEW

Fig. 1-37 General structure of a female rotifer, dorsal and lateral views. (Reprinted and adapted with permission of Macmillan Publishing Company from *College Zoology*, 10th ed., by Richard A. Boolootian and Karl A. Stiles. Copyright © 1981 by Macmillan Publishing Co., Inc.)

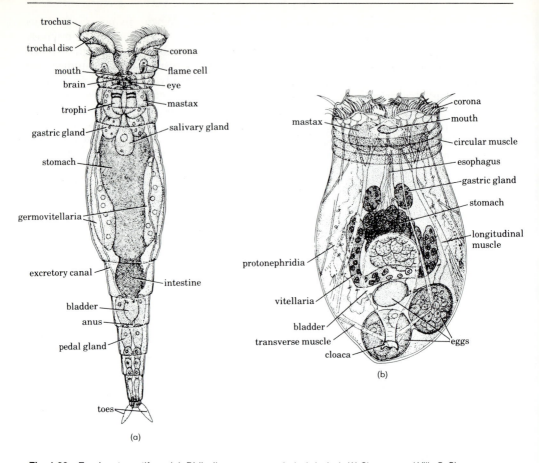

Labels (a): trochus, trochal disc, mouth, brain, trophi, gastric gland, stomach, germovitellaria, excretory canal, bladder, anus, pedal gland, toes, corona, flame cell, eye, mastax, salivary gland, intestine

Labels (b): mastax, protonephridia, vitellaria, bladder, transverse muscle, cloaca, corona, mouth, circular muscle, esophagus, gastric gland, stomach, longitudinal muscle, eggs

(a)

(b)

Fig. 1-38 Freshwater rotifers: (a) *Philodina*; (b) *Asplanchna*, ventral view of female. (Reprinted and adapted with permission of Macmillan Publishing Company from *The Invertebrates: Function and Form*, 2nd ed., by Irwin W. Sherman and Vilia G. Sherman. Copyright © 1976 by Irwin W. Sherman and Vilia G. Sherman.)

ning at the posterior end. Continue forward, cutting to one side of the midline. Cut the partitions, or septa, that hold down the body wall, and pin back the body wall. If pins are inserted through segments 5, 10, 15, and 20, the organs can be readily located (Fig. 1-47a). In a freshly killed specimen, pulsations of the swollen aortic "hearts" may be visible (Figs. 1-47b and 1-48).

To study the reproductive system, remove the esophagus to get a clear view of the seminal vesicles, consisting of three lobes on each side of the esophagus. Also locate the two pairs of small testes on the septa of segments 10 and 11.

In segment 13 find a pair of ovaries, and look for the yellowish seminal receptacles in segments 9 and 10. Crush one in a drop of water on a slide and look for masses of filamentous sperms, received from another worm during copulation. You may also want to have students observe the parasites found in earthworms; see *Monocystis* and *Rhabditis* (p. 620).

Hirudo The brightly colored rust or greenish leeches are parasitic oligochaetes about 5 to 8 cm long with a large posterior sucker. They may be found in fresh water or damp soil. While these hermaphrodites lack setae on their segments, like other annelids, each somite consists of sections of the various systems (Fig. 1-49). Somites

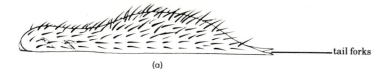

tail forks

(a)

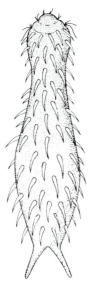

(b)

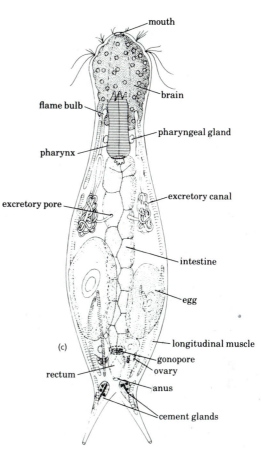

mouth

brain

flame bulb

pharyngeal gland

pharynx

excretory pore

excretory canal

intestine

egg

(c)

longitudinal muscle

rectum

gonopore

ovary

anus

cement glands

pharynx

(d)

midgut

Fig. 1-39 *Chaetonotus*, a gastrotrich commonly found in old cultures of protozoa and algae: (a) side view; (b) external anatomy; (c) internal anatomy; (d) *Lepidodermella*, another gastrotrich. (a, from D. E. Beck and L. F. Braithwaite, *Invertebrate Zoology: Laboratory Workbook*, 2nd ed., 1962, courtesy of Burgess Publishing Co., Minneapolis; b, c, d, reprinted and adapted with permission of Macmillan Publishing Company from *The Invertebrates: Function and Form*, 2nd ed., by Irwin W. Sherman and Vilia G. Sherman. Copyright © 1976 by Irwin W. Sherman and Vilia G. Sherman.)

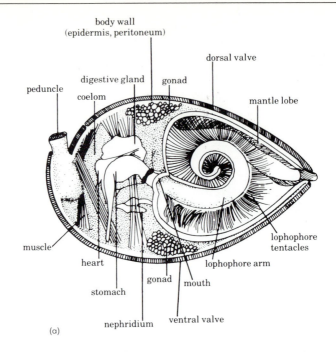

(a)

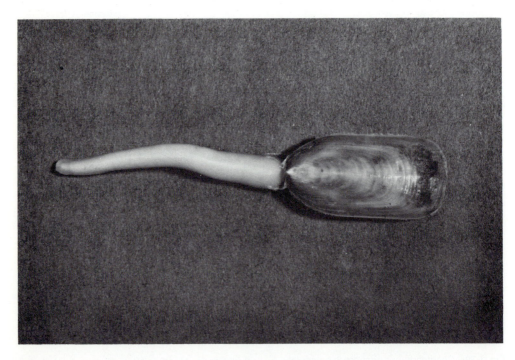

(b)

Fig. 1-40 *Lingula*: (a) internal anatomy; (b) photograph showing peduncle. (a, reprinted and adapted with permission of Macmillan Publishing Company from *College Zoology*, 10th ed., by Richard A. Boolootian and Karl A. Stiles. Copyright © 1981 by Macmillan Publishing Co., Inc.; b, Dr. Richard A. Boolootian.)

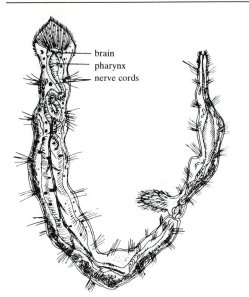

Fig. 1-41 *Aeolosoma,* an aquatic oligochaete with conspicuous orange globules found in old cultures of protozoa. Notice the asexual posterior budding. The single zooids may be 5 to 10 mm long depending on species. (Adapted from *Biology of the Invertebrates,* 2nd ed., by Cleveland P. Hickman. Copyright © 1973 by C. V. Mosby Company.)

28 to 34 are almost fused and make up the caudal sucker, which is much larger than the anterior sucker. Many ganglia are found in the posterior sucker. Blood sucked from a host is stored in gastric caeca. An anticoagulant called hirudin is produced in glands in the crop.

With the aid of Figure 1-49, dissect out the systems. Living leeches may be preserved using methods described on page 142.

Polychaetes

This class includes mainly marine, segmented worms; some live in tubes made of mucus and sand. Representative organisms, available in some regions or from supply houses, can be dissected to show the similarities within their diverse adaptations. Typical polychaetes are *Nereis virens* (*Neanthes*), *Arenicola, Amphitrite,* and *Chaetopterus.* Polychaetes have a trochophore larva (Fig. 1-50). Many polychaetes show seasonal, lunar, and diurnal cycles. Especially dramatic are the November swarms of the South Pacific Palolo worm (*Eunice viridis*); they swarm at

Fig. 1-42 *Dero,* a small aquatic oligochaete; detail of posterior end, an identifying feature of the worm. (From H. B. Ward and G. C. Whipple, *Fresh-Water Biology,* 2nd ed., edited by W. T. Edmondson, 1959; courtesy of John Wiley & Sons.)

Fig. 1-43 Posterior end of *Aulophorus,* a small aquatic worm found in old cultures of protozoa. (From H. B. Ward and G. C. Whipple, *Fresh-Water Biology,* 2nd ed., edited by W. T. Edmondson, 1959; courtesy of John Wiley & Sons.)

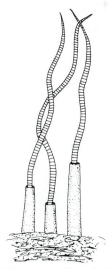

Fig. 1-44 *Tubifex,* an aquatic red oligochaete, with posterior end extended out of a mud tube. (Adapted from *Biology of the Invertebrates,* 2nd ed., by Cleveland P. Hickman. Copyright © 1973 by C. V. Mosby Company.)

low tide during the night of the last quarter moon. The Atlantic Palolo worm swarms in June and July. During this period, long extensions of sexual segments break off from male and female worms (Fig. 1-51). These reproductive segments, called epitokes, each with an eye spot on the ventral side, swim to the surface of the ocean. The waters are dense with these epitokes. Their walls rupture and sperms fertilize eggs, which develop into trochophore larvae.

Nereis   The sandworm or clamworm, *Nereis* (*Neanthes*) is a representative annelid of the Class Polychaeta. These marine worms, some 7 cm long, have a distinct head bearing tentacles and four eyes. From each segment, or metamere, extends a pair of muscular parapodia which bear setae, or bristles, and tactile receptors called cirri (Fig. 1-52).

The bright red blood of *Nereis* gives regions that are more vascular a deep red color, visible through the greenish tint of

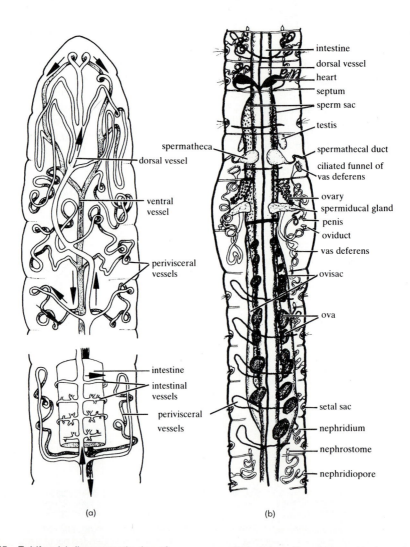

(a)   (b)

Fig. 1-45   *Tubifex*: (a) diagrammatic view of circulation in anterior region and in one segment of the intestinal region; (b) trace the reproductive system. (Adapted from *Biology of the Invertebrates*, 2nd ed., by Cleveland P. Hickman. Copyright © 1973 by C. V. Mosby Company.)

the skin. Compare the external appearance with the terrestrial earthworm (Oligochaeta).

Place a worm ventral side down in a dissecting pan of water and cut through the skin, pinning it back as you cut. Living specimens may be killed by adding Epsom salt crystals to the water. Use Figure 1-52 as a guide for dissection. There are no branchiae, but respiratory exchange takes place through the rich blood supply in the parapodia. Nephridia, paired tubes, may be found in each metamere except in the most anterior and posterior segments; the openings of the nephridia are located on the ventral side near each parapodium.

Gonads may not be visible, for they are temporary organs that develop at breeding time as proliferations from the lining of the coelom.

At breeding time, *Nereis* worms leave their sand tubes and differentiate into the heteronereis, or breeding form, which includes the normal body form (atoke) and the posterior epitoke containing gametes. Notice the parapodia and setae used for swimming by the epitokes (Fig. 1-52d).

Where possible, compare *Nereis* with other forms such as the leech (Fig. 1-49), and marine forms like *Arenicola* (Fig. 1-53), *Amphitrite* (Fig. 1-54), and *Chaetopterus* (Fig. 1-55).

Use these illustrations as guides to study the special adaptations; refer to the laboratory guides and texts in zoology for detailed studies of living and preserved forms (see end of chapter listings).

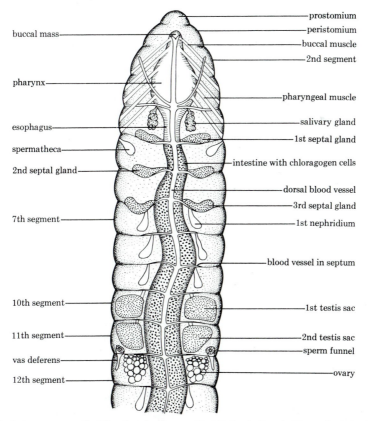

Fig. 1-46 *Enchytraeus*, a small white oligochaete; dorsal view shows organs in the anterior end. (After W. S. Bullough, *Practical Invertebrate Anatomy*, 2nd ed., London, Macmillan, 1958.)

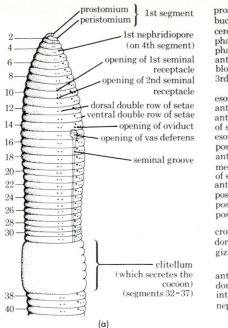

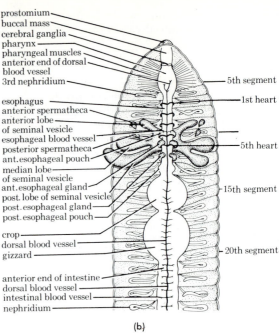

Fig. 1-47 *Lumbricus*: (a) external view; (b) general anatomy and segments. (After W. S. Bullough, *Practical Invertebrate Anatomy*, 2nd ed., London, Macmillan, 1958.)

Amphitrite   This widely distributed polychaete lives in tubes of mud or sand. It has ciliated tentacles and dorsal gills (Fig. 1-54). Dorsal gills are located at the front end of the trunk on the first three segments. These tube-dwellers are adapted to their way of life with long tentacles that catch food.

Chaetopterus   This large polychaete which may be 30 cm long shows specializations of its parapodia. Some function as fans to move water along the burrow or U-tube in which it lives. Other parapodia are modified as suction cups (Fig. 1-55).

If possible, examine living organisms in the dark to see the luminous slime. For extensive coverage of marine annelids, refer to Hickman,[28] as well as Mac Ginitie.[29]

You may want to refer to the distribution of respiratory pigments in some invertebrates (Table 1-1, p. 61).[30] Also refer to Manwell[31] and to Fox[32] for an extensive coverage.

Arthropods

This is the large phylum of jointed-legged animals with bilateral symmetry. An exosketon contains chitin, and a series of segments have pairs of jointed appendages on some, or all of the segments. There is tremendous diversity resulting from highly specialized adaptations for life on land, air, and water. The phylum is subdivided into two subphyla: a) Chelicerata, which include horseshoe crabs, spiders, and scor-

[28] Hickman, *Biology of Invertebrates*.
[29] G. Mac Ginitie, "The Method of Feeding of *Chaetopterus*," *Biol. Bull.* (1939) 77:115–118.

[30] Hickman, *Biology of Invertebrates*.
[31] O. Manwell, "Comparative Physiology: blood pigments," *Ann. Rev. Physiol.* (1960) 22:191–244.
[32] H. Fox and G. Vevers, *The Nature of Animal Pigments* (London: Sidgwick & Jackson, 1960).

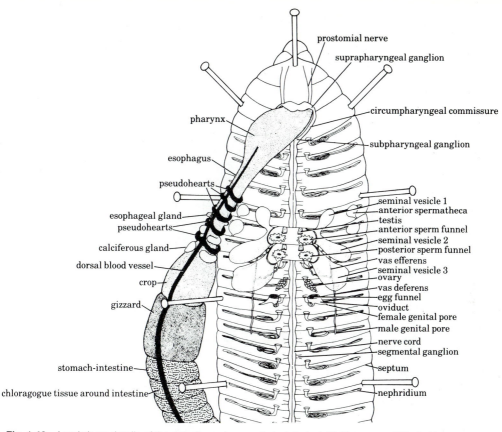

Fig. 1-48 *Lumbricus*: details of the anterior region with digestive system pushed aside. (Reprinted and adapted with permission of Macmillan Publishing Company from *The Invertebrates: Function and Form*, 2nd ed., by Irwin W. Sherman and Vilia G. Sherman. Copyright © 1976 by Irwin W. Sherman and Vilia G. Sherman.)

pions, and named for the fact their feeding structures anterior to the mouth are called chelicerae. Chelicerates lack antennae and their body is divided into a cephalothorax and abdomen; b) Mandibulata comprise both aquatic crustaceans, and the vast array of insects, as well as centipedes and millipedes. They possess mandibles and two pairs of maxillae. The crustaceans have two pairs of antennae; insects have one pair.

Crustaceans

Consider the variety among members of the Class Crustacea (Fig. 1-56). These are mainly water-living arthropods that breathe by gills, have a head and thorax (usually fused as a cephalothorax) and an abdomen. They have one pair of jaws with two pairs of maxillae and two pairs of antennae.

Eubranchipus

The fairy shrimp, *Eubranchipus*, is one of the most primitive crustaceans in the subclass Branchiopoda. Hold up a container of fairy shrimps to see how they swim on their backs, with their appendages oriented toward light (Fig. 1-57). The leaf-like appendages are adapted for respiration.

Examine a quiet form under a microscope. Fairy shrimp are about 3 cm long, pinkish, and almost transparent. Observe the flattened appendages and compound

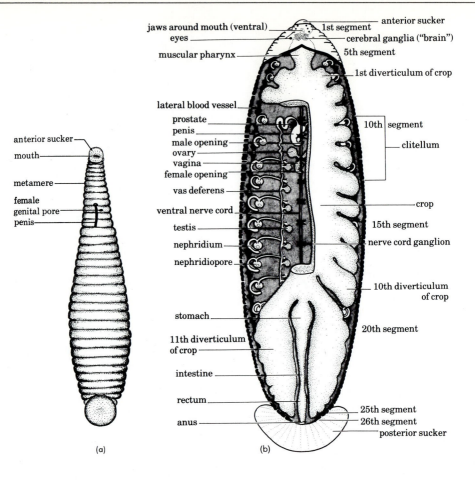

jaws around mouth (ventral)
eyes
muscular pharynx

anterior sucker
1st segment
cerebral ganglia ("brain")
5th segment
1st diverticulum of crop

lateral blood vessel
prostate
penis
male opening
ovary
vagina
female opening
vas deferens
ventral nerve cord
testis
nephridium
nephridiopore

10th segment
clitellum

crop
15th segment
nerve cord ganglion

10th diverticulum of crop

anterior sucker
mouth
metamere
female genital pore
penis

stomach
11th diverticulum of crop
intestine
rectum
anus

20th segment

25th segment
26th segment
posterior sucker

(a)

(b)

Fig. 1-49 *Hirudo*, a parasitic oligochaete: (a) external ventral view; note the large posterior sucker; (b) diagrammatic view of internal anatomy. (a, after D. E. Beck and L. F. Braithwaite, *Invertebrate Zoology: Laboratory Workbook*, 2nd ed., Burgess, 1962; b,

reprinted and adapted with permission of Macmillan Publishing Company from *College Zoology*, 10th ed., by Richard A. Boolootian and Karl A. Stiles. Copyright © 1981 by Macmillan Publishing Co., Inc.)

eyes, which are stalked, on the prominent head region. There is no carapace around the cylindrical body. The second antennae are enlarged, especially in males. The number of appendages may vary from 11 to 17 pairs, extending from as many segments of the thorax.

A period of desiccation precedes hatching, and in fact, seems to initiate hatching. The occurrence of nauplius larvae is often sporadic, since eggs are often carried to temporary pools of rainwater or melting ice where they hatch.

Artemia

The brine shrimp, *Artemia*, is another branchiopod found in saline (not marine) basins such as the Great Salt Lake. You may want to purchase dry brine shrimp eggs and hatch the nauplius larvae (instructions, p. 673). Hold up a container of brine shrimps (available from aquarium stores) and watch how they swim on their backs. Place a few in a watch glass or Syracuse dish of water in which they are found. Use Figure 1-58 to identify organs of the brine shrimp under the microscope.

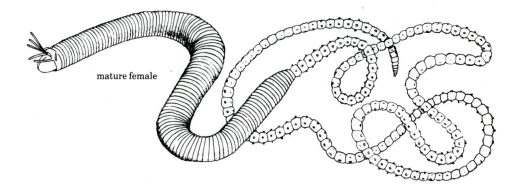

apical organ

eye

esophagus

larval nephridium

otocyst

mesenchyme

stomach

ciliated ring

blastocoel

ciliated ring

anus

anal vesicle

Fig. 1-50 Trochophore larva of polychaete worms. (Reprinted and adapted with permission of Macmillan Publishing Company from *College Zoology*, 10th ed., by Richard A. Boolootian and Karl A. Stiles. Copyright © 1981 by Macmillan Publishing Co., Inc.)

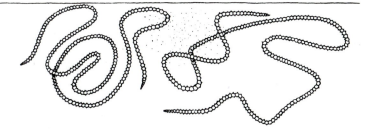

mature female

sexual segments swim to surface; eggs and sperms are discharged

Fig. 1-51 *Eunice viridis*, the Pacific palolo worm, showing reproductive segments called epitokes. (Reprinted and adapted with permission of Macmillan Publishing Company from *College Zoology*, 10th ed., Richard A. Boolootian and Karl A. Stiles. Copyright © 1981 by Macmillan Publishing Co., Inc.)

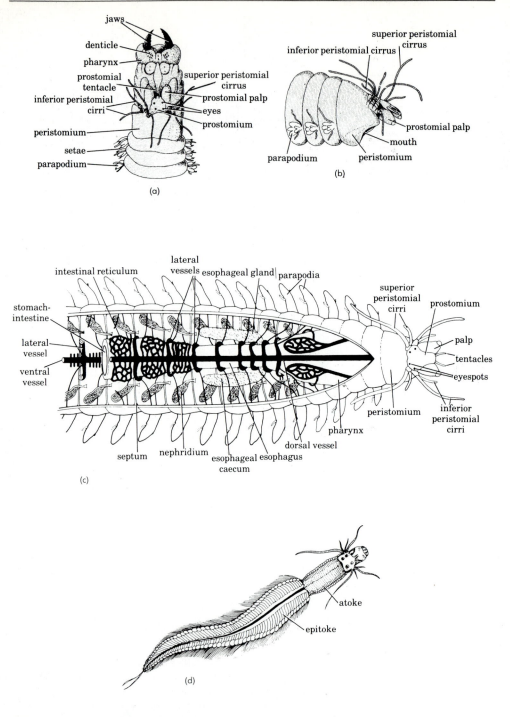

(a)

(b)

(c)

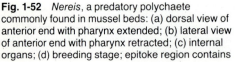

(d)

Fig. 1-52 *Nereis*, a predatory polychaete commonly found in mussel beds: (a) dorsal view of anterior end with pharynx extended; (b) lateral view of anterior end with pharynx retracted; (c) internal organs; (d) breeding stage; epitoke region contains gametes. (Reprinted and adapted with permission of Macmillan Publishing Company from *The Invertebrates: Function and Form*, 2nd ed., by Irwin W. Sherman and Vilia G. Sherman. Copyright © 1976 by Irwin W. Sherman and Vilia G. Sherman.)

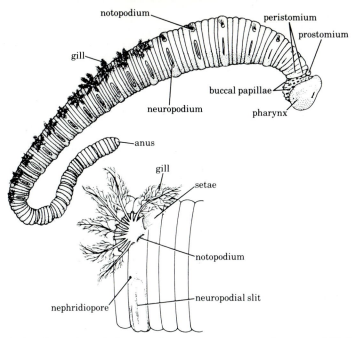

Fig. 1-53 *Arenicola*, a burrowing polychaete; note the clusters of gills or branchiae. (Reprinted and adapted with permission of Macmillan Publishing Company from *The Invertebrates: Function and Form*, 2nd ed., by Irwin W. Sherman and Vilia G. Sherman. Copyright © 1976 by Irwin W. Sherman and Vilia G. Sherman.)

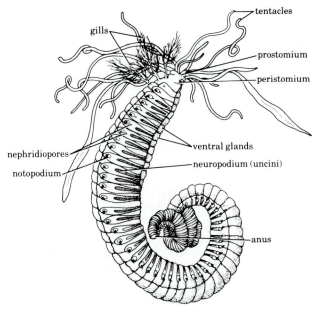

Fig. 1-54 *Amphitrite*: note the specialized prostomium modified as tentacles covered with cilia and mucus-secreting cells. Food particles are moved along a groove in the tentacles toward the mouth. (Reprinted and adapted with permission of Macmillan Publishing Company from *The Invertebrates: Function and Form*, 2nd ed., by Irwin W. Sherman and Vilia G. Sherman. Copyright © 1976 by Irwin W. Sherman and Vilia G. Sherman.)

The flattened thoracic appendages have setae that filter food particles into a ventral groove to the mouth. Sherman and Sherman suggest placing a specimen on paper toweling to dry before fixing the back of the animal in a bit of plasticene or petroleum jelly on a dry depression slide.[33]

[33] Sherman and Sherman, *Invertebrates.*

The ventral side of the abdomen should be uppermost. Add a little water, then a drop of Congo-red stained yeast suspension to study the feeding mechanism. Are there changes in pH in digestion?

Cyclops

A member of subclass Copepoda, *Cyclops* is a greenish form about 2 to 5 mm long. It

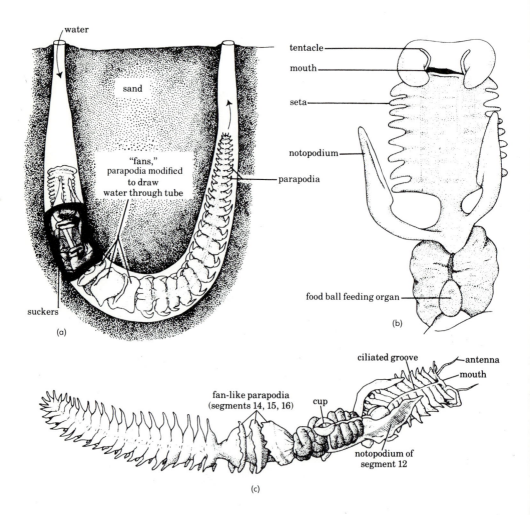

Fig. 1-55 *Chaetopterus*, a polychaete that lives in a U-shaped tube, relies on a water current created by its fanlike parapodia to obtain a food supply: (a) adult in tube; (b) dorsal view of anterior end; (c) external view. (a, b, reprinted and adapted with permission of Macmillan Publishing Company from *College Zoology*, 10th ed., by Richard A. Boolootian and Karl A. Stiles. Copyright © 1981 by Macmillan Publishing Co., Inc.; c, reprinted and adapted with permission of Macmillan Publishing Company from *The Invertebrates: Function and Form*, 2nd ed., by Irwin W. Sherman and Vilia G. Sherman. Copyright © 1976 by Irwin W. Sherman and Vilia G. Sherman.)

has one median red eye and six pairs of thoracic legs, the first four of which are biramous. The segments of the body are different sizes, a distinguishing feature in identification and classification (Fig. 1-59). Hold up a container of *Cyclops* and notice its jerky swimming motions. A pair of egg sacs is often attached to the tail. Females may be carrying capsules of sperm cells received from the male during mating. Fertilized eggs develop into nauplius larvae which pass through several molts, a stage when they are difficult to identify unless the appendages are counted.

Examine a quiet form under the micro- scope to identify the organs. Add a tooth- pick or broken cover glass to the drop of water before applying a cover glass to avoid crushing the specimen. To quiet *Cyclops*, add a crystal of Epsom salt, or mount the specimens in a drop of very thin agar.

There are also parasitic copepods. Both the body and gills of fishes may have these "fish lice" attached by suckers to the host.

Daphnia

The water flea, *Daphnia*, is a member of the subclass Branchiopoda, in order Clado-

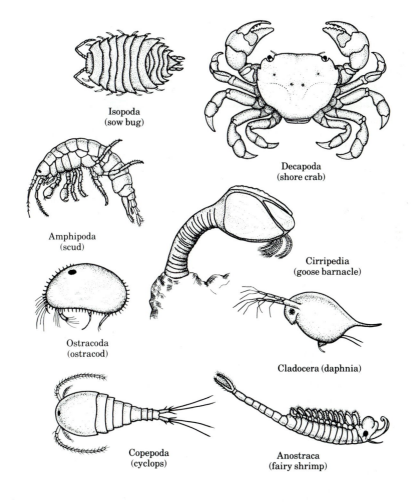

Isopoda
(sow bug)

Decapoda
(shore crab)

Amphipoda
(scud)

Cirripedia
(goose barnacle)

Ostracoda
(ostracod)

Cladocera (daphnia)

Copepoda
(cyclops)

Anostraca
(fairy shrimp)

Fig. 1-56 Variety among crustaceans. (Reprinted and adapted with permission of Macmillan Publishing Company from *College Zoology*, 10th ed., by Richard A Boolootian and Karl A. Stiles. Copyright © 1981 by Macmillan Publishing Co., Inc.)

cera. It has a laterally compressed body and large branched antennae. Occasionally abundant in pond water, *Daphnia* moves in a rapid, jerky manner. Its flattened appendages are used both for locomotion and respiration. The single compound eye is usually in active motion. A beak protrudes on the underside of the head. Except for the head, the compressed body is enclosed in a flattened transparent carapace. There is a caudal spine, and the sturdy second antennae moved by large muscles are the main organs of locomotion. Normally five (or six) pairs of appendages are attached to the thorax.

Use Figure 1-60 to identify the organs in this water flea, about 2 mm long. Mount *Daphnia* in a drop of pond water, adding a toothpick or broken cover glass to avoid crushing the forms with a cover glass.

Although there are no blood vessels, circulating blood cells may be seen as blood flows anteriorly from the rapidly pulsating heart and returns from the hemocoel and inner wall of the carapace, where gaseous exchanges occur; some exchange also occurs along the appendages. In well-aerated water, *Daphnia* have almost colorless hemoglobin; in stagnant waters, the organisms appear orange (pp. 267–68).

Sherman and Sherman suggest studying the filter-feeding mechanism of *Daphnia* by placing a dab of petroleum jelly on a dry depression slide.[34] Dry off a specimen by transferring it from a pipette onto toweling. Then use forceps to transfer the *Daphnia* to the petroleum jelly. Press down a bit so the dorsal region sticks in the jelly. Add a bit of water, then a small drop of Congo-red stained yeast suspension. Watch the motion of the third and fourth pairs of appendages, which are the main filters in water fleas. Or you may prefer to add single-celled algae such as *Chlorella* to the slide, and watch how the current from actively moving appendages carries in the algae. Notice how quickly the food tube turns green.

(a)

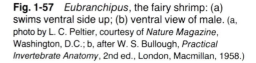

antennule
appendage
antenna
labrum

mandible
maxillule
maxilla
1st thoracic appendage
food groove
11th thoracic appendage
left penis of 12th thoracic segment
1st abdominal segment
2nd abdominal segment

7th abdominal segment
telson
anus
caudal ramus

(b)

Fig. 1-57 *Eubranchipus*, the fairy shrimp: (a) swims ventral side up; (b) ventral view of male. (a, photo by L. C. Peltier, courtesy of *Nature Magazine*, Washington, D.C.; b, after W. S. Bullough, *Practical Invertebrate Anatomy*, 2nd ed., London, Macmillan, 1958.)

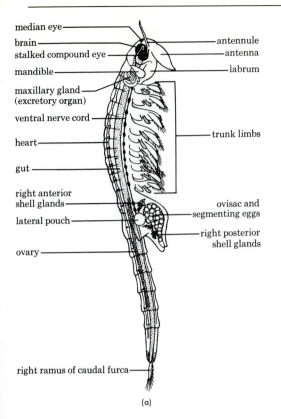

median eye
brain
stalked compound eye
mandible
maxillary gland (excretory organ)
ventral nerve cord
heart
gut
right anterior shell glands
lateral pouch
ovary
antennule
antenna
labrum
trunk limbs
ovisac and segmenting eggs
right posterior shell glands
right ramus of caudal furca

(a)

(b)

Fig. 1-58 *Artemia*, the brine shrimp: (a) side view of female; (b) larva. (Courtesy of General Biological Supply House, Inc., Chicago.)

Most *Daphnia* are parthenogenetic females; eggs develop into miniature water fleas in the brood pouch. Under adverse conditions, sexual reproduction occurs. Some eggs hatch into males, and the females then produce one or two darkly colored, monoploid eggs that require fertilization. These females are distinguished from the usual females by an altered, saddle-like shape of the dorsal valve that develops into an ephippium. At the next molt of the female, the eggs enclosed in the ephippium, or altered carapace, are left behind; they remain dormant until conditions become more favorable for development. (Also refer to reproduction in chapter 6.)

J. L. Brooks reports that overcrowding seems to initiate production of males; sexual egg production seems to occur when there is a sudden decrease in food supply for females.[35] Reported in the same source are methods of examination and preservation using a double coverslip. (Specimens are gradually transferred into full-strength glycerine jelly between two coverslips; this seal is mounted in resin on a slide.) Refer to behavior in relation to wavelengths of light (p. 333).

Cypris

Another small crustacean that bears a superficial resemblance to *Daphnia* is *Cypris* (Fig. 1-61). However, this form is more opaque. These ostracods have a body entirely enclosed in a bivalve carapace, which may be tinted greenish, orange, or yellowish brown. *Cypris* has seven pairs of thoracic appendages. Two pairs of antennae are used in swimming. There is no heart in the freshwater ostracods. Both parthenogenetic and sexual reproduction occur in this form.

Barnacles

Lepas   The goose barnacle (Fig. 1-62) is representative of an adult, sessile crustacean, a member of the subclass Cirripedia. It has a long stalk for attachment. Calcareous plates make up a carapace that protects the body. A thin mantle encloses the body. (At one time, goose barnacles were considered mollusks.) There are no

[35] "Cladocera," in W. T. Edmondson, ed., *Freshwater Biology*, 2nd ed. (New York: Wiley, 1959).

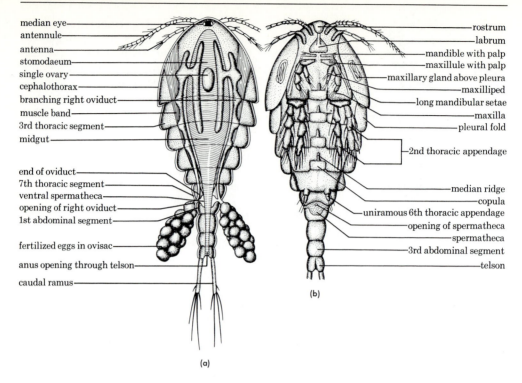

median eye
antennule
antenna
stomodaeum
single ovary
cephalothorax
branching right oviduct
muscle band
3rd thoracic segment
midgut

end of oviduct
7th thoracic segment
ventral spermatheca
opening of right oviduct
1st abdominal segment

fertilized eggs in ovisac

anus opening through telson

caudal ramus

rostrum
labrum
mandible with palp
maxillule with palp
maxillary gland above pleura
maxilliped
long mandibular setae
maxilla
pleural fold

2nd thoracic appendage

median ridge
copula
uniramous 6th thoracic appendage
opening of spermatheca
spermatheca
3rd abdominal segment
telson

(b)

(a)

Fig. 1-59 *Cyclops*: (a) dorsal view of female; (b) ventral view of female. (In both views, the 3rd, 4th, and 5th pairs of thoracic appendages have been omitted.) (After W. S. Bullough, *Practical Invertebrate Anatomy*, 2nd ed., London, Macmillan, 1958.)

antennae, no compound eyes. Six pairs of thoracic appendages, biramous and plumelike, create a current that brings plankton to the mouth region. A barnacle's orientation within its shell has been described as standing on its head kicking food into the mouth.

Sherman and Sherman suggest that respiration and the feeding phases may be observed in living barnacles.[36] Make a suspension of one part milk in three parts seawater. With a thin pipette, such as a Pasteur pipette, add the suspension near the cirri. Watch the slow, first phase in which the valves open and the limbs, or cirri, are unrolled, and create a current drawing the suspension into the mantle cavity and into the mouth. A rapid, second phase follows in which the cirri move forward and are withdrawn, the mantle

cavity contracts, and water is sieved through the cirri. Also try to feed the barnacles brine shrimps or pieces of mussel.

Goose barnacles are hermaphrodites. The fertilized eggs develop into free-swimming nauplius larvae, which, after a few molts, settle down and become sessile adult barnacles. Goose barnacles are widely distributed and may be found attached to floating seaweeds, or on ships and floating logs.

Balanus Rock barnacles do not have the stalk found in goose barnacles. They have a thick shell made of six plates that close when the animals are exposed to air at low tide. Their thin, fringed cirri create a current that moves small organisms into the mouth (Fig. 1-63). Rock barnacles are often attached to hermit crab shells, or to

[36] Sherman and Sherman, *Invertebrates*.

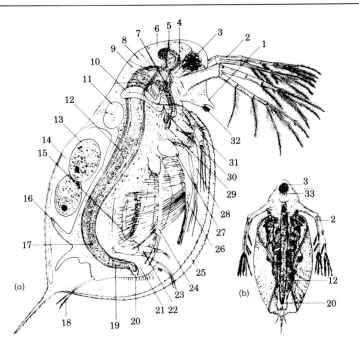

(a)

(b)

1. rostrum
2. exopodite of antenna
3. compound eye
4. supraesophageal ganglion
5. right midgut caecum
6. antenna abductor muscle 1
7. esophagus
8. antenna abductor muscle 2
9. antenna levator muscle
10. shell gland
11. heart
12. intestine
13. egg
14. brood sac
15. roof of food groove
16. median dorsal process
17. midgut, posterior portion
18. caudal seta
19. hindgut
20. anus
21. abreptor
22. trunk appendage 5, medial lobe
23. caudal furca
24. trunk appendage 4,endite
25. carapace
26. trunk appendage 3, endite
27. metepipodite
28. trunk appendage 2
29. lst maxilla
30. mandible
31. trunk appendage 1
32. chemosensory setae
33. eye muscle

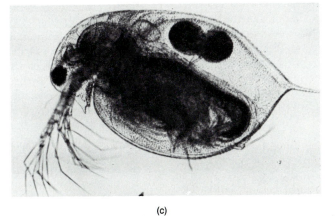

(c)

Fig. 1-60 *Daphnia*, the waterflea: (a) lateral view; (b) ventral view; (c) photograph (magnification × 40). (a, b, illustration by Carolina Biological Supply Company, © 1969; c, photograph by Carolina Biological Supply Company.)

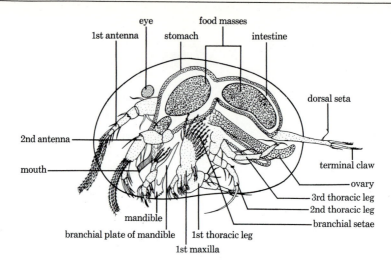

Fig. 1-61   *Cypris*, a freshwater ostracod: lateral view of female. Note the body is encased entirely in a bivalved carapace. (From H. B. Ward and G. C. Whipple, *Fresh-Water Biology*, 2nd ed., edited by W. T. Edmondson, 1959; courtesy of John Wiley & Sons.)

king crabs or other animals. They also abound on rocky surfaces.

Another member of Cirripedia is *Sacculina*. This parasitic form loses the systems found in the larva stage and degenerates into a sac attached to the abdomen of host crabs.

Gammarus

The earwig or scud, *Gammarus*, belongs in the order Amphipoda, subclass Malacostraca including a variety of forms as diverse as the sow bug, crab, freshwater crayfish, and the larger marine lobster. Another genus of amphipods includes the land-dwelling sandhoppers. *Gammarus* lives in salt and brackish water as well as in quiet lakes and ponds. It can be maintained readily in the laboratory in large battery jars of "green" water containing *Chlorella*. These laterally compressed, dark colored forms may be seen swimming upright or resting among the floating duckweeds. Use Figure 1-64a to identify scuds, and Figure 1-64b to trace their internal anatomy.

Paguras

The hermit crab, *Paguras*, inhabits empty periwinkle or whelk shells. The cephalo-

thorax protrudes from the shell, while the abdomen is especially modified, spirally twisted, to fit the shell in which it lives (Fig. 1-65). A sixth pair of appendages serves as a kind of hook that holds the shell. In some species, one pair of appendages is especially large and acts to close the mouth opening of the shell when the animal is pulled within the shell. As hermit crabs molt and grow larger, they seek larger shells. In fact, they may consume the whelk in their search for a larger shell.

Behavioral studies of living hermit crabs can be a fascinating laboratory activity (p. 351).[37] In addition, the land hermit crab, *Coenobita*, is easily maintained in the laboratory. Study its behavior as it searches for larger shells. The soft spiral abdomen takes the shape of the spiralled shell it backs into. While a nocturnal scavenger of plants and animals, *Coenobita* is active in a classroom terrarium habitat of sand, some plants for cover, and empty shells of varying sizes. For a full discussion of its physiology, refer to "Land Hermit

---

[37] There is also a well-developed lab exercise in J. Crane, Jr., *Introduction to Marine Biology*, A Laboratory Manual (Columbus, OH: Charles Merrill, 1973).

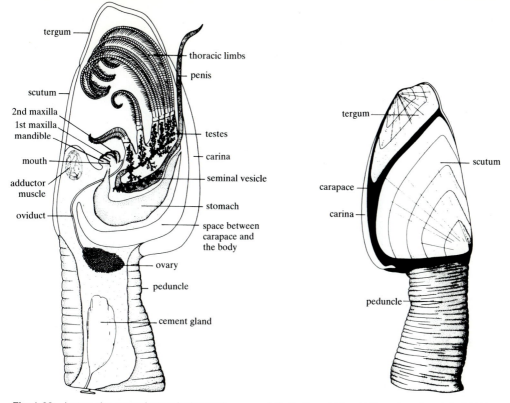

Fig. 1-62 *Lepas*, the goose barnacle: internal and external views. (Adapted with permission from *An Illustrated Laboratory Text in Zoology*, 2nd ed., by Richard

A. Boolootian and Donald Heyneman. Copyright © 1962, 1969 by Holt, Rinehart and Winston, Inc.)

Crabs" by D. James, *Carolina Tips*, March 1, 1978.

Cambarus

Crayfish are freshwater forms similar to the marine lobsters (*Homarus*). They are classified as decapods with shrimps, prawn, crabs, and hermit crabs. Eyes are stalked, and a large carapace covers the thorax; there are five pairs of walking legs. (Some sea crayfish have large antennae, no chelae, and legs that end in simple claws.)

In some sections of the country, marine lobsters (*Homarus*) may be preferred for dissection. Use Figure 1-67 as a guide. When possible, compare the anatomy of a crayfish and lobster.

Examine the external anatomy (Fig.

1-66) of a decapod (either crayfish or lobster). Notice that the body is divided into two distinct parts, a rigid cephalothorax and a more flexible, segmented abdomen.

In a dissecting pan place the specimen dorsal side up. Begin at the posterior dorsal end of the carapace, and cut forward to each side of the midline along each side of the outer edges of the thorax up to the rostrum. Now remove this top part of the carapace. In the dorsal part of the thorax find the pericardial sinus containing the angular, muscular heart. Look for the three pairs of ostia through which blood enters the heart (Fig. 1-67).

Under the heart and a bit forward are the reproductive organs. In a mature female, find the bilobed ovary (containing eggs) in front of the heart and a single

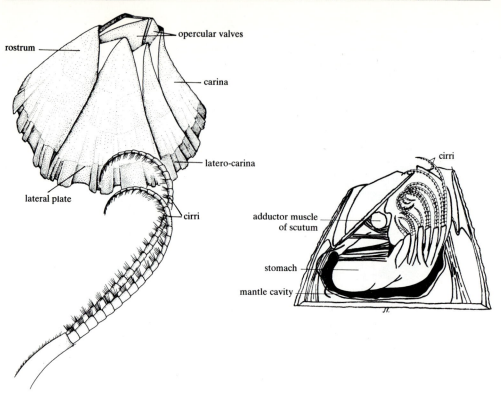

Fig. 1-63 *Balanus*, the rock barnacle: external and internal views. (Adapted with permission from *An Illustrated Laboratory Text in Zoology*, 2nd ed., by Richard A. Boolootian and Donald Heyneman. Copyright © 1962, 1969 by Holt, Rinehart and Winston, Inc.)

fused mass behind the heart. Locate the oviduct extending down on each side to the first segment of the third thoracic leg. In the male the white testis should be in a corresponding position, with two highly coiled vasa deferentia (sperm ducts). If these are filled with sperms, mount some in a drop of water, tease the ducts apart, and examine the unusual star-shape of the sperm cells. The vasa deferentia open on the first segment of the hindmost thoracic leg.

If a living specimen is available, study blood cells, clotting, and phagocytosis. Sherman and Sherman describe techniques.[38] Snip off the tip of the antenna, or a small leg, and transfer a drop of blood to the slide. Add a drop of mineral

oil to delay cell changes and apply a coverslip. Prepare a second slide without mineral oil. Under the microscope, examine how the cells disintegrate. Add a drop of 1 percent iodine solution to the edge of the coverslip of the slides to stain the fibers.

To demonstrate phagocytosis in crustacean blood, inject India ink or carmine suspension into the limb joints, or directly into the heart, with a fine hypodermic needle and syringe. After an hour, obtain a drop of blood from the specimen, as described above. Examine the blood on a slide with a coverslip under high power. Compare with blood samples of *Limulus* (p. 64). Refer to Table 1-1, invertebrate respiratory pigments on page 61.

Behavior and the action of hormones, as well as color changes associated with the

---

[38] Sherman and Sherman, *Invertebrates*.

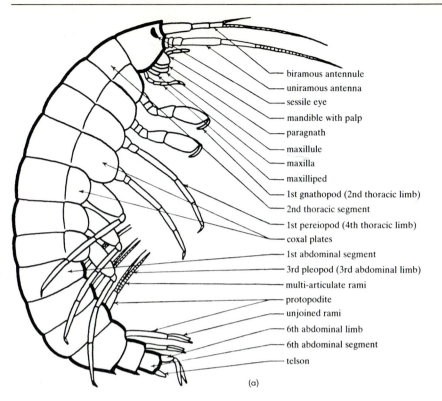

- biramous antennule
- uniramous antenna
- sessile eye
- mandible with palp
- paragnath
- maxillule
- maxilla
- maxilliped
- 1st gnathopod (2nd thoracic limb)
- 2nd thoracic segment
- 1st pereiopod (4th thoracic limb)
- coxal plates
- 1st abdominal segment
- 3rd pleopod (3rd abdominal limb)
- multi-articulate rami
- protopodite
- unjoined rami
- 6th abdominal limb
- 6th abdominal segment
- telson

(a)

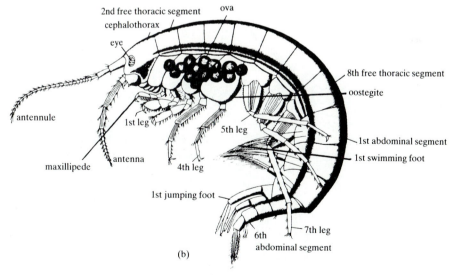

2nd free thoracic segment   ova
cephalothorax
eye
8th free thoracic segment
oostegite
antennule
1st leg
5th leg
1st abdominal segment
1st swimming foot
maxillipede
antenna
4th leg
1st jumping foot
6th
abdominal segment
7th leg

(b)

Fig. 1-64 *Gammarus*, the earwig or scud: (a) external view of laterally compressed body; (b) internal view. (a, reprinted and adapted with permission from Macmillan Publishing Company from *Practical* *Invertebrate Anatomy*, 2nd ed., by W. S. Bullough. Copyright © 1953 W. S. Bullough; b, from T. Jeffery Parker and W. Haswell, *A Textbook of Zoology*, 5 ed., Macmillan, 1930.)

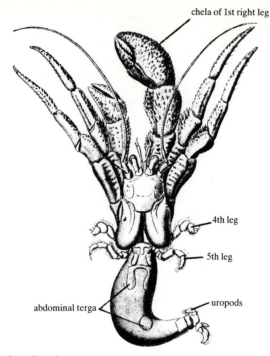

chela of 1st right leg

4th leg

5th leg

uropods

abdominal terga

Fig. 1-65 *Paguras*, the hermit crab. (Reprinted and adapted with permission from Macmillan Publishing Company from *A Textbook of Zoology*, 5th ed. by T. Jeffrey Parker and William A. Haswell. Copyright © 1930 by Macmillan Publishing Company, Inc.)

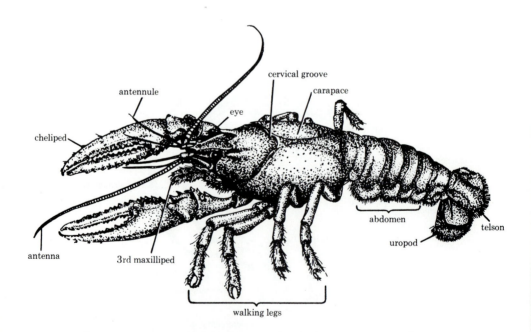

cervical groove

carapace

antennule

eye

cheliped

abdomen

telson

antenna

3rd maxilliped

uropod

walking legs

Fig. 1-66 *Cambarus*, the crayfish: external view. (From W. G. Whaley, O. P. Breland, et al., *Principles of* *Biology*, Harper & Row, 1954.)

TABLE 1-1 Oxygen-combining capacities of different bloods in some invertebrates (at 0°C. and 1 atmosphere pressure)*†

respiratory pigment	color	invertebrate and site of pigment	O_2 vol. % (in ml. O_2/100 ml of blood)
Hemoglobin	Red	Annelids (plasma)	1–10
		Mollusks (plasma)	2–6
Chlorocruorin	Green to pink	Annelids (plasma)	5–9
Hemocyanin	Blue	Mollusks (plasma)	1–5
		Crustaceans (plasma)	1–4
Hemerythrin	Red	Sipunculids (coelomic corpuscles)	1–2

* From several sources.
† Compared with the oxygen volume capacities of invertebrates, those of vertebrates range from about 4 to 30. Other factors that influence the combining capacities are the pH and activity conditions.

Source: C. Hickman, *Biology of the Invertebrates*, 2nd ed. (St. Louis, MO: Mosby, 1973).

eye stalks, are well described in Sherman and Sherman.

Insecta

The large subphylum Mandibulata comprises both crustaceans and the larger Class Insecta. Some half a million species of members of the Class Insecta have been identified. Observe the diverse variety in structure and lifestyle of orders of insects as shown in Figure 1-68. Some are beneficial, some housepests, some social insects, many are predatory and parasitic; on the whole, they are the major competitors of humans for the food supply.

Grasshopper

The larger *Romalea* or the smaller *Melanopus* may be available for study. First examine the external anatomy and especially the details of the head region (Figs. 1-69, 1-70). Then remove the wings from a freshly killed or preserved specimen and pin it with the dorsal side up in a dissecting pan containing water. Cut through each side of the top of the posterior abdomen; remove this upper part of the abdomen—the heart, which is a delicate tube, may be attached. Arteries send blood into perivisceral sinuses, and blood is collected

in a pericardial sinus found around the heart. Ostia, or openings, in the heart allow blood to enter from the sinus.

Notice the white air sacs and air tubes on each side of the abdomen. Use Figure 1-71 to follow the tracheal system.

If the specimen is a female, look for a mass of yellow sacs in the anterior end of the abdomen (Fig. 1-72). Trace the oviducts from the two ovaries to an opening in the ovipositor.

Blatta

For a study of insect anatomy some teachers prefer the oriental cockroach *Blatta* or the larger *Periplaneta*. These as well as grasshoppers are available from biological supply houses.

Examine the external anatomy; if possible, compare it with that of the grasshopper. Identify the head and its appendages, the 3 segments of the thorax, and the 11 segments of the abdomen. The end segments are not easily distinguished, for they overlap and are reduced. Locate spiracles (Fig. 1-73).

Pin the roach ventral side down in a waxed pan or on a corkboard. Remove the leathery elytra and wings, and cut into the specimen along one side so that the dorsal section can be flapped over to expose the internal organs. Students can identify the dorsal "heart," a chain of node-like swellings which extends along the length of the roach. In the vicinity of the first leg locate the white salivary glands surrounding the esophagus. Trace the food tube; find the extended crop, thick-walled gizzard, hepatic caeca, colon, and rectum. The thread-like mass is composed of Malpighian tubules. Now remove the fatty tissue and the digestive tract, and trace the tracheal tubes, which branch throughout the body. Use Figs. 1-73, 1-74 to trace the systems.

Arachnids

This class of Phylum Arthropoda includes the "living fossil," the king crab, and spiders, scorpions, mites, and ticks. There is a division of the body into a cephalothorax, to which six pairs of appendages are attached, and an abdomen. Antennae

and mandibles are lacking; the first pair of appendages is modified as nippers, or chelicerae. Book lungs are characteristic of arachnids; there may also be tracheal tubes and spiracles.

Use Figure 1-75 to explore the anatomy of a spider.

Limulus

The king crab or horseshoe crab, *Limulus*, may be found crawling through sand in search of worms or bivalves. Its body consists of a large horseshoe-shaped carapace, a smaller shield, and a telson or caudal spine. A pair of large eyes and a more anterior pair of smaller eyes are on the convex dorsal surface.

Turn the specimen over and locate the mouth at the base of the anterior walking legs; the first pair of appendages is the chelicerae (Fig. 1-76). The anus is posterior at the beginning of the long spine.

Examine the flattened appendages, the

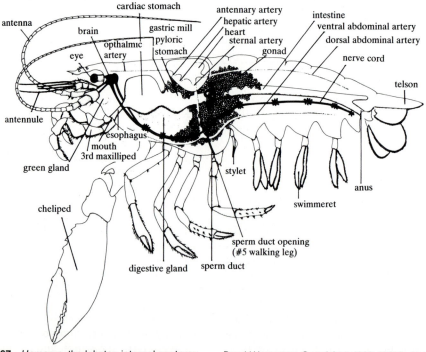

Fig. 1-67 *Homarus*, the lobster: internal anatomy. (Adapted with permission from *An Illustrated Laboratory Text in Zoology*, 2nd ed., by Richard A. Boolootian and

Donald Heyneman. Copyright © 1962, 1969 by Holt, Rinehart and Winston, Inc.)

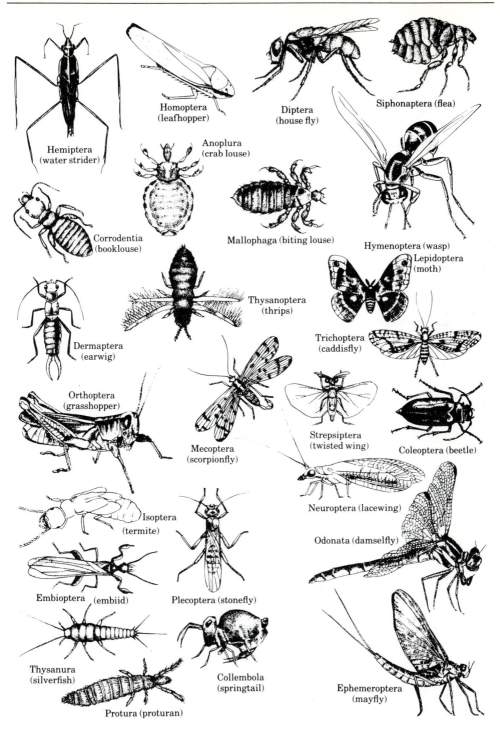

Fig. 1-68 Variety among orders of insects. (Reprinted and adapted with permission of Macmillan Publishing Company from *College Zoology*, 10th ed., by Richard A. Boolootian and Karl A. Stiles. Copyright © 1981 by Macmillan Publishing Co., Inc.)

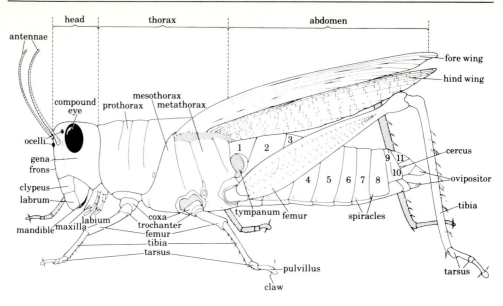

Fig. 1-69 *Romalea*, external anatomy of female. (Reprinted and adapted with permission of Macmillan Publishing Company from *The Invertebrates: Function and Form,* 2nd ed., by Irwin W. Sherman and Vilia G. Sherman. Copyright © 1976 by Irwin W. Sherman and Vilia G. Sherman.)

segmented body, and the specialized book gills composed of some 200 "pages" containing blood vessels. The bluish color of the blood is due to the copper-containing pigment hemocyanin. (See respiratory pigments, p. 61, Table 1-1.)

Sherman and Sherman illustrate how to obtain blood from *Limulus* (Fig. 1-77) as one among many activities using live specimens.[39] Draw 10 ml of blood from the heart of *Limulus* into a syringe. Defibrinate the blood by shaking it with glass beads in an Erlenmeyer flask. Then filter through cheesecloth and centrifuge. Pour 5 ml of the blood into a test tube and bubble oxygen through the blood. Fit the test tube with a one-hole rubber stopper and use a water faucet aspirator to evacuate the tube. Compare the color changes of the copper-containing hemacyanin, the respiratory pigment.

The authors also suggest tracing the path of India ink or carmine-in-saline from the heart into sinuses of the body and the gill books. Inject 10 ml of the dye into the heart of a small, almost transparent *Limulus*.

A related ancient form, the nocturnal scorpion, has a sting containing poison located at the end of the telson. Use Figure 1-78 as a guide to the external anatomy of a scorpion. Compare the anatomy with that of the king crab.

Mollusks

This is the phylum of soft-bodied, unsegmented animals with a ventral muscular foot and usually covered by a shell. Gills are protected by a mantle. Among the members of Class Gastropoda are the mollusks with one spiral shell (or none): this includes the marine whelks and abalones, marine and freshwater snails, and land slugs. Many of these forms may be obtained for dissection: either collected, or purchased from fish markets.

Members of Class Bivalvia or Pelecypoda include oysters, clams, mussels, and scallops. These bivalves have a thick foot for burrowing. Many are readily available for dissection. Squids and octopuses (Class

---

[39] Sherman and Sherman, *Invertebrates*.

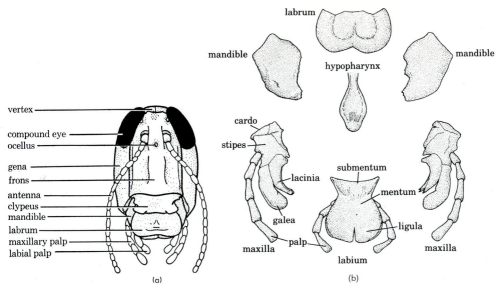

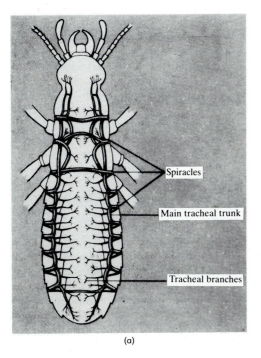

Fig. 1-70 Head region of a grasshopper: (a) head region; (b) mouth parts. (a, reprinted with permission of the publisher from *College Zoology* by R. W. Hegner and K. A. Stiles, 7th ed., © The Macmillan Co., New York, 5059; b, reprinted and adapted with permission of

Macmillan Publishing Company from *The Invertebrates: Function and Form,* 2nd ed., by Irwin W. Sherman and Vilia G. Sherman. Copyright © 1976 by Irwin W. Sherman and Vilia G. Sherman.)

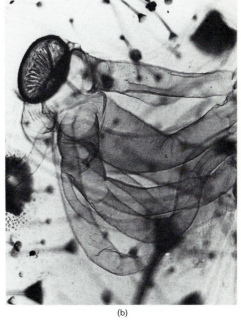

Fig. 1-71 Tracheal system of an insect: (a) air-filled tracheal tubes in relation to spiracles; (b) highly magnified view of a spiracle and attached tracheal tubes in a butterfly (× 115). (a, adapted

from *Biology* by Grover C. Stephens and Barbara Best North. Copyright © 1974 by John Wiley & Sons, Inc. Reprinted by permission of John Wiley & Sons, Inc.; b, photo by R. H. Noailles)

Cephalopoda) have a well-developed nervous system and eyes, and eight or ten tentacles. Squids may be purchased from fish markets for dissection, or from biological supply houses. Chitons are in Class Amphineura.

Compare the levels of specialization among the classes of mollusks (Fig. 1-79).

Helix

The edible land snail, *Helix*, is a nocturnal animal that feeds on fruits and leaves. It is usually available from fish markets or outdoor gardens. Use bone forceps to cut around the spirals of the shell and pull out the animal. Refer to Figure 1-80 to identify the position of the organs.

Identify the ovo-testis in this hermaphrodite. Cut into the mantle cavity through the roof behind the collar. Now locate the kidney and the heart. Cut into the pericardium and identify the thin auricle and the more muscular ventricle. Remove the shell of a living specimen and examine the beating heart, within a pericardium, that lies near the largest coil of the shell. *Helix* has a well-developed nervous system of ganglia and nerves; it is often used in studies of behavior (p. 340).

Statocysts in the foot have been shown to affect its geotaxis response.

Helix may be killed in an extended form by placing it in a corked bottle full of water for 24 hr, which causes asphyxiation.

Physa

Obtain a freshwater snail from an aquarium and compare the anatomy with that of a land snail (*Helix*). Use Figure 1-81 as a guide. Secure a snail in a watch glass with a bit of wax so that the peristome, the opening of the shell, is uppermost; add enough water to just cover the snail. Examine the shape of the foot and the rhythmic waves of motion under a dissecting microscope. Or place a snail on a glass slide where it will become attached and invert the slide over a watch glass to examine under the microscope. Look for cilia as the shape and motion of the foot are examined.

Examine the ciliated epithelial cells lining the intestine of *Physa*. First narcotize a snail in a warm solution of 1 percent magnesium sulfate. Mount the body in pond water on a glass slide. Dissect out the intestine and, with forceps, peel off the

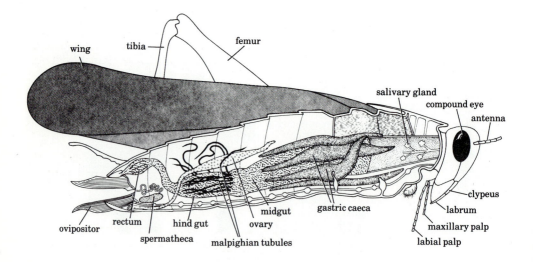

Fig. 1-72 *Romalea,* internal anatomy of a female. (Reprinted and adapted with permission of Macmillan Publishing Company from *The Invertebrates: Function and Form*, 2nd ed., by Irwin W. Sherman and Vilia G. Sherman. Copyright © 1976 by Irwin W. Sherman and Vilia G. Sherman.)

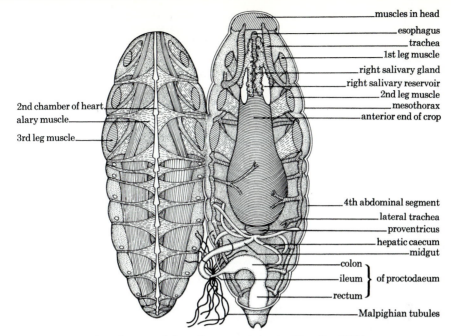

muscles in head
esophagus
trachea
1st leg muscle
right salivary gland
right salivary reservoir
2nd leg muscle
mesothorax
anterior end of crop

2nd chamber of heart
alary muscle
3rd leg muscle

4th abdominal segment
lateral trachea
proventricus
hepatic caecum
midgut
colon
ileum } of proctodaeum
rectum
Malpighian tubules

Fig. 1-73 *Blatta*, a cockroach: general dissection.
(After W. S. Bullough, *Practical Invertebrate Anatomy*, 2nd
ed., London, Macmillan, 1958.)

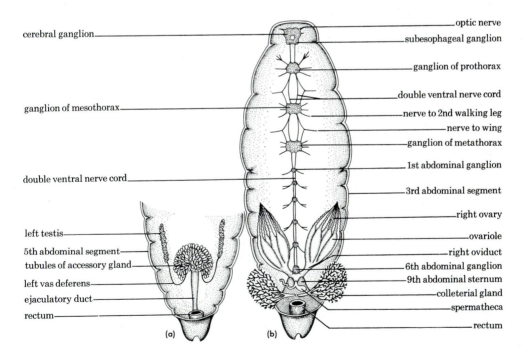

cerebral ganglion

optic nerve
subesophageal ganglion
ganglion of prothorax
double ventral nerve cord
nerve to 2nd walking leg
nerve to wing
ganglion of metathorax
1st abdominal ganglion
3rd abdominal segment
right ovary
ovariole
right oviduct
6th abdominal ganglion
9th abdominal sternum
colleterial gland
spermatheca
rectum

ganglion of mesothorax

double ventral nerve cord

left testis
5th abdominal segment
tubules of accessory gland
left vas deferens
ejaculatory duct
rectum

(a)   (b)

Fig. 1-74 *Blatta*: (a) dorsal view of male
reproductive system; (b) dorsal view of nervous
system and reproductive system of female. (After W.
S. Bullough, *Practical Invertebrate Anatomy*, 2nd ed.,
London, Macmillan, 1958.)

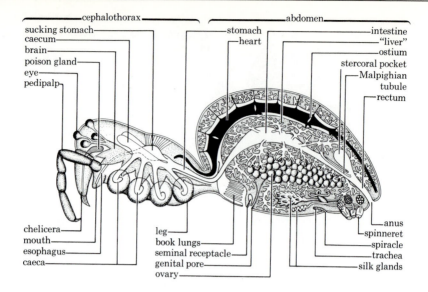

Fig. 1-75 Spider: internal anatomy. (After R. W. Hegner and K. A. Stiles, *College Zoology*, 7th ed., New York, Macmillan, 1959.)

lining. Mount in a drop of pond water on a clean slide; stain if necessary (see stains, p. 136). *Physa* and other aquarium snails are useful in studies of embryology (p. 407). Land snails are excellent for showing negative geotropism (p. 340).

Busycon

Marine gastropods, such as whelks, *Buccinum*, or *Busycon*, are available from supply houses and fish markets, and can be collected in many areas. Use Figure 1-82 as a guide in identification of organ systems. Remove the shell with bone forceps. Cut the specimen a little to the right of the middorsal area along the ciliated gill to the pericardium. Whelks have a single row of ciliated gills (ctenidia) on the left side. If living specimens are available, add a small amount of carmine powder to the water containing the whelk. Trace the current as water and red particles enter the siphon, move to the posterior end, turn toward the head end, and then leave in the region above the tentacles.

Haliotis

The gastropod *Haliotis* (the abalone) is widely distributed, feeding at night on seaweeds. The sexes are separate, and gametes are shed through pores in the shell. Break the attachment of the visceral hump to the shell and loosen the animal from the shell. Use Figure 1-83 to trace the general location of the organs. Cut into the mantle and locate the heart in the pericardium. Close to the heart locate the two kidneys. Also identify the two small auricles and a larger ventricle through which the rectum passes. There are large spaces, or lacunae, through which blood passes rather than capillaries or veins. Dissect away the heart and kidneys to locate and trace the alimentary canal.

Pelecypods

Anodonta

The dissection of this freshwater clam can be applied to other bivalves such as mussels and oysters.

Pry the two valves apart with a scalpel, or hold in one hand and tap one valve hard enough to crack. Insert a scalpel or knife blade between the mantle and valve and separate the edge of the mantle and both adductor muscles from the shell; these

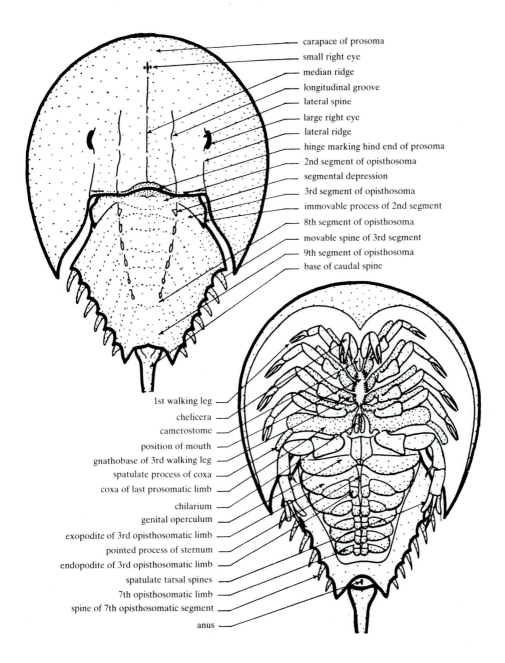

carapace of prosoma
small right eye
median ridge
longitudinal groove
lateral spine
large right eye
lateral ridge
hinge marking hind end of prosoma
2nd segment of opisthosoma
segmental depression
3rd segment of opisthosoma
immovable process of 2nd segment
8th segment of opisthosoma
movable spine of 3rd segment
9th segment of opisthosoma
base of caudal spine

1st walking leg
chelicera
camerostome
position of mouth
gnathobase of 3rd walking leg
spatulate process of coxa
coxa of last prosomatic limb
chilarium
genital operculum
exopodite of 3rd opisthosomatic limb
pointed process of sternum
endopodite of 3rd opisthosomatic limb
spatulate tarsal spines
7th opisthosomatic limb
spine of 7th opisthosomatic segment
anus

Fig. 1-76 *Limulus,* the horseshoe crab: dorsal and ventral views of female. (Reprinted and adapted with permission from Macmillan Publishing Company from *Practical Invertebrate Anatomy,* 2nd ed., by W. S. Bullough. Copyright © 1958 by W. S. Bullough. Reprinted 1960.)

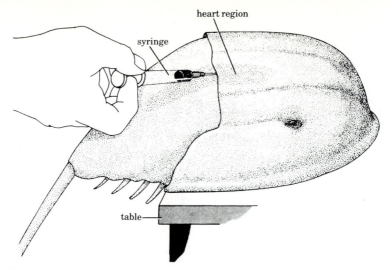

Fig. 1-77 *Limulus*: method for obtaining blood. (Reprinted and adapted with permission of Macmillan Publishing Company from *The Invertebrates: Function and Form*, 2nd ed., by Irwin W. Sherman and Vilia G. Sherman. Copyright © 1976 by Irwin W. Sherman and Vilia G. Sherman.)

muscles keep the shells closed. Cut through the large posterior adductor muscle below and behind the hinge, then cut the anterior adductor muscle. Examine the mantle, a white membrane which lines each shell. Note the posterior end, where the edges of the mantle meet to form the pigmented siphons. Also notice the papillae on the ventral siphon (through which water and food enter) and the smooth edges of the dorsal, exhalant siphon.

If a live clam or mussel is available, place it in a finger bowl of sea water. Add a small amount of carmine powder near the siphons that protrude from the posterior end of the animal. Distinguish the incurrent and excurrent siphons. Then turn back the mantle lobe and, if the specimen is freshly killed, the shimmering motion of cilia on the gills may be observed. Now add a drop of carmine suspension to the gills. Note the currents that sweep the particles, trapping them in mucus and carrying them in a ciliary stream to the mouth.

Scrape some cells from the edge of the mantle and bits of gill. Look for beats of ciliated cells on the edge of the mantle and gills under the high power of the microscope. Add Ringer's solution or the juice of

clam, then a bit of carmine powder to a piece of the living gill tissue on the slide. Note the movement of the carmine particles along the surface of the gills. Notice the cilia that create a current bringing water through the mantle cavity and incurrent siphons. Move the slide to locate gill pores into the water tubes of the gills. Notice the movement of fluid in the parallel internal water tubes of each gill. Water passes through gill pores into water tubes of the gills where oxygen and carbon dioxide are exchanged. (You may want to study the effects of different ions on ciliary motion, page 139.)

Notice in the specimen that the gills on each side unite, forming a channel above the gills that leads to the dorsal siphon. The lower inhalant siphon leads to the lower body cavity.

Now firmly grip the mantle lobe and pull the body away from the dorsal margin. Under the hinge, locate the heart. Carefully cut into this pericardial cavity, and find the yellowish ventricle which may still be pulsating.[40]

---

[40] For details and diagram of lamellibranch circulation, refer to Hickman, *Biology of Invertebrates*.

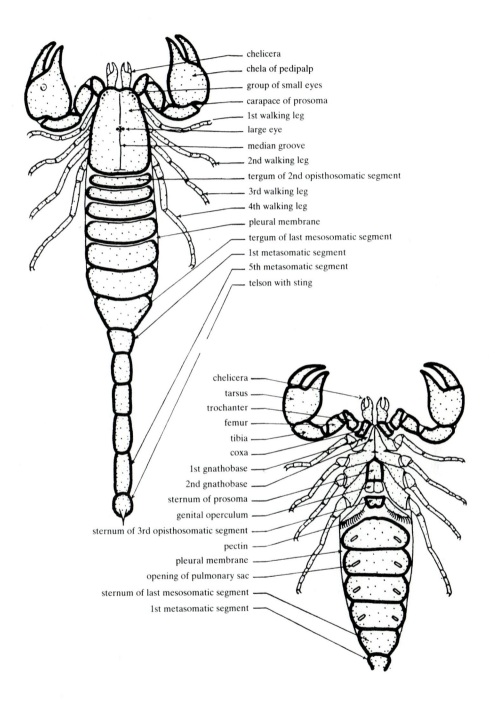

chelicera
chela of pedipalp
group of small eyes
carapace of prosoma
1st walking leg
large eye
median groove
2nd walking leg
tergum of 2nd opisthosomatic segment
3rd walking leg
4th walking leg
pleural membrane
tergum of last mesosomatic segment
1st metasomatic segment
5th metasomatic segment
telson with sting

chelicera
tarsus
trochanter
femur
tibia
coxa
1st gnathobase
2nd gnathobase
sternum of prosoma
genital operculum
sternum of 3rd opisthosomatic segment
pectin
pleural membrane
opening of pulmonary sac
sternum of last mesosomatic segment
1st metasomatic segment

Fig. 1-78 Scorpion, generalized: dorsal and ventral views. (Reprinted and adapted with permission from Macmillan Publishing Company from *Practical*

Invertebrate Anatomy, 2nd ed., by W. S. Bullough. Copyright © 1958 by W. S. Bullough. Reprinted 1960.)

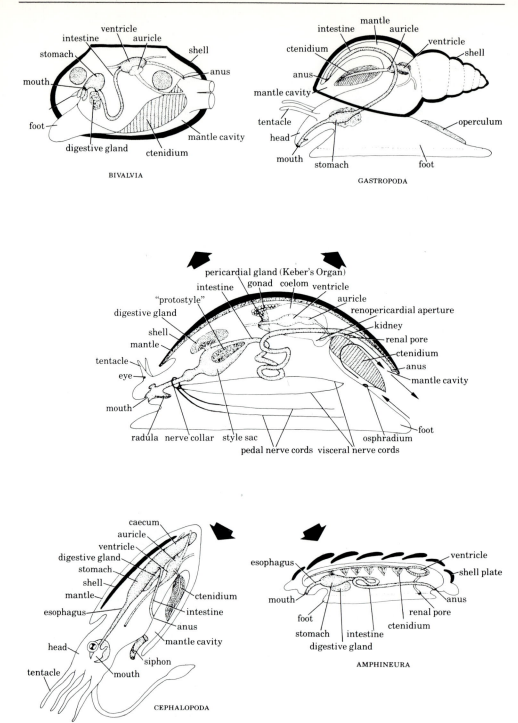

Fig. 1-79 A generalized mollusk pattern, center, with arrows pointing to specialized classes of mollusks. (Reprinted and adapted with permission of Macmillan Publishing Company from *The Invertebrates:* *Function and Form*, 2nd ed., by Irwin W. Sherman and Vilia G. Sherman. Copyright © 1976 by Irwin W. Sherman and Vilia G. Sherman.)

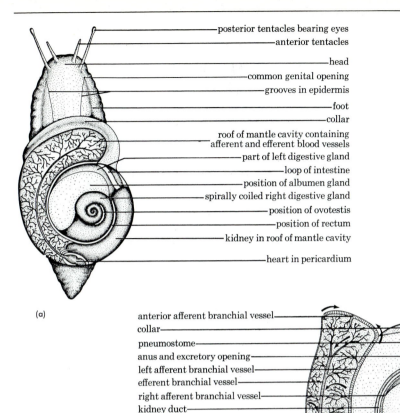

posterior tentacles bearing eyes
anterior tentacles

head
common genital opening
grooves in epidermis
foot
collar
roof of mantle cavity containing
afferent and efferent blood vessels
part of left digestive gland
loop of intestine
position of albumen gland
spirally coiled right digestive gland
position of ovotestis
position of rectum
kidney in roof of mantle cavity

heart in pericardium

(a)

anterior afferent branchial vessel
collar
pneumostome
anus and excretory opening
left afferent branchial vessel
efferent branchial vessel
right afferent branchial vessel
kidney duct
rectum

floor of mantle cavity
kidney
efferent renal vessel
branches of right afferent branchial vessel
pericardium

auricle
ventricle
aorta
afferent renal vessels

(b)

Fig. 1-80 *Helix*, a land snail: (a) dorsal view, with shell removed to show the visceral hump and the position of the organs within; (b) direction of flow of blood through the body (mantle cavity exposed). (After W. S. Bullough, *Practical Invertebrate Anatomy*, 2nd ed., London, Macmillan, 1958. Reprinted 1960.)

Locate the dark kidneys posterior to the heart, and in front of the posterior adductor muscle. The paired Keber organs are also part of the excretory system. To trace the entire digestive system, pull down the anterior part of the foot, and locate the mouth surrounded by two pairs of ciliated labial palps. In this region, just back of the anterior adductor muscle, in front of and above the foot, food strained by cilia on the gills is collected and carried to the esophagus and stomach. Behind the anterior adductor muscle, locate the darkly colored digestive gland that surrounds the stomach. From here, trace the intestine through the heart and toward the excurrent siphon (Fig. 1-83b).

Carefully pare away the muscle of the foot to expose the ovary or testis in the posterior dorsal part. In a mature female,

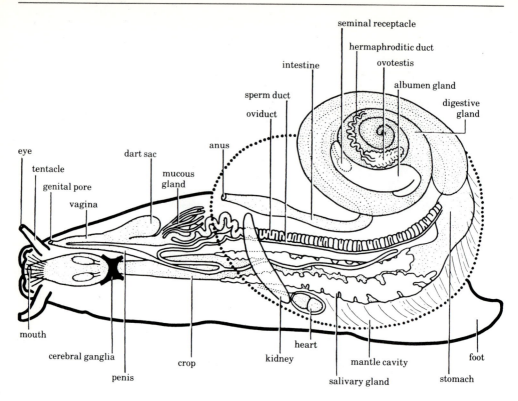

Fig. 1-81 General pattern of anatomy of a snail. (Reprinted and adapted with permission of Macmillan Publishing Company from *College Zoology*, 7th ed., by Robert W. Hegner and Karl A. Stiles. Copyright © 1959 by Macmillan Publishing Co., Inc.)

small brown glochidia larvae may be found in the lamellae of the gills. Examine these under the microscope. Many freshwater clams have this interesting parasitic stage, a glochidium, which is a modified veliger larva (Fig. 1-84). Following their release, these larvae attach themselves by fastening the valves of their shells onto the fins and gills of fish; the skin of the fish grows around each glochidium. After three to four weeks the young clam is freed and drops off to become a free-living clam. In this life cycle, freshwater clams become widely distributed as they travel distances carried by the fish host. Many marine mollusks pass through a trochophore larva stage (similar to polychaete annelids). As larvae develop, the band of cilia thickens into a rim, or velum, around the dorsal tip. This second stage of the larvae comprise the characteristic veliger larvae.

Mercenaria

Hard shell clams may be more readily available in some regions. Use the illustrations and the previous description of the dissection of *Anodonta* to study the anatomy of *Mercenaria*.

Pecten

Living specimens of scallops are available from supply houses or may be collected in marine areas. Scallops are especially well-adapted for swimming; observe their fast, clicking movement through the water. Note that one side of the bivalve is more convex than the other; also study the ridges and the wide hinge. Use Figure 1-85 to identify the main organ systems of a preserved or freshly killed specimen. Cut the adductor muscle with a knife and push the convex shell back in its hinge, then

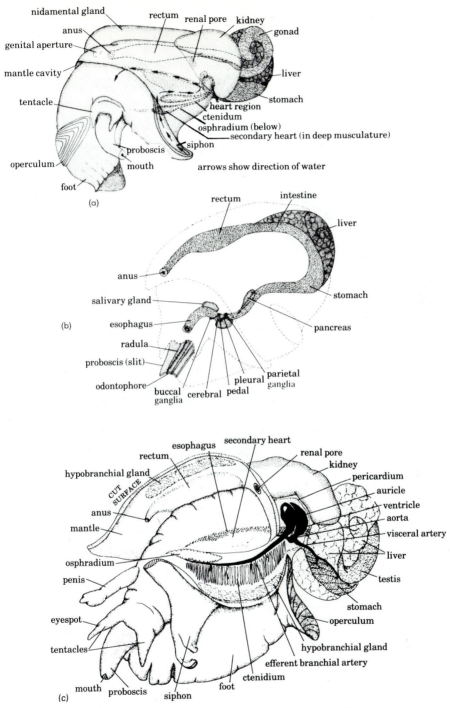

nidamental gland

anus

genital aperture

mantle cavity

tentacle

operculum

foot

rectum

renal pore

kidney

gonad

liver

stomach

heart region

ctenidum

osphradium (below)

secondary heart (in deep musculature)

siphon

proboscis

mouth

arrows show direction of water

(a)

rectum

intestine

liver

anus

salivary gland

esophagus

stomach

radula

pancreas

proboscis (slit)

odontophore

pleural

parietal
ganglia

buccal
ganglia

cerebral

pedal

(b)

esophagus

secondary heart

rectum

renal pore

kidney

hypobranchial gland

pericardium

CUT
SURFACE

auricle

anus

ventricle

mantle

aorta

visceral artery

osphradium

liver

penis

testis

eyespot

stomach

operculum

tentacles

mouth

proboscis

siphon

foot

ctenidium

efferent branchial artery

hypobranchial gland

(c)

Fig. 1-82 *Busycon*, a whelk: (a) external anatomy of female (shell removed); (b) diagrammatic view of digestive system; (c) internal anatomy of male (mantle cut back to show inside of mantle cavity). (Reprinted and adapted with permission of Macmillan Publishing Company from *The Invertebrates: Function and Form*, 2nd ed., by Irwin W. Sherman and Vilia G. Sherman. Copyright © 1976 by Irwin W. Sherman and Vilia G. Sherman.)

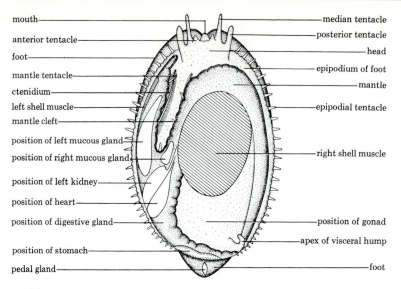

mouth

median tentacle

anterior tentacle

posterior tentacle

foot

head

epipodium of foot

mantle tentacle

mantle

ctenidium

left shell muscle

epipodial tentacle

mantle cleft

position of left mucous gland

position of right mucous gland

position of left kidney

position of heart

right shell muscle

position of digestive gland

position of gonad

apex of visceral hump

position of stomach

pedal gland

foot

(a)

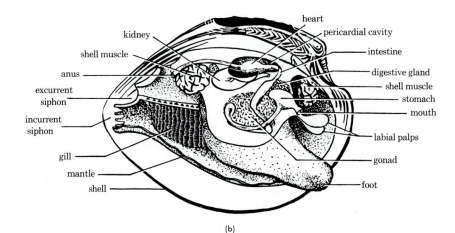

heart

kidney

pericardial cavity

shell muscle

intestine

anus

digestive gland

excurrent
siphon

shell muscle

stomach

incurrent
siphon

mouth

labial palps

gill

gonad

mantle

foot

shell

(b)

Fig. 1-83 *Haliotis*, the abalone (gastropod): (a) dorsal view, with shell removed to show position of organs; (b) *Anodonta*, anatomy of a freshwater clam. (a, after W. S. Bullough, *Practical Invertebrate Anatomy*, 2nd ed., London, Macmillan, 1958. Reprinted 1960; b, from W. G. Whaley, O. P. Breland, et al, *Principles of Biology*, Harper & Row, 1954; after R. W. Hegner, *Invertebrate Zoology*, Macmillan, New York, 1933.)

remove it. Note the highly pigmented edge of the mantle with its many "blue eyes" and tentacles. Remove the mantle in order to locate the organ systems. Directions for careful dissections are given in laboratory manuals and texts in invertebrate zoology.[41] You may also want to compare the eye of *Pecten* with that of a squid (Fig. 1-87).

Elephant's tusk shells (*Dentalium*) also may be dissected using Hickman's text as a guide.[42]

---

[41] Sherman and Sherman, *Invertebrates*.

[42] Hickman, *Biology of Invertebrates*.

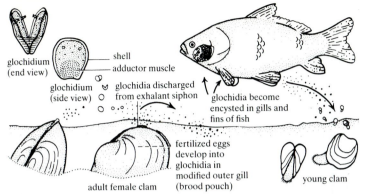

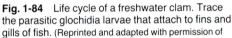

Fig. 1-84 Life cycle of a freshwater clam. Trace the parasitic glochidia larvae that attach to fins and gills of fish. (Reprinted and adapted with permission of Macmillan Publishing Company from *College Zoology*, 10th ed., by Richard A. Boolootian and Karl A. Stiles. Copyright © 1981 by Macmillan Publishing Co., Inc.)

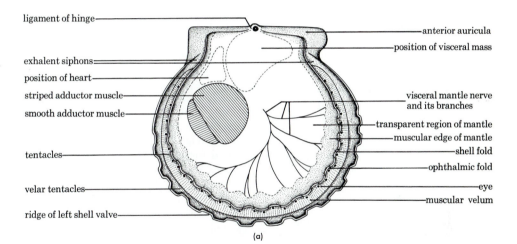

(a)

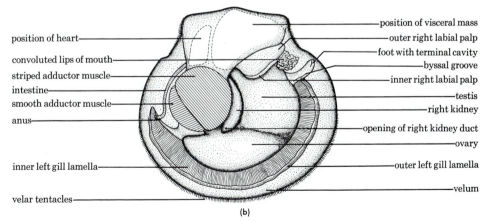

(b)

Fig. 1-85 *Pecten*, a scallop: (a) right shell valve has been removed to show the right mantle; (b) internal anatomy, after removal of right mantle lobe. (After W. S. Bullough, *Practical Invertebrate Anatomy*, 2nd ed., London, Macmillan, 1958. Reprinted 1960.)

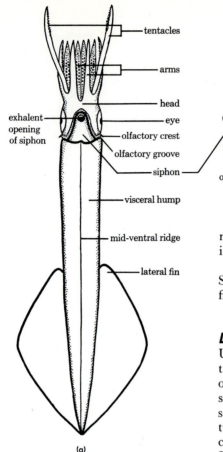
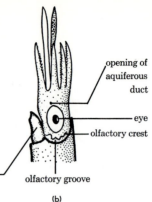

Fig. 1-86 *Loligo*, the squid: (a) ventral view, showing external anatomy; (b) lateral view of head region. (From W. S. Bullough, *Practical Invertebrate Anatomy*, 2nd ed., London, Macmillan, 1958, reprinted 1960.)

Cephalopods

Members of the Class Cephalopoda include *Sepia* (cuttlefish), *Loligo* (squid), octopus, and *Nautilus* (pearly nautilis). Squid and cuttlefish have ten arms and a reduced shell (decapods) while the octopus has eight arms and no shell. These forms have a well-developed head region with prominent eyes. There is a well-developed nervous system. In fact, the giant axons of the squid have been used in studies of conduction of nerve impulses. Note the color change in living specimens due to responses of melanophores to changes in light.

As a group, cephalopods are predatory. Sexes are separate, and the fertilized yolk-filled eggs hatch without a larval stage.

Loligo

Unlike most mollusks, squids have retained the bilateral symmetry of some other phyla. The "foot" of this mollusk has shifted forward to become ten arms with suckers and a funnel (Fig. 1-86). There are two horny beaks for tearing the prey caught by the long tentacles with suckers. Jets of water from the mantle cavity are ejected from the funnel using the jet propulsion principle, that is, water spurted forward causes the squid to be propelled backward, and vice versa. Note the two large eyes; dissect the eye of the squid and compare it with a pecten's eye (Fig. 1-87), as well as a mammalian eye as examples of convergent evolution.

After a study of the external anatomy, cut to one side of the mid-ventral line (use Fig. 1-88a as a guide for dissection).[43]

Echinoderms

This phylum comprises marine forms that have spiny skins and radial symmetry,

[43] For excellent coverage of the life cycle and embryology of cephalopods, refer to Hickman, *Biology of Invertebrates.*

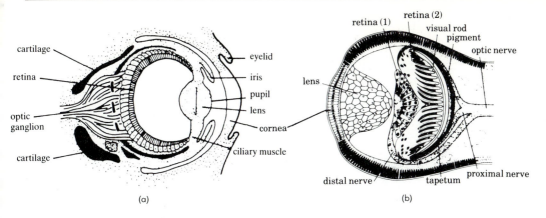

Fig. 1-87 Diagrammatic views of (a) eye of squid and (b) eye of scallop. (a, reprinted and adapted with permission of Macmillan Publishing Company from *College Zoology*, 10th ed., by Richard A. Boolootian and Karl A. Stiles. Copyright © 1981 by Macmillan Publishing Co., Inc.; b, reprinted and adapted with permission of Macmillan Publishing Company from *The Invertebrates: Function and Form*, 2nd ed., by Irwin W. Sherman and Vilia G. Sherman. Copyright © 1976 by Irwin W. Sherman and Vilia G. Sherman.)

although the larvae are bilateral. They move by means of tube feet and have an intricate water-vascular system. There are no special respiratory or excretory organs; a complex nervous system is present although there is no cephalization. The sexes are separate, and the swimming bipinnaria larvae are of the type found in starfishes (Fig. 1-89). The eggs of echinoderms have been well-studied in experimental embryology. The phylum has many fossil remains; they are an important link as the invertebrate group closely related to vertebrates. Examples of echinoderms are the attached sea lilies (Crinoidea), starfish or sea stars (Class Asteroidea), serpent stars (Ophiuroidea), sea cucumbers, (Holothuroidea), sand dollars and sea urchins (Echinoidea).

Asterias

A live starfish provides the opportunity to observe its tube feet in action, one of the unique features of echinoderms. Observe a live starfish in a marine aquarium as it "walks" and "climbs" a vertical wall using its tube feet with suckers for attachment. Place one in a shallow glass container and view its surface under a dissecting microscope or with a hand lens. Notice the spines, and at their base, the clawlike,

stalked pedicellariae that remove debris. Stroke the top, or aboral, surface with a camel's hair brush, then pull the brush away quickly. Examine the bristles for attached pinching pedicellariae. Mount some in a drop of glycerine on a slide, or use a hand lens. Notice the jaws and muscle attachments.

In a living specimen, observe the action of cilia on the outer surface of the dermal branchiae, or gills, which are extensions of the coelom and lie between the skeletal plates. Add a bit of carmine powder or India ink near these respiratory organs and watch the cilia beat toward the tips of the branchiae.[44] Cilia are also located inside the dermal branchiae; observe their action by injecting a suspension of carmine powder into the coelom of a living specimen. Count the number of dermal branchiae over an area of several square centimeters, and estimate their density on the surface of the animal.

Locate the pigmented eyespot on the tips of each arm. Study the "map" of a starfish (Fig. 1-90). Locate the madreporite, a part of the water-vascular system; note that it lies closer to two arms. Moving

[44] W. Curtis, and M. Guthrie, *Laboratory Directions in General Zoology*, 4th ed. (New York: Wiley, 1948).

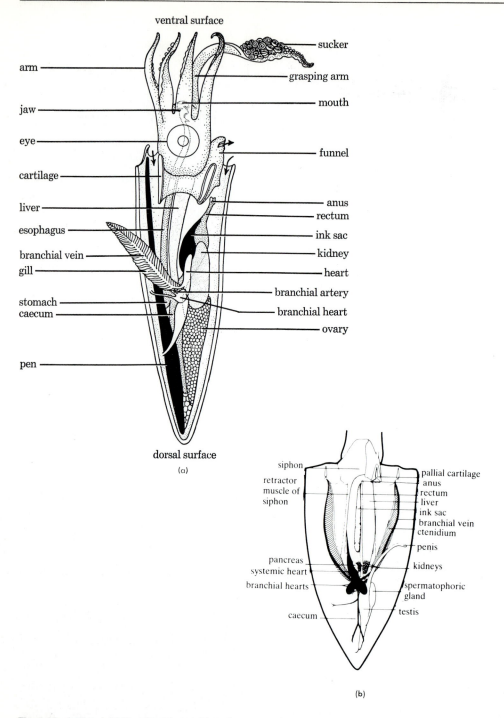

ventral surface

sucker

arm

grasping arm

jaw

mouth

eye

cartilage

funnel

liver

anus

esophagus

rectum

branchial vein

ink sac

gill

kidney

stomach

heart

caecum

branchial artery

branchial heart

ovary

pen

dorsal surface

(a)

siphon

pallial cartilage

retractor
muscle of
siphon

anus
rectum
liver
ink sac
branchial vein
ctenidium

penis

pancreas

systemic heart

kidneys

branchial hearts

spermatophoric
gland

caecum

testis

(b)

Fig. 1-88 *Loligo:* (a) internal anatomy of female with body wall and arms removed from right side; (b) internal anatomy of male, ventral view. (a, reprinted with permission of the publisher from *College Zoology* by R. W. Hegner and K. A. Stiles, 7th ed.,

© The Macmillan Co., New York, 1959; b, adapted with permission from *An Illustrated Laboratory Text in Zoology*, 2nd ed., by Richard A. Boolootian and Donald Heyneman. Copyright © 1962, 1969 by Holt, Rinehart and Winston, Inc.)

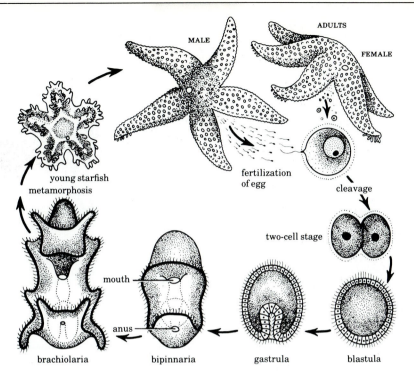

ADULTS

MALE

FEMALE

fertilization
of egg

cleavage

two-cell stage

young starfish
metamorphosis

mouth

anus

brachiolaria   bipinnaria   gastrula   blastula

Fig. 1-89 Life cycle of a starfish showing bilateral symmetry in later larval stages and radial symmetry in the young starfish. (Reprinted and adapted with permission of Macmillan Publishing Company from *College Zoology*, 10th ed., by Richard A. Boolootian and Karl A. Stiles. Copyright © 1981 by Macmillan Publishing Co., Inc.)

clockwise from the madreporite, locate the anus between the next arm. For convenience sake, these two arms comprise the bivium. The remaining three arms, the trivium, can be dissected to examine internal organs without disrupting the attachment of the madreporite to the rest of the water-vascular system.

Turn the starfish to its ventral side, the oral side, where the mouth is located. Notice the arrangement of spines around the mouth, and the ambulacral grooves with tube feet that end in suckers. Also note that the tip of each arm has a tube foot, a tentacle, without a suction disc. The eyespot is at the base of this tentacle.

Live starfish may be narcotized and killed in a small amount of water to which some crystals of magnesium sulfate (Epsom salts) have been added. When placed upside down, the tube feet usually

remain extended within the ambulacral grooves. When they no longer respond to touch, the starfish may be stored in 70 percent alcohol. In preserved specimens, part of the stomach may protrude from the mouth.

To examine the internal anatomy of the starfish, use scissors to cut across an arm of the trivium and along the sides of each arm. Remove the aboral skeletal plates from the top of the arm, but leave the central disc intact. Dissect in a pan of water. Locate the conspicuous digestive glands, the paired hepatic caeca (Fig. 1-91). When these are lifted or removed, look for the paired gonads, reddish ovaries, or testes that may be yellowish to white. Observe the ball-like ends of the tube feet called ampullae.

Cut the specimen in half and locate the mouth and baglike stomach. Carefully

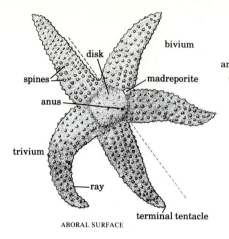

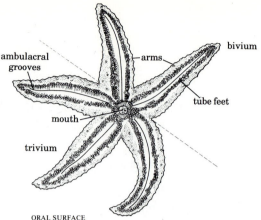

ABORAL SURFACE

ORAL SURFACE

Fig. 1-90 *Asterias*, a starfish. External views: dorsal or aboral surface (left); ventral or oral surface (right). (Reprinted and adapted with permission of Macmillan Publishing Company from *The Invertebrates:*

Function and Form, 2nd ed., by Irwin W. Sherman and Vilia G. Sherman. Copyright © 1976 by Irwin W. Sherman and Vilia G. Sherman.)

separate tissues from the central disk; parts of the digestive system are attached. Ducts leading from the pyloric part of the stomach branch into pyloric caeca containing digestive enzymes. Remove the pyloric caeca from one arm to reveal the two retractor muscles connected to the cardiac regions of the stomach; they pull in the stomach when it has been everted.

Under these muscles, locate the tubes of the water-vascular system (Fig. 1-92). Below the madreporite find the calcareous stone canal. Water enters the madreporite, then moves along the calcareous stone canal into the ring canal around the mouth to the ampullae, and then to the tube feet. Tiedemann's bodies are believed to produce the amoeboid corpuscles found in the water-vascular system.

Blood vessels and the nervous system are more difficult to trace, although you may find a nerve ring around the mouth.

Arbacia

The purple sea urchin, *Arbacia*, may be available for study in eastern regions; the purple *Strongylocentrotus* is more readily found on the West Coast. The five-part radial symmetry is especially apparent in forms from which the spines have been

removed. If the five rays of a starfish were turned backward and upward, the globular shape and symmetry of a sea urchin would result.

Look for the tube feet among the long spines; these tube feet are moved by muscles attached to the ambulacral plates. On the oral surface, examine the jaws composed of five long toothlike radii that form a structure known as Aristotle's lantern (Fig. 1-93).

Dissect a specimen by carefully cutting a circle around the globe; remove the upper half. The main systems are similar to those of the starfish. Figure 1-94 may serve as an introductory guide in locating the main organs. If available, compare the anatomy of a sand dollar with that of a sea urchin.

Refer to the class activity on fertilization and early cleavage in sea urchin eggs (p. 407).

Cucumaria

Sea cucumbers, members of the Class Holothuroidea, have unusual powers of regeneration. They eviscerate through the cloaca when attacked, and the saclike body crawls off to regenerate internal

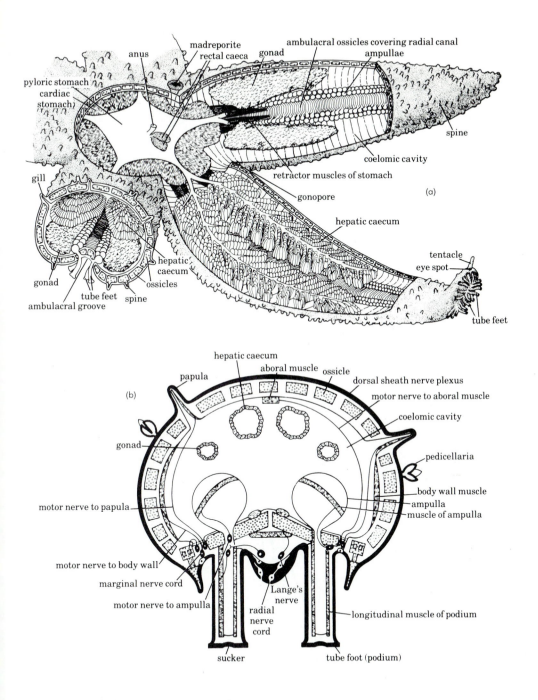

Fig. 1-91 *Asterias*: (a) internal view; top surface view shows one ray from which hepatic caeca have been removed to show gonads and ampullae; (b) diagrammatic cross section of a ray shows main nerves. (Reprinted and adapted with permission of Macmillan Publishing Company from *The Invertebrates: Function and Form*, 2nd ed., by Irwin W. Sherman and Vilia G. Sherman. Copyright © 1976 by Irwin W. Sherman and Vilia G. Sherman.)

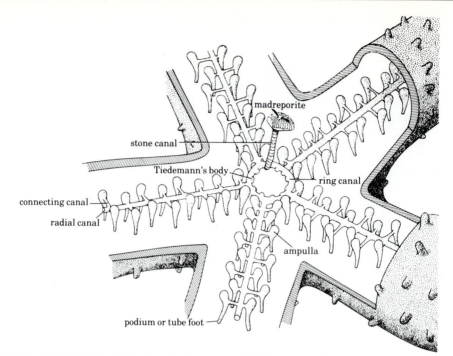

Fig. 1-92 *Asterias*: diagrammatic view of water vascular system. Trace the path of water from the madreporite to the ampullae and tube feet. (Reprinted and adapted with permission of Macmillan Publishing Company from *The Invertebrates: Function and Form*, 2nd ed., by Irwin W. Sherman and Vilia G. Sherman. Copyright © 1976 by Irwin W. Sherman and Vilia G. Sherman.)

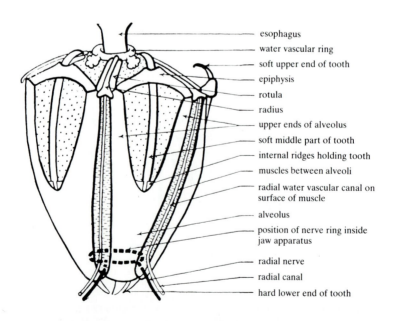

Fig. 1-93 Sea urchin: side view of chewing jaw (Aristotle's lantern). (Reprinted and adapted with permission from Macmillan Publishing Company from *Practical Invertebrate Anatomy*, 2nd ed., by W. S. Bullough. Copyright © 1958 W. S. Bullough. Reprinted 1960.)

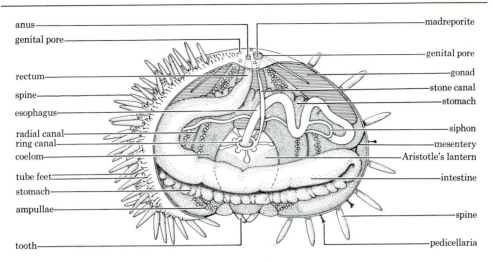

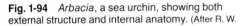

Fig. 1-94 *Arbacia*, a sea urchin, showing both external structure and internal anatomy. (After R. W. Hegner and K. A. Stiles, *College Zoology*, 7th ed., New York, Macmillan, 1959.)

organs from the oral region within three weeks.

Locate the mouth among the ten enlarged tube feet that form many-branched tentacles (Fig. 1-95). After studying its external appearance, lay a sea cucumber on its side and carefully cut along the length of the animal to expose the organs; pin back the body wall, and make a transverse cut through the body wall to flatten it. If you dissect in a pan of water, the branches of the two respiratory trees float freely. Distinguish the coils of the digestive tube from the tubes of the one ovary or testis suspended from the dorsal wall mesentery.

After removing the respiratory trees and cloaca, trace the water-vascular system using Figure 1-95 as a guide. Begin with the madreporite and water-vascular ring.

If living specimens are available, studies can be made of "breathing" (cloacal changes) and of hemocytes in the water-vascular system. Refer to Sherman and Sherman for detailed activities.[45]

Chordates

Members of Phylum Chordata are characterized by a dorsal nerve cord, a noto-

[45] Sherman and Sherman, *Invertebrates*.

chord, and pharyngeal gill slits at some time in their life cycle. Generally, three subphyla are recognized: Urochordata (tunicates), Cephalochordata (*Amphioxus*), and the typical Vertebrata. It is only the free-swimming larval stage of the tunicates that have the characteristic chordate structures. *Amphioxus*, only about 5 cm long, is often studied as a representative type of early chordate.

Amphioxus
Examine under low power prepared slides of a young lancet for a study of its total appearance: a fishlike shape with no specialized head or limbs. Find the notochord and locate the myotomes and gill slits. Reproductive organs are not apparent in these young forms that are mounted on slides.

If possible, also examine preserved mature specimens (Fig. 1-96).

Squalus
The basic chordate pattern is found in this cartilaginous fish, the dogfish shark, *Squalus*. Related forms are skates and rays. These vertebrates can be collected, or obtained from supply houses and some local fish markets.

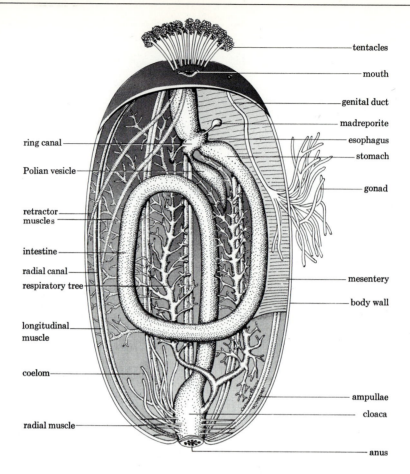

tentacles

mouth

genital duct

madreporite

esophagus

stomach

gonad

ring canal

Polian vesicle

retractor muscles

intestine

radial canal

respiratory tree

mesentery

body wall

longitudinal muscle

coelom

ampullae

cloaca

radial muscle

anus

Fig. 1-95 *Cucumaria*, a sea cucumber: internal anatomy with body wall divided along the middle of the dorsal surface. (After T. J. Parker and W. A. Haswell, *A Textbook of Zoology,* London, Macmillan, 1930, and R. W. Hegner and K. A. Stiles, *College Zoology,* 7th ed., New York, Macmillan, 1959.)

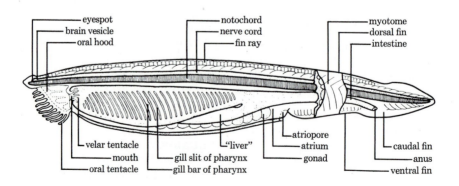

eyespot
brain vesicle
oral hood

notochord
nerve cord
fin ray

myotome
dorsal fin
intestine

velar tentacle
mouth
oral tentacle

gill slit of pharynx
gill bar of pharynx

"liver"

atriopore
atrium
gonad

caudal fin
anus
ventral fin

Fig. 1-96 *Amphioxus*, the lancet: internal anatomy. (Reprinted with permission of the publisher from *College Zoology* by R. W. Hegner and K. A. Stiles, 7th ed., © The Macmillan Co., 1959.)

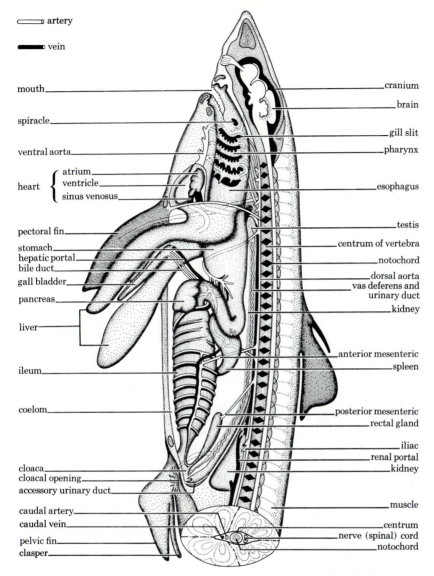

artery

vein

mouth

spiracle

ventral aorta

heart { atrium
        ventricle
        sinus venosus

pectoral fin

stomach
hepatic portal
bile duct
gall bladder

pancreas

liver

ileum

coelom

cloaca
cloacal opening
accessory urinary duct

caudal artery
caudal vein

pelvic fin
clasper

cranium
brain

gill slit
pharynx

esophagus

testis
centrum of vertebra
notochord

dorsal aorta
vas deferens and
urinary duct
kidney

anterior mesenteric
spleen

posterior mesenteric
rectal gland

iliac
renal portal
kidney

muscle

centrum
nerve (spinal) cord
notochord

Fig. 1-97 *Squalus*, the dogfish: internal anatomy. 7th ed., New York, Macmillan, 1959.)
(After R. W. Hegner and K. A. Stiles, *College Zoology,*

As a guide in dissection, use Figures 1-97, 1-98, and a textbook such as Boolootian and Stiles, *College Zoology.*

Fish

Hold the fish in one hand, begin the incision near the anal region, and cut forward for an inch or so. Then lay the fish in a waxed pan or on a corkboard, and cut forward along the midline. Be careful not to injure any organs. Find the heart in a pericardial sac beneath the pharynx (Fig. 1-99). Blood moves from the auricle, into the ventricle, into the ventral aorta, along afferent branchial arteries, and into the gills. After the blood is aerated, it moves into efferent branchial arteries into the dorsal aorta. It returns through the veins into the sinus venosus, which empties into the auricle. Students may be able to trace some of these main blood vessels.

Food enters the mouth, passes along a short esophagus into a stomach, and then past the pyloric valve into the intestine. Look for the liver and the gall bladder. At the anterior end of the intestine you may find a large, red spleen.

Students should find the large air bladder in the dorsal part of the abdomen. High up in the dorsal region of the ab-

dominal cavity look for the kidneys. Trace the ureters into a urinary bladder opening, and then into a urogenital orifice, posterior to the anus. Distinguish ovaries from testes and trace the reproductive tubes into the urogenital opening.

Examine the respiratory system next. Find the paired gills, which are supported by gill arches. Notice the branchial filaments (double rows) on each gill and their rich supply of capillaries. Also note the gill rakers, which act as sieves to hold back food particles.

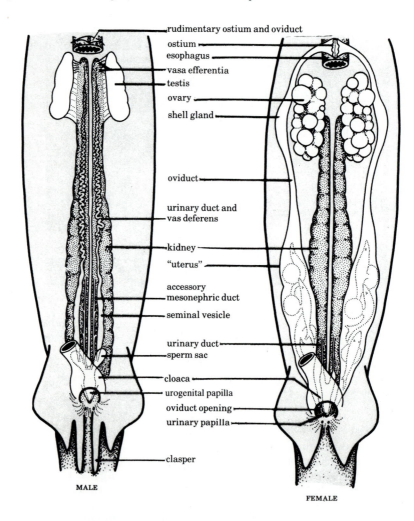

rudimentary ostium and oviduct
ostium
esophagus
vasa efferentia
testis
ovary
shell gland

oviduct

urinary duct and vas deferens

kidney
"uterus"

accessory mesonephric duct

seminal vesicle

urinary duct
sperm sac

cloaca
urogenital papilla
oviduct opening
urinary papilla

clasper

MALE

FEMALE

Fig. 1-98 *Squalus*: urogenital systems. (Reprinted and adapted with permission of Macmillan Publishing Company from *College Zoology*, 10th ed., by Richard A. Boolootian and Karl A. Stiles. Copyright © 1981 by Macmillan Publishing Co., Inc.)

A dissection of the eye of the fish will reveal a spherical lens.

Freeze-dried specimens of many organisms are available from Ward's.[46] Although expensive, they are odorless and serve as "real models" of anatomy and as useful year-round displays.

Rana

This description of the anatomy of the frog will hold for a study of frogs generally and for most toads and salamanders.

[46] Ward's Natural Science Establishment, PO Box 1712, Rochester N.Y. 14603; also PO Box 1749, Monterey, CA 93940.

Since you cannot see a beating heart or study the contrast of colors of the organs in preserved frogs, try to have at least one freshly killed (pithed, p. 348; or anesthetized, p. 142) frog to see the deeply colored auricles contract to pass blood to the ventricle, and the ventricle contract, changing from deep red to light pink as it sends blood into the dorsal aorta.

If living frogs are available, observe their breathing movements. Show heart beat (p. 265) and circulation (p. 270); and study fresh blood cells, ciliated epithelial cells (from the roof of the mouth), intestinal protozoa, and active sperm cells (see chapter 2; also page 272).

Examine the head and sense organs: ear drums, eyes, and nose. The transparent nictitating membrane attached to the inside of the lower eyelid can be raised over the eye. Cut through the angle of the jaws to identify internal mouth parts of a freshly killed frog (Fig. 1-100). Dissect out a bit of the ciliated epithelial tissue from the roof

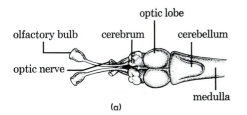

(a)

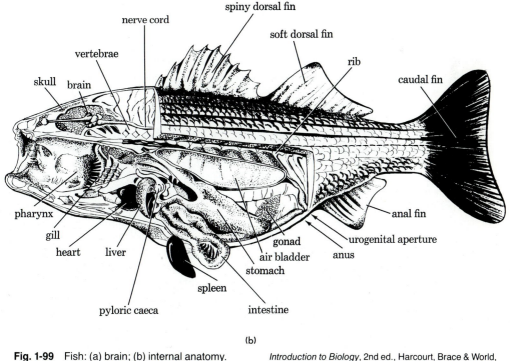

(b)

Fig. 1-99 Fish: (a) brain; (b) internal anatomy. (From G. G. Simpson and W. S. Beck, *Life: An Introduction to Biology*, 2nd ed., Harcourt, Brace & World, Inc., 1965.)

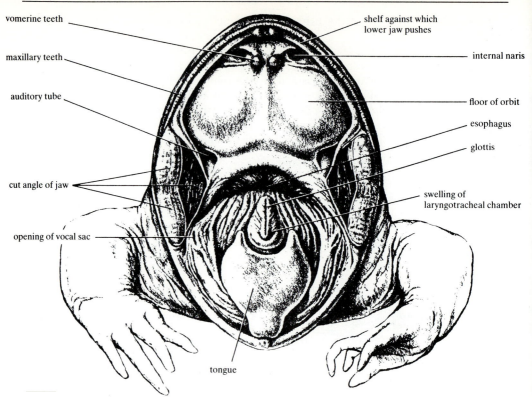

vomerine teeth

maxillary teeth

auditory tube

cut angle of jaw

opening of vocal sac

tongue

shelf against which lower jaw pushes

internal naris

floor of orbit

esophagus

glottis

swelling of laryngotracheal chamber

Fig. 1-100 Frog: internal mouth parts. (Adapted from *Laboratory Studies in Biology* by Warren Walker.

of the mouth and pharynx in a drop of frog Ringer's solution on a slide. Examine under low and high power.

Lay a pithed frog in a dissecting pan and with scissors cut off the lower jaw so that the roof of the mouth and pharynx are visible. With a toothpick, sprinkle a fine dust of carmine particles on the roof of the mouth. Then slit the esophagus down to the stomach to watch the passage of particles carried by ciliated epithelial cells. Can you time the rate of movement due to ciliary action? What happens if you wash the surface with frog Ringer's solution at 30° C (86° F) or a solution at 10° C (50° F)?

For dissection, lay a dead frog on its back in a waxed pan, and pin down the extended limbs. Use forceps to pick up the loose skin of the abdomen between the hind limbs; cut from there up to the lower jaw. Turn back the skin (notice its rich supply of blood vessels), and pin it down.

(There is some data that suggests about a third of the respiratory exchange in frogs occurs through the skin.) Now cut through the white abdominal muscles, taking care not to cut into the ventral abdominal vein or the organs beneath. At the pectoral region cut through the bony sternum. Use both hands to spread out the pectoral region. Cut away any connecting mesentery, and pin the muscles back to the pan. Make transverse incisions into each of the limbs so that the abdominal wall can be cut away or pinned down (Fig. 1-101).

Locate the heart in a transparent, tough pericardial sac (Fig. 1-102). Raise the apex of the heart and cut the thin thread of tissue that attaches the atria to the dorsal wall of the pericardium; locate the sinus venosus. Identify each organ system: respiratory, digestive, circulatory, excretory, and reproductive. Gently move the organs aside to observe their location in the body.

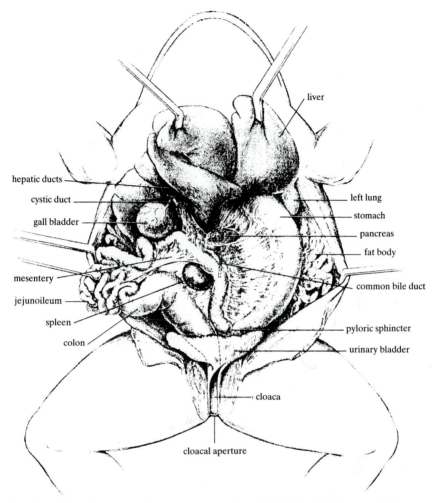

Fig. 1-101 Frog: internal anatomy. (Adapted from *Laboratory Studies in Biology* by Warren Walker.

Copyright © 1967 by W. H. Freeman, Pub.)

Note the three-lobed liver, a left lobe subdivided into two lobes, and a right lobe. Locate the large gallbladder in relation to the liver. Blow into the glottis with a fire-polished glass tube to inflate the lungs. Slit open one lung to observe the many folds that increase the inner surface. Use a blunt, flexible probe to push back into the mouth and locate the gullet. The pancreas, which looks like a yellow cord, is in the loop formed by the stomach and intestine. Open the J-shaped stomach to examine the contents, and also its lining.

If your specimen is a female distended with eggs, one ovary and an oviduct should be removed. Trace the other ovi-duct to locate the funnel (ostium at the base of the lungs) where eggs that break away from the ovary migrate (by action of the ciliated cells lining the body cavity) to enter the oviducts. Locate, at the other end of the oviducts, the opening to the exterior, via the cloaca (Fig. 1-103).

In the male, locate two oval, orange or yellow testes. Carefully examine the top surface of the mesentery which holds a testis to a kidney. Trace the fine threads of the sperm ducts, the vasa efferentia, that carry spermatozoa from the testes into the kidneys. From here, spermatozoa pass into the ureters, which also serve as vasa def-erentia. Locate the ureters leading from

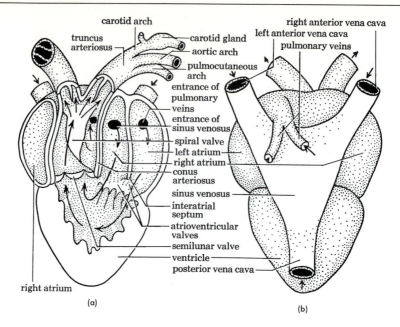

carotid arch

right anterior vena cava
left anterior vena cava
pulmonary veins

truncus
arteriosus

carotid gland
aortic arch
pulmocutaneous
arch
entrance of
pulmonary
veins
entrance of
sinus venosus
spiral valve
left atrium
right atrium
conus
arteriosus
sinus venosus
interatrial
septum
atrioventricular
valves
semilunar valve
ventricle
posterior vena cava

right atrium

(a)

(b)

Fig. 1-102   Frog: (a) ventral view of heart to show three chambers; (b) dorsal view; arrows trace the flow of blood. (Reprinted with permission of the publisher *from College Zoology* by R. W. Hegner and K. A. Stiles, 7th ed., © The Macmillan Co., 1959.)

the outer edge of the posterior part of each kidney into the cloaca where spermatozoa and urine empty. (More prominent along the side of each kidney is a Mullerian duct, a rudimentary, vestigial oviduct.)

Observe the yellow fat bodies, fingerlike masses connected to both the ovaries and testes; they are largest in summer. Find the red, spherical spleen. Insert a flexible probe through the anus to locate the urinary bladder, or inflate the bladder by blowing through a fire-polished glass tube.

In a preserved specimen, study the nervous system in fine detail (Fig. 1-104). Pare away the top of the skull with a scalpel until you find the two hemispheres of the cerebrum between the eyes. In front of the cerebrum you may be able to locate the two pear-shaped olfactory lobes that lead out into the nasal region. Posterior to the cerebrum the optic nerves may be found extending from the optic lobes. Then trace the cerebellum and extension of the medulla into the spinal cord. When you lift up the organs within the abdomen locate the ganglia and branching spinal

nerves extending from the spinal cord. Study the position of the spinal cord in relation to the spinal column. Follow the sciatic nerves into the muscles of the thighs.

Snake

If the spinal cord of a freshly killed snake is not destroyed, the snake must be fastened securely to a board. (It is best, of course, to destroy the cord.) Lay it on its back and tack down the upper jaw and the tail end. Carefully pick up the skin at the throat, and make an incision extending to the anus. Cut slowly so as not to pierce the distended air sac which fills most of the body. Pin back the skin every few centimeters. Cut through the membranes across the rib section, avoiding blood vessels (Fig. 1-105).

Use glass tubing to inflate the stomach, then the right lung. Notice the air sac extending from the lung. Look for the undeveloped left lung found in some snakes at the end of the trachea.

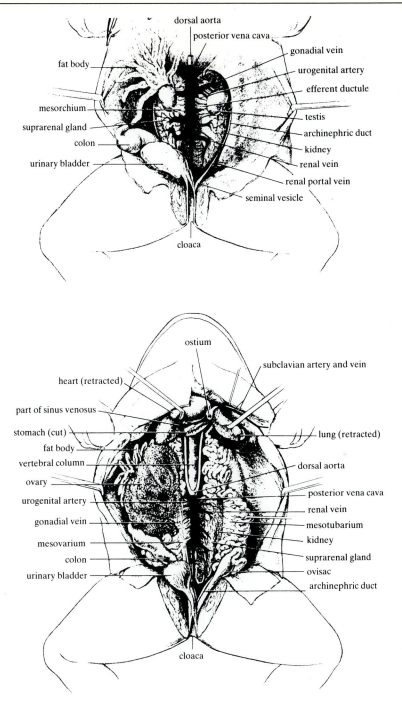

Fig. 1-103 Frog: reproductive systems, male and female. (Adapted from *Laboratory Studies in Biology* by Warren Walker. Copyright © 1967 by W. H. Freeman, Pub.)

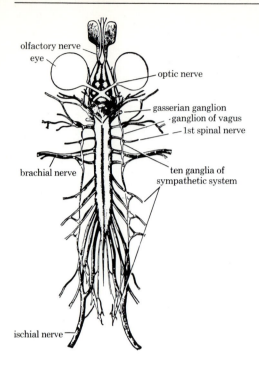

olfactory nerve
eye
optic nerve
gasserian ganglion
ganglion of vagus
1st spinal nerve
brachial nerve
ten ganglia of
sympathetic system
ischial nerve

Fig. 1-104   Frog: nervous system. (From R. W. Hegner, *College Zoology*, rev. ed., Macmillan, New York, 1926.)

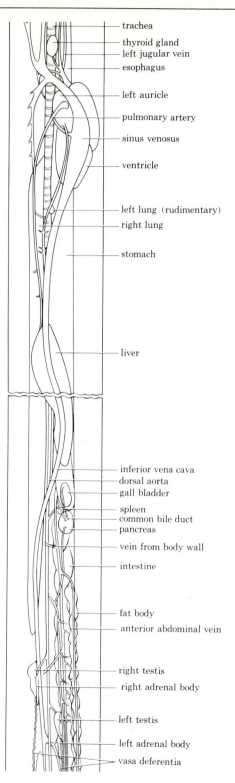

trachea
thyroid gland
left jugular vein
esophagus

left auricle

pulmonary artery

sinus venosus

ventricle

left lung (rudimentary)
right lung

stomach

liver

inferior vena cava
dorsal aorta
gall bladder

spleen
common bile duct
pancreas

vein from body wall

intestine

fat body

anterior abdominal vein

right testis

right adrenal body

left testis

left adrenal body

vasa deferentia

   The purpose of this rudimentary dissection is to make available an example of the way body form determines the disposition of the organs. Figure 1-105 is a guide for tracing organ systems.

Turtle

When preparing a fresh turtle for dissection, use a strong forceps to hold the head extended, and snip through the spinal cord at the back of the neck with strong clippers or shears. Then, with a hacksaw or coping saw, cut through the bridge which connects the carapace (upper shell) and the

Fig. 1-105 (right)   Internal anatomy of a snake. Note there is a break in the drawing. (After J. T. Saunders and S. M. Manton, *A Manual of Practical Vertebrate Morphology*, 2nd ed., Oxford Univ. Press, New York, 1949.)

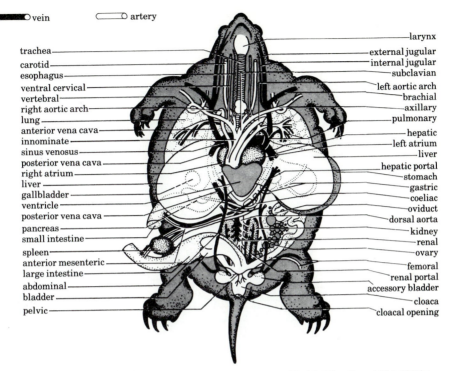

● ◯ vein ▭ ◯ artery

trachea
carotid
esophagus
ventral cervical
vertebral
right aortic arch
lung
anterior vena cava
innominate
sinus venosus
posterior vena cava
right atrium
liver
gallbladder
ventricle
posterior vena cava
pancreas
small intestine
spleen
anterior mesenteric
large intestine
abdominal
bladder
pelvic

larynx
external jugular
internal jugular
subclavian
left aortic arch
brachial
axillary
pulmonary
hepatic
left atrium
liver
hepatic portal
stomach
gastric
coeliac
oviduct
dorsal aorta
kidney
renal
ovary
femoral
renal portal
accessory bladder
cloaca
cloacal opening

Fig. 1-106 Internal anatomy of a turtle. (Reprinted and adapted with permission of Macmillan Publishing Company from *College Zoology*, 10th ed., by Richard A. Boolootian and Karl A. Stiles. Copyright © 1981 by Macmillan Publishing Co., Inc.)

plastron (lower shell) on each side. Raise the plastron, and cut away the connecting membranes; then remove the plastron.

Cut through the membrane to reveal the internal organs, and examine the conspicuous heart (Fig. 1-106). Continue to locate other organ systems.

Pigeon

Before any bird is dissected it should be dipped into hot water so that the feathers can be plucked off easily.

Stretch out the wings and legs on a corkboard or dissecting pan, and fasten them securely. Cut forward from the posterior end of the keel, or breastbone, revealing the crop (where food is macerated). Pull the crop forward to see how it is connected to the stomach, or proventriculus (Fig. 1-107). Then identify the trachea, with its cartilage rings. Find the jugular vein on each side of the neck.

Try to insert glass tubing into the glottis, and blow into it to inflate the air sacs in front of the breastbone.

Air which passes from the glottis into the trachea finally enters nine large, thin-walled air sacs, situated mainly along the dorsal sides of the body cavity. When air is exhaled, the muscles of the thorax and the abdomen contract, sending air out of the air sacs into the lungs and finally out through the trachea (Fig. 1-108).

A structure peculiar to birds is the syrinx, a vocal organ found at the base of the trachea where the trachea divides to form two bronchi.

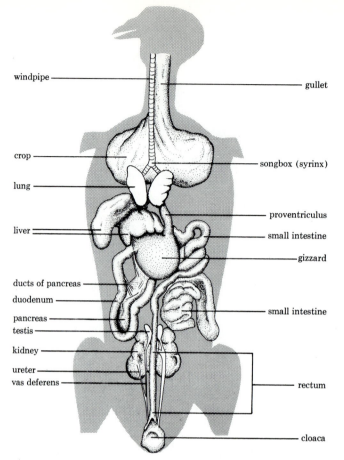

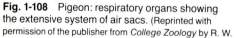

windpipe — gullet

crop — songbox (syrinx)

lung — proventriculus

liver — small intestine

— gizzard

ducts of pancreas —

duodenum —

pancreas —

testis — small intestine

kidney —

ureter —

vas deferens — rectum

— cloaca

Fig. 1-107 Pigeon: internal organs (heart has been removed). (After A. J. Thomson, *The Biology of*   *Birds*, Sidgwick and Jackson Ltd., 1923.)

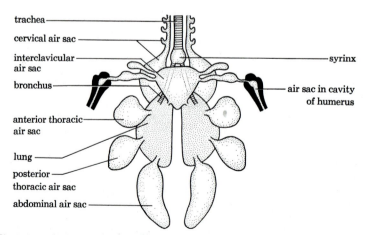

trachea —

cervical air sac —

interclavicular air sac — syrinx

bronchus — air sac in cavity of humerus

anterior thoracic air sac —

lung —

posterior thoracic air sac —

abdominal air sac —

Fig. 1-108 Pigeon: respiratory organs showing the extensive system of air sacs. (Reprinted with permission of the publisher from *College Zoology* by R. W.   Hegner and K. A. Stiles, 7th ed., © The Macmillan Co., 1959; after J. Thomson, *Outlines of Zoology*, 9th ed., revised by J. Ritchie, Oxford Univ. Press, New York, 1944.)

Now cut through the skin of the abdomen, and continue posteriorly to the anus. Abdominal air sacs may be seen if they are inflated. Locate the heart in front of the large liver. Since the heart is comparatively large, dissect it to locate the four chambers.

Also find the pink lungs attached to the back of the animal. Trace the digestive, excretory, and reproductive systems. Locate the two testes or the single ovary (the left one) in front of the kidneys. Note there is no urinary bladder in birds.

Dissect the leg to show how a bird remains perched when asleep. Show that in a relaxed position the toes grasp the perch.

Fetal pig

In some areas fetal pigs may be more readily available for dissection than rabbits, cats, or white rats. If you purchase a pig's uterus for a study of the relationship of the embryo, placenta, and umbilical cord (p. 99), there will be several fetuses available for dissection.

When specimens are purchased, estimate the approximate age of the fetus by its length: 15 to 20 cm (about 75 days); 20 to 25 cm (about 100 days); 30 cm at birth (about 115 days).

Place the pig fetus on its back in a waxed pan. Tie a cord around one forelimb, and bring it around underneath the pan to fasten back the other forelimb. Spread apart the hind limbs in the same way. Then make an incision in the midregion of the chest; cut through the skin laterally and along the medial line toward the umbilical cord. Use Figure 1-109 as a guide in dissecting around the umbilical cord. Fold back the thick skin and note the peritoneum (Fig. 1-110). Locate the liver; observe the umbilical vein that runs from the umbilical cord to the liver. Fold over and pin down the body wall. (The umbilical vein must be cut before you can fold over the part of the body wall that contains the cord stump and find the allantoic stalk, one large umbilical vein, and two umbilical arteries. The arteries have

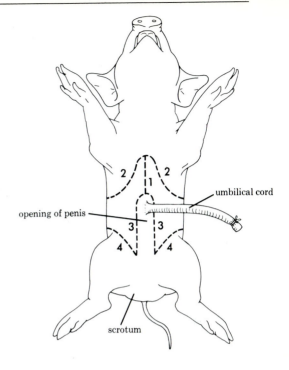

Fig. 1-109 Fetal pig: pattern of cutting lines for dissection.

thicker walls than the veins; the allantoic stalk is usually collapsed.

Identify the liver, stomach, intestine, and diaphragm. Raise the large thymus gland covering the heart. Identify all the organs now *in situ*. Carefully remove the pericardium and trace the blood vessels leading from the heart, especially the dorsal aorta. Dissect the large heart to compare the structure and size of the atria and ventricles; make a slit along the middle length of the heart.

Now trace the digestive system, including the gallbladder, bile duct, and pancreas. Also identify the large, flattened dark red spleen near the stomach. Locate the respiratory system and trace the connection from the trachea to the lungs.

Begin a careful study of the urogenital system (Fig. 1-111). Spread apart the hind limbs; feel the hard pubic symphysis of

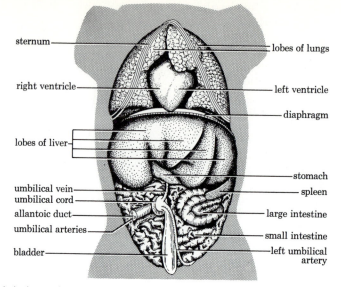

sternum

lobes of lungs

right ventricle

left ventricle

diaphragm

lobes of liver

umbilical vein

umbilical cord

allantoic duct

umbilical arteries

bladder

stomach

spleen

large intestine

small intestine

left umbilical artery

Fig. 1-110   Fetal pig: internal anatomy.
(After T. Odlaug, *Laboratory Anatomy of the*   *Fetal Pig*, Wm. C. Brown, 1955.)

the pelvic girdle. Cut through this region to examine the urogenital system in male and female specimens. In a female fetus, the urogenital opening is located below the anus; in the male, the urogenital orifice is located behind the umbilical cord. Next trace the urinary system. The large, bean-shaped kidneys lie in the dorsal part of the abdomen. From these, trace the ureters toward the urinary or allantoic bladder. Notice that one end of the bladder extends as the allantoic stalk into the umbilical cord. On top of the kidneys find the long, narrow adrenal glands. Then locate the small oval arteries, slightly posterior to the kidneys. Study the position of the Fallopian tubes and the uterus. Then cut through the muscle in the pelvic girdle, and locate the white line of fusion, the pubic symphysis. At this point, split the girdle with a scalpel to reveal the rest of the reproductive system: the urethra (a duct leading from the urinary bladder to the urogenital sinus), the vagina, and the urogenital sinus.

In the male, separate the penis from the ventral body wall posterior to the umbilical cord. In older specimens the testes may have already descended into scrotal sacs.

Look for the vasa deferentia, the thin sperm ducts from the testes, and find how they loop over the ureters. Locate the crescent-shaped epididymis enclosing the inner side of each testis, and see how each is connected with the vas deferens. Also trace the urethra from the bladder to the penis.

Finally, try to trace the nervous system, and study the structure of the eye. Cut away the skin and cartilaginous skull in a circle at the top of the head. Remove this disk to reveal the two convoluted hemispheres of the cerebrum. Observe the meninges; remove the dura mater near the cranium and the thinner pia mater that covers the brain. Locate the longitudinal cerebral fissure, the depressions (sulci) and elevations (gyri) on the two cerebral hemispheres.

You will want to refer to several laboratory guides that are available for the dissection of a fetal pig. Several are listed at the end of the chapter.

Rat

Lay the animal ventral side up and fasten with cords as described for the fetal pig.

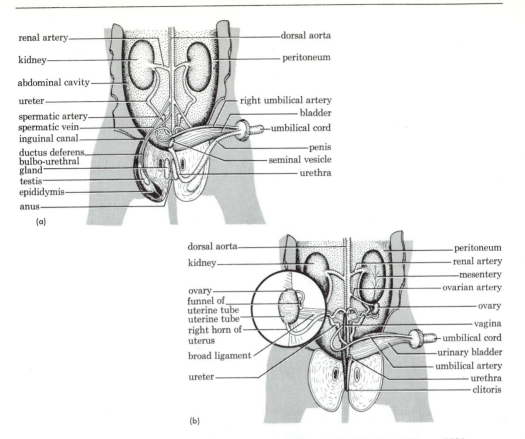

renal artery
kidney
abdominal cavity
ureter
spermatic artery
spermatic vein
inguinal canal
ductus deferens
bulbo-urethral gland
testis
epididymis
anus

dorsal aorta
peritoneum
right umbilical artery
bladder
umbilical cord
penis
seminal vesicle
urethra

(a)

dorsal aorta
kidney

ovary
funnel of uterine tube
uterine tube
right horn of uterus
broad ligament
ureter

peritoneum
renal artery
mesentery
ovarian artery
ovary
vagina
umbilical cord
urinary bladder
umbilical artery
urethra
clitoris

(b)

Fig. 1-111 Fetal pig: urogenital systems of (a) male; (b) female. (After T. Odlaug, *Laboratory Anatomy of the Fetal Pig*, Wm. C. Brown, 1955.)

Make an incision in the skin along the midline from the base of the neck to the pelvis. Fold back the skin and fasten it with stout pins to a board or dissecting pan. Carefully cut through the muscular abdominal wall up to the breastbone. When the chest is opened, the lungs and heart collapse (Fig. 1-112). Inflate the lungs to their normal size, so that the lungs fill the chest cavity and nearly encircle the heart. Study the lobes of the lungs and examine the muscles of the diaphragm.

For the most part, the rest of the anatomy can be identified from the dissection of the fetal pig (p. 97 and Fig. 1-112).

Plant kingdoms

Algae
Diversity is the key word in describing algae. They exist in freshwater, in moist soil, on tree bark, and floating in the oceans. Refer to the classification of algae among the Kingdoms Monera, Protista, and Plantae (pp. 2, 3).

Monerans
These more primitive procaryotic plant cells include both bacteria and blue-green algae; cells that do not have DNA bound within a nucleus separated from the cytoplasm, nor do they contain mitochondria or chloroplasts. The blue-greens were probably the first to survive on land; many use molecular nitrogen from the atmosphere. *Nostoc* (Fig. 1-113) and *Oscillatoria* (fig. 1-123) are two common blue-greens. (Bacteria are treated on p. 696.)

Eucaryotic algae
The eucaryotic algae of Kingdom Plantae include the red, brown, and green algae.

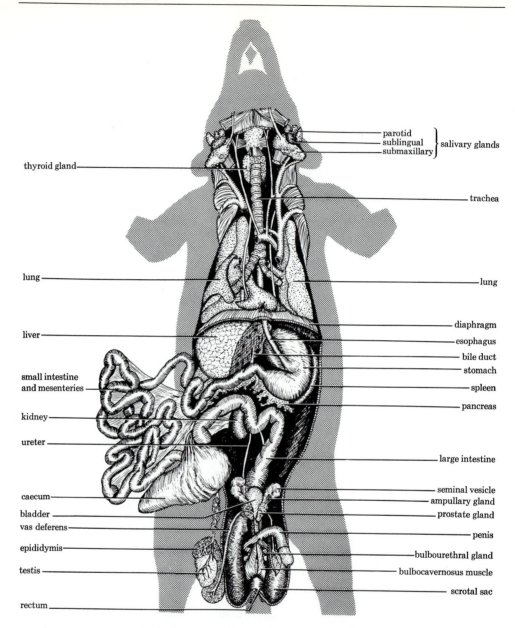

parotid
sublingual } salivary glands
submaxillary

thyroid gland

trachea

lung

lung

diaphragm

liver

esophagus

bile duct

stomach

small intestine
and mesenteries

spleen

pancreas

kidney

ureter

large intestine

seminal vesicle

caecum

ampullary gland

bladder

prostate gland

vas deferens

penis

epididymis

bulbourethral gland

testis

bulbocavernosus muscle

scrotal sac

rectum

Fig. 1-112 Rat: internal anatomy of male; the heart and part of the liver are omitted. (After M. Ulmer, R. Haupt, and E. Hicks, *Comparative Chordate Anatomy*, Harper & Row, 1962.)

Those found among the Protists include the euglenoids, diatoms, dinoflagellates, golden brown and golden greens (refer to classification, pp. 2–3).

In this introductory exploration, groups of algae are described to show variety, regardless of their classification.

Consider first, the marine red algae found floating along coastlines. Most, like *Polysiphonia* (Fig. 1-115), have a finely divided or feathery thallus. The brown algae range from small clusters of cells to large kelps. Some common green algae (Fig. 1-116) are *Chlamydomonas, Chlorella,*

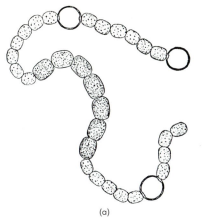

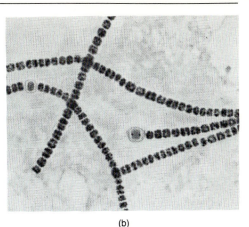

Fig. 1-113 *Nostoc*, a procaryotic blue-green alga: (a) structure; (b) photograph. (a, adapted from *The Plant Kingdom*, by William H. Brown, Ginn and Company,

Publishers. Copyright © 1935 by William H. Brown; b, courtesy Carolina Biological Supply Company.)

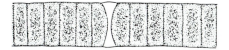

(a)

Fig. 1-115 *Polysiphonia*, a red alga. (From W. H. Brown, *The Plant Kingdom*, 1935.)

(b)

Fig. 1-114 *Oscillatoria*, a procaryotic blue-green alga: (a) structure; (b) photograph. (a, adapted from *The Plant Kingdom* by William H. Brown, Ginn and Company, Publishers. Copyright © 1935 by William H. Brown; b, Grant Heilman Photography.)

Spirogyra (a filamentous form), *Scenedesmus*, and *Pediastrum*. Among the algae that comprise a large part of marine plankton are the dinoflagellates, some of which show luminescence (Fig. 1-117). Algal blooms of dinoflagellates cause "red tide" that results in the death of large numbers of fish and other marine life. Diatoms have sculptured walls of silica (Fig. 1-118). Their golden cells make up much of the phytoplankton. One side of the cell overlaps the

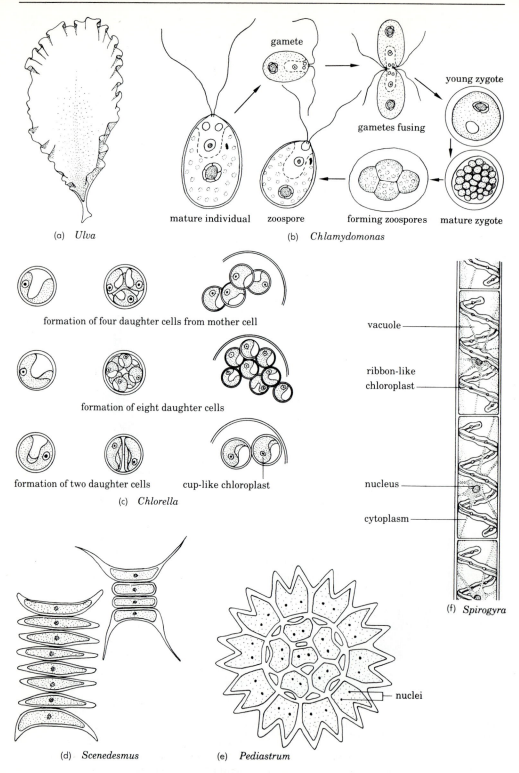

(a) *Ulva*

gamete

young zygote

gametes fusing

mature individual   zoospore   forming zoospores   mature zygote

(b) *Chlamydomonas*

formation of four daughter cells from mother cell

formation of eight daughter cells

formation of two daughter cells   cup-like chloroplast

(c) *Chlorella*

vacuole

ribbon-like
chloroplast

nucleus

cytoplasm

(f) *Spirogyra*

(d) *Scenedesmus*   (e) *Pediastrum*

nuclei

Fig. 1-116 (left)   Several green algae: (a) *Ulva*, a marine form; (b) *Chlamydomonas*, flagellated plantlike protist; (c) *Chlorella*; (d) *Scenedesmus*; (e) *Pediastrum*; (f) *Spirogyra*, a filamentous alga.

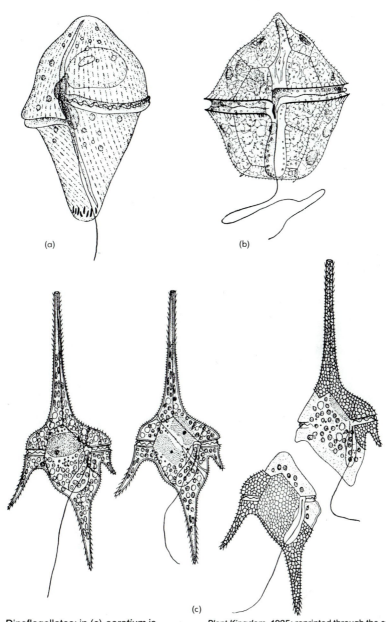

(a)

(b)

(c)

Fig. 1-117   Dinoflagellates; in (c) *ceratium* is shown dividing into two cells. (From W. H. Brown, *The Plant Kingdom*, 1935; reprinted through the courtesy of Blaisdell Publishing Co., a division of Ginn and Co.)

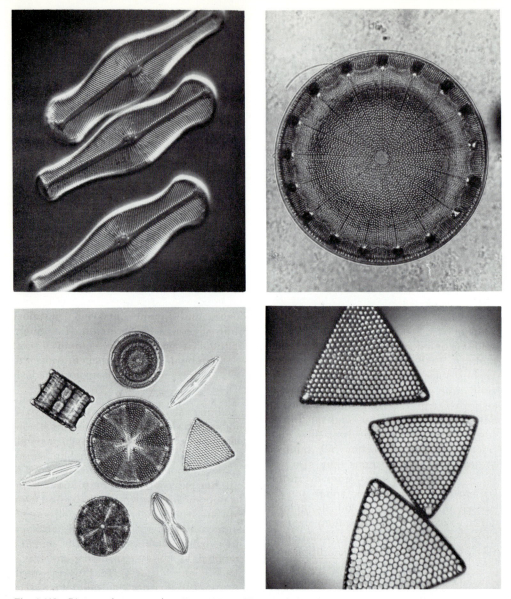

Fig. 1-118 Diatoms: four examples. (Grant Heilman Photography.)

other much like a pill box or Petri dish. Diatoms store energy in the form of oil drops which also serve to keep them afloat.

Evolutionary relationships of algae are based on the structure of the thallus. Unicellular forms are more primitive, evolving along three lines of diversifica-

tion: a) volvicine line, such as *Scenedesmus*, *Volvox*, and *Chlamydomonas*; b) syphonous line including forms that rarely have cross-walls so they are multinucleate; familiar forms are *Vaucheria*, *Hydrodictyon* (Fig. 1-119), and *Acetabularia* (Fig. 1-120); c) tetrasporine line in which cells form

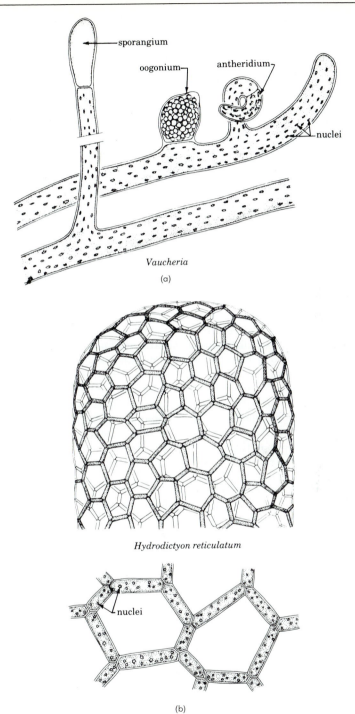

Vaucheria

(a)

Hydrodictyon reticulatum

(b)

Fig. 1-119 Freshwater algae: multinucleate algae rarely have cross-walls (syphonous line); (a) *Vaucheria*; (b) *Hydrodictyon*, multinucleate cells form a net of *Hydrodictyon*. (Adapted from *Botany: An*

Introduction to Plant Biology, 6th ed., by T. Elliot Weier et al. Copyright © 1982 by John Wiley & Sons, Inc.)

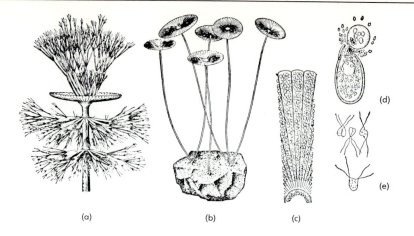

(a) (b) (c)

Fig. 1-120 *Acetabularia mediterranea*: (a) young plant with branches and early stage of umbrella top (× 5); (b) mature plant with umbrella-shaped sporangia (× 1); (c) part of umbrella top showing sporangia with aplanospores; (d) aplanospore releasing gametes; (e) fusion of gametes. (From W. H. Brown, *The Plant Kingdom*, 1935; reprinted through the courtesy of Blaisdell Publishing Co., a division of Ginn and Co.)

new cross-walls in cell division. You may want to refer to pages 460, 436 for studies of reproductive cycles among representative algae; for excellent diagrams of the evolution of alternation of generations, refer to current botany or biology texts.[47]

Cultures of algae are available from biological supply houses. Also refer to chapter 9 for culture solutions and maintenance. For microscopic examinations of algae, refer to chapter 2; also see photosynthesis, chapter 3; growth and population counts, chapter 6.

Fungi (lower and higher)

Lower fungi include many parasitic forms that cause plant diseases (potato blight), as well as saprophytic molds with hyphae, water molds like *Saprolegnia* (that commonly grows on dead flies in pond water and on radish seeds in water), and the Myxomycetes (p. 708). Higher fungi include yeasts, powdery mildews, blue and green molds (Figs. 1-121 and 1-122), bread

molds, mushrooms, rust fungi, and other parasitic fungi. For excellent coverage of reproductive life cycles, refer to current biology and botany texts. Also, refer to a sourcebook of some 227 laboratory exercises in plant pathology prepared by the American Phytopathological Society.[48]

Reproduction among these organisms is described here in chapter 6, culture media and techniques in chapter 9.

Bryophytes

Life cycles of mosses and liverworts are illustrated and described under reproduction in chapter 6; culture methods are described in chapter 9.

Lower vascular plants: Tracheophytes

Life cycles of club mosses (*Lycopodium*, *Selaginella*, and *Equisetum*) as well as ferns, are described in chapter 6, reproduction, and in the section on maintenance of plants in the classroom, page 715. Also

---

[47] See K. Norstog and R. Long, *Plant Biology* (Philadelphia: Saunders, 1976). G. Stephens and B. North, *Biology* (New York: Wiley, 1974). C. Villee, *Biology*, 7th ed. (Philadelphia: Saunders, 1977); T. Weier, C. Stocking, and M. Barbour, *Botany*, 6th ed. (New York: Wiley, 1982).

[48] A. Kelman, Chairman, Sourcebook Committee, *Sourcebook of Laboratory Exercises in Plant Pathology* (San Francisco: Freeman, 1967).

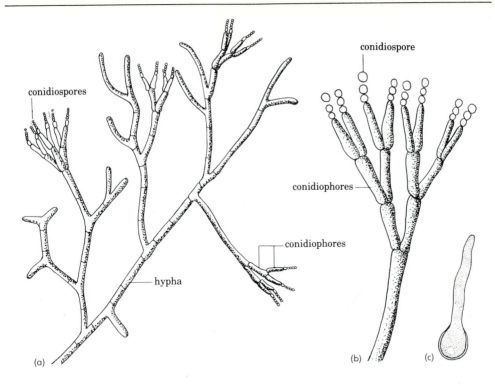

Fig. 1-121 *Penicillium*: (a) pattern of growth of hyphae; (b) conidiophores bearing conidiospores; (c) germinating conidiospore.

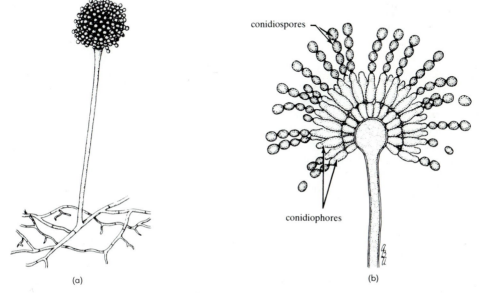

Fig. 1-122 *Aspergillus*: (a) pattern of growth; (b) enlarged conidiophores to show conidiospores.

refer to the fine botany and biology texts listed at the end of the chapter.

Seed plants: Tracheophytes

Leaves

Although leaves take many shapes, their function usually remains the same—making food for the green plant. Seven patterns of leaves are shown in Figure 1-123. Edges of leaves as used in identification are shown in Figure 1-124. Some leaves are only a few cells thick (such as elodea, or *Mnium*), while others have elaborate adaptations.

Make crude, freehand cross sections of a leaf, or purchase prepared slides. To make a freehand cross section, sandwich a piece of leaf between two halves of elderberry pith, balsa wood, or a raw potato (p. 154). With a sharp razor, slice thin sections at an angle; mount in a drop of water. Some sections cut in this way will have thin edges; under a microscope examine the cellular structure (Fig. 1-125).

Identify the upper epidermis (which lacks chloroplasts). Under this layer look for the palisade layer (or layers) filled with chloroplasts.

Study guard cells and stomata in the lower epidermis of a leaf. Peel the lower epidermis near the midvein of a lettuce leaf, or from a geranium, *Kalanchoë*, or *Tradescantia*, or an onion shoot (p. 220). Mount the lower epidermis in a drop of water, and find the numerous kidney-shaped pairs of guard cells surrounding the stomata. Also refer to a "peel" technique, page 218. Refer to Figure 3-15 and page 221 to show conditions for opening and closing of stomates.

Compare the cross section of an angiosperm leaf with that of a pine leaf (Fig. 1-126).

Specialization of leaves

Some leaves are highly specialized organs. The familiar pink and white "petals" of dogwood are modified leaves. Recall that the leaves of onions and other bulbs are adapted for

Fig. 1-123 Leaves, variety in patterns: (a) pinnately lobed leaf (horse nettle); (b) palmately lobed leaf (grape); (c) palmately parted leaf (buttercup); (d) pinnately compound leaf (black locust); (e) palmately compound leaf (horse-chestnut); (f) pinnately trifoliate leaf (desmodium); (g) palmately trifoliate leaf (wood sorrel).

food storage. Still other leaves, such as those of the Venus's flytrap, are modified to capture insects (Fig. 5-8). *Utricularia* has a leaf modified as an insect trap (Fig. 1-127). The water hyacinth has leaves containing large air spaces which keep the plants afloat. The tendrils of *Smilax*, the garden pea, and many other plants are also modified leaves, as are the spines of cactus. In *Kalanchoë* and *Bryophyllum*, the notches of the leaves are primordia of new plants (Fig. 6-110).

Stems

Make freehand cross sections of stems as described in activities concerning transport in plants, page 285. Compare cross sections of monocotyledonous and dicotyledonous stems.

Study the rise of colored liquids through the fibrovascular bundles of celery, carrots, carnations, or growing seedlings (p. 285). Still better, examine the fibrovascular bundles in the orange-flowered jewelweed, or touch-me-not (*Impatiens*), found in swampy regions in late summer. The practically transparent stems show the fibrovascular bundles leading up to the leaves. Collect the stems in August, cut them into short sections, and preserve them in 70 percent alcohol until they are needed.

Also examine the stem of a gymnosperm which is similar to the dicotyledonous stem, except that conifers have no companion cells in the phloem. The xylem of conifers contains only tracheids, which carry on conduction of water.

There may be occasion to examine stems of ferns to compare their structure with that of higher tracheophytes, seed plants.

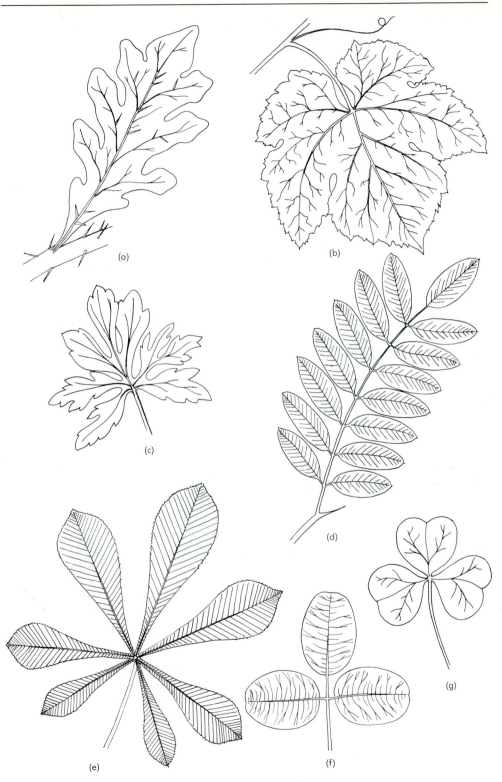

(a)

(b)

(c)

(d)

(e)

(f)

(g)

Fig. 1-124 Leaf outlines as described in keys for identification: (a) entire margin (mountain laurel); (b) serrate (birch); (c) dentate (chestnut); (d) lobed (chestnut oak); (e) cut (black oak); (f) compound (black locust). (Photos: A. M. Winchester.)

The pattern of stem growth, either alternate or opposite, is a means for identifying families of plants.

While the main function of stems is to support leaves and flowers and their fruits, many stems have specialized adaptations for other work. For instance, some stems may be storage regions, such as the fleshy stem of the white potato (tubers), the thickened stem of kohlrabi, or the rhizome of ginger. In fact, the wide spates of green which make up a cactus plant are not leaves, but stems which have taken over the process of photosynthesis. Some stems, like those of morning glory, are modified into twiners. Stems of dodder and some

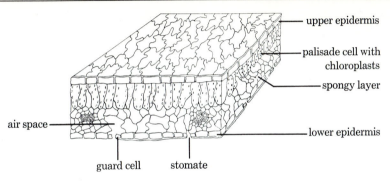

upper epidermis

palisade cell with chloroplasts

spongy layer

lower epidermis

air space

guard cell   stomate

(a)

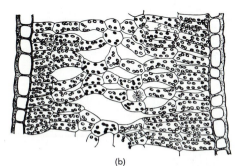

(b)

Fig. 1-125 (a) Cross section of an angiosperm leaf; (b) section of *Eucalyptus* vertical leaf; note the vertical leaf has a palisade layer on each side. (b, from W. H. Brown, *The Plant Kingdom*, 1935; reprinted through the courtesy of Blaisdell Publishing Co., a division of Ginn and Co.)

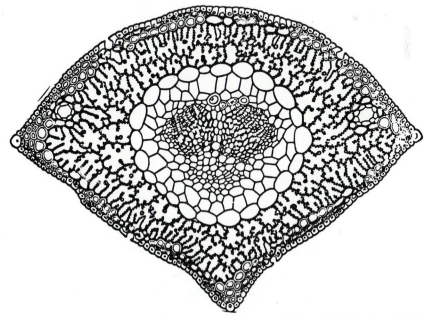

Fig. 1-126 Cross section of a pine leaf. (From W. H. Brown, *The Plant Kingdom*, 1935; reprinted through the courtesy of Blaisdell Publishing Co., a division of Ginn and Co.)

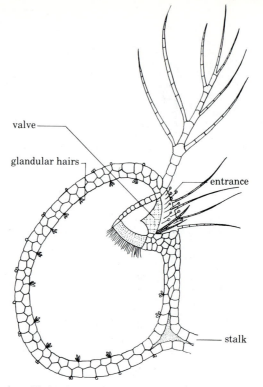

valve

glandular hairs

entrance

stalk

Fig. 1-127 *Utricularia*: leaf modified as insect trap. (From W. H. Brown, *The Plant Kingdom*, 1935; reprinted through the courtesy of Blaisdell Publishing Co., a division of Ginn and Co.)

epiphytes are specialized to absorb water and minerals, thus taking on the usual function of roots. Runners and rhizomes may also serve a reproductive function in that they can be used to clone new plants (vegetative propagation, p. 492).

Roots

Examine root hairs on growing seedlings (p. 472), or measure the elongation of young growing roots (p. 467).

Cross sections of carrots or of the large castor-bean roots will give a fair picture of root structure. Prepared slides of roots may also be purchased. There is a close similarity between the structure of mono-cot and dicot roots, although there is a distinct difference in their stem structure.

While roots serve primarily for anchorage and absorption of soil water, some roots show wide modifications. Compare the fleshy storage root of the carrot, sweet potato, beet, radish, and dahlia. Brace roots and prop roots are found in the Indian rubber plant and the strangling fig tree. New plants can be grown from some of the fleshy roots; thus these roots serve a reproductive function too. Climbing plants, such as ivy, cling by means of adventitious roots.

Representatives of other members of the tracheophytes (the gymnosperms and ferns), the mosses and liverworts (Bryophyta), and bacteria, algae, and fungi are described in other chapters where their inclusion seemed more appropriate. (Refer to the Index for specific studies.)

Flowers

The flower, as a specialized organ for reproduction, is described in chapter 6. Special adaptations of flowers, seeds, and

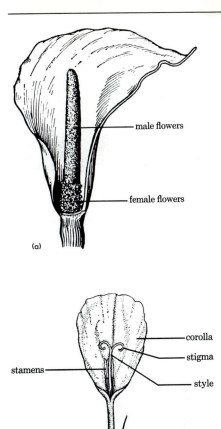

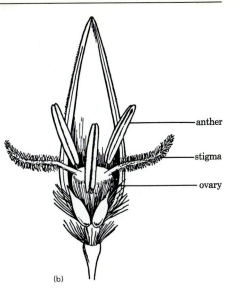

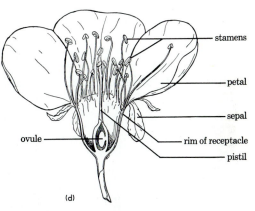

Fig. 1-128 Flowers: types of inflorescences among some families of flowing plants: (a) calla lily, showing spadix with male flowers above, female flowers below; (b) *Avena sativa* (oats) showing one flower of a panicle; (c) marigold, showing one flower; (d) sour cherry. (a, c, from W.H. Brown, *The*

Plant Kingdom, 1935, reprinted through the courtesy of Blaisdell Publishing Co., a division of Ginn and Co.; b, from R. M. Holman and W. W. Robbins, *A Textbook of General Botany*, 3rd ed., 1934, courtesy of John Wiley & Sons; d, from T. E. Weier, et al, *Botany*, John Wiley and Sons, 6th ed., 1982.)

fruits are shown in Figures 1-128, 1-129, 1-130, and 1-131.

Complete flowers have sepals, corolla, stamens, and pistils. *Incomplete* flowers lack one or more of these parts. For example, sepals and petals are missing, or reduced to hairs or scales, in the sycamore, willow, alder, grasses, and sedges. Imperfect flowers lack one of the essential struc-

tures; they may be either staminate or pistillate flowers. Willows, poplars, hop, and the date palm are some examples of plants having imperfect flowers. Some species of maple and ash bear three kinds of flowers on one tree: staminate, pistillate, and perfect. Other plants, such as maize, have staminate flowers as tassels, and ears composed of pistillate flowers (Fig. 1-131).

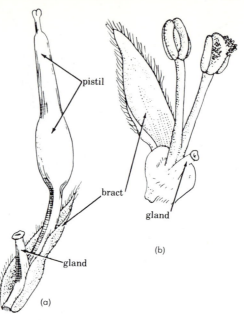

Fig. 1-129 *Salix*, the willow: (a) pistillate flower; (b) staminate flower (one or two stamens per flower). (Adapted from *Botany: An Introduction to Plant Biology*, 6th ed., by T. Elliot Weier et al. Copyright © 1982 by John Wiley & Sons, Inc.)

(a)

(b)

(c)

Fig. 1-130 *Zea mays*, the corn flowers: (a) tassel of staminate flowers; (b) exserted anthers of staminate flowers; (c) pistillate flowers with attached styles. (Grant Heilman Photography.)

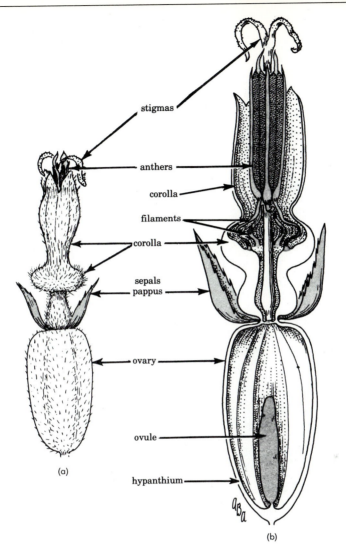

stigmas

anthers

corolla

filaments

corolla

sepals
pappus

ovary

ovule

hypanthium

(a)

(b)

Fig. 1-131 *Helianthus*, the sunflower, showing disk flowers; pollen has been shed and stigmas are raised and exserted. (a) external view; (b) enlarged lengthwise section. (T. Weier, et al., *Botany: An Introduction to Plant Biology*, 6th ed., © 1982, reprinted by permission of John Wiley & Sons, Inc.)

Flowers are identified in plant keys in a number of ways, for example, as hypogynous, perigynous, or epigynous (Fig. 1-132). Some representative types of flowers and fruits that identify common families of flowers are given in Figures 1-128 and 1-129 through 1-135. Also see Figures 6-86, 6-87, and 6-88.

With a hand lens, examine the texture and patterns of pigmentation in petals, the sticky or feathery stigma, and the powdery pollen (black in a tulip, more often yellow in other flowers) that covers the stamens.

Distinguish the outer circles of sepals and petals; look for the common basic pattern of three, four, or five, or multiples thereof. Remove the sepals, which may be green or possibly the color of the petals (as in the tulip), and then carefully remove the petals. Notice whether or not the stamens

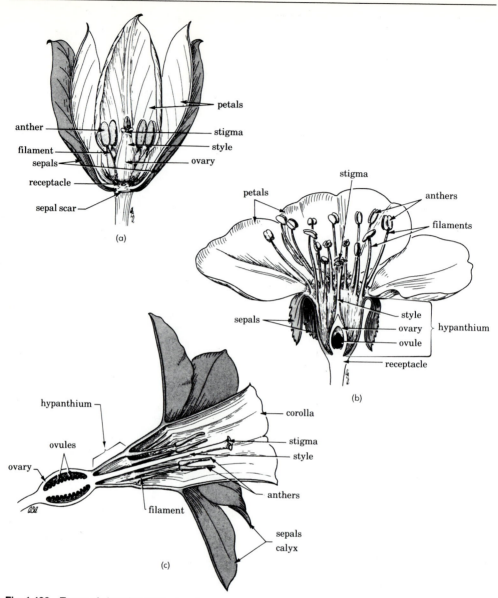

Fig. 1-132 Types of elevation of floral parts: a) hypogyny (tulip); (b) perigyny (cherry); (c) epigyny (daffodil). (Adapted from *Botany: An Introduction to Plant Biology*, 6th ed., by T. Elliot Weier et al. Copyright © 1982 by John Wiley & Sons, Inc.)

are attached to the base of the petals; this fact is used in identification of flowers.

Identify the parts of the pistil. With a razor or scalpel, make a transverse cut through the ovary to examine the number of locules, or compartments, holding ovules. Are the locules in threes or fives?

Notice how the ovules are attached to the wall of the ovary. Use prepared slides for an examination of the megaspores to be found within the ovules (Fig. 6-93).

Dust some of the pollen grains on a slide, and examine in a drop of alcohol. Methods of germinating pollen grains are

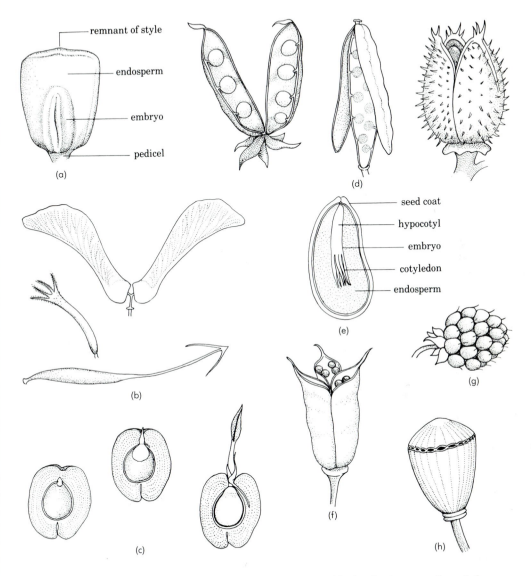

- remnant of style
- endosperm
- embryo
- pedicel

(a)

(d)

- seed coat
- hypocotyl
- embryo
- cotyledon
- endosperm

(e)

(b)

(g)

(c)

(f)

(h)

Fig. 1-133 Seeds and fruits, showing special adaptations: (a) fruit (grain) of corn (*Zea mays*), external view. The small projection at the upper end of the grain is the base of the strand of corn silk; (b), top, samara of maple (*Acer*); bottom, achenes of *Cosmos* and *Bidens* (beggar-ticks); (c) germination of coconut. The large central meat is endosperm. In the left-hand drawing the embryo is still very small; the cotyledon, which is modified as an absorbing organ, is in the endosperm, while the remainder of the embryo projects up into the husk. In the middle drawing the modified cotyledon has enlarged, while the shoot appears through the husk. In the right-hand drawing the cotyledon fills the cavity in the kernel; (d) left to right, pod of pea (*Pisum*); silique of mustard (*Brassica*); capsule of Jimson weed (*Datura*); (e) longitudinal section of pine seed; (f) three follicles of larkspur; (g) aggregate fruit of raspberry; (h) capsule of poppy. (a, b, d, after R. M. Holman and W. W. Robbins, *A Textbook of General Botany*, 3rd ed., Wiley, 1934; c, e, from W. H. Brown, *The Plant Kingdom*, 1935, Ginn and Co.)

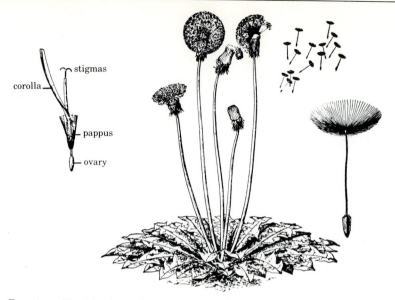

Fig. 1-134 *Taraxacum* (dandelion) showing a flower head, a ripe head with seeds and seed dispersal. To the left is a single flower showing united stamens enclosing the style ending in two stigmas; to the right is a single mature fruit (the parachute is derived from the calyx). (From W. H. Brown, *The Plant Kingdom*, 1935; reprinted through the courtesy of Blaisdell Publishing Co., a division of Ginn and Co.)

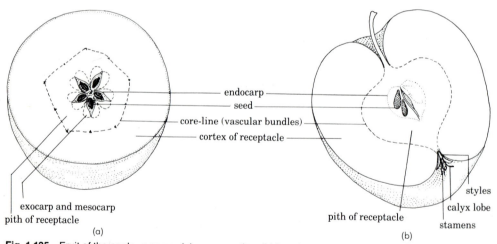

Fig. 1-135 Fruit of the apple, a pome: (a) cross section; (b) lengthwise view.

described in chapter 6. Also refer to page 455 for the preparation of a germination chamber for pollen grains. However, if the flowers look as if they have been ripe for several days, look for germinating pollen grains on the stigma. Try to cut thin slices of the stigma, or carefully crush a portion of the stigma on a slide and mount in water. Examine under low power for evidence of extensions growing out of the pollen grains; these would be pollen tubes.

Seeds and fruits

Observe the diversity in adaptation of types of seeds and fruits to enhance wide dispersal (Figs. 1-133, 1-134, 1-135, and 6-92).

CAPSULE LESSONS: WAYS TO GET STARTED IN CLASS

1-1. On occasion, bring to class an "unknown" plant or animal. Have students decide upon methods to determine relationships. Perhaps they will think of dissecting the organisms as well as examining tissues under the microscope. Also refer to ecological relationships in chapter 8.

1-2. Careful dissections of animals give evidence of phylogenetic relationships. For example, a group of four students can work together with one preserved specimen and share the task of dissection and identification. Compare organs such as the heart, brain, or other organ systems in comparative studies. Fresh calf or lamb hearts are available from local butchers; preserved materials are available from supply houses. Fish, lobsters, some snails, squid, octopus, clams, and mussels are sold in fish markets.

1-3. A phylogenetic collection of animals can be made, preserved, and mounted in similar sized jars. This collection may be used in studies of evolution as well as ecological relationships. In addition, when possible, plant on the school grounds examples of the many phyla of plants (as well as seed plants of monocot and dicot types); include shrubs and trees in the arrangement for class field trips.

1-4. Use hall cases or cabinets in the classroom to display examples of families of plants and/or animals. In early spring and fall, some teachers have exhibits of plants with identifying labels. Similarly, collections of sea shells, algae, and egg cases, make splendid exhibits and stimulate increased awareness of the biological environment.

1-5. In comparative studies, some teachers coordinate dissections of a fish, frog, or snake, especially when funds are limited. One group of students will dissect fish, another frogs, and still another, snakes. These are then stored in alcohol until the next laboratory period. Then teams of students switch specimens and compare the commonality as well as diversity among the vertebrate types.

1-6. There may be an opportunity to build and maintain a greenhouse in the laboratory or on adjoining grounds (see p. 763). Seeds can be germinated for transplanting, and, in class for studies of the effects of plant hormones, light of different wave lengths, mineral needs, and so forth.

1-7. Perhaps a botanical garden, zoological park, or museum is near enough for a class visit. Have students prepare guides in advance with questions to focus observations.

1-8. Take a vicarious field trip using 2×2 slides of a wide variety of flowering plants. Have students identify the special adaptations of flowers for pollination and for seed dispersal. (See catalogs of color slides available from biological supply houses.) Or prepare transparencies of a variety of algae, or of protozoa, or of shells—indeed, vicarious trips to the beach at low tide, to a pond or lake.

1-9. There may be space available for a nature center which may supply nearby schools with living materials such as protozoa cultures or algae, *Drosophila*, mosses, ferns, or seedlings. Such a center may serve as a place to visit for nearby schools.

BOOK SHELF

The following is a selected list of references especially relevant to the subject of this chapter. We have drawn upon some of these heavily and have referred to them in context. While some may be older references, they serve as a base for research reading.

Also consider filmstrips, loops, films, transparencies, as well as computer software, as extensions of the lesson. Refer to listings of distributors as well as producers in the appendix and explore their offerings.

Ashley, L. *Laboratory Anatomy of the Turtle.* Dubuque, IA: Wm. Brown, 1962.

Barnes, R. *Invertebrate Zoology* 4th ed. Philadelphia: Saunders, 1980.

Birky, C. W. "Studies on the physiology and genetics of the rotifer *Asplanchna*," *J. of Exp. Zool.* 155:273–88, 1964.

Boolootian, R., ed. *Physiology of the Echinodermata.* New York: Wiley, 1966.

Boolootian, R., and D. Heyneman. *An Illustrated Laboratory Text in Zoology.* 3rd ed. New York: Holt, Rinehart and Winston, 1975.

Boolootian, R., and K. Stiles. *College Zoology.* 10th ed. New York: Macmillan, 1981.

Brock, T. et al. *Biology of Micro-organisms.* 4th ed. Englewood Cliffs, NJ: Prentice Hall 1984.

Bullough, W. S. *Practical Invertebrate Anatomy.* 2nd ed. New York: Macmillan, 1968.

Calow, P. *Invertebrate Biology* (Functional Approach). New York: Wiley, 1981.

Chiasson, R. *Laboratory Anatomy of the Pigeon,* 2nd ed. Dubuque, IA: Wm. Brown, 1972.

Dales, R. *Practical Invertebrate Zoology.* Laboratory manual. 2nd ed. New York: Wiley, 1981.

Engemann, J., and R. Hegner. *Invertebrate Zoology.* 3rd ed. New York: Macmillan, 1981.

Frandson, R. *Anatomy and Physiology of Farm Animals.* Philadelphia: Lea & Febiger, 1974.

George, J. D., and J. George. *Marine Life: An Illustrated Encyclopedia of Invertebrates of the Sea.* New York: Wiley, 1979.

Giese, A. *Blepharisma.* Stanford, CA: Stanford Univ. Press, 1973.

Gilbert, J., and G. Thompson. "Alpha-tocopherol control of sexuality and polymorphism in the rotifer, *Asplanchna.*" *Science* 159:734–36, 1968.

Hall, E. *The Mammals of North America.* 2nd ed. New York: Wiley, 1980.

Hickman, C. *Biology of the Invertebrates.* 2nd ed. St. Louis, MO: Mosby, 1973.

Hickman, C., and L. Roberts. *Integrated Principles of Zoology,* 7th ed. St. Louis MO: Mosby, 1983.

Hildebrand, M. *Analysis of Vertebrate Structure.* 2nd ed. New York: Wiley, 1982.

Hill, D. *The Biochemistry and Physiology of Tetrahymena.* New York: Academic, 1972.

Jeon, K. *The Biology of Amoeba.* New York: Academic, 1972.

Jones, A. R. *The Ciliates.* New York: St. Martin's, 1974.

Jurand, A., and G. Selman. *The Anatomy of Paramecium aurelia.* New York: Macmillan, 1969.

Margulis, L., and K. Schwartz. *Five Kingdoms* (illustrated guide to 89 phyla of 5 kingdoms). San Francisco: Freeman, 1981.

Norstog, K., and R. Long. *Plant Biology.* Philadelphia: Saunders, 1976.

Norstog, K. and A. Meyerrieckes. *Biology.* Columbus, OH: Merrill, 1983.

Pennak, R. *Fresh-Water Invertebrates of the United States.* 2nd ed. New York: Wiley, 1978.

Prosser, C. *Comparative Animal Physiology.* 3rd ed. Philadelphia: Saunders, 1973.

Richardson, D. *The Biology of Mosses.* New York: Wiley, 1981.

Russel-Hunter, W. D. *A Biology of Lower Invertebrates.* New York: Macmillan, 1968.

Sherman, I., and V. Sherman. *The Invertebrates: Function and Form: A Laboratory Guide.* 2nd ed. New York: Macmillan, 1976.

Smith, R. et al. *Keys to Marine Invertebrates of the Woods Hole Region.* Marine Biological Laboratory, Woods Hole, MA., 1964.

Stephens, G., and B. North. *Biology.* New York: Wiley, 1974.

Tartar, V. *Biology of Stentor.* Elmsford, NY: Pergamon, 1961.

Villee, C. *Biology.* 7th ed. Philadelphia: Saunders, 1977.

Vinyard, W. *Diatoms of North America.* Eureka, CA: Mad River Press, 1979.

Weier, T. et al. *An Introduction to Plant Biology.* 6th ed. New York: Wiley, 1982.

Wichterman, R. *The Biology of Paramecium,* New York: McGraw-Hill, 1951.

Dissection manuals are available from W. C. Brown Co., 2460 Kerper Blvd., Dubuque, IA 52001, for the shark, perch, *Necturus*, frog, turtle, domestic chicken, pigeon, white rat, rabbit, cat, fetal pig.

A useful booklet, *Humane Biology Projects*, 3rd ed., is available from Animal Welfare Institute, PO Box 3650, Washington, D.C., 20007, 1977.

2

Levels of Organization: Cells and Tissues

The organization of life, in the biological sense, may be at the level of procaryotic or eucaryotic cells.[1] Our focus is on bacteria, protists, and many-celled organisms.

Some teachers begin the study of cells with tissue cells; others prefer to introduce the microscope with a study of motile microorganisms. Whatever the approach, there is a fundamental conceptual scheme in biology—the complementarity of structure and function. Thus, the cell is the unit of structure and function.

While many techniques and solutions are described in context in other chapters, the basic stains and techniques for preparing slides are developed in this chapter. An introduction to the biochemistry of cells is also offered in techniques relating to the action of enzymes on components of cells.

Suggested ways to introduce the organization of organisms are offered in the Capsule Lessons at the end of the chapter. As in other chapters, there are investigations that students of varying abilities may undertake wherever they seem appropriate.

Use of the microscope

In order for students to enjoy their investigations while using a microscope,

they should learn to use one correctly (Fig. 2-1). Many teachers have found the following routines useful, and students should be encouraged to practice them.

1. Carry the microscope upright by holding the arm of the microscope with one hand and supporting its weight with the other hand. Set it down gently.

2. Align the low-power objective with the tube until it clicks into place.

3. Move the mirror to find the best light, using the concave side if the microscope has no condenser. When using a lamp, use the flat, plane side of the mirror toward the lamp. Avoid having direct sunlight fall on the microscope or reflect up through the mirror to the eye.

4. Watch the low-power objective (the shorter one, usually) as it is turned down near the slide, so that the working distance of the objective from the slide can be noted. Learn to lower the objective so that it is at a working distance—about 0.6 cm from the slide. Many microscopes are adjusted to lock in this point.

5. While looking through the eyepiece, or ocular, slowly raise the barrel or tube by turning the coarse adjustment knob. Practice keeping both eyes open to avoid eye strain. Learn to move the slide with the free hand to locate the material to be examined (preparation of slides, p. 132).

---

[1] Eucaryotic cells are cells with a distinct nucleus contained within a membrane along with a variety of cell organelles. In contrast, procaryotic cells, such as bacteria and blue-green algae, are more primitive; they lack a nuclear membrane, so that nuclear matter is in contact with the cytoplasm.

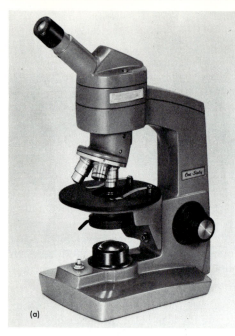

Fig. 2-1 Microscopes: (a) monocular; (b) dissecting microscope. (a, photo courtesy American Optical Co.; b, photo courtesy Polaroid Corp.)

6. Center the specimen found under low power; then slowly turn the fine adjustment knob a few degrees in each direction until the object is focused clearly. It is important that the specimen be centered before switching to high power, since field size decreases with increased magnification.

7. Switch to high power (this can be done without changing the working distance if the microscope is parfocal). Adjust the light again by moving the mirror, and use the *fine* adjustment knob only to focus the object clearly. The fine adjustment knob should be used with care. If the object was in focus under low power, students need make no more than a 10° turn in either direction to get a sharp image. Open and close the iris diaphragm until the light is comfortable to the eye.

Teachers may want to look into the advantages of a zoom optical system giv-ing an "in focus" magnification continuing from $25\times$ to $100\times$.

Much of the beginning student's difficulty results from failure to clean the lenses of the microscope. Cloudiness of the surface of the glass lenses may be removed by breathing on them and then quickly wiping them with lens paper. If this does not remove the film, wipe the surface with lens paper moistened with 95 percent alcohol and then dry the surface. Dried glycerin, blood, or other albuminous materials may be removed by lens paper moistened with water to which a drop of ammonia has been added; the lenses should then be wiped dry. Use a glass plate over the condenser when working with salt solutions or sea water; salt crystals etch the lenses and corrode the iris diaphragm. Balsam, immersion oil, paraffin, or other oily substances may be wiped away rapidly with lens paper moistened with xylene. Since lens glass is softer than ordinary glass, care should be taken to avoid scratches from dust or other fine particles. If there are black specks obscuring the view, check to see where they are by turning the ocular. If the specks also turn, remove the ocular and brush the inner glass surface with a camel's hair brush or lens paper. Do not use cloth, ordinary paper, or fingers (which are oily).

Oil immersion lens

In general, an oil immersion lens is needed to examine bacteria and other minute cells or organelles of cells. (The substage diaphragm must be opened whenever the oil immersion lens is used.) After centering the object under high power, place a drop of immersion oil on the slide, and switch the oil immersion lens in place. Slowly and carefully lower the oil immersion objective so that it just touches the oil drop. (Immersion oil has a refractive index of 1.5.) Then focus with the fine adjustment knob. (Stained slides may be examined with or without a coverslip. Wet mounts require a coverslip.)

If oil remains on the objective after a study is completed, it will harden. Use xylene to remove it. Since xylene is detrimental to the oiled parts of the microscope, immediately wipe the oil off the objective and the slide with lens paper.

Binocular (stereoscopic) microscopes

The compound monocular and binocular microscopes provide a two-dimensional image for observation; some appreciation of depth or thickness may be realized by using the fine adjustment knob. One then can observe sections, or slices, as if cut from the top, the middle, and down to the bottom of a cell.

The dissecting binocular stereoscopic microscope (Fig. 2-1b) has two oculars and gives a three-dimensional image. The true stereoscopic microscope is really two microscopes with two oculars and two objectives. The oculars can be adjusted to the user's eyes to achieve a stereoscopic effect. In the dissecting microscope, the image is not reversed or inverted as it is in

Fig. 2-2   Dark field stop.

the compound microscope. Of course, these microscopes provide less magnification since they are used to view a wide field such as a culture dish of microorganisms.

Advantages of dark field phase, or polarizing microscopes, can be gleaned from laboratory manuals such as one by Giese,[2,3] and/or college texts. Giese also describes finding the resolving power using diatoms.

Some almost transparent cells can be observed under dark field. To make a dark field stop, paste a small circle of black paper over the condenser with transparent tape (Fig. 2-2). This will remove the central beam of light so that the object is illuminated with peripheral light beams.[2,3]

Degree of magnification

With a compound microscope, magnification is expressed in diameters. To determine the total magnification, multiply the magnification of the ocular lens by the magnification of the objective lens. For low power this is generally $10\times$ by $10\times$; for high power, $10\times$ (ocular) by $43\times$ (objective). Check the numbers on the objectives of your microscope.

[2] A. Giese, *Laboratory Manual in Cell Physiology*, rev. ed. (Pacific Grove, CA: Boxwood Press, 1975).
[3] Refer also to M. Abramowitz, "Rheinberg Illumination," in *Amer. Laboratory* (April 1983), PO Box 185, Arlington, MA 02174. This technique is related to dark field, but uses different filters for "optical staining."

TABLE 2-1   Data for microscope using low and high power

eyepiece	objectives magnification	numerical aperture (NA)	total magnification	working distance	field of view
10×	10×	0.25	100×	6.3 mm	1.5 mm
10×	43×	0.55	430×	0.4 mm	0.35 mm

Working distance refers to the distance between the cover glass and the tip of the objective. The field of view refers to the area visible through the eyepiece when the microscope is in focus (see Table 2-1).

Prepare laboratory drawings with some indication of approximate scale. If a student makes a drawing that measures 4 cm, or 40,000 microns (μ), and the length of the actual object is 100 μ, the drawing is magnified 400×.

$$\text{magnification} = \frac{\text{size of drawing}}{\text{size of specimen observed}}$$

Size of cells   Find the size of a cell by first measuring the diameter of the field under low power. Lay a transparent millimeter ruler across the diameter of the field under low power. Line up the left edge of the diameter of the field with a mm line as shown (Fig. 2-3). Note that the diameter in the example is more than 1 mm. Esti-

mate what fraction of a millimeter the next space represents. In the example given, the diameter is represented as 1.5 mm, or 1500 μ.[4]

Prepare a slide of cells such as onion or elodea cells, or others. How many elodea or onion cells fit across the diameter of the field under low power? Are there 8 elodea cells? Then each cell is ⅛ of the 1500 μ, or about 200 μ in length. Next, find the width of the cells and possibly the size of the nucleus.

Diameter under high power   Switch to high power and observe that the field is now less than 1 mm. To measure the field under high power:

$$\frac{\text{magnifying power of high-power objective}}{\text{magnifying power of low-power objective}}$$

Divide this quotient into the diameter of the low-power objective field calculated previously.

magnification (ocular × objective)	field diameter (μ) (approximate)
100×	1500
430×	350
970× (oil)	150

Calculate the length of an onion (or other) cell as seen under high power.

Area of the field

Students can also measure microorganisms and cells if they know the *area* of the field of vision. If they lay a transparent plastic ruler along the diameter of the field and observe the number of millimeters

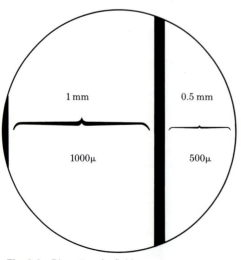

1 mm   0.5 mm

1000μ   500μ

Fig. 2-3   Diameter of a field.

[4] Most low-power fields have a diameter between 1.5 and 2.0 mm (1500–2000μ).

across the diameter, they can calculate the area by using πr^2, the formula for the area of a circle. This procedure may be used for low or high power or for oil immersion lenses. The volume of a cell can also be determined by using the formula for a sphere ($\frac{4}{3} \pi r^3$) or, if the cell is a cylinder, the formula $\pi r^2 h$. Circumference of a circle is $2\pi r$. The surface of a sphere is $4\pi r^2$ and of a cylinder is $2\pi rh$. ($\pi = 3.416$). For more careful measurements an ocular micrometer is used.

Ocular micrometer The length and width of an organism can be measured with a small micrometer scale that is inserted into the ocular of a microscope. The scale, which may be subdivided into 50 units (or, more practically, into 100 equal divisions), is engraved on a glass disk that fits into the eyepiece. (When inserting the scaled disk, place the ruled side down into the diaphragm of the ocular.) Micrometers which are permanently mounted in the ocular are less adaptable for interchanging different kinds of counting disks.

The arbitrary divisions on an eyepiece micrometer must be calibrated against some known unit, such as those found on a stage micrometer slide. Each eyepiece micrometer must be calibrated for low power, high power, or other combinations of objective and ocular, and for the specific tube length of each microscope.

The known units that are used on the stage micrometer may range from 1 to 2 mm, and be subdivided into tenths or hundredths of a millimeter. For example, a typical stage micrometer may have a distance scale of 2 mm measured off into 200 parts; each space is then 1/100 mm. The unit for measurement of microscopic objects is the micron (μ), which is 0.001 mm, or 1/25,000 in. Students soon learn that the space of 1 mm on a ruler is equivalent to 1000 μ; 0.01 mm would then be 10 μ. Each unit on the stage micrometer described above would be 10 μ.

To calibrate the eyepiece micrometer, clip the stage micrometer slide under the

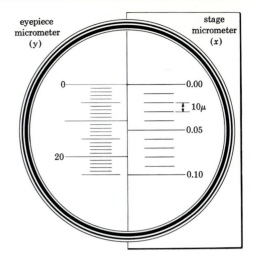

Fig. 2-4 Calibration of arbitrary divisions of an ocular micrometer against known divisions of a stage micrometer scale. Range of 20 divisions on an ocular micrometer would be the same as 80 μ; hence each ocular micrometer division equals 4 μ.

microscope; rotate the ocular micrometer, and simultaneously move the stage micrometer slide so that the scales are superimposed in one plane (Fig. 2-4). Move the slide so that the beginning of one large-scale division on the ocular micrometer corresponds to the beginning of one large-scale division on the stage micrometer. The number of small divisions on the stage micrometer slide corresponding to one large-scale division on the eyepiece micrometer can then be counted and multiplied by 10 μ—the value of each small division on the stage micrometer in this example. This will give the value of one large ocular unit in microns.

Several readings should be made and then averaged. The fine calibration of the subdivisions of the scale and variations in the thickness of the slide affect the accuracy of measurements.

Assume that one ocular micrometer division equals 10 μ (0.01 mm). When the stage micrometer is removed and a slide is substituted (using the same pair of lenses), an object that measures eight ocular micrometer divisions would be 8 × 10 or 80 μ.

Other methods The grid of a hemacyto-meter is another means for determining the dimensions of some objects under the microscope. (Fig. 4-9). However, both counting chambers and micrometers are expensive if each student needs the equipment. Clear plastic grids marked in 1 mm squares also can be placed under the slide to be examined. If still available, Bausch & Lomb makes an inexpensive thin circular disc with a measured series of sawteeth on the rim. However, the micron disk is for use with a 31-15-09 10× Huygenian eyepiece only.

Microprojection

A microprojector is highly useful in class demonstrations to show, for example, specific organisms to observe in pond water, or the behavior of a given specimen in normal versus altered environments.

One commercial microprojector is shown in Figure 2-5a. Detailed, careful directions for operation come with all commercial projectors. However, an essential requirement for efficient microprojection is a darkened room—the darker the better. If dark shades are not obtainable or do not darken the room sufficiently, you may want to devise a box with a screen, as shown in Figure 2-5b. In this box is shown a "homemade" microprojector which can also be used in the standard way. Directions for both uses follow.

Homemade microprojector

If you have a microscope and a lantern-slide projector you can prepare a microprojector at no cost. Simply remove the eyepiece lens from the microscope and the front lens from the projector. Now focus the light beam from the projector at the mirror of the microscope. Place a slide on the stage and adjust the mirror until a spot of light shows on the ceiling overhead. Then adjust the position of the projector until the cone of light just fills the mirror. (You can check this by tapping chalk dust from an eraser to show the Tyndall cone.)

Now focus on the specimen, using the low-power objective. A clear, enlarged image will form on the ceiling. If you wish, you can use a pocket mirror, clamped at a 45° angle, as shown, to throw the image forward to a screen.

If dark window shades are not available, set up the whole apparatus, with the mirror, inside a large carton. One side of the carton should be opened, leaving four flaps, and a translucent screen of tracing paper or tracing cloth attached across the opening. (The completely assembled apparatus is shown in Figure 2-5b.) The flaps of the box can be braced open to act as a shadow shield, as shown. The carton can be placed on the teacher's desk, with the screen facing the class; a large, clear image will be seen on it.

Microscope screen

When students work with microscopes, it is not always possible to know whether they all see what you hope they are seeing. An inexpensive substitute for a microprojector is a viewing device that consists of a microprojection hood that can be slipped over the eyepiece tube of a microscope. In one model, a real image is projected onto a 7.5 cm × 8.8 cm ground glass screen 17.5 cm above the microscope. A concentrated substage light source is needed, available from a filmstrip or slide projector.

Another model offers a viewing screen 10 cm × 12.5 cm that comes with an adapter and fits practically all microscope draw tubes, both vertical and inclined (Fig. 2-6). A substage light source is needed (use a filmstrip projector).

Photomicrography

The NPC MF-10 model is one model, among others, that is specially designed for classroom photomicrography (Fig. 2-7). It fits most microscopes; an adapter attaches to the microscope. There is also a focusing tube. After focusing, remove the tube and slide the camera on over the

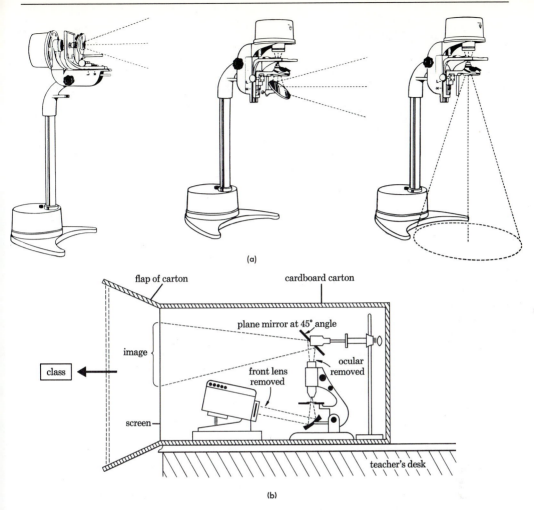

(a)

(b)

Fig. 2-5   Microprojectors: (a) Bausch & Lomb Tri-Simplex Micro-Projector; (b) homemade microprojector for use in daylight. Without the cardboard container, the same homemade projector can be used to throw an enlarged image on a screen (or, without the plane mirror, on the ceiling) in a darkened room.

adapter. The camera requires no lens; the quality of the picture is determined by the optics of the microscope. Black and white prints take seconds; color prints take a minute. The equipment may be purchased from biological supply houses. To operate, attach the MF-10 focusing tube and bring the specimen into focus. Replace the tube with the camera and shoot. The MF-10 uses regular Polaroid pack film. An instruction book accompanies the model.

First microscope lessons

Some teachers like to train students in the use of the microscope by beginning with a wet mount, or a prepared slide of newspaper print. Cut out a single letter from a newspaper (an *e*, for example) and mount it in a drop of water. Apply a coverslip. Locate the *e* under low power and then center it. Compare the direction of the letter on the slide to its position

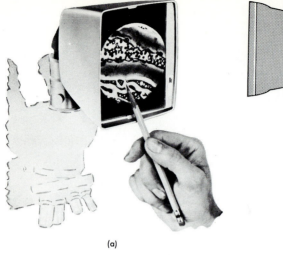

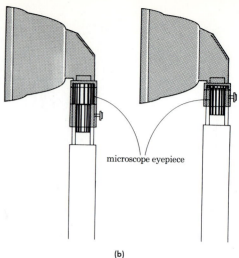

microscope eyepiece

(a)

(b)

Fig. 2-6 Microprojection hood: (a) Welch microprojection screen with (b) adapter that fits regular and long-focus microscope oculars. (Courtesy of The Welch Scientific Co., Skokie, IL.)

under the microscope (the *image*). Is it right side up, reversed, or upside down? Now center the letter and switch to high power. Move the slide a bit to the left. In what direction does the letter move under the microscope?

To practice the art of focusing, prepare a number of slides containing three crossed threads of different colors. (Mount the threads crisscross in a small drop of clear Karo syrup; seal with a coverslip. Lay flat to dry.) Find the bottom thread, then find the center and upper ones by proper change of focus.

Other teachers find that students learn quickly that the image on a slide is reversed and inverted, and immediately start with a more exciting study of *living* microorganisms. When students move the slide to follow a motile form, they quickly learn in what direction the slide must be moved. In fact, some teachers like to use rich cultures of *mixed* protists—gray *Paramecium* (Fig. 1-5), shapeless *Amoeba* (Fig. 1-13), bluish-green, trumpet-shaped *Stentor* (Fig. 1-8), dark gray, cigar-shaped *Spirostomum* (Fig. 1-10), and rose-colored *Blepharisma* (Fig. 1-7)—so that students see a variety of distinctive organisms at

the same time. Or you may want to begin with *Euglena* (Fig. 1-2), or algae such as *Chlorella* (Fig. 2-15) or with epithelial cells from the cheek (Fig. 2-21); or with *Daphnia, Cyclops,* larvae of brine shrimp, or a microscopic worm such as *Dero* or *Tubifex* (culturing, chapter 9; also see Figures 1-60, 1-59, 1-42, 1-44).

The first lessons may be made exploratory to give students some freedom in selecting the material to be used. Make several cultures available for study over several lessons and allow students to work at their own pace. Later, more careful studies of specific organisms can be made.

Teaching patterns

Consider two approaches to the study of the biology of living things—two roads leading to a common goal—the similarities and the diversity among plants and animals.

The first approach introduces living organisms from the *cell level*—an examination of single-celled forms and cells of tissues. The second approach begins with a study of the *many–celled organism.* Begin

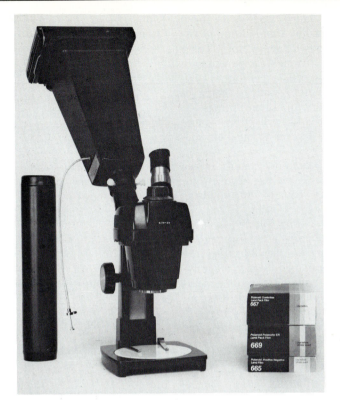

Fig. 2-7 Photomicroscopy. (From Newton Plastics, MF–10)

with the many organisms whose anatomy and physiology are briefly described in chapter 1—*Hydra, Daphnia,* an earthworm, a grasshopper, a dandelion plant. Then develop the concept from

organism ⟶ systems ⟶
organs ⟶ tissues ⟶ cells

Using single-celled organisms

Begin with a microscope study of a mixed culture—protozoa, algae, yeast cells, and possibly bacteria—to illustrate diversity among single-celled organisms. (Refer to page 132 for preparation of materials for the microscope.) Then plan to spend several subsequent laboratory periods on in-depth studies of *Paramecium,* and/or *Blepharisma, Euglena,* and others (Figs. 1-5, 1-7, 1-2). Compare adaptations for food-getting, locomotion, and responses.

From single-celled organisms, move into a comparison of many-celled organisms, or begin investigations of the organelles of unicellular organisms—mainly protists.

Using many-celled organisms

Begin microscope work with exploratory examinations of *Daphnia, Cyclops, Hydra,* or microscopic worms such as *Tubifex.* What life functions can students describe? Then move into a study of the different kinds of tissues that enable organisms to carry on their life functions. Examine tissues from the lining of the cheek, from an onion bulb, and from an elodea leaf. What is the basic uniformity in cells of plants and animals? Establish the concept of cells as "units" of organization; then examine organelles and the biochemistry of cell components, and enzymes (pp. 175–91).

When freshly killed frogs have been dissected, microscopic examination of tissues of the frog may follow (p. 150). There may also be time to look for parasites in frogs (see p. 621).

In either approach, students come to accept a basic concept: Living things are made of cells. Cells have diversity or special adaptations, depending on their specific function.

Both patterns lead into a study of the composition of cells: the nature of protoplasm, possibly cyclosis (p. 148), diffusion gradients and density of cells (p. 240), or ion antagonism (p. 153). Finally, students need to realize the difference between their concept of the cell as viewed under the average compound microscope, magnifying some 430×, and the current general concept as based on observations using an electron microscope, which may magnify some 750,000× to reveal ultramicroscopic organelles (Fig. 2-8).

Readings about cells

Some students will want to supplement their textbook by reading offprints from *Scientific American*, or using histories of science that are relevant to class work;[5] readings offer enrichment for highly motivated students. Listings of *Scientific American* offprints are available from W. H. Freeman & Company, 660 Market Street, San Francisco, CA 94104.

Students may be interested in learning how some early discoveries in science were made. For example, van Leeuwenhoek's letters provide a lucid and refreshing description of his observations of microorganisms in the 1700s. Some teachers reproduce copies for their students. A short description from Dobell's collection of van Leeuwenhoek's letters follows:

I have a very little Cabinet, lacquered black and gilded, that comprehendeth within it five little drawers, wherein lie inclosed 13 long and square little tin cases, which I have covered over with black leather; and in each of these little cases lie two ground magnifying-glasses (making 26 in all), every one of them ground by myself, and mounted in silver, and furthermore set in silver, almost all of them in silver that I extracted from the ore, and separated from the gold wherewith it was charged; and therewithal is writ down what object standeth before each little glass.

This little Cabinet with the said magnifying-glasses, as I may yet have some use for it, I have committed to my only daughter, bidding her send it to You after my death, in acknowledgement of my gratitude for the honour I have enjoyed and received from Your Excellencies.[6]

On October 4, 1723, Leeuwenhoek's daughter, Maria, fulfilled her father's request and delivered the "little Cabinet" to the Royal Society.

Leeuwenhoek described his discovery of "animalcules" in a letter dated September 7, 1674, from Delft.

About two hours distant from his Town there lies an inland lake, called the Berkelse Mere, whose bottom in many places is very marshy, or boggy. Its water is in winter very clear, but at the beginning or in the middle of summer it becomes whitish, and there are then little green clouds floating through it; which, according to the saying of the country folk dwelling thereabout, is caused by the dew, which happens to fall at that time, and which they call honeydew. This water is abounding in fish, which is very good and savoury. Passing just lately over this lake, at a time when the wind blew pretty hard, and seeing the water as above described, I took up a little of it in a glass phial; and examining this water next day, I found floating therein divers earthy particles, and some green streaks, spirally wound serpent-wise, and orderly arranged, after the manner of the copper or tin worms, which distillers use to cool their liquors as they distil over. The whole circumference of each of these streaks was about the thickness of a hair of one's head. Other particles had but the beginning of the

---

[5] Two, among many, are: M. Gabriel and S. Fogel, eds., *Great Experiments in Biology.* (Englewood Cliffs, NJ: Prentice-Hall, 1955); I. Knoblock, ed., *Readings in Biological Science*, 3rd ed. (New York: Appleton-Century-Crofts, 1973).

[6] From *Antony van Leeuwenhoek and His "Little Animals"* (p. 96), coll., trans., and ed. by Clifford Dobell. Published by Dover Publications, Inc., New York 10014, NY, 1960, and reprinted through permission of the publisher.

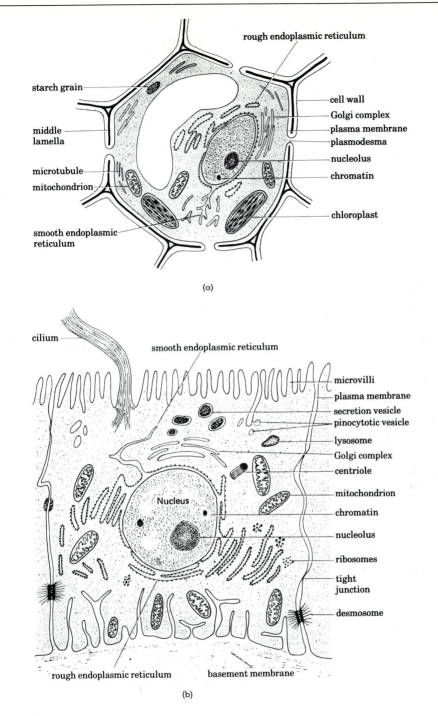

Fig. 2-8 Diagrammatic view of organelles of (a) a plant cell; (b) an animal cell. (Adapted with permission from Macmillan Publishing Company from *Biology* by Clyde F. Herreid II. Copyright © 1977 by Macmillan Publishing Co., Inc.)

foresaid streak; but all consisted of very small green globules joined together; and there were very many small green globules as well. Among these there were, besides, very many little animalcules, whereof some were roundish, while others, a bit bigger, consisted of an oval. On these last I saw two little legs near the head, and two little fins at the hindmost end of the body. Others were somewhat longer than an oval, and these were very slow a-moving, and few in number. These animalcules had divers colours, some being whitish and transparent; others with green and very glittering little scales; others again were green in the middle, and before and behind white; others yet were ashen grey. And the motion of most of these animalcules in the water was so swift, and so various, upwards, downwards, and round about, that 'twas wonderful to see: and I judge that some of these little creatures were above a thousand times smaller than the smallest ones I have ever yet seen, upon the rind of cheese, in wheaten flour, mould, and the like.[7]

In this letter, Leeuwenhoek is probably referring to the spirally arranged alga *Spirogyra*; if so, this is considered the earliest description of this alga. The animalcules are probably different kinds of protozoa; the ones with legs near the head and fins were possibly many-celled rotifers. *Euglena* is a green flagellate, and *Euglena viridis* is somewhat whitish at each end; this may be the species that Leeuwenhoek was describing in his letter to Mr. Oldenburg, who was secretary of the Royal Society in London.

Preparation of materials for the microscope

The Sourcebook is intended for teachers, and not written for students. Solutions, fixatives, and stains should be prepared by teachers since some contain toxic substances. Some solutions, percentage or molar solutions, should be prepared under a teacher's supervision to avoid errors.

Unless otherwise stated, temperature referred to is room temperature.

Only a select number of the commonly used methods for preparing slides can be described in a *Sourcebook*. Detailed guides for specialized techniques, including the preparation of materials for electron microscopy, can be found in texts such as those listed at the end of this chapter.

The techniques described here include methods for both plant and animal tissue cells and microorganisms. More extensive descriptions for preparation of stained bacteria slides, blood smears, and slides of parasites are offered in context in studies of bacteriology, hematology, and parasitology. You will find it convenient to use the Index of this reference.

Before preparing stains, solutions and/or other chemicals such as fixatives for histological work, you will want to read the safety precautions in chapter 10, pages 373–74. For example, many fixatives contain mercuric chloride (corrosive sublimate), picric acid, or formalin. It is suggested that you purchase the suggested fixatives ready-made and observe the procedures for proper use of these materials.

Temporary slides

Protozoa as well as tissue cells of plants and animals are best examined in the living state—without the artifacts that may develop in fixing and staining cells. However, fixed, stained preparations are needed to view the organelles of cells, protozoa stained to show the silver line system, cell inclusions, as well as stages in nuclear divisions (Fig. 2-26; also see Fig. 6-18). Refer to a later section in this chapter for preparing permanent slides.

Two stains suggested for temporary stained slides are acetocarmine and acetic methyl green. While these stains kill the protists, a rapid examination can be made of their stained nuclei. (For other vital stains, see page 136).

Additional preparations of temporary slides are described in other chapters; for example, see *Monocystis* from earthworms (p. 620), flagellates in termites (p. 631), parasites in frogs (p. 621), and mitochondria (p. 137).

---

[7] Leeuwenhoek, pp. 109–11.

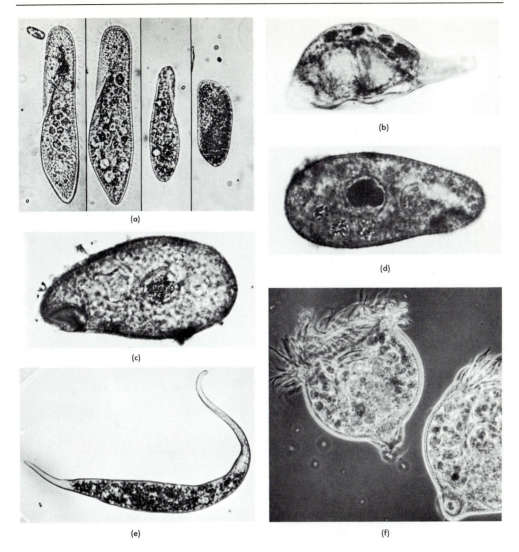

Fig. 2-9 Common ciliated protozoa: (a) four species of *Paramecium* (left to right: *P. multi-micronucleatum, P. caudatum, P. aurelia,* and *P. bursaria*; all photographs at the same magnification); (b) *Blepharisma* stained to show chain of nuclei; (c) *Tetrahymena pyriformis,* stained; (d) *Colpidium,* stained; (e) *Dileptus,* a ciliate with a long proboscis; (f) *Vorticella.* (a, c, d, courtesy of Carolina Biological Supply Co.; b, courtesy of Ward's Natural Science Establishment, Inc., Rochester, NY; e, f, photos by Walter Dawn.)

Examining pond water When collecting pond water, include some submerged leaves; possibly floating leaves of pond lilies, water hyacinths, and others; and some bottom mud. In the laboratory, divide the material into battery jars and several finger bowls (for more accessible study). Add a few rice grains or boiled wheat seeds, and store in a cool spot. Cover with several layers of cheesecloth, or loosely cap with aluminum foil. (For collecting and maintaining cultures, also see chapter 9.)

A gross examination of a large area of

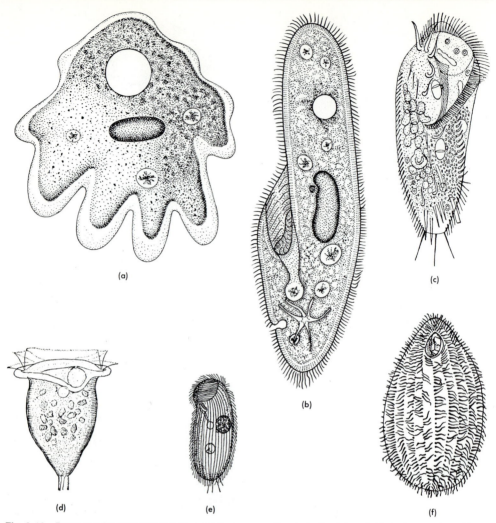

Fig. 2-10 Protozoa: (a) *Amoeba*; (b) *Paramecium*; (c) *Stylonychia mytilus*; (d) *Vorticella*; (e) *Colpidium colpoda*; (f) *Tetrahymena*

pyriformis.(c, d, e, f, from R. Kudo, *Protozoology*, 4th ed., 1954; courtesy of Charles C. Thomas, Publisher, Springfield, IL.)

the finger bowl can be made with a dissecting microscope (magnifying some 30×). Explore the contents: Focus on the surface for several species of *Paramecium*; for free-swimming stages of *Vorticella*; and for *Stylonychia* (refer to figures in chapter 1 as well as Figures 2–9 and 2–10). Focusing down, you may also find *Euplotes*, *Blepharisma*, *Colpidium*, and *Tetrahymena*. Stalked *Vorticella* may be attached to submerged leaves or to the bottom of the dish. *Para-*

mecium, *Colpidium*, *Amoeba*, and some many-celled microorganisms such as gastrotrichs, rotifers (Fig. 2-11), and crustaceans such as *Cypris*, *Cyclops*, and *Daphnia*, (pp. 50–53) may be found in the same region, feeding on bacteria around decaying leaves. Microscopic worms such as *Aeolosoma* or *Dero* may also be found. (Also refer to page 670.)

Transfer samples from *different levels* to slides for more careful examination under

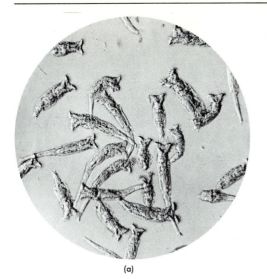

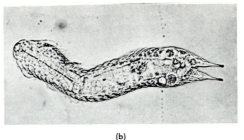

(b)

(a)

Fig. 2-11 (a) Rotifers; (b) *Lepidodermella*, a gastrotrich found among debris in collections from pond water rich in algae. Although *Lepidodermella* is about the same size as a rotifer, it lacks the whorls of cilia at its head end. (a, courtesy of General Biological Supply House, Inc., Chicago; b, courtesy of Carolina Biological Supply Co.)

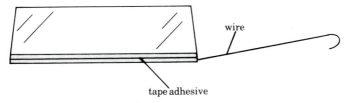

Fig. 2-12 Preparation of a slide for submersion.

low power (100×) and high power (430×). You may want to float clean coverslips on the surface as well as on the bottom of the culture dish. Remove with forceps for examination of slow-moving or sessile forms. Or immerse slides at different levels.

IMMERSED SLIDES This method was devised by A. Henrici. Obtain a sampling of algae, bacteria, and protozoa from a culture medium, pond, or lake. When microscope slides are suspended in water for periods of time, microorganisms will adhere to the slides. (Also refer to pages 578, 586.) These slides can then be quickly examined under the microscope.

Benson describes a simple procedure.[8] Tape a length of wire to the long edge of two slides with adhesive tape (Fig. 2-12). Then tape the remaining edges together

[8] H. Benson, *Microbiological Applications*, 2nd ed. (Dubuque, IA: W. C. Brown (Co., 1973).

and lower into a culture, or attach to vegetation in a pond or lake. After a week, separate the two slides, and examine the two exposed surfaces; stain if desired. Try using different lengths of wire so that "bottom" organisms may be trapped as well as those nearer the surface.

To prevent temporary wet mounts from drying, place a very small drop of mineral oil at the edge of the coverslip. A thin film will diffuse under the coverslip and enclose and seal the water drop. (Refer to page 162 for cleaning coverslips.)

Pure cultures

Pure cultures are useful for investigations into life cycles, growth and environmental factors, behavior, and reproduction. To isolate individual specimens of different genera in fresh culture medium, see culture methods, chapter 9. Refer also to regeneration after bisecting *Blepharisma* (p. 136).

Isolating single protozoans For studies of growth rate of single organisms or to obtain a clone (organisms having the same heredity), and for studies of environmental variables, refer to pages 458, 463, chapter 6 on growth.

One technique for isolating protozoans or algae is a "fast" technique. Slowly centrifuge a culture to concentrate the organisms. Pour off the culture medium and transfer the concentrate into depression slides. Use a wire loop of the type used in bacteria transfers; pick up several protozoans with the loop and transfer them to freshly prepared culture medium in deep-welled slides or Syracuse dishes (which can be stacked). Use for studies of growth and the effect of variables on growth of microorganisms.

Isolating bacteria, fungi, and some algae To obtain colonies, use the dilution techniques or streaking techniques described in chapter 9. Use standard sterile bacteriological techniques.

Dissecting protozoa Use a capillary pipette to transfer amoebae to a depression slide as suggested by Giese.[9] Examine the forms under a dissecting microscope. When a slow-moving amoeba moves under your quietly held needle, cut the amoeba in half with a rocking motion of the needle. Mount the enucleated and the nucleated halves of the amoeba in separate hanging drop preparations sealed with petroleum jelly (preparation of hanging drops, pp. 153–54).

In similar fashion, dissect *Blepharisma*. Remove the anterior hypostome (oral groove, undulating membrane, membranelles, and mouth) from the rest of the cell. Giese describes regeneration within 5 to 6 hrs at 25° C (77° F).[10]

GLASS DISSECTING NEEDLES Prepare glass needles for dissecting *Amoeba* or *Blepharisma* by heating one end of a length of a 5 mm soft glass rod in a microburner (p. 426). Touch the heated tip to the tip of a cooler glass rod and pull apart quickly to make thin glass needles. Cut the needles 1 to 2 cm long with an almost invisible tip.

To prepare dissecting needles with a tip at an angle, see page 427.

Vital stains

Stain protozoa or other living tissue cells to show specific details of cells such as macronuclei, cilia, flagella, or other organelles. Some stains, like Lugol's iodine solution, kill the cells rapidly; others, called vital stains, kill the organisms slowly. While the organisms slowly absorb the stain, they carry on their life functions for some time.

Vital stains are useful as 1 percent stock solutions; add absolute alcohol to get dilutions of 1:10,000 to 1:100,000.[11] Apply a drop of the stain to a clean slide and let dry; store sets of slides with dry stain films in covered slide boxes. When a drop of culture medium is added to a slide, the stain will slowly dissolve in the aqueous medium. Or, add a dilute aqueous solution of a vital stain directly to a drop of culture medium on a slide.

Vital stains are usually prepared in absolute alcohol solutions; basic dyes are less toxic than acidic dyes. Prepare several of the following vital stains for class use. (For additional stains, consult page 169 or the Index).

METHYLENE BLUE is used in concentrations of 1 part of stain to 10,000 or more parts of absolute alcohol (0.1 to 0.01 percent dilutions). The nucleus and granules in the cytoplasm absorb the stain.[12]

NEUTRAL RED is dissolved in absolute alcohol (1:3000 to 30,000). As an indicator in these dilutions, it is yellowish red in

---

[9] Giese, *Laboratory Manual in Cell Physiology.*
[10] A. Giese and A. Smith, "*Blepharisma* in Introductory Biology," *Am. Biol. Teacher* (1973), 35:407–19; also refer to A. Giese, *Blepharisma: Biology of a Light-Sensitive Protozoan* (Stanford, CA: Stanford University Press, 1973).

[11] Stock solutions of vital stains are available from biological supply houses as well as Eastman Kodak.
[12] Methylene blue (0.5 percent) may be effective in discharging trichocysts in *Paramecium*.

alkali, cherry red in weak acid, and blue in strong acid. The nucleus of cells stains lightly, especially the macronucleus of *Paramecium*. Also use this indicator in studies of pH of food vacuoles (p. 140).

CONGO RED   may be used in dilutions of 1:1000 of absolute alcohol. Or, as an indicator, it may be diluted in water. In the presence of weak acids the indicator turns from red to blue. (See uses of indicators in the study of change in pH during digestion in food vacuoles on page 140.)

JANUS GREEN B   prepared as a saturated solution in absolute ethyl alcohol, is a specific vital stain for mitochondria in protozoa in dilutions as weak as 1 part of the saturated solution to 500,000 parts of water.

Furthermore, mitochondria in fresh frog's blood can be stained with dilute Janus Green B (1:10,000). Prepare the dye with physiological saline solution. Add a small drop of the vital stain to a small drop of blood on a clean slide. Or prepare slides in advance so that there is a dried film of the vital stain on the slide. Then add a small drop of blood; the stain will slowly diffuse into the cells. Mitochondria stain bluish green. More permanent slides can be made by removing the coverslip before the components of the cells pick up all the stain and air-dry. Then apply balsam and seal with a coverslip. Lay the slides flat to dry. Janus Green B may also be used to stain bacterial inclusions and Golgi bodies in protozoa.

JANUS GREEN B AND NEUTRAL RED[13] Neutral red is often used together with Janus Green B. The neutral red stains the vesicles (lysosomes) near Golgi bodies without staining the Golgi bodies themselves.

While Janus Green B stains mitochondria, it is also affected by cell enzyme activity so that not all inclusions that pick up Janus Green B are mitochondria.

Prepare separate solutions of neutral red and Janus Green B. Mix them together in small amounts as needed since the combination is not stable.

neutral red stock solution

neutral red	0.5 g
neutral absolute ethyl alcohol	100 ml

Janus Green B stock solution

Janus Green B	0.5 g
neutral absolute ethyl alcohol	100 ml

First dilute the neutral red stock solution: 1 ml stock to 10 ml absolute ethyl alcohol. Then mix 3 ml of this diluted neutral red solution with 0.5 ml Janus Green B stock solution. Prepare slides with the mixture of stains and allow to dry (avoid dust).

For immediate use, add a drop of the mixture to a slide with a drop of culture of protozoa. Apply a coverslip. Mitochondria stain green. When staining tissue cells such as blood, use a small drop of blood so a thin film results. Maintain the slides in an incubator at 37° C (99° F) for 20 min.[14]

SUDAN III   stains neutral fats red. Prepare a 2 percent stock solution using absolute alcohol. When ready to use, dilute the stock solution equally with 45 percent alcohol.

LUGOL'S SOLUTION   reveals flagella and cilia, and stains glycogen reddish brown. Kudo suggests a preparation of the solution using 1.5 g of potassium iodide in 25 ml of water, then adding 1 g of iodine[15] (also see page 744).

NILE BLUE SULFATE   stains the macronucleus of protozoa green and food vacuoles blue. Prepare dilute solutions (1:10,000 to 1:15,000).

NIGROSIN   although not a vital dye, can be applied to smears of protozoa. Nigrosin can be purchased as a 10 percent solution; or dissolve 10 g nigrosin in 100 ml distilled water. Apply a drop to a drop of protozoa. Dry the slide in air or under a lamp and examine under low and

[13] G. Humason, *Animal Tissue Techniques*, 3rd ed. (San Francisco: Freeman, 1972).

[14] Humason, *Animal Tissue Techniques*.
[15] R. Kudo, *Protozoology*, 5th ed. (Springfield, IL: Charles Thomas, 1973).

TABLE 2-2   Dyes staining cytoplasm of normal living *Paramecium**

dye	minimal concentration for cytoplasm staining	percent mortality in 1 hr	hours needed for destaining cytoplasm
Bismark brown	1:150,000	0	7
Methylene blue	1:100,000	5	7
Methylene green	1:37,500	5	4
Neutral red	1:150,000	3	9
Toluidin blue	1:105,000	5	9
Basic fuchsin	1:25,000	30	9
Safranin	1:9000	30	1½
Aniline yellow	1:5500	0	1
Methyl violet	1:500,000	20	2
Janus Green B	1:180,000	40	7

* Reproduced, with adaptations, from *Introduction to Protozoology* by Reginald D. Manwell, St. Martin's Press, Inc., New York, and Edward Arnold Limited, London. Copyright © 1961 by St. Martin's Press, Inc. After G. Ball, "Studies on *Paramecium*, III: The Effects of Vital Dyes on *Paramecium caudatum*," *Biol. Bull.* 52:68–78, 1927.

high power to see the pattern of rows of cilia in the pellicle of ciliates. Mount in resin and apply a coverslip. Humason also suggests adding a small amount of carmine to the living organisms so that food vacuoles will be stained.[16] (For silver line impregnation, see page 165.)

METHYL GREEN ACETIC ACID   can be prepared by saturating a 1 percent solution of acetic acid with methyl green (p. 745 or Table 10-6) and filtering. Add 1 drop of a rich culture of protozoa to 1 drop of the methyl green acetic acid solution. The cytoplasm of the cells should remain clear or light blue while the nucleus is greenish. Trichocysts can also be seen.

ACETOCARMINE   is used to differentiate the nucleus. Add carmine powder to a boiling 45 percent solution of acetic acid until it is saturated (*caution:* work in a hood). Filter. Add 1 drop of a protozoa culture to 1 drop of the stain on a clean slide. R. D. Manwell cites the work of Ball (1927), who tried varying dosages of many vital stains on mortality of *Paramecium caudatum* (Table 2-2).

PHENOL RED   To prepare a stock solution, dissolve 1 g phenol red in 100 ml of water. An indicator in the pH range 6.8 to 8.4.

TOLUIDIN BLUE (0.1 PERCENT)   Add 0.5 g toluidin blue to enough 10 percent ethyl alcohol to make 500 ml. In dilute solutions, acts as a vital stain. Or dissolve 1 g in 0.5 ml concentrated HCl (*caution*). Add water to 100 ml. Test several dilutions (for example, 0.01 percent) of this stock solution on living material.

BISMARCK BROWN   Use in dilute concentrations as a vital stain, mainly for protozoa. Dissolve 0.5 g in 100 ml water as a *stock* solution.

A variety of microorganisms: in-depth studies of organelles

Protozoa

On a clean slide, center a small drop of culture medium of protozoa such as *Paramecium*, *Blepharisma*, *Vorticella*, or *Spirostomum*.[17] Apply a clean coverslip by pulling it along the slide at a 45° angle until it touches the edge of the drop, then carefully lower it over the drop so that the fluid is spread out evenly between the two

[16] Humason, *Animal Tissue Techniques*.

[17] Living stained cultures of protozoa and other small vertebrates may be purchased from Carolina Biological Supply Co. under the trade name Vitachrome cultures. These red-stained specimens are easy to locate and study.

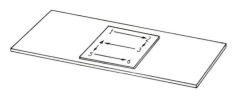

Fig. 2-13 Scanning an entire slide.

surfaces of glass. If air bubbles occur, raise the coverslip and reapply to eliminate them. Methods for slowing down protozoa, and for preventing evaporation are described on pages 141–42. Staining with vital dyes may be desired (pp. 136–38).

Examine the slide under the low-power objective. Center the slide and lower the objective to about 6 mm above the slide or until it locks in place. Then use the coarse adjustment knob to raise the tube until the protozoa are in focus. With one hand move the slide to inspect the organisms in the drop; use the other hand to sharpen the image by slowly moving the fine adjustment knob a bit to the left or right.

Scan the entire slide. Begin at the left top corner of the slide and move across to the right, back again to the left, and so on for the entire area of the slide (Fig. 2-13).

Center a specimen under low power, then switch to high power for careful examination. We will assume that a dense culture of *Paramecium* is under study (Fig. 2-10).[18] There may be *Paramecium caudatum*, or *P. bursaria* (green because of the symbiont *Chlorella*, or *Zoochlorella* in the cytoplasm). On the whole, paramecia average 135 μ. Use figures shown in chapter 1 as a guide for identification of other protozoa.

To immobilize protozoa for prolonged observation, add either a small drop of 1 percent copper sulfate or a drop of 3 percent copper acetate (from a dropper bottle) to the slide of protozoa.

To show nuclei and cytoplasmic structures in some protozoa, add a drop of dilute methylene blue (1:10,000) (p. 136) or dilute acidified methyl green (p. 138)

at the edge of the coverslip. Use lens paper or filter paper at the opposite edge of the coverslip to draw out the fluid. Notice the greenish nuclei in a light blue cytoplasm when methyl green stain is used. Compare the different kinds of nuclei in *Spirostomum*, *Paramecium*, *Pelomyxa*, and *Blepharisma*.

Vital stains may be used to show mitochondria in some protozoa. Make a film of Janus Green B and neutral red on a slide and let the stains dry (see p. 137). Then add a drop of a rich culture of protozoa such as *Spirostomum* and *Paramecium*. Look for the many mitochondria scattered in the cytoplasm.

Ciliary motion   Wichterman reports that the spiraling of *Paramecium* to the left is due to the oblique motion of cilia.[19] In a summary of findings of others, he further reports that monovalent cation salts and hydrates that were tested induced reversal of ciliary motion, except $(NH_4)_2SO_4$ and $NH_4C_2H_3O_2$. None of the bivalent and trivalent cation salts tested induced reversal, except $CaHPO_4$ and $MgHPO_4$. (Also refer to paramecia on page 153).

You may want to use the silver line pellicle staining method (p. 165).

Contractile vacuoles   Watch the alternating contraction of the two contractile vacuoles of *Paramecium caudatum*, especially evident as the fluid evaporates and the coverslip presses down on the organisms.

The pulsation rate is about once every 6 sec. You may want to have students compare this rate with the rate of once every 18 sec when paramecia are in a 0.5 percent NaCl solution. In a 1 percent NaCl solution, the rate decreases to once every 3 min. What is the rate in a 2 percent, in 5 percent NaCl solution? In addition, investigate the effect of temperature on the rate of contraction of vacuoles in *Paramecium*. There is a rise in metabolic rate as temperature rises. For example, the rate of

---

[18] Centrifuge a culture; supply concentrated cultures to classes in labeled dropper bottles.

[19] R. Wichterman, *The Biology of Paramecium*, Blakiston (New York: McGraw-Hill, 1953).

contraction of the vacuoles rises from almost zero at 1° to 2° C (34° F to 36° F) to ten to twelve times a min at 25° C (77° F).

Food vacuoles   Active cilia in the oral groove of *Paramecium* create a current of water. This can readily be observed when *Chlorella* cells are added to the slide; or by adding a dilute suspension of carmine powder, India ink, or yeast cells stained with acetocarmine (p. 138). Observe the ingestion of these materials as they are swept into the gullet forming colored food vacuoles in the cytoplasm. Watch *Chlorella* accumulate in the oral groove; when the mass reaches a certain size it is pushed into the cytoplasm in the form of a food vacuole. Carmine and India ink will be eliminated through the anal spot located on the surface, about at the level of the end of the gullet.

Observe changes in the pH in food vacuoles by adding an indicator to a source of food. Boil a small amount of yeast with dilute neutral red solution (1:10,000 parts of pond water). Add a drop to a slide containing a dense culture of paramecia; under high power note a reddish color in acid pH and a yellow tint in alkaline pH as digestion proceeds. Or prepare a small quantity of very dilute brom thymol blue (1:10,000 parts of pond water). To this add a crumb of yeast and bring to a boil for 5 min. When cool, add a drop to the slide of paramecia. At a pH of 7.6 the food vacuoles are blue, while a yellow color indicates a pH of 6.

Using another technique, add a few drops of dilute Congo red (1:1000 parts pond water) to 2 to 3 drops of evaporated milk. Add a very small drop to a slide of paramecia. Under high power, observe the gradual formation of food vacuoles that turn blue in an acid pH (acidic enzymes) becoming red as digestion is completed and the pH becomes more alkaline.

Staining *Paramecium* will reveal a macronucleus and a micronucleus. While the macronucleus contains about 400 times more DNA than the micronucleus, it is the latter structure that goes through mitotic stages in fission.

Trichocysts are barbed, harpoonlike structures released by paramecia, yet their function is not clear. Add a bit of acetic acid (or certain inks such as Parker) to the side of the slide to see the long trichocysts unleashed (Figs. 1-5, 1-6).

Further discussions of the functions of *Paramecium* and other protists are offered in chapter 1, as well as reproduction (p. 403); behavior (p. 336); vital stains (p. 136); permanent stained slides (p. 163); dissections of protozoa (p. 136); isolating protozoa (p. 136).

There may be other ciliates in pond water. Use Figure 2-9, and the section on classification of protozoa (p. 665; see also Fig. 9-5), to identify the available specimens. *Blepharisma* is a pink or rose colored form due to the presence of granules in the cytoplasm containing blepharismin, formerly called zoo-purpurin. This pigment is light-sensitive, becoming bleached in strong light; it is soluble in alcohol and acetone. The pigment stentorin, found localized in long bands of the ciliate *Stentor*, is not light-sensitive and is not readily soluble. *Stentor* is large, bell-shaped when irritated but trumpet-like in its free-swimming form (Fig. 1-8). (Also refer to chapter 1, pages 5–9.)

If amoebae are available, add a drop of yeast stained with acetocarmine (p. 138), or a small drop of dilute India ink. Add bristles or sand grains to raise the coverslip so that amoebae can move freely; observe pseudopodia in action (Figs. 1-13, 2-10).

When available, add a drop of concentrated *Paramecium* to a depression slide containing starved *Pelomyxa*. Locate specimens under low power; under high power find paramecia within food vacuoles of these giant amoebae.

Similarly, add a drop of concentrated paramecia to a slide of starved *Didinium*. Observe the rotating barrel-like *Didinium* consume paramecia through its anterior oral region. (Fig. 9-5.)

Pinocytosis   Two techniques, among many, are suggested to show pinocytosis in amoebae. Sherman and Sherman describe the appearance of channels in the pseudo-podia of *Amoeba proteus* a few minutes after placing amoebae in a solution of 0.125 *M* NaCl in 0.01 *M* phosphate buffer at pH 6.5 to 7.0.[20] The channels pinch off forming tiny vacuoles.

Giese describes the "frilled" appearance of the pseudopodia of amoebae as evidence of pinocytosis.[21] Place amoebae that have been starved for several days into a medium of 2 percent albumin dissolved in pond water. Examine under high power after about half an hour.

Laboratory strains of protozoa and algal flagellates of known genetic composition may be obtained for in-depth studies from a laboratory center or some biological supply houses. Refer to L. Provasoli, "A Catalog of Laboratory Strains of Free-living and Parasitic Protozoa," *J. Protozool.* 5:1958.

Supplementary techniques

Preventing swelling of cells   Cells that have a fairly high salt concentration may swell when mounted in tap or aquarium water. This swelling can be prevented by putting the cells in a 10 percent aqueous solution of glycerin. Then add a coverslip. On the other hand, when the mounting fluid has a higher salt or sugar concentration than the cell, the cell loses water, resulting in a shrunken or plasmolyzed state (p. 294, Fig. 4-28).

Retarding evaporation   Students may ring a wet mount with petroleum jelly to slow down evaporation. Dip the mouth of a test tube into soft petroleum jelly. Apply this circle of petroleum jelly around the drop of material on the slide, and place a coverslip so that its edges are sealed to the slide. Or add a very small drop of mineral oil to the wet mount before applying a coverslip.

Concentrating organisms in a culture
Rich cultures of protozoa such as *Paramecium* and *Blepharisma* should be available for microscopic study. Also examine flagellates and other motile algae (culturing, chapter 9). When a culture is dilute, concentrate the organisms by centrifuging 10 to 20 ml of the culture. Or pour a portion of the culture into vials or test tubes and cover all but the top third of the tubes with carbon paper. Or prepare a short length of glass tubing and insert it into a one-hole stopper in a vial filled with culture, as in Figure 2-14a; and cover the vial with carbon paper. Protozoa concentrate in the uncovered portions (the tops of the test tubes, or the tubing in the stopper, as in Figure 2-14b), since most of them are positively phototactic or negatively geotactic, or gather where the concentration of oxygen is greatest (at the top surface). Then prepare temporary mounts containing large numbers of organisms for microscopic study.

Slowing down the motility of protozoa
Protozoa, especially the ciliated forms, move too rapidly for beginners to follow under the microscope.

Most experienced microscopists simply prepare several slides ahead of time and let the fluid begin to evaporate. As evaporation continues, the weight of the coverslip is enough to impede the movement of organisms. *Paramecium, Blepharisma, Spirostomum,* and *Stentor* have conspicuous contractile vacuoles which are best observed

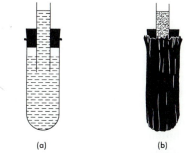

(a)   (b)

Fig. 2-14   A simple way to concentrate a culture of protozoa or flagellates: (a) fit culture vial with stopper and short glass tube; (b) cover vial with carbon paper.

[20] I. Sherman and V. Sherman, *The Invertebrates: Function and Form,* 2nd ed. (New York: Macmillan, 1976)

[21] A. Giese, *Blepharisma.*

when the slides begin to dry—the pressure of the coverslip flattens the organisms. Under a darkened field (reduce light by closing the diaphragm a bit), the rhythmic pulsations of the vacuoles may be seen clearly (p. 139).

You can also tease apart lens paper or glass wool on the slide with a drop of rich culture. (There is some danger that cilia and larger membranelles may be damaged with this method.)

The following solutions may also be used to retard motility:

GUM TRAGACANTH   In a mortar, grind a small amount of gum tragacanth to a fine powder. Add cold water to make a thick jelly. Dilute the material to the proper viscosity by placing a drop of protozoa culture and a drop of the jelly on a slide. You may need to try different dilutions.

METHYL CELLULOSE   Prepare a solution by dissolving 10 g of methyl cellulose in 90 ml of water. Add the methyl cellulose to warm water (85° C or 185° F) and stir; cool by placing the container in a water bath and bring down to 5° C (41° F) stirring constantly. Pour into tightly stoppered bottles for individual use. Add 1 drop of this syrupy solution to a drop of protozoa culture on a slide. Or ring the center of a slide with methyl cellulose and add a drop of the culture medium in the middle of the ring; then apply a coverslip. For a more dilute solution, reduce to a 3 to 5 percent solution. (Other preparations are available from supply houses.)

GELATIN   Prepare a 2 to 3 percent solution of clear gelatin by stirring in cold water and then heating gently until dissolved. Allow to cool to room temperature, and add 1 drop of this to a slide along with 1 drop of the culture containing the microorganisms. Or use a thin 1 percent agar solution.

You may also want to try a bit of isopropyl alcohol or chloretone, but these alter the structure and physiology of the organism.

Anesthetization   At times you may want to examine flatworms, a hydra, *Daphnia*, or some form having contractile tissue. Or you may want to anesthetize a large animal in order to dissect out a bit of tissue for examination, for example, ciliated epithelial tissue from the roof of the mouth of a frog (p. 150). The following materials and methods suggested for anesthetization are referred to throughout this book. Some experimentation may be necessary to meet specific needs. (Refer to safety precautions before preparing solutions, pp. 132; 728–30.)

MAGNESIUM SULFATE   A saturated solution of magnesium sulfate (Epsom salts) can be added, drop by drop, to the water containing the specimen. This procedure is useful for anesthetizing starfish and hydra; a 1 percent solution can be used to slow down some protozoa.

BUTACAINE SULFATE   A freshly prepared 0.1 percent solution of butacaine sulfate can be added, a drop at a time, to a container of protozoa and rotifers.

MENTHOL   A few crystals of menthol on the surface of the habitat water will anesthetize planarians.

URETHANE   A 1 percent solution of urethane (ethyl carbamide; *caution:* poison) may be used to anesthetize planarians and *Daphnia*, and a 5 percent solution can be used to anesthetize frogs. Mammals such as guinea pigs, rats, rabbits, and mice may be injected with 10 percent solution urethane. (Refer to humane methods, p. 175.) A preferred anesthetic is MS-222.

MS-222   A useful anesthetic for aquatic invertebrates as well as fish and amphibia (and tadpoles) is MS-222 (tricaine methanesulfonate). Animals are placed in a diluted solution (1:10,000) of MS-222. As a brief anesthetic for toads or frogs, pour enough 2 percent MS-222 into a bowl to cover webbing of the legs. Hoar and Hickman indicate that while MS-222 is more expensive than urethane, prolonged exposure to urethane should be avoided since it may be carcinogenic.[22]

COPPER ACETATE   A 3 percent solution will immobilize ciliated protozoa.

CHLORETONE   A solution of 0.1 percent chloretone can be added to water containing goldfish or frogs. One drop of chloretone of this dilution added to a large drop of culture of protozoa, hydras, or rotifers acts as an anesthetic. A frog may be anesthetized by immersion in a solution composed of 1 part of 0.5 percent chloretone and 4 parts of Ringer's solution for 15 to 30 min. (Use a 1 percent solution to kill planarians.)

ETHYL ALCOHOL (70 percent)   may be used, drop by drop, to relax and anesthetize earthworms.

ETHER OR CHLOROFORM   Spray ether or chloroform on water containing such specimens as *Daphnia* or worms, and then cover the container (*caution*). Mammals such as rabbits, rats, guinea pigs, and mice may be anesthetized by placing them in a container to which cotton saturated in ether or chloroform is added. (Quantity of anesthetic and the time factor can be altered so that these animals may be killed in this way. Refer to humane methods of handling animals on page 175.)

ANESTHETIZING A FROG   To anesthetize a frog for a demonstration, such as circulation in the webbed foot, immobilize it briefly in a covered jar with some cotton saturated with ether; the immobilization will not last very long.

A more effective method[23] is to inject a 2.5 percent solution of urethane (ethyl carbamide; *caution:* poison) into the ven-

tral lymph sac. As a guide, inject about 0.1 ml for every 10 g of the frog's weight.

There may be other occasions when you want to inject drugs of various sorts into a frog. Drugs are usually injected into the anterior lymph sac[24] in the floor of the mouth. Hold the animal in your hand so that its ventral surface is toward you. Draw the drug into the hypodermic syringe. Open the frog's mouth; avoid the tongue, and point the needle toward the floor of the mouth. When you press the needle into the skin, the needle enters the lymph sac; then inject the drug.

Algae

Soupy green water from an aquarium or culture jar standing in strong light contains a vast variety of algae. Prepare wet mounts of samples from different levels in the container (see pp. 135, 577, 586 for immersed slides). Under high power, find the bright green, cigar-shaped flagellate *Euglena* (Fig. 1-2), with its prominent red "eyespot" containing the pigment astaxanthin. Several species of *Euglena* may be found; some have a highly flexible pellicle resulting in a 'euglenoid motion', a change in shape (Fig. 1-2; also p. 4). When the population density increases, *Euglena* encysts. A dry green rim may be found around an old culture. (Refer to page 663 for culturing; also see phototaxis on page 330.)

Peranema and *Astasia* are colorless euglenoids (Fig. 1-3) found in pond water. These, as well as *Euglena*, may be stained with acidified methyl green (p. 138). Add a small drop to one side of the coverslip and apply a bit of filter paper to the opposite edge. Note that the nucleus turns a greenish hue while the cytoplasm becomes bluish.

In greenish water in a culture jar, look for small, spherical *Chlorella* (Fig. 2-15d). They are often found in the same conditions as *Euglena*. The single chloroplast almost fills the cells so that its cuplike

[22] W. Hoar and C. Hickman, Jr., *A Laboratory Companion for General and Comparative Physiology*, 2nd ed. (Englewood Cliffs, NJ: Prentice-Hall, 1975). MS-222 is available as "Finquel" from Ayerst Laboratories, Veterinary Medical Division, 685 Third Ave., N.Y. 10017.

[23] D. Pace and C. Riedesel, *Laboratory Manual of Vertebrate Physiology*, (Minneapolis: Burgess, 1947).

[24] *Ibid.*

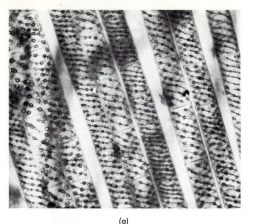

(a)

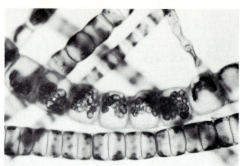

(b)

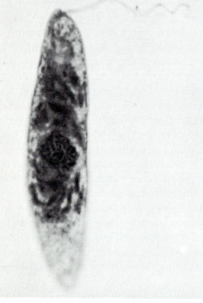

(c)

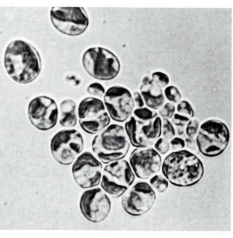

(d)

Fig. 2-15 Algae: (a) *Spirogyra*; (b) *Ulothrix*, a filamentous alga; (c) *Euglena*, a flagellate (stained); (d) *Chlorella*. (a, b, courtesy of General Biological Supply House, Inc., Chicago; c, photo courtesy Carolina Biological Supply Co.; d, photo by Walter Dawn.)

shape may not be visible unless the cells are viewed at a certain angle.

There may be desmids of rare beauty, and some filamentous algae such as *Spirogyra* or *Ulothrix* (Fig. 2-15); or you may find the moneran, *Oscillatoria*, which appears as a blue-green scum on a pond or culture.

Bacteria

Refer to Cautions in handling bacteria, page 697. Temporary slides may be made of

bacteria from several sources: decaying bean suspension, "barrel" pickle or sauerkraut juice, soil water, or water from a pond or ocean sediment. With a Pasteur pipette or a wire transfer loop, spread a *thin* film about the size of a dime along one half of a slide so that you can handle the other end during staining. Let the film air-dry; fix the *dry* film by brushing the bottom of the slide through a flame some 3 to 5 times so that bacteria will not wash off during staining. Lay the slide across the mouth of a tumbler of water and flood the

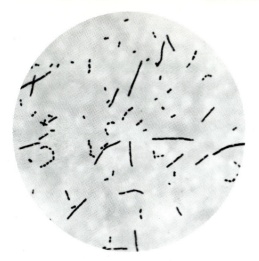

Fig. 2-16 Smear of yoghurt showing both *Lactobacillus bulgaricus* (rods) and *Streptococcus lactis* (chainlike forms). (Courtesy of General Biological Supply House, Inc., Chicago.)

slide with Loeffler's methylene blue stain for 1 to 2 min. Rinse off the stain by dipping the slide in the tumbler of water; blot dry with bibulous paper. Examine under high power and oil immersion to distinguish the shape of the bacteria.

Cocci and bacilli will be found in several preparations; spirilla are less frequent. Especially common in yoghurt are *Lactobacillus bulgaricus* and *Streptococcus lactis* (Fig. 2-16). These bacteria should be visible in the clear areas amid the dark globules of coagulated milk products. Prepare very dilute solutions of the yoghurt. (In sauerkraut or pickle juice, wild yeast cells are found in abundance.)

METHYLENE BLUE (LOEFFLER'S)   This stain keeps indefinitely and is prepared by combining the following ingredients. It also may be purchased. Prepare the two solutions and then mix; filter if necessary.

solution 1:

| methylene blue | 0.3 | g |
| ethanol (95%) | 30 | ml |

solution 2:

| potassium hydroxide | 0.01 | g |
| water | 100 | ml |

This is a simple stain for bacteria which

also shows endospore formation; cell contents stain blue while the spore remains unstained.

GRAM'S STAIN   Bacteria are classified by their ability to hold crystal violet dye. Bacteria that are "Gram-positive" have cell walls that are more sensitive to dehydration by alcohol which seems to close the pores of cell walls so that the dye is retained. Bacteria that lose the dye are "Gram-negative;" their cell walls have a higher lipid content. When the decolorizer is added, the crystal violet is washed out of the cell.

Use young cultures to determine whether bacteria are Gram-positive, that is, cultures that are 24 hrs or younger for more accurate results with Gram's stain. Place a slide with a dry fixed bacteria film across a tumbler or baby food jar of water.

1. Add drops of Gram's crystal violet stain for 1 min.

2. Wash briefly by dipping the slide in the water.

3. Add drops of a mordant, Gram's iodine solution, to fix the stain for 1 min.

4. Wash off with tap water.

5. Add 95 percent alcohol as a decolorizing solution—the most critical step; tilt the slide and add one drop at a time until no more color runs off. In this step, Gram-negative bacteria lose the violet color.

6. Counterstain the decolorized Gram-negative bacteria with Gram's safranin solution for 10 to 20 sec.

7. Wash off the excess stain and air-dry.

GRAM'S CRYSTAL VIOLET   Prepare the two solutions:

solution 1: Dissolve stain in alcohol

| crystal violet | 2.0 g |
| ethanol (95%) | 20 ml |

solution 2: Dissolve ammonium a-oxalate (0.8 g) in 80 ml of water.

Mix in *equal* parts, that is, dilute solution 1 so it is equal to the *volume* of solution 2 that is used.

GRAM'S IODINE   Dissolve 5 g $NaHCO_3$ in 100 ml water. Mix the potassium iodide and the iodine in a mortar and grind with a pestle. Add the water and the bicarb solution and mix well. Pour into individual dropper bottles for students' use. When Gram's iodine loses its color, discard and prepare a fresh solution. Maintain in dark bottles.

iodine CP	1 g
potassium iodide	2 g
sodium bicarbonate (5%)	60 ml
water	240 ml

GRAM'S SAFRANIN   Dissolve the stain in alcohol and mix, then add the water and filter if necessary.

safranin	0.25 g	
ethanol (95%)	10	ml
water	100	ml

Cultures of pure bacteria may be purchased; especially useful are *Bacillus subtilis* (rods), *Sarcina lutea* (coccus or spheres), and *Rhodospirillum rubrum* (spiral forms). These are readily available, harmless, and large enough to be seen under high power. Also available are harmless chromatogenic bacteria,[25] which are especially useful for illustrating the separation of bacteria by color of colonies. See Koch's methods for isolating bacteria on solid medium (pp. 619, 702).

The time for staining varies with the thickness of the film, the concentration of the stain, and the kinds of bacteria under study. With practice, students will learn how many minutes it takes to stain the bacterial film. Coverslips need not be used. For permanent slides, however, add 1 drop of balsam when the slide is thoroughly dry, and then apply a coverslip.

Refer to the preparation of hanging drop slides to observe motility, as found in *Bacillus subtilis* (p. 153).

Additional techniques for handling bacteria, such as cell counts, culturing, isolating bacteria of soil, water, and milk are

described (pp. 458–59); metabolic products (p. 706); nitrogen-fixing bacteria, and bacteria that cause crown gall in tomatoes (p. 619).

CRYSTAL VIOLET STAIN   This is a simple bacterial stain. Spread a loop of a 24-hr culture on the center of a clean slide and air-dry. Fix the bacteria to the slide by passing the bottom 3 to 4 times across the flame. Apply drops of crystal violet stain to cover the bacteria film. After 30 sec, rinse off in tap water; blot and air-dry. Examine the slide under oil for shape of cells and arrangement of the cells in chains, in clusters, and other identifiable patterns.

To prepare crystal violet stain, dissolve 4 g of crystal violet in 20 ml of 95 percent ethyl alcohol. In a container of 100 ml of distilled water dissolve 1 g of ammonium oxalate. Then *mix* the two solutions thoroughly and store in airtight dropper bottles or other containers.

Fungi

Yeast cells   *Saccharomyces cerevisiae* is the yeast used in bread and beer making. Transfer a drop of actively fermenting sugar medium containing yeast cells (see culturing, p. 712) to a clean slide which has a dried drop of methylene blue stain or neutral red stain. Apply a coverslip and examine under high power; observe the stain slowly diffuse into the yeast cells. Look for budding cells and whole colonies of undisturbed cells.

Budding of yeast cells may also be studied in a very dilute culture in a hanging drop preparation (p. 153); examine the edge of the drop.

Permanent slides of yeast cells may also be prepared using toluidine blue stain or Loeffler's methylene blue stain. Air-dry a *thin* film or smear, about a dime in diameter, of a yeast culture on a slide with a drop of water. Brush through a flame 2 to 3 times to fix the cells to the slide. Stain with 1 percent toluidine blue for 3 min; rinse off the stain in water and air-dry. Or apply Loeffler's blue stain as used for

---

[25] Some examples of chromatogenic bacteria are *Serratia marcescens*, *Pseudomonas aeruginosa*, and *Sarcina lutea*.

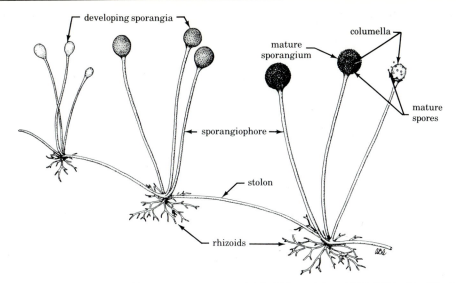

developing sporangia

columella

mature
sporangium

mature
spores

sporangiophore

stolon

rhizoids

Fig. 2-17 *Rhizopus*, bread mold. (Adapted from *Botany: An Introduction to Plant Biology*, 6th ed., by T. Elliot Weier et al. Copyright © 1982 by John Wiley & Sons, Inc.)

bacteria (p. 144). Add a small drop of clear Karo syrup and seal with a coverslip.

FAT GLOBULES   Use *S. cerevisiae* that has been grown on malt extract agar (Difco); preparation (p. 698). Mix together a loopful of yeast culture and a loopful of Sudan black B on a clean slide. After 10 min, observe the stained fat globules under high power (and under oil, if possible).

To prepare Sudan black B stain, dissolve the dye in alcohol, add 25 ml of water and mix well.

Sudan black B	0.3 g
ethanol (95%)	75 ml

GLYCOGEN   Prepare a wet mount of actively growing yeast cells. Add a drop of Gram's iodine or Lugol's solution (pp. 146, 137). Observe the dark, reddish brown areas within the cells that are much darker than iodine color.

For studies of fermentation, use side arm tubes or Durham fermentation tubes; see anaerobic respiration, page 312.

Molds   Transfer a small amount of any mold at hand to a drop of glycerin or detergent on a clean slide (glycerin and detergent reduce air bubbles). Refer to culturing methods, pages 709 and 711 for detailed studies.

To identify molds, examine the color and arrangement of sporangia under low and high power. Compare *Penicillium* (Fig. 1-121) with the conidial head of *Aspergillus niger* (black mold), and with *Rhizopus stolonifer*, the black bread mold that has large black sporangia on top of tall sporangiophores (Fig. 2-17).[26]

If possible, also examine parasitic molds (p. 618).

Examining tissue cells: multicellular organisms

Onion bulb cells

Slice a raw onion and cut the rings into 6 mm sections. With forceps or with a fingernail remove the *inner* transparent membrane and flatten it in a drop of water on a slide. Under low power, note how difficult it is to distinguish the components of these long rectangular cells. However,

[26] Germinate conidia of *Aspergillus* on Sabouraud dextrose agar, and those of *Penicillium* on potato dextrose agar (p. 710).

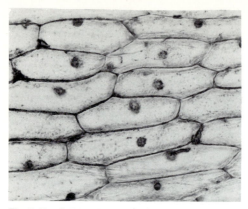

Fig. 2-18 Onion cells. (Photo courtesy Carolina Biological Supply Co.)

in reduced light, streaming of cytoplasm may be seen in fresh young onion cells. You may want to prepare a dark field stop (p. 123) for increased visibility.

Stain the cells with dilute Lugol's iodine solution or methylene blue (p. 136). Apply a folded piece of lens paper or bibulous paper to one side of the coverslip to draw out the water; add a drop of the stain to the opposite side of the coverslip, and watch how the stain flows under the coverslip. Students quickly recognize the value of using stains in examining cells (Fig. 2-18). To measure the length of an onion cell, see page 124.

NUCLEIC ACIDS IN ONION CELLS The dye toluidine blue stains nucleic acids in onion cells. Dice part of an onion bulb and peel off a small section of the transparent epidermal layer; place in a drop of $1N$ HCl on a slide. Gently heat the slide, but do not boil; add another drop of HCl if needed to prevent drying. Blot off any excess acid and add a drop of toluidine blue (p. 138). Heat gently for 1 min and blot away any excess stain. Add a drop of water and coverslip. Examine under low and high power; note the distribution of nucleic acids in the cells.

Epithelial tissue

Gently scrape the lining of the cheek with the *flat* end of a toothpick and transfer into a small drop of dilute Lugol's iodine solution (or dilute methylene blue stain) on a slide (p. 136). Stir the material in the stain with the toothpick so that the cells will not clump together; apply a coverslip and examine under low power. Center some cells and switch to high power (Fig. 2-19). Notice how the nucleus of each cell stains brown while cytoplasm granules stain a very light brown in Lugol's stain (or shades of blue in methylene blue stain). Other materials of sputum, such as free myelin globules and bronchial epithelial cells, may be found in addition to large, scattered masses of stained squamous epithelial cells. Measure several epithelial cells which are about 97μ, with a nucleus about 7μ.

Compare these epithelial cells with those found in the skin of a frog, and in ciliated epithelial cells, also with ciliated cells of a snail (p. 66), of a clam (p. 70), and of the scallop, *Pecten* (Fig. 1-85).

Elodea (Anacharis)

Observe chloroplasts in algae, such as the spiral ribbon in *Spirogyra* (Fig. 1-116) or the oval ones in *Euglena* (Fig. 1-2), or in leaf cells of the aquarium plant *Anacharis*, lettuce, or thin moss leaflets. The cells of elodea leaves are only a few layers thick; students will need to learn to focus at different levels of cells. Focus on the upper level of a cell to see chloroplasts scattered throughout. A shift in focus down to the middle of the cell shows chloroplasts forming a ring around the inner edge of cells (Fig. 3-7).

After observing chloroplasts, some teachers discuss the composition of chlorophyll using chromatograms (pp. 205, 207), and then begin a study of photosynthesis (chapter 3).

Cyclosis

The streaming of cytoplasm may be seen within several kinds of cells. Many plant cells, especially young growing cells, show cyclosis. Place germinating radish or small grass seeds in depression slides. Observe

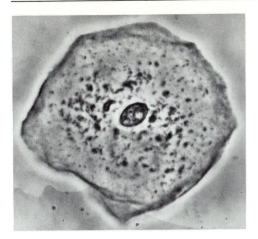

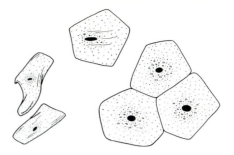

Fig. 2-19 Epithelial tissue lining the mouth. (Photo courtesy Carolina Biological Supply Co.)

the streaming of cytoplasm and vacuoles in the root hairs. Freshly picked spinach, when available, shows cyclosis in the stripped epidermis of petioles. Mount in water and examine under high power.

Many aquatic plants show cyclosis. Examine young cells of *Anacharis* (elodea), especially those from growing tips. Note the streaming cytoplasm, floating chloroplasts, the oval, dense nucleus, the central sap vacuole, and the cell wall. Focus on the cells of the midrib, as well as the cells at the edges of leaves. Keep the *uppermost* layer of cells facing upward on the slide. If necessary, stimulate cyclosis by placing leaves in warmed water, or bring a warming light near a jar of *Anacharis*. In some techniques, a bit of thiamine chloride (vitamin B_1) is added to the water. The rate of streaming may be from 3 to 15 cm per hour and as high as 45 cm per hour at a temperature around 30° C (86° F). At times, it may be sufficient to warm the slide in the palm of the hand.

In addition, mount a leaf of *Nitella*, *Chara*, or *Vallisneria* on a clean slide in a drop of water. Focus on the internodal cells of *Nitella* and *Chara*. Note the movement of the many nuclei in the internodal cells of *Nitella*. However, the chloroplasts do not move; they are fixed within the inner surface of the cell walls.

Mount several threads of the mycelium of the bread mold *Mucor* in water or glycerin (culturing, p. 709). Cytoplasm streams up one side of the threads and down the other side.

Examine the unicellular hairs on the roots of *Tradescantia* seedlings, or the staminate hairs in the flower. Mount a filament of the stamen, which has several hairs attached, in warm water on a slide. Granules may be seen moving from the strands around the nucleus along the wall to another strand leading to the nucleus. At times the direction of streaming reverses.

Streaming is studied easily in one of the large amoebae — *Pelomyxa*, sometimes called *Chaos chaos*. Mount in a drop of culture solution (p. 658). Include a small bristle or broken coverslip in the drop to avoid crushing the specimen when the coverslip is applied. Students should see the many vacuoles and the actively streaming cytoplasm, which changes from sol to gel. The clearer ectoplasm and the denser endoplasm illustrate this change.

Chromoplasts
While chloroplasts contain chlorophyll and carotenoid pigments, chromoplasts contain red or yellow-orange pigments. Examine the scrapings from the skin of a ripe tomato in a drop of water under high power.

Striated muscle in beef
In a drop of water on a clean slide tease apart a small piece of raw, lean beef.

With dissecting needles separate the fibers, then transfer a few to a slide with a dried film of dilute methylene blue stain. Add a small drop of water and apply a coverslip. Wrap the slide in several layers of toweling and gently press on the coverslip with a thumb or pencil eraser to spread the fibers. Examine under low power, center the slide, and switch to high power. Careful focusing with the fine adjustment knob and proper illumination should reveal clear cross-striations in this voluntary muscle tissue (Fig. 2-20b). Compare with a similar preparation from the thigh muscle of a freshly killed frog.

Human blood cells

Preparation of wet mounts of human blood in mammalian Ringer's solution is described on pp. 155, 156 and blood cells, p. 274. Safety precautions for drawing blood and the staining of blood smears with Wright's stain also are described.

Compare nucleated red cells of the frog's blood (Fig. 2-21d) with non-nucleated red cells of human blood.

Tissues of a frog

Epithelial tissue   A live frog placed in a jar (with about 3 cm of water) will desquamate in about 24 hr. Mount small pieces of the sloughing skin in dilute Lugol's solution or methylene blue (p. 137) on a clean slide; apply a coverslip. These mounts will last several hours if rimmed with petroleum jelly.

To examine chromatophores. Transfer a small bit of skin from the throat of the frog to a drop of water on a slide. Observe the branching cells containing melanin. Against a dark background, melanin granules move out to the branches making the skin darker. You may observe other colors, due to yellow pigment cells and to bluish tints from crystals of guanine, in some skin cells.

Observe the action of ciliated epithelial tissue. Open the mouth of a pithed frog and made an incision on each side of the mouth. Wash with frog Ringer solution. Place a bit of carmine powder on the roof of the mouth with a toothpick; examine the currents created by ciliated epithelial tissue that carry the red particles.

Students can also prepare temporary mounts of ciliated epithelial cells from the mouth of a freshly killed frog. Mount cells in 0.7 percent salt solution. What is the effect of hypotonic and hypertonic solutions on ciliary behavior and on cells (p. 155)? After careful study, show the effect of different ions on ciliary action (p. 153). Also refer to clam gill cilia (p. 70).

Dissect out small bits of the stomach, intestine, and lining of the mouth of a freshly pithed frog in separate Syracuse dishes, and add a macerating solution such as 5 percent chloral hydrate (*caution:* poison). Or place the tissue bits in 30 percent alcohol for 12 hr, or in other macerating solutions (p. 160). After 24 to 48 hrs, tease apart the tissues with dissecting needles and mount in a drop of the solution or in water. Identify the cell membrane, cilia, cylindrical cells, and columnar epithelial cells from the small intestine. Compare squamous and columnar epithelial cells.

Sperm cells   Crush the testes of a freshly killed frog (pp. 91, 93) in a Syracuse dish of isotonic frog Ringer's solution (p. 155) to dilute sperm fluid. Mount a drop with a Pasteur pipette on a clean slide and examine under high power. After a few minutes, observe the oscillating sperm heads. Then add dilute Lugol's solution to reveal details as well as flagella, although the stain kills the sperm cells.

Nerve cells   Dissect out small pieces of the spinal cord of a freshly killed frog and transfer to a clean slide. Gently press a second slide against this, mashing the tissue; hold the two slides parallel and pull apart, leaving a smear on each. Nerve cells, nuclei, cell contents, and bits of connective tissue should be treated with a dilute stain. Make permanent slides by drying these smears in air; then treat and stain them like fixed tissue (p. 165). Also

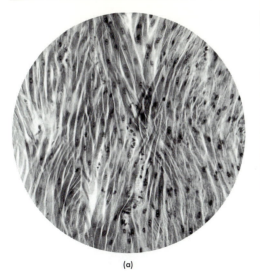

(a)

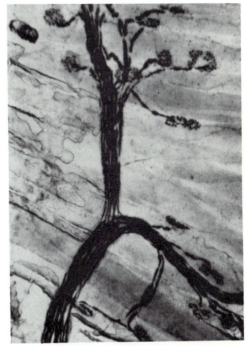

(b)

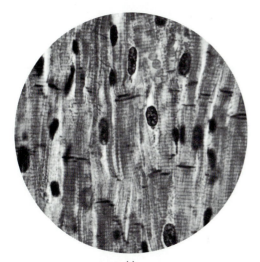

(c)

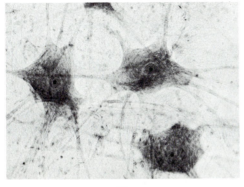

(d)

(e)

Fig. 2-20   Prepared slides of muscle and nerve cells: (a) smooth muscle as seen in section; (b) striated muscle under high power; (c) heart muscle, stained to show intercalated disks; (d) motor nerve cells; (e) motor nerve endings in muscle tissue. (Photos courtesy Carolina Biological Supply Co.)

look for mitochondria, fat globules, nuclear constituents, and intra-cellular fibrils.

Blood cells Study a wet mount of fresh frog blood in a drop of frog Ringer's solution (p. 155). Prepare blood smears and stain with Wright's blood stain (p. 272). Compare red and white corpuscles in the frog with those of human blood cells. Also stain mitochondria in blood cells (p. 273).

Muscle cells Tease apart a bit of macerated tissue from the stomach (previously described) and from the thigh of a freshly killed frog. To make the nucleus more visible, crush a bit of tissue on a slide and add a drop of acetic acid. Stain with methylene blue or Lugol's solution (p. 137). What is the difference between smooth and striated muscle cells (Fig. 2-20)?

Tissues of Hydra

Place a hydra, preferably a green one, on a slide containing pond water, and tease it apart with dissecting needles. This maceration will release the symbiotic green zoochlorellae.

nucleus

cytoplasm

cell membrane

(a)

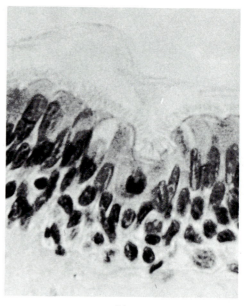

(b)

(c)

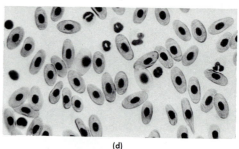

(d)

Fig. 2-21 Tissues of a frog: (a) epithelial tissue (skin); (b) ciliated epithelium; (c) macerated ciliated epithelial cells; (d) blood cells stained with Wright's stain. (b, c, courtesy of General Biological Supply House, Inc., Chicago; d, courtesy of Carolina Biological Supply Co.)

Under high power examine the large epitheliomuscle cells of the living animal. If you want to stain the cells, first fix the cells, then add stain and macerate the stained cells as follows. Prepare a solution by mixing 1 part of glycerin and 1 part of glacial acetic acid (*caution*) with 2 parts of water. Then add 2 drops of this fixative to a clean slide; add the hydra. After 3 min add 1 drop of methyl violet stain (p. 138). Leave this for a few minutes. Then draw off the stain and wash the slide in a bit of water. Now macerate the stained hydra with dissecting needles in a drop of water. Add a coverslip and examine under high power (Fig. 1-21).

To study the nematocysts of a hydra, mount a hydra in pond water on a slide. Include small bristles to prevent the coverslip from crushing the specimen. A small drop of safranin stain or a dilute acid added to the wet mount will release the nematocysts (Fig. 1-23).

Ciliated epithelial tissue and ion antagonism

Observe the effect of ion antagonism in the motion of ciliary epithelial tissue from the gills of a living clam, mouth of a living frog (p. 150), or intestine of an aquarium snail (p. 66). Dissect out several pieces of the gill of the clam in 10 ml seawater in small watch glasses. Under low power watch the free ends of the cells that show vigorously beating cilia. Add a drop of carmine suspension, and note the currents that sweep the particles of carmine.

Now transfer the pieces of gill to the following solutions in watch glasses and examine each under the microscope. First, place them into 0.9 percent potassium chloride solution; examine the motion of cilia. Then transfer the tissue into a 1 percent calcium chloride solution. Finally, return the tissue to clam Ringer's solution (p. 156). Compare ciliary motion in each solution. (Also refer to the effect of ions on the heart of a frog on page 266.)

Chromatophores

Mount scales of live fish in Ringer's solution to observe chromatophores (pigment-

Fig. 2-22   Hanging drop slide.

containing cells). To watch the contraction and expansion of these cells, draw off the Ringer's solution on the slide by putting filter paper on one side of the coverslip as a drop of chloretone (p. 143) is added on the opposite side. After examining these mounts, use the same procedure to draw off the chloretone, and add a drop of adrenalin (epinephrine) or potassium chloride to the same slide. Watch the contraction of the chromatophores. In this way, by altering the size of the chromatophores, animals such as fish, amphibians, and several kinds of invertebrates are able to simulate the coloration or the varied intensity of the shadows in their background. (Refer to chromatophores in frog skin on page 150.) Light is usually the original stimulus for the change in the chromatophores.

Additional techniques

Hanging drop slides   If a drop of water on a slide is examined under the microscope, light is reflected in several directions. To avoid this, flatten the drop with a coverslip. However, in so doing, the motility, if any, in organisms is reduced by the "crush" of the coverslip. A hanging drop preparation enables the observation of fission in protozoa, germination of pollen grains, and motility in bacteria and similar subjects. Use a clean slide and coverslip, so that the surface tension of the water is not reduced (or a drop will not form). Place a small drop of the culture medium on a coverslip. Then, with a toothpick, apply petroleum jelly to the four corners of a coverslip; or dip the mouth of a test tube into soft petroleum jelly and apply over the rim of the depression on the slide. Now invert the slide *over* the coverslip and seal the slide and coverslip in place. Quickly flip over the whole preparation; it should resemble the one in Figure 2-22 with the drop hanging into the depression of the slide.

Soft wax is especially useful in sealing coverslips in a hanging drop preparation; use 1 part petroleum jelly to 1 part of soft paraffin. Apply the melted "soft wax" with a toothpick to the four corners of the coverslip.

When motility in bacteria is under study, such as that in *Bacillus subtilis*, the cells are translucent; but a loopful of methylene blue diluted to 1:10,000 (0.01 g methylene blue in 100 ml water) increases visibility when added to the loopful of *B. subtilis* and other forms.

Cultures in depression slides Temporary hanging drops dry out quickly. A preparation which is easier to make for some studies (and more lasting) is one that uses the concavity or depression in the slide for the culture. Shallow concavity slides are for more temporary use; but the straight-walled, deep-welled slides are useful for culturing single specimens to count rate of fission, or for growing a culture from only a few specimens. However, these are thicker slides so that care must be taken in examining under high power. Straight-walled slides have a concavity that measures 3 mm deep and 16 mm in diameter. They are useful for quantitative studies such as in population counts. Place 2 to 3 small drops of the culture to be examined in the concavity; ring with petroleum jelly or soft wax (p. 153) and apply a coverslip. In addition to studies of the rate of reproduction of isolated protozoa or algae in depression slide cultures, you may want to use this means for isolating microorganisms. Use a transfer loop or add a small drop of thick methyl cellulose to the depression; spread as a thin film. Using a Pasteur pipette, or a pulled out medicine dropper, squirt a drop of culture medium onto the methyl cellulose film. Transfer single specimens that are trapped in the film to fresh medium in clean welled slides or small Petri dishes (additional methods follow).

MINI-AGAR CULTURES Place a thin agar preparation (1 percent) in the bottom of deep-welled slides and inoculate with mold spores. Algae may also be grown in deep-welled slides; streak the surface of agar with a transfer needle. In a way, these may be considered miniature Petri dishes for individual students, and they need little storage space.

The succession of microorganisms in these sealed liquid or agar film microaquaria may be examined over a period of 10 to 15 days; concepts of interdependence of organisms can be clearly developed, especially when controls are also used.

Liberation of oxygen bubbles in photosynthesis by algae, the effect of chemicals on growth, and the effects of other altered environmental conditions and pollutants on microorganisms can be traced in depression slide cultures. This technique is often referred to in specific context in other chapters (especially photosynthesis and ecology).

To retard evaporation, even in sealed slides, place the slides in plastic boxes lined with moist filter paper.

Freehand sectioning Most tissues are not rigid enough to slice into thin sections for examination under a microscope. The usual procedure is to embed tissue in some paraffin mixture. A more rapid method is to freeze the tissue with carbon dioxide from a carbon dioxide cartridge, and then to cut sections.

Nevertheless, some plant leaves and some woody stems may be prepared as temporary mounts simply by inserting them between lengths of elderberry pith, raw potato or carrot and then cutting slices.

Slice a length of elderberry (or sunflower) pith or a raw potato or carrot in half, and between the two halves sandwich a piece of a leaf, possibly of privet, or of a woody stem. Wrap these halves together tightly and soak in water for a few hours. The pith and the enclosed tissue specimen expand and become rigid enough to cut.

Then the material may be sliced at a slight angle with a sharp razor blade into thin sections. Keep the blade as well as the tissue wet, and float the sections on water

TABLE 2-3 Physiological saline solutions

invertebrates	0.75%	(7.5 g NaCl in 1 liter distilled water)
insects	0.6–0.8%	(6–8 g NaCl in 1 liter distilled water)
frogs	0.64%	(6.4 g NaCl in 1 liter distilled water)
salamanders	0.8%	(8 g NaCl in 1 liter distilled water)
birds	0.75%	(7.5 g NaCl in 1 liter distilled water)
mammals	0.9%	(9 g NaCl in 1 liter distilled water)

* G. Humason, *Animal Tissue Techniques*, 3rd ed. (San Francisco: Freeman, 1972).

so they will not curl. A better method, of course, is to insert the carrot-and-leaf preparation into a hand microtome and slice with a microtome razor blade (Fig. 2-23); the microtome, which can be adjusted for thickness produces uniformly thin slices.

Staining is optional; methylene blue, eosin, or Lugol's iodine solution may be used.

Isotonic solutions

Isotonic solutions have the same osmotic pressure as normal tissue or blood serum of organisms. In a hypertonic salt solution, for example, red blood cells shrink since the salt concentration is greater than the blood serum, while the cells swell in a hypotonic salt solution or in distilled water.

Physiological saline solutions
Humason recommends the following saline solutions for different organisms (see Table 2–3).

Other isotonic solutions may be preferred, as follows.

PHYSIOLOGICAL SALINE SOLUTION FOR COLD-BLOODED ANIMALS (GENERAL) Use as a mounting fluid in the preparation of temporary wet mounts. Prepare a 0.7 percent solution of sodium chloride in distilled water.

PHYSIOLOGICAL SALINE SOLUTION FOR WARM-BLOODED ANIMALS (GENERAL) Prepare a 0.9 percent solution of sodium chloride (dissolve 0.9 g in 100 ml of distilled water).

Ringer's solution
Ringer's solution should be prepared fresh before use. This solution is isotonic for animal tissues and may be used as a mounting fluid for living tissues and for examination of blood cells. The amounts of salts, especially sodium chloride, vary for different animals.

RINGER'S SOLUTION FOR FROG TISSUE The heart of a dissected frog will continue to beat for several hours if bathed in this buffered solution. In fact, the heart may be removed and suspended in frog Ringer's solution (p. 266). Dissolve the following salts in 1 liter of distilled water:

potassium chloride	0.14 g
sodium chloride	6.50 g
calcium chloride	0.12 g
sodium bicarbonate	0.20 g

RINGER'S SOLUTION FOR MAMMALIAN TISSUE Dissolve the following salts in 1 liter of distilled water: water:

potassium chloride	0.42 g
sodium chloride	9.00 g
calcium chloride	0.24 g
sodium bicarbonate	0.20 g

RINGER'S SOLUTION FOR BIRD TISSUE Use the same solution as for mammalian tissue, but decrease the sodium chloride to 8.5 g.

Fig. 2-23 Section razor and hand microtome for making freehand sections of plant tissue.

RINGER'S SOLUTION FOR INVERTEBRATES

Giese recommends the following proportions of salts for several invertebrates.[27]
Dissolve in 1 liter of distilled water:

	crayfish	crab	insects
potassium chloride	0.5 g	0.75 g	0.77 g
sodium chloride	10.0 g	29.00 g	9.30 g
calcium chloride	0.5 g	4.4 g	0.5 g

An additional recipe for insect Ringer's solution contains dextrose.

RINGER'S SOLUTION FOR INSECTS

Dissolve the following in distilled water; add sodium carbonate to reach pH 7.2. Bring total volume to 1 liter with water.

sodium chloride	9.0 g
potassium chloride	0.2 g
calcium chloride	0.27 g
($CaCl_2 \cdot 2H_2O$)	
dextrose	4.0 g

Earle's solution

This is another physiological solution isotonic for living tissues or small organisms. Dissolve the following salts and glucose in 1 liter of distilled water:

potassium chloride	0.4 g
sodium chloride	6.8 g
calcium chloride	0.2 g
sodium bicarbonate	0.15 g
sodium phosphate (monbasic)	2.2 g
magnesium sulfate	0.1 g
glucose	1.0 g

Locke's solution

This physiological isotonic solution may be used to mount fresh mammalian blood samples or chick embryos. It is also a component of some special culture media for protozoa. The solution should be sterilized in an autoclave. Dissolve the following salts in 1 liter of distilled water:

potassium chloride	0.42 g
sodium chloride	9.0 g
calcium chloride	0.24 g
sodium bicarbonate	0.3 g
glucose (optional)	1.0 g

[27] Giese, *Blepharisma.*

Locke's solution may be used for cold-blooded organisms by decreasing the sodium chloride to 6.5 g.

Tyrode's solution

This is a balanced physiological saline solution for mammalian tissue. Dissolve the salts in 1 liter of distilled water. Add $CaCl_2$ and shake well.

NaCl	8.0 g
KCl	0.2 g
$MgCl_2$	0.1 g
$NaHCO_3$	1.0 g
Na_2HPO_4	0.05 g
$CaCl_2$	0.2 g

Indicators

Indicators are dyes used to test the pH of a solution. As the hydrogen-ion content of the solution changes, a rearrangement of the indicator molecule occurs and a color change results. Figure 2-24 lists the range of a sampling of indicators; notice that some of the indicators commonly used in demonstrations and laboratory work are those that show a shift in pH around neutral (7). The preparation of the dye is usually given on the bottle of indicator. However, the preparation and dilution of some frequently used indicators are given below.

A convenient ColorpHast Store containing pH indicator strips ranging from 0 to 2.5 to 11 to 13 is available from MCB Manufacturing Chemists Inc., 480 Democrat Rd., Gibbstown, NJ 08027. In addition, there are times you may want to use a Universal Indicator solution (ready-made from Kodak) which is red at pH 4, shading into orange at pH 5, yellow at pH 6, light green at 6.5 and darker shades of green at pH 7. The color is deep green at pH 8 to 8.5; then blue at pH 9 to 9.3 into violet at pH 10.

Table 2-4 gives the pH values of 0.1 N solutions of common acids and bases; Table 2-5 offers the approximate pH values of a variety of substances. Also refer to the preparation of several buffer solutions on page 159, and in chapter 10.

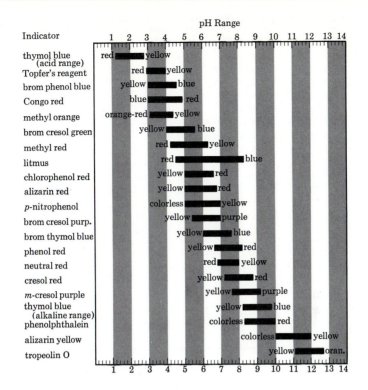

pH Range

Fig. 2-24 Range of several indicators. (Data from P. Hawk, B. Oser, and W. Summerson, *Practical Physiological Chemistry*, 13th ed., Blakiston [McGraw-Hill], New York, 1954.)

TABLE 2-4   pH values of 0.1 N solutions of a variety of acids and bases*

acids (order of decreasing strength)	pH value	bases (order of increasing strength)	pH value
Hydrochloric acid	1.0	Sodium bicarbonate	8.4
Sulfuric acid	1.2	Borax	9.2
Phosphoric acid	1.5	Ammonia	11.1
Sulfurous acid	1.5	Sodium carbonate	11.36
Acetic acid	2.9	Trisodium phosphate	12.0
Alum	3.2	Sodium metasilicate	12.2
Carbonic acid	3.8	Lime (saturated)	12.3
Boric acid	5.2	Sodium hydroxide	13.0

* Data from *Handbook of LaMotte Chemical Control Units for Science and Industry*, 13th ed., LaMotte Chemical Products Co., Chestertown, Md., 1944.

ALIZARIN RED

sodium alizarin monosulfonate	1 g
distilled water	100 ml

Indicator changes from yellow at pH 5.5 to red at pH 6.8

BROMOCRESOL GREEN

bromocresol green	0.04 g
ethyl alcohol (95%)	100 ml

Indicator turns from yellow at pH 4 to blue at pH 5.6

TABLE 2-5   Approximate pH of some common substances*

Apples	2.9–3.3	Human duodenal		Pickles, sour	3.0–3.5
Apricots (dried)	3.6–4.0	contents	4.8–8.2	Pimento	4.7–5.2
Asparagus	5.4–5.7	Human feces	4.6–8.4	Plums	2.8–3.0
Beans	5.0–6.0	Human gastric		Pumpkins	4.8–5.2
Beer	4.0–5.0	contents	1.0–3.0	Raspberries	3.2–3.7
Beets	4.9–5.6	Human milk	6.6–7.6	Rhubarb	3.1–3.2
Blackberries	3.2–3.6	Human saliva	6.0–7.6	Salmon	6.1–6.3
Bread, white	5.0–6.0	Human spinal fluid	7.3–7.5	Sauerkraut	3.4–3.6
Cabbage	5.2–5.4	Human urine	4.8–8.4	Shrimp	6.8–7.0
Carrots	4.9–5.2	Jams, fruit	3.5–4.0	Spinach	5.1–5.7
Cherries	3.2–4.1	Jellies, fruit	3.0–3.5	Squash	5.0–5.3
Cider	2.9–3.3	Lemons	2.2–2.4	Strawberries	3.1–3.5
Corn	6.0–6.5	Limes	1.8–2.0	Sweet potatoes	5.3–5.6
Crackers	7.0–8.5	Magnesia, milk of	10.5	Tomatoes	4.1–4.4
Dates	6.2–6.4	Milk, cow	6.4–6.8	Tuna	5.9–6.1
Flour, wheat	6.0–6.5	Molasses	5.0–5.4	Turnips	5.2–5.5
Ginger ale	2.0–4.0	Olives	3.6–3.8	Vinegar	2.4–3.4
Gooseberries	2.8–3.1	Oranges	3.0–4.0	Water, distilled	
Grapefruit	3.0–3.3	Peaches	3.4–3.6	(carbon-dioxide-free)	7.0
Grapes	3.5–4.5	Pears	3.6–4.0	Water, mineral	6.2–9.4
Hominy	6.9–7.9	Peas	5.8–6.4	Water, sea	8.0–8.4
Human blood		Pickles, dill	3.2–3.5	Wines	2.8–3.8
plasma	7.3–7.5				

* Data from *Handbook of LaMotte Chemical Control Units for Science and Industry*, 13th ed., LaMotte Chemical Products Co., Chestertown, Md., 1944.

BROMTHYMOL BLUE

dibromthymolsulfonphthalein	0.04 g
ethyl alcohol (95%)	100 ml

Indicator turns from yellow at pH 6 to blue at pH 7.6

CHLOROPHENOL RED

chlorophenol red	0.04 g
ethyl alcohol (95%)	100 ml

Indicator turns from yellow at pH 6 to red at pH 6.6

CONGO RED

congo red	0.5 g
distilled water	90 ml
Then add:	
ethyl alcohol (95%)	10 ml

Indicator turns from blue at pH 3 to orange-red at pH 5

CRESOL RED

cresol red	0.04 g
ethyl alcohol (95%)	100 ml

Indicator turns from yellow at pH 7.2 to red at pH 8.8

LITMUS

litmus powder	5 g
water (boiling)	50 ml

Add boiling water to litmus; let stand for 15 min and decant the liquid. Make up to 150 ml by adding 100 ml boiling water. Filter after 24 hrs. Indicator turns from red at pH 4.4 to blue at pH 8.4

METHYL ORANGE

methyl orange	0.1 g
distilled water	100 ml

Indicator turns from orange red at pH 3.1 to yellow at pH 4.4

METHYL RED[28]

methyl red	1 g
ethyl alcohol (95%)	300 ml

[28] O. Hepler, *Manual of Clinical Laboratory Methods*, 4th ed. (Springfield, IL: Charles Thomas, 1968).

Dilute by making up to 500 ml with distilled water. Indicator is red at pH 4.2 and turns to yellow at pH 6.3

NEUTRAL RED

neutral red	0.5 g
ethyl alcohol (95%)	300 ml

Dilute by making up to 500 ml with distilled water. Indicator is red at pH 6.8 and yellow at pH 8.

PHENOL RED

phenol red (phenolsulfonphthalein)	0.02 g
ethyl alcohol (95%)	100 ml

Indicator turns from yellow at pH 6.6 to red at pH 8.2

PHENOLPHTHALEIN

phenolphthalein	1 g
ethyl alcohol (95%)	100 ml

Indicator turns from colorless at pH 8.3 to red at pH 10. (For very sensitive tests, use a 0.1 percent solution.)

THYMOL BLUE

thymol blue (thymolsulfonphthalein)	0.04 g
ethyl alcohol (95%)	100 ml

Two ranges: alkaline from yellow at pH 8.2 to blue at 9.8; acid from red at pH 1.2 to yellow at pH 2.8 yellow at

TOPFER'S REAGENT

dimethylaminoazobenzene (*Caution*)	0.5 g
ethyl alcohol (95%)	100 ml

Indicator turns from red at pH 2.9 to yellow at pH 4.0

NATURAL PLANT JUICES   Natural plant juices can also be used as indicators for some purposes. For example, boil shredded red cabbage in a small amount of water and then filter. Or boil thin strips of the purple skin of eggplant in a small amount of water. Add drops of the indicator to 5 ml samples of vinegar, lemon juice, and bicarbonate of soda. Use standard pH indicators to establish the pH range of the natural plant juices.

Buffer solutions

Buffer solutions are solutions that maintain their hydrogen-ion concentration in spite of additions of quantities of acid or alkali. Many are suggested in the literature. They may be mixtures of monobasic phosphates, such as Na_2HPO_4, and a dibasic phosphate, such as NaH_2PO_4; there may be combinations of sodium and potassium phosphates. Other buffers are mixtures of weak acids and their salts: carbonic acid and sodium bicarbonate, or acetic acid and sodium acetate. Still others contain organic compounds. Buffer concentrates in 500 ml solution containers, prepared in graduated pH units, are available from MBC Manufacturing Chemists Inc., 480 Democrat Rd., Gibbstown, NJ 08027. Two buffer solutions are described here, others are described in context in chapters 3 and 10.

PHOSPHATE BUFFER $(0.2\ M)$ Humason suggests this preparation.[29] Prepare stock solutions of each of the following. For the desired pH, mix the quantities in ml as shown in Table 2-6. Check with a pH meter if necessary.

Monobasic sodium phosphate: Dissolve 27.6 g and make up to 1 liter with distilled water.

Dibasic sodium phosphate: Dissolve 53.6 g and make up to 1 liter with distilled water.

SODIUM ACETATE BUFFERS   Harrow, et al, offer this table for preparing a pH range of 3.6 to 5.6 (see Table 2–7).

Macerating tissue solutions

Caution: Refer to safety procedures in chapter 10 before preparing solutions. (Avoid putting metal forceps into the macerating fluids.)

---

[29] Humason, *Animal Tissue Techniques.*

TABLE 2-6   Phosphate buffers (0.2 *M*)

pH	monobasic sodium phosphate (ml)	dibasic sodium phosphate (ml)
5.9	90.0	10.0
6.1	85.0	15.0
6.3	77.0	23.0
6.5	68.0	32.0
6.7	57.0	43.0
6.9	45.0	55.0
7.1	33.0	67.0
7.3	23.0	77.0
7.4	19.0	81.0
7.5	16.0	84.0
7.7	10.0	90.0

TABLE 2-7   Sodium acetate buffers*

pH	0.2 M sodium acetate ml	0.2 M acetic acid ml
3.6	15	185
3.8	24	176
4.0	36	164
4.2	53	147
4.4	74	126
4.6	98	102
4.8	120	80
5.0	141	59
5.2	158	42
5.4	171	29
5.6	181	19

* Data from Harrow et al. *Laboratory Manual of Biochemistry*, 5th ed. (Philadelphia: Saunders, 1960).

NITRIC ACID   Mix 80 ml of water with 20 ml of concentrated nitric acid. (*Caution:* Pour the *acid* slowly into the *water*.) Place fresh muscle tissue from a frog or mammal in a glass dish containing this dilute nitric acid. In 1 to 3 days (at room temperature) this reagent should dissolve the connective tissue, leaving the muscle fibers isolated. Different sections of muscle may require varying periods of time for tissue breakdown. Shake the container to see the rate of maceration.

Isolate and tease apart the fibers on a slide with dissecting needles; pour off the nitric acid and wash the muscle tissue. You may want to stain the fibers with such stains as methylene blue and finally mount in glycerin or glycerin jelly (p. 161).

When it is necessary to keep specimens for several days, or indefinitely, pour off the water before staining and add a half-saturated solution of alum, prepared as follows. Add 100 g of alum to 500 ml of water and heat in an agate dish (all the alum should dissolve). Let the solution cool. Some alum will crystallize out; decant the resulting cold saturated solution. From this prepare a 50 percent saturated solution by adding 100 ml of the saturated solution to 100 ml of water. (This alum preparation is also desirable if the specimens are to be stained and mounted in glycerin.)

POTASSIUM HYDROXIDE   A weak solution of potassium hydroxide will dissolve cells; a strong solution will separate the cells but will not destroy them. Prepare a solution by warming 35 g of potassium hydroxide in 100 ml of distilled water until it dissolves (*Caution*). Let the solution cool to room temperature.

Dissect out small pieces of tissue from the leg of a frog (striated muscle), from the stomach or intestinal wall (smooth muscle), and from the heart (cardiac muscle). Place these in separate containers of potassium hydroxide solution for 15 to 30 min. On separate slides mount a piece of each kind of tissue in a drop of solution. Tease apart the tissue with dissecting needles and apply a coverslip (see Fig. 2-20).

JEFFREY'S FLUID   This solution is used for *plant* tissues. Mix together equal parts of a 10 percent solution of nitric acid and a 10 percent solution of chromic acid. (*Caution:* Do not inhale fumes or get on skin; have a well-ventilated room.) Drop thin sections of plant tissue into this fluid. Mount small bits of tissue in water on a slide to examine the rate of maceration. When the cells separate from each other fairly readily, pour off the fluid and wash the tissue in water. Mount the cells in water and apply light pressure to coverslip with the eraser of a pencil to separate them. Stain with eosin, methylene blue, or similar dyes.

Mounting small forms

Small forms may be mounted more or less permanently in several media. Organisms that contain water need to be dehydrated before they can be embedded in balsam, or the balsam will become cloudy. On the other hand, some whole mounts may be embedded in glycerin jelly or in colorless Karo syrup without dehydration.

Synthetic resins have been found to be superior in many ways to natural resins for mounting tissue on slides. Some of the natural resins often developed acidity and were variable in composition. Among the widely used synthetic resins, soluble in both xylene and hydrocarbon solvents, are Permount (from Fisher Scientific) and Kleermount (from Ward's).

Some of the aqueous mounting media may contain sugars to increase the refractive index. Many are recommended for small insects or other small invertebrates that are transferred from aqueous media or that have a high water content.

We shall describe only a few of the simpler techniques that seem practical for the classroom.

SYRUP   Small forms such as *Drosophila*, ants, fleas, mosquito larvae, *Daphnia*, *Artemia* and *Gammarus* may be mounted in a large drop of clear syrup, such as Karo.

Dehydration is not necessary. After the organism is oriented in the syrup, add a bristle so that the coverslip will not crush the specimen. Then gently lower the coverslip at a 45° angle from one side; let it sink slowly into the syrup so that no air bubbles form. Should air bubbles form, they may be broken with a dissecting needle. Then remove the excess embedding fluid with wet lens paper.

Some teachers use a medium containing fruit pectin in addition to the Karo syrup. Spread a thin layer of the following medium on a slide, and arrange small insects or worms that have been transferred from glycerin.

clear Karo syrup	50 ml
Certo fruit pectin	50 ml
water	30 ml
thymol (as a preservative)	trace

GLYCERIN JELLY   Such organisms as roundworms, insects, small crustaceans, and plant specimens may be transferred into glycerin jelly from alcohol or formalin as follows. Add glycerin to the alcohol or formalin in which the specimen is contained until 10 percent of the storage fluid is glycerin. Cover the container with a bit of gauze and allow evaporation to concentrate the glycerin. After the glycerin has become concentrated (this may take several days), transfer the specimens to a clean slide with a drop of glycerin jelly prepared in the following way. Soak 10 g of gelatin in 60 ml of distilled water for about 2 hr. Then add 70 ml of glycerin and 1 g of phenol. Heat the solution in a water bath, and then let it cool. When ready to embed specimens on a slide, soften the mounting medium by heating to about 40° C (104° F) in a water bath. The temperature should not be allowed to rise above 40° C (104° F), or the colloid will no longer solidify.

Apply a coverslip and ring it with colorless nailpolish or balsam to retard evaporation. These slides need not be stained. They must be handled with care, for the glycerin jelly melts near room temperature. Examine under the microscope with reduced light intensity.

GUM ARABIC Forms containing a high water content may also be transferred to a medium composed of gum arabic. Either of the following media offers successful slides.

1. *Farrant's medium* Dissolve 40 g of gum arabic (lumps, not powdered form) in 40 ml of warm water. Add the following ingredients and carefully stir:[30]

glycerin	20 ml
carbolic acid	0.1 g

Keep in a water bath when used. Carbolic acid is a preservative. (The medium may also be purchased from Amend Drug Co.)[31]

2. *Berlese mounting medium* Dissolve gum arabic in warmed distilled water. Add the chloral hydrate and glycerin; finally, filter.

gum arabic (lumps)	25 g
distilled water	40 ml
saturated aqueous solution of chloral hydrate (*caution*)	30 ml
glycerin	5 ml

This is especially recommended for mounting small insects.

BALSAM Scales of fish and snakes, wings of insects, hair of mammals, feathers, and similar dry specimens may be embedded in balsam. They do contain some water, so in making permanent slides it is safer to dehydrate the specimens first in xylene for several hours (see page 168); mount in a drop of balsam and seal with a coverslip.

Whole small insects must be dehydrated before mounting in balsam. For a rapid method, put them into glacial acetic acid for dehydration, then mount directly in a drop of balsam. Or put the dead insects in a carbol-xylene solution made by adding 1 part of carbolic acid crystals to 2 parts of xylene. After 1 to 2 hr transfer to pure xylene for 6 to 24 hr; then mount in balsam and apply a coverslip.

In all cases, leave the slides flat for several days until the balsam, clear Karo, or other soft media hardens; later stack the slides in a slide box.

The preparation and staining of whole chick embryos in balsam is described on page 171.

Cleaning and repairing microscope slides

Eventually, a specimen on an expensive slide may slip or the coverslip may be crushed. With some patience, these slides can be repaired.[32]

To repair a broken coverslip, soak the slide in xylene until the broken coverslip falls off by itself; avoid pushing, since the specimen may be torn. Apply fresh mounting medium and seal on another coverslip. Clear nailpolish may be used to seal the edges of the coverslip.

Protozoa may be washed off the slides in the first step in removing the broken coverslip. Draw off the xylene and collect the concentrated protozoa in the bottom of the dish.

A specimen that has slipped to an edge may be reoriented if the mounting fluid is heated slightly over an alcohol lamp. With

[30] G. Humason, *Animal Tissue Techniques*, 3rd ed. (San Francisco: Freeman, 1972).
[31] Amend Drug and Chemical Co., 83 Cordier St., Irvington, New Jersey.
[32] Also Humason, *Animal Tissue Techniques*.

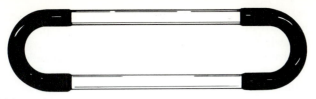

Fig. 2-25 Slide staining rack.

a sliver of a tongue depressor soaked in xylene, move the specimen in the softened mounting medium. Apply a bit more fresh medium if needed, and reapply a coverslip.

When specimens of tissue or of protozoa are faded, they may be restained. The process is laborious, since the material must be run down the series of alcohols, then destained. After restaining and dehydration, the material is moved along into fresh mounting medium.

Slide staining rack For staining a number of slides at one time, use two equal lengths of glass rods and fasten to each end short lengths of rubber tubing as shown in Figure 2-25.

Removing stains Stains on hands and glassware may be removed by treatment as suggested by Humason.[33]

> basic fuchsin—difficult to remove; use strong acetic acid in 95 percent alcohol, or dilute HCl
> carmine—strong ammonia water or weak HCl
> fast green and similar acid stains—ammonia water
> hematoxylin—weak acid or lemon juice
> iodine—sodium thiosulfate
> methylene blue—tincture of green soap or acid alcohol
> most dyes—tincture of green soap
> potassium permanganate—dilute sulfurous acid, HCl, oxalic acid
> safranin and gential violet—difficult to remove; try acid alcohol
> silver—Lugol's or tincture of iodine

[33] Humason, *Animal Tissue Techniques.*

Provide snap clothes pins for students to hold slides while staining to reduce stained hands.

Squash slide techniques

Basic techniques for making squash slides of chromosomes of *Drosophila*, *Chironomus*, and of onion root tips are described in chapter 6. Preparation of blood smears and the staining of these smears with Wright's blood stain are described on page 274; smears of bacteria are described on pages 459 and 601.

Permanent stained slides: protozoa

Concentrating protozoa A number of methods for concentrating protozoa may be used in preparing for fixation and staining. The following three methods are in common use.

1. Spread a thin film of Mayer's albumen (p. 171) on each of several clean coverslips. Then float some of these coverslips on the surface of a rich culture of protozoa; place others at the bottom of the culture jars. After 12 to 24 hr remove the coverslips with clean forceps and place them in a fixative.

2. Use a Pasteur pipette or draw out a medicine dropper in a Bunsen flame to form a capillary pipette (p. 748). Use the pipette to squirt a drop of a rich culture onto a slide or coverslip which has been spread with a film of Mayer's albumen. Quickly place the slide or coverslip into a jar of fixative.

3. This method permits the fixing of large quantities of protozoa in bulk. Centrifuge a culture slowly for some 30 sec. (Small hand centrifuges are available from biological supply houses.) Draw off the fluid quickly and pipette the concentrated culture of protozoa from the bottom into a small container of fixative. When they are to be transferred out of the fixative into alcohol, concentrate them again by centrifuging.

Note: Refer to safety precautions in chapter 10, regarding the use of toxic chemicals.

Fixation   While Schaudinn's fixative is recommended especially for fixing protozoa, others, such as Bouin's and Zenker's fixative solutions (pp. 166, 168), are also useful. (Many contain mercuric chloride so that safety precautions must be taken.)

Protozoa should stand in Schaudinn's fixative for 10 to 30 min (the average time is about 15 min at 40° C (104° F). The time requirements are different for other fixatives. Then centrifuge the protozoa and transfer them into a small amount of 50 percent alcohol. After 10 min, transfer into 70 percent alcohol to which a few drops of a concentrated solution of iodine in alcohol has been added, so that the alcohol is slightly brown. After a few minutes, transfer to 90 percent alcohol, and finally into absolute alcohol, always keeping the quantity of alcohol as small as possible.

Staining   Spread a film of albumen on slides or coverslips (see Mayer's albumen, p. 171); squirt a drop of previously fixed protozoa onto a slide or coverslip with some force so that the protozoa adhere to the albumen, and return the slides to absolute alcohol. Now the protozoa are on the slides, ready for staining. There are several suitable stains. Hematoxylin and the Feulgen nuclear reagent are chromosome stains. Borax carmine may be used for whole mounts (p. 171); or try Borrel's stain, which stains nuclei red and cytoplasm green. Many of these stains for protozoa may be found in textbooks of

protozoology (some are given here and under stains, p. 169). Delafield's or Ehrlich's hematoxylin may be used in place of Heidenhain's (p. 170).

The procedure for staining with hematoxylin is as follows. Remove the albumen-coated slides of protozoa from absolute alcohol and transfer through the series of alcohols to water. For instance, leave the slides (or coverslips) for about 2 min in each of these: from absolute alcohol to 90 percent, to 70 percent, to 50 percent, to 30 percent, and finally to water. Then place the slides in hematoxylin for 5 to 15 min so that the nuclei become stained (inspect under the microscope during the process); wash off the stain in water. Transfer the slides to ammonia water (about 1 drop of concentrated ammonium hydroxide added to 250 ml of tap water).

Next, counterstain with a cytoplasmic dye such as eosin (1 percent alcoholic solution). After 5 to 10 min transfer to water, then to 30 percent alcohol. Now follow with transfers at 2-min intervals into an upward series of alcohols: to 60 percent, 70 percent, 90 percent, and finally absolute alcohol. Next, transfer the stained slides into xylene.

The slides should be inspected under the microscope before adding balsam for final mounting. Should the nuclei be stained too deeply with hematoxylin, decolorize by placing the slides into 70 percent alcohol to which a small amount of 1 percent hydrochloric acid has been added. Then place the slides into alkaline 70 percent alcohol again to regain the blue color. (Prepare alkaline alcohol by adding 1 drop of 1 percent ammonium hydroxide to a Coplin jar full of the alcohol.) Continue into 90 percent alcohol, into absolute alcohol, and into xylene. Finally, mount with a coverslip by adding a drop of Canada balsam to each slide. Then apply coverslips. (Or add the balsam to the slide and cover with a stained coverslip, if that has been the procedure.)

Other stains   Kudo recommends some special techniques, such as using Delafield's hematoxylin of a stock solution

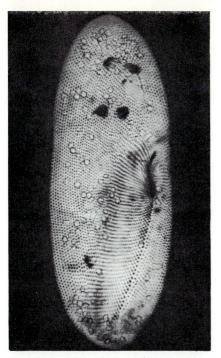

Fig. 2-26 *Paramecium* stained with silver nitrate solution to show silver line system. (Courtesy of General Biological Supply House, Inc., Chicago.)

diluted 1:5 or 1:10 in order to achieve a slow, progressive staining of protozoa.[34] Should slides become overstained, they can be decolorized in a 0.5 percent solution of hydrochloric acid in water or in alcohol. Then mount the protozoa in neutral mounting medium.

Mayer's paracarmine is recommended as a useful stain. It is prepared in slightly acidified 70 percent alcohol solution. As before, should protozoa become overstained, they may be decolorized in a 0.5 percent solution of hydrochloric acid in alcohol.

Silver line system—ciliates

Examine the silver line system of ciliates (Fig. 2-26). Use a silver nitrate solution to impregnate the ciliary apparatus and its branches on the pellicle of certain ciliates.

[34] Kudo, *Protozoology.*

For this study, use a modification of Klein's silver line system described by Kozloff.[35] Place a drop of concentrated ciliates on a slide and allow it to dry. Keep the forms from drying as a mass in the center of the slide by gently tilting the slide or stirring the drop as it dries with a clean dissecting needle. Immerse the dried films in freshly prepared 2 percent solution of silver nitrate in distilled water. After 15 min, either expose the slides in the solution to sunlight or strong artificial light; or remove the slides from the silver nitrate solution, transfer them to a dish of distilled water, and expose this to sunlight to reduce the silver nitrate. After a few minutes in light, the slides turn brownish black. Remove a slide after 10 min and inspect under low power of the microscope. Some trial and error is necessary here because of variations in the thickness of the ciliated forms, their concentration, content of saline or protein, and so on.

When the silver impregnations have been adequately reduced, wash the slides in distilled water and blot dry. Mount the films in neutral balsam or in synthetic resin media; apply coverslips.

Kozloff recommends this method for smaller ciliates, such as *Colpoda* or *Tetrahymena*, and for symbiotic ciliates. (Avoid saline solutions, since the silver chloride precipitates.)

Permanent stained slides: tissue cells

You will want to refer to safety precautions as described in chapter 10.

Histological techniques

Place single-celled organisms or bits of tissue in a fixative to kill the cells rapidly so that the cell contents are preserved and closely resemble living cells. In addition, a fixative hardens the tissue cells so that

[35] A. Galigher and E. Kozloff, *Essentials of Practical Microtechnique* (Philadelphia: Lea & Febiger, 1971).

they can be cut into thin, transparent sections for examination under the microscope.

A fixative must be washed out of the cells before they can be processed for staining. The type of fixative used determines how the cells should be washed. If a fixative contains mercuric chloride (*caution:* corrosive sublimate, toxic) or picric acid, the cells must be washed for at least 1 hr in 70 percent alcohol. On the other hand, if the fixative contains potassium dichromate, the tissues should be washed for at least 1 hr in water.

In general, the procedure for staining and mounting permanent slides is as follows:[36]

1. Fix the tissue and harden.
2. Dehydrate through the series of alcohols.
3. Clear tissue.
4. Embed in paraffin.
5. Section with microtome.
6. Dissolve paraffin with xylene.
7. Pass through the series of alcohols into distilled water.
8. Stain and counterstain.
9. Dehydrate by moving up the series of alcohols into xylene.
10. Mount in balsam.

After fixation, the next step is dehydration; then proceed with successive steps. Details for these procedures follow.

Fixation   There are several widely used fixatives that may best be purchased already prepared. (*Caution*: Many fixatives contain mercuric chloride, which is toxic, and/or formalin, which requires good ventilation.) Fixatives are listed here in alphabetical order. But first, a special note about the alcohols used in some preparations.

ALCOHOL (ABSOLUTE)   Johansen suggests a method for preparing absolute alcohol, which is, theoretically, 100 percent alcohol, with all the water removed.[37] Begin with 95 percent ethyl alcohol. Heat some crystals of cupric sulfate until only a white powder remains. Add this anhydrous form to the 95 percent ethyl alcohol. If there is still water in the alcohol, the copper sulfate will turn blue. Therefore, continue adding anhydrous sulfate until the solution no longer turns blue. Then filter the alcohol quickly into a dry stock bottle, and cork securely so that no moisture from the air enters the bottle. Apply petroleum jelly to the cork as an additional precaution. Some technicians keep a small bag of anhydrous sulfate suspended in the bottle to keep the solution free of water.

A sensitive test for water in absolute alcohol can be made by adding a few drops of the alcohol to a solution of liquid paraffin in anhydrous chloroform. If moisture is present in the alcohol, the solution becomes clouded.

ALCOHOL (ETHYL)   A 70 percent solution of ethyl alcohol is a common preservative for small forms and tissue specimens. (To dilute alcohols to lower percentages, see page 738.)

ALLEN'S FLUID   is recommended especially as an all-round general fixative.

Fix small pieces of tissue for 24 hr, and then wash in 70 percent alcohol until there is no loss of color.

BOUIN'S FIXATIVE   is excellent for general use with both plant and animal tissue; however, it is difficult to wash out of tissues before staining. Its main advantage is that specimens may be stored in it for long periods of time. Purchase a ready-made fixative.

Leave the tissue in the fixative for 24 to 48 hr; then wash in 70 percent alcohol until the color is removed. Add a few drops of lithium carbonate to the alcohol to remove the yellow pigment.

CARNOY'S FLUID   is an especially useful fixative for tissue used in chromosome studies. (*Caution:* Avoid inhaling fumes.) Purchase ready-made.

---

[36] For a full description of histological methods refer to Humason, *Animal Tissue Techniques.* Other references are listed at the end of this chapter.

[37] D. Johansen, *Plant Microtechnique* (New York: McGraw-Hill, 1940).

CARNOY AND LEBRUN'S FLUID is recommended for hard-shelled specimens, such as some arthropods, for it has high penetrability. It also dissolves fat. (*Caution*: is highly toxic and flammable.) Wash specimens well in alcohol to remove chloroform.

FAA (FORMALDEHYDE, ALCOHOL, AND ACETIC ACID) is a good hardening agent for plant tissues; in fact, plant materials may be stored in this preservative for several years. Pieces of leaf tissue should be killed and hardened in this fluid for 12 hr; thicker tissues, such as small stems and thick leaves, should be kept in it for 24 hr; and woody twigs should remain in it for a week. Tissues need not be washed after preservation in FAA. Many small animal forms can also be fixed in this fluid. The alcohol content counteracts the swelling effects of formalin and glacial acetic acid. (*Caution:* Avoid inhaling fumes.) Combine these materials:

ethyl alcohol (95%)	50 ml
glacial acetic acid	2 ml
formalin (40% formaldehyde)	10 ml
water	40 ml

FAAGO This fixative is widely used, especially for fixing small specimens such as nematodes for permanent mounting on slides. Purchase ready-made.

Let the specimens stand in an evaporating dish containing hot FAAGO until they are covered with a film of glycerin. Then add pure glycerin to the evaporating dish and let it stand for a few hours. Mount the worms or other small specimens in a drop of glycerin on a slide. Seal the coverslip with hot paraffin or with clear nailpolish.

Edmondson suggests mounting small specimens in lactophenol.[38] Transfer the specimens from the FAAGO into a mixture of formalin (5 percent) and lactophenol (5 percent). Allow the formalin to evaporate at 40° C (104° F) so that

lactophenol remains. Work in a well-ventilated room. Then mount the specimen in pure lactophenol on a slide, and ring the coverslip with clear nailpolish.

FORMALIN (10 PERCENT) Combine 10 parts of water and 1 part of commercial formaldehyde. (Treat commercial formaldehyde as 100 percent formalin.) (*Caution*: Avoid inhaling formalin fumes.)

FORMALIN Combine the following:

formalin, concentrated	10 ml
calcium chloride (10%)	10 ml
distilled water	80 ml

Remove tissue from formalin and wash in water. (*Caution*: avoid inhaling formalin; work in well-ventilated room.)

FORMOL-ALCOHOL Recommended especially for crustaceans, small insects, and for botanical specimens. Combine the following:

formalin, concentrated (*caution*)	5 ml
ethyl alcohol (70%)	100 ml
glacial acetic acid	5 ml

Fix for 1 to 24 hr, then transfer to 85 percent alcohol.

FLEMMING'S FIXATIVE is recommended especially for delicate small specimens. Purchase ready-made. Fix specimens for 12 to 24 hr. Wash out the fixative with water for 24 hr; transfer to 70 percent alcohol.

GATES' FLUID mainly used for plant tissues, is sometimes recommended to show chromosomes in root tips.

Leave specimens in fixative for 24 hr. Then wash out the fixative with running water.

GILSON'S FLUID is an excellent fixative for careful histological work as well as for small invertebrates. Purchase the fixative ready-made; it contains mercuric chloride (*toxic*) and concentrated nitric acid. Avoid inhaling the fumes. Keep steel instruments out of the fixative, or any other containing mercuric chloride (corrosive sublimate).

KLEINENBERG'S FIXATIVE is recommended especially as a fixative for chick embryos (p. 171) and many small marine organisms. Purchase the fixative ready-made; it contains picric acid.

[38] H. B. Ward and G. C. Whipple, *Fresh-Water Biology*, 2nd ed., W. T. Edmondson, ed. (New York: Wiley, 1959).

LAVDOWSKY'S FIXATIVE is a general fixative especially useful in botanical work. Purchase it ready-made. Avoid inhaling fumes of any fixatives that contain mercuric chloride; do not introduce steel instruments into the fluid.

Tissue may remain in this fixative for 24 hr. Then transfer tissue specimens into 70 percent alcohol for at least 1 hr.

NAVASHIN FIXATIVE is recommended especially in preparations of onion root tips to show chromosomes.

SCHAUDINN'S FIXATIVE is useful for preparation of protozoa (p. 164) for fixing and for staining.

ZENKER'S FIXATIVE is widely used in histological work; it is also a fine preservative for small marine forms. Fix the specimens for 4 to 24 hr. Wash out the fixative in 70 percent alcohol for the next 24 hr. (Since it contains mercuric chloride, avoid inhaling the fumes and avoid using steel instruments.)

Dehydrating, clearing, and embedding
In the fixative, tissues are killed quickly and also harden. The next step is dehydration. After water is removed, the tissues are embedded in paraffin which gives the support needed to cut thin sections for examination under the microscope. Usually tissues are transferred from the fixative into 70 percent ethyl alcohol, but the process depends on the kind of fixative in which the tissue has been killed.

Wash delicate tissues in water, then transfer gradually, first into 30 percent alcohol, then into 50 percent, and finally into 70 percent alcohol (so that diffusion currents do not distort the tissues). Keep the tissues in each solution in the alcohol series for about 1 hr, with the exception of the 70 percent alcohol (2 to 6 hr). (If the fixative contains picric acid, remove it quickly by adding a few grains of lithium carbonate to the 70 percent alcohol used in washing.) Transfer the tissues to 95 percent alcohol, and into absolute alcohol. Transfer into a clearing agent to prepare for embedding in wax. A clearing agent must be miscible with alcohol for it re-

places the alcohol in the tissues. Xylene is a rapid clearing agent. Keep the tissues in xylene for 1 to 3 hr. If xylene becomes cloudy, return the tissues to absolute alcohol. Cloudiness indicates that the tissue has not been completely dehydrated. To remove water in the stock bottle of absolute alcohol, use anhydrous copper sulfate (p. 166).

When the tissues have been cleared, they are ready for embedding in melted paraffin. Keep the wax melted in a paraffin oven at a temperature a few degrees higher than the melting point of the paraffin. Insert the specimens into square paper boats (Fig. 2-27) containing melted paraffin, and leave the boats in the oven for 2 to 3 hr. Transfer the specimens to fresh wax within this period of time. When the wax is ready to be cooled, remove the boats from the oven. Float the paper boats in a container of cold water (do not submerge), and blow on the surface of the paraffin until a film begins to form. Or, let the surface paraffin solidify first; then submerge the boat in cold water, and let the paraffin harden throughout. Remove the paper and trim the block of paraffin so that it fits on the holder of the microtome.

Sectioning and staining To attach the block of paraffin to the holder of the microtome, melt the wax at one end of the block and press it against the holder. Spread melted wax around the edges. The cutting knife and block should be set so that sections from 6 to 10 μ can be cut (1 mm equals 1000 μ). As you cut, a ribbon of wax sections forms; lift off the sections with a spatula and float on slightly warmed water which has been spread on slides already prepared with a thin film of albumen (see Mayer's albumen, p. 171). The water temperature must be below the melting point of wax. Finally, dry the slides on which the sections are floating for about 24 hr in an incubator set at 37° C (99° F). If this time is shortened there is danger that the tissues may be washed off the slides in subsequent procedures.

At this point, the prepared slides consist

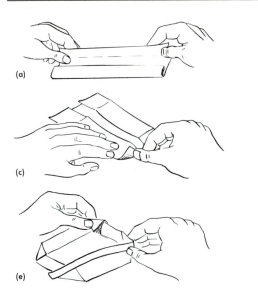

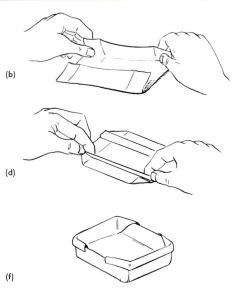

Fig. 2-27 Making a paper or aluminum foil boat in which to embed tissue in paraffin. (Gray, *Handbook of Basic Microtechnique*, McGraw-Hill, 1952.)

of slices of tissue embedded in paraffin. In the next step, the wax must be removed before the tissue can be stained. First, briefly warm the slides with the paraffin sections over an alcohol flame, and then insert the slides into a jar of xylene to dissolve the wax. After 5 min, transfer the slides to absolute alcohol for 3 min, then for about 2 min into each of the alcohols: 70 percent, then 50 percent, then 30 percent. Finally, run the slides through distilled water for 1 min. Since most stains are aqueous solutions, the sections are brought gradually through the alcohols into water.

Now the slides (with the paraffin dissolved away) are ready to be stained. There are many stains which may be used. We shall take Harris' hematoxylin, a nuclear stain, as an example of the general staining technique. First, add 2 drops of glacial acetic acid to the Harris' hematoxylin (purchase prepared). Immerse the slides in stain for 1 to 2 min. Transfer into tap water to remove acid (notice that the color fades). Wash for 5 min; then transfer slides to each of the following alcohols for 2 min: 30 percent, to 50 percent, to 70

percent, to 90 percent. To *counterstain* with eosin, a cytoplasmic stain, transfer the slides to eosin for 2 min and rinse in 95 percent alcohol. Transfer the slides into xylene where they may be left for some time. Finally, mount in Canada balsam.[39]

Stains   Many stains may be used. Certain stains penetrate tissues better after use of a particular fixative. Basic (alkaline) dyes stain acid structures within cells, and, conversely, acid dyes stain basic structures. For example, acid dyes are held by plasma, cilia, and cellulose structures. Chromosomes, centrosomes, nucleoli, cork, cutinized epidermis, and xylem tissues of plants retain basic dyes.

Some basic dyes in frequent use are safranin, hematoxylin, methyl green, gentian violet, and Janus Green B. Among the acid dyes are eosin, methyl orange, orange G, fast green, and light green. Preparations for some stains follow. For more

[39] Refer also to modifications for the laboratory: P. Mimlitsch, "Tips on Microtechnique," *Carolina Tips* (January 1970).

detailed procedures, refer to texts listed at the end of this chapter.

ACID ALCOHOL is included here since it is used to decolorize stains. (It can also be used to clean coverslips.) Prepare by adding 1 ml of 1 percent concentrated hydrochloric acid to 100 ml of 70 percent ethyl alcohol.

AMMONIATED ALCOHOL is prepared by adding 1 ml of concentrated ammonia to 100 ml of 70 percent ethyl alcohol. To prepare an alkaline alcohol, use aqueous sodium bicarbonate.

ACID FUCHSIN is useful for staining marine algae as well as many small crustaceans. Combine:

acid fuchsin	1 g
distilled water	100 ml
glacial acetic acid	1 ml
(or hydrochloric acid)	

BORAX CARMINE is prepared by dissolving 4 g of borax in 100 ml of distilled water. (Reduce the amounts if smaller quantities are needed.) Then add 3 g of carmine and boil for about 30 min. After this solution has cooled, dilute with 100 ml of 70 percent alcohol. Let stand for a few days and then filter.

Whole mounts of hydras, flukes, and embryos may be transferred into this stain directly from 70 percent alcohol. Often the slides become overstained in this dye, but they can be decolorized by placing them in acid alcohol (see above).

CARBOL FUCHSIN is a stable stain at room temperature. Prepare in two parts; combine:

basic fuchsin	1 g
absolute alcohol	10 ml

Then mix with:

phenol	5 g
distilled water	100 ml

CONKLIN'S HEMATOXYLIN is recommended especially as a stain for whole mounts of chick embryos (p. 171). Prepare as follows:

Harris hematoxylin	21 part
water	4 parts

To this add 1 drop of Kleinenberg's picrosulfuric acid (p. 167) for each milliliter of fluid.

DELAFIELD'S HEMATOXYLIN is prepared by dissolving 4 g of hematoxylin in 25 ml of absolute alcohol. Add this to 400 ml of a saturated aqueous solution of ammonium alum. Expose the solution to light for a few days in a cotton-stoppered bottle; filter. Then add the following:

methyl alcohol	100 ml
glycerin	100 ml

The stain must be ripened for two months at room temperature before it is ready for use. Finally, store in well-stoppered bottles. Wash the specimens in water before transferring them into this stain. Delafield's, Ehrlich's, and Harris' hematoxylin stains may be purchased ready-made.

EHRLICH'S HEMATOXYLIN is also ripened at room temperature for about two months. Transfer specimens from water to the stain, which is prepared by mixing:

hematoxylin	2 g
absolute alcohol	100 ml

Then add these substances in the order presented:

glycerin	100 ml
distilled water	100 ml
glacial acetic acid	10 ml
potassium alum	in excess

EOSIN Y is a plasma stain often used for whole mounts or for contrast. Make up a 0.5 percent solution in distilled water.

ETHYL EOSIN is prepared as a 0.5 percent solution in 95 percent ethyl alcohol.

FAST GREEN SOLUTION is prepared by combining:

fast green	2 g
distilled water	100 ml
glacial acetic acid	2 ml

METHYL GREEN SOLUTION is a good nuclear stain for general use. Prepare a 1 percent acidified aqueous solution by dissolving 1 g of the dye in 1 ml of glacial acetic acid. Then dilute with distilled water to make 100 ml.

METHYL VIOLET may be used to stain amphibian or human blood cells. Prepare as follows:

sodium chloride (0.7% solution)	100 ml
methyl violet	0.05 g
glacial acetic acid	0.02 ml

When human blood cells are to be stained, prepare the stain in 0.9 percent sodium chloride solution.

Other stains and reagents used in histological work At times it may be necessary to use some of the following solutions (fixatives, stains, and so forth). Many of these have been referred to in other sections of this book.

BALSAM (NEUTRAL) The slight acidity of samples of balsam is well known. This acidity is an advantage when balsam is used as the mounting fluid after acid stains, but detrimental when the basic hematoxylin stains are used. Balsam may be neutralized by adding a small quantity of sodium carbonate. Let the fluid stand for about a month. The supernatant balsam should prove to be slightly alkaline.

FORMALIN is a 40 percent solution of formaldehyde gas in water and is a stock solution.

Caution: Avoid inhaling the fumes of any formalin solutions for they irritate the mucous membranes.

Prepare a 10 percent solution of formalin, a common fixative for small forms, by adding 10 ml of stock formalin to 90 ml of water. This solution is an ingredient of many fixatives; it may also be substituted in certain fixatives for glacial acetic acid.

Sometimes a 4 percent solution of formalin is required. In such cases, add 4 ml of commercial formalin to 96 ml of water. (This is really a 1.6 percent solution of formaldehyde.)

MAYER'S ALBUMEN A dilute solution of albumen is spread as a film on a clean slide so that tissue sections or protozoa adhere to the slide throughout the many transfers from one fluid to another in fixation, clearing, and staining.

Beat the egg white, but not until it becomes stiff. Allow air bubbles and suspended matter to rise over several hours. Add equal quantities of glycerin to clear the egg white. Thymol prevents the growth of molds.

egg albumen	50 ml
glycerin	50 ml
thymol	1 crystal

(You can substitute 1 g of sodium salicylate for the thymol; they are both antiseptics.)

Store the clear fluid. It keeps 2 to 4 months without spoiling.

When the solution is ready to be used, add 3 drops of it to 60 ml of distilled water. With your finger, spread a very light film on a clean slide.

Marine algae

Temporary slides may be made of small algae in a drop of saline sealed with a coverslip. Thin slices of larger marine algae can be cut with a razor and mounted in seawater.

For permanent slides, marine algae need to be placed in FAA fixative (p. 167). Then wash in fresh water to remove the fixative. Mount thin slices of thick algae on a slide and stain the plant tissue with acid fuchsin. Add a few drops of the stain, wash off the excess with water. Blot the slide, add clear syrup (Karo) and apply a coverslip. Lie flat for 24 hr or until the syrup has hardened.

Whole mounts of chick embryo

Since a 72-hr chick embryo is the easiest to fix and mount whole on a slide, we will give a general description of the preparation of a slide of this stage (Figs. 2-28, 2-29). Then you may want to work back to the 24-hr stage. Complete directions may be found in Rugh's embryology manual, or in Humason's text.[40]

[40] R. Rugh, *Experimental Embryology: Techniques and Procedures*, 3rd ed. (Minneapolis: Burgess, 1962) and Humason, *Animal Tissue Techniques*.

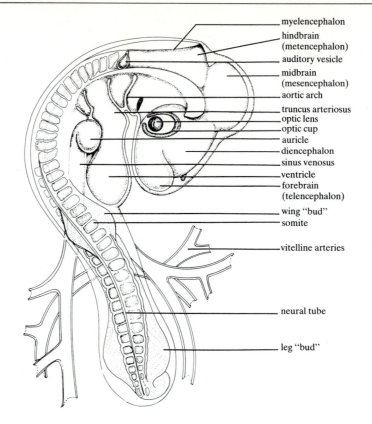

Fig. 2-28 Chick embryo (about 72 hrs). (From *Laboratory Outlines in Biology—III* by Peter Abramoff and Robert G. Thomson. W.H. Freeman and Company. Copyright© 1982.)

One way to incubate fertilized chicken eggs and to float early embryo stages in bowls of warmed saline solution is described on page 187. Break open a 72-hr egg and float the embryo in warmed saline solution. Grasp the chalazae with forceps to float the embryo to the top of the mass of yolk. With sharp scissors, cut outside the area vasculosa, and float this disc free of the underlying yolk (Fig. 6-41).

Lower a glass slide under the saline solution and float the blastodisc, yolk side uppermost, onto the slide. Flatten out the blastodisc on the slide and, to hold the embryo in position, put a ring of filter paper large enough to encircle the area pellucida over that region. Then lift the slide out of the saline solution, and gently pipette a fixative onto the slide. Use Bouin's or Kleinenberg's picrosulfuric fixative. Place several such slides in Petri dishes of fixative for 8 to 10 hr. Then transfer the slides into a graded series of alcohols in this order: 30 percent, to 50 percent, to 70 percent. Leave the slides in each alcohol for 1 hr. To the 70 percent alcohol add a bit of lithium carbonate (or ammonium hydroxide 3 percent by volume) to decolorize the bright yellow picric stain. Then transfer the specimens back to fresh 50 percent, then 30 percent alcohol, and finally into distilled water in preparation for staining.

Stain a 72-hr embryo with Conklin's hematoxylin for 5 min. For younger embryos, less staining time is needed— about 2 min for a 24-hr embryo.

Then transfer the slides into tap water for a few minutes. Follow this with transfers into 30 percent alcohol, then to 50 percent; leave the slides for ½ hr in each. Examine the slides under the microscope. Should the slides be overstained, destain them in acidified 70 percent alcohol (p. 170). Then follow with a washing in slightly ammoniated 70 percent alcohol (p. 170). The slides may remain in pure 70 percent alcohol for several hours. Finally transfer to 95 percent alcohol and follow with two changes in absolute alcohol.

When the embryos are ready for mounting, transfer the slides from absolute alcohol into pure cedar oil for clearing. After 24 hr in cedar oil, the embryos should appear translucent. Then make two transfers into xylene, ½ hr in each container. Mount the embryos in balsam with small chips of capillary tubing so that the embryos will not be crushed. Add more balsam if needed. Flame a coverslip to dry any moisture on it, and lower it into the balsam. Let the slides remain flat for several days until the specimen hardens in place. Later remove any excess balsam with xylene (Fig. 2-29).

This is one of many methods. There are several fixatives and stains used in preparing chick embryos. The blastodiscs also may be fixed in Zenker's fixative. Other stains may be preferred, and some technicians clear the mount in oil of wintergreen. Alternate techniques may be found in several specialized texts listed at the end of this chapter.

Staining skeletons of chick embryos

Rugh suggests a modification of the Spalteholz method for staining skeletons of chick embryos more than ten days old.[41] This technique may also be used for adult amphibians or small fish.

Fix specimens for two weeks in 95 percent alcohol. Then transfer the hardened specimens to 1 percent KOH. After 24 hr transfer to tap water, and, with forceps, remove most of the fleshy tissue. Transfer the specimens to 95 percent alcohol for

3 hr; change into fresh alcohol for 3 hr more. Immerse the specimens in ether for 2 hr to dissolve fat tissues (or use acetone if there is little fat). Have good ventilation.

Again transfer to 95 percent alcohol for 6 hr, changing the alcohol once within this time. Then transfer to 1 percent KOH for six days. Immerse in alizarin red "S" for 12 hr, transfer to 1 percent KOH for 24 hr and into 10 percent NH_4OH for further clearing for the next 24 hr.[42] The NH_4OH neutralizes the medium and tends to preserve the color longer. The resulting specimens should be transparent with stained skeletal elements. Store the specimens in 100 percent glycerin.

Humason offers formulas for solutions and detailed procedures.[43] For small fish, she suggests that the scales be removed and the specimens be fixed in 70 percent alcohol. Allow the specimens to remain in the fixative for several days until hard. Small embryos of birds or mammals should be fixed in 95 percent alcohol after hair or feathers have been removed. Allow to remain in fixative for three days.

After the specimens have been in fixative, rinse in distilled water and then transfer to 2 percent KOH for 4 to 48 hr, depending on the size of the specimens and how long it takes for the skeleton to begin to show through the musculature. When the specimens are clear, transfer them to alizarin red working solutions (see below for preparation). It may take from 6 to 12 hr for the skeleton to be stained red; it may also be necessary to add fresh solution.

Transfer the stained specimens to 1 or 2 percent KOH for one or two days to decolorize the soft tissues. The action can be speeded by using a lamp or direct sunlight. Next, clear the specimens in clearing solution No. 1 (described below) for two days; then transfer them to clearing solution No. 2 for one day. Immerse the specimens in pure glycerin (thymol may be added as a preservative); mount

---

[41] Rugh, *Experimental Embryology.*

[42] R. Rugh, personal communication, October 20, 1964.

[43] Humason, *Animal Tissue Techniques.*

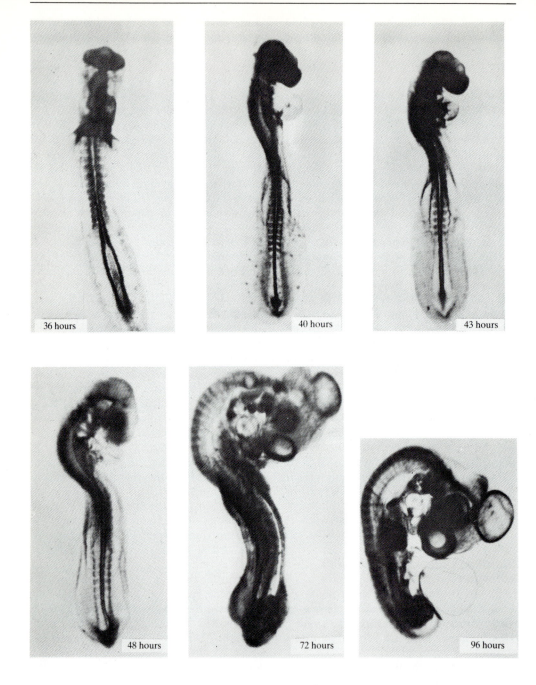

Fig. 2-29 Chick embryos: photos include 36-hr stage to 96-hr stage. (From *Experimental Embryology* by Roberts Rugh, 3rd ed., 1962. Printed by permission of Burgess Publishing Company, Minneapolis, MN, Publishers.)

them on glass slides in museum jars or store them in bottles sealed to prevent evaporation.

Humason describes the improvement of Cumley in producing specimens with greater clarity. Gradually replace the glycerin with 95 percent alcohol, then absolute alcohol. Then transfer the specimens into toluene. Prepare a solution of naphthalene saturated with toluene and immerse the specimens in this for a few days; store specimens in anise oil saturated with naphthalene. (*Caution*: combustible.)

The following solutions are needed.

ALIZARIN RED "S" STOCK SOLUTION Humason recommends the method of Hollister.[44] Prepare a saturated solution of alizarin red "S" (C.I. 58005) in 50 percent acetic acid. To 5 ml of this saturated solution add 10 ml of glycerin and 60 ml of 1 percent aqueous chloral hydrate. (*Caution*: toxic).

ALIZARIN WORKING SOLUTION Dilute the alizarin stock solution, adding 1 ml of stock solution to 1 liter of 1 or 2 percent KOH in distilled water.

CLEARING SOLUTION NO. 1 (OF HOOD AND NEILL) Combine the following:

KOH (2%)	150 ml
formalin (0.2%)	150 ml
glycerin	150 ml

CLEARING SOLUTION NO. 2 (OF HOOD AND NEILL) Combine the following:

KOH (2%)	100 ml
glycerin	400 ml

Whole mounts in glycerin, or in clear blocks of plastic, may be purchased from biological supply houses.

Humane treatment of organisms

Before students begin individual investigation of any organism, it would be well to be familiar with safety factors for teachers described in chapter 10, especially the cautions concerning working with living things.

A valuable guide for the handling, feeding, and housing of laboratory animals has been prepared by the Institute of Laboratory Animal Resources, National Research Council, titled *Guide for Laboratory Animal Facilities and Care*.[45] It includes an excellent bibliography comprising veterinary practices and husbandry. In addition, it gives the qualifications for certification for Animal Technician by the American Association for Laboratory Animal Science.

Biochemistry of cells: macromolecules

The function of complex macromolecules of cells, the various proteins, nucleotides, polysaccharides and lipids, can be examined in terms of their components. In the laboratory, students may learn some of the basic biochemical techniques of separation, extraction, centrifuging, dialysis, filtration, hydrolysis, and chromatography that are used to identify specific amino acids, lipids, simple sugars, and nucleotides.

Dialysis: different sized molecules

Wet a 15 cm length of dialysis tubing and roll the end between the fingers to open the tube. Tie off one end with string and use a funnel to add equal quantities of solutions and suspensions containing different sized molecules: 2 percent starch, 0.01 M glucose, egg albumen or an amino acid, and a chloride, such as 0.9 M NaCl. Tie off this end of the tubing; wash to remove any spillage. Immerse the tube in a beaker half-filled with water. Stir the water occasionally to distribute the diffusing molecules uniformly.

After 5 min, test samples of the water in the beaker for the presence of starch, glucose, protein, and chloride (tests described on the following pages). Again, after 15 min, test for the diffusion of mole-

---

[44] Humason, *Animal Tissue Techniques*.

[45] Public Health Service Publication #1024, 3rd ed. 1968 (Washington, D.C.: U.S. Government Printing Office).

cules to gain some notion of the speed of diffusion. To identify specific molecules, establish *the tests* for identification discussed below for pure samples of starch, glucose, protein or amino acid, and chloride.

Tests for macromolecules: starch, sugars, proteins, lipids

Unless otherwise stated, temperature refers to room temperature.

Tests for starch

There are several tests for different starches: a) microscopic examination of starch granules from various sources, such as potato, bean, arrowroot, rice, maize, pea, barley, oat, and so on; b) iodine test on a starch solution; c) diffusibility of a starch paste. Hawk[46] and Giese[47] offer test procedures for many kinds of carbohydrates such as inulin, glycogen, dextrin, cellulose, pentosan, galactan, and other pectins.

Iodine test   Place 2 to 3 ml of dilute starch solution (1 percent) in a test tube. Prepare by mixing or grinding 2 g of starch powder (preferably arrowroot) in a mortar with a little cold water. Bring 200 ml water to the boiling point and add the starch mix from the mortar stirring constantly. Bring this to a boil and allow to cool. To a sample, add a drop of the dilute iodine solution and note the blue color that results. Then heat the tube and watch the disappearance of the color which reappears on cooling of the solution. Starch granules stain blue due to the formation of an adsorption complex of starch and iodine, not due to a definite compound formed. The test solution must be neutral or a bit acid. You may want to add an alkali or alcohol to a bit of the test

sample you have of starch solution to observe this.

IODINE-POTASSIUM IODIDE SOLUTION (I_2-KI)   Dissolve the KI in 25 ml of water; grind the iodine and stir into the solution. Bring the volume to 100 ml with distilled water.

iodine	5 g
potassium iodide	10 g
distilled water	100 ml

A more *dilute* I_2-KI solution (Lugol's, p. 137) is especially useful as a stain for nuclei, cilia, and flagella as well as for testing for starch molecules in leaves or in food samples.

iodine	1 g
potassium iodide	2 g
distilled water	300 ml

Store the iodine solutions in dark bottles; pour into dropper bottles for classroom use.

Tests for hexose sugars and aldehydes of sugars

Benedict's test   In this test for the presence of glucose, pour 5 ml of Benedict's solution into a test tube and add 8 drops of the solution to be tested. Shake well. Boil for 2 min or place the tube in a boiling water bath for 3 min. When the solution cools, a precipitate forms that ranges from green to yellow to red depending on the amount of glucose present. Benedict's solution may be purchased or prepared as follows:[48]

copper sulfate	17.3	g
sodium citrate	173	g
sodium carbonate	100	g
distilled water to make	1	liter

Dissolve sodium citrate and sodium carbonate in 800 ml of warm water. Filter and pour into a glass graduate and make up to

---

[46] B. Oser, ed. *Hawk's Physiological Chemistry,* 14th ed., B. L. Oser, ed. (New York: McGraw-Hill, 1965).
[47] Giese, *Laboratory Manual in Cell Physiology.*

[48] Refer to *Hawk's Physiological Chemistry* for the preparation of quantitative Benedict's reagent and also for additional tests to distinguish monosaccharides and disaccharides.

850 ml. Dissolve copper sulfate in 100 ml of water. Pour the carbonate-citrate solution into a large beaker and slowly add the copper sulfate stirring constantly. Add distilled water to make up to 1 liter. The reagent will not deteriorate on standing.

Fehling's solution Prepare solutions 1 and 2 and store separately in rubber-stoppered bottles; also may be purchased. In testing for the presence of simple sugars, add an equal amount of each solution to a test tube of the substance to be tested and heat. A heavy yellow or reddish precipitate forms (cuprous oxide) if simple sugars are present.

solution 1:
$CuSO_4$	34.65 g
distilled water	500 ml

solution 2:
KOH	125 g
potassium sodium tartrate	173 g
distilled water	500 ml

Molisch's test Add 2 drops of Molisch's reagent to 5 ml of sugar solution in a test tube; mix well. Slant the tube and pour 3 ml concentrated sulfuric acid down the side of the tube. (*Caution.*) A layer of acid lies below the sugar. At the junction between the two layers of fluid a reddish violet color zone appears. The acid acts on the sugar forming furfural and derivatives. The test reaction is due to the presence of furfural and is not specific for carbohydrates.

Prepare Molisch's reagent by dissolving 5 g of α-naphthol in 100 ml of 95 percent ethyl alcohol. (You may prefer to substitute thymol for the α-naphthol since it does not deteriorate as quickly. In testing, add 3 to 4 drops of a 5 percent alcoholic solution of thymol to the test tube of sugar solution.) Positive tests are given by aldehydes and by formic, lactic, citric and oxalic acids.

Barfoed's test (Tauber and Kleiner modification)[49] In a test tube, add 1 ml of a 0.1 percent sugar solution to be tested (approximate percent). In a control test tube, add 1 ml of water; now add 1 ml of the copper reagent (preparation below) to each test tube. Heat both tubes for 3 min in a boiling water bath; cool for 2 min. Then add 1 ml of a color reagent such as Benedict's to each; mix well. When monosaccharides are present a blue color results. When only disaccharides are present the test sample matches the control tube. *Note:* This is not a specific test for glucose. Disaccharides will also respond under conditions of acidity, and if the sugar solution is boiled with the reagent long enough to hydrolyze the disaccharide by action of the acetic acid in the reagent. Barfoed's reaction is a copper reduction test but it differs from the other reduction tests in that the reduction occurs in an *acid* solution.

To prepare the copper reagent dissolve 24 g of copper acetate in 450 ml boiling water. Do not filter if a precipitate forms. Immediately add 25 ml of 8.5 percent lactic acid to the hot solution. Shake so that almost all of the precipitate dissolves. Then cool and dilute to 500 ml. After sedimentation, filter off impurities.

Selivanoff's test This is a resorcinol-hydrochloric acid reaction in which ketose sugars such as fructose turn bright red when heated with the reagent. Hydrolyzed sucrose yielding fructose also gives a positive test.

To 5 ml of Selivanoff's reagent in a test tube, add 5 drops of a fructose solution and heat the mixture to boiling. A positive reaction gives a red color; there may be a separation of a brown-red precipitate. This precipitate may be dissolved in alcohol giving a red color.

Hawk suggests the following comparison for the reactions of aldose and ketose sugars:[50] Place 0.5 ml portions of glucose, fructose, maltose, lactose and sucrose solutions each into five test tubes. Add 5 ml of Selivanoff's reagent to each tube, mix, and then place in a bath of boiling water. Note the time that the color first

[49] B. Oser, ed. *Hawk's Physiological Chemistry.*

[50] B. Oser, ed. *Hawk's Physiological Chemistry.*

appears in each test tube. Watch the color develop in each tube at 5-min intervals over a 15 min period of boiling.

To prepare Selivanoff's reagent, dissolve 0.05 g resorcinol in 100 ml dilute (1:2) hydrochloric acid.

Pentose sugars in plants

Leaf fragments or pure sugars may be used to demonstrate two tests for pentose sugars.

Benzidine test   Compare the reaction of glucose (a six-carbon sugar) with xylose (a five-carbon sugar). Prepare 0.25 percent solutions of each sugar. To 0.5 ml of one solution of sugar in a Pyrex test tube, add 2 ml of 4 percent solution of benzidine in glacial acetic acid. (*Caution:* Work under a hood.) Heat to the boiling point and cool immediately in cold water. Repeat the procedure with the other sugar solution. The presence of a pentose is indicated by a cherry-red color. Hexoses give a yellow or brownish color.

Bial's test   It is better to purchase prepared Bial's reagent in a concentration of 0.2%. Add 2 ml of the reagent to an equal volume of a pentose sugar. (Also run a test on glucose as described above in the benzidine test.) Heat in a boiling water bath. (Furfural is formed from the sugar and reacts with orcinol.) Now add a few drops of ferric chloride (1 percent solution). A deep green color indicates the presence of pentose.

Tests for lipids

Hawk describes many chemical tests for the presence of lipids including formation of acrolein from olive oil (test for glycerol), formation of fat crystals, saponification of bayberry tallow, iodine absorption test, and saponification number.[51] Harrow's laboratory manual also gives specific pro-

cedures for testing for lipids and other nutrients.[52]

A common elementary test is the formation of a translucent spot when a drop of fat or sample of a test substance is rubbed on unglazed paper.

Sudan III   stains neutral fat a red color. Prepare a 2 percent *stock* solution in absolute alcohol. Dilute 1 part of stock to 1 part of 45 percent alcohol immediately before use.

Sudan IV   also dissolves in, and stains fats red. Add a few grains of Sudan IV to a test tube half full of water; shake well. Then add a few drops of corn oil or other salad oils. Which region shows the dye? Test a few milliliters of milk, egg white, egg yolk, ground up seeds, among other substances, for the presence of fats. Examine under the microscope to observe the location of the red dye in the fat droplets.

Tests for proteins

It is necessary to use several tests for proteins since the reagents give a positive reaction for specific groupings in the protein molecule.[53] For example, the Biuret test depends on the peptide bond; ninhydrin. (*Caution:* toxic) tests for alpha-amino groups; and Millon reagent tests for tyrosine. In running tests, be sure to use several samples, such as dilute egg albumen, 2 percent gelatin, and water (control). Prepare egg albumen solution by beating egg white into 10 volumes of water; strain through cheesecloth. If powdered albumen is used, soak the powder in a little water for a few hours, then dilute with 10 volumes of water.

Xanthoproteic test   Pour 3 ml of a protein solution (such as egg albumen) in a

---

[51] B. Oser, ed. *Hawk's Physiological Chemistry.*

[52] Harrow et al., *Laboratory Manual of Biochemistry*, 5th ed. (Philadelphia: Saunders. 1960).
[53] Giese, *Laboratory Manual in Cell Physiology.* Also refer to *Hawk's Physiological Chemistry* and B. Levedahl et al., *Laboratory Experiments in Physiology* (see end of chapter listings).

test tube. *Slowly* add 1 ml of concentrated nitric acid and heat gently. The white precipitate turns yellow upon heating in a positive test. Cool the solution and carefully add, drop by drop, sodium hydroxide or ammonium hydroxide. An orange color indicates the presence of phenyl groups in the protein molecule (tyrosine and tryptophan). Phenylalanine yields a negative reaction.

Biuret test   Add 2 to 3 ml of a 10 percent solution of sodium hydroxide to an equal volume of egg albumen solution in a test tube; mix thoroughly. Then add, drop by drop, a 0.5 percent copper sulfate solution. Shake the tube between additions of drops. A pinkish or purplish violet color indicates the presence of proteins with at least two peptide linkages. Proteoses and peptones give a pink color; gelatin turns a bluish color.

Ninhydrin test   Pour 5 ml of a dilute protein solution into a test tube. The pH of the protein solution must be between pH 5 and pH 7. Add a few crystals of sodium acetate buffer to reach the correct pH (see preparation, p. 159). To the protein solution add 0.5 ml of a 0.1 percent solution of ninhydrin (triketohydrindene hydrate). (*Caution:* Avoid inhaling toxic fumes or eye or skin contact.) Heat to boiling for 1 to 2 min and then cool. A blue color indicates the presence of amino acids, peptides, peptones, and proteins. Carbon dioxide and ammonia are given off in the reaction. The intensity of the color produced is proportional to the amount of the protein present.

Millon test   Pour 5 ml of a protein solution into a test tube and add 3 to 4 drops of fresh Millon reagent. Heat carefully over a small flame. Egg albumen proteins form a white precipitate which turns red on heating. This reagent is also useful for testing solid proteins. In this case, dilute the reagent with 3 to 4 volumes of water. Tyrosine gives a positive reaction (Millon's is not used to test urine which contains large quantities of inorganic salts.)

Other proteins, such as secondary proteoses and peptones, turn red.

Millon reagent should be purchased ready-made since it contains mercury dissolved in nitric acid.

Hopkins–Cole test   To 2 to 3 ml of a protein solution such as egg albumen add an equal volume of the Hopkins–Cole reagent, which contains glyoxylic acid. Mix well. Slant the test tube and slowly add 5 ml of concentrated sulfuric acid down the side of the test tube. Note the reddish violet color that forms between the protein layer and the acid layer underneath. If the color does not appear after a few minutes, gently shake the tube so the fluids make contact. A positive test for tryptophan. (Gelatin gives a negative response.) Glyoxylic acid is toxic.

Hopkins–Cole reagent may be purchased ready-made.

Molecular models

To build models of organic molecules, many types of kits are available from supply houses. You may also try making your own using styrofoam balls and pipe cleaners. Recall that carbon is represented black; nitrogen as yellow; hydrogen as green; oxygen as white. Molecules of O_2 can be made by joining two white styrofoam balls with two pipe cleaners.

Testing seeds for macromolecules

Proteins

Meyer, Anderson, and Swanson suggest protein tests that may be made on slides.[54] Thick slices of lima bean seeds that have been soaked long enough to be easily cut are used. Since the directions given by Meyer, Anderson, and Swanson vary from those described above, we include here the three tests that may be more suc-

[54] B. Meyer, D. Anderson, and C. Swanson, *Laboratory Plant Physiology*, 3rd ed. (New York: Van Nostrand, 1955).

cessful when run directly on microscope slides using soft lima beans. Additional microchemical tests that may be made on slides of tissues of plants are the osazone test for sugars, the phloroglucin test for lignin, the iodine–sulfuric acid test for cellulose, or the Sudan III test for lipids.

Line up slides containing slices of lima beans on a sheet of white toweling, and run the following series of tests.

Biuret test   In this specific test for the presence of peptide linkage, proteins give a pink-to-violet color reaction. Add a drop or two of 5 percent copper sulfate solution to a thin section of soft lima bean; apply a coverslip and let it stand in a moist chamber for ½ hr. (A moist chamber may be made by lining a Petri dish with moist blotting paper or filter paper.) Remove the coverslip and wash the slice with distilled water. Blot dry with filter paper and add a drop of 50 percent potassium hydroxide solution. (*Caution.*) Does a pink or violet color appear?

Millon's reaction   To a second section of lima bean on a slide, add a drop or two of dilute nitric acid (2 percent). Let it stand for a minute and blot off the excess nitric acid. Now add a drop of Millon's reagent (purchase ready-made). Proteins give a deep red color. Since the reaction depends on the presence of the monohydroxybenzene nucleus, some compounds other than proteins may give a similar color reaction. Therefore, this is not a conclusive test for proteins.

Xanthoproteic reaction   To a section of lima bean, add a drop or two of concentrated nitric acid. A yellow color is a positive test for proteins. Since the reaction depends on the presence of the benzene nucleus, several other compounds give the same reaction. Blot the excess nitric acid with filter paper and add a drop of strong ammonium hydroxide or sodium hydroxide to the slice. Proteins turn a deep yellow or orange-brown color in this check step.

Tests for lipids, water, and minerals are described on page 178.

Principles of chromatography

Slightly different solubilities of like compounds permit such compounds to be separated in water and an organic solvent. You will recall that Michael Tswett, a Russian botanist, developed this technique. If a strip of Whatman #1 filter paper is introduced as a wick into a small sample of such a mixture or into a closed atmosphere saturated with the mixture, the compounds will migrate up or across the paper at different rates. The factors affecting the separation of pigments along the filter paper (or packed column of adsorbent, Fig. 2-30) are a combination of adsorption, ion exchange, and partition. Thus, unknown compounds can be identified by measuring the relative speed at which they migrate along the filter paper.

The distances traveled by any compound are dependent on the rate of migra-

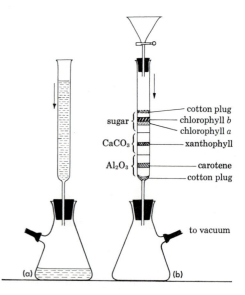

Fig. 2-30 Column chromatography: (a) column of one adsorbent; (b) use of different adsorbents and order of adsorption of pigments of chloroplasts. (After B. Harrow et al., *Laboratory Manual of Biochemistry*, 5th ed., Saunders, 1960.)

tion of molecules along a moving solvent front produced by the rise of the solvent along the length of the filter paper. As these compounds move along the filter paper, they distribute themselves between the solvent and the cellulose fibers of the paper. In general, the more soluble a compound is in water compared with the organic solvent, the slower it travels; it moves even more slowly if the pigment is also absorbed by the fibers of the paper. (Compare the results of using Whatman #1 and #3 papers.)

Since the fastest molecules will travel the greatest distance, or to the highest point along the strip in ascending chromatography (or will be located at the outer circle of the concentric rings of the circular filter paper), the relative distances can be measured, and the rate of flow, or migration, of the molecules (R_f) can be calculated.

$$R_f = \frac{\text{distance substance (solute) traveled}}{\text{distance solvent traveled}}$$

In column and paper chromatography, substances are distributed in a one-dimensional row. In some cases, compounds overlap and a second separation step is required, producing a two-dimensional chromatogram. In the latter method, developed by two British chemists, the Nobel Prize winners A. J. Martin and R. L. Synge, the procedure followed in the first separation is repeated, using a different solvent traveling at right angles to the first spread, or separation; as a result, the closely associated compounds separate out in a second dimension.

In contrast to pigments from chloroplasts or petals that can be separated and perceived, most compounds such as sugars and amino acids are colorless and need to be stained. Of course, if radioactive tracers are included in their synthesis the compounds may be identified if the chromatographic paper is placed in contact with X-ray film for a few days. (Several techniques for partitioning pigments of flowers and fruits, as well as pigments in chloro-

plasts with Whatman paper, are developed in chapter 3, pp. 207, 209.)

Chromatography: amino acids

In paper chromatography amino acids in a solution are separated and identified according to their R_f (rate of flow of molecules) and compared with standards based on temperature and type of solvents used (Table 2-8). Harrow suggests a laboratory procedure to illustrate the technique for one-dimensional chromatograms.[55] Obtain several pure amino acids such as isoleucine, threonine, and aspartic acid and prepare solutions of equal dilution. Dissolve 10 mg of the amino acid in 10 ml of distilled water; heat slightly, if necessary, to dissolve the amino acid, or add 1 to 2 drops of 0.1 N HCl. Fit each of several test tubes with strips of Whatman #1 paper (see apparatus in Figure 3-10 for partitioning of chlorophyll pigments). Make a pencil mark about 6 mm from the

[55] Adapted from B. Harrow et al., *Biochemistry.*

TABLE 2-8   Rate of flow of molecules (R_f) of several amino acids*

amino acid	R_f of 125 mm at 25° C (77° F) after 3 hr (solvent: butanol-glacial acetic-water)
Alanine	0.62
Arginine HCl	0.53
Aspartic acid	0.25
Glutamic acid	0.39
Glycine	0.49
Histidine HCl	0.81
Isoleucine	0.85
Leucine	0.86
Lysine HCl	0.41
Methionine	0.74
Phenylalanine	0.87
Proline	0.87
Serine	0.33
Threonine	0.57
Tryptophan	0.81
Tyrosine	0.53
Valine	0.82

* From B. Harrow et al., *Laboratory Manual of Biochemistry*, 5th ed., Saunders, Philadelphia, 1960.

end of each paper to indicate the place to spot the papers. Some technicians use spots, others prefer streaks of the substance to be partitioned. Use a Pasteur or capillary pipette to transfer a small drop or streak of an amino acid solution to the paper. Using separate pipettes spot the other papers. After drying, apply a second drop to the papers. When dry, insert each strip into a separate test tube; each test tube should contain 0.4 ml of butanol–glacial acetic acid solution. (Prepare by adding 250 ml of water to 250 ml of *n*-butanol; shake well and add 60 ml of glacial acetic acid.) Withdraw the bottom layer; store the butanol layer in the presence of the aqueous layer so that esterification will be reduced.

In each test tube, only the tips of the papers are immersed in the solution. Keep a record of time and rate of the solution migration until the Whatman paper is wetted to within 5 mm of the top of the paper (about 2 to 3 hrs). Then remove the papers and, with a pencil, mark the upper front; dry for about 3 min at 110° C (230° F). Under a hood, spray with a 0.25 percent solution of ninhydrin (*caution*: toxic) in water-saturated butanol. Again, dry at 110° C (230° F). As th color appears circle the spots before they fade. For each case, measure the distance that the solvent front migrated and the distance the amino acid migrated; then estimate the R_f value for each amino acid (refer to page 181 for equation).

Use Table 2-8 to identify the amino acids and to check the skill of students in measuring the R_f of the molecules in solutions. Then prepare *mixtures* of several amino acids and repeat the procedure. Or give students an "unknown" to identify.

This simple technique can be used for investigations of altered conditions of diet, drugs, or hormones on blood or urine, or to compare the contents of cells and tissues, or the composition of the food residues in seeds. (Techniques for two-dimensional chromatography are described in several readily available laboratory manuals.[56])

[56] Refer to Giese, *Laboratory Manual in Cell Physiology.*

Simultaneous chromatogram of amino acids

Instead of individual strips to make chromatograms, run a single chromatogram of several different known amino acids plus an "unknown" which may be one or a mixture of two of the known that are distributed to students. Prepare a Whatman #1 sheet (12 cm^2); lay it on wax paper and handle only the edges of the paper. In pencil, rule a line 2 cm from an edge which will be the bottom edge. Also rule a line 5 cm from the top edge of the paper which will indicate the limit of the advancing front of solvent. Mark spots in pencil equal distances across the bottom pencil line of the sheet; label these with abbreviations of the amino acid solutions to be used.

Supply students with solutions of equal molecular concentrations, such as 0.001 *M* solutions of alanine, histidine, lysine, methionine or other amino acid solutions. Use different capillary pipettes to apply spots of the amino acid solutions to the pencil marks and allow these to dry. Then apply a second set of spots and allow these to dry.

Shape the sheet into a cylinder and fasten together with staples or needle and thread but keep a gap between the edges. As a solvent, use a mixture of formic acid, water, and isopropyl alcohol in the ratio of 10:20:70. To keep the chamber saturated, line a liter container with filter paper; pour about 30 ml of the solvent into the container and allow the solvent to rise in the paper liner. Next insert the cylinder of Whatman paper with the amino acid spots into the container without touching the liner and close the container.

Allow the solvent to rise to nearly 5 cm from the top, then remove the cylinder to dry. When dry, spread out the sheet and spray the surface under a hood with ninhydrin solution. (*Caution*: Avoid inhaling, or contact with skin.) Next, warm the sheet in an oven at 80° C (176° F) for some 3 to 5 min. Then outline the spots in pencil before they fade in the light.

Calculate the R_f of the known amino acids and compare the "unknown" solution of amino acid(s) with the known amino acids on the sheet. When two amino acids are closely spaced, you may want to rotate the sheet 90° and stand the roll in a second solvent.

Separation of products of protein hydrolysis

Wald et al. suggest a comparative chromatogram study of an unhydrolyzed protein, the protein hydrolysate, one unknown amino acid, and several known amino acids (such as alanine, histidine, lysine, methionine, and/or aspartic acid) prepared on one sheet of Whatman #1 paper.[57]

A protein that is easily obtained for hydrolysis and analysis is casein in milk. Casein may be purchased or extracted from skim milk. Transfer about 0.2 g of casein into a test tube, and add 2 ml of 0.1 percent pancreatic extract dissolved in 0.1 M phosphate buffer at pH 7. Maintain in a water bath at 40° C (104° F). It is advisable to add a small crystal of thymol to inhibit growth of bacteria. The casein hydrolysate should be applied twice with a micropipette to Whatman #1 paper.

On one edge of a square sheet of filter paper, mark off dots (with a lead pencil) 1.5 cm from the edge and at an equal distance from each other. The number of dots is the same as the number of test spots to be applied. Apply different materials with fine capillary pipettes; allow the paper to dry; then reapply amino acids at least four times, and hydrolysate about six times. Label the spots with lead pencil.

Fold the paper into a cylinder, handling it only at its edges, and staple or sew it together so that the cylinder can stand in about 1 cm of solvent. Prepare 30 ml of solvent by combining 10 parts of formic acid (*caution*), 70 parts of isopropanol, and 20 parts of water. Stand the cylinder in this solvent; allow the solvent front to rise to nearly 0.5 cm from the top. Remove the

cylinder and allow it to dry. Open the cylinder and dip the sheet of paper into ninhydrin-acetone reagent (*caution:* avoid inhaling, toxic). Remove and allow to dry in hot air. Circle the colored spots.

Measure the distances from the starting mark to the front for each of the test substances and for the solvent, and prepare a table of results. Identify the amino acids in the hydrolyzed protein.

Extraction of nucleoproteins

From yeast   Wald describes a procedure for identifying nucleic acids and nucleotides in yeast cells using paper chromatography.[58] This extract of nucleic acids from yeast contains a small amount of DNA and a larger amount of RNA. Another extraction method, among several, is described in *Hawk's Physiological Chemistry.*[59]

From thymus glands   You may want to purchase cottony threads of DNA from biological supply houses, or as a special activity extract it from freshly minced thymus glands (or from spleen). In physiological saline, to which a few ml of 0.01 M sodium citrate has been added, mince the tissue and wash it several times in fresh saline solution. According to Hawk's reference, stir the tissue in 10 volumes of 1 M NaCl containing 0.01 M sodium citrate. Centrifuge at 10,000 to 12,000 rpm and separate off the supernatant liquid which contains the nucleoprotein. Now pour the solution into 6 volumes of water to reduce the NaCl content and observe the fibrous mass of nucleoproteins. Next, decant the supernatant fluid, and again, dissolve it in 1 M NaCl to purify the nucleoproteins further. Again, centrifuge at the same speed, and precipitate the nucleoproteins into 6 volumes of water. Stir the mixture with a glass rod

---

[57] From G. Wald et al. *Twenty-six Afternoons of Biology: An Introductory Laboratory Manual*, 2nd ed. (Reading, MA: Addison-Wesley, 1966).

[58] Wald et al., *Twenty-six Afternoons of Biology* (chapters 3 and 4). Separates of these laboratory activities are available from the publisher.

[59] B. Oser, ed. *Hawk's Physiological Chemistry;* also refer to Giese, *Laboratory Manual in Cell Physiology.*

(pulled out in a flame so it has a small hook) and observe the fibrous nucleoproteins adhere to the rod as the rod is twisted.[60] The nucleoproteins may be tested using the Biuret or xanthoproteic test (p. 179).

Also refer to DNA from beef pancreas in chapter 6, page 392, and making models of DNA, page 389.

From frog testes Rugh also describes an extraction method used by Mirsky and Pollister.[61] Grind a large number of testes of frogs in a small volume of $1\ M$ sodium chloride solution. To this mash, add 10 parts of $1\ M$ NaCl; a viscous mixture results due to the release of nucleoproteins. Centrifuge this mixture at 10,000 rpm and add 6 volumes of distilled water. Notice that the nucleoproteins precipitate out as fibrous matter. Pour off the supernatant fluid.

Wash the precipitate in $0.14\ M$ NaCl and redissolve the precipitate in $1\ M$ NaCl. Again centrifuge at a high speed. Precipitate the nucleoproteins again by adding 6 volumes of distilled water.

Now carefully stir with a glass rod; twist the rod and notice how the fibers of nucleoproteins wind around the rod.

Paper electrophoresis

In chromatography, separation of substances is based on the different rates of molecular movement through an adsorbing material. Electrophoresis is based on the fact that charged ions of colloidal substances migrate along paper (or other substances) toward one of the poles in an electric field. The distance through which a charged particle moves is dependent upon time and is proportional to the potential gradient at the spot where the particle is located. This last factor depends on the current, the conductivity of the solvent, and the cross-sectional area of the

solution at this point. Several zones develop if the colloidal solution is impure, each zone composed of particles having similar rates of electrophoretic migration.

Paper electrophoresis is a simple adaptation of the original principle used by Tiselius, who obtained five distinct fractions composing plasma—globulins (alpha, beta, and gamma), albumin, and fibrinogen. Later workers have obtained subdivisions of the globulins. Only small samples of colloidal material are needed in paper electrophoresis. In essence, filter paper soaked in an electrolyte is connected to a suitable source of high-voltage direct current (about 4 to 8 volts per cm). Fractions of substances are made visible by staining.

The pH at which an amino acid has no tendency to migrate toward either the negative or positive electrode is its isoelectric unit.

The electrically charged nature of a protein molecule depends on its existing pH and the composition of the amino acids. At a pH that is acid to its isoelectric point, a protein molecule will contain an excess of NH_3^+ groups and will react as a positive ion and migrate toward the cathode. When the pH is alkaline to the isoelectric point, the protein molecule moves to the anode, since it has an excess of COO^- groups and behaves as a negative ion.

Proteins differ in their isoelectric points; as a result, they have net charges of different degrees at any specific pH. In brief, in a mixture of proteins, as in blood plasma, the components migrate at different velocities in an electric field.

If the pH of the protein mixture is adjusted with a buffer to a point that is alkaline to the isoelectric points of all the components, then all these constituents of proteins will carry negative charges. When an electric current is passed through an electrolyte, these proteins will all migrate to the anode, but at different rates. The constituents are thereby partitioned out.

Using blood Stong describes a simple apparatus for the electrophoresis of

---

[60] Original procedure by A. Mirsky and A. Pollister, *J. Gen. Physiology* (1946), 30:117

[61] Rugh, *Experimental Embryology.*

blood.[62] Apply a spot of blood to filter paper and subject it to an electric current in this manner. Moisten the paper in an electrolytic solution (such as 1 tsp sodium chloride and ¼ tsp sodium bicarbonate in 300 ml of tap water). Immerse each end of the paper strip in two jars containing the electrolyte. Set the jars about 15 cm apart; support the paper by sandwiching it between two glass rectangles so that only the ends of the papers are free to extend into the solutions (Fig. 2-31).

Attach electrodes in the containers. After the components have been separated in the electric field, ninhydrin (or another suitable stain such as bromphenol blue) is applied. (See test for amino acids, p. 178.) Graphite electrodes may be used with a current of some 10 milliamperes and a voltage of about 200 volts.[63] Use pure D.C. "B" batteries, or highly regulated D.C. power supply, ±½ volt regulation.

The various components of blood that have negative charges migrate toward the positive electrode at varying rates.

Students will find some ideas for projects, such as studies of electrophorized

[62] C. L. Stong, *The Amateur Scientist* (New York: Simon and Schuster, 1960). Stong recommends the use of a rectifier from a junked radio receiver.
[63] Stong, *The Amateur Scientist*.

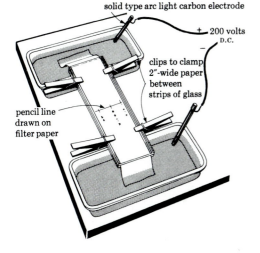

solid type arc light carbon electrode

+ 200 volts
D.C.
−

clips to clamp 2"-wide paper between strips of glass

pencil line drawn on filter paper

Fig. 2-31   Simple apparatus for paper electrophoresis. (After C. L. Stong, *The Amateur Scientist*, Simon and Schuster, 1960.)

serum of several species of tadpoles and the changes in quantity of serum albumins as metamorphosis progresses, clearly described in E. Frieden, "The Chemistry of Amphibian Metamorphosis," in *Scientific American* (November, 1963). Also read S. Clevenger's paper on flower pigments in the June, 1964 issue of *Scientific American* and E. Hadorn's paper on fractionating fruit flies. (Hadorn's technique is described here in chapter 7, page 539.) These papers suggest many avenues for possible inquiry. Also refer to *Scientific American*, October 1985, and current issues.

Enzymes and cell activities

An inkling of the speed of enzyme action—the many thousands coded by DNA in a cell—can be shown in specific demonstrations or investigations; but only a small sample can be offered, to be sure.

These catalysts, protein in composition, are active within a limited pH range and temperature and show specificity for a given substrate. A select number of activities involving enzymes in cellular respiration are offered here; the action of digestive enzymes that catalyze the hydrolysis of complex polysaccharides, lipids, and proteins are described within the context of digestion (chapter 3).

Mitochondria: centers of enzyme activity

In this technique, Janus Green B is used to stain mitochondria of celery cells, and then the dye is bleached by enzymes in the mitochondria.

Cut across two "strings" (collenchyma) of a fresh celery stalk; then make a second transverse cut 1 cm from the first. Transfer this 1-cm length, *inner surface up*, to a drop of 5 percent sucrose solution on a clean slide. With a razor blade, carefully cut away the two strings of the strip, leaving the transparent center section containing two or three layers of cells. For ease in cutting, use two razor blades as follows: Hold the two blades close to and parallel with the string. Keep the inner blade stationary and draw the outer blade along the edge of the string to make the cut. The stationary blade keeps the section from twisting. Repeat for the string on the other side. Now seal with a coverslip and look for cytoplasmic streaming in the epidermis and the uppermost subepidermal layer. Also find the green plastids, clear nucleus, and small, moving spheres and rods, which are the mitochondria. Apply filter paper to one end of the coverslip, and at the same time, add several drops of 0.001 percent Janus Green B solution to the opposite side. Watch the mitochondria stain blue; within minutes they will be decolorized by enzymes (dehydrogenases) in the mitochondria. The slides may be used for several days if they are placed in Petri dishes lined with moist toweling and stored in a refrigerator.

Also refer to staining mitochondria of frogs' blood cells using Janus Green B (p. 273).

Techniques for studying mitochondria of a living fly or bee are described in the BSCS laboratory block by A. G. Richards, *The Complementarity of Structure and Function* (Lexington, MA: D. C. Heath, 1963).

Catalase

Explore the speed of action of the enzyme catalase found in plant and animal tissues. Catalase breaks down hydrogen peroxide into water and oxygen.

$$2H_2O_2 \xrightarrow{\text{catalase}} 2H_2O + O_2$$

Students, working in groups, may use living tissues in different series of test tubes: minced raw potatoes, raw ground meat, yeast cells, fresh blood from a frog or human, or ground leaves from bean seedlings about 3 weeks old.

Fill one-third of each test tube with fresh hydrogen peroxide (3 percent). Add each tissue to be tested to a separate test tube; leave one tube of H_2O_2 as a control. Other controls may be suggested by students: boiling the tissue before adding to the test tube of peroxide, or testing the effect of

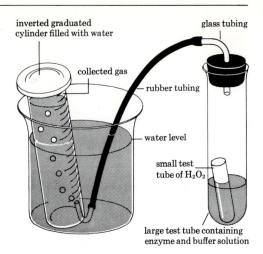

Fig. 2-32 Apparatus for collecting oxygen released from hydrogen peroxide as a result of the activity of catalase. Tip the test tubes so that hydrogen peroxide solution and enzyme-buffer solution are mixed. (After L. Machlis and J. Torrey, *Plants in Action*, Freeman, 1956.)

temperature variations on the activity of the catalase.

Note that heat is produced, and the rapid activity of catalase can be observed by the froth of bubbles.

Collection of oxygen   Some students may want to collect the oxygen gas by displacement of water, measure it, and test with a glowing splint to identify the gas (Fig. 2-32). Or use a volumeter of the type shown in Figure 3-18 to measure the quantity of oxygen produced.[64]

A quantitative technique   Study some quantitative aspects of enzyme reactions, using the technique of titration with potassium permanganate solution. In an acid solution, hydrogen peroxide decolorizes potassium permanganate within a few minutes. In a beaker, mix 1 ml of phosphate buffer (p. 159) and 9 ml of hydrogen peroxide (3 percent). Record the time at

---

[64] Also consider the technique described by Wald et al., *Twenty-six Afternoons of Biology*, using a volumeter to determine the quantity of oxygen liberated from *Serratia* cultures through the activity of catalase.

which 1 ml of catalase solution is mixed with the peroxide-buffer solution. Allow the action to continue for 10 min; then add 2 ml of sulfuric acid.

Now students can determine the quantity of hydrogen peroxide remaining in the solution after 10 min of reaction with the enzyme catalase. Count the number of drops of $KMnO_4$ needed to reach the end point where 1 drop of permanganate causes a pink color to persist. Allow a minute to elapse, and swirl between additions of drops of permanganate. Estimate the volume of permanganate needed (use a burette or graduated micropipette). What quantity of the hydrogen peroxide is destroyed in 10 minutes?

$$5H_2O_2 + 2KMnO_4 + 4H_2SO_4 \longrightarrow$$
$$2KHSO_4 + 2MnSO_4 + 8H_2O + 5O_2$$

Also titrate control solutions, that is, solutions containing the same chemicals, with water (1 ml) in lieu of the catalase solution. What is the effect of boiling on the activity of the enzyme?

Can students devise a demonstration to study the effect of metallic salts on the action of the enzyme?

Catalase activity and pH   Machlis and Torrey suggest using leaves of beans in determining the effect of pH on catalase activity.[65] Select bean leaves from ten seedlings that are about three weeks old; grind them in a mortar and add 40 ml of distilled water. Using the apparatus shown in Figure 2-32, at zero time tip the test tube so that the enzyme-buffer solution and the peroxide solution are mixed. Begin with a trial run using a buffer at 7.2 pH and let the reaction continue for 5 min, or a time sufficient to fill the 10-ml cylinder. Perform other runs using different pH buffers (3.6, 4.8, 5.4, 6.0, and so forth).

Bacterial enzymes   Refer to the action of some bacteria on starch (p. 607) and the effect of bacterial cellulase on plant fibers or filter paper (p. 606).

[65] Adapted from L. Machlis and J. Torrey, *Plants in Action* (San Francisco: Freeman, 1956).

Role of hydrogen acceptors: chick embryo

Examine the role of a vital dye such as Janus Green B in accepting hydrogen atoms in the oxidative process in cells of a developing chick embryo. Hydrogen is transferred from organic molecules to Janus Green B, thereby reducing the dye. In the oxidized state Janus B dye is green; in the reduced state it is red.

"Seat" a fertilized egg that has been incubated for 48 hr in a shallow depression of modeling clay so that it will not roll during the operation. Start to chip and carefully cut away a circle of shell from the top surface of the incubated egg. Use a forceps to lift off carefully the chipped circle of shell; with a pipette, draw off some of the more fluid albumen to expose the surface of the embryo which is lying in the center of the egg.

Place a small ring of filter paper over the egg so that the embryo is in the center. The inner diameter of the filter paper ring should approximate the diameter of the blood sinus around the embryo; the outer diameter of the paper should fit into a depression slide. Carefully cut the membrane, using the outer rim of the paper ring as a guide (Fig. 6-41). The embryonic tissue should adhere to the paper ring. Transfer this preparation into a Petri dish containing warmed (37° C, 99° F) physiological saline solution (0.9 percent sodium chloride). Several such preparations should be made. Then transfer each embryo to individual depression slides containing some warmed Janus Green B vital dye. (Prepare the solution by adding 0.1 g of Janus Green B to 500 ml of physiological saline solution.)

After about 15 min the embryo should stain green. Draw off the vital stain with a pipette and replace the fluid with warm saline. Now the stained embryo is in warm saline solution in a depression slide. Apply a coverslip and make it airtight with petroleum jelly; avoid getting the petroleum jelly under the slide near the embryo. Use low power of a microscope observe the appearance of a reddish color. Keep a record of the areas that become red and in what order in time.

Role of dehydrogenase

The removal of hydrogen in oxidation is catalyzed by a dehydrogenase which transfers hydrogen from an energy-yielding substrate to a hydrogen acceptor. Methylene blue can be used to show the action of a hydrogen acceptor. In this technique the source of the dehydrogenase is fresh, finely minced muscle from a rat.[66]

First prepare a phosphate buffer ($M/15$) at pH 7.0. Into each of a set of three test tubes pour 5 ml of the phosphate buffer, and then add 3 drops of 0.01 percent methylene blue solution. Into test tube 1 place a small amount of finely minced, fresh muscle of rat. Mix thoroughly, and quickly add a layer of mineral oil over the surface to prevent air from reaching the dye.

Prepare tubes 2 and 3 in a similar manner, except for the treatment of the muscle tissue: To tube 2 add *boiled* muscle; to tube 3 add fresh muscle that has been *washed* several times in distilled water. Place the set or sets of tubes in a water bath maintained at 37° C (99° F). Explain the changes in the tubes.

Succinic dehydrogenase: beef heart

This hydrogen transferring enzyme found in mitochondria is active in catalyzing one step in the Krebs cycle—the oxidation of succinic acid to fumaric acid.

In this procedure, beef heart is used as a source of succinic dehydrogenase.[67] In this oxidation reaction, methylene blue is reduced (and decolorized) upon acceptance of two hydrogen atoms from succinic acid. Cut up a small bit of beef heart (about the volume of a marble) into a test tube, and wash it several times with water. Prepare

sets of the following test tubes: (1) no meat, only an equivalent volume of distilled water plus 3 drops of succinic acid (0.5 M), plus 7 drops of 0.01 percent methylene blue solution, plus 9 drops of water to bring the total volume to that of the other test tubes; (2) meat, plus 3 drops of succinic acid, plus 9 drops of distilled water; (3) meat, plus 7 drops of methylene blue solution, plus 12 drops of distilled water; (4) meat, plus 3 drops of succinic acid, plus 7 drops of methylene blue solution, plus 9 drops of distilled water; and (5) boiled meat (2 min), plus 3 drops of succinic acid, plus 7 drops of methylene blue solution, plus 9 drops of distilled water.

Down the side of each tube pour a layer of mineral oil to prevent oxygen of air from entering the tubes. Stand all the test tubes in a water bath maintained at 37° C (99° F). In which tubes are there changes in color? Vigorously shake the contents of test tube 4 to show oxidation of reduced methylene blue.

At some time, students may use the technique described in the laboratory manual of Machlis and Torrey in which germinating white beans are the source of the succinic dehydrogenase.[68]

Cytochrome oxidase

In the presence of molecular oxygen, cytochrome oxidase oxidizes reduced cytochrome c. Nadi reagent, a mixture of α-naphthol and dimethyl-p-phenylenediamine (*caution:* toxic) reduces cytochrome c. These chemicals in Nadi reagent are oxidized to a blue pigment indophenol blue by cytochrome oxidase. Several techniques are described in the literature.

Using chick embryos   Wald suggests

using chick embryos of different ages.[69] Remove the chick embryo from the yolk and transfer to warm saline solution (technique, p. 187). Replace with warm Nadi

[66] Adapted from "Laboratory Experiments in Elementary Human Physiology," (1962), mimeo sheets, prepared under the auspices of the Am. Physiological Soc.; also refer to Giese, *Laboratory Manual in Cell Physiology.*

[67] From Wald et al., *Twenty-six Afternoons of Biology.*

[68] Machlis and Torrey, *Plants in Action.*

[69] Wald et al. *Twenty-six Afternoons of Biology.*

reagent and record the time. Watch for the first sign of blue color to appear and record its location at 3-min intervals. The preparation of Nadi reagent for this demonstration is given as follows. Prepare a 0.01 M solution of α-naphthol by dissolving 1.44 g of α-naphthol in 1 liter of 0.9 percent NaCl solution; heat to dissolve. Also prepare a 0.01 M solution of dimethyl-p-phenylenediamine (*caution*) by adding 1.36 g to 1 liter of physiological saline. As a buffer, two salt solutions are needed. Make up 9.5 g of Na_2HPO_4 in 1 liter of distilled water. Also prepare 9.07 g of KH_2PO_4 in 1 liter of distilled water. The phosphate buffer is a mixture of these two salt solutions in the proportion of 7.8 ml of Na_2HPO_4 solution to 92.2 ml of KH_2PO_4 solution. Nadi reagent is made by combining *equal* parts of the α-naphthol solution, the dimethyl-p-phenylenediamine solution, and the phosphate buffer.

Using potato extract   Hawk suggests adding Nadi reagent (5 drops of α-naphthol solution plus 5 drops of p-phenylenediamine hydrochloride solution —*caution*) to a tube of 5 ml of potato extract.[70] Prepare the potato extract by grating a raw, peeled potato into cheesecloth that is suspended in a beaker of 200 ml of distilled water. Use the filtered water extract in the test. In this reaction involving potato oxidase, indophenol is produced from the α-naphthol and phenylenediamine.

Phosphorylase

Phosphorylases split glycogen by introducing phosphoric acid between glycolic linkages. Refer to Wald for a demonstration using phosphorylase from a potato tuber.[71] A Waring blender and sodium fluoride (*caution:* toxic) are needed.

Phenolases in plant cells

These copper-containing enzymes in plant tissues cause the darkening of the cut surfaces of many fruits. Show the slight

[70] *Hawk's Physiological Chemistry.*
[71] Wald et al. *Twenty-six Afternoons of Biology.*

consumption of oxygen, available at the cut surfaces, in the oxidation of phenolic substances in cells. Use a cork borer to obtain several cylinders from a ripe apple and place these in a test tube. Connect the tube to a manometer; record five readings of changes in the manometer level taken at 2-min intervals.

However, since CO_2 is also given off in respiration, take a fresh set of apple cylinders and attach a manometer using a potassium hydroxide trap to absorb the CO_2. Note the effect of a copper chelator that binds the copper in the phenolase and inhibits this oxidation of phenolic substances. Place fresh apple segments in a test tube containing 0.1 M sodium diethyldithiocarbamate. (*Caution:* avoid contact with skin; eye irritant.) Compare this set of readings with those of the retreated apple segments.

Cholinesterase and acetylcholine

The activity of the enzyme cholinesterase can be measured by titration of acetic acid liberated in the following reaction.

$$\text{acetylcholine} + H_2O \longrightarrow$$
$$\text{choline} + \text{acetic acid}$$

A demonstration of this action is described by G. Cantoni.[72] Acetylcholine and cholinesterase are available from pharmaceutical supply houses. Use cresol red solution as an indicator, and titrate with 0.004 N sodium hydroxide.

Luciferase

The emission of light by many invertebrates—fireflies, marine dinoflagellates such as *Noctiluca*, and the ostracod crustacean *Cypridina*, and also by luminous fish, bacteria, and fungi is known as bioluminescence and is the result of the oxidation by an enzyme on some substrate.

Luciferin, a heat-stable substrate, is oxidized in the presence of oxygen, water, and the heat-sensitive enzyme luciferase. For

[72] Paper by G. Cantoni, "Laboratory Experiments in Biology, Physics, and Chemistry." Demonstrated for members of the National Science Teachers Assoc. by the staff of the National Institutes of Health, Bethesda, MD, March 16, 1956.

Fig. 2-33 *Photobacterium fischeri:* motile, Gram-positive, slightly curved rods; aerobic, facultative; luminescent. (Photographed in total darkness using only light emitted by the bacteria.) (Courtesy of Carolina Biological Supply Co.)

luciferin to luminesce, ATP and some divalent ions (Mn^{++} or Mg^{++}) are needed in addition to oxygen and the enzyme. A summary of reactions that result in luminescence:

$$\text{luciferin-ATP} \xrightarrow[\text{O}_2 \text{ Mg}^{++} \text{ H}_2\text{O}]{\text{luciferase}} \text{luciferin}$$
$$+ \text{ ADP} + \text{P} \quad + \text{ ADP} + \text{P}$$

Cultures of luminous bacteria (such as *Achromobacter fischeri*) may be purchased from supply houses (Fig. 2-33). Some fish have a specific, complex organ—a photophore—that is regulated through nerve and hormone control; others have only symbiotic bacteria which inhabit an organ that is set below each eye of the fish. The luminous organ in the firefly is the lantern, which is regulated by the nervous system. It is a fat body supplied with oxygen from small tubules of the tracheae.

Luciferin is soluble in water and in higher alcohols. Quick drying of organisms with acetone to remove water is the usual method for preserving luciferin. Refer to Giese for a comprehensive treatment of chemical composition and methods of extraction of luciferin from *Cypridina*, luminous bacteria, and fireflies.[73] (Giese also discusses the extraction of luciferase and the chemistry of bioluminescence.)

Bioluminescence has evolved several times so that some variations are found in the luciferin-luciferase systems. For example, in *Balanoglossus*, peroxide is used instead of oxygen and the enzyme is a peroxidase.

Observe the brilliant blue light that results when a bit of powdered marine crustacean, *Cypridina*, is placed in the palm of the hand and a few drops of water are added. The powder contains both luciferin and luciferase. The results are more effective in a darkened room. Obtain small quantities of pulverized Japanese marine crustaceans in vials of ½-dram quantities from supply houses (Appendix).

Investigations may be planned to study the step-by-step reactions and the rate of luminescence in a solution containing luciferin and luciferase. Investigate the concentrations of enzyme, pH, and water, and the presence of heavy metals or drugs. The effects of these variables follow the usual patterns characteristic of enzymatic reactions. At what time should oxygen be added to the system to affect maximum intensity of luminescence? If luciferin and luciferase are mixed before oxygen is added, maximum intensity occurs 0.02 sec after oxygen has been added to the system. However, if oxygen is added separately to the enzyme and to the substrate, and then they are mixed, maximum intensity is reached after 0.6 sec. This seems to indicate step-by-step reactions (see equations in Giese).

Luminous bacteria Crane suggests a method for obtaining luminous bacteria by allowing fish, squid, or shrimp to rot in a pan with some seawater partly exposed to air.[74] A ring of luminous bacteria forms around the body of the decaying organism, which can best be seen in the dark. Culture techniques using agar and broth media are described.

Decaying wood logs also show luminescence in the dark. Keep samples moist.

[73] Giese, *Cell Physiology*, Chapter 13.

[74] J. Crane Jr., *Introduction to Marine Biology*, A Laboratory Text. (Columbus, OH: Charles Merrill, 1973). Also refer to Giese, *Laboratory Manual in Cell Physiology*, for activities with luminous bacteria.

Then test the effect of reducing the supply of oxygen or altering the temperature on luminescence. Also examine small samples under the microscope.

Special investigations in cell physiology

Several laboratory manuals in cell physiology give procedures for special studies. For example, Giese[75] offers investigations in active transport in protozoa, in the goldfish kidney; properties of succinic dehydrogenase of some in-

vertebrates; properties of myogenic heart muscle of invertebrates; effect of RNAase on *Blepharisma* regeneration; induction of colorless mutants in *Euglena gracilis*; preparation of cytochrome oxidase; utilization of various carbohydrates by *Tetrahymena*; colorimetric microdetermination of nitrogen in biological materials.

Nerve-muscle preparations and investigations are described in Levedahl[76] as well as many physiochemical studies. Wald's *Twenty-six Afternoons of Biology* describes several investigations that are available as offprints.

---

[75] Giese, *Laboratory Manual in Cell Physiology.*

[76] B. Levedahl, et al. *Experiments in Physiology,* 8th ed. (St. Louis, MO: 1971).

CAPSULE LESSONS: WAYS TO GET STARTED IN CLASS

2-1. Plan several laboratory lessons to explore cultures of living organisms. Students can fill the blackboards with drawings of what they see. Then develop the basis of classification of protozoa: whether the forms are ciliated, where cilia are, or whether forms are flagellated. Where is the oral or mouth region located? Or are the organisms many-celled—such as rotifers, worms, and gastrotrichs? If several kinds of *Paramecium* are present, develop the idea of genus and species.

2-2. Introduce the parts and the manipulation of a microscope by having students examine a variety of protists in a drop of pond water. Or use a culture of *Tubifex* worms; since they can be seen with the naked eye, students can locate the organisms easily on their slides. Examine under low power. Which way is the slide moved to follow a motile form, or to examine the length of a worm? Why is an image centered under low power before switching to high power?

2-3. Begin a study of cells by having students examine epithelial tissue from the lining of the cheek under low power. When they are ready to examine the slide that they have prepared (stained with dilute Lugol's solution as described in this chapter), guide them to use the low power of the microscope. Then examine the slide under high power; students will begin to ask questions about the nature of the nucleus and the granular texture of the cytoplasm.

2-4. You may wish to start by asking how

cells in the shoot of an onion or garlic compare with those in the bulb. Prepare slides and then diagram both kinds of cells. How are the cells alike? How do shoot cells differ from those in the bulb? What function do they serve in the life of the plant? How are they specialized to carry on their functions?

2-5. At some time, you will want to use a block of modeling clay to represent a cell. Insert a marble in the clay. What does the marble represent? The clay? Enclose the clay in a sheet of saran wrap or other clear plastic. What does this layer represent? As an outer boundary of a cell what special functions might it have? Is this a typical plant or animal cell? Now cover the plastic with a sheet of paper. If this is a plant cell, what does the outer layer represent?

Prepare a gelatin block to which a bit of phenolphthalein solution had been added. Now enclose this in a membrane or dialysis tubing. Bring a solution of ammonium hydroxide close to this gelatin block, or suspend the block over the NH_4OH. How do students account for the reddish color in the gelatin? What is the nature of diffusion? Initiate a discussion of the nature of exchange of materials through a membrane. What materials in the environment of a cell diffuse into the cell? Do all materials diffuse readily out of cells?

2-6. Show a replica of Leeuwenhoek's microscope. Some students may be asked in advance of this lesson to prepare a talk to the class about Leeuwenhoek's findings. Read from his letters to the Royal Society in London. Compare the structure and differences in magnification of modern microscopes with Leeuwenhoek's. What effect did Leeuwenhoek's discovery of a new world of microorganisms have on the "science" of his day? What advances could be made with his new tool? What were the contributions of Hooke, Brown, Schleiden and Schwann, Dujardin, and others, to the accumulating knowledge of cells?

2-7. Use several models and/or charts of cells and protozoa such as those in Figures 2-9; 2-10; 2-20; and 2-21. Have students use the models to identify parts and then to describe the functions of the organelles of cells.

2-8. Use cells such as *Spirogyra* or *Anacharis* (elodea) in a wet mount. How could one determine that a cell has depth and is not two-dimensional?

2-9. How small is a cell? How can cells be measured in microns—fractions of a millimeter? Have students measure cells of onion, cheek cells, or others.

2-10. Use vital stains and indicators to study ingestion and digestion (change in pH in food vacuoles) in microorganisms. Refer also to activities given in chapter 3.

2-11. Which would live longer in a depression slide, single cells of the green alga *Chlorella* or *Scenedesmus*, or single-celled animals like *Paramecium*? If the slides were sealed with petroleum jelly, could students watch the slides at intervals during the day? Introduce the many life functions of plant and animal cells; begin to build facts into concepts, concepts into conceptual schemes—for example, organisms are dependent on one another, and on their environment.

2-12. As an introduction to a microscope lesson, you may want to preview some of the films and film loops in cell biology. See listings from Encyclopedia Britannica, Coronet, McGraw-Hill, Thorne, Indiana University, and others.

2-13. How has the electron microscope changed our concepts of the components of cells? Use offprints of papers from *Scientific American* and the text–atlas of scanning electron microscopy by R. Kessel and R. Kardon, *Tissues and Organs* (San Francisco: Freeman, 1979). Also refer to books that combine several readings from *Scientific American* such as *The Living Cell* and *From Cell to Organism*. Both have introductions by Donald Kennedy. Send for listings from W. H. Freeman & Co., 660 Market Street, San Francisco, CA 94104. These readings may be useful in future discussions, for example: What is the structure of mitochondria? What is their role in the cell's activities? How are pigment molecules arranged in grana of chloroplasts? What is the site of enzyme production and of protein synthesis? How are chromosomes or DNA molecules replicated when cells reproduce? What do we know of the genetics of microorganisms? Also refer to *Molecules to Living Cells*, another set of readings from *Scientific American*.

2-14. Prepare onion root tip smears or salivary gland smears of *Drosophila* larvae to see chromosomes. You may want to show a BSCS filmstrip *Smear and Squash Techniques*.

2-15. Begin a lesson by showing dry yeast in a package. Pour about 10 ml of fresh hydrogen peroxide (3 percent) into one large test tube. Sprinkle dry yeast into the test tube. Observe the tremendous activity and the bubbles pouring out of the test tube. How can the gas be collected and identified? Have students devise an apparatus for collecting a gas (see Fig. 2-32). Give the formula for hydrogen peroxide (H_2O_2). What kind of substance is present in yeast cells that could break down a toxic substance so that it would not accumulate in cells? In subsequent lessons, collect the oxygen liberated, or measure the amount of hydrogen peroxide left in the test tube after 10 min (by titration with $KMnO_4$, as described on page 187). The amount of peroxide not destroyed is in inverse relationship to the activity of the enzyme catalase.

Also use slices of potato or raw meat, as described in this chapter.

2-16. At some time, in preparation for laboratory work, show the short BSCS techniques film *Paper Chromatography*. Show how to partition the components of hydrolyzed proteins as a laboratory technique that may be used for some long-range investigation.

2-17. Plan several laboratory lessons in which students can devise their own methods of testing the amount of enzyme in relation to the reaction produced. Study methods of dilution—what is the effect of an enzyme such as amylase diluted some 10,000 times? What is the effect of pH on enzyme activity? What kinds of substances inhibit the activity of enzymes?

2-18. Show that molecules of different sizes travel through a membrane at different rates. Dialysis is a method of separating molecules of different sizes. Prepare a suspension of several compounds—amino acids, glucose, xylose, sodium chloride, and starch. Pour this into a length of dialyzing tubing (prepared as shown in Figure 3-28).

Students can become familiar with the standard reagents used in tests for the identification of several cell components by testing the water in the beaker. Use Benedict's test; Molisch's reaction; Lugol's iodine reaction; benzidine reaction or Bial's reaction; Biuret, Millon, and xanthoproteic tests. These and others are described in this chapter.

2-19. What is the effect of radiation on tissue cells? On microorganisms? Look into A. Giese, *Cell Physiology* and current texts.

2-20. Many activities with the microscope are suggested throughout the book in discussions of such topics as growth, reproduction, modes of nutrition, behavior, photosynthesis, and respiration. What is the next problem for study leading out of the study of tissue cells and unicellular organisms?

2-21. At times, teachers use free, exploratory laboratory periods to introduce some of the major topics to be studied in the year ahead. In these periods students might examine with the microscope the structure and behavior of such organisms as *Daphnia* (notice especially its heartbeat), *Tubifex* worms, and brine shrimps and their larvae (observe ingestion).

Questions continually arise about facts students learn from books and from their firsthand observations. For example, what conditions affect the size of chromatophores in the skin or scales of fish? What kinds of microorganisms can be found in a handful of soil? What is the effect of different ions on the action of cilia?

2-22. Plan to have several microscope lessons to accompany the dissection of the frog. Prepare slides of tissues of the frog. Also hunt for worm parasites in the lungs, for protozoa parasites in the contents of the intestine (refer to pages 621, 623).

BOOK SHELF

The following are only a few of the books that are pertinent to the work discussed in this chapter.

Altman, P., and D. Dittmer. *Biology Data Book*. 2nd ed. Bethesda, MD: Federation of American Societies for Experimental Biology, 1973.

Arnow, L. *Introduction to Physiological and Pathological Chemistry*. 9th ed. St. Louis, MO.: Mosby, 1976.

Blaker, A. *Handbook for Scientific Photography*. San Francisco: Freeman, 1977.

Bold, H., and M. Wynne. *Introduction to Algae*. Englewood Cliffs, NJ: Prentice-Hall, 1978.

Conn, H., M. Darrow, and V. Emmel, *Staining Procedures Used by Biological Stain Commission*. Baltimore: Williams & Wilkins, 1973.

Copenhaver, W., R. Bunge, and M. Bunge. *Bailey's Textbook of Histology*. 17th ed. Baltimore: Williams & Wilkins, 1978.

Davidsohn, I., ed. *Todd Sanford Clinical Diagnosis by Laboratory Methods*. 15th ed. Philadelphia: Saunders, 1974.

De Robertis, E., F. Saez, and E. De Robertis. *Cell Biology*. 6th ed. Philadelphia: Saunders, 1975.

Devlin, T. *Textbook of Biochemistry*. New York: Wiley, 1982.

Dyson, R. *Cell Biology: Molecular Approach*. 2nd ed. Newton, MA: Allyn & Bacon, 1978.

Freifelder, D. *Molecular Biology*. New York: Van Nostrand, Reinhold, 1983.

Galigher, A., and E. Kozloff. *Essentials of Practical Microtechnique*. Philadelphia: Lea & Febiger, 1971.

Giese, A. *Cell Physiology*. 5th ed. Philadelphia: Saunders, 1979.

Giese, A. *Laboratory Manual in Cell Physiology*. rev. ed. Pacific Grove, CA: Boxwood Press, 1975.

Head, J. *Students' Collection of Electron Micrographs*. Burlington, NC: Carolina Biological Supply Co.

Hepler, O. *Manual of Clinical Laboratory Methods*. 4th ed. Springfield, IL: Charles Thomas, 1968.

Humason, G. *Animal Tissue Techniques*. 4th ed. San Francisco: Freeman, 1979.

Jahn, T., and F. Jahn. *How to Know the Protozoa*. Dubuque, IA: W. Brown, 1949.

Jones, E. and R. McClung. *Basic Microscopic Technics*. Chicago: The University of Chicago Press, 1966.

Kessel, R., and R. Kardon. *Tissues and Organs*. Text-atlas of scanning electron microscopy. San Francisco: Freeman, 1979.

Kudo, R. *Protozoology*. 6th ed. Springfield, IL: Charles Thomas, 1977.

Lerman, et al. *General Biology Laboratory Manual*. Dubuque, IA: Kendall/Hunt, 1975.

Lillie, R. *Conn's Biological Stains*. 9th ed. Baltimore: Williams & Wilkins, 1977.

MacKinnon, D., and R. Hawes. *An Introduction to the Study of Protozoa*. New York: Oxford University Press, 1961.

Meglitsch, P. *Invertebrate Zoology*. New York: Oxford University Press, 1972.

Needham, J. *Chemistry of Life: Eight Lectures on the History of Biochemistry*. New York: Cambridge University Press, 1970.

Oser, B. L., ed. *Hawk's Physiological Chemistry*. 14th ed. New York: McGraw-Hill, 1965.

Price, F. *Basic Molecular Biology*. New York: Wiley, 1978.

Schipper, L. *Lecture Outline of Preventive Veterinary Medicine for Animal Science Students*. 6th ed. Minneapolis: Burgess, 1982.

Stryer, L. *Biochemistry*. 2nd ed. San Francisco: Freeman, 1981.

Toporek, M. *Basic Chemistry of Life*. St. Louis, MO.: Mosby, 1981.

Wolken, J. *Euglena: An Experimental Organism for Biochemical and Biophysical Studies*. 2nd ed. New York: Appleton-Century-Crofts, 1967.

Addendum: Refer to *Kodak Biological Stains* (1982) for indicators and color charts, stains, and products for microtechniques, Rochester, NY: Eastman Kodak Co. Dealers in most states. Also refer to *Scientific American*, Cumulative Index, for issues from 1948; many older issues are classics. Available as offprints from W. H. Freeman, 660 Market St., San Francisco, CA 94104.

In addition, refer to *Scientific American: Molecules to Living Cells* (New York: Freeman, 1980), a compilation of classic papers in book form, and the *Handbook of Chemistry and Physics*, CRC Press Inc., 2255 Palm Beach Lakes Blvd., West Palm Beach, FL 33409.

Also send for catalogs from distributors of films and 2 × 2 slides (addresses are given in the Appendix).

3

Levels of Energy Utilization:
Producers and Consumers

In considering the source of energy for plants and animals, we need to begin with the remarkable photosynthesizing ability of green plants. You may want to describe some of the many hypotheses and elegant experimentation that have been underway since Van Helmont probed the growth of a willow twig 300 years ago.

An excellent reference that develops the logic behind hypothesis, prediction, and experimentation in research of photosynthesis, as well as in other areas of biology, is *A Study of Biology* by J. Baker and G. Allen, 3rd ed. (Reading, MA: Addison-Wesley, 1977). You may also wish to refer to Leonard Nash's case study of photosynthesis that describes the many methods of scientists in piecing together evidence of the photosynthetic process.[1] *Scientific American* is another source of important classic as well as current papers.[2] For example, see Arnon's classic paper describing photophosphorylation (November 1960) and Bassham's paper describing the role of radioactive tracers unraveling the steps in sugar-making as illuminated chloroplasts synthesize glucose from water and carbon dioxide in as short a time as 30 sec (June 1962).

This chapter is concerned with the specific conditions under which photosynthesis occurs. Many demonstrations and investigations describe photosynthesis and digestion in plants. Also described are the many devices used by consumers to capture or to ensnare their food organisms. In addition, many of these investigations apply the methods of scientists — building models (some symbolic) to reveal underlying concepts, developing hypotheses, testing hypotheses, and experimental design.

Basic laboratory demonstrations make up the body of this chapter, along with investigations that enable students to see for themselves. An additional block of investigations using *Chlorella* aims at a more advanced level of work and somewhat more sophisticated laboratory studies and experiments. This series of problems emphasize an investigative approach by individual students or by groups working either in the laboratory or at home.[3]

Naturally, in accordance with the basic notion that teaching is a personal invention, we expect that a teacher will select demonstrations that are applicable to his or her class situation.

Photosynthesis as a process

It is estimated that 90 percent of all photosynthesis is carried on by marine and

---

[1] J. Conant and L. Nash, eds. *Harvard Case Histories in Experimental Science*, vol. 2. Reprint (Cambridge, MA: Harvard University Press, 1957).
[2] Offprints available from W. H. Freeman & Co., 660 Market Street, San Francisco, CA 94104.

[3] Also see *Research Problems in Biology: Investigations for Students* (New York: Anchor Books, Doubleday, 1963, 1965). Prepared under the direction of American Institute of Biological Sciences (BSCS).

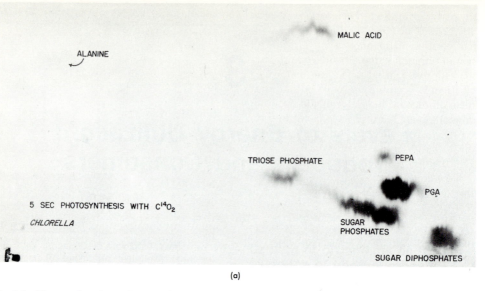

ALANINE

MALIC ACID

TRIOSE PHOSPHATE

PEPA

PGA

5 SEC PHOTOSYNTHESIS WITH $C^{14}O_2$

CHLORELLA

SUGAR
PHOSPHATES

SUGAR DIPHOSPHATES

(a)

Fig. 3-1 Three radioautographs reveal compounds containing radioactive carbon produced by *Chlorella* in light after (a) 5-sec exposure; (b) 10-sec exposure; (c) 30-sec exposure. (From J. Bassham, "The Path of Carbon in Photosynthesis," *Sci. Am.*, June 1962, p. 95.)

freshwater algae, and the remaining 10 percent by cultivated and wild land plants. Put another way, some 200 billion tons of carbon are fixed annually by photosynthesizing plants of the oceans and fresh waters and by land plants. The efficiency of the process is remarkable, since only 1 to 5 percent of the light energy is absorbed by chloroplasts. Chloroplasts absorb light in the red and blue regions of the spectrum, and reflect green light waves.

The overall process of photosynthesis can be represented by the following chemical equation:

$$6CO_2 + 12H_2O + 673,000 \text{ cal} \longrightarrow$$

low energy   from energy
level   of sun

$$C_6H_{12}O_6 + 6H_2O + 6O_2$$

high energy
level

Since the source of oxygen is water, at least 12 molecules of water must take part in the reaction for each molecule of hexose that is formed. The product, a six-carbon sugar, has built into it some 673,000 cal per mole of hexose. This energy is then available to those organisms that can reverse the photosynthetic process in biological oxidation (oxidation is discussed in chapter 4).

The equation given for photosynthesis is deceptively simple. In 1905, Blackman demonstrated that there were two distinct phases in photosynthesis, only the first phase requiring light and chlorophyll. The first phase, finally explained by Hill in 1937 and named after him, is a photochemical phase in which the energy of light is captured by chlorophyll molecules and is used to split molecules of water in a photochemical lysis of water. All the oxygen resulting from photosynthesis comes from the photolysis of water molecules:

$$2H_2O + \text{light energy} \xrightarrow{\text{chlorophyll}[4]}$$

$$4 H^+ + 4 e^- + O_2$$

ATP and $NADPH_2$ are two high-energy molecules that store the light energy that is converted to chemical energy.

Energy from sunlight is therefore used in three processes: a) production of ATP;

[4] Photosynthesizing pigments in chloroplasts absorb light of long red and orange wavelengths and short blue and violet wavelengths. The pigments are found in regions on the outer surface of membranes in chloroplasts.

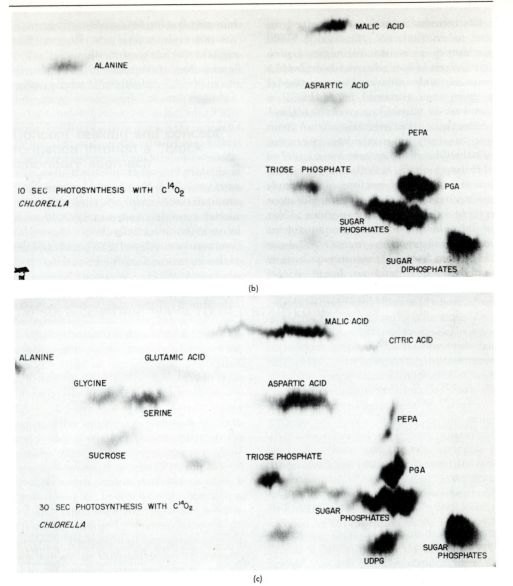

(b)

(c)

b) photolysis of water molecules; and c) use of electrons (e^-) in the fixation of carbon dioxide, the so-called dark reaction.

In the second phase, the dark reaction, there is a reduction of carbon dioxide:

$$CO_2 + 4\,H^+ + 4\,e^- \xrightarrow[\substack{\text{stroma of} \\ \text{chloroplast}}]{}$$

$$CH_2O + H_2O$$

Recall that the high-energy molecules involved in this reduction, or fixation of carbon dioxide, are produced in the light reaction (Fig. 3-1). In carbon dioxide fixation, or the Calvin cycle, three ATP and two $NADPH_2$ molecules are needed for the reaction, which occurs in stroma of chloroplasts.

$$CO_2 + 2\,NADPH_2 + 3\,ATP \xrightarrow[\substack{\text{stroma of} \\ \text{chloroplast}}]{}$$

$$CH_2O + 2\,NADP +$$

$$3\,ADP + 3\,P_1 + H_2O$$

At low light intensity, the photochemical reaction is the limiting reaction in photosynthesis. The rate at which chlorophyll molecules absorb and transmit quanta of light is not increased by raising temperature. In order to increase the overall rate of photosynthesis, however, two factors must operate: (1) light intensity must be increased and (2) temperature must be increased to optimum.

It has been shown that intact chloroplasts isolated from cells carry on three distinct functions: (1) Hill reaction (p. 196); (2) photosynthetic phosphorylation that occurs in the absence of oxygen, the process in which adenosine monophosphate is converted into adenosine triphosphate (ATP); and (3) the dark reaction, or the fixation of carbon dioxide in at least one cycle (the Calvin cycle).

Additional evidence links protein synthesis more closely to assimilation of carbohydrates. Bassham has found that alanine shows up labeled by C^{14} as rapidly as does a carbohydrate.[5] Some 30 percent of carbon taken up by algae in Bassham's studies was incorporated directly into amino acids.

Evolution of photosynthesizing systems

Our present understanding of the chemical evolution of organic substances is the result of work of many scientists in different disciplines: Haldane, Oparin, Bernal, Fox, Urey, Miller, Calvin, and Arnon, among many others. According to the hypothesis based on their ideas, methane and other hydrocarbons were probably formed by chemical abiogenesis during the thousands of millions of years before the origin of life on earth.

Those who are concerned with these theories hold that hydrocarbons could have reacted with reducing gases of the atmosphere—such as water vapor, ammonia, and hydrogen sulfide—when driven by ultraviolet radiation of the sun or by electric discharge, such as results from a discharge of lightning.

In the laboratory, successful attempts have been made to reproduce these conditions and, in fact, to produce complex organic molecules. Miller passed spark and silent discharges for a week through mixtures of methane, ammonia, hydrogen, and water vapor.[6] He obtained, along with some of the original gases, carbon monoxide, carbon dioxide, nitrogen, and several amino acids (glycine, alanine, and about 25 other amino acids in smaller amounts).

Calvin and his group[7] have experimentally confirmed methods suggested by Haldane and Oparin for duplicating the first production of organic molecules from inorganic materials in primeval seas. (Calvin's group also irradiated two-carbon substances and has obtained succinic acid, a four-carbon metabolite of living cells.)

With no living organisms to consume them, these organic molecules might have accumulated, and, as Bernal suggested, they might well have become adsorbed on clays in the seas' bottom, forming clusters at the bottom or along the edges of the seas. In such close contact, polymerization of amino acids and nucleotides could have occurred. Oparin called the swarms of organic materials that precipitate out of solution "coacervate droplets." (These droplets can be duplicated in the laboratory by mixing solutions of gum arabic and gelatin; coacervates may also be formed by mixing together various proteins such as casein, egg or serum albumin, or RNA and other materials.[8]

Recall that Cyril Ponnamperuma and others found organic compounds in the meteorite that fell in Australia in 1969.

---

[5] J. Bassham, "The Path of Carbon in Photosynthesis," *Sci. Am.* (June 1962); also refer to Bassham, "The Control of Photosynthetic Carbon Metabolism," *Science*, 172 (1971).

[6] S. Miller, "Production of Some Organic Compounds Under Possible Primitive Earth Conditions," *J. Am. Chem. Soc.* (1955) 77:2351–60.

[7] M. Calvin, "Chemical Evolution and the Origin of Life," *Am. Scientist* (1956) 44:248–63.

[8] Work of B. de Jong, as described by A. Oparin in *Life: Its Nature, Origin, and Development,* trans. A. Synge (New York:Academic, 1962).

Anaerobic heterotrophs   Since the quantity of molecular oxygen in the atmosphere was not sufficient to support life, the first living things must have been anaerobic heterotrophs, obtaining their food materials ready-made from the organic "soup" of the primeval seas. This hypothesis stating that early organisms were anaerobic heterotrophs is based on two criteria relating to known metabolic function: (1) contemporary living things have a metabolism based on the use of ready-made organic molecules; and (2) they also obtain energy through metabolic pathways based on anaerobic degradation of organic molecules (usually glucose), not by using free oxygen.

The metabolism of autotrophic organisms is based on the same chemical pathways as the metabolism of heterotrophs. It would seem that a special autotrophic mechanism—now apparently accepted as the common one, photosynthesis—has been added as a superstructure on a heterotroph foundation.

With the evolution of the photosynthetic mechanism, greater quantities of molecular oxygen became available in the atmosphere. In time, respiration became a more dominant process than the anaerobic fermentation that had first evolved. This new pathway in respiration also seems a superstructure that supplements a phase of anaerobic respiration in contemporary organisms. Oparin summarized the virtually universal chain of reactions underlying lactic acid and alcoholic fermentation in microorganisms, glycolysis in animals, and similar respiration in higher plants. In subsequent biological selections in the evolution of early organisms, the three major coenzymes of living things—DPN, ATP, and coenzyme A—developed to supplement the catalytic activities of the protein enzymes. These enzyme systems may first have developed some billion years ago as metabolic pathways in primitive organisms having fermentation.

A central point in Arnon's thinking was that photosynthetic phosphorylation is compatible with the premise that emergence of photosynthetic organisms was achieved by their acquiring a pigment system capable of decomposing water by means of light. He surmised that the early capacity for utilizing light energy was probably more closely related to the synthesis of ATP than to the assimilation of carbon dioxide (photosynthesis). Arnon's reasoning was based on the close structural association in both chloroplasts and bacterial chromatophores of the phosphorylating activity with the chlorophyll pigment system. In contrast, those enzymes responsible for assimilation of carbon dioxide can easily be dissociated from the chlorophyll pigment system.

A further advance in the evolution of photosynthetic systems might well have been the conversion of only a part of the captured energy into ATP, with the remainder used to generate a reductant for carbon dioxide assimilation. Some mechanism had to be developed to prevent the regeneration of water after it had been split, so that free hydrogen would be available for the fixation of carbon dioxide. This would have to have been accomplished by diversion of the OH radical; such a mechanism was possibly dependent in the beginning on an external hydrogen donor, as it still is today among the photosynthesizing bacteria.

In a later stage of evolution, in higher green plants, this mechanism was converted into an enzyme system for the liberation of molecular oxygen. In this way, photosynthetic phosphorylation by chlorophyll (independent of molecular oxygen) could occur before oxidative phosphorylation by mitochondria which need an oxygen supply. It is also interesting to note that after the evolution of a separate mechanism, independent of oxygen, for generating ATP in light, green plants also share with the nongreen organisms (consumers) the evolution of an oxygen-dependent synthesis of ATP by oxidative phosphorylation in mitochondria. This is the basis of Oparin's assumption that photoautotrophic organisms evolved later than heterotrophic forms.

It is important for students to recognize that many aspects of photosynthesis have not been fully unravelled. In order to demonstrate that discoveries in science are the result of a collaboration throughout history—one scientist building on the work of others—some teachers use a case history approach such as Conant developed at Harvard University.[9] Some teachers duplicate materials drawn from original sources so that students may read the contributions of Aristotle, who guessed that plants took in materials through their roots; of Van Helmont, who sought the source of the increase in weight of a willow twig in the seventeenth century; and of Priestley, who described "good" and "fixed" air in the eighteenth century. You may also want to study the elegant work of Jan Ingenhausz, elucidating the role of light in the photosynthetic process with the evolution of oxygen, and the work of Senebier (intake of carbon dioxide by a plant), DeSaussure (intake of water), and Sach (test for starch in leaves) in the nineteenth century. Studies in the twentieth century include Blackman's law of limiting factors in photosynthesis, Kamen's investigation with heavy oxygen ($O_2{}^{18}$), and Calvin's work with radioactive carbon (C^{14}).

Many fine papers may be culled from M. Gabriel and S. Fogel, *Great Experiments in Biology*.[10] Melvin Calvin's acceptance speech for the Nobel Prize is also available for reading and analysis by students.[11] Offprints of many papers from *Scientific American* are also available. These readings provide the "reach" that is greater than most students' "grasp."

Modes of nutrition

Organisms that manufacture their energy-rich organic matter from inorganic sources are producers, or autotrophs. The driving energy for the involved syntheses may be from chemicals or from sunlight. On this basis, autotrophs are grouped as chemosynthetic autotrophs and photoautotrophs.

Chemosynthetic autotrophs manufacture organic metabolites from carbon dioxide, and nitrate or ammonia, deriving energy to do so not from light, but from the oxidation of an inorganic chemical. They lack photosynthesizing pigments and form a small group that includes iron bacteria; the colorless sulfur bacteria; *Nitrosomonas*, the aerobic bacteria that oxidize ammonia to nitrite; and *Nitrobacter*, which oxidizes nitrite to nitrate compounds. (Also see investigation, p. 608.)

Photoautotrophs convert light energy into chemical energy for their metabolism. This is the main group of autotrophs, and includes the higher plants, algae, colored sulfur bacteria, and the pigmented nonsulfur bacteria.

Photoautotrophs fall into one of three subgroups characterized by these generalized reaction[12]:

1. Green plants:

$$CO_2 + H_2O \xrightarrow{\text{light}} (CH_2O) + O_2$$

2. Pigmented sulfur bacteria:

$$CO_2 + H_2S \xrightarrow{\text{light}} (CH_2O) + S$$

3. Pigmented nonsulfur bacteria:

$$CO_2 + \text{succinate} \xrightarrow{\text{light}}$$
$$(CH_2O) + \text{fumarate}$$

Van Niel used the following generalized formula for all these types in a light reaction:

$$CO_2 + H_2A \xrightarrow{\text{light}} (CH_2O) + A$$
$$\text{(hydrogen acceptor)}$$

Evidences of photosynthesis

Note: You may want to review several safety precautions concerning laboratory activities in this area as described in chapter 10.

---

[9] Conant and Nash, *Experimental Science*.
[10] M. Gabriel and S. Fogel, eds. *Great Experiments in Biology* (Englewood Cliffs, NJ: Prentice-Hall, 1955).
[11] M. Calvin, "The Path of Carbon in Photosynthesis," *Science* (1962) 135:879–89.

[12] These equations are not balanced since not all reactants and products are indicated.

Formation of sugar and conversion to starch

Simple sugars are the first products of photosynthesis. Most green plants, namely most dicotyledons, support further reaction in the leaves where the sugar is converted to starch. However, some green plants do not form starch in their leaves. Some marine algae, fleshy leaves of the onion, leaves of corn, sugar beets, and many other monocotyledonous plants (especially members of the Liliaceae) may be tested to show the presence of stored glucose rather than starch. In corn, the path of glucose from leaves may be traced to the seed, where glucose is stored. In sugar beets, glucose formed in the leaves is transported to the beet root to be stored as cane sugar. However, in the bean, and in most dicotyledonous plants, the conversion of glucose to starch continues in the leaves.

Sugar in green leaves Sprouted onion bulbs grown in light show the presence of simple sugars in the shoots. Cut 3 cm lengths of the green shoots into a Pyrex test tube; add about 10 ml of either Benedict's or Fehling's solution (p. 176). Then bring the contents of the tube to a boil over a Bunsen burner or alcohol lamp. A change in color from blue, to shades of green, to reddish orange indicates the presence of simple sugars. This color reaction (as a test for simple sugars) may be checked by boiling pure glucose solution or molasses in Benedict's solution.

Test for cane sugar in plants Before you test for the presence of cane sugar in a plant, you may want to demonstrate the sugar test (using pure cane sugar—sucrose). To a small quantity of cane sugar in a test tube add about 15 ml of distilled water. Next, add 1 to 2 ml of a 5 percent solution of cobalt nitrate (prepared by adding 5 g of cobalt nitrate to 100 ml of distilled water). Then add a small quantity of strong sodium hydroxide solution, prepared by adding 50 g of sodium hydroxide sticks (*caution:* caustic; use tongs) to

100 ml of distilled water. A violet color reaction is a positive test for cane sugar.

Note: A weak solution of cane sugar (sucrose) may be converted to glucose by adding a few drops of concentrated hydrochloric acid, and boiling gently for a minute or so. (Saliva will also change cane sugar, $C_{12}H_{22}O_{11}$, to glucose, $C_6H_{12}O_6$, on standing.) Neutralize the hydrochloric acid with sodium carbonate until no further effervescence occurs. Then test with Benedict's or Fehling's solution for the presence of glucose.

Role of light in photosynthesis

The need for light, the intensity of light, and the efficacy of specific regions of the spectrum can be studied by measuring: (1) rate of formation of starch; (2) changes in pH of the medium of aquatic organisms as carbon dioxide is absorbed in the presence of light; (3) number of bubbles of gas produced as oxygen is liberated, or distance a trapped bubble moves in a volumeter; and (4) rate of growth of cells, especially among algal forms.

Each of these methods for testing the rate of photosynthetic activity is described below. Where possible, examples will be given using land plants such as geraniums; single-celled water forms such as *Chlorella*, filamentous algae, and *Spirogyra*; or many-celled water plants such as *Anacharis* (elodea).

Need for chlorophyll in starch-making To show that starch is made in green regions of a leaf, select healthy leaves from a variegated coleus showing white regions, or from a silver-leaf geranium which has been placed in the sunlight for several days. Make a diagram of the pattern of green and white regions of the leaves. First, boil the leaves for a few minutes in water in a small Pyrex beaker on an *electric* hot plate. This preliminary step softens the leaves by breaking down cell walls. Next, transfer the softened leaves to a beaker half full of isopropyl alcohol, and warm on the hot plate. If a hot plate is not available,

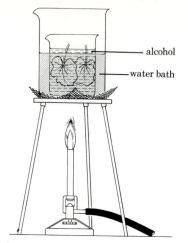

Fig. 3-2 Water bath: alternate apparatus for removing chlorophyll from green leaves when an electric hot plate is not available.

precautions must be taken to avoid igniting alcohol fumes. Alcohol may be heated in a water bath (Fig. 3-2) with the alcohol in a small beaker or long test tube standing in a larger Pyrex beaker of boiling water. (An electric immersion heater may also be used with a thermometer to make a fireproof alcohol-heating device.)

Chlorophyll is soluble in alcohol, and the pigment should be extracted completely from the leaves in about 5 min. For thick leaves, decant the alcohol and replace with fresh alcohol. Then heat again until all chlorophyll is removed. Wash the blanched leaves; spread them flat in open Petri dishes, and cover with a dilute aqueous iodine solution such as Lugol's solution (p. 176). After 2 to 5 min, rinse off the excess iodine solution. Hold the leaves up to the light to show the blackish areas which indicate the presence of starch. Note the lack of starch in the original white portions of the leaves. It may take as long as 15 min for all the iodine to react with the starch in the leaf.

As a control, establish the Lugol test for starch by first adding iodine solution to starch paste in a test tube.

Testing the role of light in starch-making
Select green geranium plants that have

been in the dark for at least 24 hr. Test some of the leaves for the presence of starch using Lugol's solution; leaves originally free of starch will serve as one control. Cover several leaves of the plants with aluminum foil or carbon paper; or cut thin disks from cork and pin to the top and bottom surfaces of several leaves (Fig. 3-3). To avoid denuding the plants of their leaves, only a portion of each leaf need be covered; the exposed portion serves as a control. Place the plants in sunlight or artificial light from a 75-watt bulb about 2 ft from the plant.

After about 24 hr, remove two or more covered leaves and an equal number of leaves exposed to the sunlight (be sure to identify them); place them in boiling water, extract the chlorophyll from the leaves in heated alcohol, and finally add diluted iodine, as already described, to test for starch. Only the exposed leaves or areas should show the dark blue-black indicative of starch formation.

The same results can be achieved in 15 min using a 500-watt lamp, 3 ft away

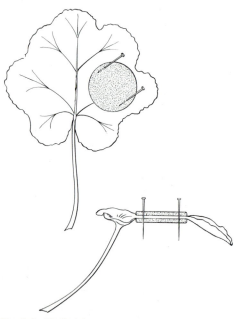

Fig. 3-3 Test of the role of light in starch making in green plants. Part of a leaf is covered with two disks of thin cork, carbon paper, or aluminum foil.

from the plant. Make certain that the plants are not overheated. Use blue cellophane to arrest part of the heat.

Starch grains   Examine the shape and size of starch grains, since these are characteristic of specific plant species. In fact, starch grains are used to identify many plants. For example, in the potato the starch grains are irregularly oval in outline, and consist of alternating dark and light lines in the grains. Compare these with starch grains of other plants (Fig. 3-4).

Prepare a mount of starch grains by pressing a clean slide over the cut surface of a raw potato or by lightly scraping the cut surface of a raw white potato with a scalpel; mount the scrapings in water, and apply a coverslip. Then allow a drop of dilute Lugol's solution (p. 176) to run under the coverslip as filter paper is placed near the opposite edge of the coverslip to draw up some of the water.

Starch-making in algae   It is especially difficult to identify the small grains of starch produced by green algae. When dilute Lugol's solution is applied, the resulting color may appear dark brown or blackish.

Whitford[13] suggests digestion of starch as a means of distinguishing algae of the

[13] L. A. Whitford, "On the Identification of Starch in Fresh-Water Algae," *Turtox News* (1959) 37:62–63.

Chlorophyceae (which contain chlorophyll *b* and starch) from Xanthophyceae (which lack both starch and chlorophyll *b*).

Place a small quantity of algae on a nearly dry slide. Place a second slide over the algae and macerate the cells (so that enzymes may penetrate the cells faster) by pressing the slides between the fingers. After the algae are thoroughly ground, remove the top slide and apply a freshly prepared diastase solution. Then apply a coverslip. Digestion of the free starch is rapid.

Wavelength and photosynthesis

Evidence of a relationship between photosynthesis and chlorophyll is obtained by comparing the absorption spectrum of chlorophyll and the action spectrum of photosynthesis (Figs. 3-5, 3-6). Chlorophyll absorbs wavelengths of light, especially in the red and blue regions of the spectrum; these are also the regions in which the highest rate of photosynthetic activity occurs. You may want to refer to Engelmann's experiment at some time.[14] (Refer to nature of chloroplasts, p. 206, and partitioning of pigments of chloroplasts, p. 207.)

[14] When Engelmann exposed slides of filamentous green algae and motile, aerobic bacteria to the spectrum of light, he found bacteria clustered at the red end mainly, and, to a smaller degree, at the blue end of the spectrum. This action spectrum demonstrated where oxygen was given off by the algae.

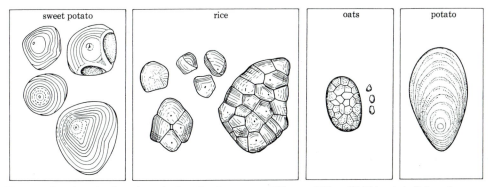

Fig. 3-4   Starch grains from four plants. (After C. Wilson and W. Loomis, *Botany*, rev. ed., Holt, Rinehart and Winston, 1957, and T. Weier et al., *Botany: A Laboratory Manual*, 2nd ed., Wiley, 1957.)

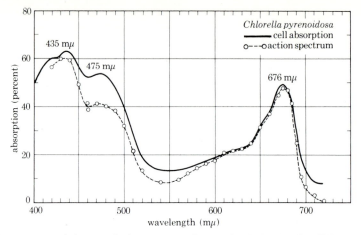

Fig. 3-5 Action spectrum of photosynthetic activity in *Chlorella pyrenoidosa*. (Adapted from M. B. Allen, ed., *Comparative Biochemistry of Photoreactive* *Systems*, Academic, 1960; from F. T. Haxo, "The Wavelength Dependence of Photosynthesis and the Role of Accessory Pigments," unpublished.)

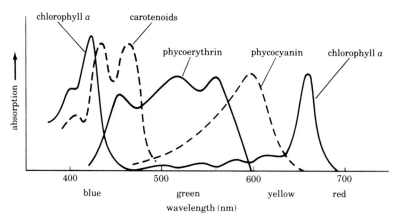

Fig. 3-6 Absorption spectra of some photosynthetic pigments show the amounts of light absorbed at each wavelength. Chlorophyll *a* has two absorption peaks. (Adapted with permission from Macmillan Publishing Company from *Biology* by Clyde F. Herreid II. Copyright © 1977 by Macmillan Publishing Company, Inc.)

In describing light falling on a plant, three factors must be known: (1) intensity, (2) quality, and (3) duration of illumination. Intensity refers to the number of quanta falling on a given surface area. Quality describes the kind of wavelength making up that light. For example, plants growing under fluorescent light get more blue than red wavelengths of light, while light from tungsten lamps is rich in red and poor in blue light. The third factor, duration, is the exposure time.

In terms of photons, or quanta of light, the greatest radiant energy value is in the violet wavelength region of 3900 to 4300 angströms (Å). The red end of the visible spectrum at 6500 to 7600 Å is the region where photons have the least radiant energy value.

Chlorophyll *a* is found in green plants in lamellae of chloroplasts. Chlorophyll *b*, also found in green algae and higher plants, and some carotenoids, are accessory pigments that extend the absorption range of chlorophyll *a* in the far red end of the spectrum, so that total absorption of chlorophyll occurs in the red and blue-violet ends of the spectrum. This is known

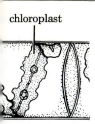

chloroplast

C. L. Wilson and W. E. Loomis, *Botany*, [Holt, Rinehart and Winston], 1957.)

eplacement of a methyl group 3 by a formyl group; chloro- n has two less atoms of the d one more atom of oxygen:

hyll *a:* $C_{55}H_{72}O_5N_4Mg$

(blue-green color)

hyll *b:* $C_{55}H_{70}O_6N_4Mg$

(yellow-green color)

e some ten different chloro- n, but in the higher green he green algae the most com- are chlorophyll *a* and *b* (see 07–209). e chlorophylls found in those sizing plants that liberate ox- are other groups of chloro- bacteriochlorophyll found in acteria; and (2) two kinds of orophyll found in the green ria. These organisms do not en in photosynthesis. ds are pigments found among gae and higher green plants. nd orange pigments of these ain four main carotenoids: lutein, violaxanthin, and and one or more of the xantho- has been postulated that end to protect light-sensitized from self-destructive oxida- ill reaction (p. 196) may be arotene is removed from the f chloroplasts. Carotenoids of izing organisms are of two ydrocarbon carotenoids, or 1 (2) oxidized derivatives, the .

r make cell counts of green suspension is e may be a clearly e green color. These vth that result from oplasm, an indirect nthesis has occurred. culture to a slide or eal with a coverslip. ells in five fields (see

Sterile dishes con- ient solution can be t and covered with eed the cooling, ster- the dishes, or streak agar has solidified. d size of clones that der different-colored e and aluminum foil. thods.)

lasts As an addi- may want to show 2,6-dichlorophenol- (0.1 percent) by illu- . The demonstration by Giese.[15]

ls of leaves: Anach- common plants used plasts is a healthy g. 3-7). Mount a leaf m water and examine

y Manual in Cell Physiology, A: Boxwood Press, 1975).

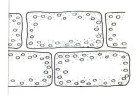

ls of *Anacharis* shows alls. Cell *A* is viewed near g the layer of chloroplasts e other cells are viewed ing distribution of l walls.

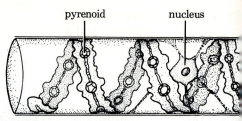

Fig. 3-8 Diagram of *Spirogyra* showing pyrenoids. When the cells are stained with Lugol's solution, pyrenoids appear black, indicating starch content. (From rev. ed., Dryde

under low and high power. When the leaves from young, growing tips are examined, the chloroplasts may appear to be moving in the cytoplasm of the cells. In reality, it is the cytoplasm that is circulating (cyclosis), and the chloroplasts are being carried in the moving "stream" (p. 148, chapter 2).

Chloroplasts in algae: Spirogyra In *Spirogyra*, the chloroplast is spiral-shaped (Fig. 3-8). Mount one or two threads in a drop of water. One species of *Spirogyra* has one spiral chloroplast; another fairly common species has a double spiral chloroplast. When Lugol's iodine solution (p. 176) is added to a wet mount, the pyrenoids along the chloroplasts stain darkly, showing that they contain starch.

Role of chlorophyll in photosynthesis

Composition of chloroplasts The chloroplast is the photochemical apparatus; it contains several pigments including chlorophylls and carotenoids.

Through the pioneer work of Willstatter and Fischer, the chemical nature of the chlorophyll molecule was established at the beginning of the twentieth century. Chlorophylls are magnesium porphyrins. A porphyrin molecule is composed of four basic nitrogen-containing pyrrole rings linked together by carbon and hydrogen atoms. In the center of the molecule is an atom of magnesium linked to an atom of nitrogen in each of the four pyrroles. Chlorophyll *b* differs from chlorophyll *a*

due to the
in position
phyll *b* th
hydrogen a

 chloro

 chloro

There a
phylls kno
plants and
mon forms
also pages

Besides t
photosynth
ygen, ther
phylls: (1)
the purple
chlorobiocl
sulfur bact
liberate ox

Caroten
the green a
The yellow
plants con
β-carotene,
neoxanthin
phylls. It
carotenoids
chloroplast
tion. The l
inhibited if
suspension
photosynth
types: (1)
carotene; a
xanthophyl

carotene $\qquad$ $C_{40}H_{56}$ $\qquad$ (orange-red)
xanthophylls $\quad$ $C_{40}H_{56}O_2$ $\quad$ (light yellow)

The central thought is that all photosynthesizing plants that produce molecular oxygen contain chlorophyll *a*, while chlorophyll *b* and associated carotenoids vary in their distribution among plant cells. Furthermore, β-carotene is almost as universal as chlorophyll *a* in those plants that evolve oxygen.

The number of molecules of chlorophyll per chloroplast seems to be the constant $\sim 1 \times 10^9$ for many photosynthesizing cells.

Among the higher plants there may be some 80 to 1000 chloroplasts within a cell. Some algae, such as *Chlorella*, have only one chloroplast that fills the cells. Besides the many pigments, enzymes, and energy-rich molecules in the chloroplasts, they also have their own DNA, RNA, and ribosomes. Chloroplasts may often be seen dividing in algae and mosses. Plants growing in shade have more and larger chloroplasts than those in bright light.

Each chloroplast is covered by a double membrane and consists of a basic colorless stroma where the enzymes active in the dark reaction (Calvin cycle) are found. Here, the high energy molecules of ATP and $NADPH_2$ produced in the light reaction are used in the reduction of carbon dioxide to form carbohydrates. In summary, three ATP molecules and two $NADPH_2$ molecules are utilized for every one carbon dioxide molecule changed to glucose.

$$6\,CO_2 + 12\,NADPH_2 \longrightarrow$$
$$C_6H_{12}O_6 + 6\,H_2O + 12\,NADP$$

Chlorophylls and other pigments active in photosynthesis are found in the stacks of green disks, or grana, connected by the many folding membranes, the lamellae or thylakoids, that extend along the chloroplast. The photosynthetic pigments seem arranged in special layers found on the inner and outer surfaces of the membranes (Fig. 3-9).

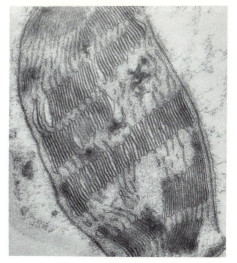

Fig. 3-9   Electron micrograph of a chloroplast from tobacco (magnification 13,700 ×). (Grant Heilman Photography.)

Analysis of chloroplasts

Pigments of chloroplasts may be separated by several extractions in a separatory funnel using different solvents, or by adsorption on a column of an adsorbent such as magnesium oxide powder. A third, more rapid and simple procedure for the classroom is the partitioning of pigments on filter paper to form a chromatogram.

Separation of pigments   Many solvents or combinations of solvents for extracting chlorophyll are described in the literature; several use acetone for the initial extraction of pigments. Further separation is based on differences in solubility of the pigments; for example, chlorophyll *a* is soluble in petroleum ether, while chlorophyll *b* is more soluble in methyl alcohol.

An extract of chloroplast pigments may be prepared by mixing 3 g of powdered leaves in 50 ml of 80 percent acetone (*caution:* toxic fumes) for about 5 min; pigments soluble in acetone will be extracted.[16] Leaf powder may be prepared

---

[16] These techniques are adapted from several described in the literature, especially from B. Meyer, D. Anderson, and C. Swanson, *Laboratory Plant Physiology*, 3rd ed. (Princeton, NJ: Van Nostrand, 1955) and also from Giese, *Laboratory Manual in Cell Physiology*.

in advance and kept on hand for several weeks under refrigeration, but it tends to deteriorate after a time. Leaves of spinach, thin leaves of certain trees and shrubs, or bean or squash leaves can be dried thoroughly in an oven and stored in a closed container. Or begin with 15 g of fresh spinach leaves; grind in a mortar with fine, clean sand in 50 ml of 80 percent acetone. Allow to stand some 10 min, grind again, and add more acetone. A deep green color results. To separate the sand from the extract, filter through a Buchner filter using several layers of filter paper to obtain an acetone extract of the chlorophylls, carotenoids, and other breakdown products of chloroplasts.

FILTER PAPER CHROMATOGRAPHY   For a *crude* separation of carotenoids from chlorophylls, each student may place a 3 cm wide strip of Whatman # 1, or preferably the thicker, # 3 filter paper as a wick into a small container of a sample of the acetone extract of chloroplasts. Within 5 min the vivid yellow band will separate from the greens of the chloroplasts. Be sure to have the room well ventilated.

A more careful partitioning of pigments can be done by hanging a strip of Whatman # 1 or # 3 paper in a closed container with a saturated atmosphere of solvent (Fig. 3-10). (Or use a disk with a narrow segment cut and turned down into the solvent as a wick as shown in Fig. 3-10.) Cut strips of Whatman paper, narrow

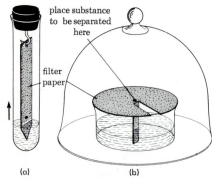

place substance to be separated here

filter paper

(a)   (b)

Fig. 3-10   Paper chromatography techniques: (a) strip filter paper method; (b) circular filter paper method.

enough to insert into test tubes without touching the sides of the tube, and long enough to fasten to a cork or stopper so that the cut tip of the paper will be only slightly immersed in the solvent to be added.

With a *pencil*, place a dot a few millimeters from the tip of the paper strip to mark the place where the spots or streaks of pigment solution will be added; use a millimeter ruler to mark the top level of the paper where the advancing front of the solvent should be stopped so it does not run off the strip. In this way, you can determine the rate of migration of molecules of different substances from the starting and advancing front distances (use R_f formula, p. 209).

Now apply a drop or streak of the leaf extract with a micropipette or Pasteur pipette a few millimeters from the tip of the strip as marked in pencil. Allow to dry; apply a second time; and after drying, apply a third drop or streak so that a *dark green stain* results.

Immerse the tip of the strip into a test tube containing about 3 cm of the solvent (a mixture of petroleum ether and acetone [9:1]; *caution:* both are toxic). Fasten the strip so the pigment spot is *not* immersed in the solvent; use a curved pin or thumb tack to secure the strip, and close the tube or flask.

When circular filter paper is used, place this over a Petri dish or finger bowl so that the wick dips into the solvent; cover with a similar dish or small bell jar. Apply the spot of pigment to the center of the filter paper so that, as the solvent rises by capillarity along the wick, the pigments spread out in concentric circles along with the advancing front of the solvent, all equidistant from the center (Fig. 3-10).

Within 10 min students should see the bands of different colors as the molecules migrate at different rates along the advancing solvent front. Remove the filter paper when the solvent reaches the marked advancing front. Before the colors fade, circle the different bands with pencil, and label them. *From top to bottom:* the fastest moving molecules are the carotenes (bright yellow); xanthophylls (dull yellow); chlorophyll *a* (dark blue-green); chlorophyll *b* (dull yellowish green).

Chlorophylls *a* and *b* are the most abundant; yellow pigments and chlorophyll *b* transfer their energy to chlorophyll *a*.

Measure the distance from the starting point on the filter paper to the advancing front. Then measure the distance traveled by the different solutes or pigments and use the formula:

$$R_f = \frac{\text{distance solute traveled}}{\text{distance solvent traveled}}$$

If a solute has an R_f of 0.5, this means the solute moved 50 percent of the distance the molecules of the solvent traveled. Compare the rates for the different pigment bands.

Some experimentation with solvents may be necessary. Benzene is also recommended, using a solution of equal parts of petroleum ether and benzene. (Be sure to have good ventilation.)

CYLINDER CHROMATOGRAM   Instead of strips of Whatman filter paper, use a sheet of Whatman # 1 paper about 10 × 10 cm. Roll into a cylinder and staple the sides so the edges do not touch. Avoid handling the paper so that perspiration from the hands does not affect the chromatogram. In pencil, mark the starting front and the final advancing front.

Stand the cylinder into a small dish containing a few ml of the pigment extract; allow the pigment extract to move up about 1 cm. Remove the cylinder and let it dry. Repeat this some two or four times so that a dark band of pigment extract appears on the cylinder. Now pour 5 ml of a solvent (9 parts of petroleum ether to 1 part of benzene) into a wide mouthed jar or beaker so that the entire cylinder is contained in a saturated atmosphere. (Have good ventilation.) Cover with a glass plate. Be sure to place the pigmented end of the filter paper cylinder into the solvent. Remove the cylinder when the solvent reaches the marked advancing front. Transfer the cylinder to a second

solvent (2 parts petroleum ether to 1 part benzene). Allow this fresh solvent to reach the marked advancing front of the solvent, then remove and dry. Remove the staples and examine the separated bands of pigment. The chromatogram should show color bands. *From top to bottom:* most rapid, orange carotenes (bright yellow alpha carotene, dark orange beta carotene) and under these, two bands of yellow orange xanthophylls; dark, blue-green chlorophyll *a*; the slowest, yellow-green chlorophyll *b*. Circle the bands with pencil before they fade; label the colors.

After learning the technique, you may find students will want to compare the rate of development of pigments in growing leaves of seedlings; or compare the development of pigments in leaves of *buds* of maples on the lawn or street with the pigment content of full-grown leaves as they develop over a period of weeks. Compare the pigments in the varieties of Japanese maples, as well as the Japanese variety of the Norway maple, and the purple beech. Pigments of flowers also can be tested with different solvents (below). Which solvents are best for carotenoids? For the anthocyanins of violets? (Refer to S. Clevenger's paper, "Flower Pigments," *Scientific American*, June 1964.)

Red and blue anthocyanins and yellow flavonoids are water soluble, dissolved in cell sap in vacuoles. Carotenes, which are located in plastids, should not be confused with yellow flavonoids.

Chromatograms of anthocyanins

What pigments make up the color of autumn leaves? Or of the red and purple tulips, fruits, and leaves? Of red cabbage? Of purple eggplant?

In a mortar, grind together about 5 to 10 g of flower petals or red leaves of houseplants or other materials in a few milliliters of 1 percent HCl and sand. Use a pipette to transfer the liquid into a test tube. Add more HCl to the mix and grind until the red pigment is extracted.

Prepare a chromatography paper strip or cylinder as previously described. As a solvent in the test tube or flask use a mix in the ratio of 5 parts of distilled water to 4 parts of *n*-butyl alcohol to 1 part of glacial acetic acid. Shake the mix; let it stand until it separates into two distinct layers. Use the *upper* layer of the mix as the solvent to prepare the chromatograms. Mark the starting front and the final advancing front of the solvent in pencil. It may take about 24 hr to get a clear separation of color bands.

Phycoerythrin is the more abundant pink-red pigment in red marine algae, and phycocyanin is the bluish pigment; both are water-soluble. The pigments of algae are difficult to extract, but some success can be gained by chopping the algae in water and letting this stand at room temperature. After a few days the algae rot and the water becomes tinted with pigments. Filter the mix to obtain the water-soluble pigments. Then use the methods for preparing a chromatogram as described for anthocyanins (also refer to page 595).

Chemical extraction of chlorophyll

Separatory funnel   The pigments of the acetone extract of chloroplasts may be separated more tediously by chemical extraction in a separatory funnel. As a first step, separate two solutions, one containing chlorophyll *a* and carotene and the other containing chlorophyll *b* and xanthophyll.

Into a separatory funnel pour 60 ml of petroleum ether (*caution:* explosive; avoid flames); add 40 ml of green acetone solution (see p. 207). Carefully rotate the separatory funnel; hold the stopper with one hand and the stopcock with the other hand; then slowly rotate and invert the funnel. Open the stopcock to release the gas pressure. (Work in a well-ventilated room.) Close the stopcock and again slowly rotate the funnel. Add 80 ml of distilled water down the side of the funnel; rotate and invert several times. Open the stopcock again to release the gas pressure.

A separation of two layers should be visible—the upper layer, a deep green color, is soluble in petroleum ether and shows fluorescence. Draw off the bottom layer (an acetone-water layer) and discard. Then wash the petroleum ether layer again with 40 ml of distilled water; rotate the funnel and again note the separation of two layers. Draw off and discard the lower water layer. Repeat this procedure three or four times to obtain a clean green petroleum ether solution.

Now this solution is ready to be separated into the two sets of pigments. To this petroleum ether solution, add 50 ml of 92 percent methyl alcohol solution (*caution:* toxic fumes). Rotate in the funnel; pour the two layers that appear into separate bottles. The lower layer, the methyl alcohol layer, contains chlorophyll *b* and xanthophyll. The upper, petroleum ether, layer contains chlorophyll *a* and carotene. (How do chromatograms of these differ from each other?)

The acetone (containing some water) is a solvent for lipids. The chlorophylls are further separated (as is the carotene from the xanthophyll) by means of their different solubilities in a series of solvents. A subsequent step, using methyl alcohol and potassium hydroxide solution, saponifies the methyl and phytyl alcohol groups of the chlorophyll pigments so that they are water-soluble.

FURTHER SEPARATION   Into a separatory funnel pour 50 ml of the methyl alcohol solution (containing xanthophyll and chlorophyll *b*), and add 50 ml of ethyl ether (*caution*).[17] Mix slowly by rotating the funnel. Add 5 ml of distilled water along the side of the funnel, and rotate; repeat four or five times until two layers separate out. Discard the lower, methyl alcohol layer.

Now two solutions are on hand: an ethyl ether solution and the earlier petroleum ether solution (the upper layer originally obtained in the separatory funnel). Obtain

---

[17] Technique described by Machlis and Torrey, *Plants in Action* (San Francisco: Freeman, 1956); see also Giese, *Laboratory Manual in Cell Physiology.*

two large test tubes; pour 30 ml of the petroleum ether solution into one test tube, and 30 ml of the ethyl ether solution into the second test tube. Down the side of each test tube carefully pour 15 ml of freshly prepared 30 percent methyl alcohol-potassium hydroxide solution (*caution*). Shake the tubes and observe for 10 min. Now add 30 ml of distilled water to each tube and shake until the two layers separate in the tubes. Prepare the methyl alcohol–KOH solution by diluting 92 ml of absolute methyl alcohol with water to 100 ml. To this 100 ml of methyl alcohol, add 30 g of **KOH** (*caution*).

These pigment layers may be separated into clean test tubes and used in studies of the absorption spectrum of each of the four pigments. Also examine each solution of pigments in reflected and transmitted light to observe evidence of fluorescence (p. 212). Place the tubes against a black background and use a strong beam of light from a projector or some other source.

COLUMN ADSORPTION CHROMATO-GRAPHY Assemble the apparatus shown in Figure 2-30. The Buchner filter flask is to be connected to a mild vacuum; the chromatographic tube is of Pyrex glass, about 20 cm long with an outer diameter of 3 cm. Insert glass wool into the tube, which has first been inserted into a rubber stopper. Clamp the tube to a ring stand and use a large-bore glass funnel in the packing process. Pack magnesium oxide powder into the tube, tap the side of the tube, and after each addition of a small quantity of powder, pack the column with a packing rod (the rod should have a diameter that just fits the bore of the tube so that the powder is evenly packed). Fill the tube to within 5 cm of the top. Then transfer the tube to the Buchner filter flask and apply gentle suction to aid in packing the powder.

Prepare a petroleum ether solution of pigments of chloroplasts (as described on page 207). Pour this deep green solution down the side of the tube so that the magnesia becomes wet; this should nearly fill the tube. Then apply to the vacuum source so that the solution is drawn down

through the length of the packed column. Different pigments are partitioned along the column due to differences in adsorptive abilities. As the extract moves along the column, add a second solution to help spread out the pigment bands so that they become easily distinguished. Prepare a 1:1 mixture of petroleum ether and benzene; pour this down the packed column. Find the following distribution of pigments, from *top to bottom:* chlorophyll *b*, a yellow-green layer; the dark green chlorophyll *a*; the xanthophylls; and the carotenes—a dark orange-yellow band of B-carotene, and the lowest layer, the bright yellow zone of A-carotene.

Several modifications of this technique may be tried. Some references suggest using a slurry of dry cornstarch in petroleum ether. In another technique,[18] several different adsorbents are used: a bottom layer of cotton, then a layer of aluminum oxide, a layer of calcium carbonate, a layer of powdered sugar, and another layer of cotton (Fig. 2-30b). Wet the column by running in petroleum ether from a dropping funnel, keeping a gentle vacuum to establish a filtrate flow rate of about 2 drops per sec. Pour in the petroleum ether extract of chloroplast pigments. Develop or spread out the pigment bands by adding a 4:1 mixture of petroleum ether and benzene (*caution*).

When dried, the column can be cut between the colored zones and the pigments dissolved in ethyl ether to which a little methyl alcohol has been added. Using a spectroscope, examine the absorption spectrum of each of the chloroplast pigments. Use chlorophyll solution as an absorption filter.

On occasion, you may also want to use a technique based on the electrostatic attraction of compounds or molecules—electrophoresis (p. 184). Here, too, the migration of complex biological compounds along a moist filter paper strip can be traced. The advantage of electrophore-

[18] Adapted from B. Harrow et al., *Laboratory Manual of Biochemistry*, 5th ed. (Philadelphia: Saunders, 1960).

sis over paper chromatography is the greater speed of molecular migration and the wider separation of compounds.

Using a spectrometer A leaf looks green because it reflects green wavelengths and absorbs others. Observe the part of the spectrum absorbed by chlorophyll extract by holding a small vial of an alcohol extract from spinach or geranium leaves halfway across the slit in the back of the spectrometer. Use a good light source. Locate the normal spectrum and the spectrum showing wavelengths not absorbed by the chlorophyll extract. As a control, also view the solvent alone through the spectrometer.

In place of a spectrometer, use concentrated light from a 35 mm projector with a lens placed to focus the beam on a tube of chlorophyll extract. Secure the test tube to a ring stand. Then fasten a prism to the ring stand to bend the wavelengths; use a white screen as a back drop. Note the wavelengths absorbed by the extract and compare with a test tube of distilled water in place of the extract.

Fluorescence Distinguish the color of light transmitted or reflected through a chlorophyll extract. Hold a beaker of the extract in a beam of light; observe the greenish color of transmitted light. Note the blood-red color of reflected light.

If the crude extract is further separated into chlorophyll *a* and chlorophyll *b* (as described earlier), observe the blue-green color of a solution of chlorophyll *a* and the yellow-green color of chlorophyll *b* in transmitted light. Chlorophyll *b* looks brown-red in reflected light, while chlorophyll *a* gives a blood-red color.

Examine chromatograms of chlorophyll pigments under ultraviolet light (*caution:* avoid looking into the light). Note the fluorescence of the pigments.

Absorption of carbon dioxide in photosynthesis

The rate of photosynthesis exceeds that of respiration in green plants by a factor anywhere from 10 to 100 times. In photosynthesis, only the dark phase is thermochemical and affected by temperature. Respiration, on the other hand, is a chemical reaction the rate of which increases some 2 to 4 times for each 10° C (50° F) rise in temperature within the optimal range. Further, the rate of respiration is affected by several factors in the environment, including the supply of oxygen and a decrease in available nitrogen, as well as hereditary factors.

The study of exchange of gases is difficult under classroom conditions since photosynthesis masks respiration during periods of light. The carbon dioxide that green cells liberate in respiration is quickly assimilated in light for photosynthesis; conversely, the oxygen released in photosynthesis is used by green cells in respiration.

In short, to study respiration, green plants must be maintained in darkness. The "apparent rate" of photosynthesis is that obtained before one makes corrections for the simultaneous, underlying process of respiration. The "true rate" of photosynthesis is established after corrections are made for respiration. This can be accomplished by keeping some samples of plants in the dark before, and after, a period of light. Then the mean value of the two dark periods can be used as the value applying during the light period.

Using indicators

Both Osterhaut and Haas (1919) used the resulting change in alkalinity as a means for measuring the rate of photosynthesis of water-living plants such as *Spirogyra* and *Ulva*. Changes in alkalinity of the suspensions in which the experimental plants were maintained were compared against a series of buffer solutions of known alkalinity. A correction could be made for the observed rate of photosynthesis as compared with the amount of respiration that had occurred. Although they did not recommend this method for accurate research, it may be useful in the classroom.

To illustrate, assume that a pH indicator (such as phenolphthalein or brom thymol blue) has been added to a suspension containing some water-living plant or algae. Compare the color of the suspension with a series of known buffer solutions (p. 217). Suppose this suspension containing an indicator held at the end point is now placed in light for 20 min. If the suspension is now found to be more alkaline, we assume that the gas has been absorbed that previously maintained the culture at a lower pH. Our assumption, based on accumulated evidence in the domain of science, is that this gas is carbon dioxide. Thus we proceed, assuming that photosynthesis is an on-going process in light and that in this process a gas, carbon dioxide, is rapidly absorbed and removed from the medium. Similarly, the acidity of the culture medium should increase in darkness. At this time, the carbon dioxide constantly liberated in respiration can be measured since photosynthesis has been stopped.

The quantity of carbon dioxide absorbed should equal the amount of oxygen liberated in photosynthesis, since the photosynthetic quotient (PQ) is 1, that is,

$$PQ = \frac{\text{volume of } O_2 \text{ liberated}}{\text{volume of } CO_2 \text{ absorbed}} = 1$$

The change in quantity of carbon dioxide is easier to measure, especially with indicators, since a change is quickly produced in 0.04 percent of a substance present in the medium. In photosynthesis, when carbon dioxide is rapidly absorbed, the acidity of the medium should decrease. (In respiration, the acidity should increase quickly.)

Indicators can be used only with media that have not been buffered, since buffered solutions can resist relatively large changes in pH. Indicators are usually considered weak organic acids or bases whose molecules give one color in the nonionized state, while their anions or cations give a different color.

Phenolphthalein is a useful indicator in titration of weak acids (such as carbonic)

and changes color in a slightly alkaline medium. Its pH range is from 8.3 to 10. (Also refer to the use of brom thymol blue on page 214.)

Absorption of carbon dioxide by algae
Cultures of *Chlorella* to which the indicator phenolphthalein is added at the start give dramatic results in 5 to 20 min. Stock cultures of *Chlorella* vary in alkalinity, showing an increase in pH as cultures grow older and remain in light (permitting active photosynthesis). There are also variations in pH of the solutions depending on the stage in the growth cycle of these algal cells. Some preliminary testing is needed to find cultures that are close to the end point for phenolphthalein. Test 10-ml samples of culture media by adding 1 or 2 drops of 0.1 percent phenolphthalein solution. If all cultures on hand are quite acidic, raise the pH by adding a small pinch of calcium carbonate.

Prepare several test tubes or Erlenmeyer flasks (125 ml) with measured amounts, such as 50 ml, of deep green suspensions of *Chlorella*. To each tube add 1 to 3 drops of the very dilute phenolphthalein solution (0.1 percent) so that the slightest pink tinge appears and then disappears on shaking the tube. Maintain several such tubes or flasks of equal cell density in light; keep the others in darkness or cover with aluminum foil. In bright light there is a deepening pink color within 5 to 20 min (depending on light intensity and cell density). Compare results with the controls kept in darkness. Other algae, or the seed plant, *Anacharis* (elodea), may be used, but *Chlorella* is especially recommended.

When stock cultures of *Chlorella* are limited, cultures may be prepared *without* the indicator. Prepare the tubes or flasks as already described, but omit the indicator solution; keep some tubes in light and some in darkness. At intervals up to 12 hr, add drops of indicator to *samples* drawn from the tubes or flasks. After 12 hr as little as 2 drops of indicator may turn the 10-ml samples of suspension to pink-rose (more alkaline). The addition of some 20 drops of indicator may not change the color of the samples from those suspensions maintained in darkness. The latter become more acidic over a period of hours, due to failure of carbon dioxide to be absorbed in darkness, and also to the liberation of carbon dioxide in respiration.

Similar materials may be used in the study of respiration (see chapter 4). If those suspensions to which the indicator was added (so that they became pink-rose in light) are now shifted to darkness, the color fades within 10 min, and the cultures become colorless in 20 min. So sensitive is the test—if cell density and dilution of the indicator are appropriate—that students can observe changes in the color of the indicator in the tubes as the day advances. There is a fading as daylight begins to diminish; as light appears the next morning, the color deepens from pink to rose again. Students can follow this over several days in school or at home by keeping a log of the time intervals.

Some students may want to measure the quantity of carbon dioxide that has been absorbed out of solution. The basis of colorimetry is given on page 216, buffer solutions on page 159, and a method of determining micromoles of carbon dioxide on page 215.

In the laboratory or at home, students may investigate the relative amount of photosynthetic activity of *Chlorella* under red, blue, green, and orange wavelengths of light (p. 222). Also refer to the series of investigations beginning on page 230.

Absorption of carbon dioxide by a water plant The fact that carbon dioxide is absorbed in the presence of light is indirect evidence of its use in the process of food-making. One of the simplest methods of demonstrating this uses the absorption of carbon dioxide from water by the seed plant *Anacharis* (elodea).

USING BROM THYMOL BLUE WITH ELODEA The procedure to be described is a modification[19] of a titration method adapted originally by H. Munro Fox. In this method the absorption of carbon dioxide from water by green plants in light is revealed through the change in the color of an indicator. (The use of phenolphthalein in studies with *Chlorella* is described above.)

Brom thymol blue[20] is the indicator used; it is blue in alkaline solution and yellow in acid medium. It has a fairly narrow range in pH, from 6.0 (yellow) to 7.6 (blue), so that slight changes in hydrogen-ion concentration show up quickly Thus, a slight increase in acidity, as when carbon dioxide is added to the solution, will change the blue color (alkaline) to yellow. When carbon dioxide is absorbed, as in photosynthesizing plants, the yellow color (due to carbon dioxide in solution) is changed back to blue.

Prepare a 0.1 percent *stock* solution of brom thymol blue by dissolving 0.5 g of brom thymol blue in 500 ml of water. To this add a trace of ammonium hydroxide (1 drop or so per liter) to turn the solution *deep blue*. In a beaker dilute the 0.1 percent *stock* solution with aquarium water in which elodea has been growing. Be certain that the solution is a deep blue. If it is not, turn it blue by adding just enough of the dilute ammonia. Then breathe through a straw into the indicator solution until the color *just* turns yellow. Now pour the brom thymol yellow into large test tubes. Prepare these demonstrations, and cork the tubes: (1) one set with no plants, exposed to light; (2) another set, each tube with a sprig of a vigorously growing elodea plant, each having an end bud, also exposed to light; and (3) a series like (2), except that each test tube is covered with aluminum foil or kept in the dark.

Tubes containing plants placed in sunlight should show a change from yellow to blue within 30 to 45 min (depending upon the intensity of sunlight). On cloudy days, place the series of tubes near a 75-watt bulb. Results similar to those in sunlight will take about 45 min. Since the (1) and (3) tubes show no color change, the color

---

[19] P. F. Brandwein, "A Method for Demonstrating the Use of Carbon Dioxide by a Plant," *Teaching Biologist* (1938) 7:76.

[20] Dictionaries usually list this indicator as "bromothymol blue"; we are following common scientific usage.

change in the tubes in the (2) series may be taken as indirect evidence of the absorption of carbon dioxide by green plants in the presence of light.

Phenol red, another indicator with a narrow range of pH (6.8 acid to 8.4 alkaline), may be substituted for brom thymol blue or phenolphthalein (described earlier). As the carbon dioxide is absorbed, the color changes from yellow (acid) to red (alkaline).

USING BROM THYMOL BLUE WITH CHLORELLA   Since brom thymol blue has such a narrow range above and below neutrality, it is obviously a good choice to indicate sensitive, slight changes from neutrality into either acidity or alkalinity. Observe how the rate of absorption of carbon dioxide changes with the intensity of light. Have students compare suspensions of *Chlorella* of different cell densities with the rate of absorption of carbon dioxide in light (or use phenolphthalein, p. 213).

The green color of the suspensions of *Chlorella* often obscures the color changes from blue to green or yellow; therefore, the indicator is best used with samples of the medium after the cells have settled. On the other hand, if the cultures are not shaken, algae settle quickly so that a color change would be apparent in the clear fluid in test tubes or flasks.

Acidify some dilute brom thymol blue (preparation, above) by blowing into it with a straw until it just reaches the turning point and remains yellow. If too much carbon dioxide is added, the time required for algae or any water plants to absorb it out of solution and show a change in color will exceed a class period. In bright light, photosynthesis is the ongoing process, and tubes of algae containing the indicator show the change to blue as the medium becomes more alkaline. This is one demonstration showing that photosynthesizing cells absorb carbon dioxide in light.

Or count the small number of drops of brom thymol yellow needed to turn a sample of the culture blue. Compare with a control in the dark. Why is it that samples from the dark do not turn blue, but remain the yellow of acidified brom thymol?

You may want to apply these techniques using indicators to test the effects of an altered environment on photosynthesis (and on respiration, too). Some investigations are suggested in the proposed block of activities on page 230.

Measuring micromoles of carbon dioxide

Suppose several flasks of suspensions of *Chlorella* or other algae, or of water plants such as *Anacharis*, are on a ledge exposed to sunlight and we now plan to determine the concentration of carbon dioxide in the medium preliminary to some other investigation. Instead of a visual inspection (colorimetry, p. 216), students may devise a quantitative measure of the amount of carbon dioxide. Since carbon dioxide is absorbed by sodium or potassium hydroxide, a titration method can be used to calculate the amount of carbon dioxide in a given volume of suspension containing an indicator. In this procedure, alkali is added, drop by drop, to the medium. Equivalent amounts of equal concentrations of alkali and of acids react; the amount of carbonic acid present (carbon dioxide in water) will be equal to the amount of hydroxide that is added to the fluid that is tested.

$$2NaOH + H_2CO_3 \longrightarrow$$
$$Na_2CO_3 + 2H_2O$$

Use a freshly prepared 0.04 percent solution of sodium hydroxide (0.4 g to 1 liter of distilled water). The value in using this percentage solution is that each milliliter can combine with 10 micromoles of carbon dioxide.

Determine the content of carbon dioxide in the medium containing the algae, such as *Chlorella*, by pouring 100 ml of medium into a flask, and then adding 5 drops of 0.5 percent phenolphthalein indicator. To this add, drop by drop, the alkali—0.04 percent sodium hydroxide. The amount should be measured to 0.1 ml; use a graduated milliliter pipette, a hypodermic syringe, or a graduated burette. Add several drops, then shake the test flask. Continue this until a pink color appears; then carefully add 1 drop at a time to reach the end point where the pink color remains. Record the number of milliliters (in tenths) of alkali used to reach this end point. This is the amount of alkali equivalent to the carbonic acid in 100 ml of unknown medium.

Compute the number of micromoles of carbon dioxide in an unknown by multiplying the number of milliliters of sodium hydroxide by ten. (Recall that each milliliter of alkali will combine with 10 micromoles of carbon dioxide.) Students will also need to compute the number of micromoles of carbon dioxide in a control medium such as Knop's solution.

In specific studies of photosynthesis, the number of micromoles of gas in the control equals the original amount of carbon dioxide in the containers at the start of the experiments. Subtract the amount obtained from the flask in light, in order to find the amount of carbon dioxide absorbed in the specific process of photosynthesis under study.

Furthermore, if the amount of carbon dioxide in the control is subtracted from the amount found for the flask kept in the *dark*, students will also have an estimate of the amount of carbon dioxide produced by green cells when respiration is the dominant on-going process.

For a clear, cogent description of a technique using radioactive barium carbonate in a study of carbon dioxide fixation, refer to Wald.[21] Teachers should not attempt these demonstrations however, unless they have received special training in handling radioactive materials.

Colorimetry

On occasion there may not be time to make titrations in the laboratory. In such cases color comparisons can be made, which provide a rough estimate of pH of the unknown. The pH of the unknown solution can be found by adding a measured volume of an indicator such as phenolphthalein to 10-ml samples of the unknown. Add the same amount of indicator to a series of buffer solutions (10 ml each) of known hydrogen-ion concentration in steps of 0.2 pH units (or other units, as desired; see Table 3-1). Then match the color of the unknown against the series of labeled buffer solutions to determine the pH of the unknown. Use test tubes of the same diameter and material. Color is matched in a comparator block, or other device that may be made in the laboratory. The block should contain a lamp of specific wattage for uniform comparison tests.[22]

Simple colorimetry is based on a visual matching of colored solutions. This is different from photometry or spectrophotometry, in which a photoelectric cell is used. While all colored solutions absorb some light, the relationship between the intensity of the incident and transmitted light depends on the thickness of the layer of solution that light passes through (Lambert's law), and also on the amount of monochromatic light the solution absorbs (Beer's law).

Colorimetric apparatus for a research laboratory is expensive; yet, knowing the basic principles involved, you may want to prepare a set-up to allow light from a tungsten lamp to pass through some standard test tubes of known pH.

A wide range of pH test paper for each full unit or for each 0.5 unit can also be purchased.[23] Then students can devise rapid methods (to be performed in one class period) to match colors of solutions when more accurate titration is not possible. Determine the pH of the initial suspension used in demonstrations to show increase in carbon dioxide in respiration

---

[21] G. Wald et al., *Twenty-six Afternoons of Biology*, 2nd ed. (Reading, MA: Addison-Wesley, 1966).

[22] Kits are available from supply houses such as LaMotte Chemical Products Co., Chestertown, MD and Hach Chemical Co, Loveland, CO 80539.
[23] LaMotte Chemical Products; Hach Chemical Co.

(chapter 4) by using pH paper or by matching (after an indicator has been added) against the standard test tubes in the student-made colorimeter.

Buffer solutions

The preparation of a series of buffer solutions may be undertaken by groups of students. Refer to A. Giese, *Laboratory Manual in Cell Physiology*,[24] for preparation of other buffers; also G. Humason, *Animal Tissue Techniques*.[25] A recipe for preparing 0.2 percent *M* phosphate buffer is given in chapter 2; other buffers are described in chapter 10.

Ready-made buffer solutions in 500 ml containers in a series of pH ranges are available from MCB Manufacturing Chemists, 480 Democrat Rd., Gibbstown, NJ 08027.

In *Plants in Action*, Machlis and Torrey give a series of phosphate buffers at intervals of 0.2 pH which may be useful in the laboratory (Table 3-1). Also refer to the preparation of a 0.2 *M* phosphate buffer using only sodium salts (p. 159).

Prepare these two *stock* solutions:

[24] Giese, *Laboratory Manual in Cell Physiology*.
[25] G. Humason, *Animal Tissue Techniques*, 3rd ed. (San Francisco: Freeman, 1972).

TABLE 3-1 Buffer solutions*

pH	$M/15$ Na_2HPO_4 (ml)	$M/15$ KH_2PO_4 (ml)
5.6	10.0	190.0
5.8	16.5	183.5
6.0	25.0	175.0
6.2	36.0	164.0
6.4	53.5	146.5
6.6	74.5	125.5
6.8	99.0	101.0
7.0	122.0	78.0
7.2	143.0	57.0
7.4	161.0	39.0
7.6	172.5	27.5
7.8	182.5	17.5
8.0	189.0	11.0

* From L. Machlis and J. Torrey, *Plants in Action* (San Francisco: Freeman, 1956).

(1) $M/15$ dibasic sodium phosphate (9.47 g of dry Na_2HPO_4 dissolved and diluted up to 1 liter of distilled water) and (2) $M/15$ monobasic potassium phosphate (9.08 g of dry KH_2PO_4 dissolved and made up to a liter of dstilled water). Mix these two stock solutions as indicated in Table 3-1.

Absorption of carbon dioxide by a by a land plant

Plants such as geranium or coleus may be used to show absorption of carbon dioxide through leaves. The amount of carbon dioxide in the air reaching the plant may be modified, removed, or increased to demonstrate a change in the rate of starch-making.

First, *reduce* the amount of carbon dioxide around the plant in the following manner. Place a healthy geranium plant on a tray or large sheet of glass together with an open, wide-mouthed jar or beaker of solid KOH or pellets or sticks of NaOH (*caution: caustic; use tongs*).[26] Cover with a bell jar; seal the bottom of the jar to the sheet of glass with petroleum jelly to make it air-tight (Fig. 3-11). The hydroxide removes the carbon dioxide from the enclosed jar. Then set up a similar demonstration, but in this one omit the hydroxide. Place both preparations in moderate sunlight. After several days, the leaves may be tested for their starch content. The plant with the carbon dioxide absorbent will show little or no starch.

A more elaborate means for reducing the amount of carbon dioxide in the air is shown in Figure 3-12. Begin with a geranium plant which has been in the dark for at least 24 hr and gives a negative starch test. An aspirator draws air through bottles containing barium hydroxide or strong soda lime in solution or pellet form. In this way the air that enters the bell jar lacks carbon dioxide. Later, test the leaves for the presence of starch.

[26] Solid KOH or NaOH should be handled with tongs; for safety and first-aid procedures, see chapter 10.

Fig. 3-11 Apparatus for reducing the amount of CO_2 in the air surrounding a green plant. The beaker contains pellets of KOH or NaOH.

The effect of an *increased* amount of carbon dioxide in air on the rate of photosynthesis can also be shown (Fig. 3-13). Select a geranium plant which gives a negative starch test, and enclose it in the bell jar. Connect a delivery tube from a carbon dioxide generator (or a flask containing dry ice) into the bell jar. (Leave the clamp open while carbon dioxide is entering the bell jar; notice the safety tube on the generator.) Later, test these leaves for starch. There should be a significant increase in the amount of starch produced in the leaves of the plants grown in an atmosphere rich in carbon dioxide. However, this is a somewhat unsatisfactory demonstration because the relative difference in starch content is difficult to detect unless a comparison of the dry weight of the two sets of leaves is made.

Stomates, the air passages of leaves

To show that stomates serve as air passages in the leaves, students may press the petioles of leaves between their fingers while the blades of the leaves are immersed in hot water. The air in the leaves will expand in hot water and escape as air bubbles through the stomates.

Floating water plants will show the stomates on the upper surface only, while

TABLE 3-2 Number of stomates per sq mm*

Leaves with no stomates on upper surface

	lower surface
Balsam fir	228
Norway maple	400
Wood anemone	67
Begonia (red)	40
Barberry	229
Rubber plant	145
Black walnut	461
Lily	62
White mulberry	480
Golden currant	145
Lilac	330
Nasturtium	130

Leaves with few stomates on upper surface

	upper surface	lower surface
Swamp milkweed	67	191
Pumpkin	28	269
Tomato	12	130
Bean	40	281
Poplar	55	270
Bittersweet	60	263

Leaves with stomates more nearly equal on both surfaces

	upper surface	lower surface
Oats	25	23
Sunflower	175	325
Pine	50	71
Garden pea	101	216
Corn	94	158
Cabbage	219	301

* From B. Duggar, *Plant Physiology* (New York: Macmillan, 1930).

most leaves will show them on the lower side. In the sunflower, clover, daffodil, and grasses, bubbles of air can be seen to escape from both surfaces of the leaves, indicating that stomates are found on both surfaces (Table 3-2).

The epidermal layers may be peeled off the leaves and mounted for microscopic

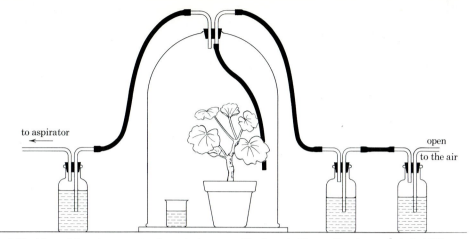

Fig. 3-12 More elaborate apparatus for reducing the amount of CO₂ in air surrounding a green plant. The three bottles and beaker contain Ba(OH)₂ or soda lime.

examination to confirm these observations (see Fig. 3-14). The fleshy leaves of *Peperomia*, types of *Crassula*, and lettuce allow easy removal of the lower epidermis (see also page 220).

You may also "lift off" the lower epidermis layer of a leaf to examine guard cells and stomates. Lay a leaf flat and brush an area of the lower epidermis about 2 cm square with acetone and quickly press down with a plastic coverslip. Mount on a slide and examine the number of stomates and the position of the guard cells. (*Caution:* Have good ventilation when using acetone and no flames.)

Entry of carbon dioxide into leaves through stomates   At another time, you may want to show that stomates are the regions through which carbon dioxide enters the leaves of land plants. Select thin-leaved plants, such as geranium or coleus, which have stomates on the lower surface of the leaves. (Other useful plants are listed in Table 3-2; the average number of stomates on the upper and lower epidermal layers is indicated.)

Paint the lower surface of leaves of geranium with benzene. If the plant has been in sunlight, benzene enters the open stomates, resulting in the appearance of translucent areas. If some leaves had been

covered, their stomates are likely to be closed, producing no translucence.

In another method, use petroleum jelly. Select two plants, geraniums for example, which have been kept in the dark overnight, or until the leaves give a negative starch test. Coat both sides of several leaves with melted petroleum jelly to close the stomates. On other leaves, coat the upper surface only; on still others, the lower surface only.

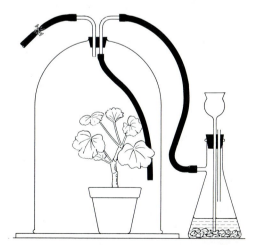

Fig. 3-13 Apparatus for increasing the amount of CO₂ in the air surrounding a green plant. Dilute acid and marble chips (CaCO₃) react in the flask to produce CO₂. Note the safety tube in the flask.

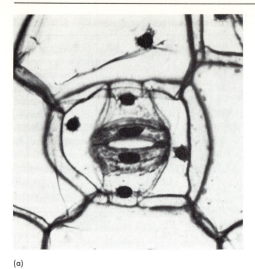

(a)

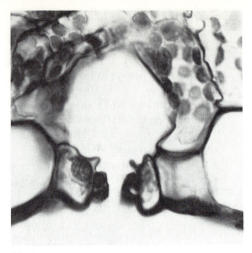

(b)

Fig. 3-14 Epidermis from leaf of *Sedum*:
(a) Showing stomates and guard cells; (b) side
view of stomates and guard cells. (Courtesy
Carolina Biological Supply Company.)

After the plants have been in light for several hours, wipe away the film (or remove it with ether or carbon tetrachloride). Boil the leaves in water, heat in alcohol until the chlorophyll is removed, and finally test for starch. Note that where petroleum jelly has clogged the stomates and prevented the entrance of carbon dioxide, starch-making has been arrested. The success of this demonstration depends on very careful technique, for example, complete clogging of the stomates. Through the use of radioactive carbon, it has been found that carbon dioxide is absorbed through the upper epidermal region in some leaves that lack stomates in this upper layer.

Examining stomates under the microscope Use leaves such as *Bryophyllum, Sedum, Echeveria, Sempervivum, Kalanchoë, Peperomia,* "hen and chicks," *Tradescantia, Zebrina,* crisp lettuce leaves, or Boston fern for this demonstration. Tear the leaf toward the main vein. Then with forceps pull off strips of the thin membrane that is the lower epidermis, and mount in water; examine under the microscope (Figs. 3-14, 3-15).

Especially desirable for the purpose of examining stomates is the epidermis of certain species of *Tradescantia* whose cell vacuoles contain a pink pigment. The guard cells bounding each stomate contain chloroplasts and can be easily located in the pink background. In the houseplant *Rhoeo discolor,* guard cells are green and epidermal cells are a brilliant purple.

Where the epidermis is difficult to pull off, blanch the leaves in boiling water for a few minutes. Then the epidermis may be easily loosened. In this condition, as well as in the fresh state, the cells may be stained with methylene blue or dilute Lugol's iodine (pp. 136, 137).

Where a study of the other layers of cells in leaves is planned, make freehand cross sections of leaves. bed on page 000.

Turgor in the guard cells surrounding stomates How do guard cells regulate the size of stomates? Strip off the lower epidermis of a plant (see above) that has been standing in bright sunlight. Mount a bit of this in a drop of water. Examine the guard cells under high power. Which side—the inner or the outer side—of guard cells has thicker walls? Now place a drop of 0.4 *M* CaCl$_2$ or NaCl solution (solutions, p. 746) near the edge of the coverslip, and draw off the water from the opposite end of the coverslip with

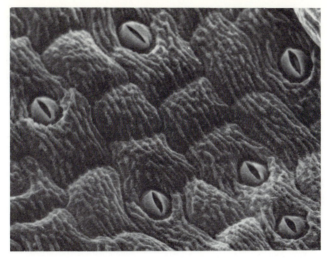

(a)

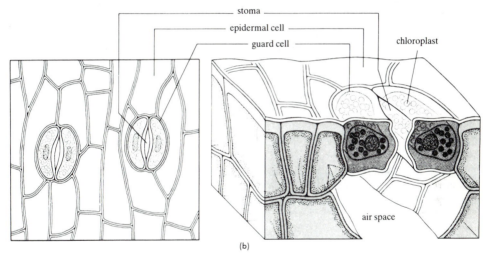

(b)

Fig. 3-15   Open stomates (magnification 260 ×): (a) scanning electron micrograph of a living leaf; (b) diagram of surface and cross-sectional views.

((a) Photo courtesy J. Heslop-Harrison. (b) From *Biology,* by Grover Stephens and Barbara Best North, 1974 by John Wiley & Sons. Reprinted with permission of John Wiley & Sons, Inc.)

filter paper. Examine the guard cells and the size of the stomates again. Notice the reduced size of the stomates as the guard cells lose turgor because of removal of water by the sodium or calcium chloride.[27] (For an explanation of plasmolysis and

diffusion, with additional examples, see page 294.)

It may be possible to return the tissue to water and note the increase in turgor of guard cells and in the size of the opening of stomates (that is, if the cells have not been killed).

Also prepare cross sections of a leaf by placing a leaf between a split section of pith or turgid carrot or potato segments. Stain the sections of leaves by mounting them in 1 percent solution of safranin.

---

[27] Guard cells are partially plasmolyzed by a 0.5 M sucrose solution so the stomate is closed. Or use drops of 0.3 M sucrose solution (preparation, p. 746). Note that a 0.03 M sucrose solution straightens the walls of the guard cells without plasmolyzing them.

Oxygen evolved in photosynthesis

No demonstration satisfactory for use in a school classroom has yet been found to show the evolution of oxygen during photosynthesis. It would be a contribution if a suitable demonstration, simple enough to be used by students, were devised. However, until one *is* devised, the following demonstrations may be useful.

Rate of liberation of gas bubbles in light as evidence of photosynthesis

Counting the bubbles of liberated gas as a method for estimating the amount of oxygen evolved during photosynthesis dates back in the literature to 1837. However, there are many inaccuracies in the method, as described by Miller.[28] He cites such variables as differences in size of air bubbles and different oxygen content of bubbles. Yet, for students, there is some value in the following two demonstrations. In flasks of rapidly photosynthesizing cells, bubbles of gas can be seen rising in the cultures; in fact, in strong sunlight, clumps of *Chlorella* rise to the surface rather quickly as they are buoyed up by a froth of bubbles. This is often apparent in a pond where masses of *Spirogyra* are brought to the surface by air bubbles.

Counting bubbles in a slide of algae   The rapid appearance of bubbles in media under controlled conditions can be used as a rough indication of rate of photosynthetic activity of algae. Students may prepare their own series of slides to compare the numbers of bubbles produced under altered environmental conditions.

Use depression slides or deep-walled concavity slides. Place 3 to 5 drops of a dense culture of *Chlorella* (deep green) in the concavities so that they are completely filled. Then slide coverslips over the fluid in each concavity so that no air bubbles are trapped. For these short-range studies it is not necessary to seal the slides with

[28] E. Miller, *Plant Physiology* (New York: McGraw-Hill, 1938).

petroleum jelly. With a hand lens, make sure each slide contains no air bubbles at the start.

Place slides in sunlight or about 2 to 3 ft from a 100-watt bulb; inspect every 3 min and count the number of bubbles that appear over a 15-min period. Compare with slides maintained in the dark. Also prepare some microaquaria with medium of varying cell density. Use dilution techniques (p. 458). Is there a relationship between cell density and the number of bubbles produced? What is the effect of using cellophane coverslips of different colors—green, red, blue? (See also page 224.)

Slides that are left for an hour or more are often useless, as small bubbles may coalesce into one large air bubble.

Counting bubbles of gas from a cut stem   Bubbles of gas escape from the cut stems of elodea (*Anacharis*) sprigs placed in bright sunlight. On the basis of evidence from research, we know this gas is oxygen. (See the list of possible inaccuracies, above.) To show bubbles, invert a growing tip of a young elodea plant 7 to 8 cm long into a test tube or beaker containing aquarium water to which about 2 ml of a 10 percent solution of sodium bicarbonate has been added for every 100 ml of aquarium water (which has been boiled to drive off dissolved gases and then cooled). The bicarbonate will provide an additional source of carbon dioxide. Tie the sprig to a glass rod immersed in the container to hold it in place (Fig. 3-16).

Expose the plant to bright light or a 100-watt bulb. Count the number of bubbles of gas escaping from the cut stem per minute. Compare the rate of production of gas in relation to light intensity by moving the test tube or beaker varying distances from the light source (15 cm, 22.5 cm, 30 cm, 37.5 cm, 45 cm). Use a light meter to check different intensities of light. Compare this with a control placed in relative shade or in darkness. If few bubbles escape, recut the stem and crush it a bit. Or use a needle to make small perforations in the stem for gas to escape.

Fig. 3-16 Apparatus to show sprig of *Anacharis* (elodea) liberating bubbles of gas in light.

Plot data on a graph and in a table to show the number of bubbles counted at varying distances.

Wavelength and oxygen production Prepare test tubes of growing tips of elodea in aquarium water adding a pinch of bicarbonate of soda. Wrap two test tubes in green cellophane, then two in blue, and two in red cellophane, and maintain all in light. Plan additional controls using clear cellophane. Where are there more bubbles produced?

While it is difficult to test for the presence of oxygen, we assume that the gas liberated in sunlight is oxygen. This assumption is based on substantial experimental evidence with other plants, and on the research with manometric techniques, and semimicro analysis.

Fogg reminds us that it is necessary to make comparisons for equal numbers of quanta since photochemical effects of light quantities remain the same although energy per quantum varies with wavelength.[29]

[29] G. E. Fogg, *The Metabolism of Algae* (New York: Wiley, 1953).

Oxygen production–pyrogallic acid In a qualitative study, demonstrate that elodea sprigs produce oxygen in light using a 10 percent solution of pyrogallic acid, which turns brown when it combines with oxygen (*caution*).

Prepare the sprigs of elodea as above in aquarium water to which a pinch of bicarbonate of soda has been added. Place two tubes in light and two in darkness for 2 to 4 hr. Then add 5 ml of freshly prepared pyrogallic acid. Which tube shows the brownish color indicating oxygen production? Plan a control tube without elodea. Does pyrogallic acid turn the liquid brown? (There may be a slight color change due to O_2 in the air.) Compare results in all the tubes after 30 to 60 min.

Also compare the rate of O_2 production in test tubes of elodea wrapped in cellophane (green, blue, red, clear) as well as in carbon paper. Place tubes in light for 2 to 4 hr; then add pyrogallic acid. In which tubes does the liquid turn brown rapidly—in wavelengths of red, green, or blue? Why is there no change in tubes wrapped in green cellophane?

Change in volume of a gas

Quantitative measurement of volume of oxygen A modification of the previous demonstrations using a graduate 1 ml pipette may give more accurate "counts" of volume of a gas produced. Invert the freshly cut stem of elodea into the wide end of the pipette and secure the pipette upside down to a ring stand. The end of the pipette should be about 2 cm into the water. At the narrow end of the pipette attach thin tubing. Suck a column of water up into the pipette and close the tubing with a screw clamp. Mark the level of the water. Use a 150- to 200-watt light bulb shielded on three sides so that light is focused on the elodea plant. Place the light 15 cm from the plant and take readings at 1 min intervals for about 10 min. Record the rate at which oxygen is released and displaces the water in the column. This indicates the rate of photosynthesis. Compare results with different groups of students.

Does the amount of light affect the rate of photosynthesis? Repeat the procedure, but position the plant *closer* to the light: 7.5 cm instead of 15 cm. Tabulate results.

Effect of increase in CO₂ Does an increase in available carbon dioxide affect the rate of photosynthesis? Repeat the procedure described but substitute 2 percent sodium bicarbonate solution in place of the water. Record data and compare with results using water.

Dense young cultures of *Chlorella* also may be used, but it may take longer time intervals to show appreciable changes depending on the growth rate and photosynthetic rate.

Use a thermometer to check temperature effects if the whole apparatus is placed in a water bath.

In addition, use the same apparatus to show the effect of wavelength on rate of photosynthesis. Cover the tubes of elodea or *Chlorella* with different colors of cellophane. Which wavelengths are absorbed or correspond to the action spectrum in photosynthesis?

"Funnel" method for collecting liberated gas In this method the escaping gas is

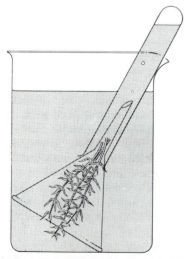

Fig. 3-17 Funnel method for collecting bubbles of gas liberated by sprigs of *Anacharis* (elodea).

collected to be tested and identified. Fill a large beaker or battery jar with aquarium water to which sodium bicarbonate solution (see technique just discussed) has been added. Cut the basal ends of five to ten sprigs of fresh elodea, and arrange them in a glass funnel so that their cut stems lie within or near the stem of the funnel. Force the wide mouth of the funnel to the bottom of the beaker (Fig. 3-17). Then cover the upper stem of the funnel with a test tube completely filled with water. Finally, set the preparation in bright light for a number of hours.

The water in the test tube should be displaced, more or less completely, by bubbles of gas rising from the cut stems of the elodea. Test the gas collected in the test tube for oxygen by inserting a glowing splint. Ganong criticized this method on the grounds that there is not enough oxygen produced to cause the splint to flare up.[30] However, the probability of getting a positive test for oxygen is increased if the first 1 to 2 cm sample of gas, which usually has a good deal of air in it, is discarded. If the water has been boiled and then cooled, most of the dissolved gases will have been removed.

Change in volume of a gas: volumeter

We have described some techniques to measure the exchange of gases in photosynthesis (and in respiration) by means of indicators, if buffered solutions are not used in the media. However, there are other methods possible for measuring volume of a gas besides those already described (pipette and funnel methods, p. 223). Students may observe changes in the volume of gases in a closed system, such as a volumeter (Fig. 3-18; also Figures 3-18, 3-19; also 4-45, 4-46, 4-48). A change in volume, due perhaps to the liberation of oxygen during photosynthesis, will affect

[30] W. F. Ganong, "The Erroneous Physiology of the Elementary Botanical Textbooks," *School Sci. Math.* (1906) 6:297.

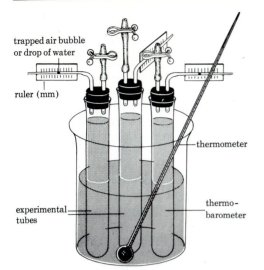

trapped air bubble
or drop of water

ruler (mm)

thermometer

experimental
tubes

thermo-
barometer

Fig. 3-18   Volumeter to demonstrate changes in volume resulting from exchanges of gases in photosynthesis. The apparatus may also be used in studies of respiration. The thermobarometer acts as a control that records changes caused by temperature or barometric pressure. (After R. E. Barthelemy, J. R. Dawson, Jr., and A. E. Lee, *Innovations in Equipment and Techniques for the Biology Teaching Laboratory* [BSCS], Heath, 1964.)

pressure in the system, and will cause a trapped air bubble or a trapped drop of fluid to move along a measured length of capillary tubing of known bore. Again, we assume the identity of the gases liberated in photosynthesis (or respiration).

In the research laboratory both photosynthesis and respiration in cells can be measured precisely with manometers and Cartesian divers.[31] In fact, Warburg first used manometric applications in 1919 in studies of photosynthesis in *Chlorella*. He was able to measure the increase in pressure in a closed system as oxygen was given off by algae; the pressure of carbon dioxide could be held constant with the use of buffers. Since oxygen is less soluble in water than carbon dioxide, the increased volume of oxygen raised the pres-

sure in the system and shifted the level of mercury in a capillary U-tube.

Methods using student-made manometers with *Chlorella* or other water-living plants are not as accurate in the classroom laboratories. However, student-made manometers made of narrow-bore capillary tubing can be used effectively to show the change in volume of *carbon dioxide* produced by rapidly growing seedlings and growing yeast cells. (Refer to chapter 4, page 331 for methods for measuring the rate of fermentation or anaerobic respiration.)

Using a volumeter   Measuring a change in the volume (in a volumeter, Fig. 3-18) is easier than measuring a change in pressure of minute amounts of gases. The volume of oxygen given off in photosynthesis, per unit time, is some 10 to 30 times the volume of oxygen that is used in respiration.

A volumeter provides some notion of a semiquantitative test if a millimeter ruler is used to measure the distance the drop or bubble travels along the capillary arm. Such factors as barometric pressure and temperature must be kept constant by placing all the tubes in a water bath, and setting up a control using an equal volume of distilled water in place of the suspension of cells such as *Chlorella*. This is a thermobarometer; maintain the same volume of gas above the medium in this control as in the experimental tubes.

When a gas is absorbed in the closed system of the volumeter, the drop or bubble should move *toward* the experimental tube; when a gas is liberated (such as oxygen in photosynthesis) at a faster rate than one is absorbed, the bubble should move *away* from the tube, due to the increase in volume within the tube. Students will recall that carbon dioxide is more soluble in water than oxygen. In the light, quantities of oxygen gas are liberated, changing the volume, so that students can readily gather data on oxygen production in photosynthesis.

Prepare several replicas of the apparatus shown in Figure 3-18 using large test

[31] Descriptions of chemical methods and of volumetric and manometric techniques are available in these and many other texts: A. Giese, *Cell Physiology*, 5th ed. (Philadelphia: Saunders, 1979), and Giese, *Laboratory Manual in Cell Physiology*.

tubes or Erlenmeyer flasks of 125 ml capacity. All the joints must be air-tight. Draw out capillary tubing of 4.5 mm diameter over a Bunsen flame and allow it to cool. (Refer to the section on manipulating glass tubing, p. 748.) Then snap off the tubing at a region with a bore sufficient to permit entry of a hypodermic needle. The needle is used to inject a colored fluid into the tubing to obtain a trapped air bubble or drop. The thinner the tubing, the greater the distance the drop or air bubble will move, and the more sensitive the reading will be. To increase visibility, red ink may be used to color the fluid, but nontoxic brom thymol blue is more satisfactory in the event that some of the fluid should be drawn into the culture medium. Add a trace of detergent to the fluid to permit free flow in the tubing.

After the capillary tubing has been inserted into the rubber stopper, loosely apply the stopper to the test tube or flask. Then add the colored fluid as a drop, or trap an air bubble between drops of fluid. If a hypodermic needle is not advisable, use a medicine dropper and apply fluid to the wider bore of the tubing that is inserted into the stopper. Then deftly apply the stopper to the container so that the drop is positioned about midway along the horizontal arm of the capillary. If a two-hole stopper is used with tubing and a clamp, pressure can be equalized more easily by opening and closing the clamp. Finally, fasten a plastic ruler under the bubble with plastic tape so that the distance in millimeters can be recorded.

Place all volumeters and the control (thermobarometer) in a water bath maintained at 25° C (77° F). Corrections may be made for changes found in the control due to pressure and temperature. Allow the apparatus to remain some 5 min in sunlight to reach equilibrium before taking readings. Then take readings every 5 min for a period of about 20 min.

Using a compensating flask   A variation of this volumeter is shown in Figure 3-19. The extra space that this piece of apparatus requires in the laboratory is justified by its sensitivity. When the apparatus is placed in bright light, a trapped bubble moves rapidly from the experimental flask toward the compensating (control) flask. In sunlight, air in the closed system is warmed and expands; the use of a compensating flask equalizes this effect so that changes due to heat may be ignored in readings.

Narrow bore teflon tubing may be substituted to make a more flexible set-up. Apply petroleum jelly to make all joints air-tight.

Designing investigations   Using some of these techniques, students may devise various investigations. We recognize a teacher's role is to ask the right questions to stimulate possible approaches. For example, some variables that students might study using volumeters or indicators are: the effect of light on rate of photosynthesis; the effect of light of varying intensity; the effect of light of varying wavelength; the effect of flashing light as opposed to continuous light; the effect of varying temperature; and the effect of varying quantities of carbon dioxide available to the cells. Students may also see how the law of limiting factors applies in these investigations. (For further activities, see pages 230–32.)

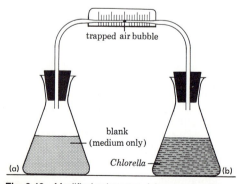

Fig. 3-19 Modified volumeter: (a) compensating (control) flask; (b) experimental flask containing *Chlorella*.

Other methods for determining oxygen content   The volumeter described above (Fig. 3-18) may also be used to absorb oxygen from the atmosphere in the closed system. Measurements may be made of the amount of oxygen liberated in photosynthesis by absorbing the oxygen with pyrogallol and taking readings of the movement of a trapped air bubble.

Add bicarbonate to the suspensions of *Chlorella* (p. 224). This will maintain a constant level of carbon dioxide in the solution and in the air above the solution. The bicarbonate solution liberates carbon dioxide as it is needed; as carbon dioxide is produced by cells, the solution absorbs the gas. As a result, the exchange of carbon dioxide does not affect the pressure in the system; only changes in oxygen affect the pressure or create substantial differences in volume.

Apply freshly prepared 2 percent solution of pyrogallol to a cotton-tipped applicator and place it in the flask so that the cotton stands above the medium (Fig. 4-46). Insert applicators with oxygen absorbent into flasks containing a suspension of *Chlorella* (or other algae) and into the controls. Also prepare controls without the pyrogallol for comparison in interpreting readings. Maintain at a constant temperature, and take readings every 3 min for a 20-min period after the apparatus has come to equilibrium.

A substantial part of the 20 percent content of the atmosphere is absorbed at the start by the pyrogallol solution, and the resulting decrease in volume of the confined gases must be considered in subsequent readings. As photosynthesis is renewed in the light, quantities of gas are produced in the closed system, as shown by the movement of the trapped air bubble in the capillary.

Some preparations of pyrogallol are alkaline (containing potassium pyrogallate), and the oxygen absorbed forms an insoluble precipitate. Since there is an excess of hydroxide in this preparation, some carbon dioxide will also be absorbed. This factor must be taken into consideration in the readings.

Prepare potassium pyrogallol solution in two parts[32]. (1) dissolve 1 g of pyrogallate in 5 ml of water and (2) dissolve 5 g of potassium hydroxide in 25 ml of water. Combine both parts immediately before use (*caution*).

Winkler method   This method for determining the amount of dissolved oxygen (D. O.) in a pond, lake, or aquarium is described in chapter 8, and in many other sources. There are many modifications; refer to more rapid, prepared solutions available from LaMotte and from Hach (see Appendix).

Evolution of oxygen by land plants

Many variations of the following method have been described. The best-known technique is probably the one in which a plant, such as a geranium, and a burning candle are placed under a bell jar. The bell jar is then sealed with petroleum jelly to a glass plate. The candle flame is eventually extinguished, presumably because the oxygen has been exhausted, and the demonstration is then allowed to stand overnight. The fact that the plant has produced oxygen may be tested for in several ways, for example, by inserting a glowing splint or by having a cigarette lighter ignite a candle inside the bell jar. A general criticism of this method is offered by several investigators.[33] It has been shown that a

[32] Adapted from R. E. Barthelemy, J. R. Dawson, Jr., and A. E. Lee, *Innovations in Equipment and Techniques for the Biology Teaching Laboratory* (BSCS) (Boston: Heath, 1964), p. 82; Giese, *Laboratory Manual in Cell Physiology.* Also see W. Andrews, ed. *Environmental Pollution* (Englewood Cliffs, NJ: Prentice-Hall, 1972); O. Lind, *Handbook of Common Methods in Limnology*, 2nd ed. (St. Louis, MO: Mosby, 1979), and many other ecology and biochemistry texts.

[33] B. C. Gruenberg and N. E. Robinson, *Experiments and Projects in Biology* (Boston: Ginn, 1925); A. Raskin, "A New Method for Demonstrating the Production of Oxygen by a Photosynthesizing Plant," *Sci. Educ.* (1937) 21:231; Ganong, "Erroneous Physiology of Botanical Textbooks"; J. Glanz, "The Infamous Candle Experiment," mimeo (author, 87–81 145 St., Jamica, NY); and H. Alyea and F. Dutton, eds. *Tested Demonstrations in Chemistry*, 5th ed., *J. Chem. Educ.* (Easton, PA, 1962), p. 18.

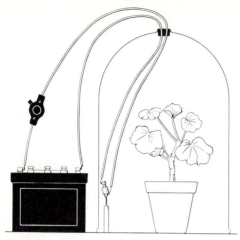

Fig. 3-20 Apparatus to show green plants giving off oxygen. An automobile cigarette lighter wired through a switch to a 6-volt battery is used to ignite a candle. The control apparatus for comparison is the same except that it does not contain the plant (or may contain different types of plants).

flame in an enclosed space will not consume all the oxygen. "An ordinary flame will not burn, as a rule, over about 3 percent of the oxygen from a confined space before it goes out."[34] The presence of residual amount of oxygen has been demonstrated. A piece of lens paper, cotton, or a match head can be ignited by a spark from a spark gap within the bell jar after the candle flame has been extinguished. However, the candle enclosed with a geranium plant under a bell jar, used with a control, can be valuable to show the difference in quantity of oxygen evolved by green plants.

An electric hot wire in the form of an automobile cigarette lighter with a 6-volt storage battery may be used to ignite a candle in this method. Prepare the apparatus shown in Figure 3-20, but do not connect the battery. Use a healthy geranium plant. Place the demonstration on a glass plate so that petroleum jelly can be used to seal the rim of the bell jar to the glass. Set up the control in the same way, but omit the green plant. Cork the opening

in the top of each bell jar. When the demonstrations have been in the sunlight for several days, connect the storage battery in series to the wires hanging outside of the bell jars. This battery supplies current for the cigarette lighter.[35] Then compare the burning time of the candle inside each of the enclosed jars. This demonstration is usually not very successful; but you may be able to turn a failure into a success by assigning a student the investigation of either explaining why it did not work or designing modifications so that hopefully it may.

Growth of organisms as further evidence of photosynthesis

When dilute Lugol's iodine solution is added to a slide of *Chlorella* that has been maintained in bright light, the cells appear much darker than *Chlorella* cells that have been kept in darkness. This difference is especially noticeable when dense cultures are used in preparing slides, which causes a darkening of the entire field.

Both initial growth and subsequent reproduction of cells are evidences of photosynthesis, for without the production of high-energy organic molecules, cells could not produce new protoplasm. The methods for measuring growth are applicable in studies of photosynthesis. (See the development in the chapter on growth, chapter 6.)

If quantities of *Chlorella* are available, students might use uniform weights (grams per liter of suspension) to compare the original weight of the culture with a possible increase in weight after photosynthesis has progressed for several weeks or a month. Or if graduated centrifuge tubes are used, students may centrifuge the media in flasks and measure increase in volume of cells per volume of liquid. Or

---

[34] Ganong, "Erroneous Physiology of Botanical Textbooks."

[35] F. Vaurio in "Photosynthesis Apparatus," *Sci. Educ.* (1938) 22:309, comments on the standing joke about cigarette lighters and suggests a 6-volt transformer, copper wire, and a nichrome wire loop to ignite the candle, or the use of a storage battery or several dry cells.

compare the dry weight of experimentals and controls. (Also refer to pages 466–67.)

Techniques describing the role of digestive enzymes in both plants and animals are on pages 245, 248.

Concept seeking and concept formation through a "block laboratory approach"

Students cannot learn all there is to know through inquiry, interpreted as laboratory research. Nor can they—through laboratory work—develop all knowledges necessary for future life. Inquiry does not mean only laboratory or experiment; it relates to general ways of seeking knowledge through personal effort, through expenditures of energy as a result of individual initiative.

The purpose of this "block" or "resource unit" is to help teachers to develop certain of the pervasive conceptual schemes in biology by means of many investigative approaches. Students need to know science as *process* (investigative approach or inquiry) as well as to evaluate the *product* of science. The products are the conceptual schemes that characterize biology and make it a special discipline. A discipline can become such only if there is an ordering of, or disciplined approach to its understandings (the product), and its efforts to enlarge its understandings (the process). Biology, and science in general, lends itself to this discipline by learning through inquiry—this method of concept seeking and of concept formation.

An *organism*, any organism, provides the focal point in the study of biology. For in its study are embodied all the understandings we have of life and of the nature of living in the biological sense. In the study of a single organism in biology, the discipline emerges in coherence.

For example, the green alga *Chlorella* fulfills the requisites for an experimental organism for studying the biological processes. This autotrophic green plant is easily maintained, its environment can be controlled, it has uniform physiological properties and is nonpathogenic, and it offers opportunity for study as an individual organism or as a population. *Chlorella* has been the tool of the research laboratory; as a result, there is much well-established literature concerning it. For these reasons, *Chlorella* is used as a focus for the activities to be suggested for laboratory and classroom.

Perhaps we need to clarify a bit what we mean by the diverse methods used by scientists in seeking knowledge. Brandwein suggests that we confuse the scientist's "concise report" of his work as given in his paper with the "manner of his work."[36] Conant distinguishes two orders of operations: (1) varied methods by which broad working hypotheses and grand conceptual schemes originate in the minds of individuals; and (2) fairly fixed patterns of testing and confirming these hypotheses.[37]

In short, experimentation by itself does not produce science. Conant continues with his operational definition of science: "As a first approximation, we may say that science emerges from the other progressive activities of man to the extent that new concepts arise from experiments and operations, and the new concepts in turn lead to further experiments and observations."[38]

Brandwein also reminds us:

What we have tried to indicate is that problem solving begins even before the problem is stated; it begins perhaps unconsciously. The specific problem may be clarified when the scientist begins the empirical or the experimental investigative portion of his work; *but a great deal of his work goes on before he states his problem* in a formal way. The scientist solves problems; but this is not to say, we repeat, that he begins his investigation with a stated prob-

---

[36] P. F. Brandwein, "Elements in a Strategy for Teaching Science in Elementary School," The Burton Lecture, in J. Schwab and P. F. Brandwein, *The Teaching of Science* (Inglis and Burton lectures), (Cambridge, MA: Harvard University Press, 1962).

[37] J. Conant, *On Understanding Science* (New Haven, CT: Yale University Press, 1947).

[38] Mentor edition of *On Understanding Science* (New York: New Am. Library, 1951).

lem. It would seem, from our observations of the way scientists work, that the problem filters out of a mixture of observations, flashes of insight ("eurekas"), vague, muddled dissatis-factions, tentative essays into empiricism, read-ing, consultation with others, thinking (con-scious or unconscious), mental scanning, and so forth. Eventually the problem does filter out, and an attempt to state it clearly is made. . . .

. . . We want, for our purposes, to distinguish between the way of the scientist in problem solving (a creative act) and the routine of the usual problem doing of the student taking an established course, in which the solution is foreordained, because there is a rigid schedule (perhaps 40 minutes) and standard equipment. Do we want our students to engage in problem solving or in problem doing? We think both.[39]

A list of some aspects and approaches to inquiry that reveal diversity of methods, yet a unity of purpose, in the efforts of scientists might well include the use of:

— the "educated guess," that is, of hypothesis

— speculation

— the "leap in the dark"

— experimental design

— special laboratory techniques (and invention)

— literature (the meaning of fact)

— criticism (confirmation)

— reports

— theory

Each of the learning situations that follow is created by asking a question that alters the conditions so as to cast doubt on the validity of the conceptual scheme. Whether the investigation is limited to a week or to a year depends clearly on what the individual student or the class as a group does.

[39] Brandwein, Watson, and Blackwood, *A Book of Methods.*

"Preludes to inquiry"—suggested investigations using Chlorella as the experimental organism (for individuals or groups)

1. *Chlorella* can be seeded into an agar medium so that colonies or clones form. Is there a limit to the size of a colony of *Chlorella?* How do these col-onies compare with colonies of *Rhizo-pus,* or of *Saccharomyces?*

2. If light is necessary for photosynthe-sis, will twice as much light double the rate of photosynthesis?

   This inquiry is meant to illustrate different levels of investigation. It is not necessary to go to the laboratory to do an "experiment" at all times. Library research is also a valid method of a scientist, leading to ex-perimental design.

   Students can inquire into what has been learned about the nature of light. They may devise an investigation to duplicate the kinds of thinking that they have read about in references. Here they will learn, no doubt, that there is all the difference in the world between reading the concise report of a scientist and going through exten-sive experimentation with its endless replication to ensure statistically valid interpretation of experiments and controls before one can reach a tenta-tive conclusion.

3. In using the "historical" approach to an inquiry, a teacher may describe the elegant experiment that Engelmann[40] designed to show the production of oxygen by plant cells in certain re-gions of a microspectrum. Or students may read Engelmann's paper.[41]

   What was known about photo-synthesis at the time? What was the

[40] Described in W. P. Pfeffer, *The Physiology of Plants,* 2nd ed., Vol. 1 (Oxford: Clarendon Press [Oxford Univ. Press], 1900).
[41] Reprinted in Gabriel and Fogel, *Great Experi-ments in Biology.*

climate of thinking, the grand conceptual scheme used to explain how green plants made food?

Special problems for investigation often arise from focus on such a productive paper. For example, did Engelmann influence the thinking of van Niel, or of Hill? What would the next steps be after Engelmann's work? How did Hill's experiments with isolated chloroplasts affect van Niel's hypothesis about the origin of oxygen in photosynthesis? How did Arnon's hypotheses modify the whole conceptual scheme of photosynthesis?

4. Perhaps begin by asking what quantity of carbon dioxide is in the air. If 0.04 percent is present, what effect will 1.0 percent have on the rate of photosynthesis?

5. What part does *Chlorella* play in the economy of a pond? (If *Chlorella* did not exist, what other organism might take its place in the pond?)

6. What is the spectrum necessary for development of chlorophyll in *Chlorella*? Is the same spectrum necessary for photosynthesis?

Note how one aspect of inquiry—hypothesis—arises out of this problem posed to students. Many hypotheses are offered in class: (a) the wavelength has no effect on formation of chlorophyll; (b) certain wavelengths are more effective than others in stimulating the formation of chlorophyll; (c) photosynthesis is independent of wavelength of light; (d) chlorophyll formation and photosynthesis are related processes; (e) formation of chlorophyll and the process of photosynthesis are independent; and (f) certain wavelengths of light are more effective than others in photosynthesis.

7. In the lowest zone of the littoral region are found red algae; green algae live at the highest region; and in the middle zone are the brown algae.

Engelmann offered a theoretical explanation for the chromatic adaptation of algae.[42] Yellow light predominates in sunlight, but blue-green light is transmitted to the greatest depth in clear seawater. Pigmentation of algae is complementary to quality of the light of the position which the algae occupy.

Here students have an opportunity to test a theory. What predictions can be tested?

8. A mutation was found in *Chlorella* that resulted in its inability to synthesize sugar although it did have chlorophyll. How could such an organism be used in designing an experiment to test some factors in photosynthesis?

9. Other mutations—cells that lack chlorophyll—have been found in *Chlorella*. How might these organisms be used in the study of the light and dark reactions in photosynthesis? How could students demonstrate that the organism is the product of heredity and environment?

10. A student observed that when the growing tip of *Anacharis* was introduced into a microaquarium containing *Chlorella*, the algae increased in number to a greater extent that did controls lacking the tip of *Anacharis*. Methods for measuring growth of microorganisms are described in chapter 6.

Clue: Growing tips are a rich source of auxins. This "prelude" offers a student an opportunity in hypothesis formation, and is fruitful of predictions that are susceptible to testing in the laboratory.

11. Is the usual biological succession of organisms in a micro-lake (microaquarium) affected by altering the light intensity? Could such alterations (if any) affect the survival of *Chlorella*? Or of other organisms in the micro-lake?

12. If some substance altered the rate of growth of *Chlorella* (or any one auto-

[42] From Pfeffer, *The Physiology of Plants.*

troph) in a pond, would there be effects on the other organisms—producers, consumers, or decomposers—in the pond?

Students might devise experiments using gibberellic acid, thiamine chloride, kinetin, or auxins.

13. What deficiency symptoms occur in *Chlorella* maintained in a culture lacking phosphates? Or lacking nitrates? Or lacking sulfates? Students may prepare Knop's solution (or others) and vary the chemical ingredients. Is the concentration of chlorophyll in cells affected? Is the rate of photosynthesis affected?

14. What is the effect of a change in the concentration of metallic ions on *Chlorella*? How do the ions of magnesium affect the concentration of chlorophyll in *Chlorella*?

Clue: It has been reported that cell division stops in *Chlorella* in a magnesium-deficient medium, with the subsequent formation of large cells.[43] In this case, the method of measuring optical density (turbidity, p. 464) is misleading since there is no growth. It has further been shown that manganese, zinc, copper, and cobalt inhibit growth and synthesis of chlorophyll in *Euglena,* while magnesium, molybdenum, and boron stimulate synthesis of chlorophyll. Do these metallic ions have similar effects on *Chlorella*?

Maintaining uniform genetic material
A technique used in bacteriological studies may be used for refining the design of some experiments with *Chlorella*. What kinds of problems could be posed, and solutions sought, if one knew a simple method of maintaining *Chlorella* colonies of the same heredity, that is, uniform genetic material?

Lederberg developed a simplified technique, replica plating, for continuing colonies of bacteria of the same heredity—

[43] Finkle and D. Appleman, "The Effect of Magnesium Concentration on Growth of *Chlorella*," *Plant Physiol.* (1953) 28:664–73.

colony cloning. A piece of velour is used to cover a cylinder of wood or other material and held securely in place. The diameter of the cylinder and velour cover is that of a Petri dish.

When colonies are grown on a Petri dish, each colony theoretically has arisen from one organism. All the cells in that colony must have the same heredity (barring mutations); such a colony is a clone.

When Lederberg and his assistants inverted a Petri dish containing colonies over the velour and pressed down lightly, a pattern of the clonal growth was formed on the velour. Now a series of fresh, sterile Petri dishes of agar could be inverted and touched to the "pattern" on the velour. When incubated, new clones formed at the points of contact with the original colonies. If several dishes were placed on top of each other, the same pattern of distribution of clones could be matched. What problems relating to *Chlorella* might now be *testable* with this method?

(Other "preludes" to investigations are offered in chapter 8, relating to "ecological studies" in which photosynthesis is the continuing thread.)

In several of the investigations presented here, emphasis is placed on the role of the altered environment in the *selection* of organisms, that is, the evolution of organisms. It becomes apparent that the organism is the product of its heredity, its interrelations with other organisms, and its environment. Adaptations of an organism may be explored, and the possibility of genetic continuity of such adaptations may be tested. Is it possible, for example, to grow *Chlorella* in highly alkaline medium? Or can *Chlorella* become adapted to an environmental change in salinity, or pH, and pass this on to successive generations?

Finally, succession of organisms leads to the problems inherent in the evolution of communities of organisms.

Developing skills in understanding the reports of scientists

Along with the skills of observing, demonstrating, and planning experiments, students need practice in another skill, the ability to read and interpret the writings of scientists. Many of the experiments or demonstrations that we have described here are a part of the history of biology and the fascinating story of how green plants make their food supply. In some classes, you may want students to read small sections of the original papers of some of the historic bench marks of the workers in this field: van Helmont, Ingenhousz, Priestley, Sachs, or Blackman, among the hundreds who have contributed in the area. Readings might also include the classic work of Calvin, Arnon, Fox, Urey, Wald, and Oparin.

In fact, some of the experiments may be elicited "afresh" from students as they think through the need for large numbers of experimental plants and for controls. In this way, students will come to appreciate the design of experiments.

At times, you may want to use a reading taken from a science journal or textbook, then prepare suitable questions based on the reading. This gives students practice in interpretation and reasoning.

Such a reading may be used as an introduction to a new topic; it might raise questions or give data related to the classwork; or it might introduce the need for experimental design in undertaking laboratory work. This is also one of the many ways a teacher can identify students with a high level of ability and interest in science. Such students read these passages with greater facility, appreciation, and understanding. In some classes, readings may be included on an examination as a test for understandings and application in a new view.

Here, for example, is an historical reading that one teacher assigned to students: Jan van Helmont's statement of his classic seventeenth-century experiment. Following the reading is a list of questions based on it that the students were to answer.

I took an earthen vessel, in which I put 200 pounds of earth that had been dried in a furnace, which I moistened with rainwater, and I implanted therein the trunk or stem of a willow tree, weighing 5 pounds, and at length, 5 years being finished, the tree sprung from thence did weigh 169 pounds and about 3 ounces. When there was need, I always moistened the earthen vessel with rainwater or distilled water, and the vessel was large and implanted in the earth. Lest the dust that flew about should be co-mingled with the earth, I covered the lip or mouth of the vessel with an iron plate covered with tin and easily passable with many holes. I computed not the weight of the leaves that fell off in the four autumns. At length, I again dried the earth of the vessel, and there was found the same 200 pounds, wanting about 2 ounces. Therefore 164 pounds of wood, bark and roots arose out of water only.[44]

Students were asked to answer the following questions:

1. How long did Helmont continue his experiment?

2. What was he trying to find out?

3. What was the gain in weight of this 5-pound willow stem?

4. How did van Helmont account for the increase in weight?

5. What would he have found if he had included the weight of the leaves over four years?

6. What kind of control would you suggest to find out whether new plant tissue grows from water alone?

7. How do we explain the increase in weight of the plant nowadays?

You may find readings for similar purposes in scientific journals such as *Science*, a publication of the American Association for the Advancement of Science (1515 Massachusetts Avenue, N.W., Washington, D.C.); *American Scientist* (Society of Sigma Xi, 345 Whitney Avenue, New Haven, Conn.); *Scientific American* (415 Madison Avenue, New York, New York 10017); offprints of papers from *Scientific*

---

[44] From L. Nash, ed., *Plants and the Atmosphere*, "Harvard Case Histories in Experimental Science," 5 (Cambridge, MA: Harvard Univ. Press, 1952).

American are available from W. H. Freeman, Market Street, San Francisco, California; and *The Sciences*, New York Academy of Science, 2 East 63 Street, New York, New York 10021.

The two-volume set of the Harvard Case Histories in Experimental Science is a valuable source of readings in the history of science. Also, you may want to look into the readings edited by M. Gabriel and S. Fogel in *Great Experiments in Biology*; J. Peters (ed.), *Classic Papers in Genetics*, and J. Knoblock, *Readings in Biological Science*. These are rich sources of reading materials (see *Book Shelf* at the end of the chapter).

Tests of reasoning

You may also want to plan tests based on observation and reasoning rather than upon strict recall. By way of example, three sets of questions follow:

INVESTIGATION 1:   A girl takes 5 test tubes containing brom thymol blue and puts elodea plants into three of them (Fig. 3-21). All the tubes are put in the dark. The next day she finds that the tubes containing the elodea (tubes 3, 4, and 5) had turned to yellow. (Brom thymol blue changes to yellow when enough carbon dioxide is added.)

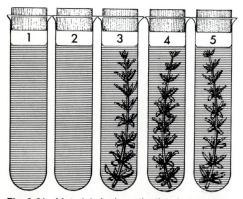

Fig. 3-21   Materials for Investigation 1. (Drawing by Marion A. Cox. From *Experiences in Biology*, by Evelyn Morholt and Ella Thea Smith, copyright, 1954, by Harcourt, Brace & World, Inc., and reproduced with permission of the publisher and the artist.)

1. What is the *best* explanation of the girl's results?

   a. Brom thymol blue turns yellow in the dark.

   b. The plants gave off oxygen in the dark.

   c. The plants gave off carbon dioxide in the dark.

   d. There is insufficient evidence for any explanation.

2. Why were tubes 1 and 2 necessary or not necessary?

   a. They were not necessary because we know the action of brom thymol blue.

   b. They were needed to match colors.

   c. They were necessary to show whether any other factors change the color.

   d. None of these.

3. What should the girl do to change the brom thymol yellow in tubes 3, 4, and 5 back to blue again?

   a. Put the test tubes in the light.

   b. Put a goldfish into the tubes.

   c. Blow into the brom thymol yellow.

   d. None of these.

4. What process, going on in the green plants in the dark, accounts for the color change?

   a. assimilation

   b. oxidation

   c. photosynthesis

   d. transpiration[45]

INVESTIGATION 2:   In the closed system of flasks and tubing shown in Fig. 3-19, there is a trapped air bubble in the tubing.

1. If the green algae (*Chlorella*) in flask (b) give off a gas and there is no change in flask (a), the drop of water in the tubing will move along the tubing due to which the following?

   a. increase in volume of gases in flask (b)

   b. decrease in volume of gases in flask (a)

   c. decrease in volume of gases in flask (b)

   d. increase in temperature of gases in flask (b)

2. In which direction will the trapped air bubble move along the tubing if the apparatus is in light?

   a. toward flask (b)

[45] Investigation 1 adapted from *Experiences in Biology*, new ed., by Evelyn Morholt, copyright, 1954, © 1960, by Harcourt Brace Jovanovich and reprinted with their permission.

b. away from flask (b)
c. oscillate between the flasks and take a central position
d. unpredictable

3. What is the gas that is liberated if the apparatus is in light?

a. carbon dioxide
b. nitrogen
c. oxygen
d. water vapor

4. If the apparatus were placed in darkness, what would the trapped air bubble probably do?

a. remain stationary
b. move toward flask (b)
c. move toward flask (a)
d. contract due to a change in pressure

5. Phenolphthalein is an indicator that turns from rose to pale pink to colorless when the quantity of carbon dioxide increases in the medium.

If equal amounts of phenolphthalein were added to the flasks (a) and (b), after several hours of darkness we would likely find that

a. flask (b) would be colorless
b. flask (b) would be rose
c. flask (a) would be rose
d. flasks (a) and (b) would be colorless

6. The main on-going process in algae maintained in the dark is

a. transpiration
b. oxidation
c. photosynthesis
d. growth

7. The color of the indicator would be markedly changed if the apparatus were

a. maintained at a higher temperature
b. kept agitated (stirred)
c. placed in the light
d. kept in darkness for several hours longer

8. Suppose the cells in flask (b) were yeast cells rather than *Chlorella*. What would you predict in the following situation?

a) In darkness, there would be a greater liberation of

a. carbon dioxide
b. nitrogen
c. oxygen
d. water vapor

b) This would be due to the fact that

a. photosynthesis goes on faster than respiration

b. respiration is the on-going process
c. an appreciable quantity of alcohol is formed
d. respiration goes on faster than photosynthesis

9. Were the culture of yeast cells placed in light there would be an increase in the quantity of

a. carbon dioxide
b. nitrogen
c. oxygen
d. water vapor

10. Yeast cells are considered

a. symbionts
b. autotrophs
c. heterotrophs
d. free living

11. Algae like *Chlorella* are considered

a. symbionts
b. autotrophs
c. heterotrophs
d. free living

INVESTIGATION 3:   Assume a welled microscope slide with pond water. Consider this pond water as a microaquarium—a micro-lake. In this micro-lake are amoebae, motile *Euglena* that are green and have flagella, green algae called *Chlorella*, and two species of *Paramecium*, some many-celled animals called rotifers, and some bacteria.

1. The primary food source, the producers, are

a. rotifers and *Chlorella*
b. *Euglena* and *Chlorella*
c. *Paramecium* and *Amoeba*
d. *Euglena* and bacteria

2. The usual role of the bacteria in this micro-lake is to act as

a. producers
b. decomposers
c. symbionts
d. parasites

3. Paramecia and rotifers show a preference for *Chlorella* as food. All things being equal, the limiting factor which determines the increase in population of paramecia in this micro-lake is the number of

a. paramecia
b. *Chlorella*
c. *Euglena*
d. amoebae

4. If two species of paramecia live in this micro-lake, a special problem may arise. This problem is competition

    a. for food between the two species of paramecia and the amoebae

    b. for food between the two species of paramecia and rotifers

    c. between the two species of paramecia

    d. between paramecia and bacteria

5. If *Euglena* should die off

    a. populations of both species of paramecia would also die off

    b. neither population of paramecia would die off

    c. population of one species of paramecia would die off

    d. population of *Chlorella* would die off

6. If the micro-lake is kept in sunlight or near a light source, many bubbles of gas become visible. The gas is probably rich in

    a. ammonia

    b. carbon dioxide

    c. hydrogen

    d. oxygen

7. Which organisms are liberating this gas in light?

    a. rotifers

    b. bacteria

    c. *Chlorella*

    d. amoebae

8. Of the following pairs of processes, the one which is a pair of *opposite* processes is

    a. photosynthesis-growth

    b. respiration-photosynthesis

    c. ingestion-digestion

    d. oxidation-locomotion

9. If red cellophane were placed over the micro-lake in the light, more bubbles of gas would occur than if green cellophane were used. This is because green algae

    a. absorb light of red wavelengths

    b. absorb light at the green end of the spectrum

    c. reflect red light

    d. carry on respiration faster in green light

10. The flagellates, *Euglena*, become bleached when subjected to streptomycin. The structures which are disrupted are the

    a. mitochondria

    b. chloroplasts

    c. cell membrane

    d. vacuoles

11. These bleached *Euglena* can live in the darkness (or in light) provided the following is added to the micro-lake:

    a. starch

    b. glucose

    c. lipids

    d. minerals

12. Of the following pairs of terms, the pair that may best be applied to *Euglena* is

    a. heterotroph-autotroph

    b. heterotroph-symbiont

    c. autotroph-parasite

    d. autotroph-symbiont

13. The fact that populations of organisms in a micro-lake may change over a few weeks is due to the principle called

    a. adaptation

    b. food chain

    c. succession

    d. growth phases

Consumers among organisms

This next section is concerned with ingestion and digestion in a variety of heterotrophs—both invertebrate and vertebrate—that require "ready-made" molecules of nutrients. In the context of enzymes, digestion in plants is also developed.

Ingestion

Ingestion by protozoa, by small freshwater invertebrates (including insects), and by some amphibia may be observed firsthand in the laboratory. These organisms are easily maintained in the laboratory and thus made available throughout the year. The cultivation and maintenance of these organisms are described in chapter 9. Where facilities favor support of a marine aquarium (p. 674), ingestion among small echinoderms, squids and other mollusks, crayfish, also may be studied. Many suggestions for investigations are offered throughout the context of this chapter.

Protozoa

Paramecium   First ring a slide with methyl cellulose. Place a drop of a concentrated culture of *Paramecium* and a small drop of a thick suspension of *Chlorella* in the center of the ring of cellulose; apply a coverslip. Under high power, watch the formation of green food vacuoles, and especially, the food mass in the oral groove. Or add a small pinch of carmine powder to a thick culture of *Paramecium* (or other ciliate). Prepare wet mounts of the mixture (pp. 138, 139, Figs. 1-5, 2-9) and watch how cilia in the oral groove create a current of water. Note also the formation of a food ball in the oral groove and the passage of the dark red "food" mass into the cytoplasm, where a food vacuole is formed. (You may substitute a drop of India ink for the carmine powder. In this case, the food vacuoles will appear black.)

Observe digestion within the food vacuoles using indicators. Stain a few drops of milk or cream with a few grains of Congo red (indicators, p. 158), then add this to an equal volume of a culture of *Paramecium* or *Stentor* or other ciliates in a Syracuse dish. Prepare wet mounts and examine the ingestion of red butterfat globules as food vacuoles. In an hour, observe the red fat globules in the food vacuoles become bluish as acid is secreted into the vacuoles, and finally change to red as digestion continues with less acid secretions.[46] Congo red is blue at a pH of 3 and orange-red at about pH 5.

A boiled suspension of Congo red and yeast may be preferred. Add a pinch of Congo red powder to a thick suspension of yeast that has been mixed with water. Bring to a gentle boil for 5 min. After the suspension has cooled, use a toothpick to transfer a small quantity to a drop of pond water containing *Paramecium*. Watch cyclosis of food vacuoles in the cytoplasm. Note the change to bluish-green indicating the more acidic contents of the food vacuoles, then back to an orange-red in about an hour (Fig. 3-22).

Or add a drop of *Paramecium* to a drop of dilute neutral red (0.02 percent) on a slide and seal the coverslip with petroleum jelly

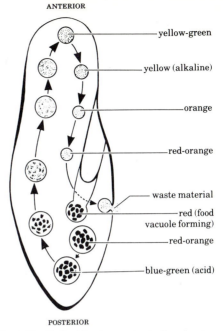

ANTERIOR

yellow-green

yellow (alkaline)

orange

red-orange

waste material

red (food vacuole forming)

red-orange

blue-green (acid)

POSTERIOR

Fig. 3-22 Path of food vacuoles in *Paramecium caudatum*. Trace the change in pH within food vacuoles containing yeast cells stained with Congo red. Digestion is first acid, then alkaline. (Reprinted and adapted with permission of Macmillan Publishing Company from *College Zoology*, 10th ed., by Richard A. Boolootian and Karl A. Stiles. Copyright © 1981 by Macmillan Publishing Co., Inc.)

for a prolonged examination. At pH 6 and below, neutral red is deep red, at pH 7 the color is rose, and at pH 8 the indicator turns yellow.

Kudo also recommends the use of indicators to render food vacuoles visible and to test their pH.[47]

Dissolve any one of the following indicators in 100 ml of distilled water: 50 mg of neutral red, 100 mg of phenol red, 150 mg of litmus, or 50 mg of Congo red. To 1 drop of the culture on a slide add 1 drop of the indicator solution (other indicators, chapter 2, pages 138, 156).

An interesting but unsolved problem may be examined by some students as an investigation. Do protozoa continue to

[46] E. Lund, "The Feeding Mechanisms of Various Ciliated Protozoa," *J. Morph* (1941) 69:563.

[47] R. Kudo, *Protozoology*, 6th ed. (Springfield, IL: Charles Thomas, 1977).

take in carmine particles which cannot be digested, or do they "learn" to select their food types? One method of attack might be to count the number of food vacuoles formed during measured time intervals. Students may think of other creative approaches. We have found this a useful investigation.

Pelomyxa Prepare several slides of thick cultures (see chapters 1 and 2) of *Stentor, Paramecium, Blepharisma,* or *Chilomonas* (Figs. 1-7, 1-8, 2-9). To each of these slides add the large amoeba *Pelomyxa,* sometimes called *Chaos chaos* (Fig. 3-23). These may be purchased from a biological supply house (Appendix). Watch how a food cup is formed as the amoeba's pseudopodia close around the prey. Food vacuoles of different colors form, depending upon the food source (red with *Blepharisma,* blue-green with *Stentor*). At times you may see the captured prey moving within the vacuole. Seal coverslips with petroleum jelly to prevent evaporation of the water. Study the change in size of the food vacuoles of the carnivorous amoebae over a period of several hours.

Blepharisma Some *Blepharisma* grow to be "giants," are cannibalistic, and contain deep red food vacuoles of ingested, smaller *Blepharisma.* Mount several drops of

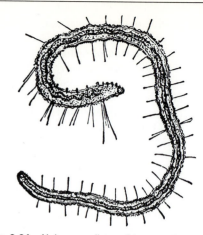

Fig. 3-24 *Nais,* a small annelid worm. (Reprinted with permission of The Macmillan Co., New York, from *Invertebrate Zoology* by R. W. Hegner. Copyright 1933 by The Macmillan Co., renewed 1961 by Jane Z. Hegner; after Leunis.)

Blepharisma culture for microscopic examination.

Trumpet-shaped *Stentor* (Fig. 1-8), mounted in a rich culture of *Blepharisma,* will soon show pink food vacuoles within the blue-green body.

Ingestion: some invertebrates

For maintaining these invertebrates in the laboratory, see chapter 9.

Mount the microscopic annelid worm *Tubifex* (Fig. 1-44) or *Nais* (Fig. 3-24), or small crustaceans such as *Daphnia* or *Artemia* in a thick suspension of *Chlorella* or Congo-red stained yeast (p. 237) for examination under the microscope. Watch green algae accumulate in the food tube; or watch Congo-red stained yeast cells change color in the food tube.

The action of tentacles and stinging organs, such as those in *Hydra* (Figs. 1-21, 1-23), makes an interesting study (p. 152).

Examine digestion in living hydras. Since hydra is an almost transparent (two-layered) organism, observations can easily be made by coloring brine shrimp or their larvae, and feeding them to starving hydras. Add some washed *Artemia* (brine shrimp) or larvae to a suspension of carmine powder; then feed to hydras in a

Fig. 3-23 *Pelomyxa,* a large amoeba. Contrast size with that of slipper-shaped *Paramecium.* (Photo courtesy Carolina Biological Supply Company.)

watch glass of water. Notice the movement of tentacles toward the raised hypostome and the ingestion of the captured shrimp. Next transfer the hydra to a shallow depression slide of water and apply a coverslip. Watch for the colored food particles ingested by the gastrodermal cells that line the coelenteron. Maintain the specimens at 17° to 21° C (63° to 70° F). (For hatching brine shrimp eggs, see page 673.)

Show the role of glutathione in the feeding response in hydras as described by Loomis.[48] Body fluids of prey attacked by nematocysts of hydras are said to contain reduced glutathione, which stimulates a feeding response in *Hydra*. According to Loomis, hydras reject dead food in which reduced glutathione becomes oxidized.

Since glutathione is also found in yeast cells, add dry yeast to some hydras in a watch glass and observe the feeding response. Or extract reduced glutathione from yeast cells by grinding dry yeast with a little water and then centrifuging and decanting the liquid portion which contains the glutathione. Now dip a bit of dead food or a minute bit of filter paper into the chemical; place in a watch glass or depression slide next to hydras; examine the ingestion response under the microscope.

Planaria worms For a good examination of the branches of the gastrovascular system of planaria worms, feed light-colored *Dugesia* (planaria) carmine powder mixed in a liver paste in a small watch glass of pond water. Within 24 hr, observe the colored digestive system. Prepare a "coverslip sandwich"[49]; place the worm on one coverslip, ring this with petroleum jelly, and apply a second coverslip. With forceps transfer this "sandwich" to a microscope slide. Examine dorsal and ventral views by turning the sealed coverslips on the slide.

[48] W. F. Loomis, "Glutathione Control of the Specific Feeding Reactions of *Hydra*," *Ann N.Y. Acad. Sci.* (1955) 62:209–28.
[49] I. Sherman and V. Sherman, *The Invertebrates: Function and Form: A Laboratory Guide.* 2nd ed. (New York: Macmillan, 1976).

(Also use Figure 1-31 as a guide to examine the ganglia and nerve cords and other organ systems in this "planaria sandwich.")

You may also want to place some planaria (unfed for several days) in Syracuse dishes along with such annelid worms as *Dero* or *Aulophorus* or with *Artemia* (brine shrimp). Observe the action of the proboscis of planaria during ingestion under a dissecting microscope or hand lens (see also page 23, Fig. 1-31).

Ingestion and digestion in Daphnia Feed *Daphnia* young cultures of *Chlorella* and observe the green color of the gut as food is consumed. (Students may distinguish female *Daphnia* by the curved rear gut, while in smaller males the gut is a straight tube.) Digestion of food goes on, and insoluble matter leaves the gut within a half hour at a temperature of 20° C (68° F).

Have students observe filter feeding in *Daphnia* as well as changes in the pH of the gut contents during digestion. Immobilize a few *Daphnia* in dabs of petroleum jelly in the well of a depression slide so that some lie on their back, and others on their side. Then observe the rapid action of the appendages between the carapace as well as the change in pH of food. Dip a toothpick into a boiled, Congo-red–yeast mix (preparation, p. 237) and add a trace to the slide. Observe the color of the gut and note the gradual change in pH; bluish in acid pH about pH 3 to 4, and reddish-orange as the pH becomes less acid.

Ingestion: some vertebrates

Demonstrate the rapid, lashing tongue movements of frogs, toads, and salamanders by placing one of these animals in a small container. Then empty a bottle of fruit flies (or earthworms) into the container. Prepare a similar demonstration using chameleons, which also feed upon fruit flies.

In some classes, this may be the time to introduce broad notions of modes of nutrition among consumers, or heterotrophs, in ecosystems. Refer to pages 617–37 for demonstrations of holozoic nutrition, saprophytism, and types of mutualism. Students will probably also ask about insectivorous plants (Fig. 8-56 and p. 635).

Digestion

The food tube   Use indicators to show the fate of food through the digestive tube in microscopic many-celled organisms.

Show changes in pH, as revealed by changes in color of the indicator as food moves along the food tube. Prepare wet mounts of rotifers, *Tubifex*, *Dero*, *Nais*, or the crustacean *Daphnia* (culture methods, chapter 9). Methods using *Chlorella* and Congo-red yeast cells have been described (p. 237).

When *Chlorella* is added to a culture of almost transparent, many-celled organisms such as *Daphnia* or rotifers, the digestive tract becomes conspicuously green. Or add a drop of a Congo-red yeast suspension to a drop of thick culture of rotifers, *Cyclops*, *Daphnia*, or some other microorganism on a slide. Seal with petroleum jelly and examine for the next 15 to 30 min. Watch the change in the food tube as the color changes from red to blue (blue at pH 3.0 and orange-red at pH 5.0).

A *more graded series* of changes in acidity may be traced with a very dilute solution of neutral red (0.02 percent). Add a drop of a thick suspension of microorganisms to a drop of the indicator and seal with petroleum jelly. At pH 9 the indicator is yellow; at pH 8 orange; rose at pH 7; deeper red below pH 6.

What change in color is visible as the food mass moves through the pharynx, stomach, and intestine of the microscopic multicellular organisms?

Also demonstrate the path of food along the digestive tract of frogs. First, pith to destroy the brain and the spinal cord (Fig. 5-12). Then pin down the frog on a dissection board and remove the skin and muscular layer to expose the digestive tube (dissection, chapter 1). Keep the organs bathed in frog Ringer's solution (p. 155).

Note the slow, rhythmic contractions along the length of the intestine and rectum. These contractions will be more apparent if the frog has been recently fed.

You may also want to demonstrate the ciliary action along the sides of the jaw and roof of the mouth of the frog. Use the same pithed frog; dissect away the lower jaw and the floor of the mouth. Then remove the viscera, leaving as much of the esophagus as possible. Sprinkle carmine powder, fine particles of cork or filings of lead or iron on the anterior part of the roof of the mouth. Watch how the particles are carried along the roof and through the esophagus in a current created by ciliated epithelial cells. The particles reappear in the cut part of the esophagus and move into the coelom, since the stomach has been cut away. (Keep the tissue moist throughout.)

Also prepare slides by mounting bits of ciliated epithelial tissue from the roof of the mouth or the sides of the frog's jaw in Ringer's solution (Fig. 2-21). Under high power look for the rhythmic ciliary motion. You may, if you wish, slow down the motion with ice. Place the slide on an ice cube. Seal the coverslip with petroleum jelly if you plan to study the motion over some time. After a study of cilia, apply a vital stain to bring out details (pp. 136, 150).

Compare the adaptive devices for food trapping among several organisms—small invertebrates as well as vertebrates—through dissection using illustrations in chapter 1 as guides.[50]

The need for digestion: diffusion through a membrane

We have studied ingestion among several heterotrophs and compared their digestive systems. What happens to the food

---

[50] Also refer to textbooks such as G. Stephens and B. North, *Biology* (New York: Wiley, 1974); C. Villee, *Biology*, 7th ed. (Philadelphia: Saunders, 1977); R. Boolootian and R. Stiles, *College Zoology*, 10th ed. (London: Macmillan, 1981); C. Hickman, *Biology of the Invertebrates*, 2nd ed. (St. Louis, MO: Mosby, 1973).

they consume? Why is this hydrolysis process—digestion—necessary? Food material must be made soluble; that is, it must be in a form that can pass through membranes. The process of diffusion may be shown in many ways. You may have used one or another of the following methods to show diffusion of molecules from place to place.[51]

Diffusion Drop a small piece of copper sulfate or several potassium permanganate crystals into water and leave them undisturbed for several days. To do so, insert a few dry crystals into a section of glass tubing. Hold a finger over the top of the tubing as you insert the other end into water. When the tubing is at the bottom of the container of water, release the finger; the crystals will fall to the bottom. Each day observe the molecules of copper sulfate as they move from the region of greatest concentration and become distributed throughout the water solvent.

To demonstrate diffusion of gases, release illuminating gas or the volatile substances in perfume, oil of wintergreen, and so on, in a room. Note that molecules move generally from the point of greatest concentration to a region of lowest concentration. This may also be demonstrated in liquids by carefully pouring red ink or a dye along one side of a container into water.

Another striking demonstration of diffusion is the movement of a gas, such as ammonia, in another gas (air). First, show the effect of alkaline ammonium hydroxide on an indicator. For example, in a test tube, add a drop of ammonium hydroxide to a few ml of phenolphthalein. Now wet a circle of filter paper with phenolphthalein and insert it into the bottom of a large test tube. Now invert the test tube over a bottle of ammonium hydroxide. Students should be ready to explain the rapid change of color of the filter paper. (You may also want to use litmus paper.)

[51] Refer to N. Unwin and R. Henderson, "The Structure of Proteins in Biological Membranes," *Sci. Am.* (February, 1984).

phenolphthalein in water

NH$_4$OH

Fig. 3-25 Apparatus to show diffusion of a gas through a membrane. A twirling red stream is visible in the test tube as molecules of ammonia evaporate and diffuse through the membrane.

Diffusion through a membrane In another impressive demonstration, show how molecules diffuse through a membrane. In large test tubes, or cylinders of water, add a few drops of a 1 percent solution of phenolphthalein in alcohol (preparation, p. 159). Cover the mouth of the tube with a dialyzing sheet, or a wet goldbeater's membrane or wet cellophane (do not use the kind that covers cigarette packs, since an additional protective agent is added). Fasten with a rubber band and invert over a bottle of ammonium hydroxide (Fig. 3-25). Since phenolphthalein turns deep red in an alkaline medium (ammonia, in this case), a stream of red molecules in motion will be visible. Watch a red stream twirl rapidly up through the test tube.

Diffusion of iodine through a membrane Pour dilute Lugol's iodine solution (p. 176) into a large test tube; cover with a wet goldbeater's membrane or use dialyz-

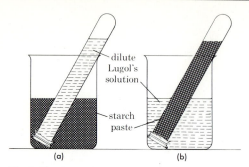

Fig. 3-26 Apparatus to demonstrate diffusion through a semipermeable membrane. Molecules of Lugol's iodine solution pass through the membrane, whereas starch does not. The characteristic blue-black color appears in the starch suspension but not in the Lugol's solution in (a) the beaker and (b) the test tube.

ing sheeting and secure with rubber bands. Invert the test tube into a beaker containing a 1 percent starch paste (Fig. 3-26). Then prepare another test tube, but this time, pour starch paste into the test tube and fasten the wet membrane over the mouth of the tube. Invert this test tube into a beaker half full of dilute Lugol's iodine solution. Note that molecules of the iodine solution pass through a membrane in either direction, whereas molecules of starch do not diffuse. Starch is insoluble; it needs to be digested in order to diffuse through membranes (Fig. 3-27).

Diffusion of different sized molecules through a membrane Fill several large test tubes with glucose solution, molasses (dilute), corn syrup, or honey. Into several other test tubes pour starch paste (1 percent). Cover each test tube with a wet goldbeater's membrane or dialysis sheeting. Fasten with a rubber band and invert each tube into an individual beaker containing a *small* volume of water. Within 15 min, test for starch by adding dilute Lugol's solution to the water containing the inverted tubes of starch paste. Then test the water in the other beakers for glucose with Benedict's solution (p. 176).

It may be desirable at this time to prepare additional test tubes containing

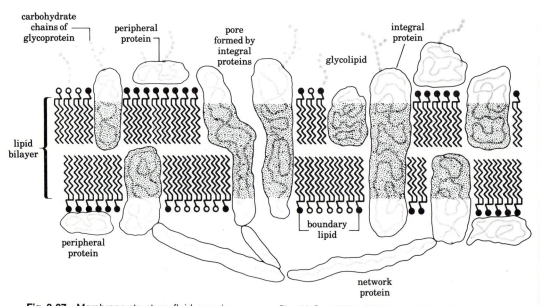

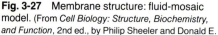

Fig. 3-27 Membrane structure: fluid-mosaic model. (From *Cell Biology: Structure, Biochemistry, and Function*, 2nd ed., by Philip Sheeler and Donald E.

Bianchi. Copyright © 1980 by John Wiley & Sons. Reprinted by permission of John Wiley & Sons, Inc.)

starch paste to which saliva has been added. Demonstrate with dilute Lugol's solution that, while starch has not diffused, some simple sugar has passed through the membrane (test with Benedict's solution). Where did the simple sugar come from? Both the starch paste and the saliva should be tested for simple sugar before saliva is added and the test tube prepared.)

Solutions and colloids   Does a colloidal solution like raw egg white (albumin) diffuse through a membrane? Do true solutions diffuse? Instead of test tubes (as used above) you may prefer to partially fill 10 cm lengths of dialysis tubing. Fasten the tubing at one end with string. Use a funnel to pour the following solutions into the dialysis tubing: glucose solution (or dilute Karo syrup), sodium chloride solution, and raw egg white. Now seal this end with string. Rinse the outside of the tubing and immerse in a beaker with a small amount of water. You may want to fasten each end of the tubing to a tongue depressor and suspend into the water. In ½ hr, test samples of the water for possible diffusion of molecules through the membrane. Test a 5 ml sample for NaCl by adding a drop of silver nitrate. A positive test gives a white silver chloride precipitate.

Use Benedict's solution to test for diffusion of glucose molecules. To a second 5 ml sample of the water add an equal volume of Benedict's and heat to boiling. Test for albumin by adding a few drops of nitric acid to a 5 ml sample of water. (Nitric acid coagulates albumin.) Has the colloid diffused? What is meant by a semipermeable membrane?

Particle size: solutions and colloids Compare the particle size of a substance in a true solution with that of a colloidal solution; recall that living matter is mainly colloidal. Into separate beakers place a NaCl solution; egg albumin and water (1:1); distilled water. Now shine a strong concentrated light beam from a lamp, such as a laboratory lamp, through the substances in each beaker. Observe the Tyndall

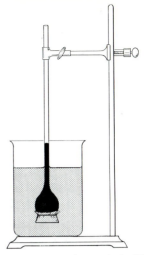

Fig. 3-28   Apparatus to demonstrate diffusion through a membrane. Water in the beaker diffuses rapidly through the membrane causing the level of molasses in the thistle tube to rise.

effect; the strong beam of light is clearly visible due to the reflection of the light from the surface of the moving colloidal particles. (The random movement of colloidal particles is known as Brownian movement.) How does this compare with a beam of light shining through a true solution?

Osmosis and osmotic pressure   Fill the bulb of a thistle tube with heavy molasses while holding the finger over the tube opening. (Instead of covering the tube opening with a finger, some teachers connect a short piece of rubber tubing with a clamp.) Next cover the bulb with a wet, semipermeable membrane (goldbeater's) or dialysis tubing (cut open) and immerse in a beaker of water (Fig. 3-28). Soon the level of the liquid inside the thistle tube will rise. This special diffusion of molecules of water through the membrane is called osmosis.[52] As the water level rises in the tube, its pressure (hydrostatic pressure) will eventually halt the further upward diffusion of water. Some teachers

[52] *Osmosis* should not be confused with *diffusion*. Osmosis is a special case of diffusion where pure water passes through a membrane permeable only to the water.

connect extensions of thin glass tubing with rubber tubing to the stem of the thistle tube. You may get a considerable rise of fluid in the tube due to the diffusion of water molecules into the thistle tube. Also refer to demonstrations of osmotic pressure using a raw egg or a raw potato (Figs. 4-29, 4-30). Giese offers many activities concerning movement of water across membranes and osmometers.[53]

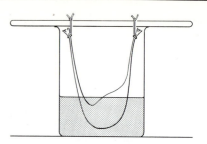

Fig. 3-29 Apparatus to demonstrate diffusion through the intestine of a frog. Intestinal casing or dialysis tubing may be substituted. The intestine is filled with molasses and suspended into water; molasses diffuses into the water and water diffuses through the membrane into the molasses.

Permeability of cells to ionized dyes
Use concentrated cultures of *Paramecium*. On a clean slide, make two heavy circles with a China marking pencil, thick enough to form a boundary. Now add a drop of neutral red solution to one circle; to the other circle add a drop of phenol red. The pH of the dyes should be neutral. Now use a wire transfer loop to add a loopful of paramecia to each drop of dye. Apply two coverslips; examine under high power. Which dye is taken up by the cytoplasm? Which is ingested in food vacuoles? Neutral red is weakly ionized; phenol red is strongly ionized. Do ions enter a cell as readily as uncharged particles? Which dye entered the membrane of the paramecia?

Or try using a yeast suspension. Shake up equal volumes (a few ml) of yeast suspension with each dye in separate test tubes. Examine the color of the mass of settled yeast cells.

Diffusion through the intestine of a frog   Dissect out the stomach and intestine of a freshly killed frog (dissection, p. 89). Clean out its contents by rinsing in Ringer's solution. Be certain the intestine is not perforated. Tie one end of the intestinal tube with cord; into the other end pour a solution of molasses diluted with Ringer's solution. Then tie this end off with cord and wash off the intestinal tract to remove excess molasses. Now suspend the intestinal tract in a beaker of water, as shown in Figure 3-29, by tying the ends to a wood splint or tongue depressor. Within a half-hour, test the water in the beaker with Benedict's solution for

glucose. For rapid, positive results use a small quantity of water in the beaker.

Intestinal casing purchased from a butcher or slaughterhouse may be used instead of the frog's intestine; or use dialysis tubing. Absorption of other soluble food products through the intestine can also be demonstrated using intestinal casing (p. 175).

Diffusion through a collodion membrane   Use large, clean, thoroughly dry test tubes. Into each test tube pour about 5 ml of collodion from a stock bottle. Rotate the test tube so that the collodion completely lines the test tube; then pour the collodion back into the stock bottle. Allow the film of collodion to harden slightly. Then loosen one edge of the film along the mouth of the tube with your fingers. Along this edge let water run down between the glass and the film of collodion so that more of the film is loosened (Fig. 3-30). Rotate the tube. By careful manipulation you can remove the collodion film intact. Test the bag of collodion for leakage by filling it with water. Then let it soak in water for a short time. Repeat the procedure until several collodion membranes are made.

You may fill these cylindrical bags with starch paste (as a control), glucose, or a 50 percent solution of corn syrup or molasses. Close the tops securely by twisting and tying with cord. Wash off the overflow materials and immerse the bags in small

[53] Giese, *Laboratory Manual in Cell Physiology*.

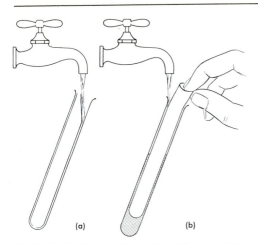

Fig. 3-30 Making a collodion bag: (a) pour water between the top of the collodion film and the test tube; (b) remove the film from the tube.

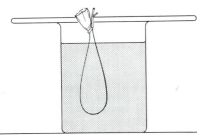

Fig. 3-31 Apparatus to demonstrate diffusion through a collodion membrane. Fill the bag with starch suspension, glucose, or solutions of corn syrup or molasses and suspend in water.

beakers of water as shown in Figure 3-31. After 10 min or so test samples of the fluid in the beakers at intervals with Benedict's solution for molecules of glucose diffused into the water.

Chemical digestion in plants and animals

The major converter of light energy is the green plant, the autotroph, which is also the source of food for animals and some plants. As heterotrophs, animals convert the ready-made organic molecules into the unique functional and structural proteins controlled by the DNA of the organism.

Digestion must take place within green plants as well as in animals since storage carbohydrates must be converted into a soluble form for transport. In this section we describe controlled investigations—both quantitative and qualitative—examining the role of enzymes, and their specificity in the digestion of starch, complex sugars, lipids, and proteins. The need for digestion was established earlier (p. 242); tests for specific nutrients are described in the context of chapter 2 (pp. 176–179).

Digestion in plants

Germination of seeds and amylase activity Soak quantities of barley seeds layered between moist rolls of paper toweling; stand in a small amount of water in a beaker. Maintain seeds at stages of germination: 1 hr, 12 hr, 24 hr, 36 hr, and 48 hr. Have groups of students grind, in separate mortars, ten seeds from each stage of germination in 5 ml of 0.1 M sodium acetate buffer (pH 5). Pour the extracts into separate test tubes and test samples for amylase activity. The disappearance of starch molecules shows the time of activity of amylase in germinating seeds. To 5 ml of 1 percent starch suspension add 1 ml of the extract and place in a warm water bath for 3 to 4 min. Then add a drop of dilute Lugol's solution. At which stage of germination of barley seeds is amylase activity first observed? Krogman describes the light-controlled activity in the phytochrome reactions that regulate the synthesis of amylase activity in germinating seeds.[54] Also refer to the role of gibberellins and the amylase activity of the aleurone layer in seeds (p. 476).

Extraction of diastase from seedlings To extract amylase from seedlings, thoroughly grind 25 germinating barley seeds (about five days old) with about

[54] D. Krogman, *Molecules, Measurements, Meanings,* A Laboratory Manual (San Francisco: Freeman, 1971).

three times as much water and a small amount of sand. Filter the preparation after it has been standing for about half an hour.

In the meantime, prepare a thin starch paste (0.1 percent) by dissolving 0.1 g of arrowroot starch in a little cold water, and adding this paste to boiling water; use 100 ml of water in all. (Stir to avoid lumps.) If it is easier to weigh out 1 g of starch, prepare a 1 percent solution with 1 g of starch and 100 ml of water as above; dilute with water to make a suspension (0.1 percent).

Add 5 ml of barley extract to an equal amount of cooled starch paste; let this stand in a warm place. Heat some barley extract to boiling, cool, and add to another test tube containing starch paste. This serves as a control since boiling destroys the activity of the enzyme. At 2 to 3 min intervals, test samples from each tube for starch with Lugol's solution. Starch disappears gradually as the diastase, extracted from the germinating barley seeds, completes the hydrolysis or digestion of arrowroot starch. The test tube containing starch paste and boiled barley extract does not show this reaction; starch paste remains in the test tube. You may prefer to test a series of dilutions of the amylase on the starch suspension.

Is there a difference in the quantity of the enzyme present during the first few days of germination and during the second week of germination, or is the same quantity produced at all times? Students might design an investigation such as this: Test dried seeds, then test seeds after one day of germination, two days, and so forth. Use the procedure already described for extracting diastase, as long as uniform quantities of water and seeds are used. Refer to the role of plant hormones in growing seedlings (p. 486).

Seedlings on starch agar film   You may prefer to vary the demonstration of the enzymatic hydrolysis of starch by germinating seedlings in the following way. Into several sterile Petri dishes pour a thin layer of starch agar (5 g of agar plus 7.5 g of arrowroot starch brought to a quick boil, with constant stirring, in 500 ml of water).

After cooling, scatter on the surface of the agar some radish, mustard, or barley seeds that have been surface-sterilized in Clorox (p. 720) and soaked overnight. Let the seeds germinate for two to three days, then remove the seedlings and pour a thin film of Lugol's solution over the surface of the agar; tilt the dishes to spread the iodine evenly. What is the explanation for the clearer areas that do not stain blue-black with iodine? How could students test whether simple sugars are present in these clearer areas?

In a simple technique, without using agar, test notebook paper with iodine to see whether there is starch in the sizing used on the paper. If so, split germinating seedlings lengthwise and place them on wet notebook paper in Petri dishes. Within hours, flood the paper with Lugol's solution to observe the clear areas around the seedlings where starch has been digested. Maintain some dishes for several days to observe wider areas where digestion has occurred. What control should be used to determine whether substances other than enzymes produce the same effect?

Starch hydrolysis on a slide   The concept of digestion as the breaking down of a substance by an enzyme may be demonstrated using a commercial preparation of a plant amylase, diastase. This may be used as a demonstration, but it is also an effective student laboratory activity.

Prepare a 0.1 percent solution of arrowroot starch (see above). While the starch paste is cooling, prepare a 0.1 to 0.2 percent diastase solution in water. (The enzyme is available from supply houses, and it is always ready to use; simply dissolve a small quantity in a large amount of water.)

Each student needs a clean glass slide or a small, square piece of glass. Add drops of several liquids to each slide in this order: (1) 2 separate drops of dilute starch suspension; (2) 1 drop of very dilute Lugol's solution (light brown in color) to each drop of starch paste so that both drops become very light blue in color; and (3) 1 drop of the diastase solution to only 1 drop of the combined starch and Lugol's solution. The other drop on the slide acts as a control throughout this demonstration.

Within a few minutes notice how the blue color of one drop disappears. Why? Why does the other drop remain blue? This experience demonstrates how rapidly an enzyme acts in splitting starch paste.

Also refer to hydrolysis on a slide using saliva on page 249.

Starch hydrolysis: a demonstration
As a class demonstration, using two *large* test tubes, prepare the same dilutions of the starch paste, diastase, and Lugol's solution. Turn starch paste in each of the two test tubes a *very* pale blue with *very* dilute Lugol's solution. Then add to *one* tube a volume of diastase solution equal to one third the volume of solution already in the tube. Invert the tube. The blue color should disappear almost immediately, depending on the quantities used and the temperature. Why doesn't the control test tube lose its blue color?

Some teachers prefer to add the enzyme to the starch paste, wait a short time, and then add iodine solution. This fails to produce a blue color, since starch is no longer present in the solution. This method is also described on page 248 in a demonstration in which students test their own saliva.

This demonstration can be used as practice in designing further experiments. For example: (1) How could you show that this same change in starch goes on within the body (use saliva instead of diastase)? (2) How could you test what the starch has been changed into? (3) How dilute a solution of the enzyme can digest the starch paste (p. 250)? (4) Is digestion independent of temperature, or would digestion go on faster at body temperature? (5) Does it make any difference whether large particles or finely chopped or chewed particles of food material are acted upon by an en-

zyme? Does boiling the enzyme affect its activity?

Students may design similar investigations using pancreatin (p. 254).

A quantitative study: enzyme activity
Demonstrate the speed of enzyme activity with diastase (preparation, p. 245) and the effectiveness of very dilute amounts of an enzyme and its action on a given substrate. A dilution technique is described for salivary diastase or amylase (p. 250).

Effect of temperature on enzyme activity
As a quantitative study, have students design a procedure for determining the effect of temperature on the action of plant diastase (or salivary amylase, p. 249). Does the amylase act as readily on a dilute starch suspension at 4° C (39° F) or 38° C (100° F) as it does at 18° C (64° F), approximately room temperature? Also refer to page 250 for detailed activity using salivary amylase.

Invertase
Show the enzymatic action of invertase from yeast on sucrose:

$$C_{12}H_{22}O_{11} + H_2O \longrightarrow$$

sucrose

$$C_6H_{12}O_6 + C_6H_{12}O_6$$

glucose   fructose

Mix half a package of yeast with 40 ml of distilled water. Let stand for 20 min, then centrifuge for 5 min at 500 × g to obtain a crude solution of the hydrolytic enzyme invertase.

Add 1 ml of the crude invertase solution to 5 ml of a 2 percent solution of sucrose, and incubate for 1 to 2 hr at 37° C (99° F). Prepare a control containing similar quantities of enzyme and sucrose, but first heat the enzyme extract to boiling, then cool. Prepare another control containing 2 percent sucrose alone, and another with enzyme alone. Initially, test a sample of the sucrose solution with Benedict's solution or a Clinitest tablet to show a negative glucose reaction.

After incubation, test 2 ml samples of the substances in the test tubes with 5 ml of Benedict's solution. Heat in a water bath for 5 min. In which test tube does a cuprous oxide precipitate form, indicating a positive test for *glucose* (an aldose)?

Also test each solution for fructose (a ketose), by adding a pinch of resorcinol and 6 drops of 25 percent HCl in a test tube; heat to boiling. In this Selivanoff test a red color indicates the presence of fructose. Run the test first on 1 ml of 0.5 M solution of fructose to demonstrate the color reaction. Avoid acid hydrolysis of the sucrose by running the test on a control tube containing a solution of sucrose alone. (See also other tests for sugars on pages 177 and 178.)

(You may also want to refer to a comparison of anaerobic and aerobic metabolism in yeast cells, chapter 4.)

Protein digestion in plants   Fresh pineapple juice contains the protein-splitting enzyme bromelin. (Another proteolytic enzyme, papain, may be purchased as a product derived from papaya.[55]) Prepare a mash of *fresh* pineapple slices and collect the juice. Add an equal volume to two test tubes containing some chopped hard-boiled egg white. As a control, boil a similar volume of the pineapple juice and add this to chopped egg white in two other test tubes. Add 2 to 4 drops of toluene to each test tube to inhibit spoilage. Later, notice the ragged, translucent digested bits of egg white in the test tubes containing fresh pineapple juice.

Repeat similar activities using papain.

Digestion in animals

Enzymes   Introduce the biochemistry of digestion—the step-by-step enzyme reactions in the hydrolysis of starch into simple sugars. Some chemical equations may be useful in developing concepts of enzymes in photosynthesis and the role of the

respiratory enzymes in glycolysis, the Krebs cycle, and the cycles of ATP formation in oxidation. Some actions of enzymes are also described in chapter 2 in cell biology.

Below are several ways to demonstrate that starch is changed into reducing sugar through the action of salivary amylase.[56]

Digestion of starch by saliva in a test tube   Collect saliva in a large test tube. Salivation can be increased by chewing soft paraffin or clean rubber bands. Filter the saliva.

Prepare two or three test tubes each containing 5 ml of a 0.1 to 1 percent arrowroot starch suspension and 5 drops of saliva. Roll the test tubes between the hands to ensure thorough mixing. Heat in a water bath at 37° C (98.6° F). Use a thermometer for a more accurate study (Fig. 3-32). A clean tin can may be substi-

[56] Carnivorous and herbivorous animals lack salivary amylase; however, rodents and pigs do have a salivary amylase.

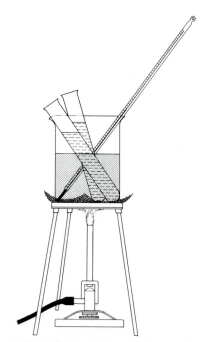

Fig. 3-32   Apparatus to demonstrate digestion of starch by saliva. Test tubes of starch and saliva are heated in a water bath at 37°C (98°F). Or substitute a hot plate for the Bunsen burner.

[55] This product is sold as a meat-tenderizing agent.

tuted for a glass beaker as a water bath. At timed intervals, remove samples of the saliva and starch mixture and test with Lugol's solution for the presence of starch. Changes will occur gradually until the blue-black reaction no longer occurs. At one point, a reddish reaction may be observed as the intermediate dextrins form. Finally, the color remains the light brown tint of the dilute iodine. Now the starch has disappeared. What has happened to it?

At this point, test the mixture for a reducing sugar with Benedict's solution (or Fehling's reagent). A change from blue to green to reddish-orange gives evidence of the presence of varying amounts of reducing sugar (maltose or glucose).

Students will question the need to demonstrate whether the original starch suspension or the saliva contained glucose (or maltose). Test both the starch paste suspension and a sample of the saliva with Benedict's solution (two controls). Other controls might have duplicates of the starch plus saliva mix.

Soluble starch results first in the reaction of salivary amylase on starch; the opaque nature of the starch suspension disappears. The activity of the salivary amylase can be plotted against time by measuring the amount of light transmitted by a test tube containing starch and very dilute Lugol's (as an indicator). Stong gives a fine description and illustrations for making a photometer using a projector for a source of light and a silicon solar cell.[57] Or make the photometer described on page 464.

Also test the effect of pH on the activity of ptyalin, and the effect of temperature. Both of these factors are critical for the activity of specific enzymes (normal pH range of saliva is 6 to 7.9.)

Salivary digestion on a slide In this laboratory activity students will need a small stock vial of dilute starch paste, a dropper bottle of dilute Lugol's solution,

a pipette, a microscope glass slide, and an empty vial.

Have each student now add 10 drops of dilute starch paste from a stock vial to an empty vial. Put 1 drop of the starch paste on the extreme left end of a slide. To this add 1 drop of iodine (Lugol's) to check the starch content. Now have students add their saliva to the 10 drops of starch paste in the vial, and warm the vial by rolling it between the palms of the hands. At 2-min intervals, put successive drops from the vial along the length of the glass slide. Test each drop with dilute iodine solution.

Gradually, the blue-black color fails to appear as successive drops of the starch-saliva suspension are tested. The reddish color of erythrodextrins may appear, and finally, there is no color change. (Starch and dextrin intermediates do not generally appear.) A few tests may show no change in the starch reaction. People have different amounts of salivary amylase in their saliva. In fact, a very small percentage of the population seems to lack the salivary amylase. This is believed to be an inherited trait.

Use Benedict's solution to test the remaining saliva–starch mixture of the vials for reducing sugar. As controls, saliva as well as a sample from the original starch paste stock should be tested for reducing sugar. The main advantage in this procedure is that students have the opportunity to test their own saliva, and each set of "experiments" acts as a control for the others.

Effect of temperature on enzyme activity
Demonstrate the action of salivary amylase under different conditions. Committees of students may prepare six test tubes, each with about 5 ml of a 0.5 percent arrowroot starch suspension prepared in 0.25 percent NaCl solution. Leave two tubes at room temperature. Chill two tubes in a beaker of water to about 5° C (41° F); place two more in a water bath at 40° C (104° F).

Add an equal quantity of saliva to one of the pair of test tubes at each of the three different temperatures. To each of the

[57] C. L. Stong, "The Amateur Scientist," *Sci. Am.* (January 1963), pp. 147–54.

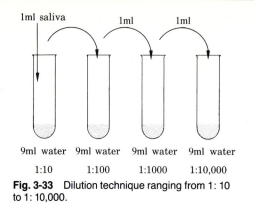

Fig. 3-33 Dilution technique ranging from 1:10 to 1:10,000.

other matched tubes add an equal quantity of saliva which has first been *boiled* for 10 min. Test all the tubes for the presence of reducing sugar.

A quantitative study: effect of dilution of salivary enzyme   Here students observe the minute quantity of enzyme required to hydrolyze starch. Dilute filtered saliva in the following manner. Into each of four test tubes pour 9 ml of water. To test tube 1, add 1 ml of saliva; shake the tube thoroughly (Fig. 3-33). With a clean pipette transfer 1 ml of this diluted solution from tube 1 into tube 2. (Tube 1 is now a 1:10 dilution; tube 2 is a 1:100 dilution.) With a clean pipette transfer 1 ml of fluid from the tube 2 into tube 3; continue until tube 4 will have a dilution of 1:10,000. Of course, further dilutions may be made as time and materials permit.

To each of the tubes of saliva of different dilutions add 1 ml of a dilute starch suspension composed of 0.5 percent starch prepared in 0.25 percent sodium chloride solution. (The sodium chloride speeds amylase digestion.) Incubate all the tubes in a water bath maintained at 38° C (100° F).

Test samples from each test tube after 1, 5, 10, and 15 min for the amount of starch present, using Lugol's iodine solution, and/or test for the presence of simple sugars using Benedict's solution. Be sure to test a control (dilute saliva), as well as the starch suspension, for the presence of simple sugars. Repeat in 30 min. Note that

time is a factor in the hydrolysis of starch. Given more time, more digestion of starch will occur. Therefore limit the period of testing to 15 or 30 min. Record the time it takes for digestion to occur so that semi-quantitative data is obtained. Note the quantity of sugar produced in a given period of time. Using the same technique, students may test the effect of pH on reaction time of this hydrolysis.

Specificity of enzymes   At this point, you may want to suggest an independent investigation to extend the work on enzymes (see procedure above). In so doing students will have the opportunity to learn how to design simple investigations with controls. Students may plan to investigate the effect of the amylase in saliva on starch, sugar, fat, and protein, as well as on foods containing many nutrients (for example, milk). Similarly, investigations may be devised to discover whether pepsin reacts with starch, sugar, fat, and protein. SPECIFICITY OF AN ENZYME   Develop the following simple plan into a class activity. Does saliva act only on starch? Dilute about 24 ml of saliva with an equal amount of water. Prepare eight test tubes so that each contains about 6 ml of the diluted saliva. Suppose we set up two test tubes for each kind of nutrient so that one may act as a check for the other. To the first set of two test tubes add 10 ml of 1 percent starch paste. To the second set of two tubes add 10 ml of a freshly prepared 3 percent sucrose solution. To the third set of tubes add 2 drops of olive oil with 10 ml of 1 percent sodium carbonate. To the last pair add a pinch of casein (or chopped hard-boiled egg white) with 10 ml of 0.2 percent hydrochloric acid. Now put all the test tubes in an incubator or in a 40° C (104° F) water bath for half an hour. Then test the starch paste and sucrose solutions for reducing sugar with Benedict's solution. Also test the casein-containing tubes for peptones (using the Biuret test, page 179).

Digestion of a cracker in the mouth   This procedure may be used as a demonstration or an individual laboratory exercise. Collect saliva into a clean test tube; then chew slowly on a piece of soda cracker (which was tested previously to ascertain that reducing sugar was absent, or present in only small amounts). Use a cracker that is all white; tan or brown patches may be indications of dextrins.

Place this chewed mixture into a clean test tube. While this part of the activity is under way, test the saliva for reducing sugar with Benedict's solution in case sweets were eaten previous to this activity. Also test the cracker for the presence of starch (as another control).

Within 5 min, test the chewed cracker with Benedict's solution; a color change from green to orange will be observed. Digestion goes on more rapidly here at body temperature than in a test tube.

You may want to divide the chewed cracker mixture so that half could be tested for reducing sugar and the other half for starch (should a student ask if all the starch had been changed to simple sugar). In most cases some starch remains. The quantity varies inversely with reaction time.

Clinitest tablets may be used in testing the end product of salivary digestion. Clinistix strips may be used to distinguish glucose from maltose.

Tests for proteins

A GENERAL TEST FOR A PROTEIN   You may want to try a technique for testing the component elements of proteins:[58] carbon, nitrogen, hydrogen, and sulfur. In a dry test tube, heat some powdered egg albumin (see preparation below). Suspend a strip of wet red litmus paper in the tube. Across the mouth of the test tube place a strip of filter paper which has been moistened in lead acetate solution. If hydrogen and nitrogen are present, ammonia fumes are produced, and the litmus will turn blue. The presence of sulfur is indicated by the blackening of the lead acetate paper. If the powder chars, you know that carbon is present. If the powder and the test tube

are dry at the beginning of the test, you can see drops of moisture condense on the sides of the test tube, which indicates the presence of hydrogen and oxygen (as water).

PREPARATION OF ALBUMIN   If you need albumin solution for immediate use, beat the white of an egg with about 8 volumes of water. Strain through cheesecloth, and filter if necessary.

To prepare powdered egg albumin, beat the egg white with about 4 volumes of water, and filter. Then evaporate the water from the filtrate on a water bath at about 50° C (122° F). Powder the residue in a mortar.

SPECIFIC TESTS FOR PROTEINS   Specific tests for proteins are described in chapter 2, page 179.

Gastric digestion

Consider the early investigations of Spallanzani in 1783, who described *chemical* digestion in the stomach, and who also found that the stomach contents were acidic. Then recall the observations of Dr. Beaumont in 1822 on his patient, Alexis St. Martin, who sustained a gastric fistula as a result of a gunshot wound. Dr. Beaumont could observe and collect samples of gastric juice from the stomach of St. Martin. The following activities demonstrate some aspects of gastric digestion.[59] Both temperature and pH can be shown to affect salivary as well as intestinal digestion (Table 3-3).

Digestion by pepsin   Separate the albumin, or white of an egg, from a hard-boiled egg; egg white is almost pure protein. (Or substitute small pieces of fibrin in which case digestion proceeds faster.) Dice the egg white into very small pieces and place equal amounts into four test tubes. (Replicas should also be prepared.) Number the tubes; then pour these substances into the tubes:

---

[58] *Hawk's Physiological Chemistry.*

[59] Also refer to a paper in *Scientific American*, January 1972 by H. Davenport, "Why the Stomach Does Not Digest Itself."

TABLE 3-3   pH range of some body fluids

body fluids	pH
saliva	5.6–7.6
gastric juice	1.5–8.4
pancreatic juice	7–8
bile	5.6–8.5
small intestine	5.8–7.6
large intestine	7.5–9
blood	7.4
lymph	7.4
plasma	7.39
cerebrospinal fluid	7.4
urine	4.6–8 (depending on diet)
sweat	3.8–6.5
tears	7.5

Data compiled from W. Spector ed., *Handbook of Biological Data* (Philadelphia: Saunders, 1956).

tube 1: 5 ml of a pepsin solution (0.5 percent pepsin)[60]

tube 2: 5 ml of 0.4 percent HCl

tube 3: 5 ml pepsin–HCl mix (add 2 drops HCl to pepsin solution)

tube 4: 5 ml of pepsin solution plus 2 drops of 0.5 percent sodium carbonate.

Maintain all the tubes in an incubator or water bath for ½ hr. Test the contents of each tube for proteoses and peptones using the Biuret test (p. 179). Pour 3 to 4 drops of the Biuret solution down the side of the tilted test tube. A positive test gives a pink-violet color. Or leave the tubes in an incubator for 24 hr and note the disappearance of the protein test substance. Does the enzyme pepsin act faster in an acidic or alkaline environment? Is there an optimum acidity for pepsin digestion?

OPTIMUM ACIDITY FOR PEPSIN ACTION Prepare three test tubes containing these substances:

tube 1: 3 ml of pepsin solution plus 3 ml 0.4 percent HCl (acidity about 0.2 percent HCl or pH 1.3)

tube 2: 3 ml of pepsin solution plus 1 ml of 0.4 percent HCl plus 2 ml water (acidity about 0.067 percent HCl or pH 1.8)

tube 3: 3 ml of pepsin solution plus 4 drops or 0.2 ml of 0.4 percent HCl plus 3 ml water (acidity about 0.013 percent HCl or pH 2.5)

To each test tube add an equal amount of chopped egg white (or small piece of fibrin); maintain at 40° C (104° F). Hawk's text gives the optimum acidity in this demonstration as pH 1.8, although it is not accurate since some acid combines with proteins in digestion.[61]

Effect of temperature on pepsin activity

Is the rate of enzyme activity affected by changes in temperature? If so, how? Place 10 ml of a pepsin–HCl mix into each of eight or more sets of test tubes. To each of the eight tubes, add equal amounts of finely diced hard-boiled egg white or albumen solution (preparation, p. 251). Fibrin may also be used and gives more rapid results. Immerse at least 2 tubes in ice water (record the temperature); place a second set of 2 tubes at room temperature; a third set of tubes in an incubator or water bath at 40° C (104° F). Boil the contents of a fourth set of two tubes for a few minutes; then cool and place in the water bath. Which seems to be the optimum temperature? What is the effect of boiling on an enzyme?

Quantitative test for the rate of pepsin activity: Mett tubes

While digestion goes on more slowly in this technique, there is opportunity for students to use quantitative measures and graph the combined results of the class. Explore the effect of temperature or pH on the rate of

---

[60] Or prepare by dissolving 750 mg pepsin USP in 100 ml of 0.1 *N* HCl, as suggested in *Hawk's Physiological Chemistry*, 14th ed., B. L. Oser, ed. (New York: McGraw-Hill, 1965).

[61] *Hawk's Physiological Chemistry.*

pepsin digestion. Prepare Mett tubes by cutting 25 cm lengths of narrow bore (2 mm) glass tubing; draw egg white into the tubes. Seal each end in a flame. Now coagulate the egg white by immersing the tubes into boiling water. When cool, cut off the ends; use a file to cut off small Mett tubes 2 cm long.

Place the Mett tubes in Petri dishes of gastric juice and/or its separate components; incubate the Petri dishes. Or test the effect of different temperatures or pH on pepsin activity. Have students collect data, measuring in millimeters the amount of digestion that occurs in specific time intervals; then graph the results.

Action of protease on gelatin Separate the two halves of several gelatin capsules purchased from a druggist. To the larger half of each add a drop of a food dye to observe the results. Fill several with water and close the capsules; immerse in a beaker of warm water (beaker *A*). To the remaining half-capsules containing a drop of food dye, add a protease solution such as pepsin (or pancreatin) and close the capsules. Immerse these into beaker *B* containing warm water. What qualitative difference is observable? What happens in the stomach to such a capsule containing medication?

Antipepsin action of some worms Obtain intestinal worms, such as *Ascaris*, from a slaughterhouse; grind a few worms in a mortar with sand. To this add some 0.9 percent sodium chloride solution; mix well and filter.

Into each of two test tubes pour 10 ml of a solution of pepsin and hydrochloric acid (described earlier), along with some fibrin or egg white. To one test tube add 4 ml of the worm extract, and to the other tube add 4 ml of water. Keep both tubes at 37° to 40° C (99° to 104° F). Study the rate of digestion over the next 24 hr. Note that the worm extract inhibits digestion of proteins.

Action of rennin on milk Demonstrate the coagulating action of the enzyme rennin. Crush a Junket tablet (rennet) and add it to a test tube of milk. Warm, but do not boil, the milk. Notice how rapidly the milk solidifies in the test tube. Compare this with a duplicate test tube in which the milk and tablet have been boiled for 5 min. Notice the destructive action of heat on enzyme activity. Some teachers prefer to crush a Junket tablet in water first before adding this to milk.

Digestion in the small intestine
The following demonstrations show digestion of proteins, carbohydrates, and fats by specific enzymes in the small intestine.

Digestion by trypsin To show that proteins are digested by trypsin in the small intestine, use commercial pancreatin powder which contains the enzyme trypsin. Demonstrate digestion and the optimum pH for the activity of trypsin. Half fill six test tubes with water; add chopped hard-boiled egg white or albumin solution (preparation, p. 251). To each of the six tubes add a pinch of pancreatin. Set two tubes aside; to two others, add 2 drops of phenolphthalein solution and then a 0.5 percent solution of sodium carbonate, drop by drop, until the first pink color appears. This will result in a pH close to 8.3 (alkaline). To two other test tubes add 5 ml of a 2 percent boric acid solution, which will result in a pH close to 5. For more rapid results, use a water bath kept at 40° C (104° F), or keep the test tubes at the same temperature in an incubator.

Or you may want to use casein of milk instead of egg white as the substrate and a more carefully prepared "pancreatic juice." Prepare 0.25 g of pancreatin in 50 ml of water; neutralize using 0.05 *M* sodium bicarbonate. Test with litmus. To this add 5 g of casein and 50 ml of 0.1 *M* sodium carbonate. Preserve with toluol and incubate at 38° C (100° F). (*Caution: Avoid flames.*)

Compare the action of a dilute solution of pancreatin with the action of such plant proteases as papain or another meat tenderizer, or fresh pineapple juice, which contains the enzyme bromelin (p. 248). Incubate casein, albumin, or hard-boiled egg white with the plant proteolytic enzyme in a water bath or incubator maintained at 38° C (100° F).

Amylase activity of pancreatic juice Test the activity of pancreatic amylase on arrowroot starch. The amylase of saliva is the same as pancreatic amylase.

Prepare a pancreatin solution as described above. Then prepare two test tubes, each containing about 10 ml of 0.5 percent starch suspension; add 5 ml of "pancreatic juice" to each. Keep the tubes in a water bath at 40° C (104° F) for about half an hour. Test the contents of one test tube for simple sugar with Benedict's solution or Clinitest tablets; test the other tube for starch with Lugol's solution. Is there an amylase in pancreatin? Recall that the digestion of starch begins in the mouth and is completed in the small intestine.

By following the pattern of demonstrations already described for salivary digestion, you may want to show the effect of temperature on pancreatic digestion of carbohydrates.

Invertase in the small intestine Demonstrate the activity of invertase as sucrose is hydrolyzed to glucose and fructose. Test the end product with Benedict's solution (other tests, for carbohydrates, pages 176–78).

Emulsification of fats In a test tube shake together a small quantity of water with a few drops of olive oil; an emulsion forms. On standing (for a short time), the oil separates from the water. To another test tube containing about 10 ml of water add 5 drops of a 0.5 percent sodium carbonate solution. To this add 3 drops of olive oil and shake well. On standing, this emulsion does not separate as quickly (since an alkaline medium exists). To a third test tube of water add a 5 percent solution of bile salts such as sodium taurocholate (or add a soap solution).

Then add a few drops of olive oil and shake well. Compare this more permanent emulsion with the preceding transitory ones. The size of the oil drops may be compared by placing a drop of fluid from each test tube on a slide for examination under low power of the microscope. Students may readily observe the small size of the oil drops in an emulsion. Elicit these observations, and the significance of the small size of the drops as a preliminary step in fat digestion by lipase in the small intestine should be apparent.

BILE SALTS AND SURFACE TENSION   To demonstrate that bile salts lower surface tension, add 5 ml of water to one test tube; to another tube, add 5 ml of bile suspension. Then sprinkle powdered sulfur on the surface of each test tube. What is the value of reduced surface tension?

Hydrolysis of fat by lipase in the small intestine Use olive oil in this hydrolysis by lipase action. Prepare three test tubes containing 10 ml of water and a few drops of olive oil. To two of these tubes add 2 ml of a soap solution or 5 percent bile salt solution (sodium taurocholate). Then prepare a pancreatin solution. Dissolve the following ingredients in 100 ml of distilled water:

pancreatin	1	g
$CaCl_2$ (1% solution)	1	ml
K_2HPO_4 (0.2 M solution)	25	ml
NaOH (0.2 M solution)	11.8	ml

Add 5 ml of this pancreatin solution to one of the test tubes with soap solution or bile salts and to the one without this solution. Thus we have three test tubes: (1) a tube containing water, olive oil, soap solution, and pancreatin solution; (2) a control containing water, olive oil, and soap solution; and (3) another control containing water, olive oil, and pancreatin solution.

At the start of this demonstration, check the pH with litmus or phenolphthalein, hydrion paper or a pH meter. The original solutions must be neutral, not acid, for the objective is to show that fat digestion results in the production of fatty acids. Demonstrate the formation of acid with litmus paper, or other indicator, after the test tubes have remained in a water bath at 40° C (104° F) for at least half an hour.

HYDROLYSIS OF CREAM BY PANCREATIC LIPASE Compare the effect of lipase and bile salts in pancreatic digestion. Prepare pancreatin solution as already described. Pour equal amounts of cream into three test tubes. To tube 1 add pancreatin solution; to tube 2 add an equivalent amount of pancreatin solution, plus a pinch of bile salts; to tube 3 add an amount of water equal to the solution used, plus a pinch of bile salts. Maintain all the tubes in a water bath at 37° C (99° F). Prepare replicates. After a few seconds, begin to test the rate of hydrolysis by adding 5 drops of blue litmus solution to each test tube. Prepare litmus solution by adding 1 g of blue litmus powder to 100 ml of water; show the action of blue litmus by adding drops to a small quantity of dilute acid.

Is pancreatin effective in hydrolysis of fats in cream? What is the apparent role of bile salts? The effect of temperature on the activity of enzymes? Are there inhibitors of any sort? How dilute may the enzyme preparation be and still hydrolyze fats? Students might devise demonstrations to test the validity of their predictions; then perform their demonstrations later in the laboratory.

LITMUS MILK TEST USING PANCREATIN You may want to show that pancreatin contains digestive enzymes which split fats into fatty acids and proteins into amino acids. Litmus can be used as an indicator to show the resulting acid reaction after digestion. First, prepare litmus milk powder: to 50 parts of dried milk powder add 1 part of blue litmus powder. Then dissolve 1 part of prepared litmus milk powder in 9 parts of water.

To each of two test tubes add 10 ml of litmus milk powder solution, and to one test tube add a pinch of pancreatin powder. To the other test tube add only water or boiled pancreatin solution. If the tubes are kept in a water bath, observe the change in litmus to pink, indicating the presence of acid.

Absorption in the small intestine

We have described demonstrations illustrating diffusion of substances through a membrane (pp. 242, 244). Although the need for digestion was established in diffusion demonstrations, the following activity serves as a summary of many related enzyme activities in the food tube.

Fill a length of sausage casing or dialyzing tubing (without holes, even pinpricks) with a variety of foods and pancreatin solution.[62] Be sure to include sources of the common nutrients. For example, include molasses, egg white and yolk, salt, bread or a cracker, and some milk. (Mix these together in a mortar first, if necessary.) Tie each end of the tubing with string and attach each end to a support as illustrated in Figure 3-29. Wash the outside of the tubing to remove traces of nutrients, and insert into a beaker of water. Place in a water bath at 40° C (104° F).

The next day test 5 ml samples of the water in the beaker for reducing sugar, simple proteins, salt, and fatty acids. Use Benedict's solution or other reagents to test for simple sugar (p. 176); the Biuret test for simple proteins (p. 179); litmus paper for acids. Add 1 or 2 drops of silver nitrate to a water sample; the formation of a white precipitate indicates the presence of a chloride.

Minerals, vitamins, calories

While few consumers stop to test the nutrient content of their food, students should understand the value of nutrient testing in a science laboratory. For example, iodine may be added to solutions of all nutrients: glucose, casein, egg albumin, corn or olive oil, and starch. This establishes the specificity of the iodine test. Similarly, tests for glucose, simple proteins, and lipids

---

[62] Dialyzing tubing made of unseamed cellophane may be purchased in rolls. When cellophane is substituted for semipermeable animal membranes or for dialyzing tubing, the nonwetting coating on the cellophane covering of packaged products needs to be removed. Immerse cellophane for 3 to 5 min in 95 percent ethyl alcohol (or in isopropyl for some 10 min). Under a faucet of warm water, wash and wipe off the film.

establish a standard for testing all nutrients. A variety of tests for nutrients such as carbohydrates, fats, and proteins are described in chapter 2.

Examining phosphates and magnesium in muscle of frog

Hawk's text also describes Hürthle's method. Use a dissecting needle to tease apart a small bit of muscle of frog on a glass slide. Then expose the slide to ammonia vapor for a few minutes. Apply a coverslip and examine under the microscope to find large amounts of ammonium magnesium phosphate crystals distributed throughout the muscle fiber.

Vitamin content of foods

We shall first deal with activities using animals to show the results of deficiencies, and then go on to demonstrate chemical tests for vitamin content of foods.[63]

Vitamin B₁ requirements of pigeons

When healthy young pigeons are fed a diet deficient in thiamine (B_1), they develop polyneuritis (beriberi). Students who maintain pigeons at home may try this investigation, or it may be carried on in school.

Separate cages are needed for the experimental and the control pigeons. The cages should be kept free of droppings, or pigeons will consume this along with food. Feed all the birds equal amounts of white polished rice, finely cracked egg shells, and water.

One group gets, in addition to this diet, a mixture of barley, hemp seeds, yellow corn, and some fresh vegetables such as shredded cabbage or lettuce.

In about ten days the pigeons without the "extras" will develop a paralytic condition due to the impairment of nerve-muscle connections. Now add the "extras" to the diet and watch how quickly the pigeons recover their health and activity.

Quantitative chemical test for vitamin C

Ascorbic acid, vitamin C, is a strong reducing agent with the ability to reduce indophenol, iodine, silver nitrate, and methylene blue (in the light). A simple, quantitative test for the presence of vitamin C in foods may be done as a class demonstration. Even better, it may serve as an excellent laboratory experience—a controlled experiment involving simple measurement.

Indophenol (2,6-dichlorophenol indophenol) is an indicator which is blue in color and is reduced in the presence of ascorbic acid or juices containing ascorbic acid.[64] Prepare a 0.1 percent solution of indophenol in water by dissolving 1 g of indophenol in 1000 ml of water.[65]

The test should be demonstrated with pure ascorbic acid first and later repeated with several fruit juices. Crush a vitamin C tablet in water. In this sensitive test, the ascorbic acid should be diluted enough so that the number of drops of ascorbic acid added to the indophenol may be counted accurately. If the ascorbic acid is too strong, 1 drop will immediately bleach the indophenol so that a quantitative test is difficult to make. Add about 10 ml of the indicator to a test tube. To this add dilute (1 percent) ascorbic acid, drop by drop; count the number of drops needed to change the color from blue to colorless. (Disregard the intermediate pink stage.) Record the number of drops needed to bleach the indicator.

Now use diluted fruit and vegetable juices—canned, fresh, and frozen. Test the juices beforehand; try to dilute them so that students will need to add about 10 to 20 drops of juice to a given quantity of indophenol. Dilute them all equally, so that the ascorbic acid content of canned

---

[63] You may want students to read W. F. Loomis, "Rickets," *Sci. Am.* (December 1970).

[64] Indophenol can be purchased from the Chemical Division of Eastman Kodak Co., Rochester, NY, or from biological supply houses.

[65] A BSCS laboratory activity recommends the use of a 10 to 20 percent solution of Lugol's iodine as a substitute for indophenol. The color of iodine bleaches until the end point is reached. See *Teachers' Guide to High School Biology: Green Version—Laboratory, Part II*, BSCS, Univ. of Colorado, Boulder, 1961 (experimental use, 1961–62).

juices may be compared with that of fresh juices. In comparative tests the quantity of indophenol used should be standardized.

Some juices may be boiled, or left exposed to air for several hours and then tested again. The data collected may then be compared with the original readings.

Bicarbonate of soda, which is often added to vegetables to preserve their green color while cooking, may be added to samples of the same juices. Note the loss of ascorbic acid under these conditions. (The larger number of drops of juice needed to bleach a given quantity of indophenol is a measure of the small amount of ascorbic acid remaining in the juices.)

There may be errors in the titration of vitamin C, due to the presence of other reducing compounds formed during the heat-processing of food or while food substances are stored. Refer to *Hawk's Physiological Chemistry* for procedures correcting such errors, as well as means of determining the ascorbic acid content of whole blood, urine, or plasma, or plant tissues; colorimetric methods are included.

The text also suggests biological methods for assaying vitamin C—experiments with guinea pigs fed on standard scurvy-producing diets.

Vitamin requirements of plants

In further studies, you may want to focus on the vitamin needs of plants—the vitamin requirements of *Euglena*: the use of B_{12} and thiamine by *Euglena*; also the use of *Neurospora* and other microorganisms, mainly bacteria, in microbiological assay of vitamin and amino acid content of foods. Many microorganisms require water-soluble vitamins. Two possible investigations using thiamine, which is readily available, follow.

Thiamine and growth of Phycomyces

Show that *Phycomyces blakesleeanus* (having a much-branched mycelium with many nuclei) requires thiamine for growth. Prepare a thiamine-deficient medium for the mold: To 1 liter of distilled water, add 100 g of glucose, 10 g of asparagine, 15 g of KH_2PO_4, and 5 g of $MgSO_4 \cdot 7H_2O$. Adjust the pH between 4.0 and 4.5.

Now prepare a solution of thiamine chloride and dilute to varying concentrations: 5 mg per liter, 2.5 mg per liter, 1.0 mg per liter, 0.5 mg per liter, 0.1 mg per liter. Pour an equal amount of medium (in milliliters) into 12 test tubes. Leave 2 as controls, and to pairs of test tubes add 1 ml of the different dilutions of thiamine chloride solution. In short, there should be 2 tubes, or replicates, of each dilution and 2 controls. Loosely stopper with cotton plugs, or cover with aluminum foil, and sterilize for 15 min at 15 lb per sq in. pressure. Inoculate each tube with an equal amount of mycelia of *Phycomyces blakesleeanus*; incubate or let stand for two weeks, and compare the rate of growth in each test tube. Can the vitamin requirement for growth of these molds be estimated? Is there growth in the controls? In assay work, the mat of mold would be weighed; possibly a rough index might be made. Does one dilution show twice the growth of another dilution? Does the mat weigh about twice as much?

Deficient strains of Neurospora

Students may repeat, on a small scale, the procedures developed in 1941 by Beadle and Tatum for identifying biochemical mutations after irradiation of the red mold *Neurospora crassa* (Fig. 7-17).

Normal strains of this red mold synthesize the requirements for growth, provided a minimal diet is supplied: a source of carbon in the form of a carbohydrate (for this mold is a heterotroph); some inorganic salts, including nitrate; and one vitamin, biotin. Some mutants have lost the ability to synthesize a specific vitamin or amino acid; apparently the enzyme is missing due to the loss or modification of a gene. As a result, these vitamins or amino acids must be added to the mineral medium to support growth.

Purchase a normal strain of *Neurospora* and a deficient strain—possibly a thiamine-deficient strain that cannot make

TABLE 3-4   Approximate energy expenditure in a variety of activities*

		male		female	
activity	time (hr)	rate (Cal/min)	total	rate (Cal/min)	total
Sleeping	8	1.1	540	1.0	480
Sitting, driving a car, bench work	6	1.5	540	1.1	420
Standing, or limited walking	6	2.5	900	1.5	540
Walking, purposeful, or outdoors	2	3.0	360	2.5	300
Occupational activities involving light physical work; spasmodic week-end swimming, golf, picnics†	2	4.5	540	3.0	360
			2880		2100

* Adapted from *Recommended Dietary Allowances*, 6th ed., *Report of Food & Nutrition Board* (Public, 1146), National Academy of Sciences–National Research Council, Washington, D.C., 1964.

† Week-end swimming, tennis, and so forth may use 5 to 20 Cal per min for a limited time.

its own thiamine, or a strain unable to make pyridoxine and choline, or arginine.

Transfer a bit of the normal and the deficient strains to separate sterile test tubes or Petri dishes containing a minimal medium; also transfer each strain to individual test tubes that have the minimal medium plus some of the required thiamine (or arginine or choline, as the case may be).

The preparation of a minimal medium for *Neurospora* is suggested in the *Teachers' Guide* to the BSCS Blue Version.[66] To 1 liter of water add 15 g of sucrose, 5 g of ammonium tartrate, 1 g of ammonium nitrate, 1 g of monopotassium phosphate, 0.5 g of magnesium sulfate, 0.1 g of calcium chloride, 0.1 g of sodium chloride, 10 drops of 1 percent ferric chloride, 10 drops of 1 percent zinc sulfate, and 5 mg of biotin. Sterilize in an autoclave for 10 min at 15 lb per sq in. Dehydrated *Neurospora* culture agar is available from Difco Laboratories (see Appendix). Media for assays of pyridoxine and choline are also available. (Agar may be added to the liquid medium given above.)

Devise ways to test for the quantity of choline or thiamine or other essential substances in foods by using specific deficient strains of *Neurospora*. Or provide students with an unknown mutant strain to identify. See also genetics of *Neurospora* (p. 522).

A well-developed series of laboratory activities is described in the BSCS laboratory block *Microbes: Their Growth, Nutrition, and Interaction* by A. S. Sussman (Boston: Heath, 1964).

Mineral requirements

Effects of minerals such as nitrates, phosphates, and sulfates may be demonstrated using growing seedlings or algae such as *Chlorella* (pp. 464, 478).

Reports in the literature indicate rotifers live longer without calcium in their cells. Is this correct?

A single substance may, in low concentration, act as a growth factor; in higher concentration as a growth regulator or inhibitor; in still higher concentrations, as an aging factor.

[66] *Teachers' Guide* to *High School Biology: Blue Version—Laboratory, Part II*, BSCS, Univer. of Colorado, Boulder, 1961 (experimental use, 1961–62).

TABLE 3-5   Recommended daily dietary allowances (RDAs)*

(Designed for the maintenance of good nutrition of practically all healthy persons in the United States.)

sex-age category	age years from	to	weight kilo-grams	pounds	height centi-meters	inches	food energy calor-ies	protein grams	minerals cal-cium milli-grams	phos-phorus milli-grams	iron milli-grams	vita-min A interna-tional units	thia-min milli-grams	ribo-flavin milli-grams	nia-cin milli-grams	ascor-bic acid milli-grams
Males....	11	14	44	97	158	63	2,800	44	1,200	1,200	18	5,000	1.4	1.5	18	45
	15	18	61	134	172	69	3,000	54	1,200	1,200	18	5,000	1.5	1.8	20	45
	19	22	67	147	172	69	3,000	54	800	800	10	5,000	1.5	1.8	20	45
	23	50	70	154	172	69	2,700	56	800	800	10	5,000	1.4	1.6	18	45
	51 +		70	154	172	69	2,400	56	800	800	10	5,000	1.2	1.5	16	45
Females...	11	14	44	97	155	62	2,400	44	1,200	1,200	18	4,000	1.2	1.3	16	45
	15	18	54	119	162	65	2,100	48	1,200	1,200	18	4,000	1.1	1.4	14	45
	19	22	58	128	162	65	2,100	46	800	800	18	4,000	1.1	1.4	14	45
	23	50	58	128	162	65	2,000	46	800	800	18	4,000	1.0	1.2	13	45
	51 +		58	128	162	65	1,800	46	800	800	10	4,000	1.0	1.2	12	45

*Adapted from *Recommended Dietary Allowances*, 8th ed., National Academy of Sciences–National Research Council, Washington, D.C., 1974.

TABLE 3-6   Caloric values of some "snack" foods*

	average serving	calories
Fruits		
Apple	1 3-in.	90
Banana	1 6-in.	100
Grapes	30 medium	75
Orange	1 2¾-in.	80
Pear	1	100
Candies, etc.		
Almonds or pecans	10	140
Cashews or peanuts	10	60
Cheese crackers	10 2-in. diameter	220
Chocolate bars, small size		
Plain	1 1¼-oz	190
With nuts	1	275
Caramels		
Plain	1 ¾-in. cube	35
Chocolate-nut	1 ¾-in. cube	60
Potato chips	10 2-in. diameter	110
Sweets		
Ice cream		
Plain vanilla	1 ⅙-qt serving	200
Chocolate and other flavors	1 ⅙-qt serving	230
Ice cream soda, chocolate	1 10-oz	270
Sundaes, small chocolate-		
nut with whipped cream	1 average	400
Midnight snacks		
Chicken leg	1 average	88
Ham sandwich	1 ½-oz ham	350
Hamburger on bun	1 3-in. patty	500
Peanut butter sandwich	1 2 tbsp peanut butter	370
Mouthful of roast	1 ½ × 2× 3 in.	130
Brownie	1 ¾× 1¾ × 2¼ in.	300
Beverages		
Chocolate malted milk	1 10-oz	450
Milk	1 7-oz	140
Soft drinks (soda, root beer, etc.)	1 6-oz	80
Tea or coffee, straight	1 cup	0
Tea or coffee, with 2 tbsp cream and		
2 tsp sugar	1 cup	90
Desserts		
Pie		
Fruit (apple, etc.)	1 average serving (⅙ pie)	560
Custard	1 average serving (⅙ pie)	360
Lemon meringue	1 average serving (⅙ pie)	470
Cake		
Iced layer	1 average serving	345
Fruit	1 thin slice (¼-in.)	125

* Adapted from "Caloric Values for Common 'Snack' Foods," published by Smith, Kline, and French Laboratories, Philadelphia.

Calories

At this point, or later in a study of respiration (chapter 4), you may want to develop a discussion of the need for calories in the diet. There are several approaches to this work, involving the use of charts and/or duplicated material.

For example, you may reproduce a chart, as in Table 3-4. After a study of the chart, students often ask why young men and women have a higher caloric requirement than older men and women. For what kinds of activities does a person need more calories? (See Table 3-4.) Why are some people overweight or underweight?

Ask students to keep a list of all the foods they have eaten during three school days. Then, in class, distribute charts which give calories, mineral content, vitamin content, and the amount of proteins in many foods. Many cereal food companies have pamphlets available listing contents of foods.

Have students check the number of calories, proteins, vitamins, and minerals in the foods they have consumed. Then total the list and divide by three to get a daily average. What number of calories is needed on the basis of body weight?

Students might also compare their own diet with the recommendations given in Table 3-5. Are they getting an adequate diet? A balanced one? Plan a diet for a week for a person who needs to gain weight. Similarly, suggest a diet for a person who wishes to lose weight. Why should dieting of any kind be undertaken with the assistance of a doctor? Nibblers will be interested in the caloric contents of some common snack foods (Table 3-6).

CAPSULE LESSONS: WAYS TO GET STARTED IN CLASS

3-1. In the last few minutes of a lesson, ask questions or show a demonstration; these are motivations for an assignment. For example: How rapidly do enzymes work? To a tall cylinder of *very dilute* starch suspension stained *very* light blue with a drop of Lugol's, add about 10 ml of diastase solution and invert the corked cylinder. Why does the blue color disappear? What is "there" instead? How dilute could you make the enzyme and still get the same effect? Does boiling affect the enzyme? The answers to these questions may serve as the next day's lesson.

3-2. At the beginning of class, display several test tubes in a rack near a light source. Use equal volumes of dense suspensions of *Chlorella* to which just enough phenolphthalein has been added so that they are at the turning point (just *slightly* pink). Use white paper as a background to increase visibility. After 10 min observe that the medium becomes a deep rose color as it becomes more alkaline (due to the absorption of CO_2 from the medium). Of course, some preliminary experimentation is needed to discover the correct cell density, light intensity, and dilution of indicator (pp. 156–59) to practice timing.

Or use sprigs of *Anacharis* with phenolphthalein or brom thymol blue (pp. 158, 159).

3-3. Or begin by having students *design* an investigation to show that green plants use carbon dioxide in light (techniques, p. 213).

3-4. In a laboratory activity, have students prepare a microlake in a welled slide containing a given density of *Chlorella*. Prepare several and cover with index tabs of different colors—red, green, blue, orange. Make sure there are no air bubbles. Expose to light and compare the number of bubbles on each slide after 5 min, then 10 min. What process is going on? Do bubbles form in the dark? If yeast cells are used instead of *Chlorella*, are the observations the same? Are the explanations similar? (Other suggestions are described in this chapter.)

3-5. Use a roll of cellophane, a plastic bag, or a clear plastic box to model a cell. Elicit information from students concerning the structures typical of all cells. As they suggest structures, build the cell. For instance, if they suggest a nucleus, insert a nucleus (made of modeling clay or a wooden bead). Small balls of green clay or beads might be chloroplasts, or a ribbon of green blotting paper the spiral chloroplasts of *Spirogyra*. How are autotrophs different from heterotrophs?

You may have models of cells made by former students to show new students. Encourage model-making of plaster of Paris or papier-mâché (model-making, p. 751). Making models can be a club project or a student project. An attractive collection of visual aids may be developed in a short time.

3-6. Suggest a possible investigation. Rotifers from older mothers show a shorter life span. How would students design an investigation?

3-7. Lessons may be devised by starting with a demonstration suggested in this chapter. At times, perform the demonstration in silence and then elicit responses through the use of questions.

3-8. Keep a log of successful lessons. As these are revised each year (in the light of variations in classes and students) you will have a valuable piece of action research.

3-9. Begin by testing a silver-edged geranium leaf for starch. Why doesn't the white portion contain starch? From here, you may have students design investigations to test the importance of light, of carbon dioxide, and so forth, as factors in food-making.

3-10. Introduce van Helmont's experiment with a willow twig, and proceed from there into a study of photosynthesis.

3-11. Begin with laboratory work. Have students examine cells of the onion bulb and onion green shoot under the miscroscope (Fig. 2-18). Compare these with elodea cells. Lead into a discussion of the function of chloroplasts.

3-12. Do green plants give off oxygen? Collect a tubeful of the gas which bubbles out of the cut stem of elodea twigs in the light (described on page 222).

3-13. Perhaps you may prefer to begin with a film or filmstrip on photosynthesis. Check those available from catalogs (see Appendix).

3-14. Using a crude acetone extract of chloroplasts of spinach, have students prepare chromatograms of the pigments, as described in this chapter. (*Caution:* Have good ventilation with a minimum concentration of acetone fumes.)

3-15. Use molecular model kits to build basic concepts of linkage of carbon to carbon and carbon to hydrogen, to nitrogen to show the structure of carbohydrates, proteins, and lipids.

3-16. Develop the equation for photosynthesis. As part of a case history have students report on the use of heavy oxygen and radiocarbon to learn the source of oxygen liberated in photosynthesis. Develop the role of ATP in the cycle of photosynthesis.

3-17. In a laboratory activity, prepare chromatograms of anthocyanin pigments in flowers, in red maple leaves, etc., using the technique described in this chapter.

3-18. Are chlorophyll chromatograms from leaves of one tree the same in early spring as compared with mature leaves? And with leaves in the fall?

3-19. In one test tube seal a snail, in another, a green water plant such as *Cabomba* or

Anacharis (elodea) along with a snail (p. 588). After developing the idea of their interdependence, you might compare the snail's dependence on plants with people's dependence on green plants. What is an autotroph? A heterotrophic organism?

3-20. Have two students stretch out some 30 ft of clothesline to simulate roughly the length of the digestive tract. How is food moved along the tract? What changes take place along the way? How does food material get out of the small intestine into the blood stream, and finally into the cells?

3-21. You may want to begin with a film which reveals many of the ways animals capture their food. Have you seen some of the films and filmstrips in the latest film catalogs?

3-22. Food-making by plants and food-taking by animals, along with digestion of insoluble materials, may be part of a broad study of interdependence among living things, as developed in chapter 8 (especially those concerning food webs and chains in an ecosystem). You may want to begin this study of digestion with activities described in that chapter.

3-23. In a laboratory lesson, have students dissect frogs so that they may trace the path of food through the body (see chapter 1). Compare this with the digestive system of humans (as shown in charts, or use a manikin). Further, how do frogs' parasites get their food?

3-24. You may want to begin a lesson with a comparison of the kinds of heterotrophic modes of nutrition: holozoic, predator-prey, parasitism, and saprophytism (see chapter 8). From here, develop a discussion of the need for digestion and the uses of food in the body.

3-25. Begin with a demonstraction of moldy fruit and discuss how molds get their food supply. How are they able to live on fruit or bread? Now you are into a discussion of interrelations among plants and animals and the dependence on green plants as the food source. Refer to soil microfauna and microflora. How do they get their food? What is the difference between intracellular and extracellular digestion?

3-26. As a laboratory lesson, have students test the action of their own saliva on starch paste, as described in this chapter. Or have one or two students demonstrate how saliva changes the starch in starch paste or a chewed cracker into soluble form.

3-27. Why do plants need a starch-splitting enzyme? Demonstrate the change from starch to sugars in germinating seedlings. Use nongerminating seeds as controls.

3-28. Have a committee of students report on the classic experiments in digestion done by Spallanzani, by Réaumur, and by Dr. Beaumont. Lead into a discussion of the role of enzymes in making food materials soluble so that they can diffuse through membranes out of the digestive system.

3-29. Have a student report on the way termites get their cellulose food supply digested by flagellates. If possible, examine living flagellates from the intestine of termites (p. 63).

3-30. You might plan a laboratory lesson to observe epithelial cells such as those in the lining of the mouth. Directions are given for scraping the lining of the cheek to get epithelial cells for mounting in Lugol's solution on page 148.

3-31. If you have not already demonstrated diffusion of molecules through membranes, you may want to begin a study of digestion with either starch suspension or glucose in dialyzing tubing. Fasten each end to a tongue depressor or splint, and suspend over a beaker of water. Or use a starch and saliva mixture (or diastase and starch). If you do not have this cellophane tubing, use test tubes across the mouth of which wet goldbeater's membranes are stretched as described previously (Fig. 3-26). Test the water in the beakers to find out which molecules pass through a membrane. Why didn't starch go through the membrane? Now you are ready to discuss digestion.

3-32. Ask students to list all the food they consumed the previous day. Develop the notion of nutrients (and balanced diet) and lead into a discussion of the "basic seven," then into tests for nutrients. Discuss animal experimentation and the use of chemical tests for the nutrients in foods.

3-33. At some other time, begin with a laboratory lesson testing the vitamin C content of some fruit juices (fresh and canned) with indophenol (described in this chapter).

3-34. Occasionally, have a supervised study lesson. Have you tested the reading and comprehension level of your students by having them read a section in their text? Or use supplementary readings to discover their interests. With slow readers you may want to try a chapter in a general science textbook.

3-35. Discuss the significance of Federal supervision of food and drugs. As an aid in this study have students examine and paste into their notebooks labels from food packages and from empty medicine bottles. What may be the dangers in using additives?

3-36. Compare the diets of an average individual in the United States, Mexico, India, the Arctic, Spain, South America, China, Greece, and Germany. Compare the amount of proteins, minerals, and vitamins as well as the other nutrients contained in these diets. Develop a discussion of what makes an adequate diet.

3-37. Examine what makes a good or a poor meal. For example, if you had black coffee with a sugared doughnut for breakfast, how could you improve this meal?

For lunch you had spaghetti, rolls and butter, custard pie, and black coffee with sugar. What's wrong with this lunch? What nutrients are in excess? Which nutrients should be added—and in what foods?

Ask students to plan a good diet, and develop a discussion of the kinds of food needed for a good diet. What is the function of proteins in the body? Of vitamins?

3-38. Prepare a demonstration to show which components of gastric juice digest proteins (described in this chapter). Better still, students may design the demonstration themselves; further, they may investigate factors that affect the activity of enzymes in a quantitative study.

3-39. With a photometer, introduce techniques for measuring growth of microorganisms. Plan several studies relating to growth (see chapter 6). Which minerals are critical for stimulating and maintaining growth?

3-40. Show a film describing how radioactive tracers are used in plant nutrition. Which free films are available from the U.S. Atomic Energy Commission? What has been learned about nutrition in plants as a result of using tracers?

BOOK SHELF

Anthony, C., and G. Thibodeau. *Textbook of Anatomy and Physiology*, 11th ed. St. Louis, MO: Mosby, 1983.

Borek, E. *The Atoms Within Us*, 2nd ed. New York: Columbia University Press, 1981.

Brooks, S. *Integrated Basic Science*, 4th ed. St. Louis, MO: Mosby, 1979.

Cameron, J. and J. Skofronick. *Medical Physics*. New York: Wiley, 1978.

Danks, S., E. H. Evans, and P. Whittaker, *Photosynthetic Systems*. New York: Wiley, 1983.

De Coursey, R. *The Human Organism*, 4th ed. New York: McGraw-Hill, 1974.

Eckert, R. and D. Randall. *Animal Physiology*, 2nd ed. San Francisco: Freeman, 1983.

Esau, K. *Anatomy of Seed Plants*. New York: Wiley, 1977.

Ford, J. and J. Monroe. *Living Systems*, 3rd ed. New York: Harper & Row, 1977.

Gabriel, M. and S. Fogel, (eds.) *Great Experiments in Biology*. Englewood Cliffs, NJ: Prentice-Hall, 1955.

Giese, A. *Cell Physiology*. Philadelphia: Saunders, 1979.

Herreid, C. *Biology*. New York: Macmillan, 1977.

Hickman, C., Jr., L. Roberts, and F. Hickman. *Biology of Animals*. St. Louis, MO: Mosby, 1982.

Holldal, P. (ed.) *Photobiology of Microorganisms*. New York: Wiley, 1970.

Knobloch, I. *Readings in Biological Science*, 3rd ed. New York: Appleton-Century-Crofts, 1967.

Lind, O. *Handbook of Common Methods in Limnology*, 2nd ed. St. Louis, MO: Mosby, 1979.

Lyons, A. and J. Petrucelli. *Medicine: An Illustrated History*. New York: Abrams Inc., 1978.

McClintic, J. *Human Anatomy*. St. Louis, MO: Mosby, 1983.

Nelson, G. and G. Robinson. *Fundamental Concepts of Biology*, 4th ed. New York: Wiley, 1981.

Nilsson, L. *Behold Man*. Boston: Little, Brown, 1973.

Noland, G. *General Biology*, 11th ed. St Louis, MO: Mosby, 1983.

Norstog, K. and R. Long. *Plant Biology*. Philadelphia: Saunders, 1976.

Saigo, R. and B. Saigo. *Botany*. Englewood Cliffs, NJ: Prentice Hall, 1983.

Schmidt-Nielsen, K. *Animal Physiology*, 3rd ed. Cambridge U. Press, 1983.

Singer, S. and H. Hilgard. *The Biology of People*. San Francisco: Freeman, 1978.

Stephens, G. and B. North. *Biology*. New York: Wiley, 1974.

Toporek, M. *Basic Chemistry of Life*. St. Louis, MO: Mosby, 1981.

Tweney, R. et al. *On Scientific Thinking*. New York: Columbia University Press, 1981.

Vander, A., J. Sherman and D. Luciano. *Human Physiology*. 3rd ed. New York: McGraw-Hill, 1980.

Villee, C. *Biology*, 7th ed. Philadelphia: Saunders, 1977.

Weier, T., C. Stocking, and M. Barbour. *Botany*, 5th ed. New York: Wiley, 1974.

Williams, T. *A Biographical Dictionary of Scientists*, 3rd ed. New York: Wiley, 1981.

4

Utilization of Materials: Building of the Organism

An organism, whether a plant or an animal, draws materials from its environment and returns other materials to the environment in the processes of maintenance, growth, repair, and reproduction.

Recall how well Walter Cannon summarized the needs of a "single cell" compared to those of cells in many-celled organisms:

Each cell has requirements like those of the single cell in the flowing stream. The cells of our bodies, however, are shut away from any chances to obtain directly food, water, and oxygen from the distant larger environment, or to discharge into it the waste materials which result from activity. These conveniences for getting supplies and eliminating debris have been provided by the development of moving streams within the body itself—the blood and lymph streams.[1]

The substance of this chapter concerns (a) transport of materials in animals and in plants; (b) energy flow from basic glucose molecules to ATP molecules in cells; (c) respiration and excretion in the maintenance of homeostasis.

Transport or circulation in animals

There are many good opportunities for individual and group investigations of heartbeat and circulation of blood in

several invertebrates and vertebrates. Activities relating to the heart, circulation, blood cells, and blood typing are described in this first section.[2]

Recording arterial pulse

Students learn easily to take their own pulse. Place the index and middle finger of one hand on the inner side of the other wrist. About 3 to 4 cm above the wrist joint, near the thumb, feel the pulse in the radial artery. Then students may want to take each other's pulse. They should count the pulse for several 15-sec periods, find the average, and multiply this by four to get the average pulse rate per minute.

Next, compare this normal pulse rate with the rate *after* exercise. Students might hop in place, do knee-bending exercises, or run up a flight of stairs. Count the pulse again. (*Note:* When teachers select students for any exercise they should be apprised first of their students' medical records. These may be obtained from the school's health education department.) Also refer to the effect of increased activity on the rate of carbon dioxide production, described on page 301.

The beating heart

Heartbeat in the frog Anesthetize a frog with MS-222 or other substances

---

[1] W. Cannon, *The Wisdom of the Body* (New York: W. Norton, 1932).

[2] Teachers must, of course, select those techniques that are best suited to their classroom situation.

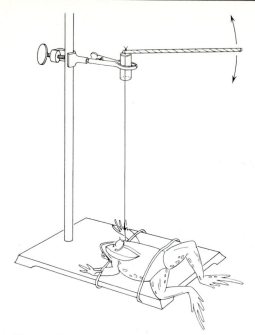

Fig. 4-1 Preparation for demonstrating heartbeat in a freshly killed frog. The tip of the ventricle is pierced by a bent pin which is tied to a thin thread attached to a straw supported by a hollow glass tube.

(page 142), or pith a frog (Fig. 5-12). Then open the frog to expose the heart. Remove the pericardial sac, and bathe the heart throughout the demonstration in frog Ringer's solution (p. 155). The heart will continue to beat for several hours.

For class demonstration you can amplify the contractions of the heart by making a lever arrangement (Fig. 4-1).

ALTERING THE HEATBEAT: TEMPERATURE The effect of changes in temperature may be shown by bathing the frog's heart with water warmed to 40° C (104° F), then with iced water. Graph the number of beats at room temperature, at 40° C, and with ice water. How long does it take for the heartbeat to be restored to "normal" at room temperature?

ALTERING THE HEARTBEAT: CHEMICALS You may also want to test the effect of epinephrine (adrenalin), acetylcholine, and serotonin in concentrations of 1 : 10,000. If you find 1 : 1000 concentrations are the only ones available, dilute to 1 : 10,000; draw 0.1 ml of epinephrine into a 1 ml syringe and fill the syringe to 1 ml with isotonic 0.9 percent saline solution. Record the normal heartbeats. Add 3 drops of the substance to be tested; wash the heart with isotonic saline solution before applying another substance.

EFFECT OF IONS ON HEARTBEAT You may want to repeat, in a simple way, the classic work of Sydney Ringer in the 1880s in determining how ions affect the heartbeat of a frog. Prepare the following solutions: (a) Ringer's solution; (b) excess calcium ions added to Ringer's (add 1 ml of 10 percent $CaCl_2$ to 99 ml of Ringer's); (c) excess potassium ions (add 1 ml of 10 percent KCl to 99 ml of Ringer's); (d) isotonic sodium chloride.

Determine the effect of ions of calcium, potassium, and sodium on an excised heart. Suspend the heart of a freshly killed or pithed frog in a beaker of Ringer's solution by means of a bent pin (like a fishhook) and thread attached to a tongue depressor. After the initial shock wears off, count the number of heartbeats in 1 min. Count the number in 2 to 3 min.

Now transfer the preparation into a beaker of Ringer's with the *excess calcium* ions; again count the beats for 2 to 3 min; return the heart to Ringer's and count the beats; transfer the heart to the Ringer's solution with *excess potassium* ions; count the beats for 2 to 3 min. Return the heart to Ringer's solution; count the beats. Finally transfer the heart to isotonic *sodium chloride* solution (0.7 percent).

Have students observe that the heart stops contracting in excess potassium chloride ion solution; in calcium chloride solution the heart contracts again. The ions in Ringer's solution are balanced to produce the normal contraction of the frog's heart.

Some teachers prefer to transfer the excised heart of a frog into separate beakers of the solutions of ions. First count the number of times the heart beats in a minute in Ringer's solution. Now transfer the heart into a beaker of 0.7 percent sodium chloride solution. Record the number of beats now; keep the solutions at room temperature. You will notice that the heartbeats are weaker and slower. Next replace the heart in Ringer's solution and note the recovery in rate of heartbeat.

Next, transfer the heart into a beaker of 0.9 percent potassium chloride solution. Again count the heartbeats until the heart muscle stops contracting. If the heart is now transferred into a 1 percent calcium chloride solution, you will find the heart contracting again. Replace the heart in Ringer's solution and watch the return to normal rate of heartbeat.

Heartbeat in the clam In many areas students may get hard-shelled clams from a fish market. Open a live clam by gently cracking one shell; then cut the anterior and posterior adductor or shell muscles with a sharp knife. Remove the mantle around the viscera and observe the beating heart in the pericardial cavity, which lies below the dorsal hinge ligament. Bathe the heart in clam juice or clam Ringer's solution (p. 155). Count the contractions of the ventricle. What is the effect of change in temperature? What is the effect of adrenalin (epinephrine) or of several types of drugs?[3]

Heartbeat in embryo snails Some students may have the snail *Physa* (or other egg-laying snails) in their aquaria at home. Tease apart a developing egg mass in a drop of aquarium water on a microscope slide and examine under low power (without a coverslip). Find a beating heart in an embryo in the early stages when the shell is still transparent. (Also note the periodic muscular contractions of the body.)

Heartbeat in Daphnia *Daphnia* (water fleas) may be obtained from aquarium stores or from lakes and ponds (culturing, chapter 9).

Estimate the average rate of heartbeats by counting the rapid heartbeats of several *Daphnia*. Depending on the species, the heartbeat may vary from 230 to more than 300 times per minute.[4] The heart is a small, muscular sac located dorsally behind the large eye (see Figure 1-60 for anatomy). There are no blood vessels. The rate of heartbeat is more rapid in smaller males, but males are rare specimens among the parthenogenic *Daphnia*.

Since the heartbeat is very rapid this effect may not be simple to detect. Students will need a stopwatch and a mechanical hand counter for these counts.

A strobe disk made of cardboard with a single slit and turned by a 300-rpm motor can be used in front of the light source to slow the heartbeat's apparent motion.

Place small dabs of petroleum jelly in the well of a depression slide. Gently embed *Daphnia* on each bit of jelly to immobilize them. Add aquarium water in which they were maintained. First count the "normal" heartbeat, then test several variables. For example, increased light intensity or increased temperature of the water speeds the heartbeat. A lack of oxygen, starvation, or cooling of the water slows the heartbeat. Acetylcholine (1:10,000) also slows the heartbeat of a myogenic heart such as that found in *Daphnia*. What is the change in heartbeat when a tranquilizer such as chlorpromazine is used? Or a stimulant such as Dexedrine?

What is the effect of epinephrine (1:10,000), or of alcohol, tea, coffee, etc.? Wash off the first substance tested with aquarium water before adding the second test substance.

You may find that some *Daphnia* are almost colorless while others are reddish in color. *Daphnia* lacks red blood cells; the iron compound, hemoglobin, is dissolved in the blood. (They do have white blood corpuscles—the type that Metchnikoff studied.) When maintained under well-

[3] Refer to W. Hoar and C. Hickman, Jr., *Laboratory Companion for General and Comparative Physiology*, 2nd ed. (Englewood Cliffs, NJ: Prentice-Hall, 1975).

[4] Rate of heartbeat varies with species as well as with environmental conditions.

aerated conditions their blood is almost colorless, while in poorly aerated water *Daphnia* gains more hemoglobin and becomes more deeply colored. After about ten days, the blood can change from deep color to almost colorless if placed in well-aerated water. In another ten days, if placed back into poorly aerated water, they gain more hemoglobin.

Heartbeat in the earthworm   Secure several live earthworms. Suspend them, anterior end down, for a few minutes or twirl them around so that the blood is concentrated into the anterior end. Anesthetize the worms by placing them in a shallow dish with enough water to cover them. To this add a small amount of alcohol, sufficient to immobilize the worms but *not* to kill them. Next wash off in water, and pin each worm to a waxed dissection pan or a sheet of cork. Make a lengthwise cut from the clitellum to the prostomium (tip of the head), cutting to one side of the mid-dorsal line (dissection, chapter 1). Cut through the body wall but avoid cutting into the viscera. Watch the beating of the five pairs of aortic loops or "hearts." With a hand lens or a binocular microscope observe the circulation of blood.

Heartbeat in aquatic worms   Mount any one of these annelid forms which may be obtained from an aquarium store or a lake or pond: *Dero, Nais, Aulophorus,* or *Tubifex* (culturing, chapter 9; figures in chapter 1). Add a bristle to the mount before the coverslip is applied. Under low and then high power, observe, especially in *Tubifex*, the single pair of aortic arches or loops, and note the movement of the dorsal blood vessel which pumps blood. The worms may be anesthetized first by placing them in a drop of water to which a drop of very dilute chloretone has been added. Other methods for anesthetizing small organisms have been described on page 142.

Dissecting a heart   Students may get an untrimmed heart of a cow, sheep, or hog from a butcher; or purchase from a biological supply house. Locate the atria or auricles, ventricles, aorta, venae cavae, and pulmonary vessels (Fig. 4-2). Insert a soft metal probe, a pipe cleaner, or use a fire-polished section of a glass rod as a probe, to explore the point of origin of blood vessels which lead into and leave the chambers of the heart. Note the semilunar valves in the blood vessels, and determine the direction of blood flow.

Dissect the chambers of the heart. Find a diagonal deposit of fat along the lower two-thirds of the heart to be dissected. Use this as a guideline that marks the wall between the two ventricles. Make an incision into each of the two ventricles. Spread the heart open and use a pipe cleaner or a glass rod to probe the position of the bicuspid and tricuspid valves which separate the ventricles from the auricles (atria).

If possible, direct a stream of water (through tubing from a faucet) through the ventricles to examine the operation of the valves.

Compare the thickness of the walls of the atria and ventricles. Note especially the thickness of the left ventricle through which blood is pumped to the entire body. Look for the coronary arteries from the aorta and trace their paths as they spread over the heart. Students might cut cross

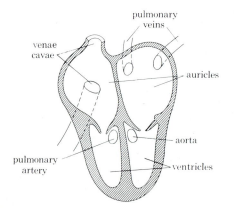

Fig. 4-2   Schematic diagram of the mammalian heart.

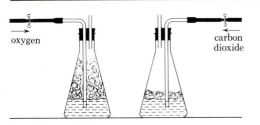

Fig. 4-3 Apparatus for demonstrating the effects of O_2 and CO_2 on oxalated blood.

sections of the aorta and a vena cava to compare the elasticity and thickness of the walls.

Oxygenated and deoxygenated blood

Get several pints of fresh blood from a slaughterhouse, and add 1 part of sodium oxalate to 9 parts of blood to prevent it from clotting.

Prepare two flasks, each with a short piece of glass tubing and a right-angle bend of glass tubing inserted in a rubber stopper. A piece of rubber tubing leads from each bent tubing—to an oxygen generator for one flask and to a carbon dioxide generator for the other (Fig. 4-3).

Note the froth and the brilliant red color of oxygenated blood and the darker red color of blood rich in carbon dioxide.

Clamp the delivery tubes and reverse the attachments so that oxygen now flows into the flask of deoxygenated blood. Watch the change to bright red as oxyhemoglobin is formed in the blood.

In lieu of a generator of carbon dioxide, the gas may be supplied by a small piece of dry ice or by chalk dust added to a carbonated beverage. Oxygen may be supplied (if no generator is available) by adding 1 ml of hydrogen peroxide to 15 ml of citrated blood.

Blood vessels When possible have students get lengths of arteries and veins from a butcher and bring them to class. Compare their thickness in cross section and the difference in elasticity. Cut along the length of a vein, and open it to show the cuplike valves. Hold it upright and try to fill it with water to show the action of the valves. Valves in lymph vessels may also be studied as a prepared slide (Fig. 4-4).

DEVICES TO ILLUSTRATE BLOOD VESSELS At times we may establish a concept with remarkably simple models. Because they are immediately contrived before students, they have additional impact.

Have you ever used this? Pull apart cheesecloth or gauze. This can represent the appearance of capillary beds. Or separate out the central strands of a coarse hemp cord. Dip one unfrayed end of the cord into red ink, the other into blue. Isn't this a standard representation of arteriole, capillaries, and venule?

Remove the insulation from the center section of a 10 cm length of electric wire. Spread apart the individual fine copper wires in the center. Imagine that the separated wires represent capillaries. Then the ends might represent an artery and vein.

Show capillary circulation around an air sac in the lungs by wrapping strands of cord around a Florence flask. Or with a glass-marking pencil draw "capillaries" on the flask. And, in similar fashion, show circulation in a villus of the small intestine. Wrap cord around a small test tube. Or

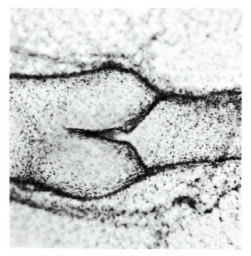

Fig. 4-4 Photo of valve in lymph vessel as shown on a prepared slide. (Courtesy of General Biological Supply House, Inc., Chicago.)

Fig. 4-5 Demonstration of blood circulation in the tail of a goldfish.

draw capillaries with a glass-marking pencil. Now insert this into a larger test tube. This will serve as a villus with an outer epithelial layer, and the smaller test tube within the larger one represents a lacteal.

Circulation in capillaries

Circulation in the tail of a goldfish

You may want to show circulation in the capillaries, arterioles, and venules in the almost transparent tail of a small goldfish or killifish. Wrap the body of the goldfish in wet absorbent cotton so that only the tail is exposed (Fig. 4-5). Place it in a Petri dish with a small amount of water. Cover the tail with a glass slide to hold it flat and expanded. Examine under low and high power of the microscope.

Watch the pulse and the swiftly moving blood in the arterioles. Contrast this with the more slowly moving blood which circulates in the opposite direction in the venules. Many criss-crossing capillaries are visible. Look for blood cells floating in the blood stream. Irregular black patches of pigment in the skin and scales (chromatophores) may be visible in the specimens (see p. 153).

Have three or four students set up several similar small group demonstrations so that everyone may observe circulation within a 15-min period. Then return the fish to the aquarium. At times the fish flip their tails vigorously, and the demonstrations must be adjusted. When desired, the fish may be anesthetized with chloretone or MS-222 (p. 142). However, circulation is very sluggish in anesthetized fish.

You may want to test the effect of a vasoconstrictor such as epinephrine (adrenalin) on small blood vessels and the effects of a vasodilator such as histamine (procedure and preparation of solutions, page 738).

Circulation in a frog

IN THE GILLS AND TAIL OF A TADPOLE When very young tadpoles are available, students may observe circulation of blood in the external gills. Later the gills are internal as epithelial cells of the skin grow over them. Nevertheless, in these older forms circulation can be observed in the almost transparent tail.

Mount a tadpole on a depression slide in a drop of aquarium water. Omit the coverslip or add a bristle to prevent the coverslip from crushing the specimen. Watch the circulation in capillaries under the low and high power.

IN THE FOOT The webbed skin between the toes of the frog is rich in capillaries. Wrap a live frog in wet cheesecloth, paper toweling, or absorbent cotton, and hold fast in a plastic bag. Lay the animal on a frog board made of cork or stiff cardboard, so that only one foot is extruded (Fig. 4-6). Or you may want to anesthetize the frog by injecting it with about 5 ml of a 2.5 percent solution of urethane (p. 142). Inject into the lymph sacs under the loose skin along the sides and back of the frog. Frogs can be anesthetized without injection by putting them into a solution of chloretone. Add 1 part of a 0.5 percent solution of chloretone in water to 4 parts of Ringer's solution. Keep the frog in this solution for about half an hour or until quiet; avoid longer immersion.

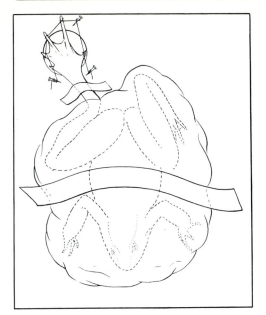

Fig. 4-6 Demonstration of blood circulation in the webbed foot of a frog. The frog is wrapped in wet gauze in a small plastic bag and secured on a corkboard with masking tape.

Whether the frog is untreated or anesthetized, wrap it in wet cheesecloth and a plastic bag; pin it to the board by fastening it with masking tape. Punch a hole in one corner of the corkboard and spread the web of one foot over this hole; pin in place or wind string around the toes and secure it with pins to keep the toes immobile (Fig. 4-6). Keep the frog's skin wet during the demonstration.

Place the board on the stage of the microscope and focus on the webbed foot. Observe capillaries, arterioles, and venules and the rate of blood flow. Notice the direction of the flow of blood and the pulsations of the blood vessels. Estimate the diameter of the capillaries (p. 124). Under high power compare the movement of red and white corpuscles. Notice the appearance and disappearance of capillaries. Under high power, look for white blood cells moving through capillary walls (diapedesis).

EFFECT OF CHEMICALS ON CAPILLARIES: WEBBED FOOT Watch the action of a vasoconstrictor on small blood vessels by adding a drop of dilute epinephrine (0.01 percent) to the webbed foot of a frog. Then wash the web and test the effect of a vasodilator such as histamine (1:10,000) solution. Again wash the web. Test whether the following are dilators or constrictors of capillaries: lactic acid (0.1 g per liter), acetylcholine (0.05 g per liter), nicotine solution, ethyl alcohol (95 percent), and sodium nitrite solution (0.1 g per liter). Each solution should be washed away before the next solution is applied.

EFFECT OF TEMPERATURE ON CAPILLARIES: WEBBED FOOT Place a small piece of ice on the web and observe the effect on circulation. Then alternate with warm water, 35° to 40° C (95° to 104° F), and observe the results.

CAPILLARIES IN THE TONGUE Anesthetize a frog by placing it in chloretone or a dilute solution of MS-222 (p. 142). Wrap the body in wet toweling, cheesecloth, or absorbent cotton; and place the animal, dorsal side up, on a corkboard. Spread the unattached tip of the tongue over the hole in the corkboard and secure with pins. Or cover the tongue with a strip of cellophane and fasten to the corkboard with pins.

Place the board on the stage of the microscrope, keep the tongue well lighted, and observe under low power a section revealing blood vessels of different sizes. Clamp the board to the stage of the microscope with a rubber band or masking tape.

Keep the tongue wet with frog Ringer's solution. Watch the pulsations in the arteries, the opposite direction of blood flow in arteries and veins, and the capillaries connecting the larger blood vessels.

Then under high power try to distinguish red and the larger white corpuscles. Estimate the diameter of the capillaries by using red corpuscles as reference. In the frog, these cells average 22 μ in length, 15 μ in width, and 4 μ in thickness.

Compare these effects on circulation: Apply slight pressure to the tongue by stroking it with a dissecting needle. Next apply, drop by drop, a small quantity of dilute urethane to a small area of the tongue. Watch the changes in the capillaries. Then tie a string around a portion of the tongue to close off circulation for a short time. Observe the effect and watch the changes as the string is released. Also apply a small amount of adrenalin (0.01 to 0.1 percent) to a small area of the tongue, and watch the changes that the adrenalin produces. What is the effect of adding a drop of warm water, or a drop of ice water?

CAPILLARIES IN THE LUNG   Anesthetize (p. 142) or pith (Fig. 5-12) a frog. Then pin the frog to a corkboard or a waxed dissection pan and dissect to expose the lungs (dissection, chapter 1). Inflate the lungs by blowing through fire-polished glass tubing inserted into the glottis of the frog. Keep the inflated lungs bathed in frog Ringer's solution. Examine the inflated air sacs and the rich network of capillaries in a bright light using a hand lens or a dissecting microscope. Observe circulation of blood cells in the capillaries.

A small section of inflated lung can be examined under a compound microscope. Spread a thin section over the hole in a sheet of cork similar to the one described above. Be sure to keep the tissue moist with frog Ringer's solution.

CAPILLARIES IN THE MESENTERY   Also try spreading the capillary-rich mesentery supporting the intestine across the opening in the cork sheet for observation—another dramatic demonstration of circulation.

CAPILLARIES IN THE URINARY BLADDER McNeil et al. recommend a study of the peripheral circulation in the urinary bladder of the frog.[5] In this preparation, blood vessels are not obscured by pigment cells, and since the vessels are confined to a single plane, complete circuits can be observed through arterioles, capillaries, and venules without refocusing the microscope.

A frog is prepared for the demonstration by a preliminary injection of 5 ml of frog Ringer's solution into the dorsal lymph

[5] C. W. McNeil et al., "The Use of the Urinary Bladder of the Leopard Frog in the Demonstration of Peripheral Circulation," *Turtox News* (August 1958) 36:8.

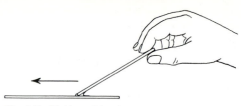

Fig. 4-7   Making a blood smear.

sinus. This pretreatment results in a distension of the bladder. A half-hour later the frog is anesthetized by an injection of 3 to 6 ml of 5 percent urethane (p. 142) into the dorsal lymph sinus. When the frog is anesthetized (withdrawal reflex has ceased), a 2-cm incision is made parallel to the midline and a bit dorsal to the ventral edge of pigmentation. Start the incision about 1 cm anterior to the hind limb. The distended urinary bladder can be extruded through the incision and prepared for study under a dissecting microscope with transmitted light. (If the bladder is not distended, Ringer's solution may be inserted through the cloaca.)

McNeil et al. also suggest observation of vasodilator effects of histamine (1 or 2 drops of 0.001 histamine) or of 1 drop of 1 percent acetic acid. Wash off the chemicals between applications. Also apply 1 drop of 0.001 M solution (or more dilute) epinephrine to show vasocontractor effect on arterioles.

Blood cells: animals

Blood cells of a frog   Mount a drop of fresh frog's blood on a clean slide. Touch one end of a second slide to the drop; hold the second slide at a 30° angle to the first slide. Push the second slide along the first so that a thin film of blood is spread the length of the slide (Fig. 4-7). Let the slide dry. Examine without a coverslip, or later stain with Wright's blood stain (p. 274); mount the slide in balsam with a coverslip if you want a permanent slide.

A drop of blood may be examined in a drop of frog Ringer's solution as a wet mount instead of a dry smear. Look for red cells that are oval, nucleated disks.

**MITOCHONDRIA AND AMOEBOID MOVE-
MENT OF WHITE CORPUSCLES** In this
technique, described by Bensley and Bensley, vital dyes are added to blood cells on a
slide.[6] As the stains are absorbed, the cells
gradually die. However, the structures
within the cells may be examined. The
following stain technique also stains mitochondria as tiny green organelles.

Prepare a number of clean slides with a
coating of the stains described as follows:
First prepare a *stock* solution of neutral red
in absolute alcohol by adding 100 mg of
neutral red to 10 ml of absolute alcohol.
Then make a *dilute* solution by adding 10
ml of absolute alcohol to 4 ml of the stock
solution. To this dilute solution add 1 or 2
drops of a saturated solution of Janus
Green B in distilled water.

Now support a chemically clean, dry
slide on a Syracuse dish, and flood the
slide with the diluted neutral red solution
containing Janus Green B. Allow the excess stain to flow off by tilting the slide into
the Syracuse dish. Set aside to dry. Prepare a supply of slides with a dried drop of
stain. (Protect them from dust.)

Mount a drop of fresh frog's blood on a
slide with the dry film of stain. Apply a
coverslip. Examine under low power and
under high power. As the dyes slowly
dissolve, notice that the mitochondria of
the cells stain green and the cell inclusions
become red. Amoeboid movements may
be studied. As the cells gradually die, their
nuclei absorb more of the stain.

PHAGOCYTOSIS BY LEUKOCYTES Observe the manner in which leukocytes of
frog's blood ingest materials. Add a small
amount of India ink or carmine powder to
some physiological saline or Ringer's solution for frogs (p. 155).).

Anesthetize a frog or use a pithed frog;
insert a hypodermic needle into a dorsal
lymph sac and draw off a small quantity
of lymph. On a clean slide add a drop of
lymph to a drop of saline plus ink (or
carmine particles). Add a bit of broken
coverslip or a piece of straw before placing
a coverslip over the preparation to prevent the coverslip from flattening the cells.
Seal the edges of the coverslip with petroleum jelly.

Inspect the slide under low power and
then switch to high power and refocus the
light to look for leukocytes engulfing particles suspended in the fluid.

If part of a webbed foot is injured by
applying a hot needle, you may shortly
observe a gathering of white cells in the
area.

Blood of invertebrates Techniques for
studying blood of *Limulus* and other invertebrate blood pigments are described in
chapter 1.[7]

Blood cells of a mammal Occasionally
you may want students to compare the red
blood cells of a frog with those of a mammal. Blood may be obtained from a
slaughterhouse; usually ox blood is available. To prevent clotting, add 0.1 g of
potassium or sodium oxalate for every
100 ml of ox blood. Or you may prefer to
add 1 part of 2 percent sodium citrate
solution to 4 parts of blood. Both the
oxalate and citrate are anticoagulants.
Other suggested preparations of anticoagulants follow.

ANTICOAGULANT 1 FOR BLOOD Add
1 ml of this solution to 10 ml of blood. To
prepare this solution, dissolve the following in 100 ml of distilled water:

potassium oxalate	2 g
NaCl	6 g

ANTICOAGULANT 2 FOR BLOOD Add
200 mg (0.2 g) of sodium citrate to 10 ml
of blood. Then mount a drop of the ox
blood as a wet mount (or prepare a blood
smear as for frog's blood, above). Both
white and red cells are visible. Use petroleum jelly on the mount to prevent
evaporation (p. 141).

---

[6] R. Bensley and S. Bensley, *Handbook of Histological and Cytological Technique* (Chicago: University of
Chicago Press, 1938).

[7] Also refer to I. Sherman and V. Sherman, *The
Invertebrates: Function and Form*, 2nd ed. (New York:
Macmillan, 1976).

Blood cells: human

Note: Whenever activities call for drawing samples of blood from students, the following procedure is strongly advised. (Also refer to safety precautions, chapter 10.)

1. Obtain consent notes from parents if your community or school requires them. A note might be obtained for the term or year to cover such activities as making blood smears and typing blood.
2. Apply alcohol to the area of the finger to be pricked. Allow the alcohol to evaporate.
3. Use individually wrapped sterile, disposable lancets.
4. After puncturing the skin and obtaining blood, clean the finger with alcohol again. If bleeding continues, have the student apply pressure by clenching a piece of sterile cotton or a square of gauze.

Unstained cells   Have students examine their own unstained blood cells in a drop of distilled water, and also in a drop of isotonic solution such as Ringer's Locke's, or Tyrode's (pp. 155,156).[8] Apply a coverslip to each drop on the one slide. Under high power, compare the shape of the red cells in distilled water with cells in an isotonic solution. The osmotic content of red blood cells is about 0.9 percent NaCl. Now examine blood cells in a hypertonic solution to show crenation of red cells. Introduce a drop of a 3 to 10 percent sodium chloride solution under the coverslip of the isotonic demonstration. How are red cells affected?

White blood cells   White corpuscles may be examined more readily after red cells have been destroyed. Prepare the slide as described above; add a drop of acetic acid at the edge of the coverslip so that it diffuses under the coverslip. Note the clear view of white blood cells that remains as the red cells are lysed.

---

[8] You may want to mention that van Leeuwenhoek described red blood cells of humans in 1673. In 1879, Paul Ehrlich developed some staining techniques.

Blood smears   Puncture the finger and place a drop of blood about one third from one end of a clean slide or touch the end of the slide to the finger puncture. Place a second slide to be a "spreader" slide with its narrow end at an angle of 30 to 45° on the first slide as shown in Figure 4-7. Now back up the upper slide to the drop of blood until the blood spreads along the narrow end of the top slide forming a uniform layer due to capillarity. *Push* the upper slide rapidly, but evenly, toward the opposite end of the bottom slide so that the blood is dragged across the slide forming an even, unrippled thin film. Let the slide air-dry. The greater the angle of the top slide, the thicker the film that is formed. A good film should be smooth and very thin toward the end of the smear with no wavy surfaces.

Chemically clean glassware is required in preparing good blood smears. Use new slides and coverslips, or place them in a beaker of 95 percent alcohol, wipe *dry*, and flame over a Bunsen burner before use.

WRIGHT'S BLOOD STAIN   Lay the dried blood smear slide across a Syracuse dish, or support it across a wide cork in a Petri dish. Cover the dry blood film completely with Wright's stain for 1 to 2 min to fix the blood film. Next, add about 6 drops of a phosphate buffer solution, drop by drop, tilting the slide to mix the solutions, until a metallic, greenish scum forms on the surface of the stain. (Prepare the buffer solution by adding 6.63 g monobasic potassium phosphate and 2.56 g dibasic sodium phosphate to 1 liter distilled water resulting in a pH 6.4.) Continue to add the buffer until you have diluted the stain by half. Let this stand for 2 to 5 min. Then rinse in distilled water.

After the first application of buffer to the original stain, place the slides in fresh buffer for 2 to 3 min, or until the stain is lavender pink. Then stand the slides on edge to dry, or blot dry with bibulous paper. The staining time varies with each batch of slides and different bottles of Wright's stain and buffer; prepare trial slides to determine staining time. Inspect

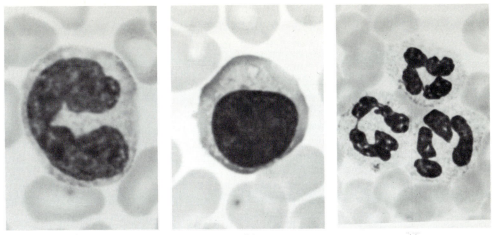

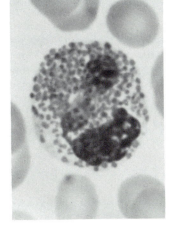

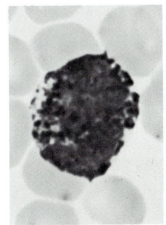

Fig. 4-8 Blood cells: Photos of five kinds of leukocytes. (a) monocyte; (b) lymphocyte; (c) netrophils; (d) eosinophil; (e) basophil.

Background ghosts are red cells and thrombocytes. (Copyright Taurus Photos, New York.)

under the microscope without a coverslip or balsam. If the stain is too dark, decolorize by returning the slides to the buffer. (For best results, purchase Wright's stain and buffer already prepared.)

Since Wright's stain contains methylene blue (alkaline stain), and eosin, which is a red, acid stain, the pH of the buffer solution affects the color of the cells. The more acidic the buffer results in cells which appear more red; blue cells indicate a buffer of higher pH. Buffers of pH 6.4 to 6.8 give best results.

Examine the blood smear under high power to locate stained leukocytes. Erythrocytes will be pink—Wright's stain stains leukocytes and platelets. Granules in basophils should be deep blue; in eosinophils, bright red; and in neutrophils, lilac (Fig. 4-8). Large monocytes will appear to have a few lilac granules, and the many lymphocytes should show a solid blue nucleus that almost fills the cell with a rim of pale blue cytoplasm. Platelets are clusters that stain blue. (If erythrocytes stain blue or green, the stain is probably too alkaline, or the slide is overstained. Excess acidity results in the whole stained smear appearing too red.)

Details of cellular structure are best studied under an oil immersion lens (p. 123). Add immersion oil directly to the stained, dried blood film.

GIEMSA STAIN   This is a fine stain for both blood smears and bacterial smears. A *stock* solution of Giemsa stain may be purchased, or prepared as follows. Dissolve 0.5 g of Giemsa powder in 33 ml of glycerin (this will take 1 to 2 hr). Then add 33 ml of acetone-free absolute methyl alcohol. To use the stain, dilute it by adding 10 ml of distilled water to *each milliliter* of stock solution of stain; buffer to pH 7.2 with phosphate buffer (p. 159).

Let the blood smear air-dry, then fix the film with 70 percent methyl alcohol for 3 to 5 min. Air-dry or blot dry with filter paper. Next apply dilute Giemsa stain for 15 to 30 min. Finally wash the slide in distilled water and air-dry.

Blood cell counts

When you introduce students to a technique using a hemacytometer to count white and red blood cells, you may want to use oxalated blood in the counting chambers. We have found it expedient to teach the technique by using young cultures of *Chlorella* or yeast cells in lieu of blood cells. Many practice sessions are needed for students to gain skill in distinguishing the corner grids on the chamber for counting white cell and the center, smaller field for red cells counts (Fig. 4-9). Consult pamphlets that accompany a hemacytometer for specific instructions for its care. Recall that descriptions of Thoma pipettes and types of diluting fluids to use, such as Hayem's (p. 277), are provided in texts in hematology and government manuals.[9] When white cells are to be counted, acetic

    [9] *Clinical Laboratory Procedure—Hematology* AFM 160-51; TM 8-227 -4; Departments of the Air Force and Army, Medical Service, (December 1973); available from Supt. of Documents, U.S. Government Printing Office, Washington, DC 20402.

acid is added to the diluting fluid to destroy red cells.

Using a hemacytometer   A brief description of a hemacytometer follows with methods for counting cells. You may want to modify the technique to measure the growth of cells in a population of yeast or *Chlorella* (growth, p. 463).

There are two counting chambers on a hemacytometer slide. Each chamber is separated by a trench or moat. Each chamber has a volume of 0.9 mm^3. The etched grids on the chamber measure 3 mm × 3 mm × 0.1 mm (depth). Notice there are nine squares, each 1 mm × 1 mm × 0.1 mm with a volume of 0.1 mm^3. The four corner squares are used in counting white blood cells under low power (marked W in Figure 4-9). These outer squares consist of 16 smaller squares 0.25 × 0.25 × 0.1 mm. The entire center square for counting red blood cells is visible under lower power. It is subdivided into 25 squares 0.2 × 0.2 × 0.1 mm with a volume of 0.004 mm^3. One square, or 1/25 of the center squares, can be seen in the field under high power.

Wet the supports to hold the coverslip. Load the chambers and allow 3 min for cells to settle. Under low power, locate the nine squares and center the central square. Switch to high power to count *red blood cells*. Count the cells in the four corners and center square—a total of 80 (16 × 5) small boxes marked R in Figure 4-9. Begin at the left of a small box in the top row. Then in the next row move, snakelike, from right to left, and so forth. To avoid counting cells twice, count cells in the boxes and those that are on lines to the left and top line of each small box. Add the cells in each of the five boxes and total the results. Make similar counts in the second chamber as a check. Refer to the example in Figure 4-10 for counting cells in *one* of the five small boxes. Many technicians count red cells in one counting chamber and white cells in the other or use the second chamber to check their counts.

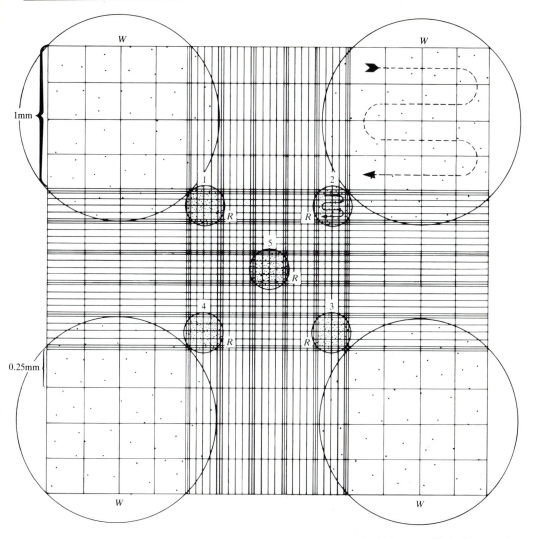

Fig. 4-9 Hemacytometer counting chamber (3 × 3 mm): Four corner boxes marked *W* are used in counting white blood cells under low power; five center boxes marked *R* are used in counting red blood cells under high power. Under low power the entire etched center region is visible; under high power one of the *R* regions (1/25) occupies one field.

White cells   These are counted in the four corner squares, each 1 mm^2, and subdivided into 16 boxes (Fig. 4-9). The cells are routinely counted in the same way as red cells. Variations of more than 20 percent indicate the suspension is not well mixed, and new counts should be made. Under low power, the dispersion of cells across the entire chamber can be checked. Other sources of error may be due to agglutination of cells or evaporation of the fluid in the chamber. For white cells, the area counted is 4 mm^2 since four corner squares are counted, each having an area of 1 mm^2; the depth of the chamber is 0.1 mm, so that the volume counted is 4 mm^2 × 0.1 mm or 0.4 mm^3.

Hayem's solution   This solution is used as the diluting fluid in preparing blood for red cell counts. It is often used as a stain for blood smears when 0.05 g of eosin is

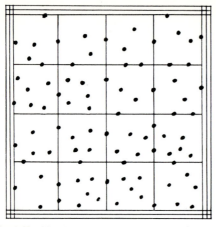

Fig. 4-10 Method for counting red blood cells in one of the five small squares. Start at the upper left to count the 16 boxes; count cells that touch lines on the upper border and left side of the box to avoid counting cells twice. The above figure has a total of 88 cells:

$$4 \to 3 \to 3 \to 4$$
$$\downarrow$$
$$7 \leftarrow 7 \leftarrow 5 \leftarrow 5$$
$$\downarrow$$
$$4 \to 5 \to 6 \to 9$$
$$\downarrow$$
$$4 \leftarrow 8 \leftarrow 7 \leftarrow 7$$

added to the solution indicated here. Before making the smear, mix 1 part of blood to 100 parts of this stain. Then make the blood smear on a clean slide (p. 274).

Weigh out these salts and add to 100 ml of distilled water:

$HgCl_2$	0.25 g
Na_2SO_4	2.5 g
NaCl	0.5 g

Differential blood counts: Human blood values After students learn to distinguish the five kinds of white cells, they can tally the cells using blood smears stained with Wright's blood stain. Examine under oil immersion; follow the standard technique in counting 200 cells. Since heavy white cells tend to accumulate at the sides of the smear, have students tally the five basic types of leukocytes in a scan of a thin part of the film as shown (Fig. 4-11). For example, if a student finds

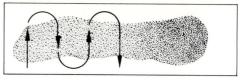

Fig. 4-11 Technique for scanning the thinnest film of a blood smear.

40 lymphocytes among the 200 counted, this would represent $\frac{1}{5}$ or 20 percent of the cells. What percent of cells are neutrophils?[10] Tables 4-1, 4-2, and 4-3 give some general information about human blood cells and other blood values.

Hemoglobin content of human blood A comparison of color may be used as a simple, but somewhat inaccurate, method for the determination of the amount of hemoglobin in blood.[11] One scale, the Tallquist scale, is made of strips of paper of different shades of "redness" of blood. Each strip is perforated in the center. A drop of blood from a finger puncture (p. 274) is placed on special white absorbent paper provided with the Tallquist scale, and the resulting color is matched against the color scale. The approximate percentage of hemoglobin may be read from the scale. Students may be able to bring to class demonstration material of this type, for doctors use several of these scales for rough estimates of hemoglobin content.

Hematocrit After centrifuging a given volume of blood (for example, 10 ml), the volume of packed red cells is estimated by reading a calibrated hematocrit tube. For example, if packed cells have a volume of 5 ml, the hematocrit reading is 50 percent. Refer to specialized references in hematology for the Wintrobe method as well as several other techniques.

---

[10] Refer to M. Ingram and K. Preston, Jr., "Automated Analysis of Blood Cells," *Sci. Am.* (November 1970).
[11] Heme is the porphyrin group containing iron; hemocyanin of snails and crustaceans has a porphyrin group containing copper and is blue in color. (Also refer to pigments, Table 1-1, chapter 1.)

TABLE 4-1 Human blood values

formed elements (45%) (plasma 55% of blood volume)	diameter microns*	number/mm³
Erythrocytes		
women	7.5	4.5–5 million
men	1 μ m thickness in center	5–6 million
Leukocytes		5,000–10,000
neutrophils	9–12	3,000–7,000 approx. 75% of wbc
lymphocytes	6–12	1,000–3,000 approx. 25% of wbc
eosinophils	10–14	100–300 approx. 5%
basophils	8–10	0–100 approx. 0.5%
monocytes	12–15	200–400 approx. 8%
Platelets	3	300,000–400,000

* Micron (micrometer μm) is one thousandth of a millimeter, 0.001 mm.

TABLE 4-2   Human blood values

Hemoglobin	15 g/100 ml blood
Hematocrit	45 percent (percentage of blood cells by volume; mainly red blood cells)
Blood pH	7.4
Red blood cells	—life span 120 days
	—1/3 of cell is hemoglobin
	—fragility: 0.44 to 0.35% NaCl
	—contain carbonic anhydrase which catalyzes the reversible CO_2 to bicarbonate conversion. Also contain glycolic system that generates energy.
	—When the count is low, or oxygen-carrying ability of the blood is reduced, the kidneys secrete erythropoietin which reaches bone marrow and stimulates the production of red blood cells.
Bleeding time	1–3 min; other studies, 2 to 4 min
Circulation time (arm to lung)	4–8 sec
Circulation time (arm to tongue)	9–16 sec
Coagulation time (venous)	6–10 min; other studies, 10–30 min
Clot retraction time	2–4 hr
Sedimentation rate	
men	0–9 mm per hr (Wintrobe)
women	0–20 mm per hr (Wintrobe)

Plasma has fibrinogen and can clot.
Serum lacks fibrinogen and does not clot.
Polymorphonuclear white cells or granulocytes are made in bone marrow; lymphocytes and monocytes are produced in reticular tissues of the spleen and lymph nodes.

TABLE 4-3   Average arterial blood pressure

age (yrs)	systolic	diastolic
15	110	70
20	120	80
30	130	85
40	140	90
50	145	90
60	150	90

Hemin crystals in blood   Obtain blood from a finger (technique, p. 274). Mount 1 drop on a slide, and allow to air-dry. To the dried blood add a crystal of salt, a drop of water, and finally a drop of glacial acetic acid. Apply a coverslip. Gently heat one end of the slide until bubbles escape. Under high power locate the reddish brown rhombic crystals of hemin.

Coagulation time   The coagulation time is the time it takes blood to begin to form fibrin, that is, to clot. The average coagulation time varies from 2 to 8 min, depending on the method used and the amount of blood. Here are two ways in which you may show how rapidly blood clots.

USING A GLASS SLIDE   This is a convenient method. Obtain a few drops of blood from the finger (technique, p. 274), and place on a clean slide. (Note the time at which the blood was taken.) Slowly pull a clean needle through the drop of blood at ½-min intervals. When a fine thread of fibrin can be pulled up by the point of the needle, coagulation has begun. Record the time interval between the flow of blood and the formation of fibrin—this is the coagulation time.

USING CAPILLARY TUBING   For this method, pull out narrow bore capillary tubing into capillary pipettes of even bore (1 to 2 mm) over a wingtop Bunsen burner. Clean the finger with alcohol, puncture the finger (p. 274), and fill a 15 cm length of pipette by placing it near the blood exuding from the finger. About 2 to 3 drops of blood are needed to fill the capillary tube. Blood rises in the fine

tubing by capillarity. At ½-min intervals break off small sections of the capillary tubing. When a red thread of fibrin forms at the broken edge, coagulation has begun. Record the time intervals as in the slide method above. In this method, coagulation time is 2 to 6 min.

Bleeding time   Obtain blood from a finger puncture (technique, p. 274). When a drop of blood appears, blot it on filter paper. Measure the diameter in millimeters (probably about 10 mm). Every 20 sec, repeat the procedure, using the same filter paper to blot the blood. Measure the *decreasing* diameter of the blood "blots". Calculate the total bleeding time in minutes and compare results among students in class. Normal bleeding time is between 1 to 3 min.

Fibrin clot and plasma   Try to obtain fresh blood from a slaughterhouse or meat-packing house or "out-of-date" whole human blood from a hospital blood bank. If possible, keep it in a vacuum bottle. Some of the blood may clot in transit, but it may be possible to get fresh blood to class. Pour some into a beaker and whip it with straws or small sticks. Fibrin will form shortly. When some of the blood is allowed to stand undisturbed, a clot also forms. As the clot contracts from the sides of the container, the clear, straw-colored plasma is visible.

Refer to journals or college texts for current ideas concerning coagulation of blood and the role of thromboplastin from platelets and injured tissues.[12]

Typing blood

A-B-O blood typing   Students may type their own blood in the classroom. Labo-

---

[12] For example, R. Doolittle, "Fibrinogen and Fibrin" in *Sci. Am.* (December 1981) describes how the surface of platelets acts as an assembly line for initiating protease thrombin to convert fibrinogen to fibrin.

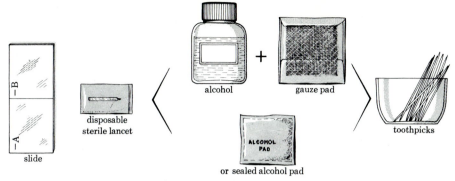

Fig. 4-12 Materials needed for A-B-O-AB blood typing.

ratory space is not essential for this activity.

Divide the class into five or six groups and supply each group with a tray or box containing the materials illustrated in Figure 4-12. Students should have access to a glass slide, which they divide in half with a glass-marking pencil. There are also individual disposable needles or lancets; toothpicks; and a bottle of alcohol or—better still—individual, sterile, disposable alcohol pads. These alcohol pads wrapped in aluminum foil eliminate the need for special precautions to keep gauze or cotton sterile, and there is no hazard in having alcohol spilled from tilted bottles.

First, demonstrate the technique for the students. Divide a slide in half with a glass-marking pencil. At the left, place 1 drop of anti-A serum; on the right, 1 drop of anti-B serum. Make this a uniform practice to avoid confusion. Since the serum is expensive, it is advisable for the teacher or one student to put the 2 drops of serum on each student's slide.

Blood serum should be refrigerated when not in use; however, it cannot be stored indefinitely. Dyes are added to distinguish the two serums. Powdered serum is also available for purchase and is accompanied by explicit directions for preparing solutions. But it is easier to purchase the anti-A and anti-B serums already prepared in solution.

Demonstrate how to obtain blood. (Request parental consent slips if necessary before distributing lancets.) Use an alco-hol-saturated pad to clean the skin; remove the sterile lancet from its wrapping, and make a quick, painless jab of the finger. With one end of the toothpick, transfer a drop of blood to the drop of anti-A serum and stir it a bit. Now with the *other end* of the toothpick transfer a second drop of blood into the drop of anti-B serum. Tilt the slide back and forth a bit and note the immediate reaction (unless the blood is type O). Students may want to check their observation by examining the slide under a dissecting microscope or hand lens. Check the results against those illustrated (Fig. 4-13).

Students readily understand the principle upon which blood typing is based. The four blood groups are classified on the basis of the presence or absence of the two agglutinogens, A and B, found in the red blood cells. Both of these antigens are present in type AB blood and both are absent in type O blood. However, when type A blood is added to type B blood it clumps. Therefore the plasma of type B blood contains a substance, an agglutinin, which clumps "A" cells, and this is known as anti-A serum (see Table 4-4). In summary, type B blood has a "B" substance, or agglutinogen, in the red blood cells and an anti-A agglutinin in the plasma. Obviously people of B blood type could not live with anti-B in their own plasma, for it would clump their own blood.

People of A blood type have "A" antigen in their red blood cells and anti-B in their plasma. (The anti-B serum which

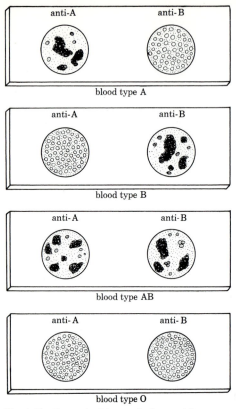

Fig. 4-13 Four possible results from adding a drop of "unknown" blood sample to each drop of serum on a slide.

TABLE 4-4   Summary for blood typing

blood group	antigen in red blood cells	antibodies in plasma
A	"A"	anti-B
B	"B"	anti-A
AB	"A" and "B"	no antibodies present
O	none present	both anti-A and anti-B present

roughly 44 percent (the percentage varies among racial groups within a population). Type A follows with about 39 percent (note there are several subtypes); type B is found in some 13 percent of occidentals; type AB is the rarest in the western hemisphere among the white population, about 4.5 percent. (We repeat: these figures are approximations.) Table 7-5 also gives some indication of the several subtypes of these four groups. Figure 4-14 includes distribution of the Rh factor among a sampling of 100 donors.

A description of the genetics of blood types and other factors in blood is given in chapter 7.[13]

[13] You may want interested students to read E. Zuckerkandl, "The Evolution of Hemoglobin," *Sci. Am.* (May 1965).

you use in blood typing in class has, of course, been extracted from blood of a person of type A).

When students find that a drop of their "unknown" blood is placed into a drop of anti-A serum and clumping occurs, the blood must be type A, as shown in Figure 4-13. Similarly, if the unknown blood clumps in anti-B serum, the blood belongs to group B. Should the unknown blood clump in both drops of serum, anti-A and anti-B, the blood belongs to type AB. O type blood, which lacks the two antigens in the red blood cells, will not clump in either serum.

After students have typed their blood, compare the frequency of each blood group in the class (see Tables 7-2, 7-3, 7-4). Type O is the most frequent, very

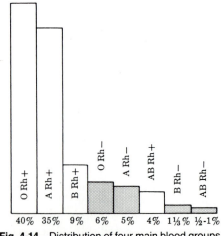

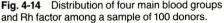

Fig. 4-14 Distribution of four main blood groups and Rh factor among a sample of 100 donors.

Rh blood typing Slide or tube techniques are used for determining Rh blood factor as in A-B-O blood typing, but false positive reactions often occur unless directions are carefully observed for using a *specific* anti-Rh typing serum.

Slides are prepared as for A-B-O typing, but the slides need to be maintained at about 37° C (99° F) and not above 47° C (117° F). Use a slide warmer or a microscope lamp. Place 2 drops of anti-Rh serum at the center of the slide. Adjacent to this, but not touching, place a drop of blood. With a toothpick mix the 2 drops and agitate by rocking the slide, keeping the slide warm. Clumping is macroscopic and occurs almost immediately (in about 30 sec) and is completed within 2 min if there is to be clumping. If the test blood clumps in anti-Rh, the blood is Rh positive; if there is no clumping, the test blood is Rh negative.

Be certain to follow the directions. When some serums are used, the blood is not to be washed with saline; in other preparations, blood is to have saline added. Oxalated blood is further specified in some techniques.

Typing serum, with specific directions or manuals of directions, are available from blood-donor services as well as biological supply companies. (see Appendix).

(See inheritance of blood factor Rh in chapter 7.)

M-N blood types Through a pair of alleles, *M* and *N*, three possible genotypes are inherited: *MM*, *NN*, and *MN*. In testing for these antigens, a slide is divided in half with a glass-marking pencil as previously described. A drop of anti-M serum is placed on half of the slide, a drop of anti-N serum on the other end. To each drop of serum on a slide add a drop of 5 percent suspension of test cells in normal saline. Cells must be washed twice in saline before use in M-N typing.

If test cells clump in anti-M, the blood type is M; clumping in anti-N indicates blood type N, and clumping in both serums indicates blood type MN. Clumping should occur within 5 min at room temperature.

The blood stream

Homeostasis Blood maintains, within a small range of limits, a constant concentration of glucose and salts, a constant pH and temperature (Table 4-5). The role of the kidneys and of hormones in maintaining a stable internal environment is described on pages 321, 323.

Hormones You may prefer to discuss the role of hormones circulating in the blood stream in this context of homeostasis. We also describe hormones and their functions in the study of behavior, chapter 5.

Radioactive tracers are used to trace the rapid uptake of iodine from the blood stream by the thyroid gland. In essence, radioactive iodine is combined into an iodide compound which is added to a sugary solution and fed by pipette to a white rat or rabbit three times daily for two days. Over the next day or days a high concentration can be located only in the region of the thyroid gland when a Geiger counter is moved over the body.

In most studies, however, the experimental animals must be sacrificed and assays made of microincinerated tissues. Have students report on Nobelist Rosalyn Yalow's elegant technique.

Body defenses In some classes, there may be the opportunity to study the immunological reactions of blood more thoroughly. For example, interested students may want to report to the class on the role of plasma cells, a specialized group of white blood cells (plasmocytes) that produces antibodies. Refer to current texts for an illustrated description of the complement-fixation test and for the latest information on immunological models. For a comprehensive account, written in clear, simple language see *Immunology Research*,

TABLE 4-5 Blood chemistry*

Glucose (0.09–0.12%)	90–110 mg per 100 ml
Nonprotein nitrogen	25–30 mg per 100 ml
Urea nitrogen	12–15 mg per 100 ml
Uric acid	1.5–3 mg per 100 ml
Creatinine	1–2 mg per 100 ml
Calcium	10–12 mg per 100 ml
Chlorides (NaCl)	0.45–0.5%
Carbonate content (CO_2)	53–77 vol. %
Alkali reserve (RpH)	8.5
Hydrogen-ion concentration (pH)	7.4

    * D. Pace, B. McCashland, and C. Riedesel, *Laboratory Manual for Vertebrate Physiology*, 3rd ed. (Minneapolis: Burgess, 1964).

prepared by the National Institute of Allergy and Infectious Diseases, Bethesda, Maryland 20014.[14] Also refer to M. Zucker, "The Functioning of Blood Platelets," *Sci. Am.* (June 1980), and J. D. Capra and A. Edmundsen, "The Antibody-Combining Site," *Sci. Am.* (January 1977).

Blood chemistry You may want to show the action of the enzyme catalase of blood in decomposing hydrogen peroxide (chapter 2). Table 4-5 gives a very brief summary of some components in blood; also refer to specialized texts in hematology and biochemistry.

Biomedical laboratory techniques

A variety of introductory courses are offered to high school and college students who plan careers in nursing, medicine, and/or biology. Activities include laboratory techniques involving blood, blood cells, and the role of kidneys and hormones in maintaining homeostasis. Readings and reports are required; identification of kinds of leukocytes; differential counts, and film loops of techniques. An outline follows of a course given over many years that students have found helpful in preparation for

advanced work. In addition, students also learn whether this area is to their liking, and thereby avoid waste of time and money in pursuit of an unrealistic goal.

Course outline: Medical laboratory techniques (one semester)[15]

    I. *Skills: Microscope*

Grow cultures of microorganisms; identification of five protozoa and algae; scanning techniques for making approximate counts; measuring cells in microns; skills in observation, interpretation, drawing; keeping a log.

    II. *Skills: Metric*

Use of gram and triple beam balances; liters and milliliters and use of graduates and calibrated pipettes; centimeter–millimeter conversions; temperature conversions.

    III. *Blood and Blood Cells*

Composition and origin of blood cells; identification of white blood cells (5 types) using Wright's blood stain; differential counts; comparison of student slides with commercial slides; counting cells with hemacytometer; hematocrit; coagulation time; Tallquist scale; film loops of specialized techniques and use of computers; blood disorders: dietary, genetic and environmental complex such as cancer; readings and reports to share with class.

[14] Available from Supt. of Documents, U.S. Government Printing Office, Washington, DC 20402; publication # DHEW NIH 73-529.

[15] A companion course in bacteriology, parasitology, and a variety of techniques is described in chapter 8, page 637.

IV. *Homeostasis and Hormones*

Hormones and feedback mechanisms; ovarian and pregnancy cycles; dissection of pig uterus; dissection of fetal pig, tracing circulation and complete anatomical dissection including brain; biological clocks; central nervous system and autonomic system.

V. *Homeostasis and the Kidneys*

Structure and function of nephron; hemodialysis; disorders of kidney; kidney clearance tests; composition of urine (chemical and microscopic); use of dip stiks, pregnancy tests, PKU, and so on.

VI. *Human Anatomy and Physiology*

In-depth studies of all systems of the body: structure and function in health and disease.

VII. *Career Information in Health-Related Fields*

Sources for career information; scrapbook to gain literacy in reading scientific journals.

Transport in plants

For heterotrophs, or consumers, ready-made organic molecules are ingested or absorbed, and then digested. But in autotrophic green plants, the raw materials—water, minerals, and gases—must reach cells. These materials also move in a stream—the transpiration stream in the xylem of trees. Further, the products of photosynthetic activity and assimilation must also be distributed along another channel moving in an opposite direction—the assimilate stream—carrying manufactured materials from leaves downward through plant tissues. Only in rapidly growing regions of a plant are the two streams moving in the same direction.

Movement of water in plants

How does water rise in plants? The generally accepted explanation of water transport is the transpiration-cohesion-tension theory. Simply stated, this theory implies

that a water column rises in the xylem tubes due to the cohesive property of water. Thus, a column of water, or transpiration stream, is continuous from roots to leaves. Since evaporation of water occurs through stomates of leaves during transpiration, there is a loss of water from the top of the water column of the plant. A tension is set up through which the column of water is lifted up through the plant to the leaves.

Both the rise of water in special tubes or vessels of the plant, and the occurrence of transpiration in leaves may be demonstrated. Study the lifting power of transpiration in plants. A related but seemingly independent action, that of root or exudation pressure, is also described here, as well as water absorption by seeds.

The techniques and materials in this section deal with those processes in plants commonly classified as conduction, transpiration, translocation, and diffusion. The exchange of gases (carbon dioxide and oxygen) through stomates of leaves or through the membranes of algal cells is described under photosynthesis (chapter 3).

Transpiration stream

Fibrovascular bundles Examine the fibrovascular bundles of jewelweed (*Impatiens*). Since the stem is transparent, the bundles may be seen by holding the plant up to light. These plants are easily transplanted and maintained in the laboratory. In fact, they can be grown from seed. However, the seeds must first be subjected to freezing temperatures for about two weeks, then dried for another two weeks. Germination and rapid growth take place in moist or wet soil. Seeds or cultivated specimens of *Impatiens* may be purchased from florists.

When the plants are to be used for demonstrating conduction of water, remove the plants—roots and all—from the soil. Then wash the roots and immerse them in a colored solution such as dilute fuchsin, phenol red, eosin, methylene blue, or red ink in water. The path of the colored solution may be traced along the fibrovascular bundles.

It may be convenient to collect jewelweed in season. Cut the stems in 6 to 7 cm sections and preserve them in a mixture of equal parts of 95 percent alcohol and glycerin. Then, the stems will be available throughout the school year to show xylem ducts. Cut the stems into longitudinal sections to reveal these ducts. Thin cross sections of stems may also be made by bundling a few stems tightly; then section them by hand with a razor blade; stain for examination under a microscope. Mount in water, and observe parenchyma cells and epidermal cells, as well as the spiral markings of the xylem ducts. Directions for the preparation of temporary and permanent stained slides are given in chapter 2.

If jewelweed is not available, other stems may be used effectively. For example, the cut stalks of fresh celery, cornstalks, or bean seedlings, to suggest a few, may be immersed in water colored by adding a few crystals of eosin, fuchsin, or red ink. After a few hours the dye will have moved up to the stems and colored the fibrovascular bundles and the veins of the leaves. Shoots of beans or the leaves of celery or lettuce show red venation (Fig. 4-15). Cross sections of the stems will show the red-colored fibrovascular bundles.

Cross sections of stems   Sandwich a length of soft stem of tomato, sunflower, or bean seedling, or of a coleus plant between two pieces of raw potato or carrot. Carefully make thin slides with a scalpel. Mount a section of stem in a drop of water and apply a coverslip. Some sections should be thin enough to examine specialized tissues of the stem. Add a drop of dilute safranin to the slide to stain xylem tubes.

Spiral conducting tubes   Examine the spiral conducting tubes in rhubarb stems. Cut rhubarb stems into very thin slices (2 to 3 mm thick) and heat in a beaker or saucepan with enough water to cover the specimens. Boil for a few minutes to soften the stems. Then add glycerin equal to the volume of boiled water and stems and transfer a small drop to a slide and add a coverslip. Locate the masses of tubes under low power and examine the spiral nature of the conducting tubes. Compare these tubes with strings of asparagus to observe the pitted conducting tubes of asparagus stems. (Use the same procedure to prepare the asparagus stems.)

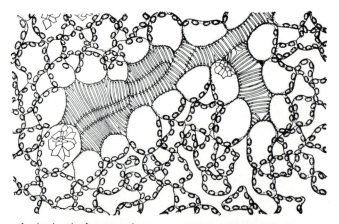

Fig. 4-15   Endings of veins in a leaf, as seen in a section cut parallel with the epidermis. (From W. H. Brown, *The Plant Kingdom*, 1935; reprinted through the courtesy of Blaisdell Publishing Co., a division of Ginn and Co.)

Root hairs To study root hairs, grow soaked seeds of radish in covered Petri dishes for a few days. Or pin a soaked seed to the cork of each of several small vials. Add a few drops of water before closing the vials. Each root will grow down into the vial if the vials are placed upright. Examine root hairs with a magnifying glass without removing them from the vials. In this way the root hairs will not dry out as many students examine them during the day (Figs. 4-16a, b, c).

Evaporation of water through leaves (transpiration)

There are many techniques for showing that water is given off by leaves, specifically, through the stomates in the leaves.

Using collodion to locate the stomates Spread a thin film of collodion over the upper surface of some leaves and on the lower surface of other leaves of a healthy plant. Do not remove the leaves from the plant. After several hours note that collodion remains transparent over the regions of the leaves which remain dry, whereas moisture due to transpiration turns the film of collodion an opaque, whitish color. (Also refer to the use of benzene, page 219; another technique for "peeling off" stomates with acetone is described on page 219.)

The different patterns of distribution of stomates may be discovered by examining upper and lower epidermal layers under the low power of a microscope (see Table 3-2; also Figures 3-14, 3-15).

root hairs

zone of differentiation

region of elongation

meristematic zone

root cap

(a)   (b)   (c)

Fig. 4-16 Radish seedlings: (a) photograph of root hairs (23 ×); (b) zones of young root; (c) region of elongation. (a, Photo courtesy of Carolina Biological Supply House; b and c, T. Weier, C. Stocking, M.

Barbour, T. Rost, *Botany: An Introduction to Plant Biology*, 6th ed., © 1982, reprinted by permission of John Wiley & Sons, Inc.)

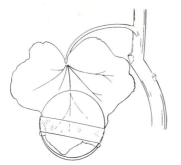

Fig. 4-17 Preparation for demonstrating transpiration. Water vapor released by stomates found on the lower surface of the leaf condenses on the lower watch crystal.

Using watch crystals to observe release of water[16]

Fasten small watch crystals to both sides of several different types of green leaves so that part of each leaf is sandwiched between the crystals (Fig. 4-17). These can be fastened with clear tape. Seal the glass edges on both sides of the leaves with a film of petroleum jelly. Since the leaves are still attached, photosynthesis continues and the leaves are not injured. Water vapor, released by the stomates in transpiration, condenses on the cool surfaces of the watch crystals and forms drops of moisture. Locate which surface of the leaf releases more water. Then compare, by microscopic examination, the relative number of stomates in the upper and lower epidermis of the different leaves studied (see Table 3-2).

Using cobalt chloride paper to demonstrate loss of water

Transpiration may be shown in leaves on a plant in yet another way. Place a square of dry cobalt chloride paper (blue) on the upper and lower surfaces of the leaves. (Cobalt chloride paper is blue when dry but turns pink when moist.) Then fold a strip of clear plastic over the paper (across the leaves) to hold it in place and to protect it from moisture in the air. Fix both with a clip as shown in Figure 4-18. Which surface of the

leaves shows a greater degree of change in color of the paper?

Cobalt chloride paper is easy to prepare. Soak strips of filter paper in a 3 percent cobalt chloride solution. The water solution is red, and the filter paper is red when wet. Dry the strips of filter paper; then store the prepared blue paper in a closed container. Before using the paper, it may be necessary to dry it (to turn it blue) in an oven or in a test tube over a flame, since the paper turns pink in the presence of minute amounts of moisture in the air.

Loss of water from excised leaves

Under a small bell jar, or in a transparent plastic bag, place a handful of leaves from geraniums, maple, *Sempervivum*, or others; let stand in bright sunlight. Use another small bell jar without leaves to serve as a control. Within 15 to 30 min, look for moisture resulting from transpiration on the inner surface of the glass or plastic bag.

Water loss in a whole plant

In this method, a whole plant is *weighed* periodically for a week. Grow sunflower or other rapidly growing plants (which have an active rate of transpiration) in small plastic pots. Cover the surface of the soil with rubber sheeting or plastic to prevent evaporation.

Weigh the whole plant in its container at the beginning of the preparation. Weigh the plant or plants at 24-hr intervals for a week. *Note:* Students might arrange to test *many* plants in this way so that they have a

Fig. 4-18 Alternate preparation for demonstrating transpiration. Water vapor released by stomates on the lower surface of the leaf turns blue cobalt paper pink.

---

[16] From W. J. V. Osterhout, *Experiments with Plants* (New York: Macmillan, 1905).

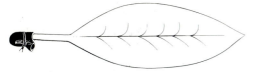

Fig. 4-19   Leaf of rubber plant with petiole sealed with rubber tubing to show loss of weight due to transpiration.

basis for comparison. (If the plants are watered during the experiment, the weight of the water added should be considered.)

Loss of weight in a leaf due to transpiration   Remove at least two leaves, approximately equal in size, from a rubber plant (*Ficus elastica*).[17] When the flow of latex has stopped, slip the petiole of each leaf halfway through a 3 cm length of tightly fitting rubber tubing. Fold over the excess of rubber tubing (Fig. 4-19), and wire each securely to avoid evaporation from the petiole.

Coat the lower epidermis of one or more leaves with petroleum jelly; coat the upper epidermis of the other leaves. Now weigh each leaf with its attached tubing and mark each leaf for future identification. Hang the leaves by means of the wire in a dry room or outside the window. After a few hours, weigh both leaves again and compare the rate of transpiration; continue to weigh leaves several times during the day. Remember that the stomates in *Ficus elastica* are found in the lower epidermis. Other succulent leaves may be substituted in this demonstration.

Potometer method for measuring water loss   Connect the bottom ends of two burettes with a short rubber tube, as shown in Figure 4-20. Now fill the whole apparatus with water and plug one burette with a one-hole stopper through which a woody stem has been inserted. Seal the plant in place with gum or modeling clay. Invert a beaker over the other burette to prevent evaporation of water. The water

level in this burette should be marked at the start of the demonstration so that as transpiration continues the amount of water absorbed may be measured by the change in the water level.

The method above is a modification of a true potometer method. Where a potometer is available, a quantitative measurement of transpiration may be made. Place a potometer filled with water into a small beaker of water (Fig. 4-21). Keep the bottom of the tube about 1 cm below the surface. Hold a leafy stemmed plant under water (to avoid air bubbles) and cut the stem near the bottom. Insert this stem into

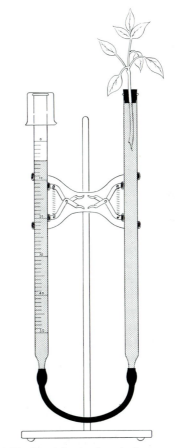

Fig. 4-20   Improvised potometer for measuring water loss in transpiration. As the plant absorbs water, the level in the burette at the left will change; the change can be measured from the calibrated markings.

[17] F. Darwin and E. H. Acton, *Practical Physiology of Plants* (New York: Cambridge Univ. Press, 1925).

a one-hole rubber stopper and fit this tightly into the potometer bowl. Support the potometer with a stand and clamp. No air should remain in the potometer.

After 15 to 20 min introduce the indicator bubble into the potometer tube by raising the end of the potometer out of the beaker and waiting (for a few minutes) until a large bubble appears at the opening. Then lower the end into the water again and measure the rate of movement of the bubble in the transpiration stream. Calibrate the tube, or attach a millimeter ruler with plastic tape to measure the distance the air bubble moves.

A simple U-tube to measure water loss
This method[18] for measuring the amount of transpiration is simpler than those previously described. Place the cut end of a leafy shoot into a one-hole stopper, and fasten this securely in one side of a water-filled U-tube. Plug the other side of the tube with a stopper through which a right-angle glass bend extends as shown (Fig. 4-22). Attach a millimeter ruler to the side arm to measure the distance the bubble moves. Clamp the preparation to a stand. Wait 15 min before taking readings. As transpiration from the leaves occurs, measure the movement of water in the extended delivery tube by measuring how far the air bubble moves.

Compare the rate of transpiration under a variety of conditions: in light in still air; in light in a stream of moving air (as from an electric fan); in the dark, with both moving air and still air. Is the rate of transpiration different if the temperature is lowered some 10° C? Or raised 10° C?

Lifting power of transpiration

Mechanics of the lifting power of leaves
A porcelain cup (called an atmometer cup) is used to simulate evaporation of water from leaves. Evaporation from the

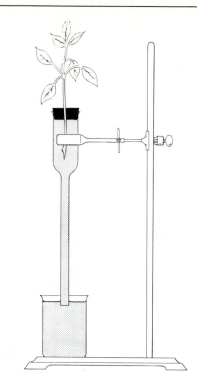

Fig. 4-21 Potometer method. An air bubble is introduced at the bottom of the tube and its rate of movement up the tube is measured.

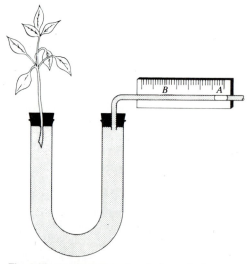

Fig. 4-22 U-tube potometer. As transpiration occurs, the air bubble in the tube will recede a measurable distance from *A* to *B*.

[18] G. Atkinson, *A College Textbook of Botany* (New York: Holt, Rinehart and Winston, 1905).

cup causes the water to lift up through thin tubes by capillarity. First, boil the porcelain cup in water to remove all air bubbles. Then insert a 75 cm length of 5-mm glass tubing into a one-hole stopper to fit the atmometer securely. Fill the atmometer and glass tubing with cooled, boiled water so that no air bubbles appear. Cover the end of the tubing with one finger and invert into a container of mercury (Fig. 4-23a). (*Caution:* wash hands thoroughly.) Or use the method shown in Figure 4-23b, which uses water only. In either method, clamp the preparation to a stand. Then use a fan to create an air current; direct it at the porcelain cup to speed evaporation. In (a), the mercury will soon rise in the tube and continue to rise higher than normal atmospheric pressure would lift it. The additional rise of the column of mercury is the result of a tension set up by rapid evaporation of water through the porous cup. In (b), the amount of water loss can be read from the calibrated burette. A similar demonstration using living plants may be developed.

Simple U-tube to show lifting power
To prepare the apparatus shown in Figure 4-24, fit a leafy shoot, airtight, into a stopper. Add some water to this arm of the U-tube; to the other arm add mercury so that the level is equal in both arms of the "U." (*Caution:* avoid handling or inhaling.) Within the hour, students may see the mercury level displaced in the direction of the tube containing the shoot.

Exudation pressure (root pressure)
You may have noticed that when a plant is cut off at the ground level water exudes from the cut stem because of root pressure. In this demonstration, select any plant, but potted plants such as the geranium may be more convenient in the classroom. Cut off the stem close to the soil level, and with a short rubber tube, attach a long glass tube to the rooted part of the stem. Clamp the tube to a stand to hold upright (Fig. 4-25). Now add a small amount of water so that the stem remains moist.

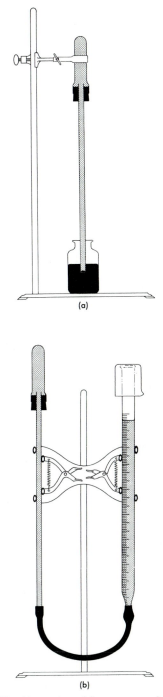

(a)

(b)

Fig. 4-23 Atmometers: (a) porous porcelain cup atmometer with water and mercury; (b) porous cup atmometer with water-filled burette from which the amount of evaporation can be measured.

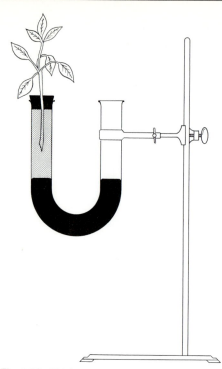

Fig. 4-24   U-tube apparatus demonstrating the lifting power of the transpiration stream.

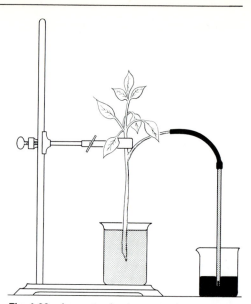

Fig. 4-26   Apparatus for demonstrating movement of water in a woody twig.

Palladin's method to show conduction of water up a stem[19]   Both the ascent of liquids and the possible effect of transpiration on the rise of fluids in plants may be studied in this method (Fig. 4-26)..

Select a plant with a forked stem or side branches; a branch of a bush or a tree with a side shoot will do nicely. Remove the leaves from the side shoot and attach the shoot to rubber tubing which leads to a glass tube. Be sure all joints are watertight. Then fill the glass tube with water and immerse it in mercury (*Caution*). For the best results, the shoot should be thin and the glass tubing of narrow bore.

Guttation   When transpiration is retarded, and there is free intake of water through the roots, exudation of water from the leaves occurs. This can be shown best by growing cereal seeds under a small bell jar. In this moisture-laden air, transpiration decreases, and *drops* of water appear on the leaves in a process known as guttation.

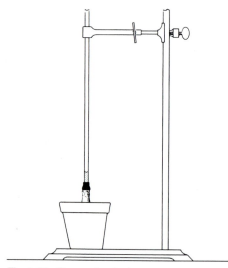

Fig. 4-25   Preparation to demonstrate exudation pressure of roots.

Within a short time (10 min or so) water begins to rise in the tube.

[19] V. Palladin, *Plant Physiology* (New York: Blakiston [McGraw-Hill], 1926).

Movement of molecules from cell to cell

A distinction needs to be made between osmosis and diffusion. *Diffusion* refers to the movement of any molecule. *Osmosis* is the passage or diffusion of *water* molecules through a semipermeable membrane. The direction of movement (in nonionized substances) is from a region of higher to one of lower concentration of the specific molecules under study. Thus, diffusion may refer to the distribution of ammonia molecules through the air as well as to the passage of glucose through a membrane.

Membranes of cells   Cells are separated from each other by their membranes. Organic molecules within cells may not be able to diffuse out of cells; smaller molecules may diffuse in or out of cells. Hence, the membranes are semipermeable. Further, membranes also have the characteristic of selective permeability; un-charged particles diffuse more readily than charged ones, and fat-soluble molecules seem to diffuse regardless of size. You may want to review the lipid-protein structure of cell membranes (Fig. 3-27), and an activity on diffusion of charged particles on page 244.

The movement of water in and out of living cells is more complex than indicated in these activities using "models." For example, permeability to water may vary with the physiological state of cells, as well as problems offered by the overall positive charge of cell membranes.

Osmosis   What conditions affect the rate of osmosis of water through a membrane? Prepare the demonstration shown in Figure 4-27. Use dialysis tubing that can be secured to calibrated thin pipettes. Tie one end of a length of dialysis tubing with string and fill it with 1 *M* sucrose; secure the tubing with string to a pipette. Similar-

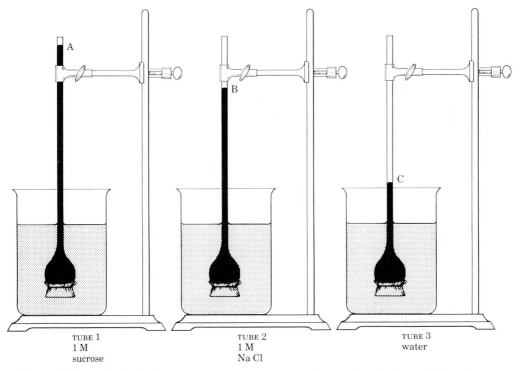

TUBE 1
1 M
sucrose

TUBE 2
1 M
Na Cl

TUBE 3
water

Fig. 4-27 Demonstration to show osmosis of water across membranes. Observe the effect of concentrations of particles on the opposite side of the membrane on the rate of osmotic flow. (Levels of fluids in the columns are labeled *A, B, C.*)

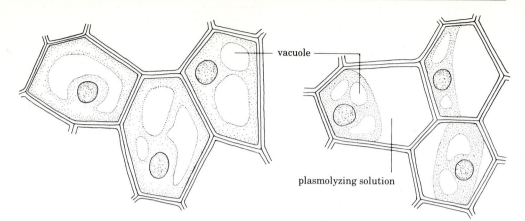

Fig. 4-28 Plasmolysis of epidermal cells of *Zebrina*. Concentrated plasmolyzing solution initially surrounds cells.

ly, prepare two more demonstrations. Fill tube 2 with 1 *M* NaCl and secure this to a calibrated pipette; complete tube 3 with water. Immerse all dialysis bags into beakers of water and hold each upright by a clamp to a ringstand.

Observe how rapidly the NaCl column rises, yet the sucrose column catches up in some 15 to 20 min and comes to rise higher in the tube than the NaCl column. Why does the flow of water, osmosis, stop in the NaCl column? Do students recognize that sucrose molecules are too large to diffuse in the opposite direction? However, NaCl ionizes in water so that there are about twice as many Na$^+$ and Cl$^-$ ions in the NaCl solution. Unlike the sucrose molecules, these smaller ions diffuse across the membrane so that when equilibrium is reached there is no further rise in the calibrated tube.

Osmotic pressure of cells Select leaves of *Zebrina*, or elodea that are in the same or associated whorls near the growing tip. Show the effect of placing cells in hypertonic solutions. As water diffuses out of cells, the cell contents shrink and cells become plasmolyzed (Fig. 4-28).

Observe the average osmotic pressure of the cell sap of the cells. Place the leaves in separate Syracuse dishes, each containing one of the following concentrations of sucrose: *M*/2, *M*/3, *M*/4, *M*/5, *M*/6, *M*/7,

and *M*/8. Place similar leaves in Syracuse dishes each containing one of the following concentrations of potassium nitrate: *M*/2, *M*/3, *M*/4, *M*/5, *M*/6, *M*/7, *M*/8, *M*/9, and *M*/10.

Under the microscope look for evidence of plasmolysis. Students can determine which concentration of sucrose and of KNO$_3$ is isotonic, that is, the concentration that just fails to cause plasmolysis in most cells.

Also observe the effect of varying concentrations of NaCl solutions on elodea cells. Elodea cells normally require a 0.9 percent concentration of NaCl within the cells.

Cytoplasm in color Cells containing anthocyanin in the cytoplasm provide a vivid observation of plasmolysis. Strip epithelial tissue from rex begonia, cells of the red onion bulb, or tissue from the lower epidermis of some varieties of *Tradescantia* that have the purple pigment on the under surface of the leaves.

Some teachers use descriptions of the experiments of Stephen Hales on water transport as described in *Vegetable Staticks* in 1727.[20]

---

[20] You may want to read an offprint of *Scientific American* (May 1976) by I. Bernard Cohen to review the work of Stephen Hales, the first to measure pressure of blood, and sap in plants.

Osmosis through a raw potato The *living* cells of a potato (or a carrot) may be used to show the passage of water through the semipermeable membranes which surround the cells. Use an apple corer to remove a center cylinder of a raw white potato. Do not plunge the corer through the entire length of the potato, but leave about 2 cm thickness at the bottom. Slowly pour a concentrated sucrose solution into the core, and close with a one-hole rubber stopper through which a piece of glass tubing has been inserted. Place this in a beaker of water; use a clamp to hold the tube upright. Then seal the stopper with modeling clay, gum, or melted paraffin to prevent leakage (Fig. 4-29).

Osmosis through a raw egg This classic technique may be of interest to some teachers, although the preparation may be time consuming. Carefully crack and re-

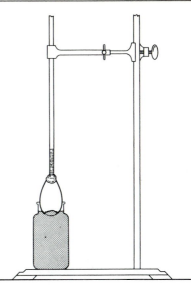

Fig. 4-30 Apparatus for demonstrating diffusion of water through the membrane of a raw egg. Water diffuses through the exposed shell membrane at the blunt end, up through the egg, out the hole in the top, up the tube, carrying with it some of the egg contents.

move the shell at the blunt end of an egg to expose the underlying semipermeable membrane. Or dissolve the shell with dilute HCl. (The cut edge of the shell should be smooth to avoid puncturing the membrane.) Invert the egg and fit it into the neck of a jar or bottle of water (Fig. 4-30). Now carefully break through the other end of the egg to make a narrow hole. Insert a length of narrow glass tubing into the egg contents. Clamp the tubing in place, and seal it to the egg with modeling clay, gum, or melted paraffin. Replace the water level in the bottle if it falls below the level of the egg membrane.

Or try a much simpler technique.[21] Stand an egg in the mouth of a bottle. Fasten a short length of glass tubing to the pointed end of the egg with sealing wax. Reach through the glass tubing with the end of the handle of a deflagrating spoon to make a small hole in the tip of the egg.

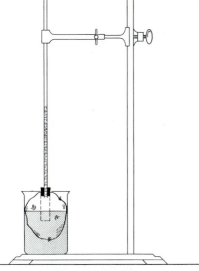

Fig. 4-29 Osmometer for demonstrating movement of water through the cells of a raw potato. The cavity in the potato is filled with sucrose solution; water diffuses through the membranes of cells of the potato to the cavity and rises in the tube, carrying with it some of the sucrose.

[21] Described by A. Applegate, Bloom Township High School, Chicago Heights, Ill., personal communication, June 1958.

By means of a short length of rubber tubing, connect a longer section of glass tubing and support this on a ringstand. Crack the blunt end of the egg so that water from the container can diffuse through the membrane; this reduces breakage of the membrane since the shell does not have to be chipped or dissolved with acid. Observe the contents of the egg rising in the glass tube.

(For a demonstration of selectivity in the absorption of ions in nutrient solutions, see hydroponics, pages 478–81; also see charged ions and paramecia (p. 244).

Thistle tube model   At times, you may want to use the thistle tube demonstration as a model of the cell sap of a cell (Fig. 3-28). Use a 20 to 30 percent *sucrose* solution in the bulb of the thistle tube to represent molecules that cannot diffuse out of the cell. (However, this model does not show changes in turgidity.) Just as water molecules enter the thistle tube, so water molecules enter root hairs and diffuse from cell to cell along a diffusion gradient. Refer also to some simple demonstrations of diffusion described earlier (pp. 242–45).

While models offer an oversimplification, you may want to refer to studies of Donnan equilibrium in a physiology text.

Active transport requires a supply of energy from ATP to pump molecules against the diffusion gradient from a region of lower concentration to one of a higher concentration. Describe the accumulation of potassium in the cells of some water plants, such as *Nitella* and *Chara*; discuss root pressure in plants, the action of vacuoles in protozoa, and the nephric organs in marine fish.

Absorption of water by seeds: Imbibition and swelling

Demonstrate the rapid absorption of water by colloids in seeds. Fill a tumbler one third full of dry beans or peas, and the rest with water. Within a few hours observe that the swollen seeds almost fill the tumbler.

Imbibition pressure of seeds   Make a cardboard box mold about 2 to 3 cm in depth. Fill it with a thick paste of plaster of Paris; drop in a handful of dry oats, wheat, or peas, and allow the paste to harden. After an hour or so, cut away the cardboard box and place the plaster block in water.

The small capillary spaces in the plaster block absorb water so that water is carried to seeds. The block bursts as a result of the swelling of the seeds due to imbibition.

Respiration in plants and animals

While glucose molecules are the major source of energy for most plants, and for animals, the ultimate source of this energy is sunlight, captured by photosynthesizing green plants. While energy is neither created nor destroyed, it can be converted into many forms available for the work of cells. Yet the conversion is not 100 percent efficient; only about 54 percent is useful energy available for the work of the cell through respiratory pathways. The rest is lost as heat, resulting in an increase in entropy. In an ecosystem, this increase in entropy is constantly counter-balanced by the continuous input of energy from sunlight trapped by the producers, the green plants.

However, living cells are open systems that exchange materials with their environment, grow, repair, and reproduce themselves, and thereby survive in a steady state with low entropy. They do not exist in a true thermodynamic equilibrium. The evolution of complex enzyme systems in cells has enabled them to maintain a steady state.

Reversing the arrow in the generalized equation for photosynthesis gives a summary equation for oxidation, or respiration —that is, the steps involved in releasing energy in the form of ATP from the energy-rich glucose molecules.

$$C_6H_{12}O_6 + 6\ O_2 + 6\ H_2O$$

$$\xrightarrow[\text{photosynthesis}]{\text{respiration}} 6\ CO_2 + 12\ H_2O + \text{energy}$$

Of course, these reactions are not reversible. The similarity is deceptive—the by-products are comparable, but the metabolic pathways are different. (Note that the respiratory quotient, R.Q., is 1: One molecule of CO_2 is evolved for each molecule of O_2 used in the oxidation of glucose. When fatty acids are oxidized, the R.Q. is less than 1; substances richer in oxygen than glucose have an R.Q. greater than 1.)

Cellular respiration refers to the oxidation of glucose or other energy-rich organic molecules. Recall that in oxidation, molecular oxygen is added to a compound, or hydrogen atoms or electrons are removed from a substance in a sequence of reactions (dehydrogenations). When electrons or hydrogen atoms are added to a compound (or oxygen is removed), the process is a reduction reaction. As a result, when one substance is oxidized another substance is reduced. In general, the addition of electrons to an organic compound stores chemical energy (hydrogen dissociates into H^+ and e^-).

Oxidation of glucose in living cells occurs both in anaerobic and aerobic respiration: two molecules of pyruvic acid result, and hydrogen is removed or transferred to a hydrogen acceptor. These intermediate steps are common to both glycolysis and fermentation. However, the fate of pyruvic acid depends on whether anaerobic or aerobic respiration occurs.

Energy pathways: Three sequences

Consider the energy pathways leading from a molecule of glucose to the formation of ATP as comprising three sequences: a) glycolysis of glucose to pyruvic acid (in the cytoplasm); b) the Krebs or citric acid cycle (in the mitochondria); c) the electron transport chains (also in the mitochondria). Incredibly, the complete oxidation of glucose to CO_2, H_2O, and energy occurs in about one second. The briefest of reviews follows: For a more in-depth study, please refer to the textbooks listed below, particularly that of Herreid, and of Stryer.

Glycolysis pathways   This anaerobic sequence occurs in the cytoplasm of cells of procaryotes and eucaryotes. Glucose is first phosphorylated, receiving a phosphate from ATP. As glucose-6-phosphate, it cannot diffuse through cell membranes. In over 11 sequential steps controlled by specific enzymes, a glucose-6-phosphate molecule is split, yielding two, 3-carbon units (pyruvic acid), and two ATP molecules.

Krebs cycle   The 3-carbon units enter the Krebs cycle. The major energy supply of cells comes from the aerobic respiratory pathway in the oxidation of pyruvic acid. This series of reactions is also called the tricarboxylic acid cycle, or citric acid cycle. Energy from this cycle is stored in molecules of ATP with the release of CO_2. The 3-carbon pyruvic acid units resulting from glycolysis enter the mitochondria where pyruvic acid dehydrogenases are located. The enzymes for this citric acid cycle occur in the *matrix* of the mitochondria (Fig. 4-31). (In procaryotes, the process is in the cytoplasm.) In a step-by-step sequence initiated by specific enzymes, the 3-carbon pyruvic acid is oxidized to form 2-carbon acetyl-CoA, along with 4-carbon oxaloacetic acid forming a 6-carbon molecule of citric acid. (Lipids and amino acids can also enter the Krebs cycle to provide energy.)

Electron transport sequence or cytochrome system   Located on the many inner folds of the mitochondria (called *cristae*) are the many enzymes involved in the transport of hydrogens in the electron transport sequence—a series of oxidations and reductions of the intermediate products in the citric acid cycle that phosphorylate ADP to ATP. Oxygen is used in the last transfer of hydrogen to yield water. Included in the cytochrome system are a series of cytochrome molecules that have a porphyrin ring containing iron: coenzyme NAD (nicotinamide adenine dinucleotide) is reduced to form two molecules of $NADH_2$; $NADH_2$, in turn, loses its elec-

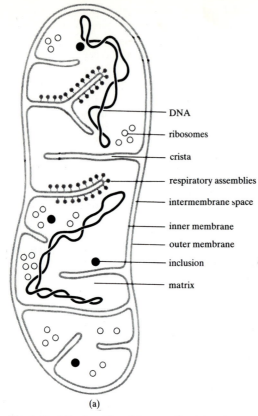

(a)

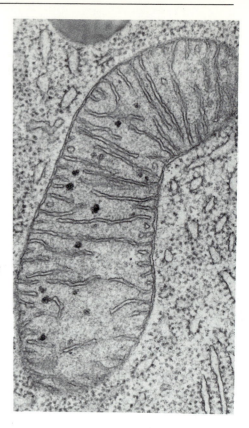

Fig. 4-31 Mitochondrion: (a) generalized drawing; (b) scanning electron micrograph. (From *Cell Biology: Structure, Biochemistry, and Function*, 2nd ed., by Philip Sheeler and Donald E. Bianchi. Copyright © 1980 by John Wiley & Sons. Reprinted by permission of John Wiley & Sons, Inc.)

trons to coenzyme FAD (flavin adenine dinucleotide), another acceptor in the oxidation-reduction sequences, to finally form water from the union of hydrogens with molecular O_2. (The electron transport system is not involved in fermentation which is anaerobic.)

Along with ATP, some carrier molecules are produced, $NADH_2$ and $FADH_2$, as already described. More oxaloacetic acid is also produced, which combines with a 2-carbon acetyl-CoA molecule from the glycolysis of glucose, and the citric acid cycle begins again. This is a continuous cycle for the breaking down of acetyl groups. Recall that CO_2 results when acetyl groups are oxidized in the citric or Krebs cycle; H_2O is produced as the last cytochrome enzyme passes hydrogens to molecular O_2.

A total of 38 molecules of ATP result from the complete oxidation of one glucose molecule: 36 in the pathways that involve glycolysis, Krebs cycle, and the electron-transport or cytochrome systems, and 2 molecules of ATP from anaerobic respiration. (Fermentation produces only 2 ATP molecules from each glucose molecule.) In summary, 38 molecules of ATP result from the complete oxidation of 1 molecule of glucose:

6 ATP from pyruvic acid to acetyl coenzyme A

24 ATP from Krebs cycle and respiratory sequences transferring hydrogens

6 ATP from glycolysis-respiratory reactions

2 ATP from anaerobic respiration

Some 70 enzymes control the sequences along the pathways in the oxidation of glucose in the mitochondria. (Note that chloroplasts also have specialized sites for their enzyme systems involved in the light reactions on the folded membranes or lamellae with stacks of grana. Enzymes of the dark reaction are in the stroma.)

For excellent coverage of biochemistry refer to C. Herreid, *Biology* (New York: Macmillan, 1977); I. Sherman and V. Sherman, *Biology, A Human Approach*, 3rd ed. (New York: Oxford University Press, 1983); L. Stryer, *Biochemistry*, 2nd ed. (San Francisco: Freeman, 1981); B. Alberts et al. *Molecular Biology of the Cell* (New York: Garland, 1983).

Respiratory enzymes and mitochondria

The role of respiratory enzymes may be demonstrated using some of the techniques presented in chapter 2. Also refer to anaerobic respiration (fermentation) in yeast cells, pages 310–13.

Mitochondria, obtained from preparations of celery stalks (see p. 185), may be examined under the microscope. Mitochondria may also be obtained from a living fly or bee, as described in a BSCS laboratory block.[22]

Quickly cut off the head, then carefully open the thorax of a live fly. Remove a bit of flight-muscle tissue to a slide and tease it apart in physiological saline solution (p. 155) to which a saturated solution of Janus Green B has been added. Apply a coverslip and examine under high power; look between the muscle fibers to locate hundreds of floating mitochondria, which are round or slightly oval. With Janus Green B the mitochondria should appear green or bluish green.

Oxidation in living cells

In the presence of oxygen the indicator methylene blue is blue in color. When oxygen is removed methylene blue is bleached and becomes colorless. This change in color may be observed in the study of oxidation in living cells. Here are two methods described in the literature.

In the first method saturate an isotonic solution (0.7 percent sodium chloride) with methylene blue.[23] Then inject about 2 to 3 ml into the dorsal lymph sac of a living frog. Within 1 hr after injection, place the frog under anesthesia (p. 142), dissect, and examine the *organs* for the presence of methylene blue.

Since the cells use oxygen rapidly, the methylene blue will be bleached or reduced. When cells are exposed to air again, or when the cells die, the blue color returns. In this way it is possible to estimate the speed at which oxidation stops in the different organs of the frog (or the rate at which tissues die). Other demonstrations are in chapter 2.

In another study methylene blue is reduced in the *muscle* tissue of a frog.[24]

You may want to show the bleaching of methylene blue by adding a few drops of sodium hydrosulfite. In the presence of hydrogen peroxide or in shaking so that oxygen is added, methylene blue is oxidized back to the blue color.

Dissect out two sartorius muscles from the leg of an anesthetized or freshly killed frog (p. 142). Stain these *muscles* with methylene blue–Ringer's solution and then mount each muscle strip on a slide. Apply a coverslip; use petroleum jelly around the rim to prevent evaporation. Within a short time, the center region of the muscle tissue may be examined; the methylene blue is colorless (reduced).

As the coverslip is raised, exposing the muscle tissue to air, the blue color will return. When the coverslip is replaced, the methylene blue will again be reduced to the colorless form in the muscle. One of the slides may be steamed (over a beaker of boiling water) until the tissue is killed.

---

[22] A. G. Richards, *The Complementarity of Structure and Function* (BSCS laboratory block) (Boston: Heath, 1963).

[23] D. Pace and C. Riedesel, *Laboratory Manual for Vertebrate Physiology* (Minneapolis: Burgess, 1947).

[24] R. Root and P. Bailey, "A Laboratory Manual for General Physiology," 2nd ed., mimeo (1946), City College, New York.

After this slide has cooled, lift off the coverslip on both slides. Note any difference in the rate of reappearance of the blue color.

Oxidation-reduction illustrated   The fact that oxidation-reduction is a reversible process may be demonstrated by mixing a solution of thionine dye with a solution of ferrous sulfate.

In the presence of intense light, the color of the dye disappears within a second. In the dark, the color reappears immediately. In the presence of light the ferrous iron reduces the dye to the colorless form, and is itself oxidized at the same time to ferric iron; in the absence of light, the process is reversed. What is the applicability of this to living tissue? Additional activities demonstrating enzyme action are described in chapter 2.

Interrelations: O_2–CO_2 cycles in animals and plants

Teachers begin in many ways, and they vary their approaches with individual classes during the day. Indeed, as we have reason to state at different times, teaching is a personal invention and the procedures vary with the individual. One teacher may begin with the concept of the interrelationship of living things using the oxygen–carbon dioxide cycle. This may be demonstrated in a sealed test tube or in a "microaquarium" in a depression slide. Another teacher may begin with a comparison of exhaled and inhaled air of the students themselves, or of changes produced in an enclosed atmosphere by plants, seeds, insects, or a mouse. Several of these approaches will be described. Students and teacher can plan together the direction in which to move.

Making a microaquarium
You may want to focus on a large aquarium in the classroom (preparation,

p. 679). What is the need for green plants? From where do animals get their oxygen supply? Or you may want to prepare microaquaria that students can monitor over the next month.

In a test tube   Begin by asking students to predict what might happen if a snail were sealed alone in water in a test tube? Seal a snail in one tube (Fig. 4-32a); into another place a snail and a green aquarium plant (Fig. 4-32b). You may prefer to "seal" each tube with a rubber stopper; cover with tape or wax to prevent an exchange of gases.

However, the demonstration seems to be more effective when the test tubes are sealed in a flame. Hold the ends of a large, soft-glass test tube (not Pyrex) and rotate it in a Bunsen flame. As the central region of the test tube softens, remove it from the flame and pull it out quickly making a 3 to 4 cm-long section (Fig. 4-33a). Prepare several of these and set aside until cool (3 min or so). Then add aquarium water to each test tube (not more than half full). To some of the tubes add green aquarium plants and a snail. Into others place a snail without the plants; into other tubes, only the plants.

To seal each test tube, hold the tube at a 45° angle and heat the narrow part of the tube until the glass softens (Fig. 4-33b). Then remove it from the flame and pull it out quickly to form a narrow constriction

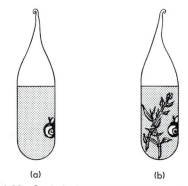

(a)   (b)

Fig. 4-32   Sealed microaquaria: *Physa,* a snail (a) in sealed tube of aquarium water; (b) with green aquarium plant in sealed tube of aquarium water.

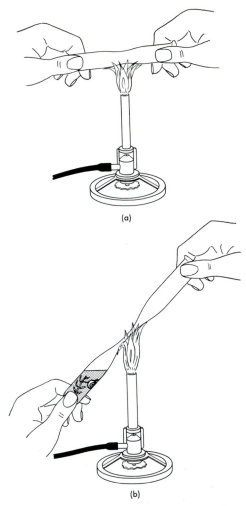

Fig. 4-33 Making a sealed microaquarium: (a) constrict an empty soft glass tube by heating and pulling; (b) after placing materials in the cool tube, make the constriction very narrow. Allow expanded gases to escape before sealing the tube.

for gases to escape. Place the center of this strand in the flame again in order to seal the tube. Be careful not to heat the liquid as you heat the glass tube, since the liquid will boil and expanded gases will cause the tube to crack. Stand the sealed microaquaria in a jar or test-tube rack in good light but not in direct sunlight.

Plastic bags also may be used instead of test tubes.

In a welled slide The oxygen-carbon dioxide cycle also may be studied using a

sealed welled slide.[25] Into the well, place the growing tip and a small segment of an aquarium plant such as *Nitella* in several drops of a rich culture of mixed protozoa and possibly some microscopic worms (*Nais, Dero*) and a crustacean, *Daphnia*. Allow some space for several air bubbles to be trapped in the microaquarium as you apply the coverslip. Seal the coverslip to the slide with petroleum jelly to prevent evaporation. Also prepare controls without plants.

Keep the microaquaria in moderate light, not direct sunlight. Examine the slides under a dissecting microscope or under a compound microscope or with a microprojector. In which slides do the worms and/or protozoa live longer?

Teachers who culture their own protozoa during the year know that cultures thrive better when algae such as *Chlorella* are introduced into the same culture dishes. Examine this relationship by preparing several microaquaria either in welled slides or as hanging drops (preparation, p. 154). Introduce six to eight paramecia to each slide in a drop of Brandwein's solution A (p. 656). To half of the slides add a drop of a thriving culture of *Chlorella*. (Culture methods for microorganisms are in chapter 9.) Maintain the slides in moderate light. During the next two to three weeks, which slides show vigorously active and reproducing paramecia? Why is this so?

Respiration: Release of CO$_2$

Carbon dioxide released in exhalation Have students breathe through a straw or plastic tube into a beaker of limewater or water containing an indicator (such as brom thymol blue). When limewater is used, it becomes cloudy (as calcium carbonate is formed).

Since carbon dioxide added to water results in a weak acid (carbonic acid),

[25] M. Rabinowitz, "A Balanced Microaquarium," *Teaching Biologist* (1938) 7:5.

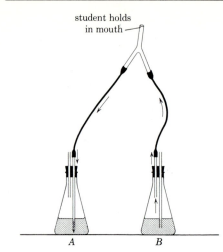

student holds
in mouth

A   B

Fig. 4-34 Apparatus for comparison of inhaled and exhaled air. Student inhales (through the mouth) air from flask B, exhales into flask A. An indicator in the water will show the difference between inhaled and exhaled air.

several indicators may be used to show the increase in acidity of the liquid. For example, pour dilute brom thymol blue (slightly alkaline and light blue in color) into a beaker.[26] The addition of carbon dioxide will turn this solution a light yellow or straw color (see Fig. 4-34) as it becomes more acidic.

Other times you may try adding phenolphthalein made slightly alkaline with sodium carbonate (or sodium hydroxide). Breathe into this, and the pink medium becomes colorless as the acidity increases. Phenol red may also be made slightly alkaline with sodium carbonate. When exhaled air is bubbled through this indicator, it turns from a red to a yellowish color (as the acidity is increased). For other indicators, see Figure 2-24.

Short-range test paper can be used by students to make determinations of pH at home. Liquid indicators are not necessary.

Effect of exercise on production of carbon dioxide   In this demonstration, six

---

[26] Dissolve 0.1 g brom thymol blue in 1 liter distilled water. If the solution appears greenish, add 1 to 2 drops of dilute ammonium hydroxide until it turns light blue.

volunteers exhale into water containing an indicator (alkaline); keep records of the time (in seconds) needed to decolorize the indicator. Then the same volunteers exercise vigorously for 1 min—by doing knee bends or by running around the hall—and at a signal breathe into a second beaker or container of the same indicator; timekeepers record the time (in seconds) that it now takes to decolorize the indicator. Students should be ready to explain why the time factor has been reduced by half (or thereabouts). (Rate of breathing, p. 320.)

Some teachers have students titrate these solutions to determine the carbon dioxide content (or increased acidity).

The solutions are prepared as follows. Add 3 to 5 drops of phenolphthalein to 100 ml of tap water. (Prepare the solution of indicator by adding 1 g of phenolphthalein powder to 40 ml of ethyl alcohol; add this to 120 ml of distilled water.) Now make the solution slightly alkaline by adding a few drops of 0.04 percent NaOH solution (4 g to 1 liter of water) until a rose color appears and persists. Test the solution by breathing into a sample; have enough NaOH so that it takes at least 30 to 45 sec for a change in color to occur.

Prepare two similar flasks with straws for each of the volunteers taking part in the activity. Either cork or add an extra drop of NaOH to the second flask of each pair, since it will otherwise be exposed to air before the student uses it.

Check students' health records so that volunteers for special class activities such as this one, involving vigorous exercise, are not overstrained.

At a signal have each volunteer exhale into the first flask of the pair that each has at hand. How many seconds are needed to bleach out the pink color? (The solution becomes colorless when the pH shifts to the acid side.) After the 1-min exercise session have students breathe into the second flask of the pair. Record the time (in seconds) needed to bleach out the pink color.

Now titrate the solution in each flask to determine the number of micromoles of carbon dioxide in the flasks of each volunteer. Each milliliter of 0.04 percent solution of NaOH will combine with 10 micromoles of CO_2. (A mole of CO_2 weighs 44 g [C = 12, O_2 = 32]; a micromole is a millionth of a mole.)

With a graduated pipette or burette tube add, drop by drop, a 0.04 percent solution of NaOH to the flasks containing exhaled air. Swirl the flasks as each drop is added. How many drops are needed to reach the end point and maintain the pink color? How many milliliters were used? To get the number of micromoles of carbon dioxide exhaled in a given time, multiply by 10 the number of milliliters of 0.04 percent NaOH used to reach the end point.

Comparison of exhaled and inhaled air Since there is a small percentage of carbon dioxide in air, students may question the conclusion that limewater (or some other indicator) turned milky because of the presence of exhaled air. The change may indicate only that carbon dioxide was present in the air.

A more valid way to show that exhaled air contains more carbon dioxide than inhaled air is shown in Figure 4-34. To make this device, connect a glass Y-tube to two sections of rubber tubing. Then fit one short and one longer piece of glass tubing into each of two two-hole rubber stoppers which fit two Erlenmeyer flasks.

Attach the rubber tubing to the *short* section of glass tubing in one flask and to the *longer* piece in the other flask.

Into each flask pour 100 ml of water to which alkaline brom thymol blue (p. 302) has been added so that the water is colored light blue. Compare your apparatus with Figure 4-34. Note that only one tube to which the rubber tubing is attached extends below the level of the fluid. Insert a short section of a straw into the mouth piece of the Y-tube. Have students inhale and exhale continuously without removing the lips from the straw. Watch the change in color of the indicator in flask *A*, the "expirator." Phenol red, phenolphthalein, or limewater may be substituted for the brom thymol blue.

The principle upon which this "inspirator-expirator" works is relatively simple. During inspiration, air from flask *B* is inhaled. As air is removed from this flask the air pressure inside it is reduced, and more air is pushed into it and bubbles through the indicator. In exhalation, air is breathed into the indicator in flask *A*, and the indicator changes color rapidly.

CO_2 production in small land animals

Measure the increase in the carbon dioxide content of exhaled air of land animals. Prepare the apparatus shown in Figure 4-35; be sure that the long and short tubes

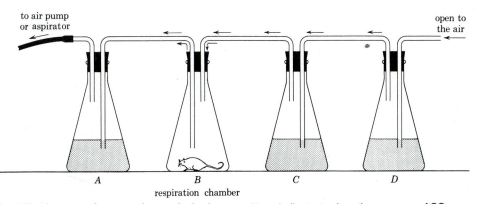

Fig. 4-35 Apparatus for measuring respiration in small animals. Flasks *A, C,* and *D* contain water with an indicator to show the presence of CO_2.

are in the right place (note that they are reversed in the respiration chamber). Pour dilute brom thymol blue, slightly alkaline phenolphthalein, or limewater into three flasks, skipping the respiration chamber. Place some fruit flies, grasshoppers, or beetles, or a small frog, toad, or white rat into the respiration chamber. (Germinating seeds can also be used; see p. 307.) Be sure to fire-polish the ends of glass tubing before inserting them into the rubber stoppers.

After connecting the apparatus, make it airtight. Connect an exhaust pump or aspirator to flask A. If an aspirator is used, begin the flow of air through the flasks by starting the siphon of the aspirator and letting the water drain into a sink or pail. As the pressure is reduced in flask A, air will be forced from the respiration chamber into this flask. If the organisms have been in the respiration chamber for a short time before the apparatus is used, carbon dioxide should have accumulated in the chamber. As this air is forced into flask A the indicator changes color. Thus, brom thymol blue becomes yellow, phenolphthalein becomes colorless, or limewater becomes milky. (The time in which this takes place will vary, of course, with the animal used—particularly in relation to its size—so that a change in the indicator may occur in an hour, or it may take several hours.)

In this same way, reduced air pressure in the respiration chamber will cause air from flasks D and C to enter flask B, the respiration chamber. Watch for a slight change in the color of the indicator in flasks C and D as the small amount of carbon dioxide in air gradually shows a cumulative effect.

After 20 to 30 min terminate the demonstration and titrate the solution in flask A with 0.04 percent NaOH (see p. 303 for details). Multiply the number of milliliters of NaOH by 10 to estimate the number of micromoles of carbon dioxide exhaled by the mouse, or by insects or other experimental animals, in a given period of time.

CO_2 production in small aquatic animals

Add enough brom thymol blue to a bottle or large test tube of pond water or conditioned aquarium water so that a light blue color results. To this add a goldfish, guppies, or several tadpoles. Then stopper the container. Within 20 min (as carbon dioxide is excreted) the blue color changes to yellow. Similar preparations, without the experimental animals, should be used as controls.

As a short review in a new context, ask students how to reclaim the yellow brom thymol so that the blue color is regained. Students should recall that green water plants carry on photosynthesis. What would happen if the plants were added to the yellow brom thymol solution and placed in strong light? In this process, carbon dioxide would be *absorbed* out of solution; thus, the liquid would become less acidic and turn back to a blue color as it becomes slightly basic.

Rate of respiration: CO_2 production and time sequence

Divide a sample of pond water into two containers. To one container, add a given weight of many of the same kind of organism (for example, test organisms might be strained *Daphnia*, a few guppies, or tadpoles). If students work in groups, samples of water may be tested for micromoles of carbon dioxide (p. 303) at different time sequences—20 min, 30 min, 1 hr, 2 hr, and so forth. At the end of the designated time, add an indicator to each container (about 3 drops of phenolphthalein solution, as used previously, to a 100-ml sample).

Titrate 0.04 percent sodium hydroxide to the end point until a pink color remains in each container. Estimate the number of micromoles of carbon dioxide by multiplying by ten the number of milliliters of sodium hydroxide needed to reach the end point. (Each milliliter of 0.04 percent sodium hydroxide solution contains 10 micromoles of carbon dioxide.)

If we let N_c equal the number of micromoles of CO_2 evolved by the control, N_e

the number of micromoles of CO_2 evolved by the experimental animals, W the weight of the animals (in grams), T the time (in hours), and R the rate of respiration (number of micromoles of CO_2 per gram of organism per hour), then we can calculate R as follows:

$$R = \frac{N_c - N_e}{W \times T}$$

In a carefully prepared series of activities for a 2-hr laboratory session, Verduin describes procedures for measuring respiration in small aquatic organisms such as tadpoles, snails, crayfish, or small fish.[27] (He also describes measurement of rate of photosynthesis using filamentous algae.)

In this technique, $0.02\,N$ sodium hydroxide and phenolphthalein are added to the pond water at the start rather than after a given time lapse as in the method already described. Then 50 ml of the stock solution, distinctly pink in color, is transferred into a 100-ml graduated cylinder. Determine the volume of the fish or tadpoles by displacement of water in the graduated cylinder. Transfer this 50-ml sample plus the organisms to one container; in another container pour 50 ml of the fluid without the animals. At the end of a half-hour titrate each solution with $0.02\,N$ sulfuric acid. Apply the same formula as given in the previous activity. In this case, the *volume* of animals (V, the milliliters) is used instead of weight.

$$R = \frac{N_c - N_e}{V \times T}$$

= micromoles of H_2CO_3 evolved per ml of organism per hr

Here N_c refers to the number of drops needed to titrate the control, and N_e refers to the number of drops needed to titrate the experimental.

The procedure recommended for measuring the rate of *photosynthesis* in this

[27] J. Verduin, "Simple Measurement of Respiration and Photosynthesis in Aquatic Organisms," *Turtox News* (1963) 41:234–37.

same laboratory period is similar to those already described in chapter 3. Verduin adds the phenolphthalein indicator to the pond water and bleaches it if necessary by exhaling into the solution with a straw. Measure 50 ml of this solution in a 100-ml graduated cylinder. Introduce a small quantity of filamentous algae such as *Spirogyra*, *Cladophora*, or *Zygnema* into this solution and determine the volume of the algae by displacement. Transfer this solution with plants into one container in the light; a control solution is also prepared with algae and placed in the dark. After 30 min titrate the experimental solution (and the control) with $0.02\,N$ NaOH until a permanent pink color appears. Since each drop of NaOH will react with 1 micromole of H_2CO_3 or bicarbonate in the water in the same way that algae remove the bicarbonate in photosynthesis, the following computation of *rate of photosynthesis* can be made:

$$\frac{N_d - N_l}{V \times T} = \begin{array}{l}\text{micromoles of } CO_2 \text{ absorbed} \\ \text{per ml of algae per hr}\end{array}$$

In this case, N_l refers to the number of drops needed to titrate the sample kept in the light, N_d to the sample kept in the dark. V refers to the volume of algae, and T to time.

A small pond might be sampled many times in the course of a day to compare the rates of respiration and photosynthesis. (Also refer to chapter 8.)

CO₂ production in aquatic plants (Anacharis)

If aquatic plants produce carbon dioxide during oxidation, they should turn a solution of dilute brom thymol blue to yellow, just as students exhaled into the indicator and turned the blue brom thymol. To demonstrate respiration in plants, the apparatus must be kept in the dark so that photosynthesis does not obscure the effect.

Prepare a 0.1 percent solution of brom thymol blue in tap water (p. 302). Add about 20 ml of this to some 50 ml of aquarium water. Set up several test tubes containing this blue solution; to each add a sprig of elodea (*Anacharis*) with a growing tip. As controls, prepare other test tubes without elodea plants. Then place some of the tubes in the dark (or cover them with aluminum foil). Let one control (containing the indicator, without a plant) remain in light. Which tubes show a color reaction? How long does it take for the solution to turn from blue to yellow as a result of oxidation?

Or use phenolphthalein and titrate by the procedure described above, or on p. 308.

CO₂ production in algae (Chlorella)

Compare the products produced by algae in light and in the dark. Other procedures described in chapter 3 using *Chlorella* in studies of photosynthesis can be used in tests of the rate of respiration.

One technique for measuring the rate of photosynthesis requires a suspension of *Chlorella* plus 0.1 percent phenolphthalein and enough NaOH (if needed) to make the suspension pink (p. 302). In the light the medium rapidly turns deep rose as CO_2 is absorbed in photosynthesis and the medium becomes more alkaline. When a similar preparation is now placed in darkness or covered with aluminum foil, a rapid bleaching of the rose color results. Prepare several test tubes with a rich (deep green) suspension of *Chlorella* plus an indicator at the sensitive end point so that slight changes in pH in light and dark can be quickly obtained. In fact, this test is so sensitive that tubes exposed to daylight will show a fading of color as daylight begins to diminish. When is respiration the dominant process?

If the stock of *Chlorella* is limited, it may be more practical to draw off samples of medium as cultures are exposed to different light conditions. Then count the number of drops of indicator needed to turn a 10-ml sample. (The pH range of phenolphthalein is from 8.3 to 10.)

Brom thymol blue, with a pH range from 6.0 to 7.6, may also be used. Because of its narrow range, it is obviously a good choice to indicate slight shifts from neutral into either acidity or alkalinity. The indicator is best used with samples of medium since the green color of *Chlorella* interferes with perception of color changes.

Also refer to measurement of micromoles of gas produced in light and dark (see p. 305).

CO₂ production in leafy plants

While green plants absorb carbon dioxide in photosynthesis, they also give off small amounts of carbon dioxide as a result of oxidation; this is masked, however, by the large amounts of the gas absorbed in photosynthesis. Therefore, a green plant must be kept in darkness to demonstrate the speed of production of carbon dioxide in oxidation.

Place a healthy, leafy green plant, such as a geranium or coleus plant, and a small beaker of limewater on a sheet of glass or cardboard. Cover this with a bell jar, and seal the glass rim with petroleum jelly. As a control, prepare a similar demonstration omitting the green plant. Then cover both bell jars with black cloth for about 24 hr. At the end of this time the limewater with the green plant should be milky (compared with the control) as a result of the production of CO_2 by the plant.

When a bell jar is not available, immerse cut stalks of plants in water in a large bottle or jar containing water (Fig. 4-36). Suspend a vial of limewater from the tightly fastened stopper. Prepare a control with branches or stalks from which the leaves have been removed, and place both containers in the dark or cover with black cloth.

CO₂ production by roots

The excretion of carbon dioxide by roots as carbonic acid can be demonstrated by using an indicator as follows. Support several vigorously growing seedlings (such as sunflower, peas, or pinto beans) on wet absorbent cotton with their roots submerged in test tubes containing water and a dilute indicator.

Fig. 4-36 Apparatus for demonstrating that leaves give off CO_2 in the dark. The suspended vial contains an indicator (brom thymol blue or limewater).

One indicator may be phenolphthalein (add a bit to the water, then introduce the small amount of sodium hydroxide needed to turn the water slightly pink). At other times, you may want to vary the demonstration by using another indicator, such as litmus solution or brom thymol blue. (Turn the litmus solution blue by adding a small amount of limewater.)

Within a few days, watch the change as the seedlings grow with their roots in the water (with indicator). The phenolphthalein solution will turn from pink to colorless, the litmus solution from blue to red, and the brom thymol from blue to yellow.

You may improve upon this demonstration if you wish by using a solid medium. Prepare some unflavored gelatin; before it solidifies, add a small quantity of litmus and limewater to tint the gelatin blue. (Or add phenolphthalein and enough sodium hydroxide to color the gelatin red.)

Fill test tubes about one-third full of the liquid gelatin. Then insert the roots of vigorously growing seedlings, such as sunflower or pinto beans, into the cool, solidifying gelatin. Loosely cork the test tubes. Within two days you should detect increased acidity. The litmus should turn red; phenolphthalein becomes colorless.

CO₂ production by germinating seeds

Rapidly oxidizing tissues, such as the cells of growing seeds, produce appreciable amounts of carbon dioxide.

Place a dozen soaked peas, beans, or corn kernels into a test tube, or fill a bottle not more than one-third full of soaked seeds. Connect this to another test tube or bottle containing limewater, brom thymol blue, or another indicator (Fig. 4-37). Where space is limited, use the apparatus shown in Figure 4-37b. In this case, a small test tube containing seeds is placed on a support inside a larger test tube containing the indicator solution.

It is advisable to disinfect the seeds first by putting them in dilute bleach (p. 470) since bacteria of decay may "confuse" the results. Then wash the seeds and soak them. Prepare controls using seeds killed by boiling.

While the previous demonstration is effective, a simple one may be used to show the rate of CO_2 production. In fact, these demonstrations may be used to show respiration in flower buds, tips of growing stems, small insects, or insect larvae. Put a grain of barium hydroxide (or limewater, brom thymol blue, or other indicator) into a test tube containing approximately 10 ml of water. Then place a small wad of cotton in the tube about 2 to 3 cm above the water level. On this cotton surface lay six germinating seeds of peas or beans, or about two dozen oat seeds, or others; seal the test tube. Prepare several such tubes

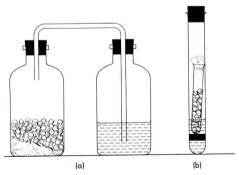

Fig. 4-37 Two types of apparatus for demonstrating that germinating seeds give off CO_2: (a) bottle at left contains germinating seeds and connecting bottle contains water with an indicator; (b) a small tube of seeds is supported inside a large tube containing an indicator.

using different seedlings, as well as controls using either boiled seeds or none at all. The rapidity with which a precipitate forms (or the color changes) is a rough indication of the rate of carbon dioxide formation.

Throughout this section a variety of indicators as well as techniques for demonstrations are given. Select from these, and use many different ways to add variety to classroom demonstrations. You can ascertain whether students have merely memorized the results or whether they understand basic principles by using demonstrations that are not described in the students' textbook or laboratory manual.

Titrating the solution You may want students to titrate the solution containing liberated carbon dioxide. Pour 50 ml of 0.2 N solution of sodium hydroxide into a flask and close with a rubber stopper. Measure a quantity of germinating pea seedlings and tie them in a cheesecloth bag; suspend the bag into the flask over the sodium hydroxide solution, holding it in position by means of a string caught in the rubber stopper.

Maintain this flask at 35° C (95° F) in an incubator, if necessary; also, prepare controls without seeds or with boiled seedlings. After two days, remove the bag of seedlings from the flask and quickly stopper again. Many variations are possible; some students may investigate how temperature affects the rate of aerobic respiration in germinating seeds. Maintain some flasks at temperatures varying from 5° to 25° C (41° to 77° F), as well as the one at 35° C (95° F).

Prepare to titrate to estimate the amount of carbon dioxide produced. First, precipitate the carbon dioxide absorbed by the hydroxide; add 5 ml of barium chloride solution (10 to 20 percent solution; *caution* in handling) to a 10 ml sample of the hydroxide from the flask. Add 3 drops of indicator such as phenolphthalein (prepared by adding 1 g of phenolphthalein powder to 50 ml of 95 percent ethyl alco-

hol and then adding 50 ml of water) to the solution to be titrated.

Now titrate with 0.1 N HCl until the rose color is bleached. Also titrate the control flasks (no seeds, dead seeds, or seeds that have been maintained at different temperatures). Then subtract the value (in milliliters) obtained from the flask with seeds from the value in the control flask and multiply by five to get the total amount of acid equivalent to the amount of carbon dioxide liberated by the seedlings.

Compare the value (in milliliters) of HCl equivalent to the liberated CO_2 in the flasks maintained at different temperatures.

Measuring the volume of CO_2 Now that the gas has been identified as carbon dioxide, you may want to measure the volume produced in respiration. Several techniques are described below. Frequently students want to find answers to questions which involve a need for effective ways to measure volumes of a gas absorbed or a gas released in a specific life activity. The suggestions described may serve as guides in planning the tools to use in an investigation.

USING WATER Place a quantity of germinating seeds such as beans, peas, or sunflower seeds, or kernels of corn, on a pad of moist absorbent cotton in a bottle or test tube. Connect the bottle to a delivery tube that runs to another bottle (Fig. 4-38). Set up as controls similar

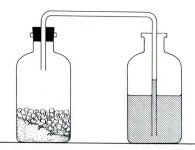

Fig. 4-38 Apparatus for measuring CO_2 given off by germinating seeds. CO_2 produced in bottle with seeds is absorbed by water in second bottle; decreased pressure in the bottle of seeds causes atmospheric pressure to push water up the tube.

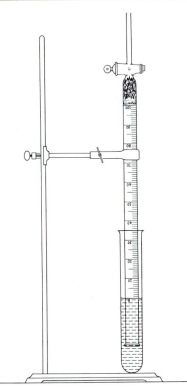

Fig. 4-39 Burette apparatus for measuring CO_2 given off by seeds. Saturated KOH solution in test tube absorbs CO_2 and the liquid consequently rises in the burette; the volume of CO_2 produced is determined by reading the level of the solution in the burette.

demonstrations using boiled seeds. Within a few hours observe the substantial rise in the level of water in the delivery tube. The rise is a measure of the volume of the carbon dioxide produced by the seeds (carbon dioxide is soluble in water).

USING A BURETTE   This is a similar technique. Place some germinating seeds into the constricted end of a 100-ml burette. Hold them in place with wet cotton. Now invert the burette into a large test tube of saturated potassium hydroxide solution (Fig. 4-39). Adjust the level of the solution to the zero mark. When the burette clamp is closed, there will be 100 ml of air in contact with the germinating seeds. Also prepare a control with dead seeds or without seeds. (*Caution:* Saturated KOH is corrosive.)

Soon there will be a difference in the rise of potassium hydroxide in the two burettes. This will be a rough measure of the carbon dioxide evolved in respiration, since carbon dioxide was absorbed by the hydroxide.

USING A RESPIRATION CHAMBER   The demonstration using a series of flasks with a respiration chamber and an aspirator described for animal respiration (Fig. 4-35) applies here as well. Fill the respiration chamber two-thirds full of germinating seeds. Then prepare a control with either killed seedlings or no seedlings.

USING A VOLUMETER   Use the volumeter shown in Figure 4-45 and the procedure described on page 315. Measure the distance the trapped drop of water or air bubbles moves. Compare the volume of carbon dioxide evolved by germinating seeds with the volume produced by dead seeds.

CO_2 production by fruits and tubers

You may want to use the apparatus shown in Figures 4-36, 4-37, 4-38 to show production of carbon dioxide in respiration of tissues of such tubers as potatoes and such fruits as apples. Place a portion of the apple or potato in a jar or tube, and line the jar with moist filter paper. Then use an indicator or limewater to show that carbon dioxide is evolved in respiration.

CO_2 production by fungi

Use a rich culture of yeast or bread mold to demonstrate the evolution of carbon dioxide.

Such forms as the edible mushroom (*Agaricus campestris*) give excellent results, because they do not readily decay. Prepare the demonstration as in Figure 4-36. Or use a container with a delivery tube extending from a one-hole stopper into a smaller container of brom thymol blue or limewater. Within an hour enough carbon dioxide is produced to change brom thymol blue to yellow (or to make limewater milky). Include a control in which the fungi are omitted.

Fermentation (anaerobic respiration)

Yeast cells   Students can devise a piece of apparatus to demonstrate that carbon dioxide is liberated by yeast cells in fermentation. A delivery tube may be carried into a beaker containing an indicator such as brom thymol blue or phenol red. Or one arm of a thin-tube manometer may be attached to a delivery tube from one test tube or flask containing a glucose-yeast solution, and the second arm of the manometer attached to a delivery tube from a flask containing only a glucose solution. Maintain both tubes in a water bath.

EFFECT OF TEMPERATURE ON FERMENTATION   Prepare duplicates of the apparatus described above and maintain them at different temperatures. Does the liberation of carbon dioxide vary with changes in temperature? Have students take readings at 3-min intervals.

An O2–CO2 cycle: yeast and Chlorella   Some teachers use this device to clarify a gas cycle: *Chlorella* liberating oxygen in light and absorbing carbon dioxide from yeast. To show this interrelationship, prepare a molasses-yeast cell mix. From this bottle, extend a delivery tube into a jar of a culture of young *Chlorella*. Prepare a control using a *boiled* molasses-yeast cell mix. After two to three days in the light, the experimental is many shades greener than the control. What gas did yeast cells liberate? In which process do *Chlorella* cells use this gas? (You may want to have students make cell counts; refer to page 462.)

Fermentation tubes   Fill the closed tubular portion of a side arm fermentation tube with a dilute solution of molasses (Fig. 4-40). Leave the bulb side partly empty. Prepare another fermentation tube in which yeast is added to the molasses. Then plug with cotton and keep the tubes in a

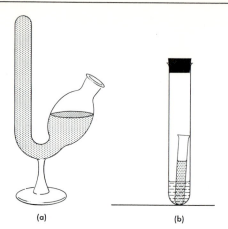

(a)   (b)

Fig. 4-40   Fermentation: (a) side arm fermentation tube containing yeast cells in molasses or 10 percent glucose solution; (b) yeast culture growing under anaerobic conditions—oxygen in outer tube is absorbed by potassium pyrogallic acid.

warm place. Within a half hour, observe bubbles of gas rise in the fermentation tube that contains yeast cells. There should be a considerable accumulation of gas in the upper arm of the fermentation tube. Now with forceps, place a small piece of potassium hydroxide in the liquid, where it will soon dissolve. Watch the liquid rise again into the tubular arm as the carbon dioxide gas is absorbed by caustic potash.

Anaerobic respiration: Methylene blue   At times, you may want to use methylene blue as an indicator of oxidative changes in yeast cells. The dye becomes colorless when reduced. Use test tubes of yeast cells in a glucose solution at varying temperatures and record the time it takes for methylene blue to become colorless. For example, add a half package of dry yeast into 100 ml of a 5 percent glucose solution and shake well. Let stand at room temperature for some 15 min. Into six test tubes, pour 5 ml of the yeast-glucose mix. Place two in a cool water bath at 15° C (59° F); two in a warm water bath at 37° C (99° F); two at room temperature. Allow a few minutes for the suspensions to stabilize at these temperatures, then add 5 drops of 0.05 percent methylene blue to each test tube and note the initial time. Record the rate of oxidation of glucose by the loss of color of the methylene blue. Plan to use controls at each temperature range— tubes of yeast suspension alone and tubes of glucose alone. Students may plot the rate of oxidation at the three temperatures on one graph using different color marking pencils. Plot the dependent variable along the ordinate (vertical axis) and the independent variable along the abscissa.

Yeast metabolism: effect of temperature and food supply

Darnell describes a simple technique to observe the effect of temperature and concentration of a food supply (glucose) on the production of CO_2 by *Saccharomyces cerevisiae* (brewers' yeast).[28] In this *two-factor* study, CO_2 production indicates the rate of metabolism. The yeast population does not reproduce since proteins are not part of the substrate. Prepare two glucose solutions: 1 percent and 5 percent. Fill a *graduated* fermentation tube with 10 ml of 1 percent glucose and 8 drops of a very concentrated active yeast suspension. (Add yeast to distilled water several hours before use and stir well.)

Add 5 ml of a specific glucose solution to the bulb of the fermentation tube; hold it horizontally so the fluid fills the long arm of the tube. Now add the 8 drops of concentrated yeast and the final 5 ml of glucose solution. Place the thumb over the opening of the bulb and invert several times to mix the contents. There should be no air bubbles in the long arm. Complete filling the fermentation tubes with yeast and 1 percent glucose, yeast and 5 percent glucose. Now incubate the yeast with different substrates. If several incubators are available, set them at 32° C, 35° C, and 42° C (90° F, 95° F, and 108° F). Otherwise, take readings at 32° C, then raise the temperature to 35° C, and use fresh preparations; finally, use a third set of fresh preparations at 42° C.

Within 5 to 10 min, production of CO_2 should be visible in the tube with 5 percent glucose at 42° C. How long does it take for the volume of CO_2 to reach 5 ml in each tube? Average the data from several groups of students.

Volume of CO_2 in yeast fermentation

Prepare a simple respirometer consisting of a small 15 ml vial or culture tube inverted within a larger test tube (Fig. 4-41). Students may need to practice this technique first using water. Fill a vial with water; cover this with a larger test tube. Then with one finger, push the smaller tube up against the bottom of the inverted larger test tube, and then quickly invert the device. There should be very little fluid lost from the vial; an air space in the smaller tube can be measured in millimeters. In the activity that follows, the CO_2 given off will displace the fluid so that students may measure the volume using a millimeter ruler at time intervals and record their data.

Fill two 15 ml vials and label their contents: vial 1 add 5 ml yeast suspension, 5 ml of glucose solution (or molasses), and 5 ml of water; vial 2 (control) add 5 ml of yeast suspension and 10 ml of water. Now bring down a test tube over each vial. Push the vial up against the bottom of the test tube and quickly invert.

Keep both respirometers in a water bath at 37° C (99° F) for half an hour or

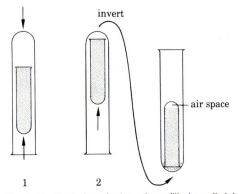

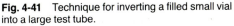

Fig. 4-41 Technique for inverting a filled small vial into a large test tube.

[28] R. Darnell, *Organism and Environment* (A Manual of Quantitative Ecology, San Francisco: Freeman, 1971).

equal volume of carbon dioxide is produced. Here, the carbon dioxide is absorbed by the hydroxide, which rises in the cylinder.

Using fermentation tubes   Place a quantity of soaked seeds (wheat, oats, or corn kernels) into the tube part of each of four fermentation tubes. Then fill this tube part with water.[30] Generate carbon dioxide and let it pass into the first tube; add oxygen to the second tube, and nitrogen (if available) to the third tube. Let the gases displace nearly all the water in each tube. Then insert a rubber stopper into the open end under water. Take out the tube, placing the corked end into a beaker containing water so that air is prevented from entering. In the control tube, the fourth tube, place seeds in a small quantity of water and stopper the fermentation tube. Within 24 hr germination will be found in the control tube and in the tube containing oxygen.

Comparing O_2 consumed by different seedlings   Is the ratio of 1:1 constant in different seeds? A small group of students might undertake this activity to compare the quantity of oxygen absorbed by seeds rich in fats with that absorbed by seeds rich in starch. In other words, is there a 1:1 ratio of oxygen to carbon dioxide when fats are oxidized? For instance, germinate soybeans and wheat seeds which have been soaked for 24 hr in advance. Prepare the apparatus shown in Figure 4-44, using soybeans (rich in fats) in one preparation; wheat seeds (rich in starch) in the second. Use colored fluid in the U-tube in each preparation and seal the bottles so they are airtight. Students should observe some differences in 24 hr. Does the level of fluid in the two arms of the U-tube manometer remain about the same in the case of the wheat seeds? If so, this indicates that the volume of oxygen consumed equals the volume of carbon dioxide given off. What

[30] L. J. Clarke, *Botany as an Experimental Science* New York: Oxford University Press, 1931).

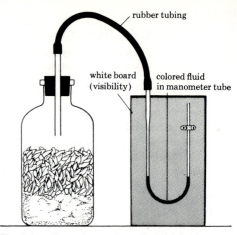

Fig. 4-44   Apparatus to show O_2 consumed by germinating seeds.

happens in the apparatus using soybeans? The additional amount of oxygen consumed is indicated by the higher column of fluid in the tube nearer the bottle.

Using a volumeter   Germinating pea seedlings, some three to four days old, consume oxygen at a rapid rate; this can be measured in a volumeter (Fig. 4-45). One molecule of carbon dioxide is formed for each molecule of oxygen absorbed so that no change in the gas volume occurs (Avogadro's law). Sodium hydroxide is added to absorb the carbon dioxide liberated in respiration.

Large test tubes or flasks may be used. Fill one tube almost to the top with germinating pea seedlings; then insert a layer of cotton. On the top of the cotton place a 3 cm layer of soda lime (sodium hydroxide; *caution*) to absorb the carbon dioxide liberated. Fill the second tube with water so that the same volume is occupied; this tube will be the thermobarometer. (You may prefer to duplicate the first tube, but using boiled seedlings.) Insert an escape tube with clamp into each two-hole stopper, and insert a 1-ml calibrated pipette in the second hole of each stopper. Or pull out capillary tubing and apply a plastic ruler as a scale (as in Figure 4-45). With a fine capillary or a hypodermic syringe insert the indicator drop or a

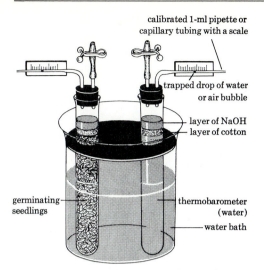

Fig. 4-45   Volumeter for measuring rate of O_2 consumption (or CO_2 liberation) by germinating seedlings.

trapped air bubble; adjust the escape clamp and set the drop or bubble at the far end of the pipette. Let the apparatus and the control reach equilibrium (place in a water bath, if needed); then take readings of the temperature and the distance that the trapped bubble or drop moves toward the flask or test tube at 1 to 2-min intervals. Subtract the readings in the thermobarometer from those of the experimental until the rate of change in the experimental becomes constant.

This is the rate of consumption of oxygen by germinating pea seedlings in the flask. What is the rate of liberation of carbon dioxide?

O_2 absorption by algae

Using a volumeter   A volumeter was used to test the amount of oxygen given off in photosynthesis of algae such as *Chlorella* (Fig. 3-18). Determine the respiration rate in the dark; let light fall on the suspension of cells and estimate the net photosynthesis.

To measure the amount of *oxygen absorbed*, the carbon dioxide must be removed from the system. A volume of oxy-

gen is simultaneously being absorbed equal to the volume of carbon dioxide released from cells (R.Q. = 1).

Use identical samples of suspensions of *Chlorella* in two flasks or tubes (Fig. 4-46). In one flask add 10 percent KOH to absorb the CO_2 in the flask and the CO_2 liberated during respiration, thereby keeping the pressure of carbon dioxide at a low level. Prepare a replica without *Chlorella* as a thermobarometer. Also prepare a control with the KOH. Use a cotton-tipped applicator that has been dipped into a freshly prepared solution of alkali—either 10 percent potassium or sodium hydroxide. If the cotton applicator is now dipped into phenolphthalein, it turns rose-colored; stand this applicator in the flask so that the cotton tip is exposed to the air above the level of the fluid. Watch how quickly the color is bleached as the alkali absorbs the carbon dioxide in the air over the fluid. Since the volume of oxygen absorbed in respiration is to be measured, the flasks must be covered so that photosynthesis ceases. Cover the flasks with aluminum foil and maintain at 25° C (77° F) in a water bath (Fig. 3-32). Take readings every 2 to 3 min for a period of

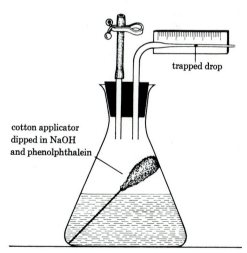

Fig. 4-46   Flask for measuring rate of O_2 consumption by *Chlorella*. Flask contains a dense culture of *Chlorella* and a cotton applicator dipped in CO_2-absorbent NaOH and phenolphthalein indicator. Large test tubes may be used instead of flasks.

15 min. (Avoid handling the flasks, and let the system come to equilibrium before beginning to take readings.)

Record in millimeters the distance traveled by the drop (or trapped air bubble) toward the experimental flasks; in the case of the controls (including the thermo-barometer), record any change in position of the drop.

Concentrated populations of cells of *Chlorella* can be studied, and by using dilutions (see p. 458), students can show the diminished rate of exchange of gases with a decrease in population density.

How might these changes in acidity affect other organisms growing in a pond? (See investigations suggested in chapter 8, The Biosphere.) Does a change in acidity of pond water affect succession of other microorganisms in the pond?

If a supply of oxygen is needed for respiration, would three times the supply of oxygen increase the rate of respiration threefold? Does gibberellic acid, or one of the auxins such as indoleacetic acid, affect the rate of respiration in plants? Is the rate of respiration the same at every phase of the growth cycle of plant cells (see chapter 6)?

O_2 shortage: Effect on seedlings

Using potassium pyrogallate to remove oxygen Soaked oat, wheat, or other small seeds or corn kernels should be placed on moist absorbent cotton in a jar or wrapped in a small cheesecloth bag, tied with string and suspended as shown in Figure 4-47. Prepare two demonstrations. Pour fresh potassium pyrogallate solution (*caution* using equal volumes of 5 percent solutions of pyrogallic acid and potassium hydroxide) into the two bottles. Then prepare two replicas as controls, using water instead of potassium pyrogallate solution. Observe the small amount of germination in the bottles which contain potassium pyrogallate. There may be a small residue of air containing oxygen trapped within the seed coats so that a little germination occurs; yet there should

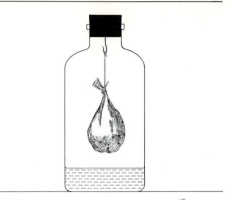

Fig. 4-47 Apparatus for demonstrating that germinating seeds require O_2. Seeds suspended over potassium pyrogallate solution should not grow, unlike those suspended over water.

be an appreciable difference in germination between the jars with water and those with potassium pyrogallate solution.

Drowning The fact that drowned seedlings stop growing is indirect evidence of the effect of the absence of oxygen. (Although this approach is described in the literature, it is not satisfactory to our way of thinking.) In this method, bean seeds or corn kernels are germinated until the root tips are about 10 mm long. Then the root tips are marked at uniform lengths with India ink. Test tubes or jars are lined with blotting paper, and the seedlings are inserted between the blotter and the glass walls. Let several jars stand with seeds drowned by water. In other jars, only moisten the blotting paper. Then cover all the jars to prevent evaporation. Measure in mm the length of the root tips after one day and after two days.

Overcrowding Soak a large quantity of beans, peas, corn kernels, or oat seeds in water after surface-sterilizing the seeds in a disinfectant (p. 470). Place some of the seeds in a small bottle, then pack the remainder of the bottle with moist sand so that every space is filled; stopper this bottle tightly. Then prepare another bottle with about the same quantity of soaked seeds. Fill this bottle only halfway with moist sand.

Compare the degree of germination of seeds in both bottles after the bottles have been exposed to the same temperature conditions for a few days.

O_2 absorption by small land animals

After students understand the techniques for measuring the consumption of oxygen (or the liberation of carbon dioxide) as previously described, they may plan investigations that focus on the effects of changes in temperature or pressure on the consumption of oxygen; the effects of treatment with nicotine, drugs, or hormones such as thyroxin and adrenalin; and the effect on breathing of animals that are quiet or in an agitated state (also see Fig. 4-35).

Using a volumeter Small insects, frogs, land snails, or mice are convenient experimental animals for this demonstration of oxygen absorption (CO_2 liberation).

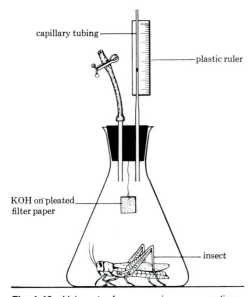

capillary tubing

plastic ruler

KOH on pleated filter paper

insect

Fig. 4-48 Volumeter for measuring consumption of O_2 by an insect or other small air-breathing organism. KOH absorbs CO_2; as O_2 is consumed, the trapped air bubble in the tubing moves toward the bottle.

Prepare a volumeter like that shown in Figure 4-48; insert thin capillary tubing (0.2 mm in diameter) into one hole of a rubber stopper and secure a plastic ruler along its length with plastic tape; or use a 1-ml calibrated pipette. Insert a short piece of glass tubing and rubber tubing with a clamp into the other hole of the rubber stopper; this is an escape tube. Fasten a square of filter paper that has been pleated to increase its surface area to the rubber stopper; or place it or cotton in a small vial that can be suspended from the rubber stopper. Soak the filter paper or cotton with 10 percent potassium hydroxide. Since KOH is caustic, avoid having the experimental animals come in contact with it. Prepare a control without the test animal; also prepare a thermobarometer. In all the preparations insert a drop of colored water containing a trace of detergent in the length of capillary tubing or pipette. Maintain in a water bath at 30° C (86° F), and wait until the flasks (or bottles) have reached the temperature of the water bath. Introduce the test animals. Take readings at 1-min intervals (unless the drop moves very slowly); also record the temperature.

Carbon dioxide liberated by the respiring organisms will be absorbed by the potassium hydroxide. Compare the experimental with the controls. Have groups of students compare their results. As oxygen is absorbed by the organisms, the change in volume within the containers will result in the trapped drop moving down into the container.

Using a manometer Hammen suggests a modification that uses the common land snail, *Helix*, for measurements in which a U-tube manometer is used instead of a volumeter.[31] (*Helix* can be obtained in fish markets.) Assemble the apparatus shown in Figure 4-49; one container accommodates the large snail (or earthworms or crayfish) along with a vial containing 20 percent

---

[31] C. Hammen, "Oxygen Consumption as a Laboratory Exercise," *Turtox News* (1960) 38:296–97.

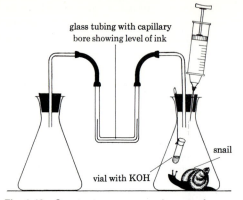

glass tubing with capillary bore showing level of ink

snail

vial with KOH

Fig. 4-49 Constant-pressure respirometer for measuring consumption of O_2 by a land snail. Both flasks are immersed in a constant temperature water bath, which may be an aquarium or a large finger bowl. (After C. Hammen, "Oxygen Consumption as a Laboratory Exercise," *Turtox News* 38:296, 1960.)

KOH to absorb the carbon dioxide that is liberated (*caution*). (Or insert a pleated piece of filter paper to increase the surface of the KOH solution.) Suspend the vial by a string as shown. Also prepare the empty container to act as a thermobarometer. Bend glass tubing and assemble the thin U-tube capillary containing colored ink. Immerse both containers in a water bath and allow the apparatus to reach equilibrium. The level of the liquid in the manometer can be adjusted by manipulating the stopper on the empty container. The syringe filled with air is inserted through the rubber stopper. Take readings of the change in the level of the manometer (using an attached plastic millimeter ruler) as oxygen is consumed by the test animal. Restore the original level by releasing the syringe to introduce air into the container. Have students take several readings, restoring the air at the end of each reading. Average the readings. The volume of oxygen that is used is the difference between these readings and the original volume. Now students can weigh the animal (or animals) to calculate the rate of respiration in millimeters of oxygen consumed per gram per hour. Some students may also calculate the difference in rate of consumption at 21° C (70° F) and

at 31° C (88° F) and plot the different rates at the two temperatures.

Production of heat by living plants

The evolution of considerable heat by seedlings during respiration may be demonstrated by preventing its escape and noting the substantial difference in temperature (Fig. 4-50). Fill a pint vacuum bottle nearly full with germinating beans, corn grains, peas, wheat, or oats on a bed of moist absorbent cotton. Insert a thermometer through a one-hole stopper and seal it in place with modeling clay to make it airtight. Prepare a control containing dead (boiled) seedlings. Unless all the seeds are disinfected (p. 470), there may be evidence of temperature change due to heat generated through bacterial decay in the control.

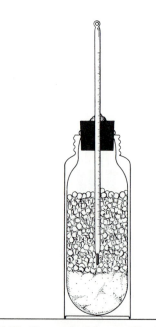

Fig. 4-50 Vacuum bottle preparation to show that germinating seeds generate heat. The thermometer in vacuum bottle with germinating seeds should show a temperature rise within 24 hr; the same setup, without seeds or with killed seeds should not.

When vacuum bottles are not available, an effective substitute can be made by using two bottles with thermometers as above and packing them into glass-wool insulation or excelsior.

Comparison of respiratory systems in animals

How are different animals adapted to permit the rapid interchange of oxygen and carbon dioxide? Have students observe how surface membranes are increased by means of different kinds of gills (as in worms, fish, and amphibian tadpoles), tracheae in insects, and book lungs in some spiders and in king crabs (see dissections, chapter 1).[32]

Among vertebrate groups, examine gills of fish, which can be obtained from a fish market. Also study the external gills of early amphibian tadpoles and the lungs of mature frogs and toads. Observe the breathing movements of amphibia by watching the external nares and throat movements.

In the freshly killed frog, the lungs may be inflated by inserting a straw or glass tubing into the glottis and blowing gently as the small lungs begin to expand. Refer to capillary circulation in the lungs as described earlier in this chapter.

Effect of temperature on breathing movements of a goldfish

Prepare several beakers, each containing a small goldfish in aquarium water. Allow a few minutes for the fish to become adapted to its environment. Have students record the temperature of the water. Observe the behavior of the mouth and gill covers as water rich in oxygen enters the mouth, flows across the capillary-rich gills, and

---

[32] Also refer to B. Heinrich, "Energetics of the Bumble Bee," *Sci. Am.* (April 1973); K. Schmidt-Nielsen, "How Birds Breathe," *Sci. Am.* (December 1971); and J. Warren, "The Physiology of the Giraffe," *Sci. Am.* (November 1974).

then out through each operculum or gill cover. Record the number of times the operculum covers beat per minute for each fish, and label the beakers to identify the fish under observation.

What happens when the temperature is raised? Is there more or less oxygen in warmer water? Place the beaker with the fish on a hot plate with a thermometer. Record the operculum beats per min for each 1° elevation in temperature up to a 5° change (no higher). (Data from three groups of students showed, for example: at 61° F, 59 beats per minute; at 68° F, 80 beats per minute; at 69° F, 109 beats per minute.) Since variations occur among fish of different size, age, and general physiology, it is advisable to pool data from the class and graph the changes in breathing rate at different temperatures.

Now remove the fish from the hot plate(s). What predictions can be made for operculum beats when the water returns to its original temperature? What happens if small bits of ice are added to the water to lower the temperature? Again, record the operculum beats per min for each drop of 1° in temperature. Lower the temperature only 5° below the original temperature.

Are some fish more sensitive to the oxygen content of water? Compare the operculum beats in young trout, killies, or other small fish with the behavior of goldfish.

Do other factors affect the rate of breathing—acid rain? Overcrowding? Many other factors may be tested. What types of adaptive behavior do the fish show aside from changes in operculum beats?

Mechanics of breathing

Prepare the bell-jar model shown in Figure 4-51 to demonstrate that breathing is mechanical. Insert a glass Y-tube into the opening of a rubber stopper which fits into the top of the bell jar. Fasten two balloons of the same size by means of string to the two arms of the Y-tube. Next cut a circle of thin rubber sheeting large enough to fit

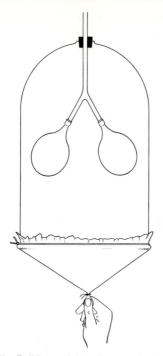

Fig. 4-51 Bell jar model to show mechanics of breathing.

over the bottom of the bell jar. In the center of the sheeting pull out a piece to enclose a small cork, and tie it off as a "holdfast." Tie the sheeting in place to the bottom of the bell jar. Check to see that there are no breaks in the sheeting.

The balloons represent lungs and the rubber sheeting the diaphragm. What happens when the rubber sheeting is pulled down? Since we have increased the volume within the bell jar by pulling down the "diaphragm," there is less pressure on the "lungs." Thus, normal atmospheric pressure pushes outside air into the "lungs." When the sheeting is pushed upward the air pressure in the interior of the bell jar is increased, the pressure on the "lungs" is increased, and the air in the "lungs" is forced out. Then the balloon "lungs" collapse.

In the body, the size of the chest cavity also changes through the action of intercostal muscles between the ribs. Students may want to design a movable model of

wood with hinges to show these changes in chest cavity.

Capacity of the lungs

A rough estimate may be made of the volume of air exhaled. Make right-angle bends in glass tubing (about 1 to 1.5 cm bore) as illustrated (Fig. 4-52). Then fill a gallon bottle about four-fifths full with water to which food dye has been added. Insert the tubing into the two-hole stopper and seal the bottle with paraffin or modeling clay.

Cover the mouthpiece with paper toweling and have students exhale into it. The amount of water displaced into the cylinder will be equal to the volume of air exhaled. Use a graduated cylinder to facilitate measuring the volume of displaced water.

Effect of excess CO_2 on the rate of breathing

Students may determine their normal breathing rate while at rest. Count the number of times breathing occurs in half a minute and then multiply by two. In general, respiration rate varies with age:

at birth	30–50 /min
1st year	25–35 /min
4–15 years	20–25 /min
adulthood	16–18 /min

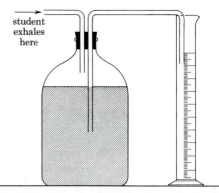

Fig. 4-52 Apparatus for estimating capacity of human lungs.

Next, have students breathe into a paper bag held closely over the nose and mouth for several minutes. Determine the rate of breathing while breathing into the bag, then again after breathing ordinary air. The increase in carbon dioxide in the air stimulates the medulla, which in turn regulates the rate of breathing. However, this activity yields very rough data. (Also refer to effect of exercise on production of carbon dioxide using indicators, p. 302.)

Study of mammalian lungs

The lungs and windpipe (called a haslet) of a cow or sheep may be obtained from a slaughterhouse or a local butcher. Or purchase from a biological supply house.

Drop the untrimmed lungs into a bucket of water and have students explain why they float. Dissect carefully to reveal the two bronchi and smaller bronchial tubes. Also examine the arrangement of cartilage rings in the trachea, or windpipe.

Many biological supply houses also have preserved plucks of dogs or sheep. These plucks, which are large and easy to handle, include heart, lungs, liver, gall bladder, pancreas, part of the diaphragm, and the duodenum. Such a specimen might serve as a fine review after dissecting a frog or rat.

Lamore recommends the use of a dog pluck and includes a labeled diagram of the pluck (Fig. 4-53).[33]

Excretion

Of CO₂   We have already described several indicators which may be used to show that CO_2 is a product of respiration (pp. 303, 304). We have also described some devices for enclosing small animals or plants in a chamber so that gaseous exchanges in the air may be tested (Figs. 4-35 and 4-36).

Of water   To show that water is also excreted by lungs during exhalation, have students breathe against the blackboard or on a mirror. Stress the differences between burning a candle and biological oxidation in cells.

Study of kidneys

Obtain an untrimmed kidney of a hog, sheep, or cow from a butcher or slaughterhouse. Study the fat capsule which envelops the kidney. Remove this sheath and slice the kidney lengthwise. Distinguish the region rich in tubules and capillaries (cortex) and the collecting funnel which leads into the ureter. Examine prepared slides to show nephrons and glomeruli. You also may want to refer to a study of nephrons (Figs. 4-54, 4-55).

Speed of action of an enzyme, urease

The enzyme urease converts urea into ammonia:

$$2H_2O + CO(NH_2)_2 \xrightarrow{\text{urease}} 2CO_2 + 4NH_3$$

Both urea and urease may be purchased from a supply house.

Dissolve some urea in water (preparation, p. 747), and divide the solution into two beakers. To each beaker add an equal number of drops of phenolphthalein so that the solutions are slightly milky. Then add a few crystals of urease to one beaker. Watch how quickly the solution changes to red (alkaline due to the presence of ammonia).

Or use a more carefully controlled procedure. Add 4 drops of 0.04 percent phenol red solution to 3 ml of 0.1 M urea. If the solution is not pink at the start, add sodium hydroxide solution (0.1 M) drop by drop. Then neutralize by adding 0.1 M acetic acid until the end point is reached and the color disappears (at pH 7).

To this solution, add pure urease and shake. Explain the rapid change. (When urease is not available, add 0.5 g of soy-

---

[33] D. Lamore, *Carolina Tips* (November 1958) 21:35.

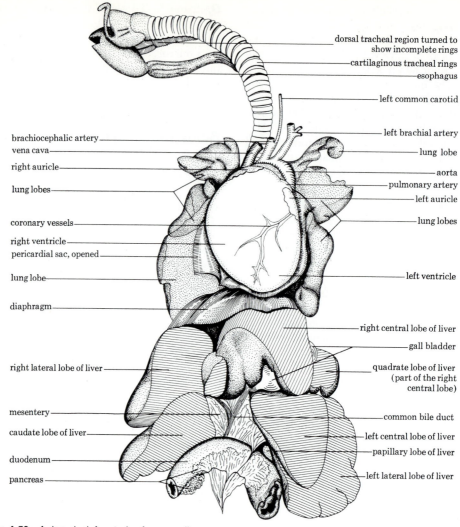

dorsal tracheal region turned to
show incomplete rings

cartilaginous tracheal rings

esophagus

left common carotid

left brachial artery

lung lobe

aorta

pulmonary artery

left auricle

lung lobes

left ventricle

right central lobe of liver

gall bladder

quadrate lobe of liver
(part of the right
central lobe)

common bile duct

left central lobe of liver

papillary lobe of liver

left lateral lobe of liver

brachiocephalic artery

vena cava

right auricle

lung lobes

coronary vessels

right ventricle

pericardial sac, opened

lung lobe

diaphragm

right lateral lobe of liver

mesentery

caudate lobe of liver

duodenum

pancreas

Fig. 4-53 A dog pluck for study of mammalian respiratory system. (After D. Lamore, *Carolina Tips* 21:35, November 1958.)

bean meal and shake. Let this stand for a while. The results are not immediate as they are when pure urease is used.)

Study of a nephron

When studying the role of the kidney in maintaining homeostasis of blood, examine the basic unit, the nephron, and accompanying glomerulus and tubules (Figs. 4-54, 4-55). Use models and diagrams. To develop the concept of the mass of capillaries comprising the glomerulus,

bunch a red pipe cleaner into a funnel and compare this to the Bowman's capsule (funnel) of the nephron. Connect tubing to the funnel to shape the rest of the nephron.

Chemical tests: Urine

Test for pH   Use litmus paper, nitrazine paper, or Labstix[34] to test the pH of urine. Freshly voided urine is usually acid with

[34] Available from Ames Co., Inc., Elkhart, IN.

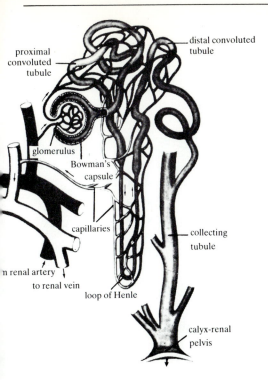

proximal convoluted tubule

distal convoluted tubule

glomerulus

Bowman's capsule

capillaries

n renal artery

to renal vein

loop of Henle

collecting tubule

calyx-renal pelvis

Fig. 4-54 A simplified nephron with blood vessels. (Adapted from *The Human Body in Health and Disease*, 4th ed., by Ruth Lundeen Memmler and Dean Lin Wood. Copyright © 1977 by Lippincott Publishers, Philadelphia.)

an average pH 6 (although the range may be 4.9 to 7.4). Diets rich in citrus fruits produce a more alkaline urine (Table 4-6).

TABLE 4-6 Chemical composition of urine

Average amount (24 hr)	1 to 2 liters depending on water intake; perspiration
pH	5–7
color	clear amber
specific gravity	1002–1030
glucose	0.1 percent
	80–120 mg/100 ml
nitrogenous substances:	g/liter
urea	25.0
uric acid	0.6
creatinine	1.5
ammonia	0.6
inorganic salts:	g/liter
chlorides (NaCl)	9.0
phosphates	2.0
sulfates (of Na, Ca)	1.5
potassium	2.0
magnesium	0.2
calcium	0.2

toxic materials; breakdown products of sex hormones

Test for glucose   Normally the glucose level in urine is 0.1 percent; in diabetes mellitus the level reaches 0.2 percent or more (300 mg/100 ml). Sugar appears in the urine when the blood sugar level reaches 180 mg/100 ml.

Test for glucose in urine by adding 8 drops of urine (0.5 ml) to 5 ml of Bene-

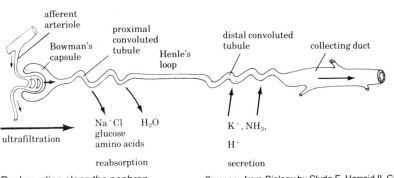

afferent arteriole

Bowman's capsule

proximal convoluted tubule

Henle's loop

distal convoluted tubule

collecting duct

ultrafiltration

Na^+Cl H_2O
glucose
amino acids

reabsorption

K^+, NH_3,

H^+

secretion

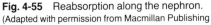

Fig. 4-55 Reabsorption along the nephron. (Adapted with permission from Macmillan Publishing

Company from *Biology* by Clyde F. Herreid II. Copyright © 1977 by Macmillan Publishing Company, Inc.)

Fig. 4-56 Microscopic examination of sediments in urine.

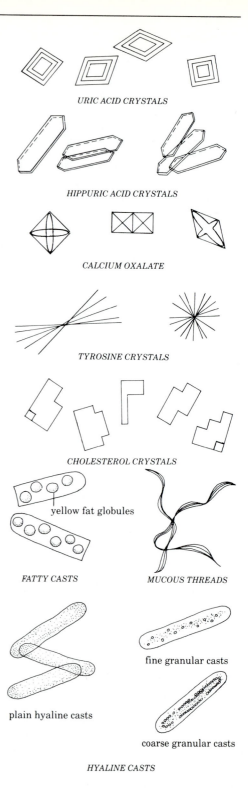

URIC ACID CRYSTALS

HIPPURIC ACID CRYSTALS

CALCIUM OXALATE

TYROSINE CRYSTALS

CHOLESTEROL CRYSTALS

yellow fat globules

FATTY CASTS

MUCOUS THREADS

plain hyaline casts

fine granular casts

coarse granular casts

HYALINE CASTS

dict's qualitative solution in a test tube. Boil for 1 to 2 min and allow to cool (or 5 min if tubes are in a boiling water bath). You may also want to demonstrate the use of commercial tablets (Clinitest or Galatest, or others) that do not require boiling.[35] Also demonstrate dip sticks that may be used to test for glucose as well as several other substances in urine.

Test for albumin There are many tests for albumin. One method is the Heller nitric acid ring test. Pour 5 ml of concentrated nitric acid into a test tube (*caution*). Tilt the tube, and use a medicine dropper to add urine so that it flows slowly down the side of the tube. Watch for the stratification of liquids with a white region of precipitated protein at the point of contact.

Some technicians suggest that the urine be diluted in this test since concentrated urine will form a white ring due to the presence of uric acid. At times, bile pigments or other substances may give a colored ring, but the albumin ring is white. See also other tests cited in medical techniques texts such as O. Hepler's *Manual of Clinical Laboratory Methods* (Springfield, IL: Charles Thomas, 1968; C. Anthony and G. Thibodeau, *Anatomy and Physiology* (laboratory manual), 10th ed. (St. Louis, MO: Mosby, 1979); E. Arthur, *Essential Human Anatomy and Physiology* (Glenview, IL: Scott, Foresman, 1977).

Phenylketonuria There is a dip-and-read test to detect this inherited but preventable mental deficiency in infants. Phenistix (available from Ames Co.) is a strip of paper impregnated with a reagent, ferric chloride (see Genetics, chapter 7).

---

[35] Clinitest tablets and Labstix are available from Ames Co., Inc., Elkhart, IN. Some dip sticks determine, in one test, the amount of protein, glucose, pH, and blood; the results can be read in seconds when compared to color tables accompanying the dip sticks.

Microscopic examination of urine

Centrifuge a 10-ml sample of urine at 1500 rpm for 5 min. Then pour off the fluid; the remaining drops of fluid will be adequate to dilute the sediment in the tube. Transfer a drop of the sediment to a clean slide and apply a coverslip. Examine under high power; scan the slide. Darken the field a bit as needed; some sediments are almost transparent.

Refer to texts describing the appearance of various crystals (uric acid, calcium oxalate, cystine, cholesterol, etc.), epithelial cells, leukocytes, erythrocytes, and many kinds of casts from renal tubules, parasites, and bacteria (Fig. 4-56).

Renal function tests and kidney clearance tests are also part of the study of kidney function.

Osmoregulators

Consider the different osmotic conditions in freshwater and marine organisms. In most marine invertebrates the internal salt concentration is the same as the seawater (isosmotic condition). Freshwater fish excrete large amounts of water in dilute urine; since salts are thereby also excreted, cells in the gills absorb salts to maintain a salt balance. Marine fish lose water, drink in large amounts of water, and produce little urine. Here the gills of these fish secrete extra salts. Refer to osmotic regulation in *Paramecium* in salt water (p. 139). Also examine the contractile vacuole system of *Paramecium* based on electron micrographs (Fig. 4-57).

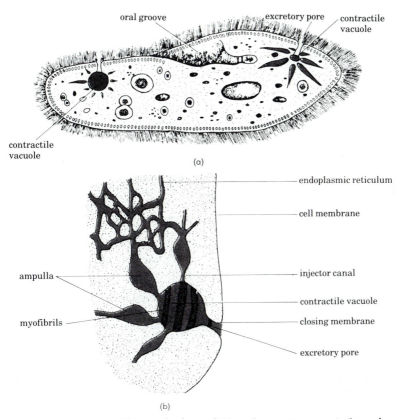

Fig. 4-57 Osmoregulators: contractile vacuoles in *Paramecium*. Electron micrographs show fluid drains from endoplasmic reticulum into ampullae, which lead to a contractile vacuole. Myofibrils probably cause contraction of the vacuole, sending fluid out the excretory pore to the surface. (Adapted with permission from Macmillan Publishing Company from *Biology* by Clyde F. Herreid II. Copyright © 1977 by Macmillan Publishing Company, Inc.)

CAPSULE LESSONS: WAYS TO GET STARTED IN CLASS

4-1. Have students take each other's pulse. Develop the idea that the beating heart pumps blood through the body by way of blood vessels. Then compare the pulse rate following some kind of exercise; this will lead to a good discussion of the role of the heart and other aspects of circulation.

4-2. Ask students to diagram their own concept of their circulatory system. Usually students begin their drawings with a heart and vessels leading in and out—and then they become "snagged." Introduce a chart of the complete circulatory system when they have reached this point. Why weren't capillaries described by early investigators who named arteries and veins?

4-3. Perhaps you may want to begin with a historical approach: Harvey and his study of circulation. What were early ideas about the function of arteries?

4-4. Bring a fresh specimen of a beef's heart for dissection to class. Elicit the functions of each chamber of the heart and of the valves. How many times does the heart beat in an hour if the average rate is 70 heartbeats per min? Develop a discussion, with diagrams, of a closed circulatory system from the heart and back again. Follow with a development of what materials the blood gains and loses on its trip through the lungs, kidneys, bone marrow, small intestine, and every cell.

4-5. Demonstrate how to prepare a goldfish for microscopic study of the circulation in its tail (see technique in this chapter). Then have several students prepare additional demonstrations so that all students may examine blood circulating through the arteries, capillaries, and veins of the tail. In which blood vessels does the blood move more quickly, and in spurts? Which are the fine connecting vessels? What are the large vessels in which blood flows the opposite way?

4-6. Set fresh celery, or bean seedlings, or carrots into red-colored water. Ask students to trace the entrance of water up into the leaves. What will the water be used for? How does the water get into the roots?

4-7. Have students examine fresh celery, jewelweed (*Impatiens*), or carrots for fibrovascular bundles. Cut very thin sections of a carrot with a razor blade; if the edges of the slices are very thin, they can be examined under the microscope. Or study prepared, stained slides.

4-8. At some point show a film on the composition of blood. Refer to available films in current film catalogs.

4-9. As a review, diagram a capillary network around one to two cells. What materials do the cells draw from the blood stream? Do all cells draw the same materials, or do some cells have "extra" special needs?

4-10. Some teachers begin with a laboratory study of the composition of the blood. (In most school systems, parental consent notes are necessary.) Study either a drop of blood in Ringer's solution or a stained blood smear. What is the effect of an isotonic solution on blood cells? Of a hypertonic solution? What is the role of different leukocytes? of platelets?

4-11. Blood typing is a very successful laboratory lesson. Have students type their own blood. Demonstrate the technique for students using the materials and safety procedures listed in this chapter. Have a student report on the distribution of types A, B, AB, and O in the total population. Compare with the findings in your class. Warn students not to consider this exercise in blood typing official, for this class work is subject to all the usual pitfalls of novices.

4-12. Begin the lesson with a question that arouses good discussion. These may be possibilities: How many know that food materials are sometimes injected into a person who is very ill? What kind of fluid is inserted into a vein? Why not into an artery? (Students might trace the path of this glucose solution through the body.)

4-13. You may want to present students with a problem. For example: A bad-tasting but harmless drug was injected into the leg of a man. In less than a minute he sensed a bitter taste on his tongue. Explain. From this elicit a discussion of circulation time, the path of circulation, and the need for circulation of blood.

4-14. Ask students what facts they know about diabetes. Some students may know a great deal about insulin injections and the precautions necessary. Here the topic of circulation is approached through the study of hormones. Of course, many other hormones might be used in this manner: role of thyroxin in metabolism; what causes a giant or a midget; a story of the adrenals and their role in stress. The role of the blood in some blood disturbances (such as types of anemia or leukemia) and some heart conditions. Elicit a discussion concerning the factors which affect the circulatory system. How is a blood clot formed? Remember that the highest death rate is caused by circulatory diseases. (Many inexpensive booklets are available from the American Red Cross or the American Heart Association.)

4-15. As a laboratory lesson have students prepare wet mounts of the water flea, *Daphnia* (usually available from an aquarium shop). Examine the beating heart found in the anterior top end of the animal. Develop a discussion of the role of the heart in circulation. Also observe blood circulating in such microscopic worms as *Dero, Nais,* or *Tubifex* worms, also available from an aquarium shop (culturing, chapter 9). What are some factors that affect the heartbeat?

4-16. Distribute vials containing seedlings with root hairs. Have students examine them with a magnifying lens. What advantage is there in the increased surface provided by the extensive root hairs? What process goes on through these cell extensions?

4-17. Provide these materials for students: several plants, cobalt chloride paper, petroleum jelly, and pins. Ask students to design a demonstration to reveal which surface of the leaves, the upper or lower surface, contains more stomates.

4-18. You may want to set up several demonstrations to show transpiration. Put them in the classroom, where they may be observed, and wait for students to ask questions. You are then addressing the topic "How is water transported in a plant?"

4-19. Sometimes you may want to include one of the demonstrations performed in class on a test with questions based on that principle or concept. For example, you might repeat a demonstration or, better still, show a modification of it (several are described in this chapter) which illustrates the same principle. Then raise five questions which test both observation and reasoning. This might be used as one test question. A sample of this type of test may be found on page 234.

4-20. Open a bottle of dilute ammonia or perfume in class for a few seconds. Have students raise hands when they can smell it. Or place a strip of wet pink litmus paper in a test tube and invert over the mouth of a bottle of ammonia. Elicit the facts that the liquid changed into a gas and the molecules diffuse among the molecules of gases in the air. Then lead into a discussion of diffusion of materials

through a membrane (p. 242). Refer to the passage of water through root hairs, and from cell to cell.

4-21. Plan a laboratory activity in which students examine the underside of such a leaf as *Sedum, Tradescantia,* or lettuce. Teach students how to strip the lower epidermis for examination under the microscope as a wet mount (see p. 147 for wet mounts).

4-22. This may be a good place to introduce the use of radioactive isotopes as tools in a study of functions of plants and animals. Students might report to the class on how radioactive iodine, phosphates, sodium chloride, and other isotopes are used (see listings at end of chapter).

4-23. At some time, students may read an offprint from *Scientific American.* The subject of the paper may be used as an introduction to a topic, as a review, or as a basis for a test of understanding.

4-24. Fill six quart-bottles with water. This is about the amount of blood in the human body. Develop the notion of a closed system of tubes for circulation of blood. Use a plaque or a chart and ask students to give their own concepts of how the blood circulates around the body.

4-25. Use a fresh haslet, a heart-lung model, or a take-apart heart model. Ask students to identify as many parts as they can. What is the advantage of the close proximity of lungs to heart? What is the advantage of thick walls in the ventricles? What is the function of the thin-walled small atria?

4-26. Begin with the inspirator-expirator apparatus described in this chapter (Fig. 4-34) and elicit an explanation for the color change in one flask from brom thymol blue to yellow. From this point on develop the need for, and the materials used in, respiration.

4-27. What effect does strenuous exercise have on breathing? (Refer to the activity on page 302.) Why is this so? From this point on, develop a discussion of the materials exchanged in respiration. Trace the path of oxygen throughout the body.

4-28. Have a student make the bell jar model as a project. Demonstrate its use, and have the class develop a description of the mechanics of breathing using the model.

4-29. Use a Florence flask to represent one air sac in the lungs. With a red glass-marking pencil draw interweaving capillaries on the glass. What materials would pass into the air sac from the blood? What would the blood gain from the air sacs? Develop a discussion of the difference between inhaled and exhaled air and the path of air into the air sacs.

4-30. Prepare a demonstration ahead of time using elodea plants and brom thymol, blue in test tubes placed in the dark (p. 305). Ask students to explain why the brom thymol blue turned yellow in the dark. This should raise many questions. Then have students set it up again as an investigation with controls. This may help clarify one difference between photosynthesis (in which carbon dioxide is absorbed) and oxidation, or respiration (in which CO_2 is liberated).

4-31. You may want to introduce the role of respiratory enzymes using a speck of dried *Cypridina* containing both luciferin and luciferase (chapter 2). Add a few drops of water. Compare this brilliant blue color with the flame of a candle, or of a burning butter candle. What is the difference?

4-32. Have teams of students dissect a kidney to trace its role in homeostasis of the blood. How are some substances reabsorbed against the diffusion gradient? Develop the role of the kidneys and the active part played by membranes.

4-33. Provide the class with jars and germinating seeds. Ask students to design an investigation, after they have established the fact that respiration may be measured, by testing either the amount of oxygen used or the amount of carbon dioxide given off.

4-34. You may want to begin a lesson with a demonstration showing that growing seedlings give off enough heat to raise the temperature above that of the room. Elicit from students a description of the process going on: respiration. What materials are used? What happens when food is oxidized? Compare photosynthesis and respiration.

4-35. Explain the importance of Dr. Rosalyn Yalow's radioimmunoassay technique which can, among other things, distinguish between human insulin and injected animal insulin. The body produces antibodies against the animal insulin that can be identified. (Recall that Dr. Yalow received the Nobel prize for medicine in 1977.)

4-36. Begin with a film on blood, respiratory enzymes, the role of the kidneys in homeostasis; also consider techniques films or film loops that show the use of a hematocrit, hemacytometer, microscopic, and chemical examination of urine. Do you have audiovisual catalogs? Have you checked sources of free or low-cost rental films?

4-37. As a laboratory activity, dissect a fresh kidney. Or use a model and trace the path of nitrogenous wastes from the kidney tubules into the ureters and urinary bladder.

4-38. Demonstrate the action of the enzyme urease on urea. Where does this occur in the body? What would happen if urea accumulated?

4-39. Begin a lesson by describing how the blockage of a ureter in a 7-month fetus was treated. While the fetus was still in the uterus, ultrasonic scans revealed one kidney was twice the size of the other. A needle was inserted through the abdominal wall of the mother. Guided by ultrasound to the exact location of the larger kidney, the needle was then used to drain the excess fluid.

Now elicit the structure of the urinary system, the kidney clearance tests, the role of the million or more nephrons in a kidney.

4-40. As a review, compare the transport, respiratory, and excretory systems of protozoa, hydra, grasshopper, worm, lobster, fish, and other vertebrates. Use models and charts or transparencies for students to help explain the various systems.

4-41. Also consider having students build their own science library and/or initiate your own reference library to offer students choices in supplementary readings. For example, obtain a listing of offprints and/or books from the following:

Careers in Science—information available from United States Dept. Labor, Bureau of Labor Statistics, reprints of *Occupational Outlook Handbook* available; write to regional offices in Boston, Philadelphia, Chicago, Dallas, New York, Atlanta, Kansas City, San Francisco.

Consider too, that the Supt. of Documents also has available a list of depository libraries which have government publications. This is valuable information for there may be times that the stock of a publication is no longer available for purchase.

New American Library, Education Division, 501 Madison Ave., New York, NY 10022.

Oxford/Carolina Biology Readers (some 100 titles; short 16 page monographs written by people active in the field for which they write).

Viking Penguin, Inc. 299 Murray Hill Parkway, East Rutherford, NJ 07073.

Scientific American, W. H. Freeman, 660 Market St., San Francisco, CA 94104.

Time/Life Books, Time and Life Bldg., Chicago, IL 60611.

BOOK SHELF

The following listing is only a sampling of texts and/or laboratory manuals in physiology.

Alberts, B. et al. *Molecular Biology of the Cell*. New York: Garland, 1983.

Arnow, L. *Introduction to Physiological and Pathological Chemistry*, 9th ed. St. Louis, MO: Mosby, 1976.

Baker, J., and G. Allen. *The Study of Biology*, 3rd ed. Reading, MA: Addison-Wesley, 1977.

Barnes, R. *Invertebrate Zoology*, 4th ed. Philadelphia: Saunders, 1980.

Borek, E. *The Atoms Within Us*, 2nd ed. New York: Columbia Univ. Press, 1981.

Calow, P. *Invertebrate Biology*. New York: Wiley, 1981.

Chen, T., ed. *Research in Protozoology*, Vol. 1. Elmsford, NY: Pergamon, 1967.

Colbert, E. *Evolution of the Vertebrates*, 3rd ed. New York: Wiley, 1980.

Dales, R., ed. *Practical Invertebrate Zoology*, 2nd ed. New York: Wiley, 1981.

Franklin, K., ed. *William Harvey: The Circulation of the Blood*. New York: E. P. Dutton, 1979.

Gabriel, M., and S. Fogel, eds. *Great Experiments in Biology*. Englewood Cliffs, NJ: Prentice-Hall, 1955.

Giese, A. *Cell Physiology*, 5th ed. Philadelphia: Saunders, 1979.

Harding, R. *Omnivorous Primates*. New York: Columbia Univ. Press, 1981.

Herreid, C. *Biology*. New York: Macmillan, 1977.

Hickman, C., Jr., L. Roberts, and F. Hickman. *Biology of Animals*. St. Louis, MO: Mosby, 1982.

Hoar, W., and C. Hickman, Jr. *A Laboratory Companion for General and Comparative Physiology*, 2nd ed. Englewood Cliffs, NJ: Prentice-Hall, 1975.

McConnaughey, B., and R. Zottoli. *Introduction to Marine Biology*, 4th ed. St. Louis, MO: Mosby, 1983.

Nelson, G., and G. Robinson. *Fundamental Concepts of Biology*, 4th ed. New York: Wiley, 1981.

Norstog, K., and R. Long. *Plant Biology*. Philadelphia: Saunders, 1976.

Norstog, K., and A. Meyerriecks. *Biology*. Columbus, OH: Charles E. Merrill, 1983.

Prosser, C. *Comparative Animal Physiology*. Philadelphia: Saunders, 1973. 3.

Schmidt–Nielsen, K. *Animal Physiology*, 3rd ed. Cambridge Univ. Press, 1983.

Sherman, I., and V. Sherman. *Biology—A Human Approach*, 3rd ed. New York: Oxford Univ. Press, 1983.

Singer, S., and H. Hilgard. *The Biology of People*. San Francisco: Freeman, 1978.

Starr, C. and R. Taggart. *Biology: the Unity and Diversity of Life*, 3rd ed. Belmont, CA: Wadsworth, 1984.

Stephens, G., and B. North. *Biology*. New York: Wiley, 1974.

Stryer, L. *Biochemistry*, 2nd ed. San Francisco: Freeman, 1981.

Toporek, M. *Basic Chemistry of Life*. St. Louis, MO: Mosby, 1981.

United States Govt. Printing Office, Supt. of Documents, *Clinical Laboratory Procedures—Hematology*. Depts. Air Force and Army, AF Manual 160–51; TM8–227–4, Dec. 1973, Washington, DC 20402 ($3.55).

Villee, C. *Biology*, 7th ed. Philadelphia: Saunders, 1977.

Wald, G. et al. *Twenty-Six Afternoons of Biology*. Reading, MA: Addison-Wesley, 1962.

Wigglesworth, V. B. *The Principles of Insect Physiology*, 7th ed. London: Chapman and Hall, 1972.

Wolfe, S. *Biology of the Cell*, 2nd ed. Belmont, CA: Wadsworth, 1981.

5

Interpretation of the Environment: Behavior and Coordination

The evolution of organisms has resulted in adaptive behavior that has survival value. Consider the many taxes among protists and small invertebrates as well as the tropisms found among plants. The advent of more specialized nervous systems has accommodated the development of reflex behavior, patterns of conditioning, imprinting, and circadian behavior. Patterns of learning, of course, are more complex (Fig. 5-1).

The investigations in this chapter describe studies of taxes in lower animals and of tropisms among plants. The interaction of plant hormones is also considered in maintaining homeostasis, growth, and differentiation. The role of hormones in differentiation in tissue cultures is described in chapter 6, p. 483.

Other investigations in this chapter show the role of animal hormones and their coordination of neural and hormonal systems. Patterns of learning, conditions conducive to learning, and the functions of the brain and sense organs complete the study of behavior. (The role of thyroxin/iodine in tadpole metamorphosis is described in chapter 6, p. 421.)

Taxes and tropisms

Adaptive behavior patterns in which animals *move* in response to a stimulus in their environment are called taxes. In 1832, De Candolle defined the bending of plants toward light as a tropism. Today, the term is used to describe *growth* responses of plants. A number of animal taxes and plant tropisms are easily demonstrated in the classroom.

Response to light: Phototaxis

Protozoa   In general, protozoa are indifferent to light, although moderate light intensity is preferred. However, some forms, such as *Blepharisma* (Fig. 1-7), are especially light-sensitive; their rose color is bleached in strong light (see p. 140).

Euglena[1]   Orienting, adaptive behavior, or taxes, are varied in *Euglena*. For example, *Euglena* shows a positive taxis toward light wavelengths that benefit its action spectrum in photosynthesis. It swims toward weak acids such as carbon dioxide in water, and to glucose. *Euglena* shows a negative taxis toward gravity (that is, it swims to the top of a vial). It also swims away from salts, extremes in temperature, and light of high intensities.

To show positive phototaxis of a population of *Euglena*, pour a concentrated culture into a Petri dish or Syracuse dish. Cover half the dish with a library card, so that light shining down on the dish falls on

---

[1] You may prefer to classify *Euglena* as a member of the plant kingdom.

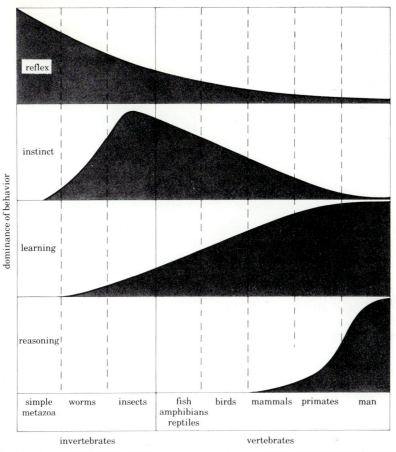

Fig. 5-1 Evolving types of behavior among invertebrates and vertebrates. In which group is reasoning the dominant pattern? (Adapted with permission from Macmillan Publishing Company from *Biology* by Clyde F. Herreid II. Copyright © 1977 by Macmillan Publishing Company, Inc.)

one half of the dish. Within 10 min, remove the cover to find the many green flagellates concentrated in the lighted half of the dish.

Students may show this phototactic response under their microscope. Cut a card to the size of a glass slide; in the center of the card, cut a narrow slit about 2 mm wide. Then prepare a wet mount of a rich culture of *Euglena* and focus under low power with good light; slip the card underneath the slide. Within a few minutes, examine the organisms that are visible through the narrow slit, and quickly remove the card. Observe the green line where most of the *Euglena* congregated where light was available through the slit.

Flagellates show a positive or directed taxis. The rate of swimming increases to about 0.18 mm per sec until it reaches the light saturation point.[2] Light is received by the photoreceptor, the enlarged base of the flagellum, and *Euglena* orients itself with its axis in line with light rays. The stigma or red eyespot, located near the base of the flagellum in the gullet, seems to be a light shield (Fig. 5-2; also Fig. 1-2).

[2] J. Wolken, *Euglena: An Experimental Organism for Biochemical and Biophysical Studies* (New Brunswick, NJ: Institute of Microbiology, Rutgers, The State University, 1961).

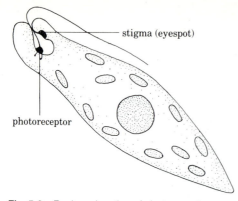

Fig. 5-2 *Euglena*: location of photoreceptor and eyespot.

You may also want to have students examine the phototactic action spectrum of *Euglena*, and observe its movement especially toward blue wavelengths of light.

Arditti and Dunn describe a technique to show that phototaxis in *Euglena* involves a response to blue wavelengths of light.[3] Prepare a cuff of black paper to fit around a corked vial or test tube so that the two ends of the cuff meet but do not overlap. On one side of the edge of the paper punch a row of holes with a paper puncher. Cover each hole with a double layer of cellophane of different colors as shown in Figure 5-3. To get far red, use one layer of red cellophane and cover with a layer of blue cellophane.

Fill the vial or test tube with a concentrated culture of *Euglena*. Fasten the black cuff with plastic tape and expose the vial to sunlight or to a light source that combines incandescent and fluorescent light for 1 to 2 hr.

Avoid shaking the vial when the cuff is removed. Where is the greatest concentration of the flagellates? For a more permanent record of the response to different wavelengths, heat the tube or vial slightly so that *Euglena* adheres to the glass. (Culture methods for *Euglena*, p. 664.)

[3] J. Arditti and A. Dunn, *Experimental Plant Physiology* (New York: Holt, Rinehart, Winston, 1969).

Planaria These flatworms are found in streams on the underside of rocks or leaves of water plants. In the laboratory they thrive best when kept in shallow, darkened jars of pond water containing some small stones under which they can hide (culturing, p. 667). Remove the cover of the containers holding planarians and watch their movement away from a flashlight as they exhibit a negative phototaxis.

Earthworms Students may keep earthworms in wet sphagnum moss in a dark box or covered dish. Or they may use a layer of humus to line the container.

Suddenly lift the cover and direct a beam from a flashlight on the anterior region of the worms. Observe the way they avoid the light by contracting or burrowing into the moss.

Fruit flies Transfer a fairly large number of adult fruit flies (*Drosophila*) from a culture bottle into a long test tube; stopper the test tube with cotton. Then cover one half of the test tube with black paper or aluminum foil. Direct a beam of light at one end of the horizontal tube and observe the attraction of the flies to the light; here is a positive phototaxis. If the position of the foil or paper is reversed and the beam of light is shifted to a new position, you will find that the flies again orient themselves in the lighted part of the tube.

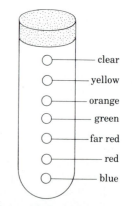

Fig. 5-3 Preparation to demonstrate phototactic action spectrum of *Euglena*.

Also refer to the effect of light on the development of animals and plants (photoperiodism, p. 375). Also investigate whether fruit flies are attracted to light of one intensity in preference to another, or to different wavelengths (see *Daphnia*).

Daphnia   What intensity of light affects diurnal migration of *Daphnia*? Use a quantity of water fleas, wide shallow containers, and a light source, the intensity of which can be varied. What happens when light of low intensity is directed at one side of the container? What reaction occurs when the light intensity is increased?

Furthermore, are *Daphnia*, *Artemia*, or other small organisms affected by the wavelength of light? Suppose that red light or green, blue, or yellow filters are used. Is the behavior of the organism the same under each wavelength of light?

How does *Daphnia* orient itself normally in a container of aquarium water? (Keep the temperature of the water constant in these observations.) In general, *Daphnia* are more active in dim light. Maintain *Daphnia* in dim light for some 15 min before exposing them to different light intensities. Place several *Daphnia* in a tall container (100 ml graduate) and illuminate the top of the cylinder with light bulbs varying from 40 to 120 watts, held at the same distance. Students should observe, that in general, *Daphnia* swim upward in jerky movements toward dim light and gradually downward as the light intensity increases. To avoid the variable of gravity, use a low or flat basin and illuminate from the side.

Besides the intensity of light, *Daphnia* also respond to light of different wavelengths. Have students observe the different behaviors in red and blue light. Smith and Baylor describe these "color dances" of *Daphnia*.[4] Under red light, *Daphnia* are calm, moving in a vertical plane called the "red dance." Under blue

light, they become "agitated" and move away from blue light. *Daphnia* move rapidly in a predominantly horizontal plane. However, when the temperature of *D. magna* was reduced from 25° C (77° F) to 10° C (50° F) the blue response disappeared. At this temperature, *Daphnia* give the "red dance" response under any light intensity or color of light.

To observe this behavior, wrap colored cellophane around light bulbs or use color filters. Many blue cellophanes also transmit red light while some red cellophanes transmit blue light (see p. 378).

The effects of temperature, drugs, and hormones on the rate of heartbeat in *Daphnia* are described on page 267; anatomy in chapter 1; digestion on page 239; culture methods on page 670.

Other organisms   Perhaps pond snails, land snails, earthworms, or *Tubifex* worms can be easily obtained. Students may use these organisms to design their own investigations of different taxes, including the effects of alternating light and dark periods, the effects of polarized light, and light of different wavelengths. How do these animal responses compare with responses of growing oat seedlings? Do seedlings bend in response to all wavelengths of light? (See pp. 364, 376.)

Ionizing radiation

Protozoa   Radiation slows down cell division among protozoa, and flagellates seem more sensitive to ionizing radiations than ciliates. For example, *Paramecium aurelia* can survive an exposure to 150,000 roentgens; yet, even in apparently normal paramecia, the effects may show up later. In general, the physiological state of protozoa and other cells affects the severity of the effects of radiations.

There is much literature on the effects of ionizing radiations on cells. There are reviews in textbooks of physiology or heredity, and current papers in journals (these are easily found in *Biological Abstracts*). Giese's textbook, *Blepharisma*, has

[4] F. Smith and E. Baylor, "Color Responses in the *Cladocera* and their Ecological Significance," *American Naturalist* (January–February 1953).

an excellent summary on this topic (see *Book Shelf* at the end of this chapter). Also refer to A. Upton's paper in *Scientific American*, "The Biological Effects of Low-level Ionizing Radiation" (February 1982).

Photodynamic sensitization

When the pink protozoan *Blepharisma* (Fig. 1-7) is maintained in moderate light or in darkness, it becomes deeply pigmented, almost rose colored (p. 140). These forms are especially sensitive to light; when exposed to intense light they quickly die. This photosensitization occurs only when oxygen is present. Students may want to compare the effect of intense light on light-grown forms with those maintained in darkness.

Further, it is possible to use color filters to determine whether light of a specific wavelength is more critical in producing death among these protozoa. Is it possible to reduce the effects of photosensitization? Is cannibalism among *Blepharisma* affected by light? (Refer to Giese's textbook *Blepharisma*, listed at the end of this chapter.)

Response to light: Phototropism

Leaves   Observe plants growing in the field. Note the mosaic of leaves formed as the leaves are oriented in relation to the angle at which light strikes. Few leaves are shaded by others. Notice the leaves of vines in particular. Indoors, healthy full-grown geranium and coleus plants show the same tilting of leaf petioles as a positive response to light from a lateral light source.

Stems   Shoots of growing seedlings show a positive phototropism more readily than fully grown plants. Soak seeds of radish, sunflower, pinto bean, oats, wheat, or kernels of corn overnight and plant them just under the surface of moist, clean sand in plastic cups or flowerpots. Students may place several of these cups of seeds under a box to exclude light. Arrange a second box covering similar cups of seeds, but with a slit on the side at about the level of the cups, and set the boxes where moderate light can enter the slit (Fig. 5-4).

Within a few days, depending on the kind of seeds used, a marked growth or bending of the stems of the shoots should be apparent in the box in which light enters from one side. In the other box, seedlings grow upright in random fashion. Variations include covering the aperture with colored cellophane to test the sensitivity of seedlings to varying wavelengths of light.

A more elaborate version of this demonstration can be constructed which provides standard conditions of humidity and ventilation. Students might install within a suitable wooden box a partition reaching to within 5 to 6 cm of the top of the box. They can also make a hinged cover for convenient handling if the box is to be a permanent piece of apparatus. Apply black paint to the inside surfaces.

Two variations are shown: In Figure 5-4a, a slit has been cut in the side of the box; in Figure 5-4b, small flashlight bulbs have been wired to the side of one compartment and to the top of the other. In this way an equal amount of light (and heat) is received by both sets of seedlings. Light the bulbs for about 5 min every half-hour or so.

Light and the bending of stems   Recall the observations of Charles Darwin and his son on plant sensitivity to light. They hypothesized that the tips of shoots were sensitive to light and that some influence was transmitted to the lower part of stems causing the bending of stems toward light. They tested their hypothesis by "capping" young shoots of canary grass and oats with caps—some made of thin glass, others of thin tin foil that were blackened inside (also see pages 364, 366).

Students may want to repeat the work of Darwin, and also some of the classic experiments of Boysen–Jensen and of F. W. Went on the production of growth

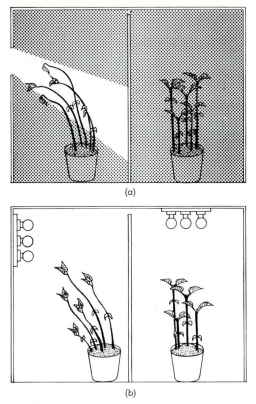
(a)

(b)

Fig. 5-4 Two demonstrations showing positive phototropism in shoots of seedlings.

Investigations using the many auxins and other plant hormones or growth-regulators are described further along in this chapter, page 364. Also refer to the effect of different ratios of growth regulators on differentiation of tissues (tissue culturing techniques, chapter 6).

Roots Roots show a negative phototropism. Soak radish or mustard seeds overnight. Spread cheesecloth across several tumblers of water and fasten with a rubber band, allowing some slack so that the cheesecloth remains wet. Then sprinkle seeds on the surface of the cheesecloth.

Enclose some of the tumblers in a darkened box. Cut a small slit in a similar box and cover replicas with this box. In this way light enters the box from one side (as described above for shoots of seedlings).

You may find it more convenient to make small black paper boxes which fit snugly over individual preparations (or use aluminum foil instead of black paper). Just cut a small slit along one side of some of the paper boxes. After the response of roots (and stems) is apparent, rotate the position of the slit 180°. Notice the change in the growth of stems and roots over the next few days.

hormones or auxins by the tips of oat coleoptiles. The tips of the coleoptiles, which are the sheaths that enclose the early folded shoots, are especially sensitive to small amounts of light when they are only a few centimeters long. Working in red light, or in the dark, students may cut off the tips, or place small caps of aluminum foil over the tips. In such cases, there is no bending toward light; compare with untreated coleoptiles (Fig. 5-26). The tip of the coleoptile produces auxin and is light-sensitive. Light (one factor among others) seems to deflect the flow of auxin to the shaded side of the shoot (Fig. 5-27). The cells on the shaded side, with a higher concentration of auxin, elongate faster and the stems bend toward light. A physiological regeneration usually occurs within 24 hr, so tips may have to be cut off again.

Response to electricity: Galvanotaxis

In an electric current, most protozoa move toward the cathode. This may be due to a direct effect on cilia (or other organelles), for the cilia on the side of the cathode beat forward, while those on the side of the anode beat backward. As a result, the organisms are turned to face the cathode. Even rhizopods put out pseudopodia on the side of the cathode.

The electrical charge on the surface of microorganisms may be demonstrated in this way. Partially fill a small U-tube or a flatsided glass container with a thick culture of *Paramecium*. Connect two dry cells in series and insert the two wires from the battery into opposite ends of the container or U-tube containing the paramecia

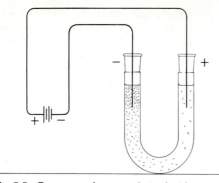

Fig. 5-5 Response of paramecia to electric current; paramecia move toward the negative arm of the U-tube.

(Fig. 5-5). Notice the clustering of the microorganisms around the negative pole; apparently the surface of the organisms is positively charged. Now reverse the current and show the migration of the microorganisms to the other electrode which is now the negative pole.

A similar preparation may be made using a Syracuse dish or an evaporating dish (even a welled slide) for examination under a dissecting microscope. This is also a rapid method for concentrating a culture.

Response to chemicals: Chemotaxis

Chemotaxis seems universal in protozoa and other small organisms; they avoid most chemicals, and exhibit a positive taxis in some weak acids. In fact, paramecia collect around a bubble of carbon dioxide in preference to a bubble of air.

Positive chemotaxis in Paramecium As food materials diffuse through the culture medium, *Paramecium* responds by swimming toward the food. Mount a drop of culture fluid containing some debris. Examine under the microscope. Protozoa will be clustered around the food material.

Negative chemotaxis in Paramecium Prepare a wet mount of a concentrated culture of *Paramecium*. Add a drop of acetic

or hydrochloric acid (1 drop of concentrated hydrochloric acid in 20 drops of water) or 1 drop of Waterman's ink[5] to one side of the coverslip. If the acid is strong enough, *Paramecium* will move away from the incoming acid. Often the trichocysts of *Paramecium* are extruded.

Another way to show negative chemotaxis in protozoa uses a bit of cotton thread soaked in dilute hydrochloric acid. This thread is placed across a drop of thick culture of *Paramecium* on a slide. Hold the slide against a dark background or view it under the microscope. Students will see a cleared area around the thread; the organisms have moved away from the acid.

What is the response when a thread dipped in hard-boiled egg yolk is brought near *Paramecium*? (Or place short lengths of millimeter glass tubing containing a variety of substances in a watch glass with protozoa cultures and view through a dissecting microscope.)

Response of Hydra to acid Mount several specimens of *Hydra* in a drop of the aquarium or pond water in which they were found. They will elongate when not disturbed. Place a bristle under the coverslip so that the organisms will not be crushed by the weight of the coverslip. Then introduce a drop of weak acetic acid to one side of the coverslip. Observe the discharge of nematocysts on the tentacles of the animals as the acid diffuses through the water. This makes an especially fine demonstration of nematocysts (p. 15). (This reaction is not considered a taxis.)

Response of planaria to chemicals Place some strips of raw liver in a culture dish of planarians and watch the rapid aggregation of these forms around the liver. In fact, you may collect planarians from a lake or brook by suspending a bit of raw liver in the water for a few hours. If planarians are present, large numbers will

---

[5] Not every ink contains enough tannic acid to give this response. Some preliminary experimentation may be necessary (also see chapter 1).

be found collected on the liver in a positive chemotaxis.

Bring a cotton applicator dipped in dilute ammonium hydroxide or acetic acid near the planarians and observe their negative taxis.

Chemotaxis in small crustaceans

Prepare 1 cm lengths of capillary tubing containing various solutions—dilute acids, dilute aspirin, salt solutions, nutrient solutions, etc. Now introduce each one separately with forceps into a Syracuse dish of a culture of small crustaceans such as *Cyclops* or *Daphnia*. Under a binocular dissecting microscope, examine the negative or positive taxes of the organisms in relation to these changes in their environment.

Capillary tubing of about 1 cm may be purchased or made by pulling out medicine droppers over a flame.

Earthworms
What is the effect of bringing a cotton applicator dipped in ammonium hydroxide near an earthworm?

Chemotropism

Pollen grains
For an investigation on how fragments of the pistils of flowers stimulate germination of pollen grains, see D. W. Rosen, "Studies on Pollen Tube Chemotropism," (*Am. J. Bot.* (1961) 48: 889–95).

Response to contact: Thigmotaxis

You may want to show how large protozoa such as *Spirostomum* or *Stentor* respond to contact or disturbance. When a drop of a rich culture of these organisms is mounted on a slide and the slide is tapped, these large organisms will contract. This is visible to the naked eye. You need a microscope, however, to see similar contractions in *Vorticella*, a stalked form (Fig. 5-6).

Students may recall that whenever *Paramecium* or other microorganisms bump

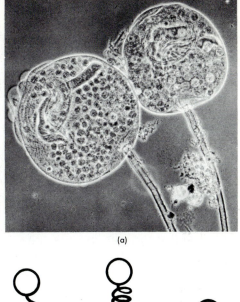

(a)

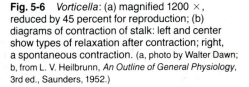

(b)

Fig. 5-6 *Vorticella*: (a) magnified 1200 ×, reduced by 45 percent for reproduction; (b) diagrams of contraction of stalk: left and center show types of relaxation after contraction; right, a spontaneous contraction. (a, photo by Walter Dawn; b, from L. V. Heilbrunn, *An Outline of General Physiology*, 3rd ed., Saunders, 1952.)

into each other, they alter the direction in which they are moving. This, too, may be considered a taxis, a thigmotaxis.

Similarly, when hydras, earthworms, insect larvae, planarian worms, and many microscopic worms are touched with a dissecting needle they contract.

Thigmotropism

Climbing roses and many vines show a positive thigmotropism, a response to touch that enables them to encircle a trellis. However, there are other cases of sensitivity in plants which probably are not tropisms. Turgor movements, which are reversible, are more rapid than growth movements or tropisms.

Mimosa shows sudden responses to touch (Fig. 5-7) as a result of loss of turgor of certain cells of the pulvinus, a swelling at the base of each cluster of leaflets. Gently tap the terminal leaflets with a pencil or finger, or pinch them with a forceps. Observe how quickly the end leaflets fold over each other. A similar response can be elicited when a lighted match is brought close to the tip of a cluster of leaflets.

This excitation of cells is followed by an action current that spreads from the tactile hairs on the leaves of *Mimosa* along the sieve tubes down through the petiole into the pulvinus. Here the permeability of the cells of the lower region of the pulvinus is affected so that the cells lose water. As a result of the loss of turgor, the stem drops.

If possible, study the response of the Venus's flytrap (Fig. 5-8). This, too, is a case of a response due to excitation created by the fly touching tactile hairs on the leaf and resulting in a wave of excitation that causes some cells to decrease their per-

(a)

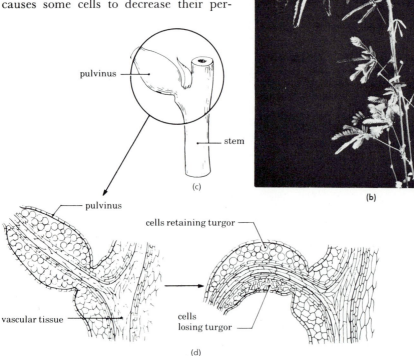

pulvinus

stem

(c)

(b)

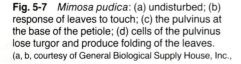

pulvinus

cells retaining turgor

vascular tissue

cells losing turgor

(d)

Fig. 5-7 *Mimosa pudica*: (a) undisturbed; (b) response of leaves to touch; (c) the pulvinus at the base of the petiole; (d) cells of the pulvinus lose turgor and produce folding of the leaves. (a, b, courtesy of General Biological Supply House, Inc., Chicago; c, d from *Biological Principles and Processes* by Claude A. Villee and Vincent G. Dethier. Copyright © 1971 by W. B. Saunders Company. Reprinted by permission of CBS College Publishing.)

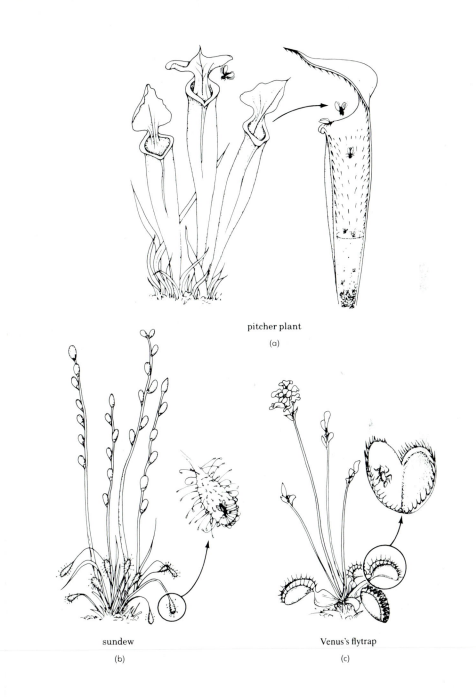

pitcher plant

(a)

sundew

(b)

Venus's flytrap

(c)

Fig. 5-8 Carnivorous plants: (a) pitcher plant; (b) sundew; (c) Venus's fly trap. (Adapted with permission from Macmillan Publishing Company from *Biology* by Clyde F. Herreid II. Copyright © 1977 by Macmillan Publishing Company, Inc.)

meability due to a change in turgidity. The protoplasm of plant cells transmits excitation accompanied by changes in electrical potential that affect the permeability of cells.

Other movements of leaves or flowers (as in *Oxalis*, and in beans, peas, and other legumes) are sometimes called sleep movements. These also rely on changes in turgidity effected by plant hormones rather than changes due to growth (tropism).

Response to gravity: Geotaxis

Paramecium   These forms show a negative geotaxis; they swim up to the top of a container. This is especially apparent when carbon dioxide accumulates in the culture medium. Use a hand lens to watch how they swim, anterior end upward, to the top of the container. You may show this by pouring a thick culture of the animals into a stoppered small test tube or vial. You may also want students to check whether light or other factors influence this response. Set up several types of controls. For example, several tubes might be prepared; some might be placed in a horizontal position, some two-thirds covered with carbon paper, and so on. Do the protozoa respond to light? To different wavelengths of light? To gravity? Their response may be useful in concentrating a culture when you plan a laboratory lesson on *Paramecium* (see Figs. 1-5, 2-9; also culturing, p. 659).

Helix   Live land snails are available from a fish market. A pound of snails will be sufficient for many investigations in behavior. Place some snails in a tall battery jar (with a cover, or they may soon be crawling up the walls). Why do they quickly move upward? What happens when the jar is placed on its side? (Also test the effect of other stimuli: a bristle near the antennae; a bit of lettuce; a lighted match; ammonium hydroxide near, but not touching.)

Pond snails may also be used, but their responses are much slower.

Magnetotaxis

Some aquatic bacteria show a magnetotaxis, orienting themselves along magnetic field lines. These bacteria synthesize small crystals of magnetite (or lodestone) that lie in a single or sometimes double chain or cluster lengthwise within the bacterial cell. This magnetic response has been found in many kinds of bacteria found in mud (cocci, spirilla, and bacilli) in fresh and saltwater environments. A possible value of this taxis is to orient these bacteria downward toward muddy sediment.

In 1975, Blakemore observed that bacteria from mud of brackish bogs and marshes always swam in one direction and congregated at the north edge of the drop under the microscope.[6] In the presence of a bar magnet near the slide, these bacteria swim in a north-seeking direction. When an electric coil is used and the current is reversed, these magnetotactic bacteria shift 180° in a new response. In the southern hemisphere, these magnetotactic bacteria swim toward the south-seeking pole.

Geotropism

Geotropism in stems   Healthy shoots of *Tradescantia* are excellent materials for demonstrating the negative response of shoots to gravity. Tropic responses can be observed within 1 hr. Clamp three test tubes containing shoots of *Tradescantia* to a ringstand in the positions shown in Figure 5-9. Be sure to seal the one-hole stopper of each tube with modeling clay, gum, or melted paraffin to avoid water leakage. (Germinating seedlings may be used in place of the shoots of *Tradescantia*, but the

---

[6] R. Blakemore and R. Frankel, "Magnetic Navigation in Bacteria," *Sci. Am.* (December 1981).

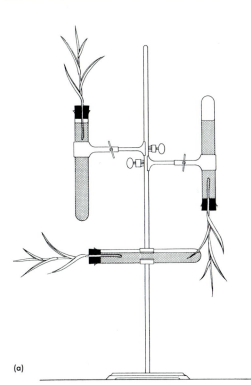
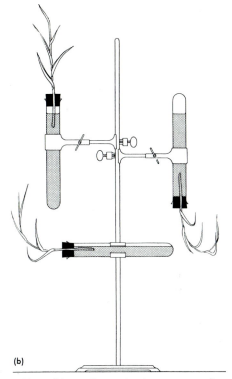

Fig. 5-9 Geotropism: response of *Tradescantia* shoots to gravity: (a) sprigs are inserted into stoppered tubes of water clamped in various positions; (b) negative geotropic responses after about 6 hr.

response will not be as rapid.) Seedlings may be grown in plastic cups of sand or vermiculite (some vertical and some horizontal). You may want to place some controls in the dark to eliminate light as a possible factor in their response.

Another way to use seedlings when fast-growing shoots such as *Tradescantia* are not available is to line several large test tubes (about 25 cm long) with blotting paper and insert a plug of absorbent cotton into the bottom of each tube. Wet the blotting paper thoroughly. Between the glass and the blotter arrange the small soaked seeds of oats, radishes, or wheat. Fill the tubes again with water; pour off the excess. If you attach string or wire to both the top and bottom of each of the test tubes, you have a loop through which the tubes can be suspended in different positions from hooks around the classroom. When the test tubes hang at a slant, the growth of

stems shows a negative geotropic response. Change the position of the tubes and watch for the change in direction of the growth of stems. In addition, set up replicas covered with aluminum foil or hang some in the dark.

This demonstration should stimulate questions about the nature of these responses. (Refer to the role of auxin, p. 364.)

Geotropism in roots Some of the preparations made to show negative response of stems to gravity are useful here. For example, when using seedlings in place of *Tradescantia*, students will notice that the roots respond positively to the stimulus of gravity. Regardless of their original position, roots bend and respond to gravity.

Also have students observe the growth of roots of seedlings planted between the glass tube and blotter and those suspended from hooks in the classroom. The stems show a negative, the roots a positive geotropic response.

If students previously germinated seeds in plastic cups of sand to show negative geotropism in stems, the seedlings may be used to demonstrate the direction of root growth. After three or four days of growth, loosen the seedlings from the moist sand by shaking them gently, and look for the bending of roots toward gravity.

Here are some other techniques that give variety to demonstrations.

USING A PETRI DISH Select soaked seeds of beans, oats, wheat, or peas which have begun to germinate so that a few centimeters of roots are visible. Then fasten seeds to a blotter which has been cut to fit the bottom of a Petri dish (Fig. 5-10). Thread fine wire through the cotyledons of the seeds to fasten them to the blotter; arrange the seeds so that the roots face the compass points; wet the blotter; and cover the Petri dish. Prepare several of these Petri dishes. Fasten the top and bottom with masking tape. Then stand the dishes in various positions by inserting one edge into a bit of modeling clay.

When roots show a positive response to gravity, shift the position of the dish

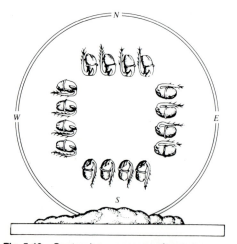

Fig. 5-10 Geotropism: response of roots to gravity. Germinating seedlings attached to a moist blotter lining a Petri dish which is held upright by a ball of modeling clay.

and watch the change in the direction of growth of each root. Similar dishes may be kept in the dark to show that light is not the active stimulus.

USING A BATTERY JAR In a variation of this demonstration, neither pocket gardens nor Petri dishes are required. Pin seedlings to a corkboard and place the board in a dark, covered battery jar lined with wet blotting paper. Add a small amount of water to the jar to keep the blotting paper wet to maintain a moist environment. Turn the corkboard from time to time to show the persistent tendency of roots to grow downward as a positive geotropic response.

USING A CLINOSTAT When roots are constantly revolving they show no response to gravity. Instead they continue to grow in the directions in which they were originally placed. Make a homemade clinostat by attaching the preparation described above, in which seedlings were wired to a blotter in a Petri dish, to the minute-hand shaft of an old alarm clock (in an upright position) so that a continuous revolving motion is produced. A small, low-geared electric motor can also be used. (See page 367 for an explanation of auxins and responses of roots.)

At another time a small committee of students may try to fasten seedlings to a water wheel. A much faster rotation may be effected with a water wheel and roots fail to respond to gravity in this condition.

REGION OF SENSITIVITY IN ROOTS Again place seedlings with roots about 1 cm long in the directions of the compass points as shown in Figure 5–10. Then cut off the tips of the roots with a razor. Prepare replicas in which the root tips are left intact. Keep all the preparations in the dark. Students will notice that the roots of the tipless seedlings do not show the normal geotropic response (until a physiological regeneration of the root tips occurs). Thus we demonstrate that it is the tips of roots that receive the stimulus, and an uneven growth of one side of the root results in a bending. Compare this with the effect of auxin on bending in stems

(Fig. 5-9). Which tissue is inhibited by auxin, that from the stems or from the roots? Also refer to the region of greatest growth in young roots (See Fig. 4–16).

RESPONSES AGAINST OBSTACLES   At some time, demonstrate the persistent response of roots to gravity, even when obstacles exist. For example, roots grow into solidified agar.

Line the sides of several finger bowls with sheets of cork. Into the bowls pour a layer of agar prepared by boiling together 1 part of agar with 4 parts of water. Pin several germinating seedlings, that have a primary root about 1 cm long, around the cork lining so that the roots are in a horizontal position and almost touch the agar. Add a film of water to the surface of the agar to cover the roots. After the finger bowls have been covered and sealed, if necessary, with petroleum jelly, put some in the dark and others in the light. After a day or so, depending upon the kind of seedlings used, notice how the roots curve and grow into the agar. Here roots respond to gravity and illustrate an expenditure of energy, for the agar offers resistance.

Hydrotropism

There are several ways to show responses of roots to the stimulus of water. You may want groups of students to try several methods.

Pocket garden   Use two squares of glass to fashion a pocket garden. Place two thicknesses of blotting paper between the glass squares so that a clear channel remains in the center (Fig. 5-11). Along the center arrange small seeds, such as radish or mustard, which have been previously soaked. Then plug the ends of the row of seeds with cotton so that they will not fall out. Fasten the glass plates together with rubber bands. Now stand one end of the preparation in water until one blotter becomes soaked. Then attach a strip of filter paper to the edge of this wet blotter to form a wick. Rest the pocket garden horizontally on a finger bowl of water, and

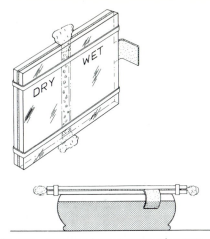

Fig. 5-11   Hydrotropism: response of roots to water. Soaked fast-growing seeds are placed in the channel of the glass-blotter-and-rubber-band pocket garden. Garden is kept moist by blotter wick extending into a finger bowl of water.

immerse the wick as shown. Be sure the other blotter remains dry. Within a few days observe the roots of the germinating seeds growing in the direction of the wet blotter rather than toward the dry blotter. Since the preparation is in a horizontal position, gravity is not a stimulus to interfere with the responses.

Hanging gardens   Shape several balls of sphagnum moss (or use sponges); soak them in water and then hold each together with cord. Insert soaked seeds of oats, corn grains, bean, radish, or wheat seeds at different locations in the balls: top, sides, bottom. Hang these balls from hooks in the classroom and keep moist.

Within a few days (depending on the kind of seed) the seeds will sprout. Students will see that roots grow *into* the sphagnum rather than downward in response to gravity. That is, the roots show a positive hydrotropism. The stimulus of water is greater than the stimulus of gravity. Should students suggest that the roots grow inward as a negative response to light, have them devise a plan for a control (the same preparation growing in the dark) so that they may refute their own suggestions.

A summary demonstration of several tropisms

You may want students to prepare the demonstrations that were first described by Coulter over 100 years ago.[7] (These are especially good for a "new" view of the topic.) Prepare balls of sphagnum moss with soaked seeds and devise investigations to test whether light, gravity, water, and so forth, affect the growth of stems and roots. Use fast-growing seedlings such as oats, barley, radishes, or corn grains.

Moisten masses of sphagnum moss and insert soaked seeds into the moss. Then shape the moss into balls about 10 cm in diameter. Tie each mass together with string so each retains its shape. Prepare some seven to ten balls of moss, so that the following demonstrations can be prepared. Where possible, prepare duplicates.

Suspend one sphagnum ball in the classroom so that it receives light from all sides. Place a second ball about 2 to 3 cm over water in a glass or plastic tumbler. Prepare a third in the same way over a tumbler, but cover the glass with black paper or aluminum foil. Then cut a thin slit in the paper or foil so that light enters from one side.

Insert a fourth wet ball part way into another tumbler as though the ball were a stopper. Thus one half of the ball is outside the tumbler and the other half is inside. Then prepare a fifth ball in the same way, but place the tumbler in a horizontal position. With two more balls (the sixth and seventh) prepare setups similar to the fourth and the fifth, but cover the tumbler with black paper so that light is excluded as a factor in the response.

Mount the eighth ball on a spindle, such as a knitting needle, which runs through its center. Then attach the spindle to the minute hand of an upright alarm clock so that the spindle forms a continuation of the minute hand.

Watch the direction of the growth of

[7] J. M. Coulter, "The Influence of Gravitation, Moisture, and Light upon the Direction of Growth in the Roots and Stems of Plants," *Science* (1883) 2:5–6.

shoots and roots. On the basis of these demonstrations, students may be able to generalize about the direction of tropic responses in plants. A discussion in class of the role of plant hormones may follow (pp. 364, 371, 372); also refer to tissue cultures, page 483.

Responses in higher animals

Reflexes in the frog

A living frog is an excellent specimen for the study of reflexes.

Blinking and other reflexes Hold a live frog securely in one hand by grasping the hind legs. Bring a blunt probe or a glass rod *near* one eye to elicit a blinking response. At times the response is so strong that the eye may be pulled into the throat region. Observe also the lids covering the eyes.

Or dip a cotton-tipped toothpick into ammonium hydroxide and bring it near the eye of the frog.

Touch the "nostrils" and watch the response. Stroke the throat and belly regions. When these regions are first stroked in males, watch the distinct clasping reflex in the forelimbs. When the frog is placed on its back and stroked longer, students will not fail to notice the quieting, almost hypnotic effect on the frog. Stroke the back and sides of a male frog to make it croak. Watch the extension of the vocal sacs and note the closed "nostrils" which prevent the escape of air.

Other responses Place a normal frog in water. Notice the resting position. Watch the normal breathing movements. Suddenly tap the jar and watch its reactions. Now watch the motion of the limbs in swimming.

Perhaps a student will demonstrate changes in the size of chromatophores in the frog's skin in response to light. Put one frog in the dark (several would be preferable) and set others in the light for several hours. Which frogs are darker in color? These responses are due to responses of the nervous system as well as the effect of a hormone, intermedin, from the intermediate lobe of the pituitary. (See also chromatophores, pp. 150, 153.)

Show the feeding responses of a live frog by dangling earthworms or mealworms before it. Note the position of the eyes in swallowing.

"Scratch" reflex   Demonstrate the scratch reflex in this way. Grasp the head and hold the forelimbs securely in one hand. Wash off any mucus present on the back of the frog. Now touch its lower back with a cotton-tipped applicator or glass rod dipped in dilute acetic acid. Notice how the hind leg attempts to brush off the irritant. With a stronger acid, a more violent response occurs. In fact, the animal may try to use both legs to brush off the irritating substance. Wash off the acid with water. Which part of the nervous system is the center for this reflex, the brain or the spinal cord? Refer to the similar response of a spinal frog (below).

Locating responses   Which region of the nervous system is responsible for these reflexes? If we destroyed only the brain of a frog, would the blinking reflex or the scratch reflex remain in this "spinal frog"? What if we also destroyed the spinal cord? Do any reflexes remain in this "pithed frog"? For destroying (pithing) the brain and spinal cord of a frog, see the photographs and description in Figure 5-12. A longitudinal section of the frog's head is shown in Figure 5–13 as a guide to demonstrating the procedure.

THE "SCRATCH" REFLEX IN A SPINAL FROG
Is the brain of the frog necessary for the scratch reflex to occur? Is the spinal cord necessary? If the brain is the center for the scratch reflex, then the frog should lose this response when the brain is destroyed. Use a spinal frog (brain removed or destroyed) for this demonstration.

Destroy the brain of a frog as shown in Figure 5–12; avoid injuring the spinal cord.

In fact, some teachers prefer to cut off the entire head to be sure they have removed the brain; insert sharp dissecting scissors into the mouth and cut off the head by cutting behind the tympanic membranes (Fig. 5-13). Place a bit of cotton over the cut end. (*Note:* This should not be done in front of the class.) Whether the brain is destroyed or severed from the body, suspend the frog from a ring stand by clamping the lower jaw (Fig. 5-14).

Wash off the mucus covering, which may be quite thick, but remember to keep the skin moist. If spinal shock has occurred the animal gives no response; wait 5 or 10 min for recovery before proceeding. Now dip a glass rod into dilute acetic acid and apply it to the back of the frog. The scratch reflex takes place even though the brain has been destroyed. Wash off the irritating acid by raising a battery jar of water up to the suspended frog and immersing the frog's body in the water. Repeat.

Release the suspended frog and pith it to destroy the spinal cord as shown in Figure 5-12. Suspend the frog once more and apply acid to the body again. Why is there no response this time? The spinal cord is the center for this body reflex.

You may now want to trace the path of the stimulus in the body. For example, when acid is placed on the skin of the back, why is the response given by muscles of the hind leg? Develop a simple pattern of a reflex arc on the blackboard as a summary activity; represent sensory neurons with one color chalk and motor neurons with another color.

The effects of epinephrine (1:10,000) and acetylcholine (1:20,000) on heartbeat and the size of capillaries in a frog are described on pages 265, 271. Also compare the effect of lactic acid (1:10,000) on the size of capillaries, as well as the effects of nicotine and histamine (1:10,000).

Nerve-muscle preparations   You can readily show how a nerve transmits a stimulus to a muscle and how this muscle

(a)

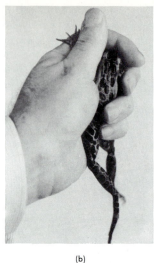

(b)

(c)

(d)

(e)

(f)

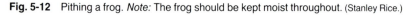

Fig. 5-12   Pithing a frog. *Note:* The frog should be kept moist throughout. (Stanley Rice.)

Spinal frog. (a) Hold the frog securely with the left hand (if you are right-handed) with its legs extended.

(b) Bend its head down by pressing its snout with the index finger.

(c) Rest the dissecting needle on a line bisecting the head.

(d) Slide the needle down until the point is at the cranial opening (foramen magnum) just below the cranium and above the spinal column (see Fig. 5-13); this is about in the middle of the posterior line of the tympanic membranes.

(e) Insert the needle without changing its direction, and move it quickly from side to side to sever the brain from the spinal cord (hold the frog firmly, for it will wriggle); tilt the needle forward, insert it into the brain case, and move it around to destroy the brain.

(f) You now have a "spinal frog": Its spinal cord is intact, but its brain has been destroyed. (*Note:* We call this a "spinal frog" for ease of reference; different authors apparently accept different terms. For example, some people refer to a "double pithed frog" when both the brain and spinal cord are destroyed.)

One reflex of a spinal frog. When the toes of a spinal frog are pinched, its legs jerk up close to its body. See the text for other reflexes to test in spinal and pithed frogs.

Pithed frog. (left) A pithed frog has not only its brain but also its spinal cord destroyed. Take a spinal frog and insert the needle into the same opening, but this time tilt it downward through the length of the spinal column, and scrape a bit. Observe the extension and relaxation of the hind legs. Notice, in the tank above, the difference between a spinal frog (swimming peculiarly) and a pithed frog (seemingly dead).

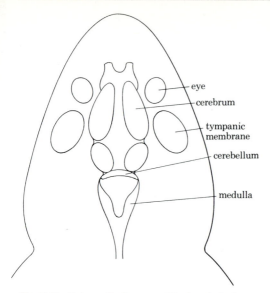

Fig. 5-13 Schematic diagram of the head of a frog (dorsal view) shows parts of brain in relation to tympanic membranes.

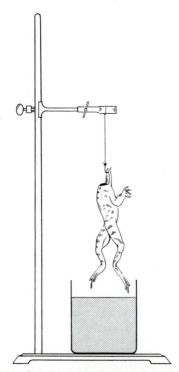

Fig. 5-14 Spinal frog, head removed, suspended from a clamp by the lower jaw for testing the "scratch" reflex. Alternately, the frog may be held in the hand, using a secure grip on the forelimbs.

tissue contracts to produce movement. Use a double-pithed frog (Fig. 5–12). Dissect the hind limb to expose the gastrocnemius muscle, or calf muscle, the largest muscle between the knee and ankle.

Either pull the skin from the gastrocnemius muscle as shown in Figure 5-15 or cut the skin around the waist of the frog and pull off the skin (as you would pull off rubber gloves). Be sure to bathe all tissues in *cold* frog Ringer's solution (p. 155). Continue to dissect the sciatic nerve from the dorsal side of the thigh, back into its origin in the spinal cord. Cut the nerve from the spinal cord and snip off, very carefully, all the branch nerves along the thigh. Transfer the nerve-muscle preparation into Ringer's solution. This preparation may now be used to simulate the discovery of Galvani, or a kymograph may be used to record muscle twitches.

Galvani's discovery If copper wires are inserted into the muscle, as shown in Figure 5–15e, and if the wires are touched to a dry cell, the current will trigger the contraction of the muscle. Or strips of different metals can be used with Ringer's solution to form a battery—a wet cell.

Prepare a "galvanic" forceps by soldering a strip of copper and a similar length of zinc (about 8 cm × 1.5 cm) at one end to form a forceps. Stimulate the nerve or muscle by placing the nerve-muscle preparation on blotting paper (soaked in Ringer's solution). With this salt solution acting as an electrolyte, touch the galvanic forceps to the muscle. Students should see the muscle twitch whenever it is touched. Now touch the nerve. What would happen if the nerve were crushed?

The surfaces of muscles and of nerves normally carry a positive charge; when dry cells or galvanic forceps are brought in contact with the tissue, negative charges are produced and conducted along the surface of the membranes; then the positive charge is restored.

When a frog's muscle is given a single electric shock, the response is a sudden twitch which lasts about 0.1 sec. Students

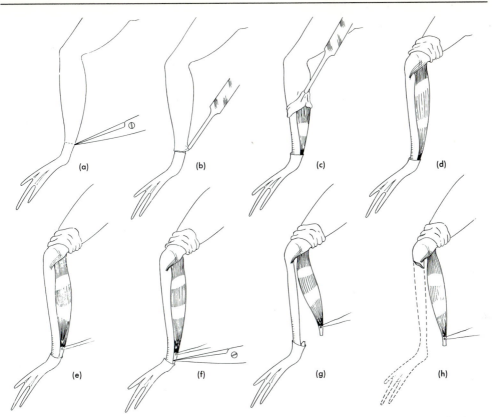

Fig. 5-15 Dissection of the gastrocnemius muscle of a frog in preparation for study: (a) the skin is carefully cut with sharp scissors near the insertion of the Achilles' tendon; (b), (c), and (d) the loose skin is rolled back with tweezers; (e) a hole is made with a dissecting needle through the tendon and a copper wire is inserted; (f) and (g) the tendon is cut below the wire; (h) the bone and other muscles are cut away. The loose skin can be pulled over the muscle to keep it from drying. (After D. Pace and C. Riedesel, *Laboratory Manual for Vertebrate Physiology*, Burgess, 1947.)

may want to read further about the separate periods of a single twitch: latent, shortening (or contraction), and relaxation periods.

There is a brief recovery period during which oxygen is consumed by muscle tissue. There must be a recovery period between stimulations or the muscle will become fatigued and the twitches cease.

However, the immediate source of energy for contraction is from ATP. After the contraction of a muscle, glycogen is broken down into lactic acid, which is oxidized in the Krebs citric acid cycle. This reaction provides energy for the resynthesis of ATP and phosphocreatine.

Using a kymograph Muscle tissue, such as a ventricle or the gastrocnemius muscle of the hind limb of a frog, may be attached to the lever of a kymograph. Trace the normal pattern of contraction of the muscle (in saline solution), especially when the muscle is stimulated electrically, or when accelerators or inhibitors are applied to the contracting muscle.

There are many models of kymographs. Essentially, the apparatus consists of a rotating drum around which a sheet of smoked paper has been wrapped. The speed of rotation can be adjusted. When the fine point of a stylus attached to a movable lever is brought near the smoked drum, the point traces a pattern on the smoked paper.

Using different kinds of muscle
Compare the contraction of striated muscle with smooth muscle of the frog. For example, cardiac muscle has a very long refractory period; each heartbeat represents a single twitch. In general, smooth muscle may remain relaxed or tightly contracted for varying periods. While the calf muscle may relax and contract in 0.1 sec, smooth muscle takes from 3 to 160 sec, and cardiac muscle may take from 1 to 5 sec.

Perhaps students can compare the smooth muscle from some invertebrates with the striated muscle from an invertebrate such as an arthropod.

Role of acetylcholine
Sympathetic ganglia liberate acetylcholine which functions at some synapses. The neurohumor or "hormone" is synthesized from acetate and choline in the presence of ATP and calcium ions. The activity of acetylcholine depends on a chemical electrogenic action, and acetylcholine is rapidly hydrolyzed by acetylcholinesterase. An idea of the effect of acetylcholine on a frog's heart may be obtained from the activity described on pages 266 and 271. Some students will want to try to reproduce the simple, elegant demonstration of Loewi. Perhaps you may want to describe the electrochemical nature of the nerve impulse. Develop the hypothesis to account for the action potential, depolarization wave, or nerve impulse in a neuron. The membrane of a neuron is polarized; negative ions are on the inside and positive ions are on the outside of the membrane (Fig. 5-16). When a stimulus is applied, the membrane becomes increasingly permeable to sodium ions so that there is an inflow of Na^+ ions; K^+ ions move out through the membrane. This causes neighboring parts of the membrane to become depolarized and the impulse is passed along the nerve. After a nerve impulse has passed, sodium ions move out and K^+ ions move in as the neuron recovers. The movement of ions

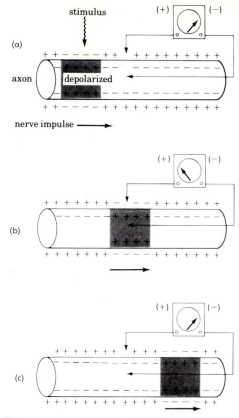

Fig. 5-16 Action potential along a neuron: (a) depolarization of the nerve cell membrane caused by a stimulus; (b) inside of neuron becomes positive to the outside; (c) nerve impulse moves along; membrane becomes repolarized to the resting state and the inside becomes negative and the outer region positive. (Adapted with permission from Macmillan Publishing Company from *Biology* by Clyde F. Herreid II. Copyright © 1977 by Macmillan Publishing Company, Inc.)

against a concentration gradient are energy dependent (active transport) and are referred to as the Na-K exchange pump. (The Na^+ concentration *outside* the neural membrane is some 10 times higher than inside the cell; K^+ ions have a concentration some 30 times higher *inside* the neuron than on the outside.)

For advanced reading that may be profitable for highly motivated students, refer to current papers in *Scientific American* as well as some older, classic papers.

Behavior in lower animals

Behavior of tube-dwelling worms

Observe the behavior of some tube-dwelling segmented worms. Gently force worms into glass tubing under water. *Amphitrite* can live in a glass U-tube for months in a marine aquarium. Record the activities of the worm on a kymograph. A laboratory manual by Sherman and Sherman offers many demonstrations of physiology of living annelids, both freshwater and marine forms.[8]

Behavior of hermit crabs

Many easy-to-do experiments on the response of hermit crabs to changes in their environment are described by Crane.[9] To remove a hermit crab from its shell, gently crack the top of the shell with pliers or a small vise. When the top of the shell is open, prod the rear of the crab and it will wiggle out of the shell. What happens when there are four hermit crabs and only three shells? What happens when larger ones are introduced in the area of smaller hermit crabs?

(Since the female broods her eggs, small yellowish masses may be found packed around the abdomen of some females.)

Adaptive behavior in fish

How does an environmental factor, such as a change in the temperature of the surrounding water, affect a fish? (Is there more or less oxygen in warmer water?) Observe the behavior of the gill cover beats in a fish as the water temperature is raised and then lowered. (Refer to page 319, chapter 4.)

Learning ("educated" responses)

Naturally, students are interested in human behavior as well as that of other organisms. Many have trained small animals. Behavior is more than a cataloguing of reflexes, instincts, conditioning, habits, imprinting, and conscious learning. Current research has raised fundamental questions about behavior, learning, and environment. Use offprints of current issues of *Scientific American* for student readings into circadian rhythms, social deprivation in young organisms, and the like.[10]

Conditioning in planarians

Students who wish to learn more about planarians might begin by reading A. Jacobson and J. McConnell, "Research on Learning in the Planarian," *Carolina Tips* (September 1962) 25:7, which describes conditioning planarian worms in a U-shaped channel cut into a plastic block. Planarians respond only briefly to bright light that is flashed at 1-min intervals. They contract when a small electric current is passed through the water. If bright light is flashed prior to the electric shock over many trials, the worms "learn" to react to the light alone; they respond to the light even when no electric shock follows. Further, the authors found that when untrained planarian worms are fed "trained" worms, they respond more frequently to the light stimulus than did the worms fed untrained planarian worms. The authors relate these findings to retention of memory in RNA. Much work has been done with planarians using a T-maze learning device (Fig. 5-17). Students may also want to read J. Best's "Protopsychology," *Sci. Am.* (February 1963), and A. Jacobson, "Learning in Flatworms and Annelids," *Psychol. Bull.* (1963) 60:74-94.

Methods for culturing planarians are described on page 667, and the techniques

---

[8] I. Sherman and V. Sherman, *The Invertebrates: Function and Form*: A Laboratory Guide, 2nd. ed. (New York: Macmillan, 1976).

[9] J. Crane Jr., *Introduction to Marine Biology*, A Laboratory Manual (Columbus, OH: Charles Merrill, 1973).

[10] Obtain a catalog of offprints from W. H. Freeman & Co., 660 Market St., San Francisco, CA.

Fig. 5-17 T-maze for training earthworms or planaria worms. Place rich garden soil and slices of apples at *A* for earthworms or bits of *Tubifex* worms or liver for planaria; at *B* place a weak acid or a weak electric current.

for bisecting and regenerating head and tail ends of these worms are described on page 497. (If any investigations involve the use of ribonuclease, this enzyme may be purchased from Worthington Biochemical Corp., Freehold, NJ or other biochemical supply houses.)

Conditioning in earthworms

With some patience, it is possible to show that "learning" of a sort takes place among earthworms. Build a T-shaped maze of wood or other materials (Fig. 5-17). At one end of the T-bar, place a rich sample of organic soil. At the other end place a piece of filter paper or cotton soaked in a dilute acid (or cover the area with coarse sandpaper). Place a hungry earthworm at the beginning of the maze and watch its trail along the length of the path. Students might keep a record of the number of times the same worm (or worms) turned toward the humus and thus avoided the acid or sandpaper. Identify the worms with India ink or food dyes so that the records of trials and errors may be identified. Repeat the maze runs daily. It may take some 50 trials before the earthworms "learn" to avoid the irritating substances and turn to the humus, where the food is located. How does learning in planaria worms compare with segmented worms?

Conditioning in goldfish

Observe conditioning in a vertebrate with a definite brain, like the goldfish. Feed the fish from one end of the aquarium regularly, and at the same time introduce another stimulus; the fish will associate the two stimuli. Flash a light each time the fish are fed, or tap on the tank. It may take three weeks or more to condition the fish. After conditioning, the fish will rise to one end of the aquarium when the tank is tapped (or when a light is on), although food may not be given. The fish now respond to a substitute stimulus (tapping or light) associated with the food stimulus.

We may consider the original response to food a reflex, an inborn behavior pattern. As a result of conditioning, the inborn response is now given to a substitute stimulus.

Learning in the white rat

Construct a rat maze using wood and window screening (Fig. 5-18). Trace out a maze and nail down strips of wood topped with screening (15 cm high) so that the animal moves along channeled paths.

Select a hungry white rat for training. Lead it through the maze several times. When it is successful in running the maze, reward it by giving it food and fondling it. In this way the rat comes to associate food and petting with a completed task.

Keep a record day to day or several times a day, as the animal (when hungry) is put through the maze. Students may compare the number of trials needed for learning the maze among several rats, some young, some older.

A group of students may study the rate of learning of rats which have a thiamine (or other) deficiency. Perhaps they may want to study the effect of drugs (caffeine) on speed of learning. (Normal controls should be used for comparison.)

Learning seems to go on faster when a reward rather than punishment accompanies the learning task at hand. What would happen if you slapped the rat each time it failed to complete the maze? (Do not try this in class, for the animal may become vicious.)

Interesting experiments on behavior

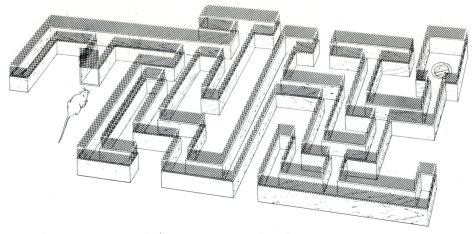

Fig. 5-18 Maze for training a white rat is constructed of wood and screening.

among protozoa, planarians, insects, and other arthropods, as well as among snails, fish, and rodents are described in M. Hainsworth, *Experiments in Animal Behavior* (Boston: Houghton-Mifflin, 1967).

Reflexes in humans

Have students observe their own reactions in the following situations.

Blinking   Have one student stand and hold a sheet of clear plastic or screening in front of the face. Ask a second student to throw cotton balls or crumpled paper balls at the student. Can the student avoid blinking even when protected?

Contraction of the iris   Have students cover one eye for a minute. When they remove the hand have them look at once into a mirror. Note the dilation of the pupil of the eye that was covered. As light strikes the eye observe the change in size of the pupil (as the iris contracts in bright light). Or darken the room for a minute. Have students observe their eyes in a mirror as the light is suddenly turned on. The contraction of the iris is marked. Also note the size of the pupil when the eye accommodates for close and distant vision.

Flow of saliva   Slice a lemon in front of the class, or describe its sour taste. Recall a favorite food by giving a detailed

description (or pictures). Ask students to describe their reflex response.

Cilio-spinal reflex   Watch the pupils of the eyes of several students when the skin of the back of the neck is pinched.

Patellar reflex (knee jerk)   Have a student sit on a chair or a table with legs crossed or freely suspended. When the subject is completely relaxed, strike a blow just below the patella bone with a rubber hammer or with the side of the hand.

Conditioning in humans

You may want to condition students. Stand at the rear of the classroom so that your students are not facing you. Ask them to try an experiment with you and to follow your verbal instructions. Direct them to mark a tally line on a sheet of paper each time you *say* "write." Use a ruler to tap on the blackboard or on a desk each time you say "write," so that the two stimuli are associated with the students' response (the drawing of a tally line).

After some 20 times of repeating these signals (about two every second) continue to rap with the ruler but stop giving the stimulus word "write." Many students will continue to draw lines at the sound of the rapping. They have been conditioned temporarily so that they draw a line in response to the stimulus (the sound of the tap). Students will vary in this experiment. Some stop immediately when you omit the oral command. Others may continue to write as many as ten extra lines when results are compared with the rest of the students. Have students explain the factors needed for conditioning. How is this type of behavior "unconditioned"?

Refer to the classic work of Pavlov in conditioning dogs.

Habits in humans

What is the value of making certain kinds of behavior habitual? Here are two methods to show that habits are time-saving and require little conscious thought.

Have students write their full names in the usual manner as often as they can in half a minute. Then have them write their names with the other hand at least five times. Record the time it takes. Why does it take so long to write with the other hand? What is one value of a habit?

Direct students to copy as quickly as possible an oral paragraph of material. Read at a fair pace and record the time it takes for students to copy the material you dictate. Then read another paragraph with one change in directions. Students are not to dot any *i* or cross any *t* in the words they copy. Select material with many words containing *i*'s and *t*'s; read at the same fair pace. Have students score the number of dotted *i*'s and crossed *t*'s.

Have a boy describe in detail how he ties a shoelace. Have someone describe in detail the kinds of houses, trees, store signs, and so forth on the way to school. Why must students stop to think, even though they pass the same way to school daily? Even though they tie shoelaces daily?

Learning in humans

Considerable time may well be spent on gaining some experimental evidence with young people. Many students lack proper study habits; they fail to get the "big idea" in their assignments, in reading, in work in class. Some of these "experiments" may help to change students' work habits.

Comparing the rate of learning under differing conditions

LEARNING SENSE AND NONSENSE RHYMES On a cardboard chart or on the blackboard list in two vertical columns the words in column *A* and in column *B*. Cover the lists so that students cannot see them.

column A	column B
thing	Its
whatever	a
a	very
as	odd
odd	thing
eats	as
its	odd
very	as
Miss T	can
can	be
Miss T	that
odd	whatever
as	Miss T
be	eats
turns	turns
that	into
into	Miss T[11]

Now uncover the list of words in column A. Direct students to memorize the list in vertical order. Record the time. Ask students to raise their hands when they have memorized the list. Allow this activity to go on for 5 min.

Now uncover column B, and have students memorize it. Keep a record of the time and the number of students who quickly raise their hands when they have memorized the words. Students readily explain that column B made sense and that column A was meaningless. From here it is not difficult to guide students toward the need for understanding main ideas so that most work and reading in

[11] Walter de la Mare, "Miss T"; reprinted by permission of the Literary Trustees of Walter de la Mare and the Society of Authors as their representative.

school does not fit into the category of "nonsense."

LEARNING UNDER DISTRACTION   Have students copy a stanza of poetry which is dictated. Better still, distribute sheets containing two stanzas of a poem.

Ask students to memorize the first stanza as quickly as possible. Students might raise their hands when they have memorized the stanza. Keep a record of the time it takes for the first hand to be raised until some dozen or so students raise their hands.

After complimenting them on their success, ask them to memorize the second stanza. However, occupy yourself by making distracting noise. Turn on a portable radio and find a "distracting" program. Also slam drawers or metal lockers from time to time. Keep a record of the time it takes for the first student to indicate success in memorizing the second stanza. Count others. Ask for an explanation of the relative slowness or even failure to memorize the second stanza. There is an opportunity here to stress the need for a quiet place for work and study.

Learning by trial and error   One of the simplest but most time-consuming ways to learn something is by trial and error. Give students a simple puzzle to solve without pictures as a guide. Use the exact shape shown in Figure 5-19 as a guide. Or use a simpler one, a larger letter *T* or *R* cut into three or four pieces. Draw it on onion-skin paper, and have a group of students volunteer to cut out sets of four cardboard or oaktag pieces from this pattern. Place four of these pieces in each one of enough envelopes for all the students in class.

Distribute an envelope to each student (note the time). Tell the students to put together the four pieces to form a perfect square with no spaces. Allow 10 min. Have successful students raise a hand; keep a record of the time. Then have these students take apart the puzzle and put the pieces back in the envelope. After 5 min, tell each of the successful students to try solving the problem again.

Why is it that most students have dif-

Fig. 5-19   Learning by trial and error. Cut four pieces of this shape and have students put them together to form a perfect square.

ficulty in solving the puzzle? Why is it that some students who successfully completed the puzzle the first time cannot repeat their performance? Elicit from students that this is hit-or-miss learning, as they say, or trial-and-error learning.

Show that when something has meaning, learning goes on faster. Have a student who was successful show the rest of the class, in detail, how to put the puzzle together. Then time the students in their performance in learning to solve the puzzle.

These experiences often help the students gain insight into some methods of improving study habits. Many who worry about their poor study habits appreciate this help. Naturally, these are suggestions only and need to be personalized through individual guidance.

What is meant by a "gestalt"? What is the effect of emotions on learning? The effect of motivation? Have students refer to recent texts in psychology of learning, as well as offprints from *Scientific American*.

The nervous system and the sense organs

The brain

Have students use a series of models to compare brains of different classes of chordates. What changes have occurred in the evolution of the cerebrum? Diagram

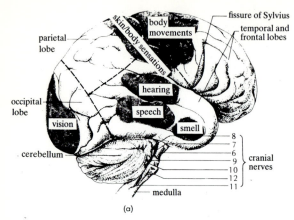

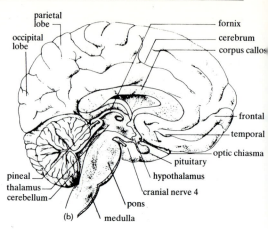

Fig. 5-20 Brain: (a) regions and special functions; (b) lateral section. (Adapted with permission from Macmillan Publishing Company from *Biology* by Clyde

F. Herreid II. Copyright © 1977 by Macmillan Publishing Company, Inc.)

the brain and associate these activities with their control centers: thinking; sneezing and swallowing; coordination of body movements and equilibrium; sight; speech; hearing; heartbeat; breathing; reflexes of the head such as blinking and change in size of the pupil (Fig. 5-20).

Dissection of a brain Obtain the brain of a sheep or calf from a butcher. This organ, as well as fetal pigs, may also be purchased from biological supply houses (See Appendix). If fetal pigs are available, dissect the brain.

When a fetal pig is used, carefully cut each occipital bone by inserting scissors into the foramen magnum and then cutting down on each side; then cut forward. Lift off the central, dorsal portion and locate the outermost tough membrane of the brain, the dura mater. Remove all the bone fragments; then cut into the dura mater and observe how the membrane separates the cerebral hemispheres from the cerebellum.

Use a take-apart model as a guide in completing this dissection. Locate the olfactory lobes, the sulci and gyri of the brain, the olfactory nerve (first cranial nerve), the cerebellum, medulla oblongata, and choroid plexus.

Lift the brain gently out of the skull, and identify the structures that can be seen from the ventral view: the optic nerves, optic chiasma, and the infundibulum, to which the pituitary may still be attached. Try, also, to find the cranial nerves and identify them.

The brain may also be cut in half sagittally, so that a median view may be examined.

Class discussions may follow from reading of papers in *Scientific American*: "Nutrients that Modify Brain Functions" by R. Wurtzman (April 1982) describes the role of tryptophan, tyrosine, and choline as precursors of neurotransmitters; S. Kety's paper "Disorders of the Human Brain" (September 1979); and in the same issue, "The Chemistry of the Brain," "The Organization of the Brain," and "The Development of the Brain." Also refer to the text by S. Springer and G. Deutsch *Left Brain, Right Brain* (San Francisco: Freeman, 1981). Two collections of papers are available as *Scientific American* Books: *Animal Behavior* (1975) and *The Brain* (1979). D. Waltz describes "learning" in computers in "Artificial Intelligence," *Sci. Am.* (October 1982).

The eye

Sheep's eye Obtain the eye of a sheep or calf from a butcher or biological supply house. Examine the eyelids and the third, or nictitating membrane. Look for the optic nerve at the back of the eye, and examine the arrangement of muscles that move the eyeball. (Have ammonia inhalants on hand, for the fixed "stare" of the eye has an eerie effect on many students.)

Locate the iris and pupil. Dissect away the lids, fatty tissue, and muscles, but leave the optic nerve. Students may readily identify the thick, white, tough sclera; the black choroid; and the whitish retina, which is probably shriveled and may have fallen into the eyeball cavity. The sclera continues along the front of the eye as the cornea. Of course, in a living animal the cornea, lens, and vitreous chamber are transparent.

Locate the large vitreous humor or area that holds the retina in place in the living animal. Find the lens and study its shape and attachment to the ciliary body. Students can also focus images with this lens.

Eye of Limulus   The arachnid eye of the horseshoe crab, or king crab, is often studied for comparisons with the eye of a mammal.

Some teachers may find the occasion to study the biochemistry of the visual pigment (red in the *Limulus* eye, formed by attachment of retinene to opsin). It may also be interesting to stimulate the eye of a living king crab with brief flashes of light. The duration of the dark adaptation, the pattern of nerve impulses of the optic nerve, and electronic recording equipment needed are well described by G. Wald et al.[12]

The ear

If possible, use a take-apart model of an ear to trace the path of sound waves. Then examine a prepared slide of the cochlea of the ear under low power. How many times is the coiled tube of the cochlea sectioned? Use high power and try to locate the

basilar membrane, and hairs; then look for the auditory nerve.

Especially try to observe fibers of different lengths in the basilar membrane. Explain how sound waves are received and transmitted to the auditory nerve.

Sense receptors of the skin

The receptors for taste, touch, sight, hearing, and smell are specialized clusters of nerve endings that keep the organism in touch with the outer environment through feedback mechanisms.

You may want students to map the location of several of the sense receptors in the skin. Note the errors in sensing the location of these receptors in the following ways.

Localization of touch receptors   Have a student keep eyes closed throughout the test. Show another student how to touch the skin of the hand of the subject with the pointed end of a soft pencil so that a mark is made on the skin. Remove the pencil, give the subject a blunt probe or toothpick and ask the subject to locate the place on the skin where the stimulus was received. Use a millimeter ruler to measure the subject's error in locating where the stimulus was applied. Try this on several students.

Here is another method for localizing these receptors. Insert two pins closely spaced in a cork, or use pointed forceps or scissors. Then gently bring the pinpoints to the surface of the hand, forearm, and fingertips. When the pins are placed closely the sensation received by the subject is that of one pinpoint. When the distance between the pinpoints is increased slightly the subject receives two sensations. Use a millimeter ruler to measure the distance between points. Record the subjects' responses; this approximates the distance between two sense receptors in the skin. Are receptors grouped more closely on the forearm or at the fingertips?

Temperature contrast   Have students immerse one finger in water at 40° C (104° F) and at the same time put a finger from the other hand into water at 20° C

[12] G. Wald et al, *Twenty-six Afternoons of Biology: An Introductory Laboratory Manual*, 2nd ed. (Reading, MA: Addison-Wesley, 1966).

(68° F); after 30 sec they should transfer both fingers into water at 30° C (86° F). What is the sensation? Use the same procedure but this time have students first immerse one finger in water at 45° C (113° F) and the other finger in water at 30° C (86° F), and finally shift both fingers into water at 10° C (50° F). Have them describe the sensation.

Temperature discrimination   Draw out several glass rods to form a blunt point. Fire-polish the ends, and allow them to cool. Chill the rods in ice water, then dry. Have students apply one to the skin of the back of the hand and also to the forearm. You may want to mark a grid on the skin about 3 cm^2 and test several regions of the skin within the grid squares. Locate the receptors that sense the stimulus of cold. Use washable ink to mark the receptors on the skin.

Repeat the procedure, but this time with rods warmed in hot water (*caution:* not too hot). Locate the receptors that sense the stimulus of heat. Mark these receptors on the skin with different colored ink. Have students diagram in their notebooks the regions that receive "cold" and "hot" stimuli.

Confusing the senses   Ask students to cross the middle finger and the index finger of one hand. Then have them roll a small pill or bean in the palm of the other hand with the crossed fingers. The students should be able to describe the sensation.

Estimation of weight   A blindfolded student might be asked to compare the weight of two graduated cylinders each holding a given volume of water. Then add small volumes of water to one cylinder; estimate at what increase in weight (1 ml of water weighs 1 g) the student perceives a difference. Shift the cylinder with increased weight from hand to hand at random during the time of the test.

Blind spot   The blind spot is the area in the eye in which there are no visual recep-

tors because it is the point at which the optic nerve leaves the retina. Wald et al describe a demonstration that each student may try.[13] Draw a small cross on a large sheet of white paper, a bit left of center. Have one student close the left eye and stare at the cross with the right eye holding the paper 30 cm from the eye. A partner of each subject should now bring a pencil point into the subject's field of vision, starting some 5 to 10 cm to the right of the cross on the paper. At what point does the pencil disappear? Mark this point on the paper; repeat, bringing the pencil from another angle. By bringing the pencil from different directions toward the cross, one can plot the boundary of the blind spot.

Chemical receptors   Map the areas of the tongue that are sensitive to salty, sweet, sour, and bitter substances. Apply the solutions with individual cotton-tipped applicators. Or apply small squares of filter paper that have been soaked in a solution to different regions of the tongue. (Use a forceps to handle the filter paper.)

Wash out the mouth between tastings. Place the test solutions in unlabeled but coded bottles. First use a solution of 2 parts of water to 1 part of vinegar. Apply it to the tip, the sides, the center, and the back of the tongue. Locate the area of the tongue that is sensitive to sour substances (Fig. 5-21).

In testing for the salt-sensitive area of the tongue, use a 10 percent solution of sodium chloride. You may want to use various concentrations of salt solutions

[13] Wald, *Twenty-six Afternoons of Biology.*

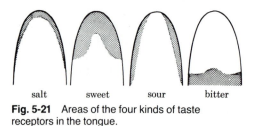

| salt | sweet | sour | bitter |

Fig. 5-21   Areas of the four kinds of taste receptors in the tongue.

(2 percent, 4 percent, 6 percent). Weigh out the salt in grams and add distilled water to make up to 100 ml. Apply it to the same regions of the tongue as before. Remember to wash out the mouth between tests.

For a bitter substance, use a weak solution of aspirin in water. Use a 5 percent sucrose solution as a sweet substance. Again, test sensitivity to sweetness by using different concentrations of sugar (0.5 percent, 10 percent, 20 percent; see page 746 for preparations of dilutions). Also test students with tap water to eliminate the factor of suggestion. Compare students' data with the areas shaded in Figure 5-21. (There are inherited differences in taste sensitivity, p. 503.)

Sense of smell Blindfold a student and have the subject hold his or her nose. Now offer cubes of raw potato, apple, and onion. Have the subject chew and identify the materials. (A solution might stimulate nerve endings associated with taste or smell or even heat.)

Coordination of responses in animals

Ciliated epithelial tissue in coordination

While ciliated epithelial tissue may have been observed earlier in studies of tissues of a frog (p. 150), here is a "new view" in the context of coordination of responses.

Ciliated epithelial tissue covers the surface of the roof of the mouth and pharynx of the frog. Examine the action of cilia. Cut away the lower jaw of a pithed frog, and lay it on its back. Place a drop of India ink or carmine powder at the anterior part of the mouth; watch the rate at which the particles move toward the rear of the mouth.

Cut off a small bit of tissue, and fold it over in a drop of saline solution. Under high power the active current set up by the cilia can be observed at the edge of the small bit of tissue. Cut down the light under high power until this current is apparent.

J. Longley describes a procedure that may readily be duplicated in class.[14] Demonstrate the movement of an object, such as a tiny glass tube, by cilia. Seal one end of a glass tubing that is several centimeters long and has a bore of about 0.5 cm. Place this over a vertically held hat pin or similar device on which it can freely rotate. Attach a light-weight pointer—a thread of glass—to the upper end of the glass tube. This may be done by bringing a heated piece of thin glass tubing close to the upper end of the tube, which is also hot; then draw out a fine thread about 4 to 5 cm long. Around the glass spindle wrap a strip of ciliated epithelial tissue (keep it moist), keeping the ends of the strip secure with threads. Cilia should rotate the glass spindle, and the pointer should be seen to move at a readily detected rate.

Coordination through hormones

How do the hormones of endocrine glands affect behavior? Most students have heard of diabetes and can describe the conditions under which insulin is given. They also can describe their feelings in "stage fright" and how this may be affected by hormones. Some students may want to report on the work of Hans Selye and describe stress and the alarm reaction. Many students are familiar with a basal metabolism test. Which is the master gland that coordinates all the other endocrine glands? What triggers the pituitary gland and the production of its many hormones? Have students describe the role of the releasing hormones of the hypothalamus in triggering pituitary hormones.[15] Describe the

[14] J. Longley, "The Activity of Ciliated Epithelia," in National Institutes of Health staff, "Laboratory Experiments in Biology, Physics, and Chemistry," mimeo pamphlet, U.S. Dept. of Health, Education, and Welfare, National Institutes of Health (Bethesda, MD: 1956).

[15] R. Guillemin and R. Burgess, "The Hormones of the Hypothalamus," *Sci. Am.* (November 1972); also B. O'Malley and W. Schrader, "The Receptors of Steroid Hormones," *Sci. Am.* (February 1976).

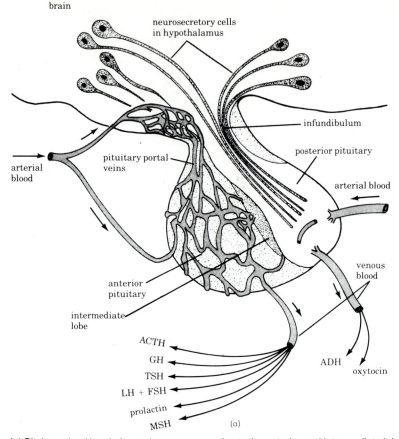

brain

neurosecretory cells
in hypothalamus

infundibulum

posterior pituitary

pituitary portal
veins

arterial
blood

arterial blood

venous
blood

anterior
pituitary

intermediate
lobe

ACTH

GH

TSH

LH + FSH

prolactin

MSH

ADH

oxytocin

(a)

Fig. 5-22 (a) Pituitary gland in relation to the hypothalamus: neurosecretory cells of the hypothalamus extend into the posterior lobe of the pituitary where they release hormones into the blood. Neurosecretory cells also release hormones in the hypothalamus, which are carried by portal veins to the anterior and intermediate lobes where they control the release of hormones. (b) Summary of hormonal control of digestion in humans. (Adapted with permission from Macmillan Publishing Company from *Biology* by Clyde F. Herreid II. Copyright © 1977 by Macmillan Publishing Company, Inc.)

role of the anterior lobe in secreting growth hormone, ACTH, thyrotropic hormone (TSH), FSH or follicle-stimulating hormone, and luteinizing hormone (LH). Describe the neural fibers in the posterior lobe and its embryological connection with the brain as contrasted with the anterior lobe (Fig. 5-22). What is the role of progesterone of the corpus luteum that is formed in ruptured ovarian follicles? What is the feedback mechanism for estrogen and FSH, and progesterone and LH?

Have students report on the role of beta blockers in reducing the risk of fatal second heart attacks as well as hypertension. Beta blockers are drugs that compete with epinephrine at receptor sites on heart muscle.

Use a manikin or a chart to locate the endocrine glands. Some organs may be purchased from a butcher: pancreas, possibly thyroid and adrenals. Plan to dissect a frog, fetal pig, rat, cat, or rabbit, to locate many of these organs (dissections are described in chapter 1). In season, a batch of tadpoles and possibly axolotyls may be treated with thyroxin or iodine (p. 421), or use pituitary glands to induce ovulation in frogs (p. 419).

There may be a time to observe the effect of epinephrine (adrenalin) in dilutions of 1:10,000, on increasing the heart-

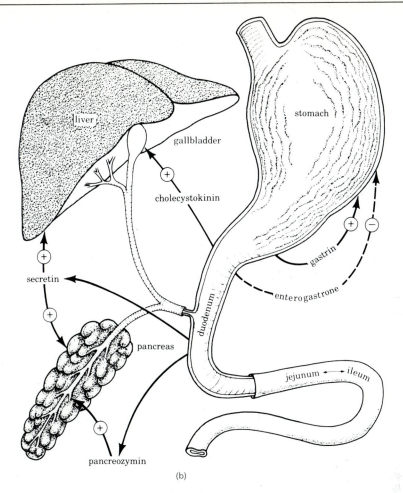

liver

stomach

gallbladder

(+)

cholecystokinin

(+) (−)

gastrin

secretin

(+)

enterogastrone

duodenum

(+)

pancreas

jejunum ← ileum

(+)

pancreozymin

(b)

beat in *Daphnia* (p. 267), as well as the effect of epinephrine on melanophores of fish scales (p. 153).

Compare these results with those that follow on the effects of hormones on a frog's heart.

Hormonal control of a frog's heart   As a laboratory activity have students observe the effects of adrenalin (epinephrine) and of acetylcholine on the contraction of the heart (technique for suspending a beating heart from a support into Ringer's solution, pp. 155, 266).

If a kymograph is available, the following preparation may be used to compare the action of several hormones and drugs. Lay a freshly killed frog on a frog board and dissect out the beating heart (cut

through the pericardial membrane to expose the heart). Keep the heart bathed in frog Ringer's solution.

Fasten a fine thread to the tip of the ventricle or pass a needle and thread through the tip of the ventricle. Attach the extension of one thread to the lever of a kymograph. Adjust the level of the frog and the position of the lever; the needle should just touch the smoked paper on the drum.

Get a reading of the heartbeat when the heart is bathed in Ringer's solution. Apply 1 or 2 drops of a solution of acetylcholine (1:10,000). Record the slower heartbeat. Then rinse the heart in Ringer's if it stops beating. Also, what is the effect on the heartbeat of a solution of epinephrine (1:10,000)? What is the role of the sinoauricular node?

Metamorphosis: Insects and frogs   How do hormones affect metamorphosis of insects? Refer to the fine monographs in the Oxford/Carolina Biology Readers series.[16] The monograph "Metamorphosis" (J. R. Tata, 1973) describes and illustrates the role of hormones in both insect and amphibian metamorphosis (also see Figs. 5-23, 5-24).

Autonomic nervous system

Pretend to give an unannounced test by distributing your usual test paper. Wait a minute; now ask students to describe their physical state. Why does the heart beat faster? Explain the causes of hyperglycemia in examination anxiety.

Have students use a diagram such as Figure 5-25 to describe the role of the autonomic nervous system and hormone coordination. Distinguish those organs innervated by the sympathetic and the parasympathetic systems.

Perhaps in a case history approach to a study of hormones, students may report on the classic work of Bayliss and Starling "The Mechanism of Pancreatic Secretion".[17] There may be time to discuss Otto Loewi's paper "On the Humoral Transmission of the Action of Heart Nerves" found in the same book. In addition, refer to an investigation using a frog's heart, pp. 266, 271.

As a summary, have students coordinate the interactions of the autonomic nervous system, central nervous system including the brain, and the ductless glands.

Behavior and drugs

Categorize the variety of drugs into (1) narcotics; (2) sedatives; (3) tranquilizers; (4) stimulants; (5) hallucinogens. Students should be familiar with the depressant action of narcotic drugs such as codeine, heroin, morphine, demerol, and methadone. Also mention the depressant action of sedative drugs such as alcohol and barbiturates.

Can students list the effects of stimulants such as cocaine, amphetamines, caffeine, and tobacco? How do tranquilizers affect the body—Miltown, Valium, Librium, and reserpine? What is the origin and effect of LSD, DET, STP, marijuana, mescaline, and psilocybrin?

Further reading   Case histories of discoveries of hormones in both plants and animals offer fine examples of experimental design. To test students' understanding, ask them to distinguish the underlying hypotheses, the problems raised, and the methods for testing, and for verifying suggested hypotheses. Again, offprints of *Scientific American* offer a readable background for case history studies.

Current information about drugs is available from many sources; one source is R. Julien, *A Primer of Drug Action*, 3rd ed. (San Francisco: Freeman, 1981). Also refer to papers, available as offprints, in *Scientific American* such as C. Van Dyke and R. Byck, "Cocaine" (March 1982).

Role of pheromones in behavior

While the honeybee communicates by means of a round dance and a tail-wagging dance to signal the location of a food supply, the firefly uses light flashes to attract its mate. Many insects also release chemicals or pheromones, that are sensed as specific sex attractants, or identify food trails, or are alarm pheromones. Recall the chemical attractants produced by slime molds (acrasin) and the pheromone of the female gypsy moth that may be detected by the large antennae of males several kilometers away if carried by wind.[18] Stu-

---

[16] Carolina Biological Supply Co., Burlington, NC.
[17] Reproduced in M. Gabriel and S. Fogel, eds., *Great Experiments in Biology* (Englewood Cliffs, NJ: Prentice-Hall, 1955).

[18] You may want to refer to D. Schneider, "The Sex Attractant Receptor of Moths," in the July 1974 issue of *Scientific American*; the receptors are on the feathery antennae of males.

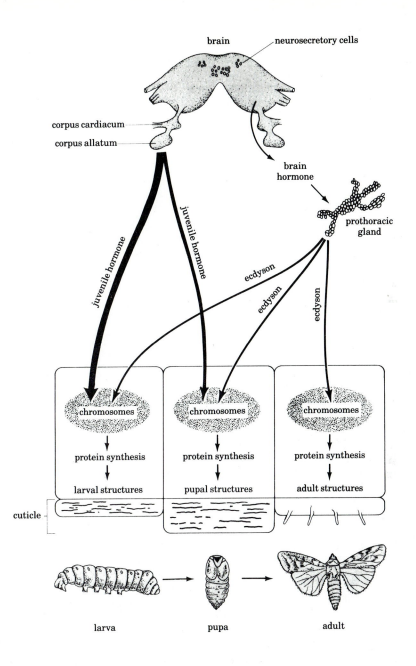

Fig. 5-23 Molting and metamorphosis in insects: Under the influence of light and temperature, a neurosecretory hormone of the brain stimulates the prothoracic gland to release ecdyson, the molting hormone. Another hormone, the juvenile hormone, released from the corpus allatum affects molting; when the secretion of juvenile hormone decreases the larva develops into pupa and adult stages. (Adapted with permission from Macmillan Publishing Company from *Biology* by Clyde F. Herreid II. Copyright © 1977 by Macmillan Publishing Company, Inc.)

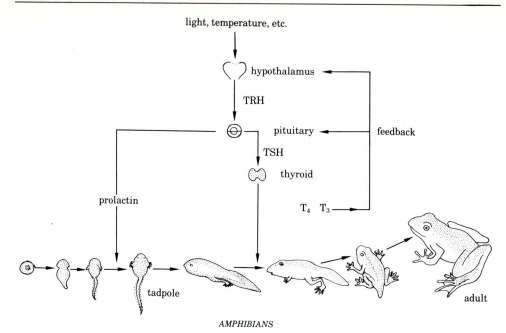

light, temperature, etc.

Fig. 5-24 Metamorphosis in amphibians: Under the influence of light and temperature, a thyroid releasing hormone (TRH) from the hypothalamus stimulates the pituitary to release thyroid stimulating hormone (TSH); this in turn stimulates the thyroid gland to secrete thyroxin which induces the metamorphosis of tadpoles into adults. (Photo courtesy Carolina Biological Supply Company.) (T_4T_3 refers to thyroxin and tri-iodothyroxin.)

dents may have read how ocean salmon find their way back to the same stream where they were hatched by means of characteristic odors that fish recognize.

Students may want to read E. O. Wilson's book, *Sociobiology* (Cambridge, MA: Harvard University Press, 1975).

Role of hormones in plant behavior

The discoveries of the many roles of plant hormones in regulating growth in response to light had their beginnings in the 1880 publication by Charles Darwin and his son. In it they describe their experiments and observations on the bending of young shoots of oats and canary grasses, and conclude that some effect transmitted from the tip to the base of stems made plant stems curve toward light (Fig. 5-26). Plan to have students read an excerpt from Darwin's "Sensitiveness of Plants to Light:

Its Transmitted Effects," in *Great Experiments in Biology*, edited by M. Gabriel and S. Fogel (Englewood Cliffs, NJ: Prentice-Hall, 1955).

Role of auxin in phototropism and geotropism in stems

Some 30 years after Darwin's observations Boysen–Jensen showed in 1911 that material was transmitted across a wound cut into the stem and that the curvature occurred in the *shaded* side of the stem when exposed to a unilateral light source. As part of a case history approach, develop the experiments of F. W. Went (1928) that led to an assay method, a quantitative method for determining the amount of growth regulator present by measuring the angle of the curvature of the stem. Tips of oat coleoptiles were placed on measured blocks of agar. These tiny blocks which had first been exposed to different numbers of coleoptile tips were then placed on

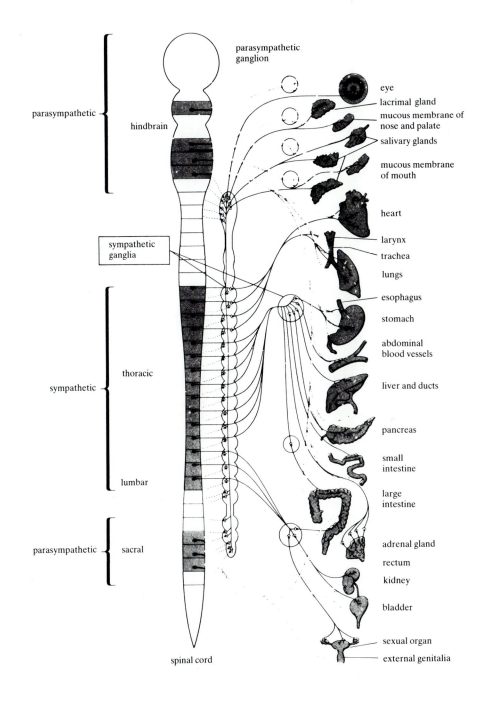

parasympathetic
ganglion

parasympathetic

hindbrain

eye

lacrimal gland

mucous membrane of
nose and palate

salivary glands

mucous membrane
of mouth

sympathetic
ganglia

heart

larynx

trachea

lungs

esophagus

stomach

abdominal
blood vessels

sympathetic

thoracic

liver and ducts

pancreas

small
intestine

lumbar

large
intestine

parasympathetic

sacral

adrenal gland

rectum

kidney

bladder

sexual organ

external genitalia

spinal cord

Fig. 5-25 Autonomic nervous system. (Adapted
from *Biology* by Grover C. Stephens and Barbara Best

North. Copyright © 1974 by John Wiley & Sons, Inc.
Reprinted by permission of John Wiley & Sons, Inc.)

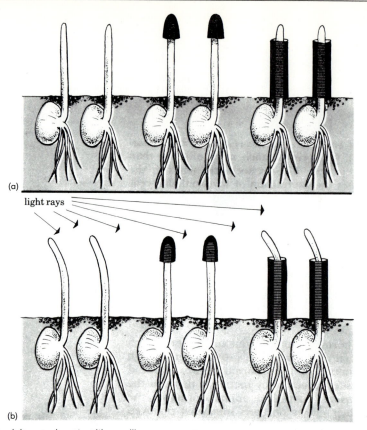

Fig. 5-26 Darwin's experiments with seedlings of canary grass: (a) some seedlings are left uncovered, some with tips covered, and some with shoots covered and tips exposed; (b) results when exposed to a lateral source of light. (From *Biological* *Principles and Processes* by Claude A. Villee and Vincent G. Dethier. Copyright © 1971 by W. B. Saunders Company. Reprinted by permission of CBS College Publishing.)

cut coleoptile tips. Different angles of stem curvature could be correlated with the quantity of growth promoting substance that had diffused into the blocks.[19]

This growth regulating substance produced in the tips of stems—showing polarity in diffusion from tip to base of stem—was called *auxin*. The flow of auxin is diverted by light as it diffuses down the stem so that its net effect, the elongation of cells, occurs on one side just below the tip of the shoots (Fig. 5-27). As a result stems bend toward light. Cells below the tip of

[19] Excerpts from papers by Boysen–Jensen (1911) and by Fritz Went (1928) are also included in *Great Experiments in Biology*.

the coleoptile respond to blue wavelengths of light and are linked with the flow of auxin.

Refer to the bending of the shoots of *Tradescantia* (Fig. 5-9). Now have students explain the curvatures in terms of auxin flow and cell elongation. When a stem is placed in a horizontal position, auxin concentrates at the bottom part of the stem; cells elongate at the bottom side of the stem and the stem turns upward.

There is evidence that amyloplasts act as "statoliths" in cells sinking to the bottom part of cells. There seems to be a sensing or "cue" in relation to gravity. Horizontal stems thus show a negative

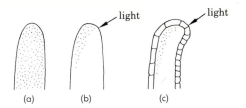

Fig. 5-27 Effect of light on flow of auxin: (a) auxin flows from coleoptile tip; (b) flow of auxin is deflected by light; (c) cells on the shaded side elongate, causing the shoot to bend toward light.

response to gravity. How the shift in auxin flow is diverted is unknown.

Role of auxin in root geotropism

Root tips also produce auxin but the effect is different from that of stems. In roots, the concentration of auxin at the lower part of the root acts to *inhibit* cell elongation. As a result the upper side of the root elongates and grows faster so that the root bends down into the soil (positive geotropism). There is some evidence that the active hormone in geotropism is not an auxin, but may be an unidentified hormone. Or possibly that the ethylene gas produced by roots may inhibit the action of an auxin, since a high concentration of auxin is known to stimulate the production of ethylene. This gas, in turn, may inhibit the action of auxin found in high concentration.

Perhaps there is a time to repeat some of the classic experiments of Boysen–Jensen, Went, and others to examine the curvature of stems, growth of roots on cuttings, and the role of plant hormones in initiating stem and root formation in tissue cultures (chapter 6, pp. 483–89).

Other effects of auxins   The auxin found in plants, indole-3-acetic acid (IAA), is synthesized from the amino acid tryptophan by the tips of stems and coleoptiles. There are several auxins that have been synthesized in the laboratory, and some are more effective in stimulating some kinds of growth. Auxins consist of a number of indolyl derivatives: indole-3-pyruvic

acid, alpha indole-3-butyric acid, naphthaleneacetic acid, and so on.

Auxins produced in meristematic tissue affect cell division as well as cell elongation. IAA stimulates cambium cell division. In fact, auxins may affect DNA replication in mitosis since it has been found that auxin-treated cells show more polyploidy.

IAA produced in apical buds inhibits the growth of lateral buds on the stem (activity, p. 368). Auxin also travels to roots and stimulates root growth (activity, p. 368). IAA-treated cuttings grow abundant roots (technique and preparation of auxins for cuttings, p. 368; also see effects in tissue cultures, p. 483). In addition, auxins applied to pistils of flowers produce parthenogenic fruits by stimulating the development of ovaries. They also have some role in sleep movements, and in the shrinking and swelling of guard cells (abscisic acid also affects turgor pressure of guard cells, p. 374).

As a whole, hormones have different effects on different parts of a plant. They give the plant "cues" about its environment. Bound to proteins, they probably act at receptor sites in cells to carry on enzyme activities. No doubt, some hormones may affect patterns of gene expression through RNA coding. (See Fig. 5-29.)

Investigations using auxins

Sensitivity of shoots to light   Germinate soaked oat seeds in paper cups of vermiculite or moist sand kept in the dark. When the coleoptiles emerge and are about 1 to 1.5 cm high, cut off about 3 mm of the tips of several with a sharp, sterile razor or the edge of a coverslip. Avoid injuring the first leaf. Then apply a bit of lanolin paste to the cut surfaces. Decapitate an additional, equal number of coleoptiles; this time apply lanolin to which 0.1 percent indoleacetic acid has been added. (Prepare this by dissolving 100 mg of indoleacetic acid in 2 ml of absolute ethyl alcohol. Add this solution to 100 g of lanolin paste and mix thoroughly so that the auxin is evenly distributed in the "carrier" paste to get a concentration of 0.1 percent or 1000 ppm. Or slightly heat the lanolin on a hot plate and stir in IAA.) Set aside a third batch of coleoptiles from which the tips have *not* been removed. Place all the shoots (that is, cut tips with lanolin, cut tips with lanolin plus indoleacetic acid, and untreated tips) near a source of unilateral light. After 24 hr students should measure the angle of curvature in all three sets of plants. Lay the plants on paper and measure the angles with a protractor. Which part of the coleoptile seems to produce a growth-regulating substance? Which part of the coleoptile seems sensitive to light? Where does the stem elongate and curve?

Growth of roots   In a similar approach, students may repeat the above technique for growing oat seedlings to show the inhibitory effect of growth-regulating substances on root growth. In this case students should cut off the 2-mm tip of several roots. Prepare untreated roots and treated roots with the lanolin paste and lanolin-plus-auxin preparation, respectively. Then set the seedlings in Petri dishes for a demonstration of geotropism as shown in Figure 5-10. Observe in subsequent examinations that auxins initiate root formation, but, in this case, after the root primordia are formed and the roots elongate, auxin inhibits further growth of roots.

Some students may also plan investigations of the inhibitory effect of terminal buds on the growth of lateral buds. Or they may plan activities to show the effect of leaves on stimulating root formation. Does root initiation proceed faster when stem cuttings have leaves?

Effect of leaves and buds on growth of roots   Select healthy coleus, geranium, or other herbaceous cuttings which have several internodes. In some, remove all the leaves; in others, remove the terminal buds and the growing tips; in a third group, remove only the lower leaves. Set aside several others as controls. Then compare the rate of root formation after planting the cuttings in moist sand. More successful results may be observed if cuttings are grown in nutrient solutions (p. 484) to which thiamine, pyridoxine, and other B vitamins are added.

Effect of IAA on rooting   Grow pea seedlings until the first pair of true leaves appears. Then cut off the radicle and any growing roots from each seedling so that stems are about the same length. Retain untreated seedlings as controls.

What is the effect of different concentrations of IAA (or other auxins) on root growth? Surface sterilize the stems by dipping them into 5 percent Clorox solution (5 parts Clorox to 100 parts water) for a few minutes. Divide the stems into three batches and soak the basal ends of the young plants for 12 hrs in various concentrations of auxin; for example, 1 mg/l, 10 mg/l, and 100 mg/l.

After 12 hr, transfer the seedlings into Hoagland's or Shive's solutions for a week or more (preparation, p. 480). Maintain these (as well as the controls) at room temperature in moderate light. Observe the rate of growth of roots on the young plants. Note the effect of the different concentrations (as well as the effects of different auxins) on root initiation. Root-forming substances made in buds and leaves are transported to the region where new growth, root growth, is stimulated.

Prepare the dilutions used here from a *stock* solution of IAA. Stock solutions of IAA, IBA (indole-butyric acid), and NAA (naphthalene acetic acid) are prepared in the same way. Investigate their effect on rooting, bud inhibition, and/or geotropism.

Preparation of stock solution of auxins: (**1 mg/ml**)   Dissolve 100 mg IAA (or other) in 10 ml of 95 percent ethyl alcohol. Add 90 ml of water to make 100 ml of *stock* solution. To prepare a *dilution* of 1 mg/l, add 1 ml of the stock solution to 1 liter of water. To make a dilution of 10 mg/l, use 10 ml of stock solution and bring the volume to 1 liter with water. To make 100 mg/l dilution, use 100 ml of the stock solution of IAA, NAA, or IBA, and bring the volume to 1 liter with water. (A dilution of 100 mg/l results in a 0.1 percent solution, or 1000 ppm.)

Effect of IAA on lateral buds

Grow pea seedlings until they are several cm tall; then cut off tips of at least two dozen seedlings. Apply IAA-lanolin paste to the cut surface of some seedlings while the controls are treated with plain lanolin paste. Prepare the paste by adding 1 ml of the stock solution of IAA (that is, 1 mg/ml) to 9 g of lanolin. IAA inhibits lateral buds; compare with untreated controls.

Auxins and the growth of stems

To show the region of the stem where growth is greatest, use fast-growing pea or bean seedlings that are several weeks old, or when a pair of first leaves are visible.[20] Prepare an auxin in lanolin paste by dissolving 100 mg of indoleacetic acid (IAA) in 2 ml of 95 percent ethyl alcohol. Mix this in 100 g of lanolin to get a 1000 ppm auxin paste.

Apply the auxin paste to the *bottom* region of the stems of some young plants; to the *middle* part of the stems of a second set of plants; at the position of the *first leaves* of a third set of plants. Set aside a fourth set as controls. Label all the sets of seedlings. For a week, examine daily the degree of curvature of the stems of the young plants (use a protractor). Notice the strong curvature of the stem in the upper part where auxin is most active.

You may want to tint the auxin paste applied to different parts of the stem by adding a pinch of carmine powder.

[20] Allow three to four weeks for pea or bean seedlings to grow about 15 cm tall; when radish seeds are used, allow about five days for radish seedlings to show rootlets and root hairs. Sprinkle on wet filter paper in fingerbowls or Petri dishes and cover with wet filter paper. Then cover with glass squares or Petri dish covers.

Geotropism and auxin (IAA)

Germinate sunflower or pinto bean seeds and grow seedlings in moist vermiculite. When seedlings are four to five days old, remove their cotyledons and cover with plastic bags to retain moisture. Maintain them in darkness for four days so that auxins do not accumulate in the upper stems. Then shift the pots of seedlings to the light and lay the pots on one side so that the stems are horizontal. After 6 to 12 hr, stand the pots so that the stems are upright again.

Now use an applicator to apply IAA in lanolin paste to the cut tips of the stems; to the controls, apply plain lanolin paste; curvatures should be visible in about 4 hr. Prepare the IAA-lanolin paste by adding 1 ml of stock solution of IAA (that is, 1 mg/ml stock) to 9 g of lanolin.

Sensitive test plants

Young bean, sunflower, and cucumber plants are useful in introductory laboratory studies since they are especially sensitive to growth-regulating substances and show dramatic results. Germinate seeds in moist vermiculite at a temperature between 26 to 30° C (79 to 86° F); allow five to six days for the seedlings to grow to a useful size. At this stage, primary leaves are some 3 to 4 cm wide.

The growth substance may be applied with a toothpick as a paste (in a lanolin carrier), or apply drops of the solution to the apical bud with a capillary pipette. Or in some cases, you may dip cut stems in a solution or in a powdered form of the auxin mixed with clay or talc (see p. 370).

Auxins: Additional preparations

While indoleacetic acid is the growth-promoting substance produced by plants, some substitutes that are less expensive and that are not quickly destroyed by the enzyme systems of plants also can be used; β-indolebutyric acid and α-naphthaleneacetic acid are effective substitutes. In fact, small amounts of the latter duplicate the entire range of activities initiated by the natural growth substances.

Rooting preparations may be applied as liquids, powders, or pastes.

LIQUID FORM Mitchell et al. suggest these concentrations for soaking stem cuttings.[21] Dissolve 1 g IAA (or other auxin) in 2 ml of 95 percent ethyl alcohol; stir this into 1 liter of water to make 0.1 percent, or into 10 liters of water to get 0.01 percent solution. The solution can be refrigerated for several months. The time needed for soaking stem cuttings varies with the kind of plant used (from 2 to 24 hr). For example, young leafy cuttings may take root quickly after soaking for some 1 to 2 hr in a 0.005 percent solution of indolebutyric acid. More woody cuttings of herbaceous plants may need to soak for 6 to 24 hr in the same dilution. Woody cuttings of trees such as dogwood or elm will root best if the leaves are still attached. They should be soaked for 1 to 2 hr in 0.01 percent indolebutyric acid.

At times you may need a more rapid method. Dip the cuttings (about 3 cm of the stem) into a stronger solution (0.05 to 0.1 percent) for about 1 sec. Prepare the 0.1 percent solution by dissolving 28 g by weight of indolebutyric acid in 30 ml of ethyl alcohol. Make more dilute solutions by adding more of the alcohol.

POWDER FORM Probably the easiest method is to dip the cuttings in a powder mixture. One method suggests rolling succulent stem cuttings in a mixture of 1 part of growth hormone powder to 1,000 to 5,000 parts of clay or talc. Stronger mixtures are needed for stems of woody plants, that is, 1 part to 500.

There are several commercial products on the market; powder seems easier to use than the liquid form, although both are available.

This method may be useful to demonstrate how growth hormones stimulate rooting. Select several cuttings of privet or willow which are about 10 to 15 cm long. Wet the cut ends of several of the stems in

water and dip them into Hormodin powder (or a preparation of auxin and clay, or talc). Shake the twigs to remove excess powder; the amount of powder sufficient to produce results is that amount which just adheres to the wet tip. Plant the cuttings, the treated ones and several untreated controls, in separate pots of sand. Examine the growth of roots each week.

PASTE FORM A paste is made by mixing indolebutyric acid or other growth-promoting substance in slightly melted lanolin. The advantage of a paste is that the substance can be placed on a localized region of the plant. First, prepare a solution (0.01 to 0.05 percent) of indolebutyric acid or naphthaleneacetamide. Use the directions given above under "Liquid Form" for the preparation of a 0.1 percent solution, and dilute this. Then mix 1 tsp of the solution in 1 tbsp of melted lanolin. The small amount of alcohol used as a solvent for the growth substance evaporates in the heated lanolin. After a thorough mixing, apply this paste with a toothpick to different parts of several plants. For example, apply a circle of paste around the stem, to the upper leaves, to the lower leaves, and to cuttings. You may add a pinch of carmine powder to increase visibility.

During the next few weeks, compare the treated parts with untreated controls. Keep both sets of plants under similar conditions of temperature and moisture. Indoleacetic acid may be used in some tests, naphthaleneacetic acid or naphthaleneacetamide in others.

Another easily prepared auxin formula suggested in the literature uses indoleacetic acid in this way (Fig. 5-28). Mix thoroughly 0.1 g of β-indoleacetic acid with 1 ml of 70 percent ethyl alcohol and then add this to 50 g of hydrous lanolin. A smooth paste should be made; heat the lanolin a bit. If you wish to have a color for identification, add a pinch of carmine powder to tint the paste red.

In an investigation, students may want to use β-naphthoxyacetic acid which is effective in stimulating the formation

[21] Refer to J. Mitchell, G. Livingston, and P. Marth, *Methods of Studying Plant Hormones and Growth-regulating Substances*, Agricultural Handbook 336 (Washington, DC: U.S. Dept. Agriculture, 1968).

(a)   (b)

Fig. 5-28 Effect of indoleacetic acid (IAA) on petioles of nasturtiums: (a) petiole before treatment; (b) petiole 1¼ hr after IAA paste was applied halfway up the petiole. (Photographs by Caroline Biological Supply Company.)

of parthenocarpic fruits (seedless fruits formed without pollination).

Other related compounds have practical applications; some show that a hormone may have different effects on different parts of a plant (refer to tissue culturing, chapter 6).

Making an agar block   Mitchell et al describe the preparation of an agar block.[22] You may want to place cut tips of coleoptiles on agar blocks in some investigations.

Prepare a 1 percent agar solution by dissolving 0.1 g agar in 10 ml of water. Prepare several 15 to 30 cm lengths of glass tubing with a 5 mm diameter. Pipette the warm agar into the glass tubing; place a finger at the lower end of the tubing while filling it with agar. Use your fingers to close both ends of a length of filled tubing and solidify the column of agar under cold water. Remove the column of agar by tapping the glass tubing. Then use a glass rod to push out the solid core of agar. Slice each core about 2 mm thick on moist filter paper in a Petri dish.

Now place cut oat coleoptile tips on agar blocks to receive the transported auxin from the tips. Or mix growth regulating substances into the warm agar so that thin "donor" blocks may be applied to cut apical tips (at the top or sides) to test the effects of growth regulators.

Other plant hormones

There are several other physiological effects caused by different auxins.[23] Some affect sexual development of flowers; for example, naphthaleneacetic acid alters the sex of flower buds, and a high level of auxin is associated with female flower bud tissue.

Some five kinds of hormones are found among seed plants. Besides auxins, there are cytokinins (or phytokinins), gibberellins, abscisic acid, and ethylene gas. Figure 5-29 gives some indication of the site of action of several plant hormones and their coordination in regulating plant behavior.

Comparing effects of several growth-promoting substances   Scatter previously soaked cucumber seeds on filter paper soaked in a growth-regulating substance in Petri dishes. Treat batches of cucumber seeds with indoleacetic acid; with gibberellin; kinetin; IAA and kinetin; and IAA and gibberellin. (Also refer to tissue culturing, chapter 6, p. 483.) Store for 24 hr in the dark. In 20 hr or less, the seed coat should be broken and the radicle visible. Measure the growth of roots in experimentals and controls in millimeters. After two to four days graph the data, the length of stems and roots in experimentals and controls.

Cytokinins   While these plant hormones were first isolated from coconut milk as the factor responsible for inducing cell division, yeast cells and other plant tissues have also been found to contain cytokinins. In addition to inducing cell division, cytokinins act together with auxins to

---

[22] J. Mitchell et al, *Methods of Studying Plant Hormones and Growth Regulating Substances.*

[23] Students interested in planning investigations may want to read issues of *Biological Abstracts* and *Annual Rev. of Plant Physiology* for current research.

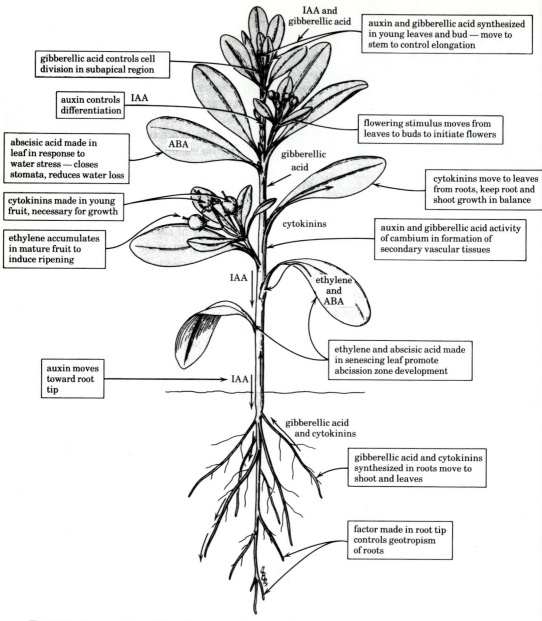

gibberellic acid controls cell division in subapical region

auxin controls differentiation IAA

abscisic acid made in leaf in response to water stress — closes stomata, reduces water loss ABA

cytokinins made in young fruit, necessary for growth

ethylene accumulates in mature fruit to induce ripening

auxin moves toward root tip

IAA and gibberellic acid

auxin and gibberellic acid synthesized in young leaves and bud — move to stem to control elongation

flowering stimulus moves from leaves to buds to initiate flowers

gibberellic acid

cytokinins move to leaves from roots, keep root and shoot growth in balance

cytokinins

auxin and gibberellic acid activity of cambium in formation of secondary vascular tissues

IAA

ethylene and ABA

ethylene and abscisic acid made in senescing leaf promote abcission zone development

IAA

gibberellic acid and cytokinins

gibberellic acid and cytokinins synthesized in roots move to shoot and leaves

factor made in root tip controls geotropism of roots

Fig. 5-29 Hormone interrelationships among the organs of a plant. (T. Weier, C. Stocking, M. Barbour, T. Rost, *Botany: An Introduction to Plant Biology*, 6th ed., © 1982, reprinted by permission of John Wiley & Sons, Inc.)

effect cell differentiation of roots and shoots (Fig. 6-108). Some students may want to learn tissue culture techniques and duplicate some of the investigations of Folke Skoog and his graduate student,

C. Miller, on stem tissue using different combinations of cytokinins and auxins. (Refer to sterile techniques, sterile box, and tissue culturing, chapter 6.)

In addition, cytokinins control the

synthesis of RNA and the separation of daughter cells (called cytokinesis). They do not show polarity as auxins do; they also seem to slow down senescence of leaves in plants, and to break dormancy of seeds and buds. (Kinetin is a synthetic cytokinin.)

Gibberellins　Of the group of some 37 different gibberellins that have been isolated, the most active is gibberellic acid (A_3). The effects of gibberellic acid in rice attacked by a fungus, *Gibberella fujikuroi*, were originally observed and described in 1898. Young rice plants grew unusually tall and died of the "foolish seedling" disease. This effect was studied more thoroughly in 1926 and now the growth substance has been extracted from the fungi and its molecular formula is known. Since gibberellins have also been isolated from meristematic tissue of higher plants, they are considered naturally occurring plant hormones. The effect of gibberellins seems to be through the production of RNA. For example, the enzyme alpha-amylase is produced from amino acids under the influence of gibberellic acid (GA) made by the early embryo in a seed—barley seeds have been well studied (Fig. 6-100). GA seems to activate the aleurone cells to synthesize the amylase that converts starch in the endosperm (also refer to investigation, p. 476).

While gibberellic acid shows some of the same effects as auxins, it is not classified as an auxin since it does not give the usual oat coleoptile test or show epinasty of leaves. Gibberellins are plant hormones that increase mitotic activity and effect the elongation of stems below the apical meristem in some plants. However, genetic composition is also a factor.

Unlike auxins, gibberellins do not show polarity of transport nor do they inhibit the growth of lateral buds or check abscission of leaves. They do induce flowering among long light period plants even when the long light period is omitted (photoperiodism, p. 375); they also retard root growth and break dormancy. There seems

to be an increase in the surface area of leaves (mitosis) which results in increased assimilation of carbon. Auxins and gibberellins may act together to provide an additive effect. GA may start the production of auxins in stem tips, so that the auxin-sensitive stages follow the production of gibberellins.

Effect of gibberellic acid on young plants　Germinate bean seeds of several different types at 25° C (77° F) in moist vermiculite; in four to five days, young plants should be visible. When plants are some ten days old, spray gibberellic acid solution (preparation, p. 374) to which a few drops of baby shampoo have been added as a wetting agent[24] on stems and leaves. (Confine the plants in a large carton as you use a hand spray to avoid spraying control plants.) Or use a cotton-tipped applicator to apply gibberellin to leaves, buds, or stems. At room temperature, in moderate light and moisture, students should observe some changes in 24 to 48 hr. Students may recognize the need to spray control plants with water and a few drops of baby shampoo. Have students record changes in controls and experimental plants and note the color of leaves; measure differences (in millimeters) in size of leaves and in the height of plants over a period of three weeks. At the end of three weeks, gently uproot the plants to observe any differences in root growth.

At this time, you may want to investigate the effect of gibberellins on dwarf plants (p. 374); also, is there an effect on growth of tissues in tissue cultures (p. 486)?

Instead of spraying young seedlings, you may prefer to have students scatter seeds on filter paper soaked in gibberellic acid in covered Petri dishes; observe the *rate of germination* among varieties of plants. On initial treatment, rapidly growing leaves are pale green but later develop a deeper color, especially if fertilizer is

---

[24] The addition of a wetting agent prevents the gibberellic acid from running off the plant surfaces.

added to the soil. (Many supply houses offer suggestions for successful demonstrations with plants.)

Refer to tissue culture techniques to investigate the effect of GA on stimulating callus formation in carrots (p. 487). Are there other factors such as wavelength or intensity of light or temperature that affect the sensitivity of plants to gibberellic acid? Further, does GA affect the rate of regeneration of such plants as *Bryophyllum*? Is the rate of growth of *Lemna* (duckweed) affected? Duckweed is a useful plant for students to work with since new leaves are easily measured, and the plant requires little care or space (see p. 490).

Dwarf plants and gibberellins   In the 1950s Phinney investigated the effect of gibberellin on dwarf plants.[25] Seedlings of dwarf bean, corn, or pea plants show dramatic changes in growth when treated with gibberellic acid. As with auxins, this is due to the elongation of cells rather than to an increase in the number of cells. The effect of GA on dwarf corn is used as an assay technique. Do gibberellins supply some substance that has been lost through mutation in dwarf plants?

Obtain tall and short varieties of seeds of peas, beans, corn,[26] marigolds, snapdragons, and/or zinnias from a seed company or biological supply house. Germinate the seeds and grow them in moist vermiculite for some two to four weeks. Then apply with a pipette 1 ml of gibberellin solution to a young apical leaf of both tall and dwarf varieties of several kinds of seedlings. For controls, apply distilled water with a few milliliters of ethyl alcohol as used in the preparation of the growth regulating solution. Repeat the application of the GA solution at two- to three-day intervals using newly developed leaves over the next several weeks. Maintain all plants in moderate light at room temperature.

Prepare a stock solution by dissolving 100 mg gibberellin in a few milliliters of 95 percent ethyl alcohol; dilute with distilled water to 1 liter. This gives a 100 μg/ml solution; prepare more dilute solutions with distilled water. When refrigerated, the solution may be stored for about two weeks. Try the effect on growth of gibberellin solutions of varying concentrations: 0.1, 1.0, 10, 100 μg/ml. Add a few drops of baby shampoo as a wetting agent.

Over two to three weeks, have students measure in millimeters the height of each plant and then graph their data. If students use different coded symbols in their graphs, they can superimpose growth results for one kind of plant—both tall and dwarf varieties—for different concentrations of gibberellin. Do dwarf plants grow as quickly as normally tall plants?

Ethylene   This gas is known to ripen fruits. It is considered a plant hormone by some investigators since it is synthesized by many plant tissues and ripening fruits from the amino acid, methionine. A high concentration of auxin stimulates the synthesis of ethylene. This may be an explanation for the inhibiting action of auxin in roots. Recall that in a horizontal root the concentrated auxin on the bottom part of the root does not stimulate the elongation of the bottom cells. Ethylene may inhibit the action of auxin, and as a result, the top cells elongate and grow more rapidly so that roots turn down to give a positive geotropic response.

Abscisic acid   Another plant hormone, abscisic acid (formerly called dormin, since it was isolated from dormant buds and cotton bolls), seems to inhibit nucleic acid syntheses and thereby slows growth. It can be said it induces dormancy. However, it is also a promoter of flowering among some short-day plants. Abscisic acid also slows down transpiration by regulating the closing of stomates. It may interfere with the action of K ions that

[25] B. O. Phinney, "Growth Response of Single-gene Dwarf Mutants in *Maize* to Gibberellic Acid," *Natl. Acad. of Sci. Proc.* (1956) 42:185–89.
[26] Several single-gene mutants of corn (*Zea mays*) are especially sensitive: anther ear–1; dwarf 5232; dwarf–1; dwarf 8201.

function in keeping high turgor pressure necessary to keep stomates open. Since a high concentration of this hormone has been found in fleshy fruits, it may also prevent seeds from germinating in the fruit (thereby having the opposite effect of gibberellin on the aleurone layer in the seed). As rain leaches abscisic acid away from fallen fruits and seeds, the seeds begin to germinate.

Height retardants

Occasionally, the growth of some plants is deliberately retarded to produce bushy, compact plants for ornamental purposes. TIBA (tri-iodobenzoic acid) is one retardant that may act by inhibiting auxin flow in stems. Another commercially available growth retardant for certain plants is Phosfon (tributyl-2,4-dichlorobenzylphosphonium chloride).[27] Phosfon causes a shortening of the internodes of the plant thereby producing a thicker stalk with deeper green leaves. It is effective on such plants as chrysanthemums, Easter lilies, seedling red and silver maple, black locust, mimosa, coleus, azalea, alfalfa, summer cypress, and some species of holly. Many retardants may block the activity of gibberellins.

Phosfon is completely soluble in water, ethanol, isopropanol, and acetone. It persists in soil so it may retard the height of plants for several years.

The substance is applied to the young growing stages of plants in a lanolin base or as an additive to soil. A spray is not used because the chemical destroys chloroplasts. Some plants, such as Black Valentine beans, show the effect of treatment in a few days; in other plants, two to three weeks are needed to show results. Plants are reduced to one-third to one-half of their normal height.

Felton and Downing suggest specific methods for retarding growth of several plants.[28] For example, plant Black Valentine beans in small pots. Just before the terminal bud unfolds use a wooden applicator to apply a circle of Phosfon in lanolin paste around the stems midway between the cotyledons and the primary leaf node.

To use Phosfon as a soil additive, dissolve 4 tsp of 10 percent Phosfon liquid in some 6 liters of water. Mix 60 ml of this solution with soil for a 10 cm pot, or 120 ml for a 15 cm pot. Plant seeds of garden annuals such as *Salvia*, or petunias in flats of untreated soil. Then transfer young plants with good leaf growth into larger pots containing Phosfon-treated soil. Set aside some pots of untreated soil with the same kind of plants.

For rooted chrysanthemum cuttings, dissolve 1 tsp of 10 percent Phosfon liquid in some 6 liters of water.

Photoperiodism

Light affects many activities of green plants—phototropism, photosynthesis, and the duration of light periods. Table 5-1 shows the action spectrum of these responses. The original work of Garner and Allard of the U.S. Department of Agriculture in 1920 was based on the amount of daylight plants received. They called these sequences photoperiods.

[28] S. Felton and C. Downing, *Carolina Tips*.

TABLE 5-1   Action spectrum of some responses in plants*

	range (Å)	peak (Å)
Phototropism (Flavoprotein)	3000–5000	4400–4700 blue light
Photoperiodism (Phytochrome)	5800–7200	6500 red light
Photosynthesis (Chlorophylls)	4000–7300	4400 blue light

* Adapted from A. G. Norman, "The Uniqueness of Plants," *Am. Scientist* 50:445, 1962.

[27] Phosfon is available from Carolina Biological Supply Co., Burlington, NC. Refer to the experiments described by S. Felton and C. Downing in *Carolina Tips* (March 1963) 26:3; a good bibliography is included.

While this notion of long-day and short-day plants persists, it is the hours of darkness that the plant "clocks" that are critical in initiating flowering. The term long-night plants would be more accurate than short-day plants. Table 5-2 lists some plants and their photoperiods; note that some plants are day-neutral.

The fact that plants flower at different times eliminates the harsh competition among seeds that might occur if all plants flowered and produced seeds at the same time. Long-day plants usually flower under constant light; they need no dark period. When a flash of light breaks the long dark period, short-day plants do not flower.

Role of phytochrome   The existence of this light-sensitive blue or bluish-green pigment in plants was announced by the U.S. Department of Agriculture in 1959, and isolated in 1960. Phytochrome is responsible for initiating the time of flowering and the germination of some seeds. It is also active in periods of light and dark sequences of photoperiods.

Phytochrome is made in leaves and must exist in small amounts; while its

bluish pigment is not apparent in albino plants, these plants respond to red and far red wavelengths. The existence of phytochrome was postulated when the action spectrum for sensitivity to photoperiodism was found to lie between 5800 and 7200 Å and to be more effective in the red end of the spectrum. This did not coincide with the absorption spectrum of any pigment that was known at the time. The research, done by Hendricks and Borthwick at the U.S. Department of Agriculture, led to a more complete explanation of how growth and development of a plant are controlled by phytochrome. Phytochrome has now been found in all vascular plants.

Reversible shift   The pigment phytochrome found in leaves controls many photoperiodic reactions of plants. Phytochrome exists in two forms; phytochrome red (P_r), which absorbs red wavelengths and phytochrome far red (P_{fr}), which absorbs far red wavelengths.

Exposure to different wavelengths of light causes a *reversible shift* of the form of phytochrome. In red light, P_r is converted into the active form, P_{fr}. In far red light, phytochrome far red (P_{fr}) is converted to P_r. In the dark, the active unstable P_{fr} form is slowly converted to the stable P_r form. In light, the conversion takes seconds.

TABLE 5-2   Photoperiods of some plants*

long-day plants	short-day plants	day-neutral plants
aster	cocklebur	azalea
beet	goldenrod	begonia
bindweed	morning glory	bluegrass
chrysanthemum	poinsettia	carrot
clover	ragweed	celery
daisy	soybean	corn
loosestrife	sweet potato	cotton
oat	tobacco	cucumber
phlox	(Mammoth)	geranium
radish		kalanchöe
spinach		pea
		nightshade
		periwinkle
		tobacco
		tomato

* A more comprehensive listing available in W. Spector, ed., *Handbook of Biological Data* (Philadelphia: Saunders, 1956).

$$\text{(stable,} \quad P_r \xrightarrow[\text{in far red light}]{\text{in red light}} P_{fr} \quad \text{(active form)}$$
$$\text{long-lasting form)} \qquad \text{(also in the dark)}$$

Apparently the long continuous dark period when P_r is formed is needed by short-day plants to flower. This is why a light flash interrupting the dark photoperiod initiates the conversion of the needed P_r to P_{fr} and inhibits flowering in short-day plants; its P_r supply has been converted to P_{fr}. On the other hand, the plant's metabolic processes may be reversed in the P_{fr} form of phytochrome. P_{fr} initiates flowering in long-day plants; it prevents flowering in short-day plants. (See Table 5-2.)

Both forms of phytochrome have an interesting effect on light-sensitive seeds, such as Grand Rapids lettuce seeds. In darkness, these light-sensitive seeds do not germinate. A short exposure to light, such as the turning over of soil, is needed for these seeds to germinate. The P_r form of phytochrome in darkness prevents germination, but the P_{fr} form initiates germination of light-sensitive lettuce seeds. P_{fr} seems to cause an activation of gibberellin.

In the woods, with a heavy canopy of green leaves, red wavelengths are filtered out by the chlorophyll so that only far red wavelengths reach plants at the floor of the wooded region. In this case, P_{fr} is converted to P_r resulting in the elongation of stems which brings the plants into better light for photosynthesis.

Phytochrome forms and germination of light-sensitive seeds

Have students show the effects of the two forms of phytochrome in red and far red light on germination of Grand Rapids lettuce seeds which are light-sensitive. First store the seeds at 20° C (68° F) in the dark. Surface sterilize a batch of seeds by immersing for 15 min in household bleach (1:9 parts water). Then soak the seeds in wet toweling for 24 hr. Scatter an equal number of seeds—about 25 to 30—into each of five Petri dishes lined with wet filter paper. Wrap Petri dishes in different

colors of cellophane. For example, for a red filter, use a double layer of red cellophane; for a blue filter, use a double layer of blue cellophane; for a far red filter, use *both* red and blue cellophane. Maintain controls in clear cellophane (see Fig. 5-30).

Label the Petri dishes and treat each dish as follows: *dish 1*: place in continuous light for four days; *dish 2*: place in continuous darkness for four days; *dish 3*: wrap in a double layer of red cellophane and expose to light for 10 min (red light), and then transfer to darkness for four days; *dish 4*: wrap in a double layer of red cellophane and expose to light for 10 min, then shift into far red light by covering the red with double wrappers of blue cellophane for 25 min; transfer to darkness for four days; *dish 5*: wrap in red cellophane (red light) for 10 min, then into far red (cover over with blue cellophane) for 25 min, then return to red light (remove the blue wrappers) for 10 min; transfer to darkness for four days. Which dishes show the greatest percentage of germination? Which were given red light as the last signal? Which form of phytochrome stimulates germination of these light-sensitive seeds? Which dishes show the greatest stem elongation? (Light in the far red range lowers P_{fr} concentration so that stems elongate faster.) Is there a survival value for seeds that receive light rich in far red—seedlings in a dark woods?

A flash of light, especially red light, will initiate germination of some kinds of

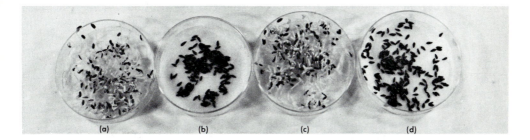

Fig. 5-30 Responses of Grand Rapids lettuce seeds to light and dark: (a) seeds kept in white light from a fluorescent tube; (b) seeds kept in darkness; (c) seeds kept in darkness except for 15-min exposure to red light; (d) seeds kept in darkness except for 15-min exposure to red light, followed by 15-min exposure to far-red light. The total length of the culture period in each case was 48 hr. (From *Plant Biology* by Knut Norstog and Robert W. Long. Copyright © 1976 by W.B. Saunders Co.)

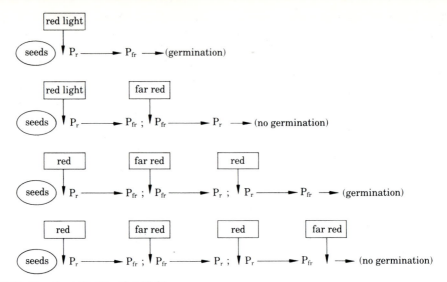

Fig. 5-31 Flowchart of action of red light.

soaked seeds since red light changes P_r to P_{fr}; apparently, metabolic blocks are removed and the seeds germinate. If exposure to red light is quickly followed by an exposure to far red, P_{fr} is converted back to P_r before the active P_{fr} has had a chance to initiate germination. As a result, seeds exposed to red light germinate; seeds exposed to red, then far red light do not germinate; seeds given red, far red, then red light germinate; seeds given red, far red, then red and again far red, do not germinate (see Fig. 5-31).

While some mature seeds remain dormant and need light to germinate, their growth may be different under an artificial light source. A mixture of fluorescent and incandescent light similar to sunlight is needed.

Filters for use in greenhouses   Filters of colored cellophane can approximate pure color. Use a fluorescent lamp with a red filter to get almost pure red light, and an incandescent filament lamp with blue and red filters to get far red light.

Incandescent filament lamps resemble sunlight in providing red and far red light; they are effective in promoting flowering and the vegetative growth of plants, even though there is some far red. If a greenhouse is equipped with fluorescent lights only, add some incandescent lamps to supply far red if needed.

Source for red light   Wrap two layers of DuPont 300 MSC red cellophane around a 15-watt fluorescent tube.

Source for far red light   Wrap two layers of DuPont 300 MSC red cellophane and two layers of 300 MSC dark blue cellophane paper around an incandescent bulb.

Far red phytochrome seems to act as a cue to seedlings that they have grown out of the darkness of the soil into light.

Students may also want to use *Lemna* (duckweed) in investigations of flowering. Use small test tubes, each containing 2 to 3 plants floating in about 15 ml of either Hoagland's or Knop's medium (pp. 469, 480, 690).[29]

[29] W. S. Hillman, "Test tube Studies on Flowering," *Bull. Torrey Bot. Club* (1962) 885:327; also W. S. Hillman, "Experimental Control of Flowering in *Lemna*. IV. Inhibition of Photoperiodic Sensitivity by Copper" *Amer. J. Bot.* (1962) 49:892.

Notice that rapidly growing weeds in open fields need only a short flash of light to activate the phytochrome of their seeds; do weeds germinate faster than the other seeds? Where space is available (for example, a greenhouse or an outdoor plot of ground) study the effects on flowering in plants when the number of hours of light and dark is altered. For example, students may grow short-day plants, such as ragweed, soybeans, cocklebur, or poinsettia, which need only about 10 hr of light in every 24 hr in order to flower. These plants need long periods of dark to flower normally; if this dark period is interrupted by even a short exposure to light, the plants remain in a vegetative stage. It is believed that the substance that controls flowering is made in the leaves and transported to the sites of flower formation (see Fig. 5-29). A long range investigation might involve learning more about the effect of covering leaves, rather than the total plant; the possible effect of covering and exposing leaves on flower development of grafted branches of shrubs or trees; or the effect of temperature on photoperiodism.

Similar investigations may be undertaken using sturdy long-day plants (light for 14 hr) such as aster, rose mallow, spinach, or barley plants. Plan to use plants that are about four to five weeks old, and expose them to changed light conditions for a week. Students should provide for controls. Also use corn, snapdragon, cucumber, tobacco (most varieties except Maryland mammoth), and tomato plants as examples of plants not affected by fixed periods of light and dark.

Photoperiods and flowering

You may want to investigate the effects of photoperiods on flowering of some short-day and some long-day plants.[30] Transfer plants from flats into small pots when they are about one month old. For long-day plants, use spinach and/or radish; for short-day plants, use cocklebur and/or soybean. (Other plants are listed in Table 5-2.)

After the plants have grown for another week in the small pots, label one pot of each type of plant according to these treatments: 8 hr light, 16 hr of continuous darkness—short-day treatment; 16 hr of light, 8 hr of darkness—long-day treatment. Set some control plants in constant light, some in constant darkness. At the start of the treatment, examine the plants for flower primordia.

Maintain the pots of plants for a week under these photoperiods. Later transfer them to normal day-night environmental conditions. Be sure all plants are maintained at the same temperature. After about four weeks, students can examine the plants to observe which photoperiods induced flowering in the short- and long-day plants.

Photoperiodism in animals

Students may rear certain Cladocera, such as *Daphnia*, in different light regimens; raise some under a light period of 12 to 16 hr and illuminate others for only 8 hr per day. Maintain several replicas, as well as controls maintained in constant light and constant darkness. Count the number of juveniles and adults at the end of a week. Replace in fresh media separating the adults from the juveniles, and again keep weekly counts of the number of juveniles.[31]

Among many male birds, the gradual increase in hours of light stimulates activity of the testes; among ferrets, the increase in illumination brings them into estrus ahead of their normal reproductive cycle. Also, among immature ducks, red

---

[30] Adapted from L. Machlis and J. Torrey *Plants in Action* (San Francisco: Freeman, 1956).

[31] Refer to A. Giese, "Reproductive Cycles of Some West Coast Invertebrates," in R. Withrow, ed., *Photoperiodism and Related Phenomena in Plants and Animals*, Am. Assoc. for the Advancement of Science, Washington, DC, 1959.

and orange light especially stimulates the pituitary gland which triggers gonad activity. Light need not reach the eye in the duck; direct stimulation of the pituitary has been achieved with a quartz rod. When direct stimulation is used even blue light is effective. Furthermore, pituitaries of illuminated ducks have been transferred to immature mice causing an increase in gonadotropic activity in the mice. In the case of ferrets, light must reach the eye; in the sparrow this is not needed. However, there is no common pattern among vertebrates. Nerve connections between the pituitary and hypothalamus are necessary for most gonadotropic activity; yet there are cases where gonadal activity seems to be the result of hormonal transmission only. (Refer to the role of the releasing factor from the hypothalamus on the stimulation of the pituitary gland in Figure 5-22).

Circadian cycles: Biological clocks

Some plants, even algae, show a 24-hr circadian cycle of physiological activity that is independent of photoperiods. Sleep movements in some plants respond to a 24-hr cycle, even in the dark. Mitotic division of cells—both plant and animal—are regulated by circadian cycles.

Circadian cycles are also found in animals and in humans. Students may record their pulse rate, heartbeat, and hours of sleep, and their periods of heightened keenness for learning and for skills. Is there a pattern?[32]

[32] Refer to the following articles in *Scientific American*: D. Saunders, "The Biological Clocks of Insects" (February 1976); S. Emlen, "The Stellar-Orientation of a Migratory Bird" (August 1976); and J. Palmer, "Biological Clocks of the Tidal Zone" (February 1975). Also refer to J. Aschoff, "Circadian Rhythms in Man," *Science*, June 11, 1965, and to current literature in *Biological Abstracts*.

CAPSULE LESSONS: WAYS TO GET STARTED IN CLASS

5-1. Pour a rich culture of paramecia or *Euglena* into a tall container or test tube. Cover it with foil so that only a few cm are exposed to light. Then uncover the tube in class and ask why the organisms are oriented mainly at the top. Or prepare a vial of *Euglena* covered with black paper with a narrow slit along the side where light enters. Remove the cover, ask students to explain the thin green streak due to the clustering of *Euglena* in the lighted area. Elicit the survival value of taxes.

5-2. Ask students if they have ever seen roots of plants grow upward. Elicit possible factors—Is the response due to light or gravity (or something else)? Share the design of an investigation to test one factor, one variable, at a time and provide for a control.

5-3. What kind of behavior could you do perfectly from birth—never had to learn? Elicit responses that develop the definition of a reflex. Then trace the path of a reflex from stimulus and receptor to the brain (for head reflexes) or to the spinal cord and back to muscles or glands.

5-4. Have a student stand before the class and hold a square of screening or clear plastic in front of his or her face. Why does the subject blink when another student throws paper balls at the subject? What is this stereotyped behavior? What are the advantages of reflexes?

5-5. Is riding a bike a reflex? Is typing or dancing? How is a habit different from a reflex? Compare the value of each type of behavior. How are new habits formed? What is the role of reinforcement? How can some habits be broken or "unlearned"?

5-6. Have students list their fears. Are these inherited patterns of behavior? What factors exist in conditioning behavior?

5-7. Begin by asking students how they would train a new puppy to answer to its name. Develop the steps in habit formation: desire, repetition, and satisfaction.

5-8. Have students list the traits they admire in other people. Then have them check themselves against this model. Develop a discussion of ways to improve oneself. What are the characteristics of a pleasant personality?

5-9. Have the class plan a field trip around the school grounds to look for examples of tropisms among plants. Students may search for examples of plants responding to environmental conditions. Give one or two examples; then set them free for 10 min to find more. Bring the class together again; all of the students will be working together to check the examples they have found.

5-10. Ask students to bring in some *Tradescantia* sprigs, and supply equipment to set up the demonstration in Figure 5-9. Have students plan the demonstration themselves. Ask what might happen if the plant were turned upside down. Why do stems grow up and roots grow into the soil?

5-11. Periodically, show a short film of a special technique that cannot be done in class—action of hormones, behavior among insects, nerve-muscle physiology, or time lapse films of plant growth. (Keep a file of film catalogs available for rental or purchase.) In general, most films serve as summaries or reinforcement of work developed in class.

5-12. A student might want to give a report of the kind of learning that W. Kohler described in his book *Mentality of Apes* (Harcourt, Brace & World, 1927). Was the putting together of two sticks to make a longer one for reaching food an example of an instinct or insight? Also compare mentality of apes with behavior in sharks, ducks, baboons, and other forms, as described in many issues of *Scientific American*.

5-13. Have students describe instincts in animals, such as nest-building, spinning of a spider's web, migration of some birds and fish, behavior in a beehive, and many others. How do these behavior patterns differ from problem-solving behavior in humans?

5-14. Start a discussion of study habits by using the activity described in this chapter in which students memorize stanzas of poetry in silence and with distractions. Or use a list of nonsense words first, and let students develop the conclusion that problems can be solved faster when one knows what the problem is, and how to tackle it.

5-15. Use the puzzle referred to in Figure 5-19. Students usually take from 5 to 15 min to solve it. After 10 min, have one student explain how to put the pieces together. Then time the class again. Why is it easier to solve the puzzle now? Keep a tally of the number of students who solve the puzzle in each case. How does this apply to doing a school assignment? How do past experiences affect our insight into new problems?

5-16. As a review lesson, you may want to show a film on the structure of the nervous system and the function of such sense organs as the eye and ear.

5-17. Begin by asking students to describe the conditions under which a diabetic must live. Or ask students how they feel when they are frightened. From these questions, elicit the uses of insulin and of adrenalin in the hormonal coordination of the body. Develop the notion of a regulator and a feedback system involving the gland and the pituitary and its releasing hormones.

5-18. Elicit an experimental design to investigate how the effects of a ductless gland are discovered. It may be necessary to describe an experiment such as that of Gudernatsch, who removed thyroid glands from tadpoles and also added thyroid extract to the diet of tadpoles. Or read together the work of Banting.

5-19. Develop a flowchart to show the coordination, the homeostasis involved in the flow of FSH, LH, estrogens, and progesterone. Develop the ovarian and pregnancy cycles.

5-20. Diagram a total plant: roots, stems, leaves, and its flowers. Develop the role of auxins, gibberellins, and kinetin in maintaining a plant in homeostasis.

5-21. Have a student describe how to knot a tie without using gestures. Why is the description slow and stumbling? Develop the notion that our habits are activities done without thinking.

5-22. Have students write their names five times. Then have them shift their pens into the other hand and again write their names five times. Why is it so much more difficult and time-consuming? Elicit from students that each letter must be thought through in spelling a word. What is the value of having a habit?

5-23. At some time, bring to class an example of some tropism described in the chapter, and then elicit explanations from the students as to the cause of this response.

BOOK SHELF

The following references give some scope of the work in behavior. Papers from *Scientific American* are available in book form or as separate offprints. Send for the W. H. Freeman catalog; also refer to the cumulative index for *Scientific American* (available in libraries).

Refer to references listed at the end of chapters 1, 2, 6, 7, 8, as well as the following:

Bentley, P. *Comparative Vertebrate Endocrinology.* New York: Cambridge Univ. Press, 1976.

Brown, J. *The Evolution of Behavior.* New York: W. W. Norton, 1975.

DeCoursey, R. and J. Renfro. *The Human Organism.* 5th ed. New York: McGraw-Hill, 1980.

Dewbury, D. A. *Comparative Animal Behavior.* New York: McGraw-Hill, 1977.

Emlen, J. *Population Biology.* New York: Macmillan, 1984.

Giese, A. *Blepharisma: Biology of Light-Sensitive Protozoans.* Stanford, CA: Stanford Univ. Press, 1973.

Gorbman, A. *Comparative Endocrinology.* New York: Wiley, 1983.

Hampden-Turner, C. *Maps of the Mind.* New York: Macmillan, 1981.

Hofer, M. *The Roots of Human Behavior.* San Francisco: Freeman, 1981.

Julien, R. *A Primer of Drug Action.* San Francisco: Freeman, 1981.

Lesher, A. *Behavioral Endocrinology.* New York: Oxford Univ. Press, 1978.

Restak, R. *The Brain.* New York: Bantam Books, 1985.

Springer, S., and G. Deutsch. *Left Brain, Right Brain.* San Francisco: Freeman, 1981.

Toates, F. *Animal Behavior.* New York: Wiley, 1980.

Wilson, E. O. *Sociobiology.* Cambridge, MA: Harvard Univ. Press, 1975.

6

Continuity of the Organism: Development, Differentiation, and Growth

We begin with the classic studies of Redi, Spallanzani, and Pasteur and their experiments to disprove spontaneous generation. Then consider the transmission of similar DNA in several types of asexual reproduction; the replication of DNA molecules; the evolution of sexuality and meiosis; laboratory studies of development and differentiation; and patterns of growth.

The transmission of DNA from generation to generation and the pattern of *appearance* of traits—that is, the realm of genetics, adaptation, and the evolution of ecosystems of the past—form the substance of chapter 7. Populations of organisms as they exist in present day ecosystems are described in chapter 8.

Classic experiments disproving spontaneous generation

Redi's experiment

Redi was among the early investigators who suspected that the familiar stories of spontaneous generation of living things were false. Recall that in 1668, he set up three jars to discover whether or not maggots and flies came from decaying meat. In season, this experiment may be duplicated (Fig. 6-1). Leave one jar open, another covered with cheesecloth (close mesh), and a third sealed with a sheet of plastic, as a substitute for Redi's parchment. Place

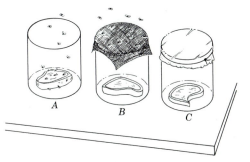

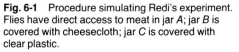

Fig. 6-1 Procedure simulating Redi's experiment. Flies have direct access to meat in jar *A*; jar *B* is covered with cheesecloth; jar *C* is covered with clear plastic.

a piece of fresh meat in each jar and set the jars out of doors overnight. You will find eggs of blowflies on the meat in the open jar and on the cheesecloth of the second jar, and no trace of flies on the jar from which the odor could not escape. In fact, at any time throughout the year you may use fruit flies and a medium of ripe banana (p. 535) in place of decaying meat. When fruit flies are used to repeat this classic experiment indoors, cover sets of half-pint bottles containing the banana medium (some covered with cheesecloth, others with cellophane or plastic, still others open) with an open-top bell jar (Fig. 6-2). Then select a thriving stock bottle of fruit flies and invert the mouth of the bottle into the opening in the bell jar. Quickly insert a stopper to close the bell jar. The flies within have the opportunity to select

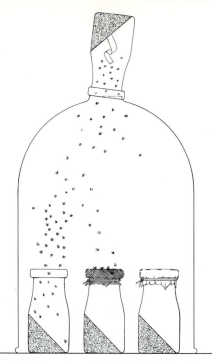

Fig. 6-2 Apparatus using fruit flies to simulate Redi's experiment. Flies are introduced into a bell jar containing bottles of banana medium; some are open, some are covered with cheesecloth, and some are sealed with plastic.

among the bottles. Within a week students should observe that eggs have been laid and larvae are burrowing in the medium in bottles that have been kept open.

Spallanzani's experiment

For a time Redi's work seemed at least to have dispelled the idea that spontaneous generation occurred in large forms. Then with the invention of the microscope and the discovery of the new world of microorganisms doubts again arose. Could such minute organisms have arisen from previously existing, equally tiny organisms? You may want to develop some of the case histories involving Needham and Spallanzani, Schwann, Pouchet, and Pasteur (to mention only a few who took sides in this controversy, which lasted for some three centuries). Pasteur finally gave evidence of

the propagation of microscopic forms from previously existing forms.[1]

Spallanzani's experiments disproving spontaneous generation of microorganisms can be repeated in class. You may recall that he boiled broth and quickly sealed the flasks. Prepare nine soft glass test tubes pulled out in the center over a Bunsen burner (Fig. 4-33). Make the constriction about 2 cm in length. As the tubes cool, prepare beef broth (p. 698) and fill the lower part of each tube with broth. Then arrange the tubes as follows:

Boil the broth in three tubes (*A*) in a water bath or double boiler and leave them unsealed.

Boil the broth in another three tubes (*B*). Plug the tubes with sterile cotton. When the test tubes are cool enough to handle, hold them at a 45° angle over the Bunsen burner and heat the constricted

[1] J. Conant, ed. *Pasteur's and Tyndall's Study of Spontaneous Generation*, "Harvard Case Histories in Experimental Science," 7 (Cambridge, MA: Harvard University Press, 1953). Also see J. Conant, ed. *Pasteur's Study of Fermentation*, 1952, case study 6 in the series. The studies have been compiled into the two-volume set *Harvard Case Histories in Experimental Science*, repr. ed. (Cambridge, MA: Harvard University Press, 1957.) Also refer to an abstract of Redi's paper in "Experiments on the Generation of Insects," in *Great Experiments in Biology*, M. Gabriel and S. Fogel, eds. (Englewood Cliffs, NJ: Prentice-Hall, 1955).

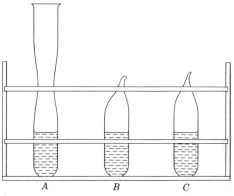

Fig. 6-3 Procedure simulating Spallanzani's experiment. The broth in tube *A* is boiled and left open to the air; broth in tube *B* is boiled and tube is sealed; broth in tube *C* is not boiled but tube is sealed.

area. Pull out the top and seal each tube (see Figure 6-3B).

Seal three tubes of broth (*C*) *without* any preliminary heating.

In a few days students should find that sets of tubes *A* and *C* contain turbid broth while the broth in tubes *B* remains clear. The tops of the sealed tubes may be broken with a triangular file and sample drops of the broth examined under the microscope under high power. Use a sterile wire loop to transfer a drop of the broth onto a clean glass slide. Spread the drop to a thin film and allow to dry. Flame the loop after making each slide. You may want to stain dry smears with methylene blue (pp. 144–45).

Modification of Pasteur's experiment

Pasteur devised many experiments to test his hypothesis that microorganisms did not arise by spontaneous generation. He designed a series of flasks like the one shown in Figure 6-4a. To simulate his experiments prepare flasks or tubes of broth with an S-shaped delivery tube (Fig. 6-4b). As the test tubes of broth are heated to sterilize the broth, water vapor collects in the trap (*A*) in the delivery tube. In this way, Pasteur permitted dust-free (bacteria-free) air to enter the tubes. The heated broth should remain clear, but the sterilized broth in the controls, open to the air (Fig. 6-4c), should become turbid after incubation for 48 to 72 hr at 37°C (99°F). Students may examine the bacteria in the broth under the microscope to confirm their observations. (Refer to staining bacteria smears, pages 144–46.)

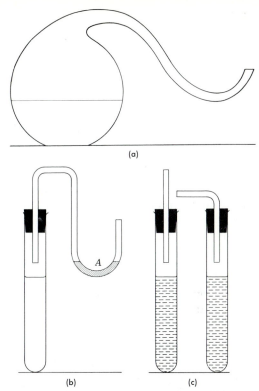

Fig. 6-4 Flasks for Pasteur's experiment: (a) one of Pasteur's flasks; (b) simple apparatus that works like Pasteur's flasks, forming a water trap at *A* to catch dust particles; (c) control tubes to be used with tube in (b).

the same DNA. Replication of DNA perpetuates the same genotype (barring mutation) in asexual methods of reproduction: fission, budding, spore formation, as well as vegetative propagation, so that the resulting clones have the same DNA.

Asexual methods of reproduction: clones

Fission   Equal divisions of an organism can be observed in cultures of protists, bacteria, and algae. Techniques for preparing cultures are described in chapter 9.

On occasion, you may want students to study the *rate* of fission in some protozoa or algae and try to discover circadian rhythms in cell division. Isolate individual

Transmission of similar DNA

Virchow extended the cell theory in stating that cells come from cells. During mitosis, the genetic information, located in triplet sequences of DNA in one cell, is replicated. When the cell divides into two or more cells, each of the resulting cells has protozoa under a dissecting microscope with a Pasteur pipette (or micropipette made of glass tubing pulled out to a fine bore in a Bunsen flame). Have students count the number of protozoa in a small quantity of medium in the well of a depression slide. In 24 hr, they may find two or four times the original number of organisms. Then isolate each of these specimens into fresh medium in welled slides.

If you lack a dissecting microscope, use a Pasteur pipette to place a row of small drops of the protozoa culture on a slide and examine each drop. In a random sampling, you should find a drop with only a few organisms. Transfer this drop to a drop of culture medium. (See isolating microorganisms, p. 458.) Examine daily and prepare a graph of the rate of fission over a given number of days (cycles of mitotic divisions, Fig. 6-8). You also may want to compare the effects of varying temperature, pH, vitamins, hormones, antibiotics, and an additional enriched medium on the rate of fission.

Use a solution of methyl cellulose or gum tragacanth (p. 142) for slowing the protozoa. Vital stains for temporary slides and staining techniques for permanent slides are described in chapter 2.

You may also want, at this time, to develop growth curves of organisms and populations of organisms (p. 461); also refer to standard plate counts, page 459.

Budding While there is an unequal division of cytoplasm in budding, the DNA complement of a yeast cell is equally transmitted to the small bud during mitosis. Prepare yeast cultures as described in chapter 9. Prepare slides to observe small buds protruding from the parent cells.

Use a vital stain (such as neutral red) to make the cells easily visible. Add a drop of a 1 percent solution of neutral red to one side of the coverslip of a slide of yeast culture. Remove the fluid in which the yeast cells were cultured by applying a piece of filter paper to the opposite side of the coverslip. Note how the volutin vacuoles and the granules of the living cells take up the neutral red stain. Dead cells, on the other hand, stain a uniform red color.

Examine also a culture of *Hydra* and look for budding: small replicas of *Hydra* may be found growing out of the stalk. Culture methods are described in chapter 9; also refer to chapter 1, and population growth, p. 455.

Spore formation Spore formation may be demonstrated readily using molds such as *Rhizopus* (bread mold, Fig. 6-5), *Penicillium* (Fig. 1-121), *Aspergillus* (Fig. 1-122), or *Neurospora* (Fig. 7-16). Under unfavorable conditions yeasts also may form spores, but these are difficult to show. While sporulation is found to some extent among protozoa (especially the malarial *Plasmodium*), examples are difficult to demonstrate unless prepared slides are used.

Prepare cultures of molds as described in chapter 9. Mount the fruiting bodies (sporangia) of the molds in a drop of glycerin or alcohol to prevent the formation of air bubbles (see Fig. 2-17).

Simulate the mechanical dissemination of spores from a mold sporangium. Almost fill a round black balloon with balls of cotton no larger than 1 cm in diameter.[2] Then put the neck of the balloon over one end of a 1 meter length of 1.5 cm glass tubing with fire-polished ends. Inflate the balloon by blowing through the open end of the glass tubing. Plug the open end with modeling clay so that the air will not escape. Sink this end into a mass of modeling clay to represent the upright "sporangium and sporangiophore" of a bread mold. The mass of clay may be considered the substrate. A network of glass tubes may be partially embedded in the clay to represent mycelia.

Burst the "sporangium" by dropping one or two drops of xylene on the balloon, or casually stroke the balloon after wetting your fingers in xylene. Students can watch the release of the cotton "spores," and notice that the cotton balls are expelled to fairly great distances around the room.

[2] P. F. Brandwein et al, "Laboratory Techniques in Biology," mimeo, 1940.

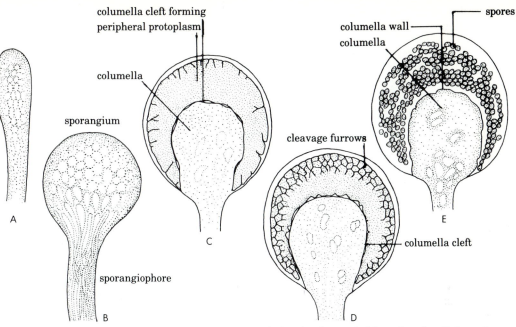

Fig. 6-5   Spore formation in *Rhizopus nigricans* (black bread mold): a sporangiophore enlarges to form a sporangium in *A* and *B*; *C* through *E* show further development of the columella with furrowing of outer region to form spores.

You may also want to study spore formation in other molds (refer to pp. 708–11), or alternation of generations in mosses or ferns (Figs. 6-61, 6-64). Students can observe dispersal of fern spores. Macerate sori (clusters of sporangia) on a dry slide without a coverslip. By the time students have focused under low power, the sporangia will be ejecting spores.

Vegetative propagation and regeneration

Extensive techniques are described on page 492 to observe the differentiation into whole new similar plants that have the same DNA (clones). Techniques for maintaining tissue cultures are also described (p. 483).

Instructional information in cells: DNA

It is interesting to note that the ideas Richard Goldschmidt developed in the 1930s, concerning the importance of the entire chromosome as a physiological unit, come close to the current models of sequences or codons in a DNA strand or chromosome. Unfortunately, Goldschmidt was not able to provide a *model* to fit his hypothesis. The model of a chromosome as a double helix of nucleotides was devised in 1953 by Wilkins, Watson, and Crick (Figs. 6-6, 6-7). Notice that the base sequences are complementary; in addition, the two strands of the helix run in opposite directions. The four different bases are two pyrimidines, thymine and cytosine, and two purines, adenine and guanine. These four bases carry the genetic information to code for the 20 amino acids. Adenine is linked to thymine by two hydrogen bonds; guanine to cytosine by three hydrogen bonds. In any organism, there are equal amounts of purines and pyrimidines. The ratio $A + T : G + C$ shown in Table 6-1, is a constant for similar DNA molecules of a given species (Chargaff's Rules).

Replication of DNA

The synthesis or replication of strands of DNA occupies a major part of the life cycle of a cell. The shortest part of this cycle is

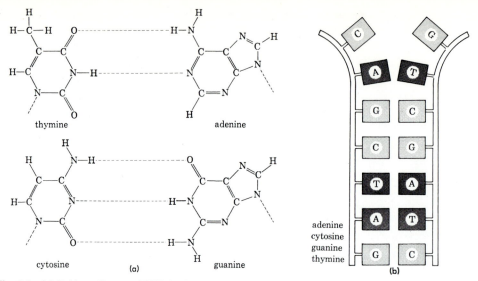

Fig. 6-6 (a) Pairing of bases of DNA: hydrogen bonds (dotted lines) link adenine to thymine, and guanine to cytosine; (b) sequence of bases: the sequence of bases in one strand of DNA determines the order in the second strand, because adenine pairs with thymine and guanine pairs only with cytosine.

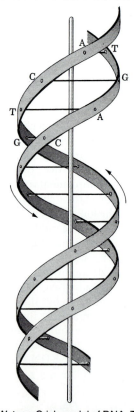

Fig. 6-7 Watson-Crick model of DNA. Two phosphate-sugar chains (shown as strips) held together by bases (shown as bars) represent the double helix form of DNA molecules.

when chromosomes are visible in mitosis—when they move apart on a spindle to opposite poles and cell division occurs. In a long G_1 stage, the cell and its organelles grow and carry on the metabolic functions of the cell. The S stage is the time for synthesis of new DNA strands; the G_2 stage is another growth period in which the cell prepares for mitosis. Stages G_1, S, and G_2 comprise the interphase stage of a cell's cycle (Fig. 6-8). In higher plants and animals the completion of a cell cycle may take some 10 to 24 hr.[3]

In replication, new complementary strands are synthesized using "old" strands of DNA as templates. Each strand or double helix "unzips" so that the two original strands attract free nucleotides to make new complementary strands. As a result, a new adenine pairs with a thymine on the old strand; a new guanine pairs with a cytosine on the old strand (Fig. 6-9). Note also that new strands form in opposite directions (Fig. 6-10).

This is the fundamental pattern of replication, called semiconservative, that Watson and Crick hypothesized for the

[3] In cleavage stages there is a brief G_1 stage; the cell cycle may take only an hour.

Table 6-1   Percentage of bases in DNA content of some organisms

species	percent purines		percent pyrimidines	
	adenine/guanine		cytosine/thymine	
bacteria	26.0	24.9	23.9	25.2
salmon sperm	29.7	20.8	20.4	29.1
sheep	29.3	20.7	20.8	29.2
human	30.4	19.6	19.9	30.1

Adapted from C. F. Herreid, *Biology* (New York: Macmillan, 1977).

new DNA strands. Each resulting DNA strand should be composed of an old strand and new strand making up its double helix.

In 1958, Meselson and Stahl traced the replication of DNA molecules by treating *E. coli* with heavy nitrogen (^{15}N). New strands of DNA incorporated the ^{15}N while old or "parental" template DNA contained ^{14}N. This observation confirmed the hypothesis of Watson and Crick.

You may also at this time want to refer to differential staining of sister chromatids at metaphase using radioactive thymidine, which becomes incorporated into the synthesis of a new strand before metaphase.

Models of DNA replication   Some students have used tongue depressors or

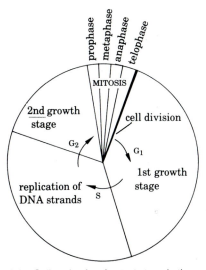

Fig. 6-8 Cell cycle: the shortest stage in the cycle is mitosis when chromosomes are visible. Periods G₁, S, and G₂ comprise the interphase period.

microscope slides, and strips of stiff plastic to make models of a double helix to show several possible triplet sequences of base pairs. An effective visual can be made by preparing cutouts in different colors representing the four nucleotides and then glueing small magnets to these cutouts. Then the magnet cutouts can be manipulated on a metal board so that replication can be shown. Separate the paired bases and insert free nucleotides to build new strands. (Later, the same cutouts can be used to show the formation of a strand of RNA, replacing the thymine base with uracil to build a single strand of RNA.) In addition, cutouts of nucleotides in different colors may be prepared and distributed in envelopes for each student to prepare possible sequences of triplets.

Transcription: DNA to RNA   The genetic information in the DNA molecules needs to pass from the nucleus into the cytoplasm. Recall that during transcription only one strand of DNA is active in forming an RNA molecule (Fig. 6-11). In place of thymine, RNA has the pyrimidine base uracil. As a result, a free uracil bonds with adenine of the original DNA strand through activation by enzymes.

Messenger RNA, transfer RNA, and ribosomal RNA are all active in protein synthesis.

Protein synthesis   While the master code DNA remains in the nucleus, messenger RNA molecules (mRNA) carry the genetic information into the cytoplasm, where the synthesis of both functional proteins (enzymes), and structural proteins (proteins that differentiate specific tis-

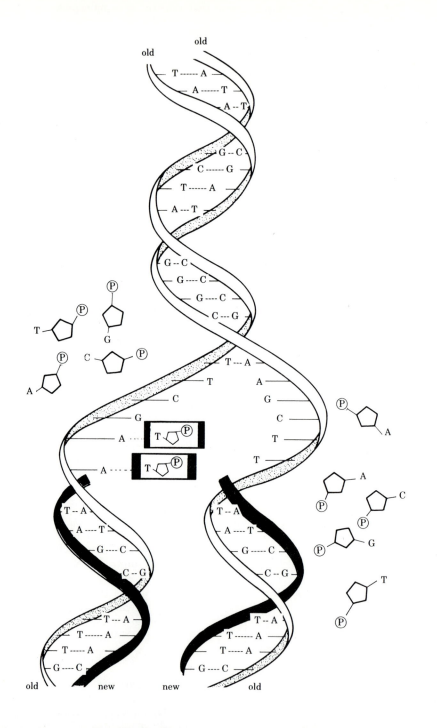

Fig. 6-9 Replication of DNA: helix unwinds and unzips; each old strand attracts free nucleotides forming complementary new strands. (Adapted with permission from Macmillan Publishing Company from *Biology* by Clyde F. Herried II. Copyright © 1977 by Macmillan Publishing Company, Inc.)

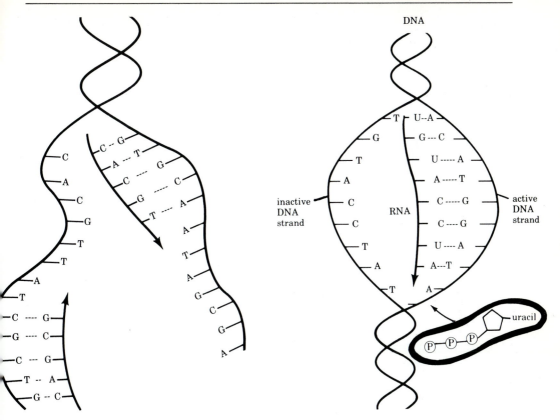

Fig. 6-10 Diagram to show new daughter strands develop on old strands in opposite directions. (Adapted with permission from Macmillan Publishing Company from *Biology* by Clyde F. Herried II. Copyright © 1977 by Macmillan Publishing Company, Inc.)

Fig. 6-11 Transcription of DNA to form RNA: only one DNA strand is active in forming an RNA strand. Note that uracil replaces thymine in the single RNA strand. (Adapted with permission from Macmillan Publishing Company from *Biology* by Clyde F. Herried II. Copyright © 1977 by Macmillan Publishing Company, Inc.)

sue cells—muscle, blood, bone, and so on) are made. An mRNA molecule may consist of thousands of nucleotides in length. Three nucleotides (a codon or triplet) code for one specific amino acid. All the codons needed to synthesize one polypeptide chain of amino acids make up a cistron. (Each cistron or blueprint for a polypeptide may be considered a "gene.") The mRNA molecules pass through pores in the nucleus to reach the cytoplasm where ribosomes (rRNA) are found—the site of protein synthesis. Ribosomes are found free in the cytoplasm and also along the endoplasmic reticulum. In the ribosomes the genetic instructions (codons) on the mRNA strands are interpreted or

translated. A ribosome separates into a larger part and a smaller part and clamps a mRNA strand. Then the ribosome moves along the single strand of mRNA "reading" the instructions to build or link specific amino acids into polypeptides (proteins) as shown in Figure 6-12. In sum:

DNA $\xrightarrow{\text{transcription in the nucleus}}$ mRNA
(double   (single
strands)   strands)

$\xrightarrow{\text{translation in ribosomes}}$ specific structural and functional proteins

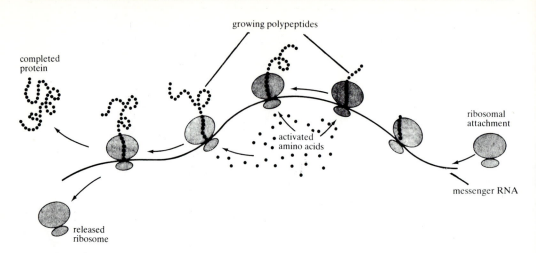

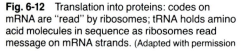

Fig. 6-12 Translation into proteins: codes on mRNA are "read" by ribosomes; tRNA holds amino acid molecules in sequence as ribosomes read message on mRNA strands. (Adapted with permission from Macmillan Publishing Company from *Biology* by Clyde F. Herried II. Copyright © 1977 by Macmillan Publishing Company, Inc.)

Transfer RNA molecules (tRNA) are formed on the original DNA template as are molecules of mRNA. The folded strands or loops of tRNA usually have the shape of a cloverleaf. There is a specific tRNA molecule for each of the 20 kinds of amino acids. Energy from ATP and many enzymes are involved in these reactions. The tRNA molecules have attachment sites for amino acid molecules found in the cytoplasm, and bring the specific amino acids to mRNA. A codon-anticodon complement forms when the codon on mRNA complements the sequence of bases on the anticodon site of the tRNA molecule. There is also an attachment site on tRNA that enables it to attach to the ribosomes.

Use Figure 6-12 to trace how ribosomes interpret the specific code on mRNA to build polypeptides from the specific sequence of amino acids held by tRNA.

There are also stop and start codons along the mRNA strand to indicate where one protein sequence begins and ends. Develop the fascinating work of Marshall Nirenberg and Severo Ochoa (1960) who established the triplet codons for the 20 amino acids (Table 6-2). As part of a case history, students might read "The Genetic Code: 11" by Nirenberg in *Scientific Amer-*

ican (March 1963). Students may also describe the hypothesis of Jacob and Monod (1961) explaining the role of an operon and promoter series that turn genes on and off.

There may be the opportunity to develop the role of enzymes and of some steroid hormones in regulating metabolic processes in cells. Describe the hormone-receptor complex that enters the nucleus where it affects DNA transcription (Fig. 6-13).

Consider using these offprints from *Scientific American* in a continuing series of studies of DNA and its control of protein synthesis: M. Nomura, "Control of Ribosome Synthesis" (January 1984); R. Dickerson, "DNA Helix and How It is Read" (December 1983); J. Darnell Jr., "The Processing of RNA" (October 1983); and L. Grivell, "Mitochondrial DNA" (March 1983).

DNA from beef pancreas Giese describes a method for extracting nucleoproteins from beef pancreas (sweetbreads).[4] Crush a small amount of beef

[4] A. Giese, *Laboratory Manual in Cell Physiology*, rev. ed. (Pacific Grove, CA: Boxwood Press, 1975).

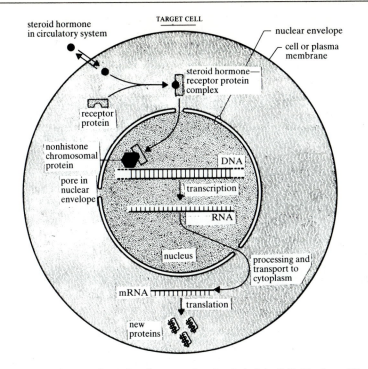

Fig. 6-13 Hormone-receptor complex enters the cell nucleus where it affects DNA transcription. (From *Cell Biology: Structure, Biochemistry, and Function*, 2nd ed., by Philip Sheeler and Donald E. Bianchi. Copyright © 1980 by John Wiley & Sons. Reprinted by permission of John Wiley & Sons, Inc.)

TABLE 6-2   Messenger RNA codons and their amino acids

listed by amino acid

Alanine—G-C-A	Glycine—G-G-A	Leucine—C-U-A	Threonine—A-C-A
G-C-G	G-G-G	C-U-G	A-C-G
G-C-C	G-G-C	C-U-C	A-C-C
G-C-U	G-G-U	C-U-U	A-C-U
Arginine—A-G-A	Glutaminyl*—C-A-A	U-U-A	Tryptophan—U-G-G
A-G-G	C-A-G	U-U-G	Tyrosine—U-A-C
C-G-A	Glutamic acid—G-A-A	Methionine—A-U-G	U-A-U
C-G-G	G-A-G	Phenylalanine—U-U-C	Valine—G-U-A
C-G-C	Histidine—C-A-C	U-U-U	G-U-G
C-G-U	C-A-U	Proline—C-C-A	G-U-C
Aspartic acid—G-A-C	Isoleucine—A-U-C	C-C-G	G-U-U
G-A-U	A-U-U	C-C-C	Terminator†—U-A-A
Asparaginyl‡—A-A-C	A-U-A	C-C-U	U-A-G
A-A-U	Lysine—A-A-A	Serine—A-G-C	U-G-A
Cysteine—U-G-C	A-A-G	A-G-U	
U-G-U		U-C-A	
		U-C-G	
		U-C-C	
		U-C-U	

* Glutaminyl is glutamic acid plus NH_3.
† Terminator codons are the last triplet in the code for protein.
‡ Asparaginyl is aspartic acid plus NH_3.

From C. F. Herreid, *Biology* (New York: Macmillan, 1977).

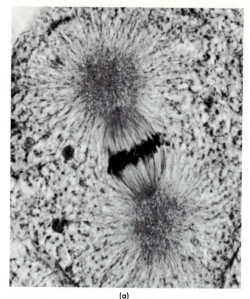

(a)

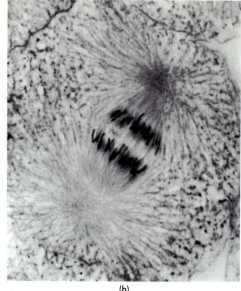

(b)

Fig. 6-14 Prepared slides showing mitosis in cleavage stages of whitefish eggs: (a) during metaphase, chromosomes line up at center of spindle (note the spindle fibers and centrosomes);

(b) during anaphase, chromosomes drawn on spindle fibers to opposite poles of the cell. (Courtesy of General Biological Supply House, Inc., Chicago.)

pancreas in a few milliliters of Ringer's solution. Then transfer 10 drops of the concentrated pancreas to a small beaker and dilute with 50 drops of 1 M NaCl. The solution becomes viscous as nucleoproteins are extracted. Then pass this extract through thin cloth toweling; tie the cloth and press out excess fluid. Transfer the extract to a small flask to swirl the extract. Now transfer a bit of the extract to several milliliters of distilled water. Long strands of nucleoproteins are precipitated. Twist the strands on a dissecting needle.

If available, frog sperm or sperm of sea urchins also may be used as a rich source of nucleoproteins. In this case, disintegrate the sperm of frogs by adding 1 M NaCl (or 3 M NaCl for sperm of marine forms). Then place 10 drops of sperm suspension into a small beaker and dilute with 50 drops of 3 M NaCl. The rest of the procedure is the same as described above for beef proteins.

Cell division: mitosis

Have students examine chromosomes undergoing mitosis in actively growing

tissues: root tips or fertilized eggs are classic examples. Under high power, examine prepared slides of onion root tips or of cleavage stages in eggs of whitefish (Fig. 6-14), sea urchin, and/or of the parasitic roundworm *Ascaris megalocephala*, which has only two pairs of chromosomes (Fig. 6-15).

Techniques to prepare chromosome squashes of *Drosophila* larvae salivary glands, as well as onion root tips, are given below. Have students compare their preparations with the mitotic figures in the prepared slides. Also consider the interval of mitosis in the total life cycle of a cell (Fig. 6-8).

Wire chromosomes   You may want to make wire chromosomes to illustrate, in a general way, how sister chromatids separate in mitosis and are pulled along the spindle fibers (Fig. 6-16). Use insulated double strand electrical wire for the models.

Have students compare their models with the stages they observe on prepared slides of onion root tip, *Ascaris*, and whitefish cleavage stages.

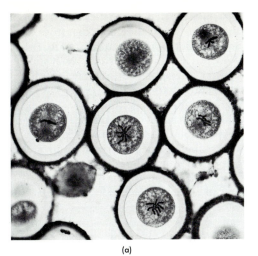
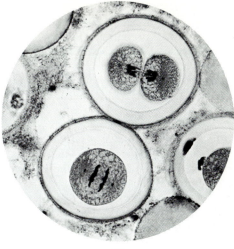

(a)   (b)

Fig. 6-15   Prepared slides showing mitosis in the parasitic roundworm, *Ascaris*: (a) large field of view; (b) two stages at higher magnification. (Photo courtesy Carolina Biological Supply Company.)

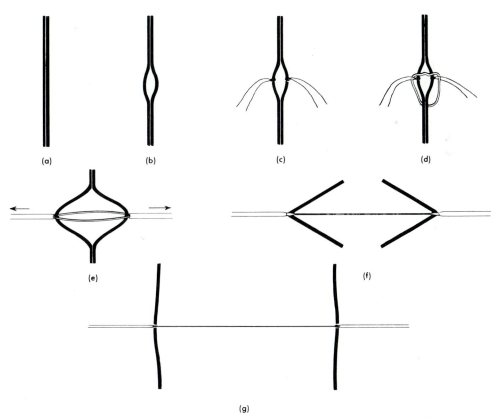

(a)   (b)   (c)   (d)

(e)   (f)

(g)

Fig. 6-16   Construction of wire model of chromosomes and spindle fibers in mitosis: (a) take a length of insulated electrical cord; (b) separate the middle portion; (c) tie a piece of string to each piece of cord; (d) slip a rubber band around the cord and through the strings; (e) and (f) pull the strings apart slowly until (g) the wire splits in two as chromosomes are drawn across spindle fibers.

Chromosome squash slides

Chromosomes of cells rapidly undergoing mitosis may be prepared using *Drosophila* larvae, larvae of *Chironomus* flies (midges), and in plant tissues, growing root tips of onion or beans. (Meiosis in young anthers may also be observed in squash slides of *Vicia* p. 401.)

Salivary gland squash Chromosomes in the salivary glands of some kinds of flies are about 100 times the size of those found in other tissue cells.

LARVAE OF DROSOPHILA Demerec and Kaufmann, and others offer detailed instructions for raising the larvae to be used in preparing these smear slides.[5] Rear the larvae at a temperature averaging 18° to 20°C (64°F to 68°F) with extra yeast added to the vials or bottles of culture medium (preparation of media, p. 535). Transfer egg-laying females to fresh vials of medium every two days so that hatching larvae are not overcrowded. Third instar larvae (the last moult) are slow-moving and begin to migrate out of the medium up the sides of the vials, or on paper to pupate. This occurs about six days after the eggs are laid. (The 14-day development period consists of eight days from the egg through three larval instars, and six days as a pupa undergoing metamorphosis.) Male larvae are easily recognized since they are larger and have more conspicuous gonads visible through the almost transparent skin in the posterior third of the body (Fig. 6-27).

Collect the slow-moving third instar larvae to prepare salivary gland slides. Transfer a large larva into a drop of insect Ringer's solution (preparation, p. 156) on a slide. Under a dissecting microscope with illumination from above, locate the black mouth hooks at the head end. Use a dissecting needle to hold down the abdomen on the slide, and with forceps pull the head end to separate it from the body. The

paired shiny salivary glands may be attached to the head; or tease apart from the white fat bodies and digestive tract that exudes from the cut end (Fig. 6-17). Now transfer the glands to a clean slide in a drop of aceto-orcein (or acetocarmine) stain (preparation, p. 399). After 3 to 5 min, transfer the glands to a second clean slide; add a small drop of fresh stain from a dropper bottle. Add a coverslip and blot with bibulous paper or toweling; gently but firmly apply thumb pressure or press on the coverslip with the eraser of a pencil, or roll a round pencil to spread the salivary tissue and remove excess stain. Locate the tissue under low power; under high power observe the banded chromosomes and centromeres. Good preparations of these temporary slides may be sealed with Duco cement or colorless nail polish.

Humason[6] and Gray[7] describe several methods for preparing permanent slides. Gray suggests placing the slide of stained chromosome smears on dry ice for about 3 min until it is completely frosted. Then raise the coverslip by inserting the point of a dissecting needle against the edge of the coverslip. With the other hand, slip a safety razor blade against the opposite edge of the coverslip. The stained chromosomes remain attached to the slide. Immerse the slide in acetone for 5 min. Remove the slide and add a drop of terpineol for 2 to 3 min to clear the preparation. Finally, tilt the slide so the terpineol runs off; add a drop of balsam and seal with a coverslip. Keep slides flat until the balsam dries.

LARVAE OF CHIRONOMUS Bensley and Bensley describe a similar technique using larvae (midges) of *Chironomus* flies.[8] The

---

[5] M. Demerec and B. Kaufmann, *Drosophila Guide*, 8th ed. (Washington, DC: Carnegie Institute of Washington, 1967); also P. Flagg, *Drosophila Manual* (Burlington, NC: Carolina Biological Supply Co., 1971).

[6] G. Humason, *Animal Tissue Techniques*, 3rd ed. (San Francisco: Freeman, 1972).

[7] P. Gray, "Permanent Whole Mount Microtechniques," Curriculum Aid. Ward's Natural Science Establishment, Inc. (undated).

[8] R. Bensley and S. Bensley, *Handbook of Histological and Cytological Technique* (Chicago: University of Chicago Press, 1938). Also refer to a similar technique using black flies, (*Similiidae*), having six chromosomes (2n) described by D. Barley, "Salivary Gland Chromosomes of Black Flies," *Turtox News* (December 1964) 42:12.

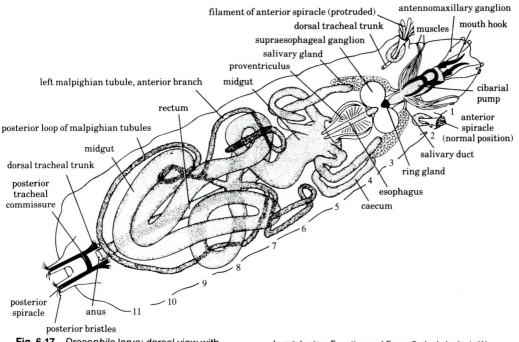

filament of anterior spiracle (protruded)
antennomaxillary ganglion
dorsal tracheal trunk
muscles
mouth hook
supraesophageal ganglion
salivary gland
proventriculus
midgut
left malpighian tubule, anterior branch
cibarial pump
rectum
1
anterior spiracle
2 (normal position)
posterior loop of malpighian tubules
salivary duct
midgut
3
ring gland
dorsal tracheal trunk
4
posterior tracheal commissure
esophagus
5
caecum
6
7
8
9
10
posterior spiracle
anus
11
posterior bristles

Fig. 6-17 *Drosophila* larva: dorsal view with tracheae, fat bodies, and heart omitted to show other internal organs. (Reprinted and adapted with permission of Macmillan Publishing Company from *The*

Invertebrates: Function and Form, 2nd ed., by Irwin W. Sherman and Vilia G. Sherman. Copyright © 1976 by Irwin W. Sherman and Vilia G. Sherman.)

larvae found in tubes in mud or in water are soft, wormlike, and often red. Collect them in the autumn or early spring, and culture them on decaying vegetation or in soil.

Mount the intact larva on a clean slide in insect Ringer's solution and pull out the head. Press between the second and third segments of the larva to release the shiny salivary glands from the body. Add a few drops of aceto-orcein stain for 10 to 15 min. Apply a coverslip and gently press with bibulous paper to smear the tissue. Under high power, chromosomes and the nucleolus are visible.

Root tip chromosome squash slides
ONION ROOT TIPS   Rapidly growing root tip cells of onion or garlic show chromosomes in many mitotic stages (Fig. 6-18). Many variations of techniques have been described. One simple technique is to rest onion bulbs on the rim of a container of water. After four to five days, when roots are 2 to 3 cm long, cut off 1 to 2 mm of root

tips. Then mount on a slide in a drop of acetocarmine stain (preparation, p. 399). (Most active mitosis seems to occur from 11:30 A.M. to 1:00 P.M.) Use forceps to mash the tips. Then heat the slide over an alcohol burner a few seconds (do not boil).

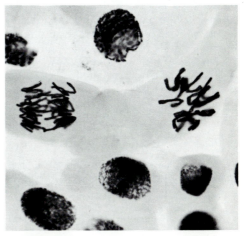

Fig. 6-18 Squash preparation of onion root tip. (Courtesy of General Biological Supply House, Inc., Chicago.)

Add another drop of stain and heat again without boiling. Apply a coverslip and over it place bibulous paper. Press firmly with thumb or pressure from an eraser of a pencil to spread the tissue. If, upon examination, the stain is too light, restain in acetocarmine and reheat.

Parker describes another technique using toluidine blue which shows an affinity for nucleic acids in the nucleus of onion root tips.[9] Sprout onion bulbs (as described above), cut off 1 to 2 mm of root tips and cover with 4 drops of 1 N HCl on a slide. Tease apart the tissue and warm the slide over an alcohol burner flame (do not boil) for 1 min. Use enough HCl to avoid drying. Blot off the excess HCl and add 0.5 percent aqueous toluidine blue (preparation, p. 158. Again, pass the slide through an alcohol flame for 1 min without boiling. Now blot off the excess stain; add a drop of fresh stain and apply a coverslip. Cover with bibulous paper or toweling and use thumb pressure or a plastic ruler to squash the preparation. These temporary slides may be sealed with clear nail polish or Duco cement.

Vena describes the Feulgen squash technique which can proceed at room temperature.[10] Sharma and Mookerjea describe a technique in which root tips are placed in a saturated aqueous solution of p-dichlorobenzene (*caution*: toxic) for 3 hr at 12° to 16° C (54° to 61° F).[11]

BEAN ROOT TIPS   Gillette describes a squash technique using the broadbean, *Vicia faba*, which has large roots and large chromosomes ($2n = 12$) while onion has 16 chromosomes.[12] Soak the seeds overnight; germinate in vermiculite. When the roots are 1 to 2 cm, cut off the root tips and fix for 2 to 4 hr in a freshly prepared solution of 3 parts absolute alcohol to 1 part of glacial acetic acid. Transfer the root tips to a watchglass or small test tube and add a mix of 10 parts of orcein stain to 1 part of 1 N HCl.[13] Heat but do not boil over an alcohol lamp. Let stand for 10 min. Transfer the stained tips to a clean slide in a drop of fresh stain and mash with a glass rod or forceps. Apply a coverslip, then bibulous paper, and apply gentle pressure. Then pass the slide through an alcohol flame several times.

Wolf describes a simpler version of Gillette's technique.[14] He recommends soaking the seeds overnight in a disinfectant, then germinating seeds in vermiculite until root tips appear. Place the root tips in a watchglass of acetocarmine and heat (not boil) for 5 min. On a slide make cross sections and squash these with a plastic ruler. Add more dye and a coverslip.

Ploidy in root tips of onion or broadbean

Colchicine suppresses cell division by preventing the formation of spindle fibers and new cell walls. As a result, chromosomes replicate but remain in the same cell, doubling the number of chromosomes (ploidy).

Fill several tumblers with water and use toothpicks to balance onion bulbs over the mouth of the tumblers so that only the bottom parts of the bulbs are immersed in water. Or soak broadbeans (*Vicia faba*) overnight and germinate. Leave some untreated bulbs and seeds in water to compare later with treated samples.

Bulbs left in the dark for a few days should have a good growth of roots about 1 cm long. Then transfer bulbs (or bean

[9] Nancy Parker, "The Four Minute Chromosome Squash," *Turtox News* (September 1968) 46:9.
[10] Joseph Vena, "Feulgen Reaction at Room Temperature," *Turtox News* (October 1967) 45:10.
[11] A. Sharma, and A. Mookerjea, "Paradichlorobenzene and Other Chemicals in Chromosome Work," *Stain Technology* (1955) 30:1–7.
[12] N. Gillette, "A Demonstration of Mitosis," *Turtox News* (March 1965) 43:3.

[13] Gillette's aceto-orcein stain is prepared by heating (not boiling) 45 ml of glacial acetic in a small flask and adding 1 g orcein stain. With a glass rod, stir and add 55 ml of distilled water. Filter and refrigerate.
[14] F. Wolf, "A Simplified Method for Making Chromosome Squashes," *Turtox News* (September 1966) 4:9. Also refer to P. Dustin's paper, "Microtubules," *Sci. Am.* (August 1980), which gives good photo of mitosis and the role of microtubules in cell division.

seeds) to containers of 0.03 to 0.05 percent colchicine prepared as a water solution. (*Caution:* Colchicine is toxic; use goggles to prevent damage to the eyes, and wear plastic or rubber gloves to avoid getting the chemical on the hands.) This step stops spindle formation and chromosomes are in late prophase.

Now return the beans or onion bulbs to water for the next 24 to 48 hr so cells resume growth and division. Once more, transfer the material to colchicine solution (*caution*). After 3 hr, cut off the 2 to 3 mm tips of the roots. Some students may stain untreated tissues while other groups prepare slides of colchicine-treated tissue.

To prepare permanent slides, place both untreated and colchicine-treated root tips in separate containers of a fixative, such as Carnoy's or Navashin's. After 12 to 24 hr, transfer the tips into separate solutions of equal parts of concentrated hydrochloric acid and 95 percent ethyl alcohol. After about 10 min, transfer the tips through 70 percent ethyl alcohol and into a drop of aceto-orcein stain on a slide. Apply a coverslip and bibulous paper and squash the tissue by pressing with the ball of the thumb in a rotary motion. Or apply pressure with the eraser of a pencil; avoid crushing the tissue.

For a more rapid technique that omits the fixative, transfer the root tips into either aceto-orcein or acetocarmine on a slide. Warm the slide, but do not boil, if acetocarmine is used. After 5 min, transfer root tips into fresh stain on clean slides. Now mince the root tips; apply a coverslip. Place the slide between sheets of bibulous paper and press down firmly with the eraser of a pencil. Examine the slides under high power. How do chromosomes on untreated slides compare with those of the colchicine-treated roots (Fig. 6-19)? What stages are most frequently observed in colchicine-treated root tip tissue? Students should observe that chromosomes in colchicine-treated tissue are shorter and thicker than untreated chromosomes, and are almost straight in one plane. Seal good preparations with clear nail polish to prevent drying.

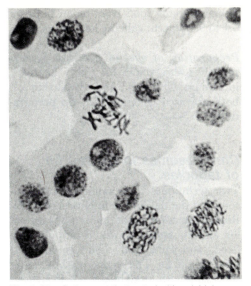

Fig. 6-19 Onion root tip treated with colchicine. (Courtesy of General Biological Supply House, Inc., Chicago.)

Preparation of stains   To prepare the aceto-orcein stain, begin by heating 45 ml of 70 percent acetic acid. The acid acts as a fixative. When the acid is hot, add about 2 g orcein. Allow the solution to cool and dilute it with 55 ml of distilled water. Apply a drop of the stain to a slide, and add the tissue. In some techniques, aceto-carmine is substituted for aceto-orcein.

To prepare acetocarmine stain, first boil 0.5 g of carmine in 100 ml of 45 percent acetic acid for about 3 min. Cool and filter the solution. In staining chromosomes, dilute 1 part of the solution with 2 parts of 45 percent acetic acid. The addition of 1 or 2 drops of ferric hydrate gives the stain a darker bluish tint.

Caffeine and onion tips   Hennessey, Martin, and Carr found that caffeine delays many stages in mitosis, and that in stronger concentration, metaphase chromosomes are especially distinct.[15] It is less dangerous to treat onion root tips with caffeine than with colchicine. To do so,

---

[15] T. Hennessey, M. Martin, and V. Carr, "A Cytological Study of the Effects of Uranium Nitrate and Caffeine on the Roots of *Allium cepa*," *Turtox News* (1951) 28:146.

condition *Daphnia* also shows cyclomorphosis—a cycle of change in the shape of the head, from a round head to a helmet head and longer tail.

Eggs of Artemia, the brine shrimp

These eggs remain viable for long periods in the dry state. They may be purchased at aquarium shops. Hatch the eggs using culture methods described in chapter 9.

Larvae may be observed in depression slides under the microscope (Fig. 6-26). Rim coverslips with petroleum jelly and slides may be used for hours; place them in a moist chamber to preserve them even longer. You may want to arrange a simple chamber by placing slides across Syracuse dishes containing water; then cover all these with a small bell jar, or place the dishes in a covered plastic box lined with moist toweling.

Metamorphosis of Drosophila

Culture fruit flies as described on pages 535–36. When flies are cultured in many small vials of medium they will lay eggs which students may examine closely with a hand lens or microscope. In fact, students may streak banana mash (or other medium) in plastic spoons or depression slides and insert them into bottles of fruit flies. Then some flies will lay eggs on the surface of these slides. With a hand lens, observe that the eggs are white, with two filaments attached to one end (Fig. 6-27). After examination, return the slides to the breeding bottles. In time, observe larval stages

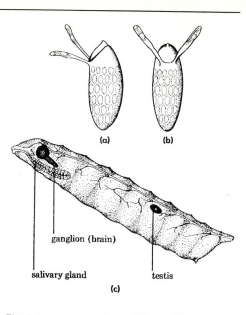

Fig. 6-27 Eggs and larva of *Drosophila melanogaster*: (a) side view of egg; (b) dorsal view of egg; (c) lateral view of third-instar larva.

and pupa formation on these slides (or spoons) and on the walls of the culture bottles. To show chromosomes in cells of salivary glands of third instar stages of *Drosophila* see preparation, p. 396. Show the entire life cycle of *Drosophila*, p. 536; also refer to heredity, p. 531.

Egg masses of praying mantis

Students may find the tan, foamy egg masses of the praying mantis, or purchase them from botanical gardens. When these masses are put into a terrarium they will hatch in early spring. Study the nymph stage of these insects which undergo an incomplete metamorphosis, that is, egg, nymph, and adult. Students may compare this with complete metamorphosis in such insects as fruit flies, beetles, butterflies and moths. (Also see chapter 9.)

Cocoons and chrysalises

Examine pupal stages of moths and butterflies by cutting open the protective coverings. Common forms include the bagworm (a small, spindle-shaped form covered with twig fragments that dangles on a stalk and the pupae of *Cecropia*, *Polyphemus*,

Fig. 6-26 *Artemia* (brine shrimp): larva stage. (Courtesy of General Biological Supply House, Inc., Chicago.)

Vanessa, and *Promethea.* Allow pupae to remain at room temperature in a terrarium with a moist sponge. Watch the colorful adults struggle to escape the pupa case at hatching. Students might trace the life cycles of the silkworm, as well as that of the honeybee. (Students may recall that male honeybees or drones are monoploid; they develop from parthenogenetic eggs.) The role of hormones in the metamorphosis of insects makes for fascinating studies (refer to Figure 5-23).

Earthworms   General directions for dissection are given in chapter 1. Holmes and Smith suggest using mature earthworms as a ready source for live sperm cells and ova.[23] Inspect the dorsal region of the worm to locate the swollen, prominent seminal vesicles (segments 9 to 12); also observe whether the clitellum is fully formed. Draw fluid with a hypodermic needle from the seminal vesicle and place in cold Ringer's solution; view this fluid in dimmed light under high power. To obtain ova, make an incision into the segments of the body containing ovaries (13 to 14). Since fertilization occurs within the cocoon, you may want to dissect out developing embryos from the cocoons. Dissect under water using a dissecting microscope.

Eggs and cleavage patterns   Introductory experimental embryology techniques are described here. However, students will find excellent diagrams of types of cleavage and fate maps of insect, mollusk, echinoderm, and vertebrate eggs in current editions of B. I. Balinsky, *An Introduction to Embryology,* 5th ed. (Philadelphia: Saunders, 1981). Students may also want to read the classic work of Holtfreter, Mangold, Spemann and others in R. Ham, and M. Veomett, *Mechanisms of Development* (St. Louis, MO: Mosby, 1980). Clear descriptions of many techniques are given in V. Hamburger, *A Manual of Experimental Embryology* (Chicago: University of Chica-

go Press, 1960), and in R. Rugh, *Experimental Embryology: Techniques and Procedures,* 3rd ed. (Minneapolis: Burgess, 1962).

Have students model cleavage stages (see model making, p. 751). Or try a technique using balloons and different colors of marking inks, showing three layers of a gastrula. Use a long balloon to bend into a gastrula stage.[24]

Eggs and sperms of snails   Maintain the aquatic snail, *Physa,* in an aquarium tank for a study of eggs, sperms, and developing embryos within the transparent shells. Dissect out the multiple-lobed ovotestis located in the uppermost portion of the spiral of a fair-sized snail. Macerate it a little in a drop of aquarium water on a clean slide. (Also refer to dissections, chapter 1.) Sperms as well as the spheroid eggs are visible under high power if the snails are sexually mature. Fertilized eggs may be found on leaves of aquarium plants or on the walls of the aquarium. Transfer eggs into welled slides or watch glasses and observe the early stages of cleavage under a microscope or dissecting microscope.

Prepared slides of cleavage stages in mollusks and echinoderms may be used to supplement these observations.

Differentiation of early embryos

Development of sea urchin embryos

Fertilization and early cleavage stages of living eggs of sea urchins can be observed and compared with prepared slides. How do bipennaria larvae of Asteroids compare with the pluteus larvae of the Echinoids (Fig. 6-28)? Early embryos of echinoderms resemble early differentiation among chordates.

To determine the sex of the sea urchins in advance of class activities, you

[23] D. F. Holmes and J. Smith, "Earthworms: A Source of Living Gametes," *Turtox News* (August 1962) 40 : 8.

[24] Described by C. Mohler, "Using Balloons as Embryological Models," *Turtox News* (April 1965) 43:4.

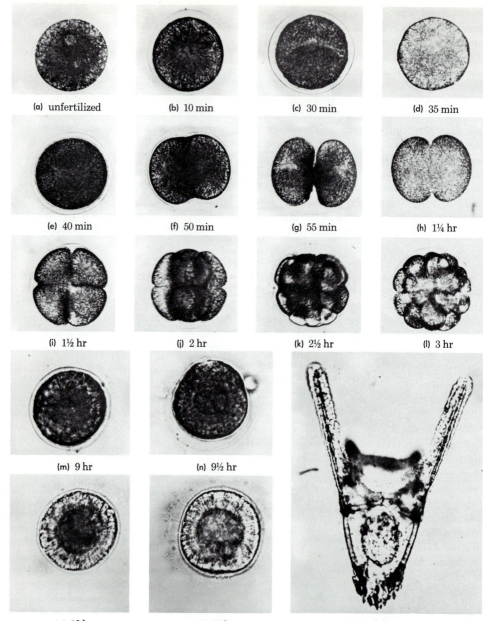

(a) unfertilized   (b) 10 min   (c) 30 min   (d) 35 min

(e) 40 min   (f) 50 min   (g) 55 min   (h) 1¼ hr

(i) 1½ hr   (j) 2 hr   (k) 2½ hr   (l) 3 hr

(m) 9 hr   (n) 9½ hr

(o) 12 hr   (p) 18 hr   (q) 2 days

Fig. 6-28 Fertilization, cleavage, and pluteus larva in *Arbacia punctulata*, a sea urchin. (Courtesy of Dr. Ethel Browne Harvey.)

may want to use a method developed by Harvey.[25] Place the oral region of a sea urchin in a glass dish and touch two electrodes (10 volts) to opposite sides of

[25] E. Harvey, "Electrical Method of Determining the Sex of Sea Urchins," *Nature* (1954) 173:86.

the animal. In less than a minute, observe whether white sperm fluid or orange eggs flow from the goniopores near the center of the dorsal or aboral surface. Label the specimens and maintain in a refrigerator at 2° C (36° F) until needed.

chordate *Amphio*:
dents to observe
Compare echino
terns of early cle
and differentiatio

Ovaries and test

smelts or perch
for dissection in
from the anal reg
Fig. 1-99). Rem
veal ovaries or
tubes, either ov
which carry the o
Use a hand lens
or ovaries. Show
caviar, cod roe, a
able as canned fo
some tropical fis

Embryology

Fish are splendid
investigations in
a complete descr
a timetable for
ment of severa
(*Brachydanio*), th
medaka (*Oryzi*
(*Xiphophorus* or *F*
He describes ted
of fish embryos
explants *in vitro*.
good discussion
addition, refer
photographs o
Japanese medak
Biological Suppl
beginner follows

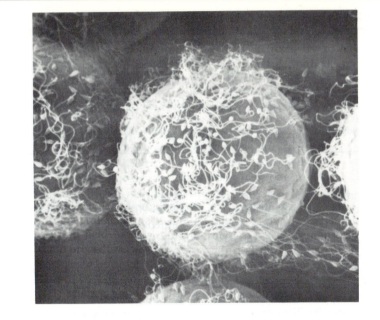

Fig. 6-29 Sea urchin egg with sperm cells as seen in scanning electron micrograph. (Photo courtesy Dr. Mia Tegner, Scripps Institute of Oceanography, University of California, San Diego.)

Shedding or spawning may also be induced in mature *Arbacia punctulata*, available on the East Coast, or in *Strongylocentrotus*, available on the West Coast. Use a syringe with a 25 gauge hypodermic needle to inject 2 ml of 0.5 *M* KCl solution[26] into the soft peristome tissue surrounding the oral region. The solution will reach the perivisceral region and stimulate shedding of hundreds of millions of eggs and billions of sperm cells within 2 min. Repeat the injection if there is no response.

Invert a shedding female sea urchin so that it rests on the rim of a 250 ml beaker of seawater. Observe the orange eggs as they flow out of the genital pores in the aboral genital plates into the beaker. When the eggs settle, pour off the water and wash the eggs three times with fresh seawater. Next make dilute suspensions of the eggs by transferring 5 drops of the eggs to a graduate cylinder; add seawater to bring the volume to 100 ml. These diluted suspensions can be maintained for 6 hr at about 5° C (41° F). To collect shedding sperms, invert a male

sea urchin over a *dry* Petri dish. Cover the dish and refrigerate until use.

To fertilize the eggs, dilute the sperm suspension to prevent polyspermy, which results in abnormal cleavage. Add 1 or 2 drops of dry sperm to 10 ml of seawater. (This sperm suspension must be used within half an hour.) First have students observe unfertilized eggs by adding a drop of the dilute egg suspension to a depression slide and a coverslip; examine the spherical eggs surrounded with jelly, and the red granules within the eggs (in *Arbacia*) due to echinochrome pigment.[27] Also try to locate the nucleus. After a rapid examination of the eggs, raise the coverslip and add a drop of sperm suspension. Note the activity of sperm cells in the presence of egg cells, as they lash about, causing eggs to rotate (Fig. 6-29). Work quickly. Within 2 to 5 min, observe the formation of the

[29] Refer to the
duction in fish in R.
Related Phenomena in
the Advancement
1959.
Also refer to W. S
Cycles, 2nd ed. (I
P. Korringa, "Rel
Periodicity in the Br
Monographs (1947)
[30] Rugh, *Experi*
[31] R. Kerchen a
ka—Its Care and I
Carolina Biological

[26] Add 37.3 g KCl to 1 liter distilled water.

[27] Procedures adapted from I. Sherman and V. Sherman, *The Invertebrates: Function and Form*, A Laboratory Guide, 2nd ed. (New York: Macmillan, 1976); P. Abramoff and R. Thomson, "Fertilization and Early Development of the Sea Urchin," *Freeman Library of Laboratory Separates* # 864 (San Francisco: Freeman, 1972); J. Crane Jr., *Introduction to Marine Biology* (Columbus, OH: Charles Merrill, 1973).

TABLE

Genus

Arbacia
Strongy
Dendra:

* P.
1972).

fertiliz
(under
brane
5 sec
This h
fertiliz
observ
eggs ir
Within
parent
egg th
togeth
60 mi
cleava
cleava
divisic
first c
the p
Table
blastu
tion ir
a blas
forme
fertiliz
tually
Abc
result:
teus l
Since
tempe
lethal
tion c
fertiliz
within
covers
or ri
evapo
labor:
In
morpl

Fig. 6-33 *Xenopus*, the South African clawed frog. (Photo courtesy Carolina Biological Supply Co.)

and early tadpoles are easy to study. Tadpoles are practically transparent so that students can observe blood vessels, blood cells, study heartbeat, and optic nerves.

Tadpoles also may be treated with thyroxin in a concentration of 1 part thyroxin to 7 million parts water. An excellent booklet, *Xenopus: Care and Culture*, is available from Carolina Biological Supply Co., prepared by D. Thompson and R. Franks, 1978.

Spring peepers In some regions, these small frogs may be more available than *Rana*. They also develop more rapidly. In addition, tadpoles may be treated with thyroxin to speed metamorphosis. Dissolve 10 mg of thyroxin in 5 ml of a 1 percent solution of sodium hydroxide and bring to a liter with distilled water. This makes a *stock* solution of 1:100,000 concentration. Refrigerate the stock solution; dilute to 1:1,000,000 before use or prepare varying dilutions.

Mexican axolotl Refer to Ward's ser-

vice bulletin on rearing of axolotl in the laboratory.[34]

Effect of temperature on developing frog eggs Have students establish the rate of cleavage at some constant temperature and then compare the rate of cleavage and differentiation of early embryos at different temperatures. Refer to Shumway's chart, Figure 6-34, and also to Figure 6-35. Maintain finger bowls of fertilized eggs in frog Ringer's solution at 15° C (59° F) as a control. Prepare similar bowls at 10° C, 20° C, and 30° C (50° F, 68° F, and 86° F). Then examine the eggs at intervals. Gastrulation begins in approximately 27 hr at 15° C; in about 72 hr at 10° C.[35]

What other environmental factors may affect the development of frog eggs? Does crowding affect development? Aeration providing extra oxygen? Drugs? Chemi-

[34] Ward's Natural Science Establishment, Inc., Rochester, NY or Monterey, CA.
[35] R. Rugh, *Experimental Embryology*, Chapter 11, "Temperature and Embryonic Development."
Also refer to F. Moog, *Animal Growth and Development* (BSCS laboratory block) (Boston: Heath, 1963).

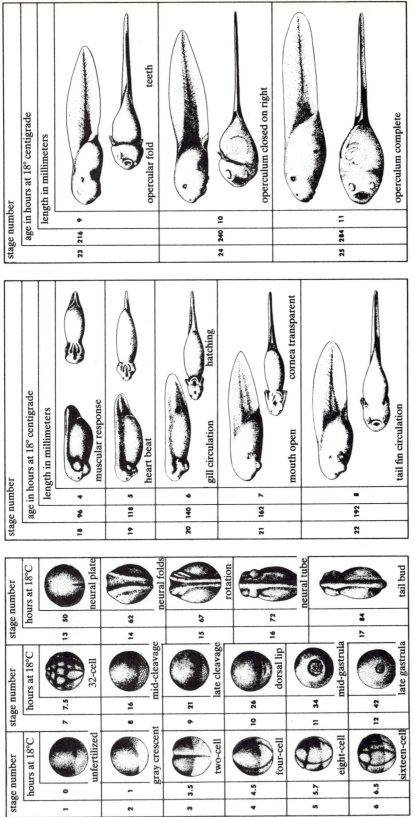

Fig. 6-34 Shumway's chart with drawings of stages in the development of *Rana pipiens*. (Adapted from *The Anatomical Record*, vol. 78, no. 2, 1940.)

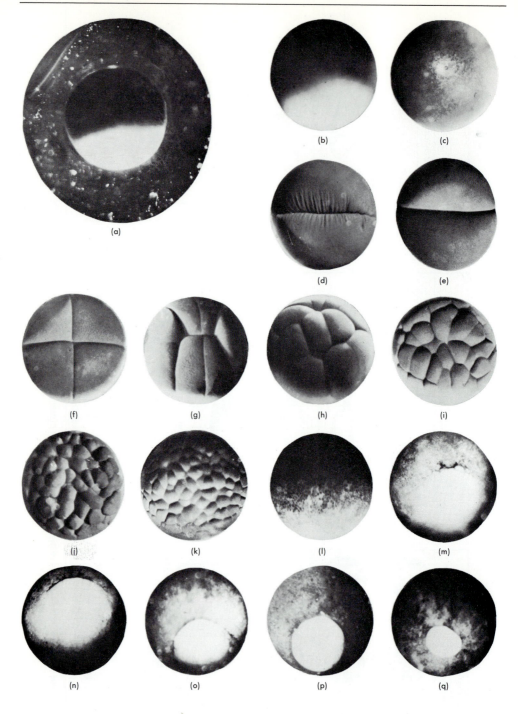

Fig. 6-35 Stages in the development of amphibian egg and early tadpole stages: (a) unfertilized egg; (b) fertilized, gray crescent; (c) polar body formation; (d) and (e) first cleavage; (f) second cleavage, 4 cells; (g) third cleavage, 8 cells; (h) fourth cleavage, 16 cells; (i) fifth cleavage, 32 cells; (j) temporary morula, 64 + cells; (k) early blastula; (l) late blastula, epiboly; (m) early dorsal lip; (n) and (o) active gastrulation; (p) and (q) yolk plug formation;

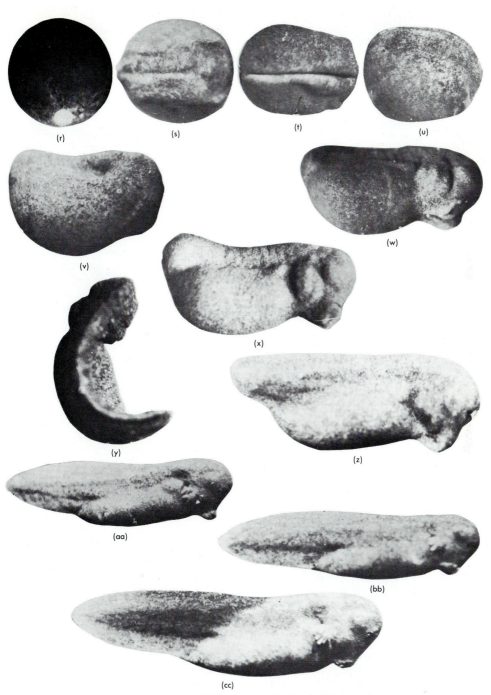

Fig. 6-35 (continued)   (r) disappearing yolk plug; (s) neural folds; (t) neural groove, rotation; (u) neurula—neural tube closed; (v) early tail bud; (w) muscular response; (x) initial heartbeat, myotomes; (y) and (z) hatching, gill development; (aa) heartbeat; (bb) gill circulation; (cc) mouth open;

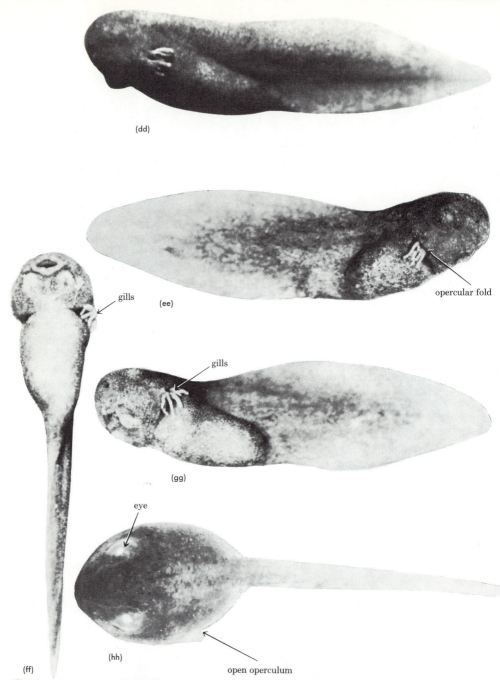

(dd)

gills (ee)

opercular fold

gills

(gg)

(ff)

eye

(hh)

open operculum

Fig. 6-35 (continued) (dd) tail fin circulation; (ee) opercular fold and teeth, right side; (ff) opercular fold and teeth, ventral view; (gg) opercular fold and teeth, left side; (hh) operculum complete. (a–cc, ee–hh, from R. Rugh, *Experimental Embryology: Techniques and Procedures*, 3rd ed., Burgess Publishing Co., 1962; dd, from *Vertebrate Embryology: The Dynamics of Development* by Roberts Rugh, © 1964, by Harcourt, Brace & World, Inc., and reproduced with their permission.)

cals? Rugh's text has photographs of several types of abnormalities that may develop

Embryos to tadpoles After the initial rapid growth of the hind limbs, major dramatic changes are visible when the forelimbs emerge and the tail is reabsorbed. Have students record these changes: growth in hind limbs in millimeters; thinning of the skin so that gill regions begin to show where forelimbs will emerge; breakthrough of limbs—the left one first—through the spiracle; absorption of the tail; reabsorption of lips and loss of horny teeth; larger eyes and mouth; formation of a tympanum on each side of the head. Note the changes in body shape from oval to angular as water loss occurs.

Tadpoles show a high mortality rate during this crucial period when there is a shift from gills to lungs. You may want students to dissect those that die to observe the lungs, and the change from a long "watch-spring" intestine to a shorter, differentiated organ system.

When students know the normal "time table" of development, plan investigations of the effect of thyroxin and/or iodine on early tadpoles (p. 421). Compare the normal development with that of tadpoles in varying concentrations of thyroxin and/or iodine. Graph data when possible. On the ordinate, place the change dependent variable, and on the horizontal axis, or abscissa, place time, concentration, or some other independent variable. Or use a histogram to show graphically the effect of different concentrations by the height of vertical lines on graph paper.

Induced ovulation in Rana pipiens

Certain amphibians can be stimulated to ovulate out of season when injected with pituitary glands. In late summer and fall, large doses are needed, but from October into spring the dose needed decreases (Table 6-5). While there is no qualitative difference between the glands in the two sexes, the female's is larger and

TABLE 6-5 Number of pituitary glands needed to stimulate ovulation

Sept.–Jan.	10 male or 5 female glands
Feb.	8 male or 4 female glands
Mar.	5 male or 3 female glands
Apr.	4 male or 2 female glands

likely contains more of the follicle-stimulating hormone. Two pituitaries from a male are about the equivalent of one from a female. You may prefer to purchase pituitary extract from a drugstore (it need not come from a frog).

Inject pituitary glands through a wide bore hypodermic needle under the skin into the abdominal cavity of mature females two days before eggs are to be obtained. A description of the technique for obtaining the pituitary glands follows.

Excising of the pituitary The pituitary gland is located just posterior to the level of the eyes of the frog (Fig. 6-36). It is necessary to cut off the upper head and jaw region of a killed frog to reach the pituitary gland. Insert sharp scissors into the mouth at right angles to the jaw, and quickly cut just back of the tympanic membranes to sever the cranium at its junction with the vertebral column. Then cut the base of the skull (the upper palate) on either side by inserting the point of the scissors through the foramen magnum (the opening into the cranium); turn the point

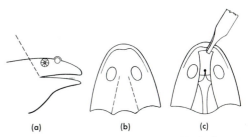

(a) (b) (c)

Fig. 6-36 Locating the pituitary gland in the frog: (a) cut off the upper jaw and the head; (b) view of upper palate: insert scissors into back of cranial cavity as far to the right as possible, and cut through bone; repeat on left side; (c) pull up flap with forceps, exposing pituitary gland (black ball). (Courtesy Carolina Biological Supply Co.)

toward the underside of the orbit. Turn back the base of the skull and hold it in position with the left thumb. Look for a small, ovoid, pink pituitary gland adhering to the turned base of the back of the skull (the ventral surface). Frequently the gland is surrounded by flocculent tissue. Remove the gland, and store it in 70 percent alcohol if it is not to be used immediately.

Mash the necessary number of whole glands in 2 ml of 0.1 percent frog Ringer's solution (p. 155) or 10 percent Holtfreter's solution.[36] Use a needle of rather wide bore to pick up this material and inject it into the abdominal cavity lymph spaces lying directly under the loose skin. Inject along the side of the animal—be careful to avoid injury to underlying organs. Rub the finger over the injected area after you remove the needle to keep the injected material from running out. Keep the injected female in a battery jar at 15° C (59° F) in about 3 cm of water. If the frogs are kept at a higher temperature, say 23° C (73° F), you will find that the eggs are often abnormal (in addition, the reserve food is too rapidly consumed).

If enough frogs are available, you may want students to dissect one of the ovulating females to study eggs in all stages of development. Cilia that line the coelom propel the eggs to the ostium. There are also cilia on the edges of the liver, and on the pericardium and mesovarium. No cilia are found in males or immature females. (Possibly, the presence or absence of cilia may be considered a secondary sex characteristic.)

Artificial fertilization of frog eggs

After 36 hr at 15° C (59° F) the eggs of the pituitary-injected female may be stripped.

---

[36] Into 1 liter of distilled water, dissolve 3.5 g sodium chloride, 0.05 g potassium chloride, and 0.2 g sodium bicarbonate. Dilute this stock solution with distilled water (1:9).

It is certain, however, that after 72 hr most of the eggs will be in the uterus. (At 15° C the eggs can be left in the body for five days after injection.) Gently hold a female and apply pressure to both sides of the abdomen and press in the direction of the cloaca. Strip the eggs in ribbons into large, dry dishes so that the eggs are well separated. Be sure that all glassware is biologically clean. Then with a *clean* pipette, wash the eggs with 0.1 percent frog Ringer's solution for a few minutes. Finally, draw off the Ringer's solution, and pipette previously prepared sperm suspension over the eggs.

Make the sperm suspension earlier so that it can stand for about 15 min, to give the sperms time to become activated. To obtain the suspension, kill two males (save their pituitary glands in 70 percent alcohol) and then dissect out the testes. Roll the testes on toweling to remove blood and body fluids; wash them in 0.1 percent frog Ringer's solution.

Macerate the four testes in 50 ml of 0.1 percent frog Ringer's solution or 10 percent Holtfreter's and let stand. This amount of sperm suspension will be sufficient to fertilize about 1000 eggs. If it is necessary to save the testes you can keep them for a few days in wet cotton at a temperature of from 4° to 8° C (39° to 46° F).

Pipette the sperm suspension over the eggs and shake the dish occasionally. Let this stand for about 20 min. Finally, wash the eggs with more 0.1 percent Ringer's and flood the dish so that the eggs float freely. After half an hour students will be able to see which eggs have been fertilized, because the fertilized eggs rotate within the jelly layers so that the black surfaces are uppermost. When the jelly is completely swollen (after about an hour), cut the clusters of eggs apart with scissors and separate them into smaller finger bowls. Put no more than 50 eggs in each finger bowl in Ringer's solution and keep at 15° C (59° F). Discard the unfertilized eggs before they begin to decay. Transfer the eggs into fresh Ringer's solution twice a week.

Artificial parthenogenesis in frog eggs

Certain eggs can be made to begin cleavage without the entrance of a sperm cell. Rugh describes a technique for stimulating unfertilized frogs' eggs to develop.[37] Segregate the female to be used from males for several days prior to the work. Boil the instruments and glassware or wash them in alcohol to be sure that they are clean and free of sperm cells. If the work is to be done at a time other than the breeding season, you will first need to induce ovulation (p. 419).

When eggs are available, parthenogenesis can be induced in this way: Strip the eggs in single file along the length of clean glass slides. Prepare several slides with rows of eggs. The slides may be kept in a moist chamber for about an hour, if necessary, until ready for use. A suitable moist chamber may be prepared by setting the slides over Syracuse dishes of water and covering with a bell jar. Now pith a nonovulating female (see Fig. 5-12). Dissect it to expose the heart. Cut off the tip of the ventricle and let the blood flow into the coelomic fluid. Next, dissect out a strip of abdominal muscle and dip it into the mixture of blood and coelomic fluid. Then streak each egg so that a film of blood-coelomic fluid remains on each egg. Now prick each egg with a glass or platinum needle within the animal hemisphere (that is, the black area), just off center. Finally, immerse the slides with the eggs on them into spring water or distilled water. Keep the eggs in Petri dishes with just enough fluid to cover the eggs. Keep them in a cool place as you examine the eggs hourly under a dissecting microscope or hand lens for signs of cleavage stages.

Speeding metamorphosis

Effect of feeding thyroxin to tadpoles

Light is necessary to stimulate the secretion of thyrotropic hormone from the amphibian pituitary. In turn, the hormone thyroxin stimulates the metamorphosis of the tadpole into an adult frog. Bullfrog tadpoles with hind limbs appearing give the best results in this study. Set up finger bowls containing different concentrations of thyroxin. Use a concentration of 1 part of thyroxin to 5 million parts of water in one series. In another set, use 1 part of thyroxin to 10 million parts of water. (The hormone is absorbed directly through the skin of the tadpoles.)

Put individual large bullfrog tadpoles or five to ten small frog tadpoles in separate finger bowls. Set aside one series as controls (use spring or pond water without thyroxin); establish two series using different concentrations of thyroxin.

Rugh suggests preparing stock solutions of thyroxin by dissolving 10 mg of thyroxin (crystalline) in 5 ml of 1 percent sodium hydroxide.[38] Dilute to a 1-liter volume with distilled water. This 1:100,000 concentration should be refrigerated until it is used. Dilute this stock solution to prepare the solutions needed for the series of finger bowls.

Feed tadpoles on alternate days with small amounts of hard-boiled egg yolk or washed canned spinach, and change the water each week to prevent fouling of the medium.

Hamburger notes that at room temperature, the first changes occur after two to three days; tadpoles in solutions of 1:100,000,000 showed changes in four days and lived for two weeks.[39]

Rugh also recommends another method using thyroxin tablets. When these are available, crush and dissolve five 2-grain tablets in 5 ml of distilled water in a mortar. Weigh out an equal amount of whole wheat flour and grind it up with the thyroxin tablets. Then spread this paste in a thin layer on glass squares and leave to dry. Finally, powder the dry mixture and

---

[37] R. Rugh, *Experimental Embryology*. Also refer to F. Moog, *Animal Growth and Development* (BSCS Laboratory Block) (Boston: Heath, 1963).

[38] Rugh, *Experimental Embryology*.
[39] Viktor Hamburger, *A Manual of Experimental Embryology*, rev. ed. (Chicago: University of Chicago Press, 1960).

store it in closed bottles in a refrigerator. Use 50 mg of this wheat-thyroxin powder per tadpole daily for one week. Feed the tadpoles, both controls and the experimental animals, with washed, canned spinach or boiled lettuce. Change the medium daily to prevent fouling.

Have students record changes in the growth of the tail, body, and limbs (in millimeters); also note changes in shape of the mouth and body contour. Transfer a tadpole to be measured into a Petri dish with a transparent millimeter ruler underneath. How does the growth rate compare with normal metamorphosis in tadpoles?

Effect of feeding iodine to frog tadpoles Since the main fraction in thyroxin is iodine, students may devise an investigation of the effects of iodine on the metamorphosis of amphibian tadpoles.

Here again, work with bullfrog tadpoles is most successful, but any species of frog, or the salamander *Necturus* may be used. No injections are necessary, since iodine is absorbed directly through the skin. Select tadpoles in which the hind limbs are just becoming visible, 1 mm or so. Set up several finger bowls with a maximum of 20 tadpoles of small frogs such as leopard frogs, or one tadpole of a bullfrog, in about 30 ml of prepared medium. Prepare several finger bowls with tadpoles in spring water or pond water as controls.

In the experimental dishes use the following medium. First, prepare a stock solution: Dissolve 0.1 g of iodine (crystalline) in 5 ml of a 95 percent solution of alcohol; then dilute to 1 liter with distilled water. This is a stock solution of a concentration of 1 : 10,000.

Dilutions as weak as 1 : 500,000 and 1 : 1,000,000 have been successful in stimulating metamorphosis. Each week change the medium to prevent fouling due to decay of food. Treat the tadpoles for 7 to 14 days. However, avoid more frequent changes, for too much iodine will result in such accelerated growth that tadpoles may die. On alternate days feed the controls as well as the experimental animals with small amounts of hard-boiled egg yolk, cooked lettuce, or spinach. Rugh recommends a square about 3 × 3 cm of spinach leaf per tadpole.[40] Remove uneaten food. Record changes in the length of tail and hind limbs and the time of appearance of forelimbs in both the experimental and the control animals. Also note the changes in the shape of the head and body. Normally, larvae of *Rana pipiens* (leopard frog) reach stages of metamorphosis about 75 days after fertilization of eggs (maintained at 23° to 25° C (73° to 77° F).

Many techniques for club work or individual investigations in experimental embryology are available in the excellent text manuals by R. Rugh in *Experimental Embryology*, and by V. Hamburger in *Manual of Experimental Embryology*. Both texts give excellent directions for preparing glass operating needles and for making transplants of tissue from donor to host early tadpoles. At what time is tissue "determined"—a future eye, gill, or limb? An introduction to this type of technique is given on page 424. Students should be forewarned that the techniques are difficult and require much practice before any degree of success can be expected—but the successes are wondrous to behold.

Differentiation in frog embryos

Early amphibian development

How does a cell in a developing gastrula, having a complement of the same chromosomes or DNA as other cells, become differentiated, developing specificity of function and structure?

Students may want to read the classic, elegant experiments of Roux, Vogt, Mangold, Spemann, and others, in Hamburger's or Rugh's texts or in *Mechanisms of Development* by R. Ham and M. Veomett (St. Louis, MO: Mosby, 1980).

With a hot needle, prick one cell of a two-celled stage of a frog or salamander

[40] Rugh, *Experimental Embryology*.

egg. What are the subsequent cleavage patterns of the eggs? Are the results the same if the "killed" cell is removed from the healthy cell?

Techniques for centrifuging eggs, for constricting early cleavage stages with ligature, and for providing a restrictive environment by partially embedding early embryos in agar are developed in Hamburger's and Rugh's texts (cited above); also refer to the formation of double embryos and of Vogt's experiments in applying vital stains to trace the "fate maps" of developing eggs.[41]

You may want to provide offprints of J. Gurdon's paper, "Transplanted Nuclei and Cell Differentiation," *Sci. Am.* (December 1968). What happens when a nucleus of an intestinal cell of a tadpole is transplanted into a frog's unfertilized egg whose nucleus has been destroyed?

Embryonic induction

Mesodermal tissue underlying ectoderm in the late blastula stage of amphibians influences or causes the differentiation of the overlying ectoderm into nervous system, eye, or other ectodermal derivatives. This is, in very elementary terms, the essence of embryonic induction. When Spemann transplanted tissue of the dorsal lip of the blastopore of a very early gastrula into a host in the gastrula stage, the transplanted section of dorsal lip caused organization and differentiation of a brain and spinal cord. Thus, the host had its own original organization plus an imposed "smaller embryo" growing on it. Spemann referred to the dorsal lip region as the "organizer" in the induction of tissues, leading to differentiation of organ systems.

Time is a factor in induction and differentiation. At a very early stage, there seems to be a faint surface etching of outlines of organs—tissue has a presumptive fate (Figs. 6-37 and 6-38). Yet, if

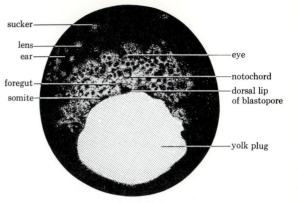

Fig. 6-37   Gastrula stage of frog showing anlage for several organs. (After R. Rugh, *Experimental Embryology: Techniques and Procedures*, 3rd ed., Burgess, 1962.)

such tissue from a presumptive, or future, eye region (anlage for eye) is transplanted to the belly region of a host in the tail bud stage (Fig. 6-35), the tissue differentiates into skin similar to the surrounding tissue. In the early gastrula, however, the presumptive fate is fixed, or determined, and it is boldly drawn so that, when transplanted, an anlage for an eye differentiates into an eye regardless of its environs.

Rugh[42] and Hamburger[43] describe specific techniques for investigations to test the organizing potencies of the dorsal lip of the blastopore. Urodele gastrulae withstand these operations more successfully than do the anuran embryos.

These techniques require considerable practice, maturity, and skill, and, although students enjoy reading these fascinating studies, only persevering students achieve success in these techniques.

Students may, however, subject eggs to different temperatures and pressures and to centrifuging. Or they may try to constrict eggs with a hair loop, possibly try to use vital stains, and to experience the "awe" of watching cleavage stages, gastru-

[41] Refer to classic papers in M. Gabriel and S. Fogel, eds., *Great Experiments in Biology* (Englewood Cliffs, NJ: Prentice-Hall, 1955).

[42] Rugh, *Experimental Embryology.*
[43] Hamburger, *A Manual of Experimental Embryology.*

lation, and formation of a neural tube in a living mass of cells. Hamburger and Rugh (previously cited) give splendid directions and illustrations for these many techniques, and for transplanting pieces of the dorsal lip and for eye and limb field operations. Other possible investigations utilize fish eggs, chick embryos, and even early mouse embryos.

To give a view of the delicate manipulations required in this work, we will briefly describe a technique for transplanting anlage for gills from an early embryo (Fig. 6-35) to an older host, such as stages (v) or (w) in Figure 6-35. All three germ layers contribute to the formation of the external gills.

Making a transplant

Select appropriate donor and host embryos. Remove the jelly and membranes as follows. Cut an egg apart from an egg mass with a pair of scissors. With jeweler's forceps, gently grasp an egg and roll it along a piece of toweling to remove most of the jelly. *Be gentle:* Do not injure the egg. Next, with a pair of finely pointed jeweler's forceps, hold onto the vitelline membrane and, with a second forceps, carefully pull or prick the membrane. When done correctly, the egg or embryo will pop out of the encasements of membrane and jelly; however, many embryos are jabbed or squashed in this procedure. Now the delicate embryos must be

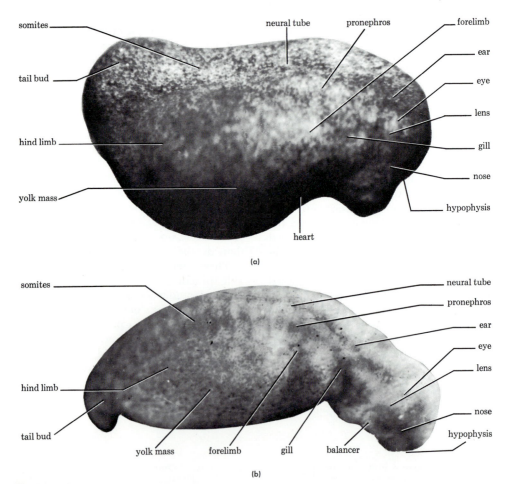

(a)

(b)

Fig. 6-38 Organ fields in amphibian embryos: (a) *Rena pipiens*; (b) *Amblystoma punctatum*.

(From R. Rugh, Experimental Embryology: Techniques and Procedures, 3rd ed., Burgess, 1962.)

handled and transferred with a hair loop (described below) or with a glass pipette of wide enough bore (see Fig. 6-40). Operating dishes are either salt cellars or Syracuse dishes embedded with plain agar, wax to which lampblack has been added to reduce the glare, or Permoplast clay. With a glass ball at the end of a glass rod, shape two depressions in the embedding material used in the operating dishes. With the hair loop or the glass pipette, transfer the donor and the host to the depressions in the operating dish (Fig. 6-39). Add enough operating solution (described below) to the dish to cover the embryos.

First prepare the host embryo to receive the transplant by cutting out a rectangular piece of tissue with a glass needle (preparation, p. 427). Remove ectoderm with the underlying mesoderm, from either the tail bud region or the belly region. In fact, plan several operations, transplanting gill anlage into different sites in different hosts. With a hair loop, cut deeply enough to accommodate the rather thick transplant. The transplant must fit snugly into this area, and the operation must be done quickly since the cells are alive and mitosis proceeds rapidly.

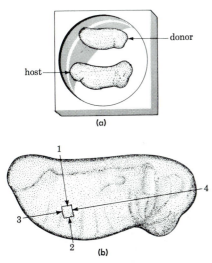

Fig. 6-39 Transplanting anlage to host: (a) donor and host in operating dish; (b) parallel cuts made to remove tissue from host so that transplanted anlage from donor can be inserted into belly or flank of host.

Using the glass operating needle and the hair loop, cut a similarly sized rectangle of tissue from the gill swelling in the donor. This gill swelling is located in line with the eye, directly below the first few somites (see Figs. 6-35 and 6-38). The tissue transplanted from the donor should include ectoderm with underlying mesoderm; transfer it, without changing its orientation, to the host. Use both the needle and the loop to insert the transplant carefully so as not to injure the tissue. Then use a coverslip bridge to hold the graft in place for about 30 min (for preparation, see directions below.) If any extra "fringes" of tissue remain around the wound after ½ hr, carefully trim them with a hair loop. Finally, transfer the host to fresh medium—which Rugh calls "growing medium" in contrast to "operating medium" (recipe, below)—and maintain in a cool room, or in a refrigerator at 10° to 20° C (50° to 68° F).

Also maintain the donors to examine the region from which the anlage for gills was removed; use the other side of the donor embryos as controls. Compare the rate of growth of the gills in the original donor with the rate of growth of the transplanted tissue in the host. Record observations, including drawings or photographs.

Solutions and operating tools

Although contamination by bacteria is not as great a danger in fast growing amphibian tissue maintained at cool temperatures, sterile procedures should be followed in handling the embryos. The agar in which the embryos are embedded is plain agar, not nutrient agar; wax or Permoplast clay may be preferred.[44]

Operating solution[45]   Operating solutions are usually more concentrated than solutions in which embryos are left to grow. A dilute solution (0.5 percent) of sodium sulfadiazine may be added to the

---

[44] Permoplast is available from American Art Clay Co., Indianapolis, IN.
[45] Rugh, *Experimental Embryology.*

solutions to inhibit growth of contaminants; it has been found harmless to embryos.

FOR URODELES Prepare quantities of the following stock solution; autoclave and store.

spring water[46]	10 liters
NaCl	70 g
KCl	1 g
CaCl$_2$	2 g

For an operating medium, dilute 2 parts of stock solution in 1 part of spring water. This is a hypertonic solution.

For longer periods for growth, prepare a more isotonic solution. Mix 2 parts of stock solution to 4 parts of spring water.

FOR ANURANS: HOLTFRETER'S SOLUTION For an operating and growing medium, use Holtfreter's stock solution. To 1 liter of distilled water add

NaCl	3.5 g
KCl	0.05 g
CaCl$_2$	0.1 g
NaHCO$_3$	0.02 g

This stock may be diluted to a 10 percent solution by adding 1 part of stock solution to 9 parts of distilled water. This solution is especially useful in the study of sperm cells in frogs (p. 413).

FOR FROG EMBRYOS For an operating medium, prepare Holtfreter's solution *double strength*; autoclave and store.

Light source A strong light source is needed for operating on early embryo stages. To reduce the heat of the lamp, place a round Florence flask filled with water to which a few crystals of copper sulfate have been added, between the light and the operating dish. Support the flask on a ring stand.

Microburner A microburner is needed to prepare the glass needles and hair loops used in the above operations. Select a 15 cm length of soft glass tubing (7 mm diameter). Pull out one end in a flame to a

diameter of about 1 mm. Make a right angle bend in the center of the length of tubing, as in Figure 6-40. Allow to cool, and insert the more pointed end through a cork supported in a burette clamp on a ring stand. Arrange the cork and tubing so that it may be tilted in several directions for ease in making glass needles. Use rubber tubing to connect the large end of the glass tubing to a source of gas. Apply a screw clamp to the rubber tubing so that you can control the amount of gas passing through the thinner tubing. Light the microburner. It should be placed fairly close to the desk or work table since students will find it easier to twirl glass tubing and make 45° glass needles with both arms supported against a desk.

Hair loops Prepare hair loops from 10-cm lengths of 6-mm-bore glass tubing. Pull out one end of the tubing in a microflame so that the opening is reduced to

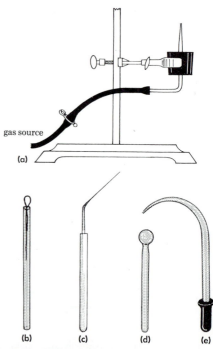

Fig. 6-40 Microburner and operating tools: (a) microburner; (b) hair loop; (c) glass needle; (d) glass ball roller; (e) transfer pipette.

---

[46] *Conditioned* tap water may be substituted.

about 1 mm; close the opening in the other end. Obtain small lengths of fine hair, preferably baby's hair, and thread each end of a 2.5-cm length into the tubing, leaving a very small loop (about 3 mm) protruding from the glass tubing. Secure the loop in place by dipping this end into warm wax (beeswax or paraffin); allow some of the wax to run up the capillary so that a good seal is made. Should an excess of wax adhere to the hair loop, touch the loop lightly to a warmed slide to remove the excess wax. Prepare several of these hair loops and store in a test tube with a cotton stopper to avoid contamination by dust in the laboratory.

Glass needles   Use 10-cm lengths of 6-mm glass rods to prepare several needle holders. In a flame, draw out the glass rod to a tapered end. Insert this tapered end into the flame again; the glass melts and balls up, forming a sphere. This is the region into which the needle (to be prepared next) will be inserted.

In the microflame, pull out lengths of 4-mm soft glass rods to a thickness of 1 mm. Twirl 7- to 8-cm lengths of this fine thread of glass in the microflame to melt the center; support your arms against the work table and pull out a bit and then quickly downward so that the apex of a triangle is formed. Quickly break this thread with a quick brush near the flame. Allow to cool. If a small ball of glass has formed at the tip, break it off with your fingernail. For pointed needles, the pulled thread of glass should be pulled apart completely. Check the point of the needles under a microscope.

Rugh suggests another method for preparing needles.[47] Pull out lengths of glass rod to 1 mm thickness (as above), and put a hook or bend at the end of each length. Then hang this hook over the metal ring of a ring stand in such a way that the glass is directly over a 20 × 60 mm glass vial containing a small bit of cotton at the bottom. At a point 2 cm from the bottom

of the hanging glass rod apply the microburner flame from one side. As the glass melts, the weight of the hanging rod will draw the end of a fine point and will cause the rod to drop into the glass vial.

Now the needles are ready to be inserted into the glass rod holders that were prepared earlier. Heat the glass ball at the tip of the holder and, when it is quite soft, insert the needle (held in a forceps) in a straight line with the holder. Allow to cool. Stand the completed needles temporarily in small balls of clay.

In use, the needles are broken very easily, so many needles should be prepared at one time. They can be stored in wide-mouthed vials containing a cushion of cotton. Stand the vials upright in a test tube holder or in an appropriate block of wood. Some workers prepare and store the proper lengths of 1-mm diameter glass rods; then they prepare the needles in a microflame as they are needed.

Glass ball rollers   A length of glass rod with a round ball at one end is needed to shape depressions in beeswax, paraffin, or agar in the operating dishes (Fig. 6-40d).

Glass transfer pipettes   Pull out several lengths of glass tubing to obtain pipettes with different sized bores. A slight twist in the bend of the pipette enables the operator to insert the pipette at an angle into a dish, thereby increasing the ease of manipulation. Twist the glass as it becomes soft by spreading or turning your hands so that the palms are upward a bit more; this will bend the soft glass. Or let the weight of the glass produce a bend; then pull out a short length. Some pipettes should have bores wide enough to pick up eggs (see Fig. 10-3); others should have smaller bores for transferring tissue to be transplanted (Fig. 6-40e).

Glass coverslip bridges   With forceps, hold one edge of a clean coverslip in a microflame until it is softened along its center line. Then lay the soft glass over a thin length of glass rod, causing a slight

[47] Rugh, *Experimental Embryology.*

bend in the coverslip. Allow to cool. The thickness of the glass rod, of course, determines the degree of bend. Only a slight bend is needed to apply pressure to a transplant to hold it in place as the graft grows together. Practice will show how much of a bend is necessary. Remember that the embryo is in a depression in the operating dish, so that often a bend is not essential.

All glass materials may be stored in Petri dishes or vials containing cotton. They may be cleaned in alcohol and then dipped in operating solutions containing 0.5 percent sodium sulfadiazine.

Development and differentiation in chick embryos

It is fascinating to watch a chick develop from a speck on the yolk of the egg cell. Fertilized eggs may be incubated in the laboratory so that students can examine developing embryos over the 3-week period to hatching. (Students can also build their own incubator.) If you stagger incubation of the eggs, you will have eggs in the incubator at different stages of development; these can all be studied in one laboratory period, or over several that you and your students plan.

You may want to candle eggs to see if embryos are developing. Place each egg over a box with a 3-cm hole at the top. Inside the box place a lamp with a 150-watt bulb. Work in a darkened room. The appearance of branching blood vessels indicates a live embryo.

Compare living embryos with prepared slides of whole mounts (preparation, p. 171); also use Figs. 2-28, 2-29 as guides. You may also want to determine the time of appearance of certain enzymes, such as cytochrome oxidase using NADI reagent (p. 188), and/or examine the role of a dehydrogenase in chick embryos (p. 187).

Examining incubated eggs   Some companies guarantee eggs up to 90 percent fertility, but seasonal variations lower this

figure. You cannot store fertilized chicken eggs in the laboratory before incubation unless you can get temperatures as low as 10° to 15° C (50° to 59° F).

Keep the incubator at 37° C (99° F) and insert the bulb of the thermometer at the level of the eggs, not high up in the incubator, for there may be a difference as great as 10° C between the two locations. Include a pan of water in the incubator to keep a uniform humidity. A relative humidity of 60 percent is optimum. Do not wash the egg surface; washing will remove a protective film which reduces bacterial infection. Turn the eggs daily to prevent adhesion of membranes. Students might mark the eggs with a pencil so that they know which surfaces to turn.

There seem to be two periods when mortality is high: the third day, and just before hatching time. Some mortality must be expected, for some hens produce fewer viable eggs. And you will want to allow for variations in handling.

After eggs have been incubated for at least 36 hr they are relatively easy to handle for examination. The number of hours of incubation is not a true index of the age of the embryos since cleavage started in the oviduct before the egg was laid. In addition, it may take up to 4 hr for a cold egg to reach the temperature of the incubator.

As you remove an egg from the incubator, hold it in the same position so that the blastoderm is floating on top of the heavier yolk.

48-hr chick embryo   Crack the egg on the edge of a finger bowl containing slightly warmed saline solution (or Locke's or chick Ringer's solution, pp. 156, 429). Place finger bowls of solution in the incubator beforehand, so that they are at the proper temperature and the embryo will not be chilled. Let the egg contents flow into the saline solution so that the embryo is completely submerged. Notice how the chick blastoderm rotates to the top position. When the embryos are larger they do not rotate as easily, but must be moved with forceps by pulling on the chalazae, the thickened albumin "ropes" on each side of the egg yolk mass.

Use a hand lens or dissecting microscope to examine the early embryo. In a 40-hr embryo you should begin to see the heartbeat and after 45-hr, a complete blood circulation should be discernible. How rapid is the heartbeat?

CHICK RINGER'S SOLUTION   Weigh the following salts and dissolve in distilled water to make up to 1 liter:

NaCl	9.0 g
KCl	0.4 g
CaCl$_2$	0.24 g
NaHCO$_3$	0.2 g

Have students locate the pairs of somites (27 pairs in the 48-hr chick); note that the embryo is about 7-mm long floating on top of the yolk. With a lens trace the blood circulating in developed blood vessels. At this stage the head end has twisted so that the head is viewed from the right side while the left side lies closer to the yolk (Fig. 2-28). Locate the amnion, vitelline arteries, and neural tube visible at the tail end. Students should recognize the twisted heart of two chambers and note the close proximity of the brain which now has differentiated into five divisions. Also note the large optic cup and the lens (Fig. 2-28). If possible, compare these dramatic changes of the 48-hr chick with the 24-hr embryo (Fig. 2-29).

Removing the embryo from the yolk   For a clearer view of the parts, remove the 48-hr chick embryo from the yolk and place it in a Petri dish, welled slide, or Syracuse dish of warm saline or chick Ringer's solution for viewing under low power of the microscope. (Be sure to replace the saline with warm saline during the study of the embryo.)

Cut a filter paper ring so that its inner diameter is a little smaller than the blood sinus around the embryo. The outer diameter of the paper ring should fit the diameter of a welled slide if you plan to use such a slide. Place the paper ring over the embryo (Fig. 6-41). Try to leave the embryo portion of the yolk exposed above the saline in the dish to avoid clouding over with yolk when you make the next cuts. With forceps hold the paper ring and the membrane that adheres to it and cut the membrane using the outer rim of the paper ring as a guide. Lift off the embryo and quickly transfer it to the welled slide or Petri dish of warm saline solution. Pipette off any excess yolk that clouds the saline and quickly replace with warm, clear saline solution. Examine under low power. Locate the major organs, as described above for the intact embryo.

72-hr chick embryo   Transfer this three-day old embryo from the yolk into warm saline solution using the technique described above and in Figure 6-41. Students should take note of the extensive differentiation of the organs. Note the formation of 36 pairs of somites; the curve of the body. At this stage the amnion covers the embryo. (You may have to jar the embryo to recognize the amnion.) Note the wing buds and the hindlimb buds, and the allantois growing out of the hindgut. At this stage most of the major organs are formed; the remaining period of incubation is used for final differentiation. Count the strong heartbeat and the well developed circulation in the vitelline arteries over the yolk and the vitelline veins that return to the small auricle, or atrium. Embryos that are older than this stage become too opaque to view through a microscope. You may want to fix these and later stages, clear, stain, and mount as whole mounts (p. 171).

Be sure to examine older embryos (remove the shell and shell membranes) in warm saline in a finger bowl. Examine the entire embryo within the chorio-allantois and note the extensive network of blood vessels that carry carbon dioxide and oxygen to and from the embryo through the surface of the porous shell. To see the amnion more clearly, carefully pull away the chorio-allantois. Rugh[48] and

---

[48] Rugh, *Experimental Embryology*.

A. Crack egg carefully.

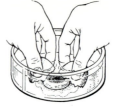

B. Hold shell under saline solution and release yolk from shell.

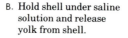

C. Place filter-paper ring over blastoderm.

D. Grasp edge of paper ring and adhering membrane with forceps, and clip the membrane completely around the paper ring.

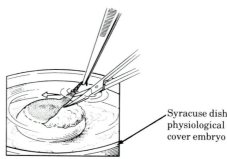

E. Lift the filter-paper ring with adhering membranes and embryo and transfer it to physiological saline.

Syracuse dish with enough physiological saline to cover embryo

F. Examine with low power microscope.

Fig. 6-41 Suggested procedure for removing early chick embryo from yolk of egg. (From *Laboratory Outlines in Biology–III* by Peter Abramoff and Robert G. Thomson. W. H. Freeman and Company. Copyright © 1982.)

Hamburger[49] describe additional investigations involving chorio-allantoic grafting. As an extensive investigation, Moog describes the role of thiourea as a thyroid inhibitor of developing chick embryos.[50] Make a 0.25 M solution of thiourea by dissolving 1.9 g thiourea in 100 ml of 0.9 percent saline and inject into chick embryos.

Differentiation in mammalian embryos

You may want to refer to several investigations: dissection of a pig uterus (p. 435), and dissection of a fetal pig (Fig. 1-109).

[49] Hamburger, *A Manual of Experimental Embryology*.
[50] Moog, *Animal Growth and Development*.

Pregnancy and delivery

Many types of uteri and mammalian embryos are available from biological supply houses (Fig. 6-42). Placental types include diffuse (pig), cotyledonary (cow and sheep), zonary (cat), and discoid (rabbit). Also available is a set of comparative uterine types: duplex, bipartite, bicornate, and simplex.

Where a permissive social and intellectual atmosphere exists in the community, a study of the development of the human embryo in the uterus may be made (Figs. 6-43 and 6-44). Human fetuses ranging from 3 to 4 months are available from supply houses. These have been cleared and mounted in plastic so that they can be examined easily in class and can be readily stored.

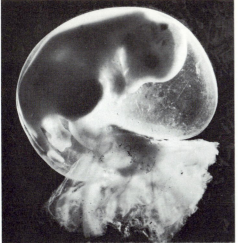

(a)

(b)

(c)

6-42 Some specimens of uteri and embryos available for study: (a) pig embryo in amnion; (b) human fetus, bisected along median line, cleared and mounted in plastic; (c) placental types (left to right): diffuse (pig), cotyledonary (sheep and cow). (a,c, courtesy of Carolina Biological Supply Co.; b, courtesy of General Biological Supply House, Inc., Chicago.)

There are numerous models of human embryology. An excellent set of models of human embryology is useful for teaching different stages of fertilization and pregnancy. There is also a set of models useful in teaching parturition.

Rugh's text tells the story of human development in splendid photographs;[51] another excellent reference is that by L. Nilsson.[52]

[51] R. Rugh and L. B. Settles, with R. Einhorn, *From Conception to Birth: The Drama of Life's Beginnings* (New York: Harper and Row, 1971).

[52] L. Nilsson, *A Child Is Born* (New York: Dell Publishing, 1977).

Trace the hormones involved in ovarian and pregnancy cycles as diagrammed in Figure 6-45. What is the role of FSH? Of LH? When is progesterone produced? Which hormones alter the lining of the uterus? Which hormones are active in maintaining pregnancy? Trace the feedbacks to the pituitary that maintain homeostasis. Compare the length of gestation among some mammals (Table 6-6). Also trace the evolution of a well-developed uterus from the platypus to the marsupials, to placentals. Make available offprints from *Scientific American* such as "The Placenta" (August 1980) in which P. Beaconfield et al describe the placenta as weighing 500 g or about ⅙ the weight of the infant. A full-term placenta measures 20 cm in diameter and is 3 cm thick in the center. There is a good discussion of the use of the placenta in testing toxicity and effects of drugs on cell division since cells of the placenta proliferate rapidly. The sad

TABLE 6-6   Length of gestation among some mammals (in days)*

mammal	gestation period	mammal	gestation period
Ape	210	Fox	52
Baboon	210	Guinea pig	68
Bat	35	Hamster	16
Bear	208	Horse	336
Bobcat	50	Kangaroo	38
Buffalo	275	Man	274–280
Camel	315–410	Mouse	19
Cat	63	Rabbit	31
Cattle	281	Rat	21
Chipmunk	31	Sheep	151
Dog	63	Squirrel	44
Elephant	624	Whale	360

* Adapted from W. Spector, ed., *Handbook of Biological Data*, Saunders, Philadelphia, 1956.

effects of thalidomide on limb buds of early embryos is described by H. Taussig, "The Thalidomide Syndrome," in *Scientific American*, August 1962.

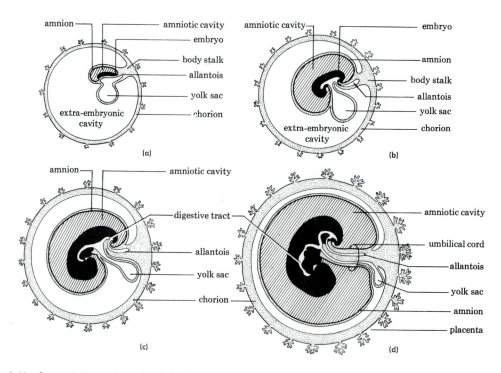

6-43 Stages in the early weeks of development of human umbilical cord, membranes, and early embryo: (a) 15 days; (b) 20 days; (c) 30 days; (d) 33 days. (After C. Villee, *Biology*, 4th ed., C. Saunders, 1962.)

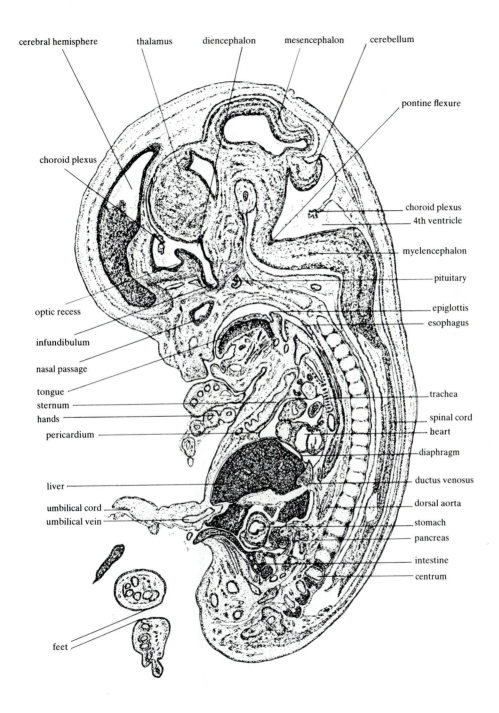

6-44 Human fetus: Longitudinal section of fetus, a little more than 8 weeks (25 mm).

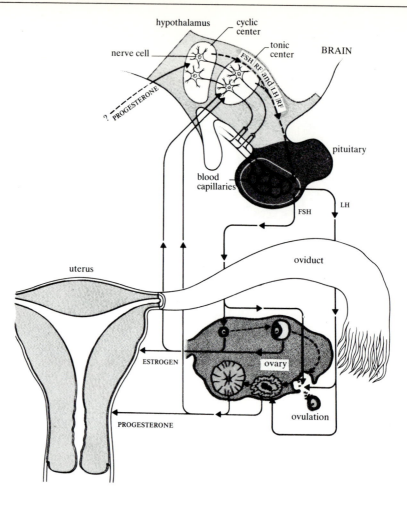

Fig. 6-45 Hormonal control of human reproduction. Hypothalamus produces releasing factors which stimulate production of gonadotropins FSH and LH by the pituitary. These hormones stimulate the growth of the egg in a follicle, ovulation, and production of corpus luteum. The sex hormones estrogen and progesterone, produced by the follicle and corpus luteum, prepare the lining of the uterus to receive a fertilized egg. A feedback loop affects the production of further releasing factors by the hypothalamus. (Adapted from *Biology* by Grover C. Stephens and Barbara Best North. Copyright © 1974 by John Wiley & Sons, Inc. Reprinted by permission of John Wiley & Sons, Inc.)

Ovaries and testes of a mammal   Dissect a freshly killed rat or other small mammal that has been etherized. (Place in a closed container with some ether-soaked absorbent cotton.) Check the reflexes to make certain that the animal is dead before you begin dissection.[53] Make an incision in the ventral body wall at the posterior end and cut forward along one side of the median line. Cut through the sternum to the shoulder or pectoral girdle. Then cut back the abdominal wall flaps and pin them back to the dissection pan (dissection, chapter 1).

In the female, identify the ovaries, oviducts, and uterus. Cut into the uterus to look for developing embryos or fetuses if the female is pregnant. Then with a blunt probe examine the uterine lining, and

---

[53] Review the section on humane treatment of animals in chapter 2, page 175.

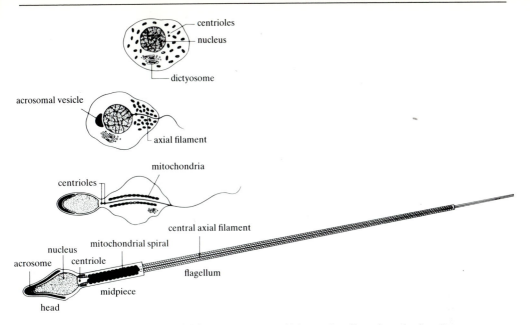

Fig. 6-46 Development of sperm cell from spermatid: The head contains the nucleus, the acrosome is produced by dictyosomes, the axial filament is made of microtubules. Mitochondria are spiral shaped around the filament forming the midpiece; other fibers form the flagellum. (Adapted from *Biology* by Grover C. Stephens and Barbara Best North. Copyright © 1974 by John Wiley & Sons, Inc. Reprinted by permission of John Wiley & Sons, Inc.)

trace the oviducts forward, and posteriorly to the vagina. You may want to show a microscope slide of a section of the ovary. In the male, dissect a scrotal sac to expose a testis. Examine a cross section of the testis; trace the epididymis, vas deferens, vesicles, and the urethra in the penis.

If there is time, students may identify other organs (dissections, chapter 1). Also refer to studies on photoperiodism and the role of the pituitary gland in the reproductive cycle (chapter 5).

Students will want to examine prepared slides of sperm cells of mammals (Fig. 6-46) and also to trace the development of the almost microscopic egg of a mammal. Compare sperm of several invertebrate and vertebrate forms (Fig. 6-47). Review oogenesis and spermatogenesis.

Dissecting a pig's uterus   If you purchase the uterus of a pig from a supply house, you will receive a bifurcated uterus with several fetuses. Carefully cut through

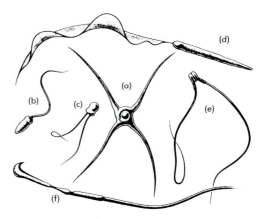

Fig. 6-47 Variations among sperm cells: (a) crayfish; (b) sea urchin; (c) toadfish; (d) toad; (e) opossum; (f) guinea pig. (Adapted from *Biology* by Grover C. Stephens and Barbara Best North. Copyright © 1974 by John Wiley & Sons, Inc. Reprinted by permission of John Wiley & Sons, Inc.)

the uterine wall to reveal the chorion and amnion. Hold up the amnion to show the fetus within, in amniotic fluid. Have stu-

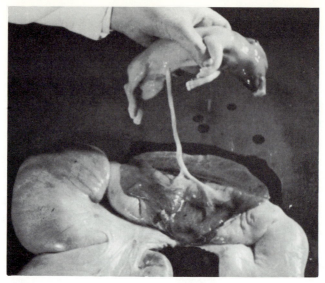

Fig. 6-48 Bifurcated uterus of pig showing one fetus that has been dissected out. (Photo from *Reproduction Among Mammals*, Encyclopaedia Britannica Films Inc.)

dents identify the placenta and umbilical cord as you gently lift out the fetus (Fig. 6-48). Fetuses can be removed and preserved in alcohol in museum jars or saved for dissection (see chapter 1).

Sexual reproduction in plants

Laboratory studies

Spirogyra In this filamentous alga, whole cells act as isogametes in conjugation. Note that the active cell moves across the "bridge" to fuse with the contents of the female, or passive cell (Fig. 6-49); a zygospore results. Examine prepared slides; methods for culturing are described in chapter 9.

Ulothrix In this filamentous alga, the contents of the cell divides in asexual reproduction into many zoospores, each with four flagella. The isogametes that unite in sexual reproduction are smaller and have two flagella (Fig. 6-50). How is sexual reproduction more specialized than in *Spirogyra*?

Acetabularia At some time, examine this beautiful branching green alga that

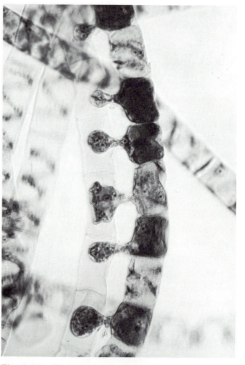

Fig. 6-49 Conjugation in *Spirogyra* (magnified 435 ×, reduced by 43 percent for reproduction). Notice the bridge along which one entire cell, as a gamete, moves into the passive cell, forming a zygospore. (Courtesy of General Biological Supply House, Inc., Chicago.)

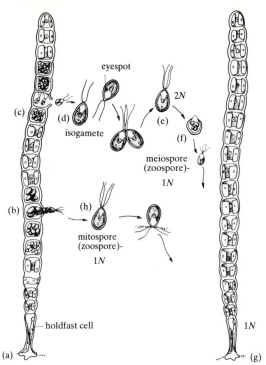

Fig. 6-50 Life cycle of *Ulothrix*: (a) entire filament (1*N*) showing (b) zoosporangia that form mitospores (or zoospores) which give rise to new vegetative filaments; cells at (c) give rise to isogametes, which fuse to form a zygote (2*N*); the zygote produces meiospores (1*N*) each of which gives rise to a new filament. (Adapted from *Plant Biology* by Knut Norstog and Robert W. Long. Copyright © 1976 by W. B. Saunders Company.)

has, when fully differentiated, an umbrellalike top composed of many sporangia (elongated cells) (Fig. 1-120). Aplanospores are formed within sporangia; after a rest period, these produce within them a number of biflagellated isogametes, which in turn, fuse to form zygospores that grow into new branched plants with umbrella tops. Some students may want to read about J. S. Hämmerling's grafting experiments which show the role of the nucleus in differentiation of umbrella tops.[54] This experiment is also described in chapter 7, page 526.

[54] J. S. Hämmerling, "Nucleo-cytoplasmic Relationships in the Development of *Acetabularia*," *Internal. Rev. Cytol.* (1953) **2**:475–98.

Ulva   This marine green alga, often called sea lettuce because of its flat form, is only two cells thick. It may be long and attached by holdfasts to rocks on the ocean floor. In asexual reproduction, the cell content divides into zoospores, each with four flagella (Fig. 6-51). Zoospores grow into new plants. Asexual and sexual plants resemble each other until sexual reproduction occurs. At this time, isogametes with two flagella are formed in the cells of monoploid sexual plants. The diploid zygote germinates into an asexual plant. *Ulva* has a diploid asexual generation with an alternate generation of monoploid sexual plants. This type of alternation of generations anticipates similar alternations that occur among land plants—mosses, ferns, and seed plants.

Fucus   Use fresh material or preserved specimens of this brown marine alga to trace its life cycle (Fig. 6-52). There is no asexual reproduction. Spermatozoids are released from antheridia and swim by means of two flagella, one projecting forward, the other backward. The mature plant is diploid.

Vaucheria   This tubular green alga, often found on flowerpots and called "green felt," is available for study of sexuality in plants (Fig. 6-53; also see culturing methods, chapter 9).

Chlamydomonas   Refer to Figure 6-54 for a study of sexual reproduction; also see heredity in *Chlamydomonas* (p. 525).

Oedogonium   This common unbranched green filamentous alga reproduces by *unlike* gametes: eggs and sperms (Fig. 6-55). These gametes are produced in special cells called gametangia, which are developed from vegetative cells of the filaments. Oogonia are oval cells within which an egg develops. Motile ciliated sperms are liberated from the antheridia; they swim to the egg through an opening, or pore, formed by the breakdown of a small area of the wall of the oogonium.

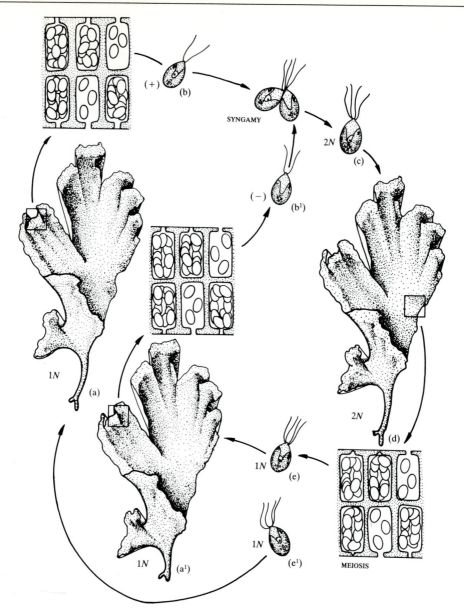

Fig. 6-51 Life cycle of *Ulva*: Flat thallus gametophytes of two strains, plus and minus as shown in a, a¹, produce plus and minus gametes (b, b¹) which unite to form a zygote (c). The resulting 2*N* sporophyte plant (d) produces meiospores (e, e¹) after meiosis, resulting in two gametophyte strains. (Adapted from *Plant Biology* by Knut Norstog and Robert W. Long. Copyright © 1976 by W. B. Saunders Company.)

Oedogonium also reproduces asexually by forming ciliated zoospores, which are transformed vegetative cells. These cells with a collar of cilia break away from the filament and swim off to take hold at another location, become elongated, and carry on cell division. Thus they form a new filament. Stain a wet mount with Lugol's solution (p. 176) to show cilia.

Fungi

In the mold, *Rhizopus*, hyphae show no visible differences between the two mating

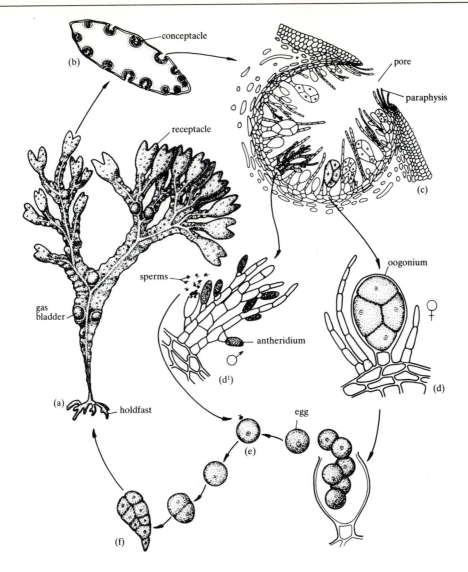

Fig. 6-52 Life cycle of *Fucus*: (a) diploid plant with receptacles shown at the tips; (b) cross section of a receptacle containing conceptacles; each antheridium (c) produces 64 sperms and each oogonium (d) produces 8 eggs; (e, f)

fertilization and developing embryo that will produce a new generation of bisexual plant. (Adapted from *Plant Biology* by Knut Norstog and Robert W. Long. Copyright © 1976 by W. B. Saunders Company.)

types, but a physiological difference exists. In conjugation, hyphae of opposite mating types (plus and minus) grow toward each other, and the tips fuse (Fig. 6-56). A heavy wall develops around the zygote resulting in a zygospore. Meiosis occurs in the zygospore so that the new sporangium and spores are monoploid. Chapter 9 describes methods of culturing.

Neurospora   Obtain sexual strains of *Neurospora* from a biological supply house. This pink mold grows well; however, it readily contaminates other cultures in the laboratory unless care is taken. If possible, make transfers in another room; handle the tube cultures as though they were bacterial cultures. Flame the cotton stoppers and mouths of the tubes; use a moist

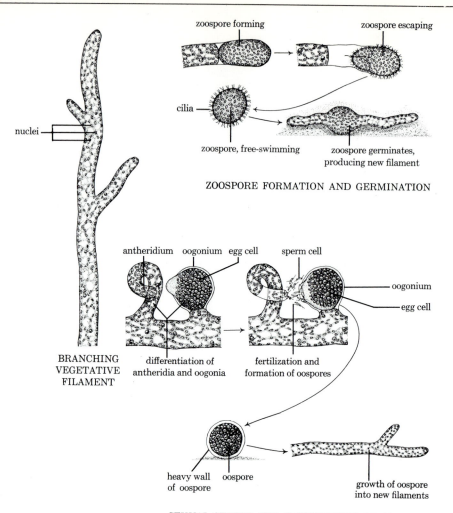

ZOOSPORE FORMATION AND GERMINATION

SEXUAL STAGES AND GERMINATION OF OOSPORE

Fig. 6-53 Life cycle of *Vaucheria*. (Courtesy of General Biological Supply House, Inc., Chicago.)

sterile loop to transfer spores from an old culture to new agar slants. Some workers use an aqueous suspension of spores to make transfers.

Prepare slanted agar medium (Difco has a prepared medium,[55] or see page 524) in wide-mouthed test tubes or small jars plugged with cotton. Perithecia become visible in about one week and mature in two weeks (Fig. 6-57; see also Figs. 7-16, 7-17, 7-18).

Crosses may be made between two strains, for example, wild type and albino.

[55] See Appendix.

Inoculate spores at opposite regions of a prepared Petri dish, or at opposite ends of the slants in test tubes. Watch the growth of the mycelia and the meeting of the mycelial fronts forming sexual spores. Also refer to genetic studies using *Neurospora* (p. 523); and nutrient deficient strains (p. 524).

Yeasts   Trace the life cycle of *Saccharomyces cerevisiae*, the yeast used in baking and brewing (Fig. 6-58). Note that asexual budding occurs in 1N cells (plus and minus strains). In sexual reproduction, a fusion or conjugation of a plus and minus

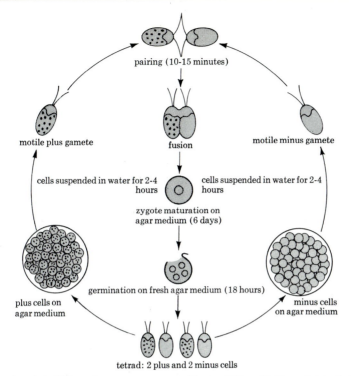

pairing (10-15 minutes)

motile plus gamete

fusion

motile minus gamete

cells suspended in water for 2-4 hours

cells suspended in water for 2-4 hours

zygote maturation on agar medium (6 days)

germination on fresh agar medium (18 hours)

plus cells on agar medium

minus cells on agar medium

tetrad: 2 plus and 2 minus cells

Fig. 6-54   Sexual cycle in *Chlamydomonas*. (Adapted from R. Levine, *Genetics*, Holt, Rinehart and Winston, 1962, and R. Manwell, *Introduction to Protozoology*, St. Martin's, 1961.)

cell results in 2N yeast cells. These undergo budding; meiosis follows with four cells developing in each ascus. Half the liberated cells are plus, half are minus strain (1N); the asexual budding stage begins.

Bryophytes

Marchantia   This flat bryophyte has a lobed thallus and is dioecious, developing umbrellalike branches that bear either antheridia or archegonia (Fig. 6-59). Antheridia are located in the upper surface of the branches, while archegonia are found on the lower surface of the umbrella-like branches. Spermatozoids swim on the surface of the flat thallus to reach the archegonia. Use Figure 6-60 to trace the life history of *Marchantia*. Also refer to culturing, p. 713; and the effect of light on inducing sexual differentiation (antheridia first), p. 714.

Alternation of generation in mosses   Compare the life history of the moss (Fig. 6-61) with that of the fern (Fig. 6-64). The gametophyte of the moss is the conspicuous stage, familiar as the green plant; the sporophyte grows on top of the gametophyte and is parasitic on the gametophyte. Examine prepared slides to observe both antheridia and archegonia developing on a gametophyte. Biflagellated spermatozoids swim to an egg in the archegonium. The gametophyte is monoploid; the sporophyte is diploid. Where did meiosis occur?

Also try to grow the spores of mosses (p. 714) and follow steps similar to those for growing fern spores (p. 716, also p. 444).

Tracheophytes

Equisetum   Among the vascular plants, examine the scouring rush (or horse-tail),

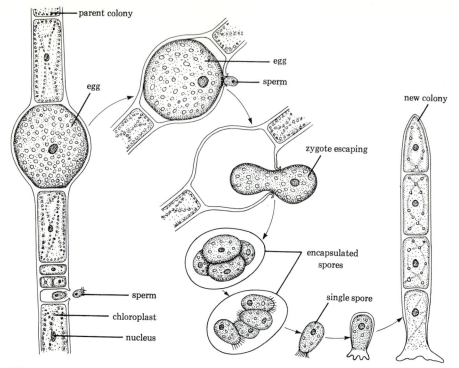

Fig. 6-55 Sexual reproduction in *Oedogonium*: Notice four meiospores resulting from meiosis in the zygote giving rise to new filaments. (After D. Marsland, *Principles of Modern Biology*, 3rd ed., Holt, Rinehart and Winston, 1957.)

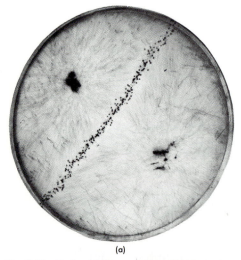

(a)

(b)

Fig. 6-56 Fusion of plus and minus gametes of molds to form zygospores: mycelia of *Phycomyces blakesleeanus* fuse, forming zygospores at the junction of plus and minus strains; (b) zygospores resulting from a similar type of fusion or conjugation in *Rhizopus nigricans* (magnified). (a, courtesy of Carolina Biological Supply Co.; b, courtesy of General Biological Supply House, Inc., Chicago.)

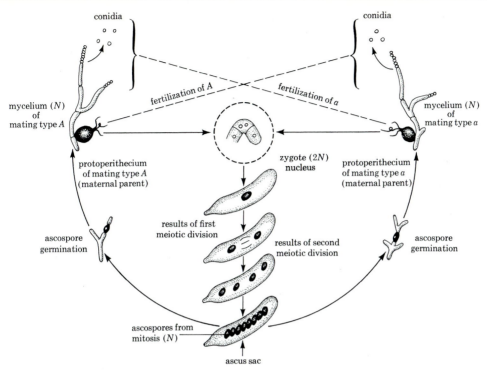

Fig. 6-57 Sexual reproduction in *Neurospora* (pink bread mold), which is an ascomycete, or sac fungus. (After R. P. Wagner and H. K. Mitchell, *Genetics and Metabolism*, Wiley, 1955.)

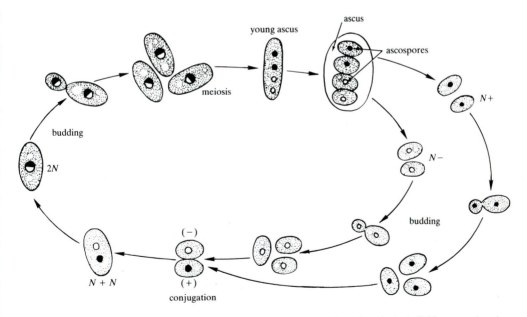

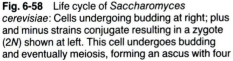

Fig. 6-58 Life cycle of *Saccharomyces cerevisiae*: Cells undergoing budding at right; plus and minus strains conjugate resulting in a zygote (2N) shown at left. This cell undergoes budding and eventually meiosis, forming an ascus with four ascospores, two of each strain (1N); asexual cycle begins. (Adapted from *Botany: An Introduction to Plant Biology*, 6th ed., by T. Elliot Weier, C. Stocking, M. Barbour. Copyright © 1982 by John Wiley & Sons, Inc. Reprinted by permission of John Wiley & Sons, Inc.)

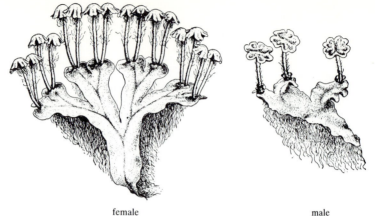

female   male

Fig. 6-59 *Marchantia*: At top left, branches of female gametophyte plant bear archegonia on their underside; at right, specialized branches of male gametophyte plant have antheridia sunken in the upper surface. (Adapted from *The Plant Kingdom* by William H. Brown, Ginn and Company, Publishers. Copyright © 1935 by William H. Brown.)

Equisetum (Fig. 6-62). Notice the branched green shoots and the upright fertile shoots with cones (sporophyte). During Carboniferous times, these plants were treelike. Figure 6-63 illustrates an extinct relative, *Calamites*. Today there is only one living genus of *Equisetum*.

Trace the life cycle in Figure 6-62. Also observe the characteristic spore of *Equisetum* (method for culturing spores, p. 717). Multiciliated spermatozoids are shown in Figure 6-62.

Alternation of generation in ferns
Collect fern spores and germinate them using the method described on page 716. Also use Figs. 6-64, 6-66 to trace the life history of a fern from the tiny gametophyte generation that alternates with the conspicuous, independent sporophyte generation (Fig. 6-65). Diploid sporophytes grow from the small heart-shaped prothallus, or gametophyte (Fig. 6-66). When the fern gametophyte is about 2 mm in diameter, transfer to a slide and look for spiral, many-flagellated sperm cells near the antheridia (Fig. 6-67). Or you may want to observe prepared slides of the many-flagellated sperm cells that swim to a large egg found in the flasklike archegonium.

The mature fern gametophyte, only about 1 cm, soon disintegrates, while the independent sporophyte generation develops a vascular system with xylem and phloem.

Antheridia and archegonia develop on the underside of a fern prothallus, with antheridia usually appearing earlier. Some bracken fern prothallia seem to grow faster than others and secrete an antheridogen that stimulates antheridia formation on nearby slower growing prothallia. Gibberellins show a similar effect.[56]

Germination chamber: fern spores
Heat a 2 percent glucose solution to which 2 percent agar is added. Dip a glass rod into the hot solution and draw it flat across the surface of a coverslip. Add fern spores to the surface as the solution cools. Dab a bit of petroleum jelly to four corners of the coverslip. Then place a depression slide over the agar on the coverslip and press to seal. Invert the slide; examine under low power to check the inoculum. Maintain at room temperature and examine in a few days for signs of germination. (The same method may be used to examine pollen grains, or a variety of microorganisms in

[56] B. Voeller, "Gibberellins: Their Effect on Antheridium Formation in Fern Gametophytes," *Science* 143 (January 1964).

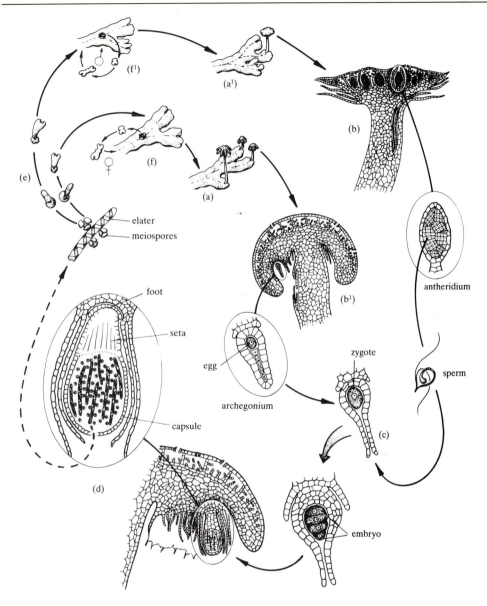

Fig. 6-60 Life cycle of *Marchantia,* a liverwort: begin at the top of the diagram with (a) and (a¹) as the female and male gametophytes; trace the production of an egg in the archegonium (b) and flagellated sperm from the antheridium (b¹); fertilization occurs, forming a zygote (c); (d) developing sporophyte in archegonium and mature sporophyte; (e) meiospores resulting from meiosis produce 1*N* gametophytes that eventually bear either archegonia or antheridia. (Adapted from *Plant Biology* by Knut Norstog and Robert W. Long. Copyright © 1976 by W. B. Saunders Company.)

garden soil, or other. Use a drop of filtered soil to inoculate the agar by streaking with a sterile loop or needle.) Also see p. 716 (agar medium) and 717.

Sexual reproduction in higher Tracheophytes

The formation of gametes after reduction division, or meiosis, occurs in the spore mother cells in the anthers and within the ovules. In seed plants, the conspicuous plant is the diploid sporophyte, while the monoploid gametophyte generation is now reduced within a pollen grain and ovule.

At this time, it may be helpful to study flowers of shrubs and trees in season as they come to blossom. If your school is located near a park, or in a rural area, observe flowering plants on short field trips. Have students collect flowering shrub or tree branches as well as dandelions, violets, crocus, pussy willow, tulip, or others. Since they bloom about the

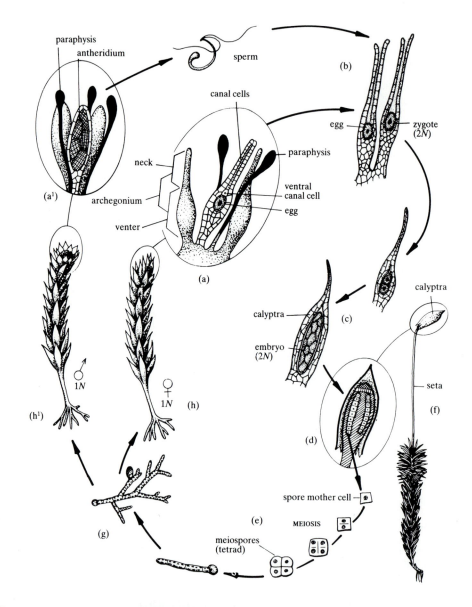

Fig. 6-61 Life cycle of a moss: alternation of monoploid sexual stage (gametophyte) with diploid asexual stage (sporophyte). Note that meiospores germinate into protonema which grow into male and female gametophytes (1*N*). (Adapted from *Plant Biology* by Knut Norstog and Robert W. Long. Copyright © 1976 by W. B. Saunders Company.)

same time, students should be able to observe blossoms of magnolia, *Forsythia*, maple, cherry, dogwood, oak, and elm.

Later, return to collect fruits and seeds for identification (Fig. 1-133).

Soon students will learn that not all flowers have petals, and that some flowers contain only stamens, while others have only pistillate parts. (Refer to Figures 1-128 through 1-134.)

Parts of a flower Tulips, daffodils, and gladioli are excellent forms to show as almost diagrammatic flowers; the pistil and stamens are clear. Students should also see a variety of flower shapes and realize that these shapes help to identify families of plants. They might learn to recognize a few such as Compositae, Leguminosae (Fig. 6-68), and Rosaceae (Fig. 6-69). (Also refer to pages 113–118). How

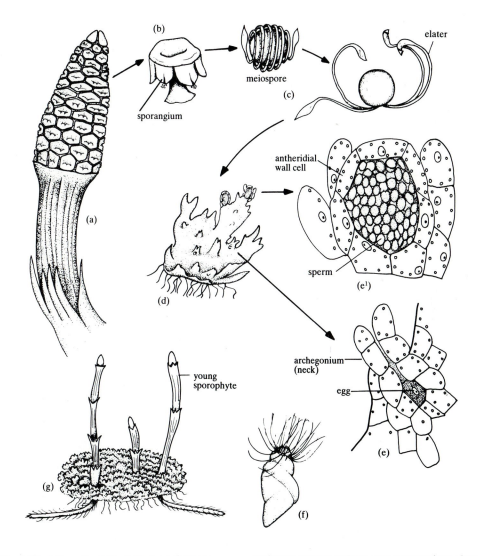

Fig. 6-62 Life cycle of *Equisetum*: sporophyte (a) produces meiospores (b) with elators (c); gametophyte (d) with differentiated antheridia and archegonia (e, e¹); multiflagellated sperm (f) (about 80 flagella) unites with an egg; gametophyte plant (g) with young sporophytes attached. (Adapted from *Plant Biology* by Knut Norstog and Robert W. Long. Copyright © 1976 by W. B. Saunders Company.)

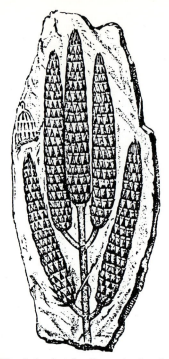

Fig. 6-63 *Calamites*, fossil scouring brush. (Adapted from *The Plant Kingdom* by William H. Brown, Ginn and Company, Publishers. Copyright © 1935 by William H. Brown.)

do flowers of willow, oat, or corn differ? (Figs. 6-70, 1-129, 1-130). Some students may show interest in classifying plants; learn how to use a flower key (study the pattern, p. 113, Fig. 1-132). Figures 1-128 and 6-70 shows the main kinds of inflorescence. Some flower keys are listed in the appendix.

Examine flowers with a hand lens or dissection microscope. Dissect out the stamens after the exterior whorls of sepals and petals have been studied. Examine anthers with a hand lens, and then dissect to show the pollen chambers. Prepare anther squash slides (p. 401). Or use a prepared slide; mount pollen grains on a slide and examine under high power (Fig. 6-71; also see p. 451, 455). Note the sculpturing and the different shapes of pollen grains from several plants. Some flowers may be identified by means of their pollen grains. On occasion, students may want to germinate pollen grains (p. 453).

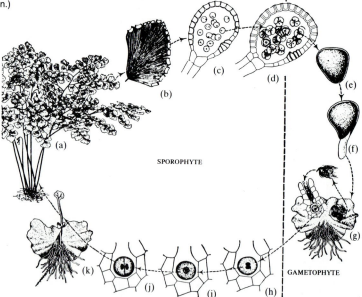

Fig. 6-64 Life cycle of a fern: begin with the sporophyte (a), showing sporangium (b) with tetrads of spores (c,d). Note the line separating the sporophyte from the gametophyte generation (a sporangium is part of the sporophyte, but a spore is the first stage of the gametophyte generation); spore germinates (e,f) and a prothallus results (g); antheridia and archegonia differentiate producing eggs and sperms; the fertilized egg (h) grows into the young sporophyte, attached to prothallus (k). (Adapted from *The Plant Kingdom* by William H. Brown, Ginn and Company, Publishers. Copyright © 1935 by William H. Brown.)

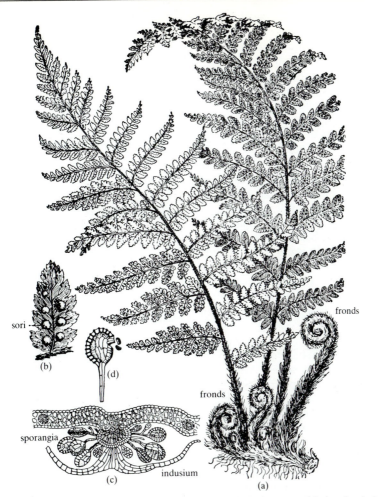

Fig. 6-65 Sporophyte of a fern showing fiddleheads (or unrolled fronds, which usually develop from (a) the rootstock the next season); underside of a pinnule showing sori (b); cross section through a sorus (c) showing indusium and sporangia; spores are liberated from a sporangium (d). (Adapted from *The Plant Kingdom* by William H. Brown, Ginn and Company, Publishers. Copyright © 1935 by William H. Brown.)

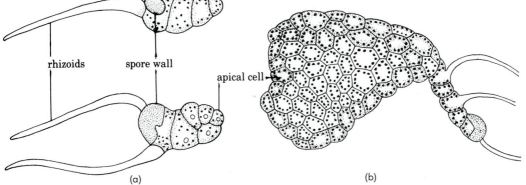

Fig. 6-66 Fern: (a) early stages in the development of a prothallus (gametophyte); (b) prothallus showing spore wall attached.

(Adapted from *The Plant Kingdom* by William H. Brown, Ginn and Company, Publishers. Copyright © 1935 by William H. Brown.)

Similarly, examine cross sections and lengthwise sections of the ovary. Locate the ovules. The number of chambers in the ovary equals (or is a multiple of) the number of petals and stamens. Floral parts are mainly in "fives", or "threes" especially among the monocots.

During the year, short field trips may be taken to show fruits and seeds on trees and shrubs around the school. Students may collect fruits and prepare a display of some of the special adaptations that facilitate seed dispersal (Fig. 6-72). Also refer to Figure 1-133.

Seeds also may be studied. Soak some large seeds such as beans, peas, or a corn grain (really a fruit containing a seed). Then dissect seeds to learn their structure. (See page 471 for studies of seeds and germination.) Monocots and dicots can be distinguished in their pattern of growth. (Germination methods, p. 470.)

Some students may also collect pine cones to study the relation of seeds to the scalelike bracts in the cones. Also study the life cycle of gymnosperms (Fig. 6-73) and compare this with the life cycle of angiosperms (Fig. 6-74).

The *Ginkgo* tree, a living fossil, may be available in parks or specially planted on some streets. *Ginkgo biloba* is the only living form of the widespread ancestral types we know only from fossil remains of the Triassic some 200 million years ago. It is dioecious; pollen trees have catkins containing microspores, and ovules are paired on short stems of the seed trees (Fig. 6-75). Microspores, carried by wind, enter the ovule where they develop into gametophytes. (Sperm cells swim through pollen tubes to reach egg nuclei.) Among seed plants, cycads and *Ginkgo* are the only forms with flagellated male gametes.

Artificial pollination   You may want to demonstrate how breeders would pollinate flowers selected for mating. Hybridization techniques of this kind are commonly used to obtain new varieties, through recombinations in the offspring of different DNA. The flower to be pollinated should

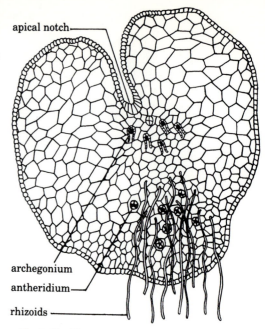

Fig. 6-67   Mature prothallus (gametophyte) of a fern, viewed from the ventral surface.

Fig. 6-68   Typical flower shape of pea family. (Adapted from *The Plant Kingdom* by William H. Brown, Ginn and Company, Publishers. Copyright © 1935 by William H. Brown.)

Fig. 6-69   Typical flower shape of rose family. (Adapted from *The Plant Kingdom* by William H. Brown, Ginn and Company, Publishers. Copyright © 1935 by William H. Brown.)

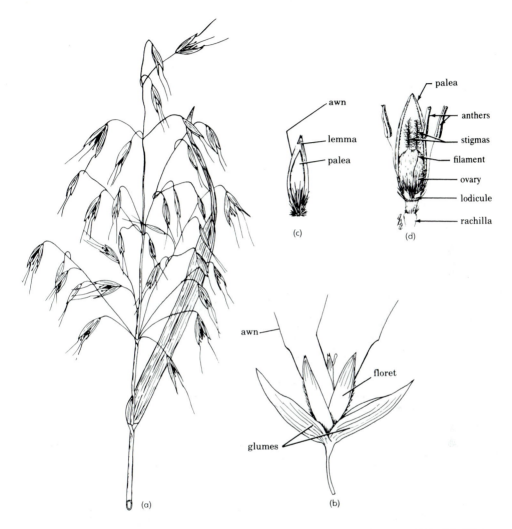

Fig. 6-70 *Avena* (oat). (a) inflorescence;
(b) spikelet; (c) floret; (d) open flower. (Adapted
from *Botany: An Introduction to Plant Biology*, 6th ed.,
by T. Elliot Weier et al. Copyright © 1982 by John Wiley
& Sons, Inc. Reprinted by permission of John Wiley &
Sons, Inc.)

be selected in the bud stage, that is, before
its own pollen becomes ripe. Open the
flower bud carefully and remove the un-
ripened stamens by cutting them out with
scissors. Wash with alcohol all the equipment you use, to prevent any foreign pollen from entering the stigma of the selected
flowers. Again cover the flowers you have
now pollinated to prevent other pollen

grains from fertilizing the ovules. Plan to
pollinate several flowers in one operation.
Students can plant the seeds that result,
and, if possible, study the inheritance of
flower color of the next generation.

Examining pollen grains Prepare slides
of many kinds of pollen grains. Have
students discover the variety of shapes and

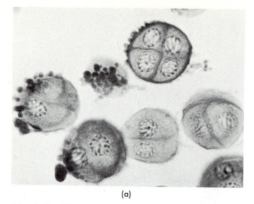

Fig. 6-71 Examples of pollen: (a) lily; (b) pine. (a, courtesy of Carolina Biological Supply Co.; b, courtesy of Ward's Natural Science Establishment, Inc., Rochester, NY.)

Fig. 6-72 Adaptations for seed dispersal: (a) cocklebur; (b) milkweed. (Photos by Hugh Spencer.)

patterns of the grains by mounting some in a drop of water or in melted glycerin jelly (preparation, p. 161). (The motion of the pollen grains in water is due to Brownian movement.) Also study the kinds of pollen in the air at particular seasons (allergy time). Lightly coat slides with petroleum jelly and suspend the slides out of doors to catch pollen. (Refer to counting pollen grains, p. 455.)

Staining pollen grains To a small amount of aniline oil add a bit of crystal violet so that a light purple tint results. Mount pollen grains in a few drops of the tinted aniline oil on a slide. Then hold the

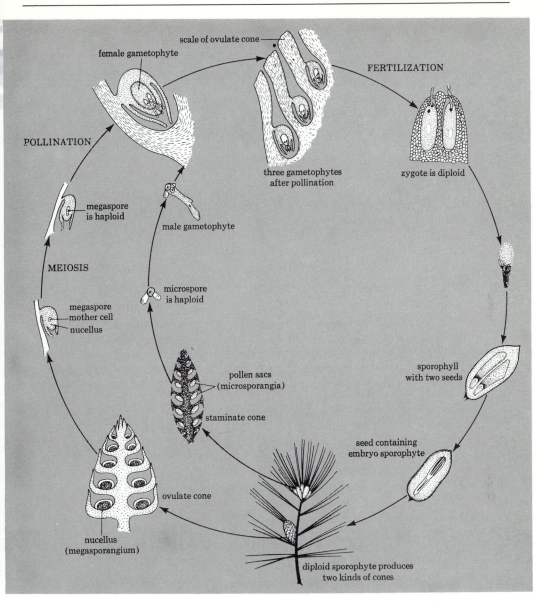

Fig. 6-73   Life cycle of gymnosperm. (After C. Villee, *Biology,* 4th ed. Saunders, 1962.)

slide over a Bunsen flame until the pollen grains stain deeply, but do not let the slide become warm to the touch.

Let the slide cool and remove the excess oil with filter paper. Wash several times by adding drops of xylene to the slide, then drawing it off with filter paper until the excess stain has been removed. Then add a drop of balsam and a coverslip.

Germinating pollen grains   There are several ways to grow pollen grains which germinate in the sticky, sugary exudate of the stigma. Students can brush ripe anthers against a ripe stigma of the same kind of flower, then crush the stigma in water on a slide. Include a few bristles or pieces of broken coverslip before adding a coverslip to prevent the grains from being

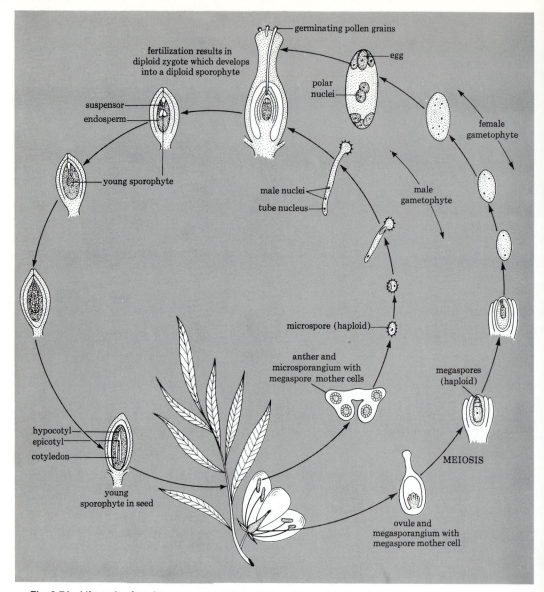

Fig. 6-74   Life cycle of angiosperm. (After C. Villee, *Biology*, 4th ed. Saunders, 1962.)

crushed; or make a hanging-drop prepara-tion (Fig. 2-22). The slides can be kept in a moist chamber between examinations. Maintain at a temperature of 21° C or 25° C (70° F or 77° F). Study pollen grains of narcissus, lily, daffodil, tulip, or sweet pea in this way. Pollen grains can also be sprinkled into a sucrose solution. How-ever, different pollen grains require differ-ent sugar content to germinate. Where

the stigma of the flower is quite sticky, pollen grains may require more concen-trated sugar solution for germination. In general, a majority of pollen grains grow in solutions varying from 2 to 20 percent; most at 10 percent. However, many com-posites require 30 to 45 percent sugar solution. The pollen grains of many cycads grow tubes in two to three days when grown in a 10 percent sucrose solution.

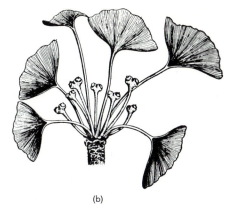

(a)   (b)

Fig. 6-75 *Ginkgo biloba*: (a) branch of male tree bearing strobili (sporophylls); (b) branch of female plant with young ovules. (Adapted from *The Plant*

Kingdom by William H. Brown, Ginn and Company, Publishers. Copyright © 1935 by William H. Brown.)

Pollen grains in agar medium   Try a more precise technique described by Johansen.[57] Boil 1 g of sucrose and 0.5 g of agar in 25 ml of water. Cool the medium to about 35° C (95° F); then add powdered gelatin (0.5 g) and stir until it melts.

Keep the solution at 25° C (77° F) in a water bath or on a hot plate. Put a thin film of this solution on clean slides. Dust the slides with pollen grains and keep them in a moist chamber without applying coverslips. Examine the slides under the microscope for growth of pollen tubes. (Try crushing anthers of *Tradescantia* in a film of this sugar solution.)

About 5 hr after pollen tubes form, mitotic figures can be seen if slides are to be made and stained. For a study of genetic differences between two types of pollen grains, see page 521 and Figure 7-14.

Counting pollen grains or other nonmotile cells   To count pollen grains or other stationary cells (*Chlorella* or yeast cells), use the following method. Mark off a section or grid 2 cm × 1 cm, an area of 2 cm² with a razor blade on a plastic slide. When pollen grains in the air are to be

[57] D. Johansen, *Plant Microtechnique* (New York: McGraw-Hill, 1940).

TABLE 6-7   Daily counts of pollen grains or other nonmotile cells

No. grains on 2 sq. cm.	*No. grains in reported count for 1 sq. cm.*
8	4
6	3
7	3.5
⋮	⋮

counted, grease one side of the marked area of the slide with a thin film of petroleum jelly. Leave the slide out of doors or on a window sill. Use a hand counter to tally all cells. Record the total number of pollen grains in an area of 2 cm². Then divide by two to get the number of pollen grains per cm² of collecting surface as the count for the day (Table 6-7).

Measurement of growth

In a general sense, growth is often identified with an increase in height or size. In a biological sense, growth is associated with an increase in the mass of the living substance. The distinction is not a simple one; an increase in height or length may accompany an increase in mass. In either sense, growth is a dy-

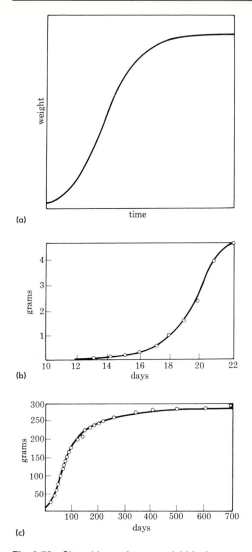

(a)

(b)

(c)

Fig. 6-76 Sigmoid growth curves: (a) ideal sigmoid curve; (b) intra-uterine growth of white rat; (c) postnatal growth of white rat. (In (b) and (c), the open dots indicate actual measured values.) (After B. I. Balinsky, *An Introduction to Embryology*, Saunders, 1960.)

growth. Plotting the increase in weight against time gives a sigmoid curve (Fig. 6-76), which is a measure of the absolute increase from starting to final weight regardless of other factors.[58] In short, a sigmoid growth curve gives a graphic illustration of absolute increase in weight and shows the greatest absolute increase in weight per unit of time in the middle part of the curve. Now, if increases in mass are measured for equal time intervals, then the amount of increase at different periods in the life cycle of one organism may be known. An absolute increase is not a true picture of the rate of growth at *different periods* of life (Fig. 6-76), nor does it offer a valid basis for comparing the rates of growth of *different organisms* (Fig. 6-77).[59] Clearly, if the same absolute increase in a given time is shown by a large and a small animal, their rate of growth is not the same; the smaller animal had to grow at a faster rate.

Growth, then, is an increase in geometric progression, an exponential process. The increase is proportional to the initial quantity of growing material (if the initial quantity is exceedingly small and if the initial size is negligible). The general formula for exponential growth is $W = e^{vt}$, where W is the weight of an animal at any time t, v is the observed rate of growth, and e is the base of natural logarithms (equal to 2.71828 . . .).

Students may plot exponential growth curves of several organisms, and of populations of organisms. Analyzing growth at the intracellular level, however, is

[58] The common practice in laboratory work is to plot time, or concentration, or another variable along the horizontal X axis (abscissa), and the dependent variables, the changes, along the vertical Y axis (ordinate).

[59] B. I. Balinsky, *An Introduction to Embryology*, 5th ed. (Philadelphia: Saunders, 1981) offers a splendid chapter on growth on the cellular and intercellular levels and on the organismic level. Balinsky also interprets growth curves and aging, as well as disproportional growth in organs. Some of the material presented here has been gleaned from Balinsky's text.

namic indicator of metabolism of cells. Furthermore, each organism seems to have a specific cycle of growth, and even providing an abundance of nutrients does not permit a continued accelerated rate of growth, since enzymes must also be present.

Students can measure the increase in weight of an organism as an index of its

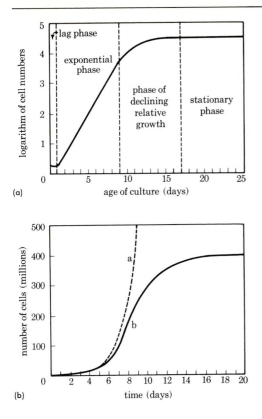

(a)

(b)

Fig. 6-77 Sigmoid growth curves of different organisms: (a) growth of *Chlorella*; (b) growth of yeast cells showing theoretical rate of multiplication with unlimited space and food (curve *a*) and actual rate of multiplication in a definite limited amount of nutrient solution (curve *b*). (a, after G. E. Fogg, "Famous Plants: 4, *Chlorella*," in M. L. Johnson, M. Abercrombie, and G. E. Fogg, eds., *New Biology*, Vol. XV, Penguin, Harmondsworth, Eng., 1953; b, after D. Marsland, *Principles of Modern Biology*, rev. ed., Holt, Rinehart and Winston, 1951.)

more complex. The overall rate of synthesis of molecules depends on the quantity of molecules of enzymes at hand, and the quantity that is available for each step in the chain of syntheses. Possibly the enzymes that control growth of protoplasm exist as only a few molecules in a cell and must be multiplied before organic syntheses in growth can get under way.[60]

[60] Balinsky, *An Introduction to Embryology.* Also read Balinsky's chapters on differentiation and regeneration.

Investigate the factors affecting the change in the height of a tree. Is this growth pattern similar to that of a definitive structure such as a leaf or a fruit? Is the growth pattern of an individual cell or a multicellular plant or animal similar to the growth of a population of cells, such as a culture of yeast cells, or of *Chlorella*, or of *Paramecium*? Does the logistic theory of population growth hold for all kinds of organisms? For some limitations, see H. Andrewartha and L. Birch, *The Distribution and Abundance of Animals* (Chicago: Univ. of Chicago Press, 1954) and more recent references.

Logarithmic growth curves

Populations of such cells are *Chlorella*, *Paramecium*, and yeasts have been carefully studied (Fig. 6-77); populations of these organisms exemplify distinct growth phases (Figs. 6-76a and 6-78). The first phase is a lag phase, in which the organisms become adapted to their medium or environment. The second phase is an exponential growth phase, in which rapid growth of viable cells occurs. When young cultures are used in studies of growth, counts of cells may be made regularly and plotted against time. The points will be

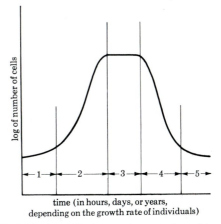

Fig. 6-78 Logarithmic growth curve: phase 1 is the initial lag phase; phase 2 is the exponential phase of rapid growth; phase 3 is the stationary phase; phase 4 is an exponential death phase; phase 5 is a survival phase.

found to lie closely along a straight line. How long can the exponential growth rate continue in a given culture medium? Are there physical conditions of density of cells per volume of medium that have detrimental effects on growth? What deficiencies in the medium may arise? What materials may accumulate in the medium?

The third phase of growth is a declining growth phase, or maximum stationary period. Cells are viable, but there is little growth. The duration of this phase may depend upon alterations in the environment. Final phases 4 and 5 are stationary or decline phases, which may lead to the death of the population (or to spore formation among cells that have this ability).

Measuring growth of microorganisms

Students may either estimate the rate of growth of individual cells (that is, the rate of fission or budding), or measure the growth of a population of organisms.

Making serial dilutions

Dilute suspensions of microorganisms to isolate and/or make cell counts. Prepare a series of small test tubes or vials of the culture medium in which the cells grow, for example, use Knop's solution (p. 480) if the cells are *Chlorella*. When an entire class is engaged in a laboratory activity, small quantities may be used. Add 9 ml of Knop's solution to each of four or more small test tubes. With a micropipette of 1-ml volume, or a calibrated micropipette, transfer 1 ml of the algal suspension from the main culture to the first of the series of test tubes. Agitate this tube, then with a clean pipette draw up 1 ml of the mixed dilution, and add it to the second test tube. Again, agitate this tube to distribute the cells, draw up 1 ml and add this to a third test tube, and so forth. (Also refer to diagrams of dilution series, p. 459.)

TUBE 1: 1 ml of algal suspension
  + 9 ml of Knop's = 1:10 dilution
TUBE 2: 1 ml of 1:10 dilution
  + 9 ml of Knop's = 1:100 dilution
TUBE 3: 1 ml of 1:100 dilution
  + 9 ml of Knop's = 1:1000 dilution
TUBE 4: 1 ml of 1:1000 dilution
  + 9 ml of Knop's = 1:10,000 dilution

Students may make counts of cells in each dilution (p. 462). Also refer to standard plate counts (p. 459). Or they may streak solidified agar in Petri dishes, or use welled slides to count colonies or clones of cells (p. 460, 465).

Isolation of individual cells

Individual cells isolated in welled slides, or in a hanging drop, may be used for studies of the rate of fission. The same technique can also be used for developing a clone, a group of organisms of similar heredity (barring mutations). Picks up individual cells with capillary tubing pulled out to a fine thread. Or use a row of small drops of the dilution on a slide. Inspect each drop for isolated cells. Or use the following method, which was originally used by Spallanzani. Place 1 small drop of culture on a slide; near it, place 1 small drop of fresh culture medium. With the tip of a capillary pipette make a connecting channel of fluid between the 2 drops. Observe under a microscope. When a *motile* cell moves into the drop of fresh medium, destroy the channel, pick up the isolated cell, and transfer to a welled slide or hanging drop. Seal the coverslip with petroleum jelly.

Nonmotile cells, such as bacteria, yeasts, or small algae, may be isolated on agar by streaking a dilute culture with a transfer loop or needle.

Each colony that forms may be considered a pure colony—a clone of cells having the same DNA. The change in the diameter of a colony may be taken as an estimate of the rate of fission of cells. (Other techniques for handling bacteria and culturing, p. 396–704.)

Bacterial population count—standard plate count

Counting the actual number of organisms on a slide, or using a hemacytometer gives the number of dead as well as live microorganisms. Turbidity studies are not accurate either since they also count the number of dead microorganisms with the living. The standard plate count, or quantitative plating method, reflects the number of living microorganisms in a given volume of solution—water, milk, soil water, raw meat, other foods, and so forth.

Quantitative plating   In this procedure, a given volume of suspension of microorganisms is diluted through three sterile water bottles or blanks and then poured into empty sterile Petri dishes; sterile nutrient agar is then added. Duplicate plates of each dilution should be made for increased accuracy unless this involves too much glassware in a given situation. Make a count of colonies and calculate the number of organisms per milliliter of original culture.

The procedure is diagrammed in Figure 6-79. Note that three small bottles (A, B, and C) each contain 99 ml of sterile water. Line up four empty sterile Petri dishes.

With a sterile 1 ml pipette transfer 1 ml of the culture to water blank A so that this bottle has a dilution of 1:100. Shake well

to dislodge clumps of microorganisms. Now with a second pipette transfer 1 ml of fluid from blank A to B. Now bottle B has a dilution of 1:10,000. Shake vigorously, then deliver one 1 ml from blank B to an empty sterile Petri dish and label this dilution 1:10,000. Label a second Petri dish 1:100,000; into this dish transfer *only 0.1 ml* from blank B.

Next carry over 1 ml from blank B into blank C and label this dilution 1:1,000,000. As before, use a clean pipette to transfer 1 ml from this dilution into an empty sterile Petri dish and label the dish 1:1,000,000. For a further dilution, 1 part in 10 million, transfer from blank C *only 0.1 ml* and label the Petri dish 1:10,000,000.

Now pour a thin film of nutrient agar maintained at 45° to 50° C (113° to 122° F) in a water bath, into the four Petri dishes. (Or use special agar medium, depending on the optimum growth requirements of the microorganisms.) When solidified, invert the dishes, and incubate for 24 to 48 hr.

For an accurate count of colonies use only those dishes that have at least 30 colonies and no more than 300. Make a tally with a mechanical hand counter. Unless you have a colony counter, place a Petri dish against a grid made of graph paper so that each colony, even the smallest, can be counted. A hand lens is useful.

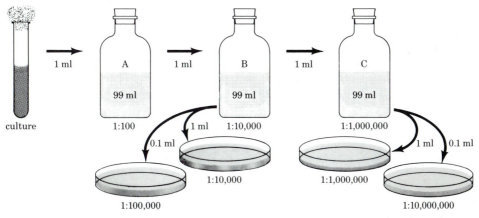

Fig. 6-79   Quantitative plate count of living organisms: diagrammatic representation of dilution and quantitative plating.

Mark off sections of the graph paper so that a count can be made without duplication.

Finally, the number of colonies counted multiplied by the dilution factor gives the number of organisms per milliliter of the original culture:

$$\begin{array}{c} \text{number of} \\ \text{colonies} \end{array} \times \begin{array}{c} \text{dilution} \\ \text{factor} \end{array} = \begin{array}{c} \text{number of micro-} \\ \text{organisms per ml} \end{array}$$

A specific example is described on page 603, for counting the numbers and kinds of microorganisms (bacteria, molds, actinomycetes) found in soil and cultured in a medium of sodium-caseinate-glucose agar available from Difco.

Growth of single protozoans in tubes

Giese describes the study of the rate of growth (fission) in protozoa using small culture tubes.[61] Use 3 mm soft glass tubing cut into 6.5-mm lengths. Seal one end with heat. Widen the other end a bit by twirling on a pointed carbon held in the opening. Inoculate the tube with a single protozoan and culture medium[62] and plug the opening with a firm cotton plug. Medium and tubes may be sterilized first for certain types of investigations.

Or study the rate of growth of single protozoa in the well of a depression slide with culture medium and seal coverslip with a ring of petroleum jelly. To reduce the rate of evaporation, place the welled slide preparation in a Petri dish.

Determine the rate of divisions and plot against time. Graph the data on the ordinate (dependent variable), and time on the abscissa (or other independent variable).

Take counts both morning and afternoon for three or more days.

Pasteur pipettes may be used to pick up single protozoans. Giese also describes the preparation of inoculating pipettes for the transfer of protozoa.[63] Over a Bunsen flame heat 12-cm lengths of 4 mm soft glass tubing until soft, then pull. Cut the cool glass by scratching with a diamond pencil at a region where a fine tip about 0.2 mm may be obtained. Press lightly to break. Fire-polish the tip by inserting quickly into a Bunsen flame. Insert cotton into the wider end of each inoculating pipette. Store them, tip down, in a test tube with a soft cotton cushion at the bottom of the test tube. These tubes may be sterilized in the test tubes if desired, but for transfer of single protozoa sterilization is not necessary.

Allow 8 to 12 days for a good clone culture to develop from a single protozoan (depending on temperature).

Growth of Tetrahymena: cell counts

Establish a thriving culture of this small ciliate that is often used in experimentation. With a sterile pipette, add 1 to 2 drops of *Tetrahymena* to sterile medium containing 1 percent proteose peptone medium (Difco). A good culture results after 24 to 48 hr in medium light at room temperature.

For studies of growth, or generation time (time for doubling of a population), the number of cells must be known at the start. Pull out medicine droppers into capillary tubes, or use capillary pipettes. Draw a sample from the culture, then seal both ends of the tube with modeling clay or melted wax. Count the number of individuals by examining the capillary tube under a dissecting microscope or compound microscope. Place the tubes in a Petri dish; make cell counts and record data. Groups of students may compare their results.

The effects of variables in the environment (such as pH, temperature, light, concentrations of nutrients, and so on) on the growth of *Tetrahymena* are possible investigations.

Growth of Euglena

You may want to investigate the effect of variables on the

---

[61] Giese, *Laboratory Manual in Cell Physiology.*
[62] Giese uses a medium made of lettuce powder from Difco, or Cerophyl (dried cereal grass) available from Cerophyl Laboratories, Inc. 4722 Broadway, Suite 259, Kansas City, MO 64112.

[63] Giese, *Laboratory Manual in Cell Physiology.*

growth pattern of *Euglena gracilis* grown on agar plates maintained in light or in liquid medium in small test tubes. Variables may include organic nutrients, salts, minimum medium, and enrichment with vitamin B_{12}. (*Euglena* is an assay organism in studies of this vitamin.)

Effect of variables on the growth of yeast cells

Many laboratory activities can be planned with a package of dried yeast cells and dilute molasses solution. Make cell counts or analyze turbidity using the methods described for *Chlorella* or other nonmotile microorganisms (pp. 458, 462, 464).

You may consider diluting a yeast-molasses suspension, then adding equal volumes to a number of small test tubes. Now freeze a tube at the start (0 hr); incubate the rest of the tubes, but remove one at 4 hr to freeze, and another at 8 hr to freeze, and so on. In this way, students have populations of yeast cells available for laboratory studies that have grown over various spans of time.

Small groups of students may count equal volumes of cells. For example, 2 drops of an ordinary pipette delivers approximately 0.1 ml; or use 0.1 ml pipettes. Count yeast cells in five fields under high power; or use a counting chamber if available (p. 276). Or devise other methods, such as scratching the surface of a plastic coverslip with a razor blade into

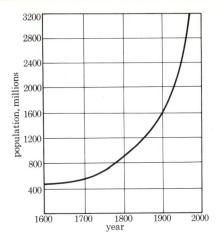

Fig. 6-81 Actual growth of human population since A.D. 1600.

quarters, then a quarter into eighths, to count cells in small areas and multiply by the total area in millimeters. Refer to method for counting pollen grains (p. 455) or bacteria (p. 603).

You may want to compare the growth rate of a yeast population with that of a population of *Chlorella* cells in light and darkness. Or study the effect of adding a protein source, such as a pea, to the yeast-molasses suspension to compare any gain in growth rate due to reproduction.

Growth of populations of microorganisms

Viable populations of cells increase exponentially: 1, 2, 4, 8, 16, 32, 64, ... (that is, 2^0, 2^1, 2^2, 2^3, 2^4, 2^5, 2^6, ... Notice in the growth curve that there are limits to the theoretical curve of "compound interest"—limits set by such environmental factors as the quantity of food. Then the rate of growth falls off so that the number of cells reaches a constant value. Populations of other organisms show similar growth curves (Fig. 6-80). Present data of the human population of the earth indicates that the same general growth curve will be approximated (Fig. 6-81).

There are several methods of measuring growth suitable for the high school or

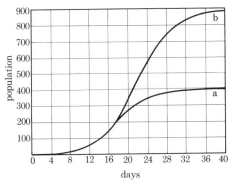

Fig. 6-80 Growth of *Drosophila*: curve *a* in a ½-pint bottle; curve *b*, in a 1-pint bottle. (After D. Marsland, *Principles of Modern Biology*, rev. ed., Holt, Rinehart and Winston, 1951.)

college laboratory.[64] One good method uses *Chlorella* as an experimental organism because it is nonpathogenic, nonmotile, easy to maintain in the laboratory, and useful in studies of photosynthesis and respiration (chapter 3). Yeast cells can also be used, but *Chlorella* is preferred because it can be inspected without the use of a microscope since the culture medium very soon begins to turn green.

Under optimum conditions, *Chlorella* has a very short lag period—about one day. It then moves rapidly into active growth for about ten days, until the rate begins to level off. In the later, stationary phase, *Chlorella* cells are large and vacuolated and have thick walls. The generation time for *Chlorella* is about 15 hr, more than twice that for yeast cells or bacteria. (Culture media for *Chlorella* are described in chapter 9.[65])

Counting cells in a population

The number of cells in a given volume of suspension must be known at the start of experiments in which growth rate is estimated (for example, under varying conditions of light, minerals, or wavelengths).

You may want to standardize your own simple ways to count cells in a population using a known dilution factor and measuring off the area of a field under the microscope (dilution techniques, p. 458). Or for some cells, such as bacteria, you may want to count cells in a bacteria smear. For more careful work, a counting chamber or hemacytometer is useful (p. 276). Cells like *Chlorella* and yeast approximate the size of blood cells,[66] so that students can count the number of cells in a carefully measured cubic millimeter.

Counting fields Plan with students a simple way to calculate an approximate change in population size of microorganisms. Mount three sample drops on a slide from a known dilution suspension, or from a specific region in the culture medium. Use *Paramecium*, *Blepharisma*, or other protozoa, or yeast cells, *Chlorella*, or *Euglena*.[67] Cover each drop with a coverslip. Now count cells in 5 fields under low power. Tally the 15 samples (5 fields × 3 drops) over a period of days or weeks depending on the rate of growth of the specific organisms used. Record the data at starting time (zero time) and number of days after inoculation of organisms into fresh culture medium. What is the lag phase? When is there exponential growth? (Refer to growth curves, p. 457.) What factors in the medium account for the decline phase? Construct a graph of your data.

There are many uses for this simplified technique to show variations in growth rate. For example, investigate the effect of light and dark as well as different wavelengths of light on photosynthesizers, of temperature, and of enriched medium (add a few drops of 3 percent nitrate or phosphate solution when test samples are photosynthesizers). Studies of the possible adaptability of *Paramecium* to increases in salt content in the medium show these organisms can adapt to small changes as measured by gain in population size. Welled slides may be used for culturing when space is limited. The area of the well can be calculated, and the size of the drop estimated. For example, depending on the bore, an ordinary pipette delivers approximately 20 drops for a volume of 1 ml (or use 1 ml pipettes).

Also consider cell mass (centrifuge) or cell volume, or turbidity as a measure of population growth (p. 466, 464).

What happens when two species of an organism, such as *Paramecium aurelia* and *P. caudatum* compete in the same ecological

---

[64] Also examine the many activities in A. S. Sussman, *Microbes: Their Growth, Nutrition, and Interaction*, BSCS laboratory block, (Boston: Heath, 1964).

[65] Urea is a good source of nitrogen for *Chlorella*, since urea does not support the growth of contaminants in the culture medium; further, it does not change the pH.

[66] Red blood cells are about 7 μ, white cells about 12 μ.

[67] When counting *Euglena*, heat the slide a bit to immobilize them.

niche? Culture each species separately, than mix equal volumes of both species in a dish of culture medium. Set up several dishes for comparison (see culturing, page 659). Which has survival value—*P. aurelia* or *P. caudatum*? In a mixed culture, *P. aurelia* continues to grow past the fourth day, while *P. caudatum* population shows a fairly rapid decline.

Have students graph the S-growth curve of each species when grown separately, and compare these curves with the populations in competition. *P. caudatum* has a higher reproductive rate, but the larger population excretes wastes into the medium. However, *P. aurelia*, the smaller form, is able to survive in the increasingly toxic medium.

Counting chamber   Use a known dilution of *Chlorella* or yeast cells in making cell counts with a hemacytometer. Mix the dilution well and insert by means of a calibrated pipette or syringe so that a film covers one counting chamber. The film spreads across the grid of the chamber and should show no air bubbles (technique, p. 276). Allow the cells to settle for about 3 min before counting. Under low power, locate the center section of 25 smaller boxes surrounded by double or triple lines (Fig. 4-9). The entire region can be seen in one field under low power. Under high power, count cells in 5 boxes (16 smaller squares in each of the boxes) using the technique for counting red blood cells.

Agitators for cultures of microorganisms   Some students may want to build agitators, which are used for keeping cultures of microorganisms stirred to facilitate distribution of nutrient materials and to promote growth. Some students have contrived an agitator from discarded parts of an old phonograph. The turntable and motor can be fastened to a wooden base, and to this may be attached wire test-tube baskets to hold a supply of test tubes or flasks.

Counting   microorganisms   (bacteria)   Umbreit describes Breed's method for counting organisms.[68] This method is especially useful for counting small cells such as bacteria in a sample, and can be done without a counting chamber or any special equipment. It has wide use for laboratory activities.

With a glass-marking pencil, mark off a 1-cm^2 on each of several microscope slides. With a sterile standardized loop, transfer a small drop of suspension of bacteria onto a slide, and spread the drop evenly over this *known*, marked-off area. The volume of fluid transferred by a given loop is about the same for each loopful and can be determined by weighing and averaging 50 or 100 loopfuls. Probably a loopful will carry 0.01 ml of fluid. This amount of fluid, then, has been spread over an area of 1 sq cm.

After the smear is dried, fixed, and stained (see p. 145), the cells are counted under an oil immersion lens. This lens usually has a field of 0.16 mm, and the area of the smear seen under the field is about $\frac{1}{5000}$ of the 1 cm^2 area over which the sample was spread. The area of the oil immersion field is $\pi r^2 = 3.14 \times 0.08 \times 0.08 = 0.02 \text{ mm}^2$.

Consider that the smear was spread over 1 cm^2, or 100 mm^2; therefore there are 100/0.02, or 5000 microscopic fields in the 1 sq cm. Umbreit continues this calculation, estimating that each organism seen in one field represents 5000 in the area of 1 sq cm. If the loop carried over 0.01 ml of the sample of the fluid, this would represent 5000 per 0.01 ml, or 500,000 per ml. It can be calculated, in this case, that each organism seen per field represents 500,000 per ml of the original sample.

Thus, in general, for any number of organisms N seen in a field, there are

$$N\left[\frac{100Y}{\pi(d/2)^2}\right]$$

organisms in the area of the slide, where d is the diameter of the microscopic field (in millimeters) and Y is the known area over

[68] W. Umbreit, *Modern Microbiology* (San Francisco: Freeman 1962).

which the total sample is spread (in square centimeters). To convert this to a general formula giving the number of organisms in 1 ml, multiply by $1/Z$, where Z is the total sample on the slide (in milliliters). Thus,

$$\text{organisms per ml} = \frac{400NY}{Z\pi d^2}$$

If, as in the preceding discussion, the sample (Z) is 0.01 ml and the given area is 1 sq cm, the equation can be reduced to

$$\text{organisms per ml} = N\left[\frac{4 \times 10^4}{\pi d^2}\right]$$

Turbidity as a measure of growth   A photometer of the type shown in Figure 6-82 may be made by students. Fairly good approximations can be obtained of changes in turbidity of test tube cultures of *Chlorella* or yeast cells, or of protozoa, bacteria, or *Euglena*. Measurements may have to be made over days or even weeks to note definite variations in relative rate of growth. (Recall, however, that in turbidity measurements both living and *dead* cells are counted.)

In the measurement of turbidity, a beam of light is passed through the culture medium, and the transmitted light is reg-istered on the photosensitive surface of a light meter (or a photoelectric cell). This, in turn, causes a current and a corresponding deflection of the needle of the galvanometer; the greater the illumination, the greater the current. A dial can be calibrated to measure foot-candles of intensity of illumination on a surface.

Since the intensity of illumination is inversely proportional to the square of the distance, when the meter is placed *exactly 1 ft away* from the lamp, the reading in foot-candles will be equal numerically to the candlepower of the lamp:

$$\frac{\text{intensity of illumination}}{\text{(foot-candles)}} = \frac{\text{candlepower}}{d^2}$$

With increased turbidity, due to the multiplication of cells in the culture medium, less light is transmitted through the suspension, which reduces the reading on the meter (foot-candles in this case).

Students will need to compare the meter reading to a standard sample of culture medium. Also, blanks, or empty test tubes, should be measured (similar test tubes must be used in all the experimentals).

If three cultures of *Chlorella* of varying density, as determined visually, are placed in the light of a 100-watt lamp, the ex-

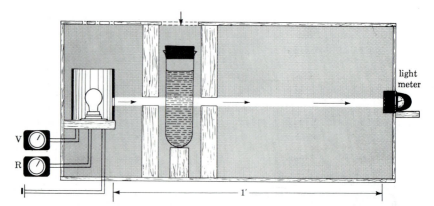

Fig. 6-82 Student-made photometer for measuring growth of microorganisms in a culture. The interior of the box must be lightproof, painted dull black, with black felt fitted around apertures. An aluminum shield on the light source concentrates light through a slit on one side, thus permitting light to pass in a direct line through the wood partitions and the culture to the light meter, placed 1 ft. away from the light source. The voltmeter is inserted into the circuit to maintain a constant voltage, and the rheostat permits variation of the current through the light bulb. Ventilation holes above the light bulb prevent the culture medium from becoming overheated.

posure meter readings might approximate the following order:

solution	reading (foot-candles)
blank tube	28
Knop's solution (control)	24
pale green suspension	20
light green suspension	12
deep green suspension	8

Counts of cells made of these three cultures in a counting chamber have given results of the following order:

solution	cell density (cells per cu mm)
pale green	12,400
light green	14,500
deep green	22,700

In this case, turbidity may be standardized in terms of the number of cells estimated in a counting chamber.

Clones of algae as a measure of growth   Students may count colonies of cells of algae in much the same way as clones of bacteria are examined in Petri dishes. A diluted suspension (see pp. 458, 459) of cells is added to cooling agar medium before the agar is plated into sterile Petri dishes. Or a measured sample of cells is poured into the Petri dish and cooled agar is then added.

The agar solution is a thin medium composed of about 1 to 1.5 percent agar prepared with Knop's solution or other culture (media, p. 658). Use mineral agar rather than nutrient agar in order to hold down the bacterial contaminants that are in the cultures of algae. (Agar melts at 100° C [212° F] and solidifies below 42° to 45° C [108° to 113° F]. Algae may be added to the agar cooled to 42° to 45° C. Some practice is needed because, on the addition of the cool algae culture, the agar will solidify rapidly so that it cannot be poured.) There is also an advantage to use Nalgene nutrient pads: cellulose pads impregnated with pre-measured quantities of dehydrated medium, with a membrane

lining the dish with grid markings for easy counting of colonies.[69]

Petri plates may be stacked and kept in moderate light if green algae are to be cultured. After about two weeks, hold the plates up to the light or against a white background to locate small pinpoint colonies as they begin to appear.

A student who has some manual dexterity may construct a colony counter using a diagonal ledge of wood with a hole in it. A light bulb is placed on the underside of the board and a Petri dish is placed over the hole (Fig. 6-83). Complete directions for making a colony counter may be available from the American Optical Co., Buffalo, NY. In the slanted side of a wooden light box, cut a hole the size of a Petri dish. Place a glass plate over the hole and rule vertical and horizontal guidelines about 1.5 cm apart on the glass. Position a Petri dish containing colonies on this glass plate, and with a hand lens, count the colonies that are within each square drawn on the glass plate. When light shines through the Petri dish, even small pinpoint colonies are observable.

In experimental work, students would inoculate the medium with a given volume of dilute algal suspension. When colonies arise, transfers may be made into small test tubes or welled slides to provide cells of the same heredity, that is, clones.

In a mixed culture of algae, *Chlorella* usually appears first in the colonies developing on agar. Prepare a slide for checking under the microscope.

The diameter of the colonies may be measured with a millimeter ruler, or the plates may be laid on graph paper and students can devise their own method for estimating numbers of colonies. Plastic Petri dishes that are already ruled for counting colonies are available.

If the colonies are evenly distributed over the surface of the Petri dish, students may count only those in a small section and, from this, calculate the clones in the total area of the sphere. Use a cork-borer

[69] Nalgene nutrient pads are available from Nalgene Company, Box 365, Rochester, NY 14602.

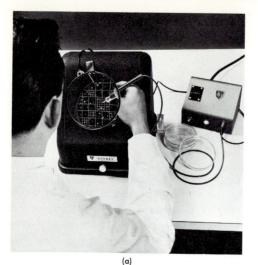

(a)

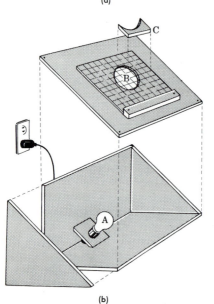

(b)

Fig. 6-83 Colony counters: (a) commercial counter with an electronic register used in microbiology laboratories; (b) plan for construction of colony counter (*A,* 25-watt light bulb; *B,* opening, the diameter of a Petri dish, covered by ruled glass; *C,* wooden block to hold Petri dish). (a, courtesy of American Optical Co.; b, after an American Optical Co. drawing.)

with a known diameter to measure off a plug or cylinder of seeded agar in the Petri dish. Then calculate the volume of this small cylinder ($\pi r^2 h$) and count the number of clones in this volume of cylinder.

From this, calculate the number of colonies in the known volume of agar plus medium used (possibly 15 ml plus 1 ml of culture). Welled slides with a straight-walled concavity usually are 16 mm in diameter and 3-mm deep.

Counting clones may be a useful measure, provided one remembers its limitations; the number of live cells is not determined. The count represents an estimate of the number of live cells that are capable of reproduction. A constant error in counting cells, particularly in methods which measure turbidity, is that dead cells cannot be distinguished from live ones.

Dispose of Petri dishes as though they were contaminated with bacterial cultures; cover with Lysol or other strong disinfectant.

Small prescription bottles with one flat side may be substituted for Petri dishes. Add 15 ml of agar medium to which a specific dilution has been added. Seed the agar medium while it is cooling at 45° C (113° F). Shake the bottle, and allow the agar to solidify on one flat side (Fig. 6-84). When the bottles are to be incubated, invert them, as is also customary for Petri dishes. Inverting the bottles prevents the condensed drops of moisture that usually accumulate on the "roof" of the container from dripping back onto the growing colonies on the surface of the agar medium.

Students can count the colonies through the glass. Suspensions of *Chlorella*, yeast, or bacteria may be seeded in the agar medium in these bottles.

Measuring growth by a gain in volume (or weight) The growth of populations of organisms over several weeks (or even over a month) can be studied by estimating the increase in volume or weight of cultures that have been maintained under a variety of experimental conditions.

Either allow tubes of a given volume of culture to stand undisturbed so that cells become packed at the bottom of a graduated tube (bulk volume), or centrifuge 10-ml samples of agitated experimental cultures, as well as the controls, in cali-

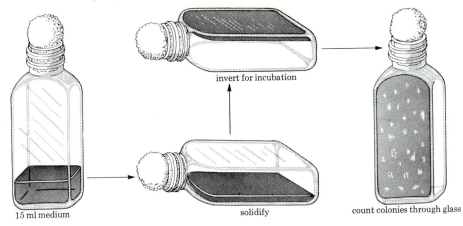

Fig. 6-84 Procedure for using prescription bottle cultures (in lieu of Petri dishes) for measuring growth of colonies of microorganisms. (After W. Umbreit, *Modern Microbiology*, Freeman, 1962.)

15 ml medium

solidify

count colonies through glass

invert for incubation

brated centrifuge tubes at a given speed for a given time. Then calculate the volume of sedimented, packed cells in relation to the volume of fluid (milliliters of algae per 100 ml of fluid). The packed volume of cells in a sedimentation tube per milliliter of culture may be determined daily. Or the cells from diluted samples of cultures can be counted, and the number of cells per milliliter of fluid can be estimated.

Wet weight may be obtained by running a sample collected from a centrifuge through a Buchner flask attached to a vacuum pump. Collect the cells on filter paper in the funnel and weigh. Or allow the filter disks to dry in a dust-free container or oven and weigh to obtain the dry weight of cells in 1 liter of fluid.

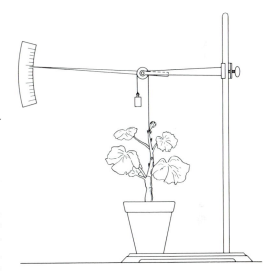

Fig. 6-85 Auxanometer for measuring vertical growth of a plant.

Measuring growth of plants

Growth in height of a total plant

Use a ruler to measure the total vertical growth of a plant. For more accurate work use an auxanometer (Fig. 6-85). Measure growth in width of stems or roots with a caliper. Also weigh the total plant; an increase in weight is an index of growth. (If a truer index of growth is desired, the dry weight of the plant should be taken.) Also refer to the role of growth regulators affecting height (chapter 5).

The growing regions of a plant

Region of greatest growth in roots
Germinate seeds of sunflower, pinto beans, peas, lima or other beans, as well as kernels of corn, in a germinating dish or

roll of toweling as described under the section on germinating seeds, p. 470. When the rudimentary roots are about 3 cm long, dry them carefully; with India ink and a pen, mark off 1 mm intervals along the length of each root taking care not to injure the tissue. Or fasten a thread on a jumbo paper clip and roll the thread on an inked stamp pad. Replace the seeds in the germinating dish and examine after 24 hr, 48 hr, and so on (Fig. 6-86). Record the data and prepare a graph to record total growth of the roots. Where is the region of greatest growth along the roots? Compare results with prepared stained slides of mitosis in onion root tips.

Growth of stems   Shoots of seedlings of beans, oats, wheat, or peas are suitable for marking with India ink. Blot the stems dry and mark intervals 1 mm apart. For curved stems (or roots) you may want to use wet string to measure the length of the stems and then use a ruler to measure the number of millimeters.

Is there a difference if the seedlings are kept in light or in the dark? Measure the difference in length between nodes.

Growth of leaves   Germinate seedlings of plants that have different shaped leaves: pinto beans, lima beans, peas, sunflower, squash, and so on. When small leaves are visible, flatten them against a glass or

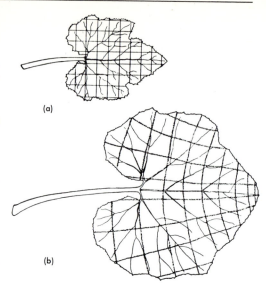

Fig. 6-87 Growth of a squash leaf: (a) leaf marked into squares with India ink; (b) same leaf after growth. Note that some squares have enlarged more than others. (From R. M. Holman and W. W. Robbins, *A Textbook of General Botany*, 4th ed., 1939; courtesy of John Wiley & Sons.)

plastic graphed square or other support marked off into squares of known dimensions. Outline the shape of the entire leaves on graphed paper. Then subsequent growth changes can be superimposed on the original outlines. Or trace equidistant marking on the leaves with India ink (Fig. 6-87). Examine the leaves from day to day; watch the changes in the pattern of the boxes. Note how growth in different planes gives pattern to different species of leaves. Estimate the surface area of each leaf (length × width).

As an interesting variation of this investigation, grow these seedlings—with marked roots, stems, and leaves—under different colored lights or different conditions of moisture and atmospheric pressure, or in media lacking specific minerals (see p. 479).

Further studies   For additional studies and activities describing patterns of growth of leaves, stems, and roots, as well as internal changes in structure, refer to

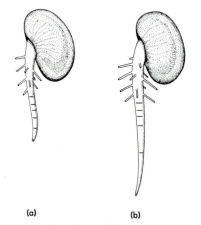

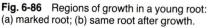

Fig. 6-86 Regions of growth in a young root: (a) marked root; (b) same root after growth.

the BSCS laboratory block by A. E. Lee, *Plant Growth and Development* (Boston: Heath, 1963). The block also describes activities on metabolism and regulation by auxins and gibberellic acid.

Refer to chapter 5 of this text for activities involving auxins and other plant hormones and photoperiodism. Some students may want, at this time, to look into Sir D'A. Thompson's classic *On Growth and Form* (2 vols., Cambridge University Press, New York, 1952). Is there a hereditary basis for the plane of growth of cells of leaves or of fruits that gives them their shape? While light is a factor in photosynthesis, light of varying intensity, duration, and in different regions of the spectrum have effects on morphology and differentiation in plants. Look into the chapters on "Cellular Basis of Growth" and "Symmetry," in E. Sinnott, *Plant Morphogenesis*, (New York: McGraw-Hill, 1960). Also refer to the genetic control of development in genetics texts.

Peculiarities in growth

The characteristic growth of twining plants is due to unequal distribution of cells, causing curvature in cylindrical organs. This pattern of growth is called nutation. Unequal distribution in growth of cells lying in one plane results in another kind of curvature—nastic curvature. If more cells are located at the lower layer of one plane, leaves or stems may curve upward giving a hyponastic curvature. In epinasty, the curvature is downward.

These growth characteristics are not regarded as tropic responses; tropisms are regarded as growth movements caused by asymmetrical conditions and resulting in a bending due to elongation of cells on one side (see tropisms, chapter 5).

Lemna

Duckweeds, genus *Lemna*, may well be useful in investigations of growth of a *whole plant*. These tiny seed plants that float on ponds and lakes consist of a leaf-frond about 2 to 4 mm long, with a threadlike rootlet. Duckweeds reproduce rapidly by producing new fronds from two pocketlike areas in the leaf. New fronds remain attached to older ones and rapidly cover a surface of a test tube, finger bowl, or battery jar when undisturbed (Fig. 6-88). See culture methods, chapter 9.

Have students record the increase in number of fronds over two weeks and graph the number of new fronds generated each day. These clones, all with the same DNA, are valuable in investigations of many variables in the environment. When the average rate of growth is known, students may investigate the effect of increase in nitrates or of phosphates to the medium. Is there a different rate of growth in light of different wavelengths—red or blue? Is there any alteration in chloroplasts under these variables? Do auxins or gibberellins affect *Lemna* (refer to p. 490)? What periods of light (or dakness) affect flowering in *Lemna* (refer to p. 375)[70]? Also refer to competition between two species.

Competition among plants: Lemna

Abiotic as well as biotic factors affect the rate of growth of plants as described in this

---

[70] W. S. Hillman, "The Lemnaceae or Duckweeds," *Bot. Rev.* (1961) 27:2, 221–287.

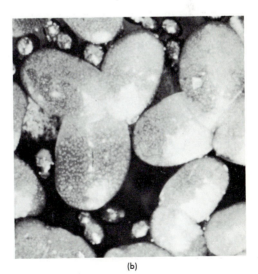

(b)

Fig. 6-88 Photo of *Lemna*, the duckweed, showing new fronds produced by old fronds.

chapter. (Also refer to chapter 8.) While many kinds of animals may exist in the same niche and make different demands on the ecological resources, green plants draw from a common, limited pool of materials: minerals, water, light, carbon dioxide, and oxygen. As a result, plants may compete for materials and interfere with one another's growth.

J. Harper describes the exponential rate of growth of fronds of duckweed (growing in 400-ml beakers of nutrient solution) as a population growth of 0.3 g per g per day.[71] Were this to continue, the fronds would cover an acre in some 50 days. But this exponential growth rate stops, and a linear growth rate results, increasing by 15 mg *per beaker* per day. In the *exponential* phase, the limiting factor is the number of fronds; in the *linear* phase of growth, the limiting factor is some environmental factor. A third or stationary phase of growth occurs in about 7 to 9 weeks, as many fronds die off as new ones grow. The limiting factor in this case seems to be light; the fronds under the thick mass growing on the surface do not get light and thus die off. (Comparisons can be made with land plants, especially young plants and the relationship between their growth rate and the amount of shading they receive.)

What happens when two species of duckweed, *Lemna minor* and *L. gibba*, are grown in the same culture? *Lemna gibba* gains dominance and *L. minor* is eliminated; this seems to be due to the presence in *L. gibba* of air sacs in the fronds, which probably raise the plants to the top of the mat of growth of duckweeds.

(Ecologists are also concerned with the condition in which one species of land plant may produce toxins that inhibit growth of other plants in their proximity.)

Germination of seeds

Surface sterilization of seeds   Because conditions favorable for germination of

[71] J. Harper, "Interference Between Plants," *The Times Sci. Rev.* (Autumn 1963) 13:7–8.

seeds also enhance bacterial and/or mold germination, the surfaces of seeds should be sterilized. Soak seeds for about 15 min in household bleach, such as Clorox (5 percent sodium hypochlorite); add 50 ml to water to make 1000 ml.

Germinating seeds   Seeds presoaked in water germinate faster. Seeds with hard seed coats impervious to water need to be scarified; either file an edge of the seed or insert a dissecting pin at one end. On a small scale, seeds may be germinated in moist vermiculite, sphagnum moss or soil, or in layers of wet toweling in Petri dishes or other containers. However, roots are often damaged in transplanting the young seedlings. This is the main advantage in using a paper roll for germinating seeds. Wet two sheets of toweling and lay both on a similar size sheet of wax paper to make three layers (Fig. 6-89). Fold back the top layer of toweling about 5 cm. Use this top sheet as a planting guide as you lay presoaked seeds in rows along the width of the second layer of toweling. (It is advisable to dip the seeds in a fungicide, such as Clorox, before this planting.) Now fold back the top toweling so it covers all the seeds. Loosely roll the three layers (toweling and wax paper) and then stand the roll in about 10 ml of water in a tall beaker. The wax paper serves to keep moisture in the toweling. Now incubate or keep in the dark at a temperature between 25° to 30° C (77° to 86° C). Allow some four to six days for germination, depending on the type of seeds. Pinto beans, sunflower, peas, oats, lima beans, cucumber, corn, melon, and millet all grow well in this technique. (Also refer to the role of phytochrome, p. 377.)

Large-scale germination   At times you may prefer to use a method that allows visibility while germinating many seeds at a time. Prepare a germinating dish with two porous clay flowerpot saucers (about 30 cm in diameter) and a smaller one that fits within as in Figure 6-90a. Wash all the porous saucers in hot water and Clorox at the start to reduce contamination espe-

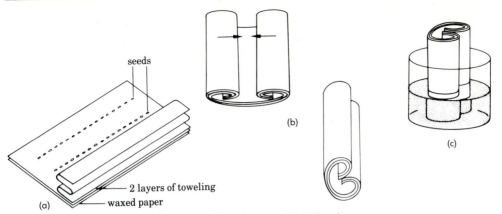

seeds

2 layers of toweling

waxed paper

(a)

(b)

(c)

Fig. 6-89 Paper roll for germinating seeds. (U.S. Department of Agriculture.)

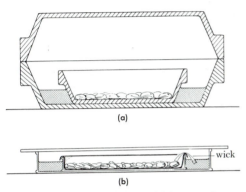

(a)

wick

(b)

Fig. 6-90 Germination dishes: (a) large-scale dish made from three porous flowerpot saucers; (b) smaller dish made from two glass dishes with blotting paper liner and filter paper wick in the inner dish.

cially by molds. Place the smaller saucer containing soaked seeds inside of one of the larger saucers; fill the peripheral "moat" with water. Now cover the whole with the other 30-cm saucer. The top cover can be raised to observe rate of germination without disturbing the seeds.

In some cases, you may want to use glass to increase visibility without removing a cover (Fig. 6-90b). Fit a small, flat dish within a larger glass dish and fill the outer moat with water. Then line the inner dish with moistened filter paper or toweling; two pieces of absorbent paper are then bent from the inner dish into the water of the outer dish to form a wick. Cover the dishes with a plate of glass or a small bell

jar. For rapid germination, cover the entire dish with heavy cloth or aluminum foil.

Examining seedlings

Presoaked seeds germinate faster since increased water content activates metabolic activity. If seedlings are to be examined for early development and to show root hairs, students may grow them in Petri dishes of moistened paper toweling. When the shoots become tall, replace the cover with a plastic tumbler. Be sure the moistened paper is enclosed within the tumbler, otherwise the paper will dry out rapidly.

Monocot and dicot seedlings   Compare the embryonic structures of a number of monocot and dicot seedling (Fig. 6-91). Remove the seed coat of a soaked dicot seed such as bean (pinto or lima bean, pea). Identify the two thick cotyledons. Separate these and locate the early embryo plant and compare with embryos of older seedlings. Identify the plumule, the tiny folded leaves, and the embryonic root. What is the function of the hilum? Of the micropyle? Sketch the embryos of early and later germinating seedlings. Have students observe the position of the embryo root (radicle) as a bulge on the seed near the micropyle. Also compare the progressive development of the hypocotyl and plumule

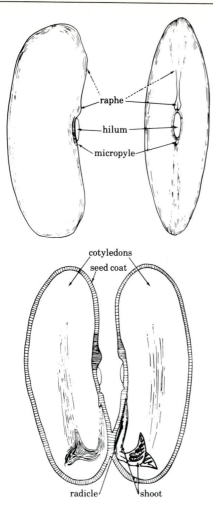

Fig. 6-91 *Phaseolus vulgaris* (bean): above, external side views of the seed; below, seed opened to show embryo. (Adapted from *Botany: An Introduction to Plant Biology*, 6th ed., by T. Elliot Weier, C. Stocking, M. Barbour. Copyright © 1982 by John Wiley & Sons, Inc. Reprinted by permission of John Wiley & Sons, Inc.)

as they differ in the growth of a bean and a pea seedling (Figs. 6-91, 6-92 and 6-93).

Next, examine a soaked corn kernel which is actually a fruit containing a seed. Cut a kernel in half lengthwise. Make the cut across from front to back of the kernel cutting through the center of the light colored area where the embryo is located (Fig. 6-94a). Locate the parts of the

embryo. Note the one cotyledon, the whitish area where the embryo lies. The rest of the kernel outside of the cotyledon makes up the endosperm. Note there is no hypocotyl. Students might sketch the parts of a very early germinating kernel and that of one some 3 to 4 days older (Fig. 6-95).

Growing seeds to show root hairs

This is a rapid method for use by students in the classroom. Students might line several small vials or test tubes with wet blotting paper and place radish seeds between the wet blotter and the glass walls of the vials. Enough for the entire class may be prepared. In a few days, students may examine the root hairs with a hand lens without disturbing their growth. Occasionally, add a few drops of water to the bottom of the vials; this will be absorbed in the blotter by capillarity.

When seeds larger than radish seeds are used, soak them first and pin the germinating seeds to the cork of a vial, one seedling per vial. When the vials stand upright, the root with the root hairs grows down into the vial.

Grow seeds of several different types of plants for comparative studies, for example, oats, wheat, and onion seeds (Figs. 6-96 through 6-98).

Students often question why all the seeds do not germinate. You may want to investigate seed viability, as well as the effects of environmental variables on the rate of germination, factors such as temperature, light and dark, red light, far red light, and so on. (Refer to pages 375–79.)

Test of viability in seeds

Viable seeds produce a dehydrogenase, which reduces tetrazolium salts; dead seeds do not. As a result, tetrazolium salts—which are colorless in the oxidized form and colored in the reduced state—can be utilized to detect and separate viable seeds from dead seeds. For example, when 2,3,5-triphenyl tetrazolium chloride

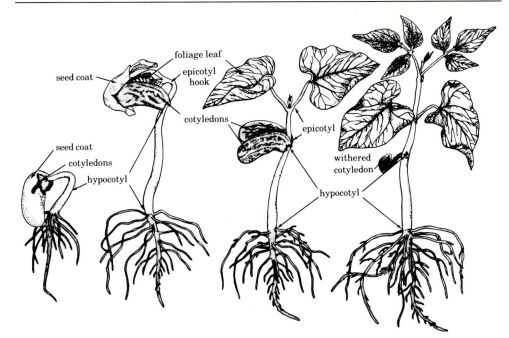

Fig. 6-92   *Phaseolus vulgaris*: stages in germination of the seed. (Adapted from *Botany: An Introduction to Plant Biology,* 6th ed., by T. Elliot Weier, C. Stocking, M. Barbour. Copyright © 1982 by John Wiley & Sons, Inc. Reprinted by permission of John Wiley & Sons, Inc.)

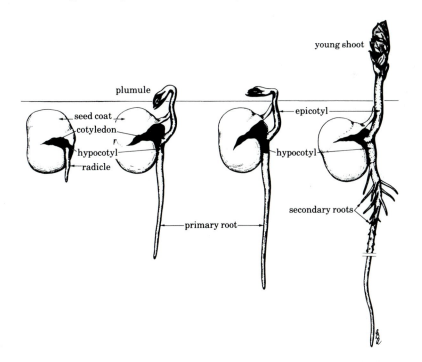

Fig. 6-93   *Pisum sativum* (pea): stages in germination of the seed. (Adapted from *Botany: An Introduction to Plant Biology,* 6th ed., by T. Elliot Weier, C. Stocking, M. Barbour. Copyright © 1982 by John Wiley & Sons, Inc. Reprinted by permission of John Wiley & Sons, Inc.)

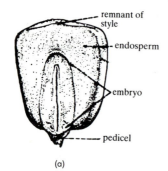

remnant of
style

endosperm

embryo

pedicel

(a)

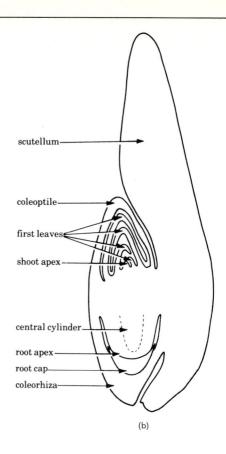

scutellum

coleoptile

first leaves

shoot apex

central cylinder

root apex

root cap

coleorhiza

(b)

Fig. 6-94 *Zea mays* (corn): (a) fruit or grain of corn showing embryo. The remnant of the style (strand of corn silk) is visible at the top; (b) lengthwise section of a developing embryo within the monocot corn kernel. (a, adapted from *The Plant Kingdom* by William H. Brown, Ginn and Company, Publishers. Copyright © 1935 by William H. Brown; b, adapted from *Botany: An Introduction to Plant Biology,* 6th ed., by T. Elliot Weier, C. Stocking, M. Barbour. Copyright © 1982 by John Wiley & Sons, Inc. Reprinted by permission of John Wiley & Sons, Inc.)

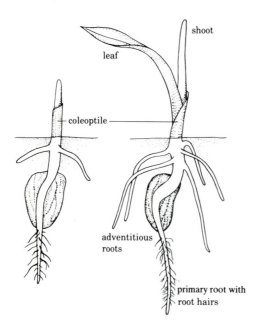

shoot

leaf

coleoptile

adventitious
roots

primary root with
root hairs

Fig. 6-95 *Zea mays:* stages in germination.

(TTC) is added to living tissues, it will be reduced by the dehydrogenase produced by the tissues and will change from colorless to red. (Other derivatives of tetrazolium may be black or blue in the reduced form.) The usual germination tests take up to ten days to perform, but the use of TTC reduces the test time to 24 hr or less.

Soak 100 barley seeds or seeds of other fast growing plants like squash, sunflower, or others in water for about 18 hr. Then germinate 50 seeds on moist filter paper in Petri dishes kept in the dark. Cut the remaining 50 seeds in half lengthwise; cut through the embryo plants as well (Fig. 6-99). Place these seeds in Petri dishes.[72] Add a 0.1 to 1.0 percent solution of TTC (*caution:* toxic) so that the seeds are just covered. (Add 0.1 g of tetrazolium chlor-

[72] As a *control,* you may want to treat similar seeds that have been boiled for about 10 min in the same manner.

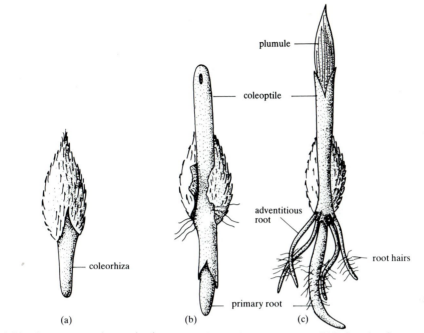

Fig. 6-96 *Avena*: stages in germination. (Adapted from *Plant Biology* by Knut Norstog and Robert W. Long.   Copyright © 1976 by W. B. Saunders Company.)

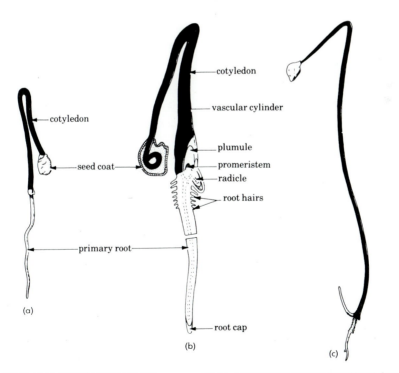

Fig. 6-97 *Allium cepa* (onion): stages in germination. (Adapted from *Botany: An Introduction to Plant Biology,* 6th ed., by T. Elliot Weier, C. Stocking, M.   Barbour. Copyright © 1982 by John Wiley & Sons, Inc. Reprinted by permission of John Wiley & Sons, Inc.)

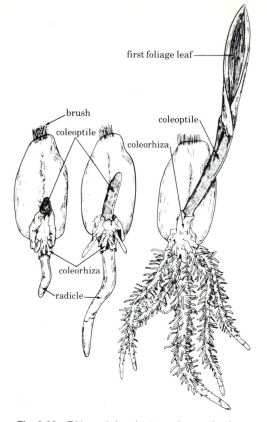

Fig. 6-98 *Triticum* (wheat): stages in germination. (Adapted from *Botany: An Introduction to Plant Biology,* 6th ed., by T. Elliot Weier, C. Stocking, M. Barbour. Copyright © 1982 by John Wiley & Sons, Inc. Reprinted by permission of John Wiley & Sons, Inc.

minated in the untreated dish with the number obtained in the rapid test using the indicator. Now apply the formula for percentage of seed viability:

$$\frac{\text{number of seeds stained red}}{\text{total number of seeds used}} \times 100$$

Synthesis of amylase in some early seed embryos

Many embryo monocots, such as corn and barley, need the starch and protein stored in the endosperm for growth. In 1960, Yomo and Paleg independently discovered that gibberellin is the key that unlocks the food supply in cereal seed endosperm. After a seed absorbs water, the embryo produces gibberellin which causes cells of the aleurone layer (the layer that covers the endosperm) to produce enzymes, especially amylases, that hydrolyze starch to sugar (Fig. 6-100).

When early embryos are removed from corn or barley, there is no production of amylase so that endosperm is rich in starch. Compare the glucose content in corn kernels with kernels from which the embryos have been removed. Or divide kernels so that a half contains the embryo and the other half lacks the early embryo. Then test both endosperm reserves for starch and for glucose.

ide to 100 ml water.) Soak for 2 to 4 hr in the dark at 20° C (68° F). The tetrazolium salt is colorless, water-soluble, but also toxic. By chemical or phytochemical action this indicator is reduced to triphenylformazan—an insoluble, bright red dye. As germination starts, chemical action changes the colorless solution into the insoluble red dye. The viable seeds may be identified by the staining of at least half of the scutellum, the whole of the shoot, and the regions where the adventitious roots will later develop (Fig. 6-99). This is a test of viability *before* germination.

Some days later, have students compare the number of viable seedlings which ger-

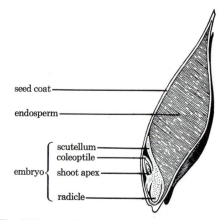

Fig. 6-99 Lengthwise section through a barley seed. (From *Plants in Action* by Leonard Machlis and John F. Torrey. San Francisco: W. H. Freeman and Co., 1956.)

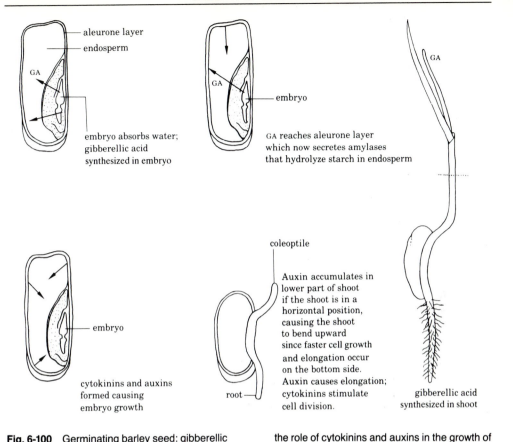

Fig. 6-100 Germinating barley seed: gibberellic acid synthesized in the embryo stimulates the secretion of amylases by the aleurone layer. Trace the role of cytokinins and auxins in the growth of the embryo.

Growth of seedlings in relation to stored food

Generally, the growth of seedlings is dependent upon the amount of stored food available to the embryos. Have students germinate several pinto or other bean seeds that have been soaked in water overnight. When the rudimentary roots, or radicles, which usually emerge first, become visible, select several seeds at this stage of germination and treat them in the following manner: (1) remove one cotyledon from several seeds; (2) remove both cotyledons from several other seeds; (3) as controls, leave the cotyledons intact on some seeds.

Suspend the seeds on a wire mesh with the roots is nutrient solution (p. 480), or plant them all in clean, moist sand. Measure their growth at regular intervals over the next few weeks and record increases in millimeters of roots and stems.

This preliminary laboratory activity may lead to further investigations which are developed in this chapter. Some questions will likely arise. Can nutrient solutions be substituted for the stored food in the seeds? What type of enzymes are synthesized? How do roots grow? What substances are responsible for the growth of roots or the bending of stems? How do leaves grow into different patterns? (Refer to page 468.) What factors affect the growth rate of parts of plants? Can we speed up the growth of roots on cuttings by using growth regulators? Can we retard the height of plants? How does the amount of light or the wavelength of light affect seed growth? (See also role of plant hormones, chapter 5.)

Effect of light and quality of light on the growth of seedlings

Light seems necessary for several of the formative processes as well as for photosynthesis. Seedlings in darkness generally grow tall and spindly, because they lack adequate supporting tissue and have long internodes. Most plants do not develop chlorophyll in the dark. Students may test this by growing some bean or pea seedlings in the dark and others in the light. In about a week the seedlings grown in the light should be sturdier, shorter (since there are short internodes), and greener than the etiolated plants grown in the dark. A few days of exposure to light should cause the development of chlorophyll in the pale leaves. Also, refer to the role of phytochrome in germination of seeds, page 377.

Mineral requirements of plants

Although green plants manufacture simple sugars during photosynthesis, certain minerals are needed for the total metabolic activities and growth of the plant. As a series of laboratory investigations, grow plants in soilless cultures (hydroponics). Show the effects on plants of a deficiency of macronutrient elements (N, K, P, Ca, S, Mg) and micronutrient elements (Fe, Mn, B, Zn, Cu).

Although the term "hydroponics" was originally applied to water-culture methods, the term has been extended to water, sand, and gravel cultures (Fig. 6-101). While detailed care is not possible in the classroom, emphasis may be given to the need for balanced solutions for the growth of plants. In our experience, these activities are successful if the following conditions are observed.

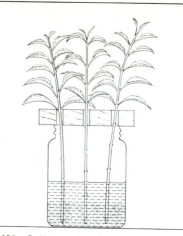

Fig. 6-101 Soilless culture for growing plants. Seedlings or cuttings supported by a perforated wooden or cork block are grown in a glass jar containing a layer of sterile sand and nutrient solution.

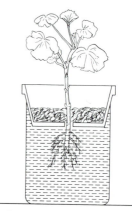

Fig. 6-102 Soilless culture for growing single plant or cutting. A porous flowerpot is immersed in a container of nutrient solution.

Provision should be made for roots to receive adequate aeration (by using an aerator pump). Glazed crocks are excellent containers; enameled pails may also be used. Metal cans or Mason jars (soft glass) may be used if a paraffin or asphaltum lining is applied to the metal or glass to prevent the solution from making contact with the fairly soluble surface. Glass jars may be painted black or covered with black oilcloth or aluminum foil to inhibit the growth of algae in the solution.

When single plants are used, plant them in wide, porous flowerpots (Fig. 6-102).

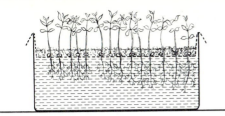

Fig. 6-103 Soilless culture for growing seedlings bedded in glass, wool, or sphagnum moss in a paraffined zinc mesh trough. The roots grow into the nutrient solution.

Wash the pot thoroughly in hot water and household bleach and rinse well. Choose a jar into which 5 cm of the pot will fit. Fill the jar with the nutrient solution and fit the pot into this. Plant the cutting in the pot in vermiculite, glass wool, or sphagnum moss. As the cutting grows, supports may be made of sections of plywood through which holes have been bored. Instead of wood supports you may find it useful to make supports for cuttings by molding paraffin sheets to fit as covers on the pot; then make holes in the paraffin by means of a hot iron rod.

Seedlings or cuttings need not be grown in solution alone; they can be grown in coarse sand cultures. For this method, use a mixture of peat moss (about 30 percent by volume) and clean sand. Peat moss (sphagnum) retains moisture and also reduces the alkalinity of the medium. Fill a small flowerpot with the sand-peat moss mixture and fit it into a container of solution. Run glass wool, placed in contact with the roots of the plants, through the drainage hole in the flowerpot, into the nutrient solution. This wick transports nutrient solution to the roots with less evaporation than would result if you watered the sand from the top of the mixture.

Although most plants can be grown by these methods in the classroom, certain plants are especially resistant to poor environmental conditions. Good results should be forthcoming with wheat, rye, corn, radish, sunflower, castor beans, peas, beans, tomatoes, and potatoes. (*Precaution:* Avoid growing these plants in a laboratory where illuminating gas or bottled gas is used since this causes wilting and loss of leaves; solutions for hydroponics, pp. 480–82.)

Effect of mineral deficiency

This is a useful laboratory activity for many groups. Students may prepare a series of test tubes, each supporting a growing seedling held fast in a one-hole stopper. Or insert seedlings into a hole in foam plastic (Fig. 6-104). In each of the containers, omit a different salt from a nutrient solution such as Knop's or Sachs' (p. 480). Later, compare each seedling's growth with seedlings grown in the total

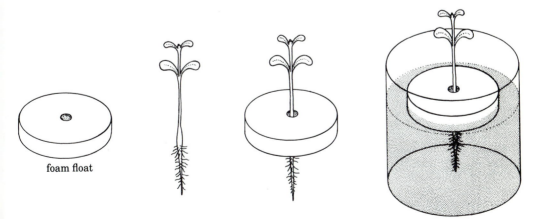

foam float

Fig. 6-104 Young seedling inserted into a hole in a disk of foam plastic which acts as a float in a jar of nutrient solution.

nutrient solution. Also test the effect of different degrees of acidity on plant growth. For example, while potassium dihydrogen phosphate, KH_2PO_4 gives the solution a slightly acid reaction, potassium monohydrogen phosphate, K_2HPO_4 produces an alkaline solution. Do these salts have an effect on growth of plants?

There are several valuable keys for identifying nutrient deficiencies in plants. In these keys, the main traits described refer to changes in chlorophyll and to types of plant growth. All these observations may be made by students. In addition, the work has a practical application in some school areas. Useful keys may be found in agricultural references.[73]

Some nutrient solutions for plants

Plants seem to have a wide tolerance for different salts, so no one solution of nutrient salts serves all plants.

Some of these solutions date back to the original ones of Knop, Sachs, and Pfeffer, developed nearly 100 years ago. Other solutions show modifications, including varying quantities of trace elements. Some trial-and-error activity may be needed to learn which of these solutions is the most fruitful in a specific schoool situation. (Other culture solutions for algae are described in chapter 9, and buffer solutions on pages 159, 741.)

KNOP'S SOLUTION   Weigh out the following salts and dissolve in 1 liter of water.

$Ca(NO_3)_2 \cdot 4H_2O$	0.8 g
KNO_3	0.2 g
KH_2PO_4	0.2 g
$MgSO_4 \cdot 7H_2O$	0.2 g
$FePO_4$	trace

SACHS'S SOLUTION   Dissolve the following salts in 1 liter of water. (The calcium phosphate is only partially soluble.)

KNO_3	1.0 g
NaCl	0.5 g
$CaSO_4$	0.5 g
$MgSO_4$	0.5 g
$Ca_3(PO_4)_2$	0.5 g
$FeCl_3$ (1% solution)	trace (1 drop)

PFEFFER'S SOLUTION   Weigh out and dissolve the following salts in 1 liter of water. (This is a slight variation of Knop's solution.)

$Ca(NO_3)_2$	0.8 g
KNO_3	0.2 g
KH_2PO_4	0.2 g
$MgSO_4 \cdot 7H_2O$	0.2 g
KCl	0.2 g
$FeCl_3$	trace

SHIVE'S SOLUTION   Dissolve the salts listed below in 1 liter of water. The iron and phosphate salts should be added first to the water, since they precipitate readily. (Enough of the minerals will remain in the solution, however.)

$Ca(NO_3)_2 \cdot 4H_2O$	1.23 g
KH_2PO_4	2.45 g
$MgSO_4 \cdot 7H_2O$	3.70 g
$FePO_4$ (0.5% solution)	1.0 ml

HOAGLAND'S SOLUTION   Weight out the following salts and dissolve in 1 liter of water.

$Ca(NO_3)_2 \cdot 4H_2O$	0.95 g
KNO_3	0.61 g
$MgSO_4 \cdot 7H_2O$	0.49 g
$NH_4H_2PO_4$	0.12 g
ferric tartrate	0.005 g

Agricultural experimentation stations may vary the ingredients of some of these well-known solutions. Yet most of the nutrient solutions contain six of the essential macronutrient elements for plant growth: phosphorus, nitrogen, sulfur, potassium, calcium, and magnesium.

One practical activity is growing seedlings or cuttings in Hoagland's solution minus the nitrogen compounds. Later test the petioles of plants for nitrates using diphenylamine sulfate reagent. (*Caution:* skin, eye irritant.) Place thin sections on a slide in a few drops of the reagent; a blue

---

[73] Dept. of Agriculture yearbooks; Another useful guide is the text by T. Weier, C. Stocking, and M. Barbour, *Botany*, 6th ed. (New York: Wiley, 1982).

color indicates the presence of nitrates.

Most workers in this field advise the addition of trace elements or micronutrient elements to all nutrient solutions. These trace elements needed in minute quantity include boron, iron, manganese, copper, zinc, and probably molybdenum.

HOAGLAND—ARNON NUTRIENT SOLUTION[74] This is a basic water-culture medium for growing plants without soil. *Stock Solutions of Major Elements* Prepare 1 M stock solutions of *each* of the following salts with distilled water.

KH_2PO_4	131.1 g/liter
KNO_3	101.1 g/liter
$Ca(NO_3)_2 \cdot 4H_2O$	236.2 g/liter
$MgSO_4 \cdot 7H_2O$	246.5 g/liter

Iron Solution Versene or Sequestrene are chelates of iron. Prepare a 5 percent solution with water. Then dissolve 200 ml of the solution in distilled water to make up to 1 liter; this stock solution is 10,000 ppm of iron. In use, dilute by adding 1 ml of stock solution to 1 liter of distilled water to get 10 ppm of iron concentration.

Solution of Micronutrients Dissolve the following trace elements together in 1 liter of distilled water to make a stock solution. Add the micronutrients in the order given to avoid precipitation:

H_3BO_3	(boric acid)	2.50 g
$ZnCl_2$	(zinc chloride)	0.50 g
$CuCl_2 \cdot H_2O$	(cuprous chloride)	0.05 g
MoO_3	(molybdenum oxide)	0.05 g
$MnCl_2 \cdot 4H_2O$	(manganese chloride)	0.50 g

Solution for use from Stock Solutions Combine the following volumes of the 4 solutions of major elements, iron solution, and solution of micronutrients in 500 ml of distilled water:

KH_2PO_4	1 ml
KNO_3	5 ml
$Ca(NO_3)_2$	5 ml
$MgSO_4$	2 ml
iron solution	1 ml
micronutrient solution	1 ml

Make up this solution to 1 liter with distilled water. In preparation for use, dilute solution to ⅓, ½, or ¾ strength.

KNUDSON'S CULTURE MEDIUM Weigh out and mix:

$Ca(NO_3)_2 \cdot 4H_2O$	1.0	g
KH_2PO_4	0.25	g
$MgSO_4 \cdot 7H_2O$	0.25	g
$(NH_4)_2SO_4$	0.50	g
$FeSO_4 \cdot 7H_2O$	0.025	g
$MnSO_4 \cdot 4H_2O$	0.0075	g

Add 1 ml of a micronutrient solution, either Trelease and Trelease, or Hoagland's. Dissolve in 500 ml of distilled water and mix well. To this add 20 g sucrose and bring the volume to 1 liter.

To solidify the medium, add 12 to 15 g of agar (1.2 to 1.5 percent) and stir while heating. Autoclave or place in a pressure cooker to sterilize. Artificial seawater may be added in place of the micronutrient solution.

TRELEASE AND TRELEASE MICRONUTRIENT SOLUTION[75] This micronutrient solution combines the elements Zn, Si, Al, I, B, Cu, Ni, Na, Li, As, Co, and Mn. Prepare a stock solution by dissolving the following salts in distilled water and bringing the volume to 1 liter.

$ZnSO_4 \cdot 7H_2O$	5.0	g
Na_2Si	2.5	g
$Al_2(SO_4)_3 \cdot 18H_2O$	10.0	g
KI	0.625	g
H_3BO_3	2.5	g
$CuSO_4 \cdot 5H_2O$	0.5	g
$NiCl_2 \cdot 6H_2O$	1.25	g
$NaCl$	2.5	g
$LiCl$	1.25	g
As_2O_3	0.625	g
$CoCl_2 \cdot 6H_2O$	1.25	g
$MnSO_4 \cdot 4H_2O$	2.5	g

[74] As described in *Growing Plants Without Soil for Experimental Use*, Publication #1251, U.S. Dept. Agriculture, December 1972. Available from Supt. Documents, U.S. Government Printing Office, Washington, DC 20402.

[75] S. F. Trelease and H. M. Trelease, "Changes in Hydrogen-Ion Concentration of Culture Solutions Containing Nitrate and Ammonium Nitrogen," *Amer. J. Bot.* (1935) 22:520.

Acidity of solutions

Have students check the pH of a solution with an indicator (as described on page 156), pH meter, or Hydrion paper. Most plants grow best within a range between 4.5 and 6.5 (acid); a pH over 7 (alkaline) may retard plant growth. Further, when the solution is slightly acid, iron salts are more soluble. Check the pH every few days, since the solution may become alkaline as nitrogen-bearing ions are absorbed. If potassium ions are absorbed rapidly, sulfate ions accumulate and produce an acid solution (see buffer solutions, pages 159, 741).

To avoid these changes in pH, some workers prefer a drip method to immersion of the plants directly in the solution. If you cannot get an intravenous drip bottle, make one by filling a bottle with nutrient solution and sealing a siphon with a clamp to the stopper. Invert the bottle, and at regular intervals release the clamp to moisten the bedding in which the plants grow. When the plants are grown without bedding, it is easier to change the solution than to use a drip bottle.

Growing seedlings in solution

Begin with a simple method that is successful with small, fast-growing seeds— especially cereal plants, which require little support. Use surface sterilized seeds (p. 470).

To prepare support for the seedlings, you may want to immerse a layer of cheesecloth or mosquito netting in hot paraffin and spread it across the top of a container of fluid. Allow a little slack in the center to dip into the solution. (Plant the seeds in this depression.) You may prefer to add sterile sand or vermiculite to the bottom of the container. This gives some support to plants rooted in the sand. Or bore holes into a foam plastic block (or cork), which can be fitted over a jar to give support to stems (Fig. 6-104). Or you can just float a sturdy seedling in the solution.

Use zinc (or galvanized iron) mesh troughs to make a more substantial support for larger plants. These supports must be given several coats of paraffin or asphaltum to prevent toxicity. You may try to use nontoxic stainless steel mesh. Shape 5-mm mesh into a trough with flanges to support it against the sides of the jar.

Most plants need a surface area of about 10 cm^2 of nutrient solution and an air space of about 5 cm^2. When large numbers of seedlings are planted in crocks, use a fish-tank aeration pump to provide a stream of air through the solutions to enhance root growth. Other specialized devices for ensuring aeration and circulation of nutrient solutions are described in texts in hydroponics.

First, cover the mesh trough with a 5-cm layer of straw, glass wool, coarse sawdust, or sphagnum moss. (You may want to test the possible toxicity of the bedding beforehand by germinating some seeds in the specific material you plan to use.) Next, plant the surface-sterilized seedlings in the bedding.

Add enough solution to the container to cover half of the roots of the seedlings. Then cover the other half of the roots with another 5-cm layer of bedding. This second layer excludes light and provides a moist air space around the roots. Keep the jars in medium light. As the plants grow, support them with stands. Change the nutrient solution weekly.

Assay of vitamin B₁₂

The flagellate *Euglena* is one of many microorganisms that is unable to manufacture its own vitamin B_{12}, even though it is an essential factor for growth. (Vitamin B_{12} is the anti-pernicious anemia factor.) Because of this dependence on an outside source of vitamin B_{12}, the rate of growth of *Euglena* may be used as an assay measure of the presence and concentration of vitamin B_{12} in a substance.

Begin with suspensions of *Euglena* of known cell density. Using a photometer, such as the student-made instrument shown in Figure 6-82, students may measure the growth of *Euglena* in each suspension. They may then prepare a graph, plotting the concentration of cells on the *Y*-axis and the growth variable (time) on the *X*-axis. Then, the concentration of vitamin B_{12} in the unknown suspension may be estimated by determining the ratio of the *Y*-intercepts of the unknown suspension and the known suspension.

For culturing *Euglena*, see chapter 9. Add glutamic acid, malic acid, and asparagine to the basic culture; culture medium available from Difco Co. (Commercially, B_{12} is a by-product of *Streptomyces griseus* from which streptomycin is obtained.)

Nutritional deficiencies in Neurospora

The use of deficiency diets in studies of heredity of the pink mold *Neurospora* is described in chapter 7 in isolating mutants.

Plant tissue culture techniques

For beginners, culturing plant tissue is more successful than the culture of animal tissues. Recall the early studies of plant tissue culturing: In 1922, W. J. Robbins grew root tips of tomato plants in an aseptic medium to which were added vitamins from yeast extract, sugars, and minerals; in 1939, P. R. White isolated root tips of tomato plants grown and subcultured into fresh medium for 30 years; in the 1960s, F. C. Steward isolated phloem cells from carrot roots. His aseptic culture medium included coconut milk, which supports growth of plant cells and contains a growth substance.[76] The phloem cells grew and organized into small "embryos" that differentiated into whole carrot plants

demonstrating that specialized individual cells had retained the potential to develop into whole plants. Terminal meristem of geranium plants also have been cultured, as have the anthers of *Datura* (Jimson weed) and tobacco; these have also given rise to whole plants. In the case of anther cells, the new plants are monoploid. These experiments have been duplicated with members of the same plant family (tomato, tobacco, and potato); this is a useful tool in genetic studies of pure line breeding.

You may also want to refer to the culturing of protoplasts, or single plant cells, from which the cell wall has been removed by enzymes—again, an example of cells that differentiate into whole plants.[77] Consider also the possibilities inherent in hybridizing plant cells of two different species. Students also may want to read current work in journals on the transferring of bacterial genes into cells of tomato and other plants to improve their nutritional value as a food crop.[78]

As an introduction to aseptic techniques and the field of tissue cultures, students may dissect out the *entire* embryo of a soaked seed to observe its growth and differentiation in sterile tissue culture medium. This may be a time to compare early embryos of plants of diverse genera; and a time to reflect on the expression of DNA in seeds, and the eggs of animals.

Then students may isolate parts of a plant—root tips, or growing meristem (bud tissue)—to investigate the potential for developing a whole plant.

Further questions arise: What is the effect of growth substances—auxins and gibberellins—on growth of lateral roots and the growth of stems? In general, auxin promotes root growth while gibberellin promotes shoot elongation (see also the effect on dwarf pea plants, page 374); kinetin promotes shoot growth. Kinetins

[76] F. C. Steward et al, "Growth and Development of Cultured Cells," *Science* (1964) 143:20.

Also refer to S. K. Pillai and A. Hildebrandt, "Induced Differentiation of Geranium Plants from Undifferentiated Callus *in vitro*," *Amer. J. Bot.* (1969) 56:1; and for additional techniques, J. P. Nitsch, "Experimental Androgenesis in *Nicotiana*," *Phytomorphology* (1969) 19:389.

[77] P. Carlson, "The Use of Protoplasts for Genetic Research," *Proc. Nat. Acad. Sci.* (1973) 70:2; also J. Shepard, "Regeneration of Potato Plants from Leaf Cell Protoplasts," *Sci. Am.* (May 1982).

[78] I. Vasil, ed. *Perspectives in Plant Cell and Tissue Culture* (New York: Academic Press, 1980).

(or cytokinins) are synthesized in root tips and reach shoots through the phloem; these cytokinins and some other growth regulators stimulate cell division and differentiation of organs in seed plants. Investigate how different ratios of indoleacetic acid (IAA) and kinetin affect differentiation of excised tobacco pith (p. 489).

In general, the apical meristem of tobacco, chrysanthemum, geranium, sunflower, and bean plants give successful results in tissue cultures. Transfer methods are similar to those used in bacteriology. Culture media should contain both macronutrients and micronutrients. Nonphotosynthesizing plant parts used in tissue culture require a source of energy and vitamins such as sugars and vitamin B complex. Since many vitamins are made in shoots of plants, the excised root tips must have vitamins added to the medium. After autoclaving, growth regulators can be added to the medium for more advanced studies. Consider the antagonistic or synergistic effects of two or more growth regulators on differentiation of a plant from a small calluslike undifferentiated excised part. Media may be purchased ready made from Difco, Fisher Scientific, Eastman Organic Chemicals, and other supply houses (see Appendix). White's basal medium (preparation follows) can be supplemented with coconut milk which contains kinetin (cytokinins). Murashige's medium (preparation, p. 485) is available from Carolina Biological Supply Co.

Further suggestions for investigations are described in classic papers by White,[79] Murashige,[80] and Paul,[81] and by Mitchell in a Department of Agriculture Hand-book.[82] Consult current abstracts and journals for new work (for example, *Annual Review of Plant Physiology*).

A student-made sterile box (described on p. 485) is useful for making transfers of tissues, since it reduces air currents carrying mold and bacteria spores. More elaborate boxes or work cubicles may be made of clear plastic sheets or plywood if permanent space is available.

Preparation of media

White's medium   Purchase from a commercial source, or prepare the following *five stock* solutions, each in 1 liter of distilled water. Add sucrose as a final step.

	solution A	
$Ca(NO_3)_2$		2.0 g
KNO_3		0.8 g
KCl		0.65 g
NaH_2PO_4		0.165 g
	solution B	
$MgSO_4$		36.0 g
	solution C	
$MnSO_4$		0.45 g
$ZnSO_4$		0.15 g
H_3BO_3		0.15 g
KI		0.075 g
$CuSO_4$		0.002 g
Na_2MoO_4		0.021 g
solution D (must be freshly prepared)		
$Fe_2(SO_4)_3$		0.25 g
	solution E	
glycine		0.3 g
thiamine		0.01 g
pyridoxine		0.01 g
nicotinic acid		0.05 g

In preparing White's solution use only small amounts of the stock solutions. For example, begin with 500 ml of distilled water; to this add 100 ml of stock solution A, and 10 ml each of solutions B, C, D, and E. After these have been thoroughly mixed

[79] P. White, *The Cultivation of Plant and Animal Cells,* 2nd ed. (New York: Ronald, 1963).

[80] T. Murashige, "Suppression of Shoot Formation in Cultured Tobacco Cells by Gibberellic Acid," *Science* (1961) 134:3474, 280, and "Plant Propagation Through Tissue Cultures," *Ann. Rev. Physiol.* (1974) 25:135–166. Also refer to M. P. Kefford and A. H. Rijven, "Gibberellin and Growth in Isolated Wheat Embryos," *Science* (1966) 151:3706, 104–105.

[81] J. Paul, *Cell and Tissue Culture,* 5th ed. (London: Churchill Livingston, 1975).

[82] J. Mitchell and G. Livingston, *Methods of Studying Plant Hormones and Growth-regulating Substances,* Agriculture Handbook #336 (February 1968). Available from Supt. of Documents, Washington, DC 20402.

by shaking, add 20 g of sucrose and make up to 1 liter by adding distilled water.

Now divide the solution by pouring 50-ml amounts into smaller flasks, plug with cotton; autoclave for 15 min at a pressure of 15 lb per sq in. These flasks may be stored for a few days in a refrigerator until transfers can be made. Since the solution contains organic compounds it cannot be stored indefinitely. Once they have mastered this tissue-culture technique, students may try to grow excised sections of roots, stems, and leaves in nutrient solutions to examine how differentiation and growth take place.

Agar may be added in the preparation of the medium if a solid medium is needed (5 to 10 g agar to 1 liter of fluid).

Murashige's medium   Prepare the following *four stock* solutions:

I. STOCK SOLUTIONS OF SALTS To avoid precipitates of salts, dissolve each of the eleven salts listed below separately in 75 ml distilled water and then combine them and bring the total volume to 1 liter with distilled water:

ammonium nitrate	NH_4NO_3	16.5	g
potassium nitrate	KNO_3	19.0	g
calcium chloride	$CaCl_2$	4.4	g
potassium phosphate	KH_2PO_4	1.7	g
manganese sulfate	$MnSO_4 \cdot 4H_2O$	0.22	g
zinc sulfate	$ZnSO_4 \cdot 7H_2O$	86	mg
boric acid	H_3BO_3	62	mg
potassium iodide	KI	8.3	mg
sodium molybdate	$Na_2MoO_4 \cdot 2H_2O$	2.5	mg
cupric sulfate	$CuSO_4 \cdot 5H_2O$	0.25	mg
cobaltous chloride	$CoCl_2 \cdot 6H_2O$	0.25	mg

II. STOCK SOLUTION OF MAGNESIUM SULFATE Dissolve in 1 liter of distilled water:

$MgSO_4 \cdot 7H_2O$	37 g

III. STOCK SOLUTION OF EDTA AND IRON Dissolve in 200 ml distilled water and keep refrigerated:

ferrous sulfate	$FeSO_4 \cdot 7H_2O$	1.1 g
sodium edetate	$Na_2 \cdot EDTA$	1.5 g

IV. STOCK SOLUTION OF VITAMINS AND GLYCINE Dissolve vitamins and glycine in 100 ml of distilled water and refrigerate:

Myo-inisitol	2.5	g
thiamine hydrochloride	10	mg
nicotinic acid	50	mg
pyridoxin hydrochloride	50	mg
glycine	200	mg

To prepare Murashige's medium for use, select stocks I, II, III, and IV in the following proportions: I. Stock: 100 ml; II. Stock: 10 ml; III. Stock: 5 ml; IV. Stock: 1 ml. Combine these volumes of stock solutions and add 30 to 40 g glucose or sucrose that has been dissolved in 200 ml distilled water. Bring the total volume to 1 liter with distilled water.

For a solid medium, add 5 to 8 g of agar for each liter of solution (0.5 to 0.8 percent agar).

The final pH of the solution should be about 6; adjust with small amounts of NaOH or HCl.

Autoclave or use a large pressure cooker for tubes of medium; refrigerate for a few days. However, like White's medium this contains organic matter, and cannot be kept too long.

Sterile work box

To reduce contamination of instruments, culture medium and living tissues, work within a confined area. Students may construct a sterile work box from a sturdy carton, heavy gauge plastic sheeting, and aluminum foil. A convenient size, depending on space available, is a box 3-ft high, 2-ft long, and 2-ft wide (Fig. 6-105). Cut the sides of the carton so that they slant toward the front thereby providing a higher rear area. Ring out two circles for openings to insert your hands. Line the box with aluminum foil or surface sterilize the carton with diluted household bleach.

Stretch the clear plastic sheeting over the slanted top and fasten securely with masking tape.

Keep the box away from drafts and insert materials needed to excise tissues from plant seedlings, or meristems or other materials inside the box. For example, along with the sterile test tubes of liquid or solid medium, filter paper cones (presoaked in rubbing alcohol) upon which to place tissue when liquid medium is used; also include sterile forceps, scissors, transfer needles, razor blades, sterile Petri dishes, a bottle of rubbing alcohol, and a container of household bleach (to surface sterilize seeds and outer plant tissues). Also include a flat aluminum tray with sterile cotton in which to rest instruments. You may need a gooseneck lamp outside the box to increase visibility. Wash hands in soap and in rubbing alcohol before beginning.

Without a work box, students may work in a draft-free area and use bacteriological techniques: flame needles, scalpels, mouth of tubes, and flasks.

Growth of excised embryos

Use seeds that germinate rapidly, for example, pinto beans, peas, sunflower, radish, oats, or morning glory. Sterilize the surfaces of the seeds by placing them for 15 to 30 min in household bleach (diluted 1:9 with water), or in a 0.5 percent solution of sodium hypochlorite to which a little detergent has been added. Then soak the seeds in sterile water to hasten germination. Be sure to observe aseptic techniques; you may want students to try to use the student-made sterile box previously described.

Excise the embryo of each seed by cutting with a sterile scalpel near the hilum, then around the seed. Remove the seed coats, separate the embryo from the cotyledons or from the endosperm of each seed. Use sterile transfer needles to place each embryo in a separate small flask or test tube of sterilized solid medium, using the medium of either White or Murashige (p. 484). If necessary, make a small hook

in the transfer needle to catch or secure the embryo tissue. Maintain the cultures in the dark at 28° to 30°C (82° to 86°F) for four to five days. Some practice may be necessary; students should prepare several tubes of excised embryos. They may compare rate of growth of a whole seed, an excised embryo, and even test the effect of the addition of growth regulators to the medium. For example, what is the effect of IAA (indoleacetic acid) or of gibberellin? Or cytokinin? (preparation, pages 489, 374). Does light of different wavelengths (red and far red) affect growth of embryos? (You may want to refer here to phytochrome and to photoperiods, pages 375–79.)

Gibberellic acid solution Dissolve 1 g of crystalline gibberellic acid in 5 ml of 95 percent ethyl alcohol. To this add a few drops of baby shampoo as a wetting agent. Bring the volume to 1 liter with warm water. This makes a solution containing 1000 ppm of gibberellic acid.

Also refer to the interaction of IAA and kinetin in the differentiation of tissue, as demonstrated by Skoog and Miller and described on pages 489–90.

Growth of excised roots of seedlings

Seeds that germinate quickly are recommended; tomato and tobacco seeds are often used in tissue culture investigations. Dissect a tomato with sterile scalpel and forceps and transfer the seeds to sterile Petri dishes lined with filter paper that has been soaked in sterile water. Dry tomato seeds or tobacco seeds may be substituted; surface-sterilize for 15 to 30 min in 0.5 percent aqueous solution of sodium hypochlorite or household bleach (diluted 1:9 water) to which a trace of detergent has been added. Germinate the seeds in the dark at 25°C (77°F) in the Petri dishes. After five to eight days, students may excise 1-cm root tips from a dozen or more of the germinating tomato or tobacco seedlings. Use aseptic techniques (and

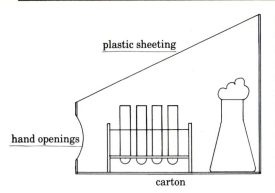

plastic sheeting

hand openings

carton

Fig. 6-105   Preparation of a sterile box.

possibly a sterile box, Fig. 6-105.) With a sterile transfer needle or loop, transfer each root tip to a separate test tube or small Erlenmeyer flask containing White's or Murashige's culture medium. (Medium may be solidified by adding 0.8 percent agar.) When liquid culture medium is used, rest the root tip on a filter paper disk or cone at the surface of the medium so it serves as a wick.

Record the amount of growth, and observe the number of lateral roots that have begun to differentiate. Measure the root lengths in millimeters by holding a plastic ruler behind the flask. Small segments of root may be subcultured in fresh medium. What is the effect if glycine is omitted, or if some of the vitamin B groups are absent from the medium? Do growth regulators affect the rate of growth of lateral roots?

Growth of excised stems

Grow tobacco plants from seeds using the procedure previously described for excised roots. Transfer seedlings to small pots of moist vermiculite and/or garden soil. The meristematic growing tips and leaves may be cultured in culture medium using aseptic techniques (see above). In this investigation, use the stems of the tobacco plants. Cut some 20-cm lengths of a stem, sterilize the bark of the stem with cotton soaked in rubbing alcohol (or dilute household bleach). Peel off the bark of the tobacco stem, and in a sterile Petri dish cut transverse sections about 4-cm long. Ex-

pose the inner layer of soft pith by dissecting away the outer layers of tissue. Now cut 20 small disks of pith 6-mm thick and transfer two disks into each of several test tubes or vials of sterilized solid culture medium. Examine the tubes after five to ten days; over the next five to ten days record the rate of growth, the kinds of differentiation of pith into stem or root parts or the formation of a callus, a growth of undifferentiated cells. What are the results after a month or so?

Tissue squash slide   After the 7th day, some students may want to remove small bits of the pith under sterile conditions and prepare a tissue squash slide to inspect the number of mitotic stages. This is an especially interesting investigation when pith cells have been grown in medium to which growth regulators have been added. For example, cytokinins stimulate cell division, auxins do not; auxins cause cell elongation, however, due to softening of cell walls. Both steps are requisites of cell division.

Growth of excised root tissue: carrot

Use a narrow cork-borer to cut round plugs of tissue (2 mm × 8 mm) from the phloem region of a carrot. Use the aseptic techniques as described previously and grow these explants in plant tissue culture medium containing vitamins and glycine (p. 484). Grow the plugs either in solid medium in Petri dishes or in test tubes of liquid medium using paper cones as wicks with the phloem plugs lying in folds of the wet paper.

Record the weight of the plugs first, then plant them in the medium after surface sterilization (dip the plugs in alcohol or in dilute household bleach (p. 470). Hocking-Curtin suggests maintaining the carrot explants at 24° to 27° C (75° to 81° F) with a photoperiod of 16 hr of light and 8 hr of dark.[83] After five to ten days callus forma-

---

[83] J. Hocking-Curtin, "Plant Tissue Culture and the Carrot," *Carolina Tips* (December 1982).

tion begins, and after five weeks the callus becomes four times the original size. Over the next two to three months, as differentiation occurs in a medium that stimulates roots and shoots, students may weigh the young plantlets and compare the increase in weight over time.

Callus forms from cambium at the cut end of root or stem tissues and gives rise to primordia of regenerating new plant tissue. You may recall that most monocots lack cambium and this limits the possibility of regeneration of lost parts; however, nodes and bases of leaves have some meristem and the possibilities for new growth of cuttings of stems and roots is better if buds are present since there is a root-forming substance in buds.

Also prepare medium to which different quantities of growth regulators have been added: 0.0, 0.5, 5.0, 25, 50, and 100 ppm. What is the effect of auxin on growth? On differentiation? What is the effect of gibberellin? Is there a synergistic effect or countereffect if auxins and kinetin are added to a medium together? You may want to use the technique on page 489,

which to some extent, duplicates the work of Skoog and Miller.

Cloning from a carrot callus   Refer to Ward's *Bulletin*, March 1983, "Here Comes the Next Green Revolution" for a review of current as well as background studies in tissue culture cloning techniques with fine color photos.[84] Excise pieces about 0.5 cm from germinating carrot seedlings and place in a medium containing the plant hormone 2, 4-D that induces callus formation. Incubate tissue at 23° C (73° F) for one to two months. Then transfer bits of the growing callus to another medium to induce differentiation; allow three months for clones to develop.

Cloning from a single cell   Describe the procedure used by Steward in culturing *single cells* from phloem of a carrot to obtain embryoids that developed into flowering plants (Fig. 6-106).[85]

[84] Ward's Natural Science Establishment, Inc. 5100 West Henrietta Rd., P.O. Box 92912, Rochester, New York, 14692:
[85] Steward, "Growth and Development of Cultured Cells."

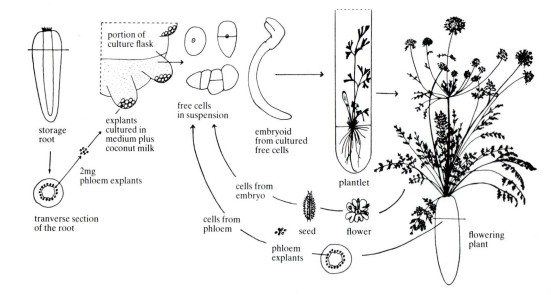

Fig. 6-106   Procedure developed by F. Steward in producing clones from single cultured cells of carrots. Explanted cells grow into embryoids and mature flowering plants. (Adapted from *Plant Biology* by Knut Norstog and Robert W. Long. Copyright © 1976 by W.B. Saunders Company.)

Interaction of IAA and kinetin

In 1957, Skoog and Miller showed the interplay of auxins and cytokinins in growth and differentiation in young plants.[86] Many similar studies revealed that plant hormones do not act alone, but in a homeostatic ratio, depending on both physiological and environmental cues (see also Figure 5-30). In tobacco, the growth patterns and the differentiation of tissues of pith explants vary depending on the ratio of auxins and kinetin added to tissue culture medium (Fig. 6-107). A high ratio of auxin to kinetin stimulates differentiation of roots; a low ratio of auxin to kinetin stimulates the differentiation into apical meristems and shoots.

Prepare the following growth hormone solutions:

IAA   Dissolve 3 mg IAA in a few milliliters of ethyl alcohol and bring the volume to 1 liter with distilled water; this makes a 3 ppm solution (1000 mg/1 = 1000 ppm). Further dilutions can be made for investigations of relative effects of growth regulators.

Kinetin   Prepare 1 mg/liter of kinetin by dissolving 1 mg kinetin in a few milli-

---

[86] F. Skoog and C. Miller, "Chemical Regulation of Growth and Organ Formation in Plant Tissues Cultured *in vitro*," in H. K. Porter ed. *The Biological Action of Growth Substances*, Symposia of Soc. for Experimental Biology (New York: Academic Press, 1957). Also refer to C. Miller, "Kinetin and Related Compounds in Plant Growth," *Ann. Rev. Plant Physiol.* (1961) 12:395–408.

ters of ethyl alcohol, or in 0.05 *M* NaOH that is slightly warmed; bring the volume to 1 liter with distilled water. Dilute further as needed.

Excise pith from tobacco stems as described above; prepare sterile medium and tubes or vials of tissue culture medium. As the agar medium cools, add the growth regulators as shown in Figure 6-107 or in other ratios of concentrations. Observe the growth patterns over one to two months (Fig. 6-108). Under aseptic conditions, small segments of roots or of stems may be subcultured into fresh medium. What happens to the callus formation in other ratios of concentration? Also refer to Murashige's paper on the effects of gibberellic acid on tobacco cells.[87]

Growth of excised apical buds

Using the techniques described, grow young seedlings of tomato plants until they are about 6 to 8 cm tall in 25 days, or use bean seeds that grow above the surface soil after three to four days for investigations in tissue culturing.

Use the excised leaf primordia of ferns, or of flowering plants (apical meristem) such as sunflower, chrysanthemum, bean, tomato, and tobacco plants. Transfer the inner leaves of the apical buds (1 to 2 mm) with sterile forceps and/or transfer loops,

---

[87] T. Murashige, "Suppression of shoot formation in cultured tobacco cells by gibberellic acid," *Science* 134:3474 (1961). Also refer to T. Murashige and F. Skoog, *Physiol. Plants* 15 (1962).

Fig. 6-107.   Tissue culture media with different ratios of IAA and kinetin show their effects on tobacco pith explants.

callus growth	callus growth	root growth	shoot growth	no growth
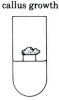			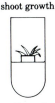	
control medium without IAA or kinetin	medium *plus* 3 mg/l IAA 0.2 mg/l kinetin	medium *plus* 3 mg/l IAA 0.02 mg/l kinetin	medium *plus* 0.03 mg/l IAA 1mg/l kinetin	medium *plus* 0.2 mg/l kinetin no IAA

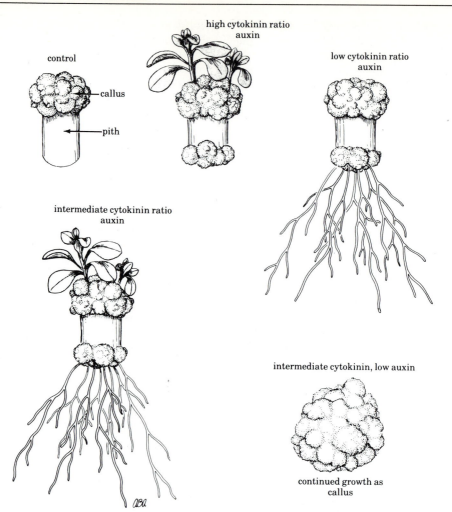

control

callus

pith

high cytokinin ratio
auxin

low cytokinin ratio
auxin

intermediate cytokinin ratio
auxin

intermediate cytokinin, low auxin

continued growth as
callus

Fig. 6-108 Diagram representing the stages in differentiation of shoots, roots, or both, with different ratios of auxin and cytokinin. (Adapted from *Botany: An Introduction to Plant Biology,* 6th ed., by T. Elliot Weier, C. Stocking, M. Barbour. Copyright © 1982 by John Wiley & Sons, Inc. Reprinted by permission of John Wiley & Sons, Inc.)

or hooked needles, to sterile culture medium solidified with agar. Students should be able to observe differentiation of small mature leaves. Some investigators suggest adding 15 percent autoclaved coconut milk or acid casein hydrolyzate to the sterile medium in concentrations of 1:1000. To avoid chlorosis of leaves, ammonium nitrate may be added to the medium.[88]

[88]T. Steeves, H. Gabriel, and M. Steeves, "Growth in Sterile Culture of Excised Leaves of Flowering Plants," *Science* (1957) 126:350–51.

Also investigate the effect of auxins and cytokinins on the growth and metabolism of leaves. Grow small isolated disks of leaves (cut with a cork borer) in sterile medium. In addition, you may want to test the effects of cytokinins in delaying breakdown of chloroplasts in excised leaves (delay of senescence).

Duckweeds and cytokinins *Lemna* shows rapid growth of new fronds in the presence of cytokinins added to culture medium but little response to auxins or gibberellins. Students may compare growth of controls and experimentals by recording the fresh weight of quantities used initially, and then comparing weight of duckweeds over a week or two. (Also refer to growth of *Lemna*, p. 469.)

For enrichment and a clear account of science reporting, have students read J. van Overbeek, "The Control of Plant Growth," *Sci. Am.* (July 1968). Students also may be interested in reading the classic papers by Skoog and Miller,[89] and by Erikson and Goddard,[90] among many others. An abridged edition is available of Sir D'A. Thompson's classic text *On Growth and Form*, J. Bonner, ed. (New York: Cambridge Univ. Press, 1961).

Animal tissue culture techniques

Growing animal tissue *in vitro* may require direct supervision from a researcher in the field. Students may read the descriptions by Rugh[91] and by Hamburger[92] of methods for culturing isolated sections of early amphibian embryos from the tail bud stage to observe differentiation of these primordia, or anlage, when grown in an embryo host or in a hanging drop. Frog embryonic tissue is readily available for studies. (See introductory description for growing frog tissue grafts, p. 498.)

In general, culture media for animal tissue are difficult to prepare and maintain, but they are available commercially.[93] Temperature must be carefully controlled for warm-blooded vertebrate tissues, which are highly susceptible to infection. Refer to texts such as *Cell and Tissue Cultures* by J. Paul.[94]

[89] Skoog and Miller, "Chemical Regulation of Growth and Organ Formation."
[90] R. Erikson and D. Goddard, "An Analysis of Root Growth in Cellular and Biochemical Terms," *Growth* 15: symposium supplement 10 (1951).
[91] Rugh, *Experimental Embryology*.
[92] Hamburger, *A Manual of Experimental Embryology*.
[93] Available from Colorado Serum Co., Laboratories, Denver, CO 80216, as well as from other suppliers of animal blood products and culture media.
[94] Paul, *Cell and Tissue Cultures*.

Dedifferentiation and regeneration

How does the seemingly undifferentiated tissue at the notches of leaves of *Bryophyllum*, or in the "eye" of the white potato become a whole new plant—a clone of the parent plant? What happens to a planarian worm when cut in half? How does regeneration occur in *Tubifex*, or in tadpole tails? DNA in cells retains its potential to develop a whole organism. But what mechanisms trigger this differentiation?

Regeneration of missing parts is more easily studied in plants than in animals. Regeneration regularly occurs in wound healing, growth of adventitious roots, tumors, "witches' brooms," fasciations, and the many forms of propagation by vegetative multiplication.

Among higher plants most vegetative multiplication results from mitotic growth of primordia or dormant buds; at times, regeneration may be possible only in meristematic tissues. Among animals, regeneration involves a reorganization and growth of cells. Cells seem to become dedifferentiated and resume the characteristics of "embryonic" tissue with instructions for developing whole new plants or animals.

Plants and animals resulting from regeneration have the same DNA as the parent (barring mutations or chromosome aberrations), since new cells result from replication of DNA in mitotic divisions.

Regeneration among plants

Regeneration in plants may occur naturally, in which case it may be considered a form of asexual reproduction. Or plants may be multiplied from isolated parts, such as leaves, stems, roots, or even buds of some desired variety of plant (usually obtained through breeding). The latter would be a purposeful, or artificial, method for propagating new plants.

Many studies of regeneration may be done at home by students as extensions of class discussions.

Natural methods in plants

Root stocks and rhizomes   Many species
of plants spread themselves by means of
vigorous underground stems which grow
out in all directions from the parent plant.
Rootlets develop at intervals and stems
shoot up at these nodes. Two hardy forms
that may be grown in the classroom and
used to demonstrate this are varieties of
snake plant (*Sansevieria*) and *Aspidistra*.
Other common forms that reproduce similarly are Solomon's-seal, calla lilies, many
ferns (especially New York, Boston, holly,
hay-scented, and polypody), Johnson
grass, and Bermuda grass.

Runners (stolons)   Some plants produce
trailing or reclining stems which take root
at their ends or at nodes; erect stems grow
from these nodes. Specimens of the hardy
strawberry geranium (*Saxifraga*) and the
spider plant (*Anthericum liliago*) are available for classroom use throughout the
year. Other useful plants are the strawberry (*Fragaria*), cinquefoil (*Potentilla*), and
Cotyledon secunda. Several aquarium plants,
such as eel grass (*Vallisneria*), also produce
long runners.

Bulbs and corms   A bulb is a compact
group of fleshy scales or leaves surrounding a small, sometimes vestigial stem. In
bulb forms like the lily the scales are narrow and loose, while in the onion, narcissus, amaryllis, daffodil, and hyacinth
the scales are continuous and fit closely.
Small bulbs, or bulblets, may develop
around the parent bulb. These small bulbs
are borne above the ground, for example,
in the axils of the leaves of the tiger lily, or
at the top of the flower stalk in the leek.
Many other bulbs divide into two or more
parts.

A corm, on the other hand, is solid
throughout. It is the swollen base of the
previous year's stem. Cormels are developed in the same way bulblets are
formed. Good examples of corms are those
of Indian turnip, or jack-in-the-pulpit
(*Arisaema*), *Caladium*, crocus, and gladiolus.

Students may grow many of these bulbs
and corms in class. For example, when
onions are partially submerged in water
(with the root end just below the water
surface), a good root system develops in a
week. The onion may be supported by the
rim of the container. When bulbs are not
placed in water they first send forth shoots,
not roots. Also use a garlic and have many
of the bulblets sprout.

Similarly, narcissus and hyacinth bulbs
may be grown in water or gravel. Narcissus bulbs grow quickly and flower in a
warm room in a month's time. Hyacinth
bulbs, however, require a preliminary
cooling treatment in a dark place for about
four weeks, and flowers appear two to
three months later. Tulip and spring crocus bulbs require a cool temperature of
about 4° C (39° F) for a preliminary "forc-
ing." Cover the bulbs lightly with sandy
loam; water them well; and keep them in a
cool, dark place for about nine to ten
weeks. The bulbs of Easter lilies, which are
rather slow to flower, should be covered
with some 5 cm of sandy loam, watered
well, and kept in the dark until the shoots
are about 5-cm long. After this, transfer
them to a warm location, where it takes
from three to four months for the lilies to
flower.

Tubers   These are thickened underground stems, in which starch is stored.
The new plants arise from rather incon-
spicuous buds in the tubers, such as the
spirally arranged leaf scars, or "eyes," in
the white potato. Cover tubers of the white
potato, dahlias, *Anthericum*, or *Caladium*
with 2 cm of wet sand, or set them in a
tumbler of water so that about half of the
tuber is covered. Growing plantlets may
be separated from the tuber with a melon
ball cutting spoon.

Fleshy roots   If sweet potatoes, carrots,
or radishes are placed in containers of
water and supported with toothpicks so
that they are not totally immersed, shoots
will differentiate into plantlets.

Commercial methods

Cuttings   A cutting is a part of the plant,
either a stem or a leaf, which has been

separated from the parent plant and which, under suitable conditions, will regenerate a complete plant.

STEM CUTTINGS   Select the tip of a healthy, vigorous young stem of geranium, begonia, or *Tradescantia* to use for a cutting. Include several nodes in the cutting, usually a piece 5-to 10-cm long, depending on the plant and the length of the internodes. When vines are used, such as bittersweet (*Celastrus*), the older parts root better.

In preparing a stem cutting, cut squarely below a node with a sharp knife. Pinch off the flower bud if there is one; cut off some of the leaves to reduce the rate of transpiration so that the cutting does not wilt. Plant the cutting about 3-cm deep in medium-grained wet sand or sphagnum moss.

For good results with fleshy-stemmed plants such as geranium and *Dieffenbachia* (which should see use more often where stem cuttings are desired), the tip of a plant is not used. Instead, cut the stem into 5- to 8-cm pieces and plant them horizontally in moist sand. Allow a cover of about 2 cm of sand.

Other plants suitable for stem cuttings are coleus, ivy (either *Hedera* or *Parthenocissus*), privet (*Ligustrum*), willow (*Salix*), golden-bell (*Forsythia*), pineapple (*Ananas*), and cactus (*Opuntia* or other similar cacti). Succulent forms usually take root within two to three weeks, while woody types often require months. (Also see polarity, Fig. 6-111.)

There are a few precautions to observe in making cuttings. Use healthy stocks and work in a place that is not excessively hot or dry so that plants do not wilt rapidly. Keep the plant cuttings in moist paper until they are ready for planting.

The planting medium must be moist to keep the plants from wilting, and well drained so that they do not rot. Pack the medium around the cuttings to keep them upright. Use coarse-grained sand or vermiculite for good drainage. (You will find that round-grained sand such as sea sand does not pack well.)

After the cuttings have been planted, water them well and cover with a plastic bag or invert a tumbler over each plant to retain moisture. Remove the covers after a few days. Keep the young cuttings sheltered from strong sunlight.

A callus forms around the cut end and adventitious roots develop in about two to three weeks. Then set the small plants into 5-cm pots. The pots should first be washed and soaked in water for several hours so that they will not immediately absorb water from the potting soil. To prepare this soil, mix 3 parts of loam (garden soil), 1 part of sand, and 1 part of humus. Crush the lumps so that the soil is smooth and fine textured. Then water the soil thoroughly about an hour before potting, and finally, fill the pots with the soil.

Slip a broad knife or spatula under the cutting, lifting the sand along with the cutting to avoid injuring the roots. Make a hole in the soil and set the cutting in place. Avoid packing the soil too tightly. If the roots are long, hold the plant in place in the empty pot and gradually add the soil. When the plants show good growth, transfer them, with the same technique, to larger pots.

In general, most cuttings, fleshy roots, and stems of herbaceous plants are less prone to bacterial or fungus infection when started in moist sand rather than plain water. Storage parts of plants containing carbohydrates are most susceptible to decay.

A special note about geranium plants. In the fall they should be cut back, repotted, and allowed a period of rest. Cut the plants back about two-thirds of their length and divide the plants so that each cutting has about three main stems. These cuttings should be repotted in a mixture of equal parts of loam, sand, and leaf mold. You may plant smaller cuttings in a large flowerpot around a partly submerged smaller pot as in Figure 6-109. Water the center pot so that equal drainage takes place with less chance for leaching of the minerals from the soil.

LEAF CUTTINGS   A new plant may regenerate from part of a leaf. Use the leaves of *Bryophyllum* (Fig. 6-110), *Kalanchöe*, *Sedum*, begonias (especially *Begonia rex*), *Sansevieria*, or African violets. Lay the

Fig. 6-109 Growing plants from stem cuttings. The smaller inset flowerpot is kept full of water to provide equal moisture to the cuttings.

on sand, many shoots appear at the regions where the veins are severed. Plantlets differentiate from the notches along the edges of *Bryophyllum* and *Kalanchöe*.[95]

Layering While layering occurs naturally especially among wild roses, raspberries, and blackberries, it may also be done purposely to propagate more bushes. Place a branch in contact with the moist soil and hold fast so that roots and shoots are produced where they contact the soil. These can be separated from the parent plant. This method is used when plants cannot be grown successfully from cuttings because they do not root readily. Layering is the commercial method for propagating such plants as magnolia, grapevines, and raspberries.

leaves of the first three plants on wet sand and hold them flat with a small stone or flat piece of glass placed on the leaf blade. Cut across the veins in a begonia leaf. For *Begonia rex* and *Sansevieria* plant upright a section some 5-cm long about 3-cm deep in sand. Look for rooting and the appearance of shoots. When a leaf of *Begonia rex* is laid

---

[95] *Kalanchöe* and *Bryophyllum* have tissue in the notches of leaves that is differentiated—called a preformed foliar embryo by E. Sinnott, *Plant Morphogenesis* (New York: McGraw-Hill, 1960).

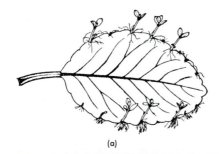

(a)

Fig. 6-110 Growing plants from leaf cuttings: (a) diagram showing small plants arising from notches along the edges of *Bryophyllum pinnatum*; (b) photograph of new plants differentiated from tissue in notches of leaf of *Bryophyllum*. (a, from W. H. Brown, *The Plant Kingdom*, 1935, reprinted through the courtesy of Blaisdell Publishing Co., a division of Ginn and Co.; b, Richard F. Trump, from R. F. Trump and D. L. Fagle, *Design for Life*, Holt, Rinehart and Winston, 1963.)

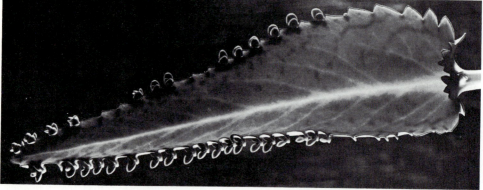

(b)

In class you may want to use *Vinca*, English ivy, and *Philodendron*. The small climbing rose and the small blackberry *Rubus villosus* are especially suitable for classwork.

Sometime you may want to use the Chinese, or pot method of layering. This is generally successful with rubber plants (*Ficus*) and castor-oil plant (*Croton*). Cut into the bark of a branch about 30 cm from the tip, where the wood is slightly hardened, and pack it around with sphagnum moss tied with raffia or cord. Keep the packing moist. Then cut a flowerpot in half lengthwise and shape each half around the wounded section of the stem. Hold it together with cord. Paper may be used instead of a flowerpot, but it is less effective. In two to three weeks, when the stem becomes rooted, cut off the branch from the tree.

Grafting   In this method of vegetative propagation the cut surfaces of two woody plants are fitted together so that their cambium layers make contact. In our experience, useful plants for classroom work in grafting are the common privet and reciprocal grafts between the potato (*Solanum*) and the tomato (*Lycopersicum*). Thriving young tomato plants furnish suitable scions to graft onto the potato stock.

Some precautions are necessary in grafting. Try only the simplest kinds of grafts, that is, stem grafts. Bud, saddle, and other types of grafts are not generally successful in the classroom. The scion and the stock should be similar in width. Remove all the leaves of both the stock and the scion when you graft. Make clean cuts with a sharp knife.

Fit the cut ends snugly and tie them together with raffia or similar material. Impregnate the binding material with beeswax or grafting wax (p. 747). Whenever possible, cut the stock about 3 to 5 cm from the soil line in order to eliminate many buds. On the scion there should be as many buds as possible, including the terminal bud.

Students may try bud grafts and tongue grafts with rose, apple, and privet. Select a stem about 3 cm in diameter. Also try grafts among a variety of cactus plants.

For studies of the effects of growth hormones on the rate of regeneration in new plant parts, see page 483.

Polarity   You may want to repeat the classic study of Vöchting, showing polarity in willow shoots. What happens when a twig is inverted and kept moist? Do roots continue to be regenerated from the original, or morphologically basal, region? What growth occurs at the original apical region? If the twigs are subdivided into many smaller sections, is polarity retained? (Refer to Figure 6-111), note however, that polarity is more complex than Vöchting anticipated (Fig. 6-112).

Students who want to investigate how "embryonic" tissue becomes differentiated, and what is meant by polarity will find lucid reading and illustrations in E. Sinnott, *Plant Morphogenesis* (McGraw-Hill, 1960) and also T. Weier et al, *Botany* (New York: Wiley, 1982). For more ad-

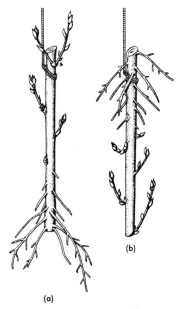

Fig. 6-111   Polarity in the willow: (a) stem suspended in moist air, producing roots and shoots; (b) stem in inverted position. (After E. Sinnott, *Plant Morphogenesis*, McGraw-Hill, 1960.)

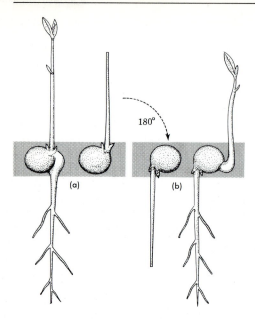

180°

(a)   (b)

Fig. 6-112 Inversion of polarity: (a) the epicotyl is decapitated from etiolated pea seedlings; (b) when seedling is inverted and placed in water, roots grow from the epicotyl and a shoot grows out from a cotyledonary bud. (After E. Sinnott, *Plant Morphogenesis*, McGraw-Hill, 1960.)

vanced reading, also refer to G. L. Stebbins "From Gene to Character in Higher Plants," in *Am. Scient.* 53 (1965).

Regeneration among animals

Some invertebrates—especially worms, crustaceans, and echinoderms—are able to regenerate missing parts. New cells are formed as a result of differentiation and growth. Consider regeneration in hydra, planaria, *Tubifex*, small *Dero* and *Aeolosoma* worms.

Regeneration in hydras   Place a well-fed hydra in a Syracuse dish containing aquarium or pond water. Cut the animal transversely with a sharp razor or coverslip. Try several cuts: one below the ring of tentacles to remove the hypostome and tentacles; one above the bud region; one below the bud region; or one at the base of the hydra. Isolate the pieces in welled slides containing aquarium or pond water, and

seal the slides with coverslips and petroleum jelly. Prepare several specimens and store them at 15° to 30° C (59° to 86° F). Diagram the types of cuts used, and record the observations made during the next week or so. Is there polarity? Which parts regenerate fastest? Are there factors that enhance regeneration—thiamine chloride, growth substances, or others?

Grafting hydras   Lenhoff reviews several techniques for making grafts on hydras. Also consider the classic work of E. B. Harvey showing induction.[96] Transplant a small portion of the mouth or hypostome of a hydra, with tentacles attached, to the midregion of a recipient hydra. The hypostome acts as an "organizer" inducing a new hydra at this region. Lenhoff suggests that shallow Petri dishes lined with paraffin be used for the operations. Further, he suggests that if the hydras are introduced a day in advance into the waxed dishes, they tend to adhere to the wax and are easier to cut. Harvey used both green hydras and albinos; the albinos resulted when green hydras were placed in 0.5 percent glycerin, which apparently releases the symbiotic algae, the zoochlorellae, from the cells of the hydras.

Hydra chimeras   An interesting demonstration of the formation of chimeras is described by Sherman and Sherman[97] and by Lenhoff.[98] Cut green hydras (*Chlorohydra*) and brown hydras (*Pelmatohydra*) into 4 or 5 pieces and *slowly* centrifuge for about 15 min so that a mass of tissue is formed at the bottom of the centrifuge tubes. Transfer the mass of tissue into separate watch glasses or Syracuse dishes of pond water. Observe the reassociation or reorganization of tissues within a day,

[96] H. Lenhoff, "Laboratory Experiments with *Hydra*," *Carolina Tips* (October 1960) 23:8. E. B. Harvey's original paper (*J. Exptl. Zool.* 7:1, 1909) may be available for students to read.
[97] Sherman and Sherman, *Invertebrates*.
[98] Lenhoff, "Laboratory Experiments with *Hydra*."

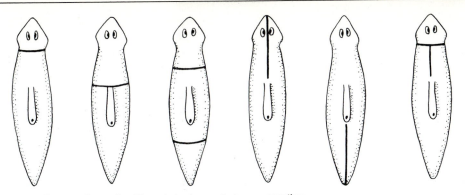

Fig. 6-113   Patterns for cutting *Planaria* to demonstrate regeneration.

and within several days, the formation of small budlike structures, which become tentacles. Chimeras of different kinds result, containing a mixture of cells from both types of hydra. Examine under a dissecting or compound microscope for details.

Regeneration in Planaria   Most planarians may be used for regeneration studies, but *Dugesia dorotocephala* and *D. tigrina* are especially suitable (culturing techniques, chapter 9). With a small brush, transfer worms to a drop of water on a glass slide, a moist cork, or on a slide resting on an ice cube. Cut the worms with a sharp razor or a coverslip in the patterns suggested in Figure 6-113. Make sketches of the kinds of cuts made, and isolate the cut pieces into Syracuse dishes containing pond water or Brandwein's solution A (p. 657). Store the slides in a covered container in the dark (or in an enameled container at 21° to 27° C (70° to 81° F). Change the water frequently to encourage rapid healing, but do not feed. Notice the light colored mass of regenerating wound tissue—a blastema—that protrudes from the cut edge. Record growth changes over several weeks. As an extended laboratory activity, study the rate of regeneration in relation to the metabolic gradient (or degree of cephalization). Which part regenerates fastest? Is there anterior-posterior polarity? Some students may also be interested in reading C. M. Child's classic work in observing the rate of oxidation using methylene

blue and planarians sealed in anaerobic conditions.[99]

Refer to papers on conditioning cited on page 351 for studies on naive heads, trained heads, and regenerated heads in studies of memory, RNA, and learning. Also refer to Best's paper on learning in planarians.[100] Some earlier writings offer techniques and ideas for research. Refer to Balinsky's discussion of regeneration;[101] to Bronsted's fine review paper;[102] and the investigations offered by Hamburger.[103]

Regeneration   in   protozoa   Discuss experiments that have been done using *Stentor, Paramecium, Euplotes,* and *Blepharisma,* among others to show regeneration of parts of the cell provided the nucleus is intact. (See p. 136.)

Regeneration   in   Tubifex   Posterior regeneration is most successful in *Tubifex,* the small annelid worm (purchase in aquarium shops). Place the specimens in Syracuse dishes and make a transverse cut with a scalpel or razor blade. Then separate the sections into individual dishes of pond water and maintain at room tem-

[99] C. M. Child, *Patterns and Problems of Development* (Chicago: Univ. of Chicago Press, 1941).
[100] J. B. Best, "Protopsychology," *Sci. Am.* (February 1963).
[101] Balinsky, *An Introduction to Embryology.*
[102] H. V. Bronsted, "Planarian Regeneration," *Biol. Rev.* (1955) 30:65–125.
[103] Hamburger, *A Manual of Experimental Embryology.*

perature (10° to 16° C or 50° to 61° F). A new anal opening is formed within 48 hr when the cut end heals. A successful regeneration takes some two to three weeks (culture methods, p. 669).

Dero, a microscopic annelid, may also be used for regeneration; also try using the earthworm, in which anterior-end regeneration seems to be more successful.

Regeneration in Aeolosoma   This small aquatic oligochaete, some 2-to 3-mm long regenerates a new prostomium rapidly (culture methods, p. 669). Lonert describes the following technique.[104] Use a pipette to transfer *Aeolosoma* to a slide with a small amount of water to slow its activity. Under low power, cut off and discard the first segment containing the prostomium. With a soft brush or medicine dropper, transfer the remaining worm to a Syracuse dish or welled slide of fresh culture medium. Prepare several specimens since the wiggling worms are difficult to cut. Examine cut worms under the microscope. Within some 15 to 20 min, observe the wound has healed. A small prostomium forms in three to four days; a fully regenerated tactile organ regenerates in two weeks.

In another set of experiments, Lonert describes cutting *Aeolosoma* at the 4th or 5th segment so that an anterior end with a prostomium results, along with the remaining posterior end. Remove any asexual buds at the posterior end that may confuse observations of results. Prepare several specimens; isolate them in separate covered dishes or welled slides. By the 3rd day, the posterior end regenerates a bulbous prostomium; and a smaller, fully functioning tactile organ is formed by the 7th day. The anterior end of the worm regenerates a new tail end by the 8th day, and in two weeks, budding of new worms from the tail end occurs. At times, Lonert found that some posterior body segments regenerated a new prostomium at both ends—yet one was better developed than the other. The confused worms lived about a week.

Regeneration and polarity in Sabella   The bilobed branchial crown of this tiny tube-living marine polychaete serves as both a food-getting and respiratory organ. The crown, about 1 mm in diameter, can be regenerated, including the positioning of the light-sensitive pigment spots located along each radiole. Some students may want to read Fitzharris's experiments in *American Scientist*.[105] Remove a worm from its tube under a dissecting microscope to observe the thorax (anterior 5 to 11 segments) and the posterior abdominal region comprising many, many segments. Note that the thoracic segments have dorsal setae and ventral hooks while the abdominal segments have the opposite—dorsal hooks and ventral setae. About two weeks after the crown and a few anterior segments are amputated, anterior regeneration occurs. Investigations of polarity, action of colchicine and neurohumors are described by Fitzharris in fine illustrations as guides for a capable student.

Regeneration of tails of tadpoles   When young tadpoles of frogs or salamanders are available, cut off the tips of their tails with a sharp razor. Refer to specific techniques in Hamburger's book.[106] Make drawings of the cuts and compare the regeneration of tails in the specimens under observation. To reduce mortality, maintain the tadpoles at a cool temperature. Keep a record of observations for a week or 12 days. (See also mitosis, using cells of tadpoles, p. 400.)

Suggestions for long-range investigations

Students who are highly motivated may be guided into long-range investigations that

---

[104] A. C. Lonert, "Regeneration in *Aeolosoma*," *Turtox News* (November 1967) 45:11.

[105] T. P. Fitzharris, "Control Mechanisms in Regeneration and Expression of Polarity," *Am. Scient.* (July–August, 1973) 61. Also refer to N. J. Berrill, *Growth, Development, and Pattern* (San Francisco: Freeman, 1961).
[106] Hamburger, *A Manual of Experimental Embryology.*

may develop into "independent research." Tissue culturing techniques require patience and practice to avoid contamination, but these techniques and problems also challenge interest and abilities of young people. The role of special hormones, photoperiodism, differentiation, and polarity are areas of biology rich in possible investigations.[107] The techniques described throughout this text and the suggested references are a useful beginning in raising problems, finding some observations that raise further questions for investigation.

For example, some students may study the causes of dormancy in seeds, and the effects of plant hormones or photoperiods in breaking dormancy. Some seeds may be filed or cut, others placed in acid, and so forth. Controls should be included in these investigations. Also consider investigations in regeneration in animals; cleavage patterns of both invertebrate and invertebrate eggs.

Some of the early, classic studies may provide some insight into the thinking and methods of early investigators. Offprints from *Sci. Am., Biological Abstracts, Ann. Rev. Plant Physiol.* are excellent sources for both investigations and background material. For example, you may want to refer to B. O. Phinney and C. A. West, "Gibberellins as Native Plant Growth Regulators," *Ann. Rev. Plant Physiol.* 11 (1960). Also refer to J. Liff, "Interaction Between Kinetin and Light on Germination of Grand Rapids Lettuce Seeds," *Plant Physiology* (1964) 39:3.

We have discussed many aspects of growth in animals and plants. Many of these investigations may be considered an extension of classwork. Topics presented in other chapters—such as reproduction, deficiency diseases, photosynthesis, and genetics—include some aspect of the work described in this chapter. The techniques and the subject areas may be coordinated effectively as the teacher sees fit into related classroom and laboratory experiences.

[107] For an early study of differentiation, see J. Bonner and M. Hoffman," "Evidence for a Substance Responsible for the Spacing Pattern of Aggregation and Fruiting in the Cellular Slime Mold," *J. Embryol. Exptl. Morphol.* (1963) 11.

CAPSULE LESSONS: WAYS TO GET STARTED IN CLASS

6-1. Plan several lessons with a microscope to show fission in protozoa, or budding in yeast cells, or sporulation in some fungi. Develop the notion that, in general, asexual reproduction results in offspring with heredity identical to that of its parents. How do these cloning methods compare with reproduction by seeds, or reproduction in higher animals?

6-2. Encourage students to bring in specimens to keep in the aquaria and terraria in the classroom, and also to maintain such specimens at home. Some may want to study germination of seeds as well as devices for dispersing seeds. Embryology offers rich opportunities for laboratory work with individual plants or animals. Have students share their investigations with the rest of the group. Reference books can be made available in the classroom.

6-3. Present students with a hen's egg (*caution:* hard-boil). What will it give rise to?

And why a chick and not a frog or a duck? Focus this introductory discussion on the notion that an organism arises from a fertilized egg and that it has a specific blueprint, its own DNA or code of heredity—half from each parent. How many eggs do animals produce? Compare codfish with chickens. Elicit differences between external and internal fertilization. What are the odds against each egg of a codfish becoming fertilized? Begin to touch upon the evolution of sexuality from protozoa, algae, yeasts, and bacteria to the flowering plants up to the mammals among vertebrates.

6-4. The questions that students raise probably lead to the next lesson: What is in a fertilized egg? What governs the genetic continuity? How does the fertilized egg develop? You may want to show one or more of some fascinating films concerning the embryology of fish, chick, and human; consult catalogs of film producers and distributors (Appendix). Combine models, films, and slides with other approaches suggested here.

6-5. At some time, plan a reading of offprints from *Scientific American* (catalog from Freeman, San Francisco). For example, plan seminars where each students gives a 10-min talk and answers questions of the class concerning such topics as hormones and reproduction; photoperiodism and its effect on flowering and on reproductive cycles of animals; role of messenger RNA in protein synthesis; replication of DNA; patterns of cleavage among eggs of mollusks, worms, echinoderms, and vertebrates; differentiation in time and space of tissues from early gastrula; polarity and regeneration; metabolic disorders; radiation.

6-6. Begin with fertile chick eggs that have been incubated for 48 hr, 72 hr, or longer. Locate the beating heart and try to trace the circulatory system; discuss the origin of blood cells, somites, anlage of eyes, limbs, and other organs and organ systems.

6-7. Collect frogs' eggs and watch their development. Consult Schumway's chart giving a timetable of development of frogs' eggs. Some students may want to test the regenerative power of the tail of the early tadpole stages (and also to examine prepared slides of chromosomes, page 400; see also Figures 6-34, 6-35 for tadpole stages).

6-8. Have students use a series of models of cleavage and describe the stages in the development of the frog or the chick. Have students arrange them in chronological order. Many students have made models of papier-mâché or of special materials. Or use commercial models to develop early stages in cleavage; then, to show what goes on within the nucleus show stages of mitosis. Compare with models showing stages in meiosis of cells in the ovaries and testes or in the ovary and anthers of flowers. Transparencies are available, and sets of slides for the microscope are useful for firsthand, close examination of details of chromosomes, and of sperms and pollen grains.

6-9. When in season, crush stigmas of flowers on a slide, and look for the growth of pollen tubes under the microscope. (Some students may try to find the optimum concentrations of sugar solutions which serve as an artificial medium for germination of pollen grains of a given flower.) Have students describe how a sperm nucleus in the pollen grain reaches the nucleus of an egg cell in the ovule so the fertilization occurs. Compare fertilization in plants with that in animals.

6-10. Have students examine prepared slides showing meiosis in testes of insects or other forms. There are also films available to compare mitosis and meiosis.

If available, examine living sperm cells in frogs. Or make chromosome squashes of salivary glands of fruit flies or onion root tips.

6-11. Develop laboratory activities from a film showing techniques for making chromosome squash slides, or methods for handling *Neurospora* or *Drosophila*.

6-12. As a summary, show color transparencies of mitosis and of DNA replication using an overhead projector. Many transparencies can be made by students. Also show films of reproduction in a variety of organisms.

6-13. A committee of students may want to report on the development of eggs of different animals and plants. One committee could describe the reproductive habits of salmon or of tropical fish or of pond snails. Have students bring in living material for examination. Fish may be purchased for dissection. In season collect frog's eggs and watch them develop into tadpoles.

Other students may raise pigeons or chickens; they can bring in eggs. If possible, incubate fertilized eggs and examine different stages of development. Compare the similarities and differences in reproduction among many groups of animals: How many eggs are laid? Are the eggs fertilized internally or externally? How long does it take for the embryo to develop? How large is the egg? What factors determine the size of the egg? How can a mammal's egg be microscopic?

6-14. Plan a field trip, in season, to collect eggs of animals. Also study flowers as trees and shrubs begin to blossom. Establish the similarity between sexual reproduction in plants and animals. What is the advantage of the diversity resulting from sexual reproduction?

6-15. Examine prepared slides of cleavage stages in whitefish eggs or mitotic figures in dividing cells in the tip of the onion root. Trace the stages in the replication of chromosomes in mitosis, where each new cell formed receives an equal amount of chromosome material. Have students explain how all the body cells are replicas of the fertilized egg. How does differentiation occur? Present a short film illustrating chromosome movements in mitosis.

6-16. Use pipe cleaners of different colors to show chromosomes undergoing mitosis; compare with stages in meiosis. Also thread a few wooden beads of different colors on the pipe cleaners to represent genes. Compare "chromosome" and "gene" with sequences of nucleotides in DNA. How are similar codons in DNA replicated in mitosis?

6-17. Begin with a handful of soaked seeds. Diagram a cross section of a cereal plant such as barley or oats and also of dicot seeds. As students dissect seeds, label the embryo, aleurone and endosperm of a monocot. What are the parts of a dicot seed?

How does the embryo plant gain food stored in the endosperm? As water enters the seed, metabolic activity begins. Develop the role of gibberellic acid and amylases; cytokinins, and auxins in the roots and shoots of the developing embryo.

6-18. At some time, clone a plant. Have students try several methods at home and bring their successes to class. Will the new plants have the same DNA? Explain.

6-19. As a research problem, a student might want to explore polarity in the regeneration of some worms or *Hydra*. Will the anterior parts of cut pieces always regenerate a "head" part, or can they grow a "tail," or posterior portion?

If a shoot of a plant is turned upside down, will a new shoot regenerate at the anterior part, or may roots grow on the portion which was originally in the most anterior (terminal) part?

6-20. In a laboratory activity dealing with population studies of yeast cells use the methods suggested in this chapter. When 1 g of dry yeast is added to 10 ml of water, the average number of cells per field under a high-power objective is about six. Refer also to Breed's method for estimating numbers of microorganisms (in this chapter).

6-21. Prepare several grids by marking off squares in India ink on rubber sheeting. On the sheet, outline the shape of a specific leaf, fruit, or fish. Stretch the sheet in several directions to show how altered patterns of the plane of growth might affect the shape or form. Elicit a discussion of ways that changes in the genetic code might alter the pattern or plane of growth or polarity; lead into a discussion of interrelated steps in the genetic control. Also refer to studies of mitotic divisions and to the effects of hormones such as auxins and of retardants such as Phosfon.

6-22. When a batch of amphibian eggs is available (they may also be purchased), students may devise a whole series of investigations, from simple, careful observations of cleavage patterns, invagination, and formation of a neural tube to plans for a study of environmental factors that affect rate of growth.

6-23. At some time, raise the following questions. Are both cells in the two-celled stage of the developing egg equally totipotent? Can each develop an entire embryo? What would happen if one cell were killed by pricking it with a hot needle? What would happen if the two cells were separated from each other by constricting them with a hair loop? Each cell might then be isolated in a separate dish and left to develop.

6-24. Ask the class how an apple tree could produce five different kinds of apples. When possible have students demonstrate how a graft is made. What is the value of vegetative propagation?

6-25. Many students may do experimental work at home; there they may try to regenerate planarian worms, limbs of salamanders, or tail fins of salamander tadpoles. Plan to have such students report to the class or at special seminars. Perhaps there is space in a hall cabinet to display their work.

6-26. As a review or summary, begin with a handful of *dried* beans and another of germinating beans. You may develop a lesson on respiration and the use of the stored food supply. Or you may raise questions: What are the characteristics of living things? Are the dried beans living? This is one way to develop the notion of enzymes and use of materials in living systems.

You may want to show tropisms suggested by students. Or you may plan a study of plant functions beginning with seedlings, which are so adaptable for class demonstrations.

6-27. To introduce the concept of a hormone as a substance made in one part of a body which produces an effect in other parts, begin with studies of plant hormones. Students may report to the class on some of the classic experiments in the field. You may also want to demonstrate the effect of feeding iodine or thyroxin to tadpoles of salamanders or frogs. Compare the action of plant hormones with hormone action in animals.

6-28. Diagram a total plant—leaves, stems, roots, and flowers. Then trace the role of plant hormones (auxins, gibberellins, cytokinins, and abscisic acid) in maintaining plant homeostasis in both growth and differentiation.

6-29. Have students grow seeds of bean, sunflower, squash, or other plants at home. Show them how to soak and place the seedlings between wet blotter and the glass walls of baby-food jars or other containers. This activity may lead into a study of growth hormones, the way plant embryos grow, or root hairs and the conduction system of seedlings.

These are the times when firsthand experience is easier to provide, and more meaningful and exciting than a film. However, there may be appropriate times for showing time-lapse photography of plant growth.

6-30. When possible, try to grow excised embryos of seedlings or pieces of roots or leaves as tissue cultures, as described in this chapter.

6-31. Develop a case history of the discovery leading to the recognition of a growth-promoting substance (auxin). Investigators would include, after Darwin, Boysen-Jensen (1910), Paal (1914), Went (1920). Went placed cut oat tips on gelatin blocks and then applied these blocks to one side of decapitated coleoptiles. The observation that cells under this region elongated was proof that there was a diffusible growth substance.

BOOK SHELF

Consult references in chapter 7 on heredity along with these references on growth and differentiation.

Abramoff, P., and R. Thomson. *Laboratory Outlines in Biology*. San Francisco: Freeman, 1982.

Alberts, B. et al. *Molecular Biology of the Cell*. New York: Garland, 1983.

Balinsky, B. *An Introduction to Embryology*, 5th ed. Philadelphia: Saunders, 1981.

Bottino, P. *Methods in Plant Tissue Culture*. Kensington, MD: Kemtec Education Corp., 1981.

Brookbank, J. *Developmental Biology: Embryos, Plants, Regeneration*. New York: Harper and Row, 1978.

Dodds, J., and L. Roberts. *Experiments in Plant Tissue Culture*. New York: Cambridge Univ. Press, 1982.

Gabriel, M., and S. Fogel, eds. *Great Experiments in Biology*. Englewood Cliffs, NJ: Prentice-Hall, 1955.

Gilbert, E., and E. Frieden, eds. *Metamorphosis*, 2nd ed. New York: Plenum, 1981.

Gorbman, A. et al. *Comparative Endocrinology*. New York: Wiley, 1983.

Ham, R., and M. Veomett. *Mechanisms of Development*. St. Louis, MO: Mosby, 1980.

Hamburger, V. *A Manual of Experimental Embryology*, rev. ed. Chicago: Univ. Chicago Press, 1960.

Hickman, C. P. Jr. et al. *Biology of Animals*, 3rd ed. St. Louis: Mosby, 1982.

Nostog, K., and A. Meyerriecks. *Biology*. Columbus, OH: Charles Merrill, 1983.

Ransom, R. *Computers and Embryos: Models in Developmental Biology*. New York: Wiley, 1981.

Rugh, R. *A Guide to Vertebrate Development*, 7th ed. Minneapolis: Burgess, 1977.

Saunders, J. Jr. *Developmental Biology: Patterns, Problems, Principles*. New York: Macmillan, 1982.

Sharp, W. et al., eds. *Plant Cell and Tissue Culture Principles and Applications*. Columbus: Ohio State Univ. Press, 1979.

Sherman, I., and V. Sherman. *The Invertebrates: Function and Form*. A Laboratory Guide, 2nd ed. New York: Macmillan, 1976.

Stewart, A., and D. Hunt. *The Genetic Basis of Development*. New York: Wiley, 1981.

Street, H., ed. *Plant Tissue and Cell Culture*. Berkeley: Univ. of California Press, 1973.

Thorpe, T. A., ed. *Plant Tissue Culture: Methods and Applications in Agriculture*. Wayne, NJ: Avery Publishing Group Inc., 1981.

Wetherell, D. *Introduction to In vitro Propagation*. Wayne, NJ: Avery Publishing Group Inc., 1981.

7

Continuity of the Organism: Patterns of Inheritance, Adaptation, and Evolution Within Ecosystems

Consider, if you will, the hundreds of thousands of replications of DNA and the translation of the code within a given organism into structural and functional proteins. Consider the questions that arise out of recent studies of the code. How does one determine the limits of a single gene? How does DNA affect the development of the organism? How does the environment favor some genotypes so that they have survival value? Consider the traditional forms of gene inheritance that may be calculated in light of Mendelian inheritance, crossover, and the like.

Further, how have genomes interacted in the past to survive sometimes tumultuous changes in the environment? These are questions that we will consider in this chapter.

DNA: Its expression in the organism

The expression of DNA in humans is considered first, followed by plant and animal genetics.

Human heredity

Since most people are interested in their own heredity, a study of heredity among humans is a good place to start. Students enjoy looking into the heredity of colorblindness, sex, freckles, musical ability, hair color, shape of ears, curly hair, blood

types, and so forth. They learn the patterns of heredity using examples from human traits as well as those from traditional plant materials—peas, corn, or *Neurospora*, and also from other animals—*Drosophila* or protozoa. They will want to study Mendel's work, and demonstration crosses using rats, peas, maize, sorghum, fruit flies. Students may also wish to model, in three dimensions, the replication of a chain of DNA molecules (p. 389).

Tasters vs. nontasters

About seven out of every ten human beings taste PTC (phenylthiocarbamide) as a salt, sweet, sour, or bitter substance. To others, PTC is tasteless. Paper soaked in this harmless chemical may be purchased from biological supply houses; or you may want to prepare it yourself.

Cut the paper into 1-cm squares. (Do not explain what kinds of tastes will be obtained; otherwise some students will imagine a taste.) Distribute the paper and have all the students taste it along with you at the same time. Then notice their reaction. Tally the results—tasters and nontasters. Ability to taste PTC paper seems to be a dominant trait.

Have students use symbols to illustrate some possible crosses, using T for taster, and tt for the recessive. Is it possible for one child to be a taster and his or her sibling a nontaster? Trace back to the parents' genetic makeup. For example,

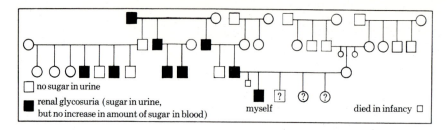

Fig. 7-1 Family pedigree of a student who suffered from renal glycosuria (excess sugar in the urine). (Drawing by Marion A. Cox. From *Experiences in*

Biology, by Evelyn Morholt and Ella Thea Smith, copyright 1954, by Harcourt, Brace & World, Inc., and reproduced with permission of the publisher and artist.)

both parents might be hybrid for the trait, resulting in a possibility of 75 percent of the offspring tasters and 25 percent nontasters. Or one parent might be hybrid and the other a nontaster; then 50 percent of the offspring could be tasters and 50 percent nontasters.

Elicit from students that there are six possible crosses whenever *one* pair of genes is under study.

Try to provide enough paper for students to take home to test their parents' reactions as well. Then students might compile all the data.

You may also purchase tablets of mannose, a sugar which tastes different to a number of people. But the genetic basis for these differences is not as well known as in the case of PTC tasting.

PREPARING PTC PAPER   Gradually dissolve 500 mg of phenylthiocarbamide in 1 liter of water. At room temperature this takes about 24 hr. Then soak sheets of filter paper in this solution and hang them up to dry. Cut small pieces, about 1-cm square, and store them in envelopes for distribution to students. (At higher concentrations, nontasters will also react.)

A family pedigree   Students may be interested in tracing the inheritance of a trait in their own family. They may study the inheritance of eye color, hair color, freckles, PTC tasting, and many other traits. They may want to model their pedigree after the one diagrammed in Figure 7-1. See also Table 7-1 listing a number of hereditary traits.

TABLE 7-1   Some human traits that are inherited

	dominant	recessive
Hair, skin, etc.	curly hair (or kinky or wavy)	straight hair
	dark hair	blond hair
	nonred hair	red hair
	early baldness	
	(dominant in males)	normal
	white forelock	self-color
	widow's peak	straight hairline
	abundant body hair	little body hair
	mid-digital hair	
	(middle segment fingers)	no mid-digital hair
	normal	absence of sweat glands
	piebald (skin and hair spotted	
	with white)	self-color

(continued)

TABLE 7-1 (_continued_))

	dominant	recessive
	pigmented skin, hair, and eyes	albinism
	black skin (several pairs of genes—cumulative)	white skin
	scaly skin (ichthyosis)	normal
	skin sensitive to slight abrasions—epidermis bulbosa	normal
Eyes, ears	astigmatism	normal
	congenital cataract	normal
	nearsightedness	normal
	farsightedness	normal
	large eyes	small eyes
	long eyelashes	short eyelashes
	brown	blue or gray
	hazel or green	blue or gray
	"Mongolian fold"	no fold
	narrow, high bridge of nose	broad, low bridge
	free earlobes	attached earlobes
	broad lips	thin lips
	tongue rolling	nontongue rolling
Skeleton	brachydactyly (short digits)	normal
	syndactyly (webbed fingers or toes)	normal
	polydactyly (more than five digits)	normal
	short stature (many genes)	tall stature
	dwarfism (achondroplasia)	normal
	midget (ateliosis)	normal
Systemic conditions	normal	alcaptonuria (homogentisic acid accumulation; "black" urine)
	progressive muscular atrophy	normal
	hypertension	normal
	normal	hemophilia (sex-linked)
	normal	sickle-cell anemia
	hereditary edema (Milroy's disease)	normal
	normal	phenylketonuria
	normal	diabetes mellitus
	taster (PTC)	nontaster (PTC)
	normal	congenital deafness
	normal	amaurotic idiocy
	normal	dementia praecox (several pairs of genes)
	migraine	normal
	blood group A	blood group O
	blood group B	blood group O
	Rh positive factor	Rh negative factor

Adapted from C. Villee, _Biology_, 7th ed. (Philadelphia: Saunders, 1977).

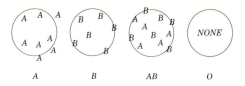

Fig. 7-2 Diagrammatic view of antigens on red blood cells: Type A blood: only A antigens; Type B: only B antigens; Type AB: both A and B antigens; Type O: no antigens.

Blood groups You may recall that Karl Landsteiner in 1900 discovered the major blood groups: A-B-O-AB. There are three multiple alleles for the antigens on the red blood cells (Fig. 7-2) which react with antibodies in plasma. Why there are antibodies is unknown. As a result, two antigens, A and B, react with two antibodies (a and b) in the plasma. Note that people with blood type O can give red cells to those of all blood types; they are universal donors. However, people with O type can safely receive only O blood because they have antibodies for both A and B antigens. Universal recipients are those of blood type AB since they lack antibodies in their plasma (Table 7-2).

About 90 percent of Caucasians have A or O type blood. In tracing the genetics of these blood types use I^A for antigen A; I^B for antigen B; use i or I^O for examples of no antigens present (O type). Gene I^A is dominant over i (I^O); gene I^B is dominant over i (I^O). AB blood type results when both genes $I^A I^B$ are present; neither dominates the other. Refer to Table 7-3 for the possible genotypes of people with the four phenotypes of blood groups. The frequency distribution of the four blood groups in the United States is given in Table 7-4. Since the discovery of A-B-O, it has been found the antigen A has two forms: A_1 and A_2. If we consider that multiple alleles are the basis for the genetics of blood types, it is clear that genes A_1, A_2, B, and O belong at the same locus. As a result, ten possible genotypes may be found from recombinations of the four alleles.

Develop the possible phenotypes of children when those of their parents are

TABLE 7-2 Four major blood groups

blood type	antigen	antibody in plasma
A	A	b
B	B	a
AB	AB	none
O	none	a and b

TABLE 7-3 Blood groups: phenotypes and possible genotypes

phenotypes blood groups	possible genotypes
A	$I^A I^A$ $I^A i$ (or I^O)
B	$I^B I^B$ $I^B i$ (or I^O)
AB	$I^A I^B$
O	ii (or $I^O I^O$)

TABLE 7-4 Blood groups (A_1-A_2-B-O)

group	frequency (percent)	phenotype and frequency		genotype and frequency	
O	43.5	O	43.5	OO	43.5
A	39.2	A_1	31.0	A_1O	25.1
				A_1A_2	2.3
				A_1A_1	3.6
		A_2	8.2	A_2O	7.9
				A_2A_2	0.3
B	12.7	B	12.7	OB	11.9
				BB	0.8
AB	4.5	A_1B	3.4	A_1B	3.4
		A_2B	1.1	A_2B	1.1

known. Students also learn quickly to recognize which phenotypes are not possible as they apply their knowledge. Also refer to paternity cases that determine who could father a child of blood type AB. Table 7-5 uses the simplified A-B-O symbols to show the inheritance of these blood groups.

At this time you may want to determine the frequency of blood groups in a class (technique, p. 280).

Inheritance of M and N blood types
Antigens M and N found on erythrocytes

TABLE 7-5 Inheritance of blood groups

Phenotype blood type of parents	Genotype possible genetic combinations of offspring	Phenotype blood type of offspring	Phenotype blood types that do not occur in offspring
O × O	OO	O	A, B, AB
O × A	OA, OO	A, O	B, AB
O × B	OB, OO	B, O	A, AB
O × AB	OA, OB	A, B	O, AB
A × A	AA, OA, OO	A, O	B, AB
A × B	OA, OB, AB, OO	A, B, AB, O	——
A × AB	AA, OA, OB, AB	A, B, AB	O
B × B	BB, OB, OO	B, O	A, AB
B × AB	OA, BB, OB, AB	A, B, AB	O
AB × AB	AA, BB, AB	A, B, AB	O

are inherited as a pair of alleles resulting in three genotypes: *MM*, *NN*, and *MN*. Natural antibodies against them are not often found in the serum of humans.

Parents who have the genotype *MM* produce 100 percent *MM* offspring; *NN* × *NN* results in 100 percent *NN*; *MM* × *NN* results in 100 percent *MN*; *MN* × *MN* gives a probability of 25 percent *MM*, 25 percent *NN*, and 50 percent *MN* offspring.

HUMAN MIGRATION There may be occasion for students to read further about the use of blood groups to trace the patterns of migration of peoples over the face of the earth. Study Table 7-6. The alleles I^A, I^B, and i represent the gene that codes for the production of the antigens A and B, and for the recessive, respectively. (Sometimes the alleles are indicated as L^A, L^B, and l, in honor of Landsteiner, the discoverer of these blood groups.) Figure 7-3 shows the distribution of the allele for blood group A.

You may want to refer to the discussion of migration of genes in population "pools" (p. 509).

Secretors and nonsecretors Some students will want to investigate A, B, and H antigens in water-soluble form in body fluids as well as saliva, tears, sweat, urine, semen, and gastric juice. Substance A is secreted by individuals of blood group A; substance B by people of blood group B. The antigenic substance H is secreted by members of group O. (Anti-H serum is used to detect H; all secretors have H in their saliva regardless of A-B-O blood groups.)

Let us assume that *S* is the gene for secretor, and *s* is the recessive allele. Secretors are either *SS* or *Ss*. This gene is inherited independently of the A-B-O antigens. Some association has been found between this *S*-gene and the Lewis blood group system.

Rh factor The Rh factor in blood was identified in 1940 in the rhesus monkey (hence its name) and later found to be present in human blood. About 85 percent of Caucasians have the Rh antigen on their red corpuscles, designated as Rh positive (Rh+). Those lacking the antigen are Rh negative (Rh−). Nearly all Chinese, Japanese, African Negroes, and North American Indians have the antigen for the Rh factor. Although there are at least eight alleles for the Rh factor, consider here only one allele: that Rh positive is dominant over Rh negative. A person with Rh-positive blood may have the genotypes *Rh+ Rh+* (homozygous) or *Rh+ Rh−* (heterozygous). In nonmatched transfusions of Rh blood, an Rh negative recipient builds antibodies against the Rh factor, or antigen, from the donor's blood. (There

TABLE 7-6 Frequencies in different populations of the alleles *i*, *I^A*, and *I^B*, giving rise to the blood groups, O, A, B, and AB*

population	number of persons tested	i	I^A	I^B
Americans (white)	20,000	0.67	0.26	0.07
Icelanders	800	0.75	0.19	0.06
Irish	399	0.74	0.19	0.07
Scots	2,610	0.72	0.21	0.07
English	4,032	0.71	0.24	0.06
Swedes	600	0.64	0.28	0.07
French	10,433	0.64	0.30	0.06
Basques	400	0.76	0.24	0.00
Swiss	275,644	0.65	0.29	0.06
Croats	2,060	0.59	0.28	0.13
Serbians	6,863	0.57	0.29	0.14
Hungarians	1,500	0.54	0.29	0.17
Russians (Moscow)	489	0.57	0.25	0.19
Hindus	2,357	0.55	0.18	0.26
Buriats (N. Irkutsk)	1,320	0.57	0.15	0.28
Chinese (Huang Ho)	2,127	0.59	0.22	0.20
Japanese	29,799	0.55	0.28	0.17
Eskimos	484	0.64	0.33	0.03
American Indians (Navajo)	359	0.87	0.13	0.00
American Indians (Blackfeet)	115	0.49	0.51	0.00
W. Australians (aborigines)	243	0.69	0.31	0.00
African Pygmies	1,032	0.55	0.23	0.22
Hottentots	506	0.59	0.20	0.19

* From E. Sinnott, L. C. Dunn, and T. Dobzhansky, *Principles of Genetics*, 5th ed., McGraw-Hill, New York, 1958; data from W. Boyd, *Genetics and the Races of Man*, Little, Brown, Boston, 1950, and A. Mourant, *The Distribution of the Human Blood Groups*, Thomas, Springfield, Ill., 1954.

are no natural antibodies against this substance in human blood.)

Another complication arises in pregnancy when a female who is Rh negative has an Rh-positive fetus developing in the uterus. (What is the father's genotype?) There is no problem with the first born (although there may be seepage of the Rh-positive blood cells from the fetus into the Rh-negative blood of the mother). The mother's antibody titer remains low, even though her blood is sensitized and produces antibodies. But in subsequent pregnancies the newly formed Rh-positive blood cells of the fetus are endangered by the antibodies of the mother. In these cases, *erythroblastosis fetalis* may result in the death of the child. However, this incompatibility can be detected before birth by amniocentesis and treated. Rh-*negative*

mothers may also receive Rh antibody serum to destroy any fetal Rh antigens in her blood and to protect the fetus in subsequent pregnancies.

You may want students to work out several possible crosses to show the pattern of inheritance of the Rh factor. Assume for purposes of simplicity that it is a trait controlled by a single pair of genes. Suppose an Rh-negative woman marries an Rh-positive man. Will all the children be Rh-positive? What are the chances, if any, for an Rh-negative child? Why are there no complications if the mother is Rh negative, and the fetus is Rh negative? Some students ask why there are no complications if the mother is Rh positive and the fetus Rh negative. Of course, the reason lies in the quantity of blood and the ability of the fetus to make antibodies.

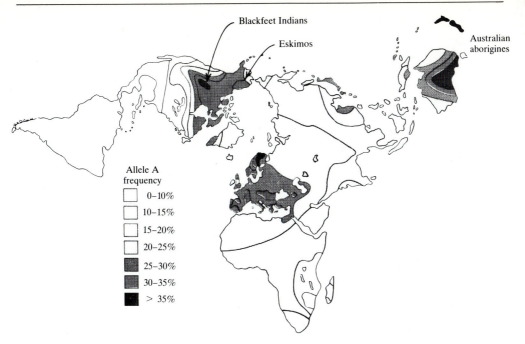

Blackfeet Indians

Eskimos

Australian
aborigines

Allele A
frequency

☐	0–10%
☐	10–15%
☐	15–20%
☐	20–25%
▨	25–30%
▦	30–35%
■	> 35%

Fig. 7-3   Distribution of the allele for blood type A.
(Adapted from *Biology* by Grover C. Stephens and
Barbara Best North. Copyright © 1974 by John Wiley &
Sons, Inc. Reprinted by permission of John Wiley & Sons,
Inc.)

What little effect would antibodies of the
fetus have on the five quarts of maternal
blood?

Adding to the complexity of inheritance
of the Rh factor is the use of two systems of
symbols: Wiener proposed the use of
Rh_0 and rh'; Fisher and Race used
alleles C and c and genes D and E. The two
systems are based on differences in interpretation of serological reactions.

Out of 100 individuals:

45 will have group O blood, and of these
39 Rh+, 6 Rh−
40 will have group A blood, and of these
35 Rh+, 5 Rh−
10 will have group B blood, and of these
8 Rh+, 2 Rh−
5 will have group AB blood, and of these
4 Rh+, 1 Rh−

Sickle-cell anemia and sickle-cell trait
An inherited disorder of hemoglobin is
transmitted by a single gene Si or Hb^s.
People who are heterozygous are said to
have the sickle-cell trait; their blood shows

normal hemoglobin as predominant in red
blood cells with a small amount of abnormal hemoglobin. When the oxygen tension
is low, the hemoglobin is altered, and the
red cells become sickle, or crescent-shaped
(Fig. 7-4). Those who are homozygous for
the gene, that is, $SiSi$ (or $Hb^s Hb^s$) suffer
from a severe form of anemia. In these
cases, about one-third of the red cells are
sickle-shaped.

Several disorders of hemoglobin can be
distinguished by electrophoretic mobility.
For example, in an electric field, normal
hemoglobin has a negative charge and
migrates to the positive pole; the "sickle"
hemoglobin has no charge and does not
migrate in an electric field.

You may recall that Linus Pauling
showed that the sickle-cell disease was due
to an abnormal hemoglobin, hemoglobin
S.[1] In 1954, Vernon Ingram discovered

[1] You may want to refer to the classic paper by
L. Pauling, et al., "'Sickle-cell Anemia'; A Molecular
Disease," *Science* (1949) 110:543–48.

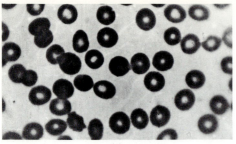

(a)

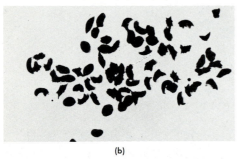

(b)

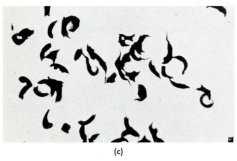

(c)

Fig. 7-4 Human red blood cells: (a) normal cells from individual of genotype Hb^AHb^A; (b) cells from a heterozygous individual of genotype Hb^AHb^S show sickling of some red cells; (c) red cells from a homozygous individual of genotype Hb^SHb^S show a high percentage of sickled red cells. (Courtesy of the Univ. of Michigan Heredity Clinic, Dr. James V. Neel, Director.)

the chemical difference between normal and abnormal sickle-cell hemoglobin. Normal hemoglobin contains a molecule of glutamic acid; the disorder is due to a substitution in one peptide group—one different amino acid out of 287. In sickle-cell hemoglobin, a molecule of valine replaces the normal glutamic acid. Several other hemoglobin alleles exist in the same locus.

Students may readily work out crosses to solve problems such as: If both parents are heterozygous for the "sickle" gene, what percent of their children may have sickle-cell anemia? How do we determine whether the 75 percent with normal hemoglobin are homozygous or heterozygous for normal hemoglobin (and have sickle-cell trait)? Students may report on the simple blood test in which a drop of blood is added on a slide with a drop of a chemical that reduces the oxygen level, such as sodium metabisulfite solution. A large number of red cells will sickle within 15 min in people who are heterozygous for the trait, that is, sickle-cell trait. A low concentration of sodium cyanate inhibits sickling of red blood cells with this type of abnormal hemoglobin. Have students read the fine paper by A. Cerami and C. Peterson, "Cyanate and Sickle-cell Disease," *Sci. Am.* (April 1975). About 99 percent of individuals with sickle-cell disease are Negroes; among Caucasians, the disease occurs among Italians and Greeks, or other groups in the Mediterranean region.

You may want to discuss the relationship between malaria and sickle type hemoglobin. Since the malarial plasmodium does not attack red cells with the abnormal hemoglobin, heterozygous individuals (Hb^AHb^S) are not likely to be killed by malaria. As a result, the frequency of the sickle gene may be as high as 40 percent in some regions in Africa. Those homozygous for the gene die in infancy; those homozygous for normal hemoglobin are susceptible to malaria. In a special environment, then, the *Si* gene has survival value, so it accumulates in the gene pool. A similar condition occurs in Asia and other Mediterranean regions where thalassemia follows a similar pattern of inheritance.

Thalassemia factor   The gene for thalassemia major, when homozygous, causes Cooley's disease, a severe anemia leading to early death of its victims. However, there is a less deadly, and more common form, thalassemia minor, or the heterozygous condition. The heterozygous forms show varying degrees of anemia in their blood picture. The frequency of the gene is high in Mediterranean countries; in certain areas of Greece and Italy, there is a frequency as high as 10 or 20 percent, and in other regions, only about 1 percent. In the United States, the gene is found mainly in immigrants from Italy and Greece.

In thalassemia, there is one amino acid in the hemoglobin sequence that is different: lysine is substituted for glutamic acid which is present in a normal hemoglobin molecule.

Deaf-mutism   Two different dominant genes are needed to produce normal hearing. Each is inherited independently. For example, *D* and *E* interact to produce normal hearing. Any other combination of genes—such as *Ddee*, *DDee*, *ddEE*, or *ddEe*—results in children deaf at birth. In this type of inheritance two "normals" may have a child who is deaf; two deaf parents may have a normal child. Most people have at least one dominant of each pair of genes, and are normal. Assume normal parents, *DdEe* × *DdEe*; could these parents have a deaf child (*DDee*)?

Albinism   There are two types of albinism. In one type, the pigment melanin is not made by the cells; in another type melanin is produced but does not accumulate in tissue cells. As a result, two albinos may have normally pigmented children if their albinism is due to different sets of genes. For example, if one parent is *AAbb* and the other is *aaBB* the offspring receives a dominant gene from each parent and the heterozygote, *AaBb*, has normal pigmentation. What are the chances for albinism to appear when two parents are heterozygous for the two genes affecting melanin: *AbBb* × *AaBb*?

Sex-linked genes

In 1902, McClung suggested that the unpaired set of chromosomes in several insects might be a determiner of sex, even

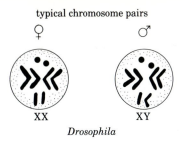

typical chromosome pairs

Fig. 7-5   *Drosophila* chromosomes.

though he had made no observations of cells of female flies. In 1905, Stevens and Wilson determined that XX and XY chromosomes were sex chromosomes distinct from the rest of the autosomal chromosomes (Fig. 7-5).[2]

It was quickly recognized that during spermatogenesis, the X and Y chromosomes segregate at meiosis so that 50 percent of the sperms carry the X chromosome and 50 percent carry the Y chromosome. Since the egg has only an X chromosome, the type of sperm—either X-carrying or Y-carrying—determines the sex of the offspring.

With this knowledge, Morgan was able to explain the results of crossing red-eyed and white-eyed *Drosophila*. The genes for eye color in *Drosophila* are sex-linked since they are located on the X chromosome. Among humans, the well-known sex-linked traits are hemophilia and red-green color blindness.

Hemophilia   Hemophilia is a sex-linked disorder resulting from a recessive gene located on the X chromosome. People with hemophilia lack a specific globulin in their plasma so that their blood does not clot normally. While hemophilia has been traced in some pedigrees of royal families (Fig. 7-6), the condition occurs in about 1 in 25,000 males and in 1 in 625 million females. A female would need to be homozygous for the gene (X^hX^h) to have hemophilia (Fig. 7-7). Women can be

---

[2] Among birds, sex chromosomes are alike in the male, called ZZ; the female is ZO.

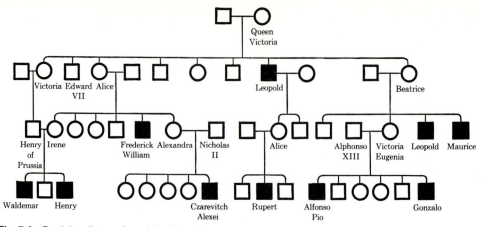

Fig. 7-6 Partial pedigree of royal families of Europe showing transmission of the gene for hemophilia. (From L.H. Snyder and P.R. David, *The Principles of Heredity*, 5th ed., Heath, 1957.)

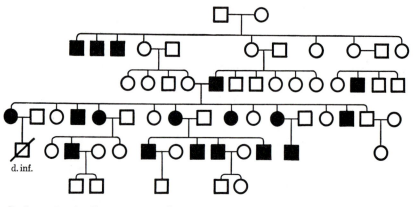

Fig. 7-7 Pedigree showing the occurrence of hemophilia in women. (From L.H. Snyder and P.R. David, *The Principles of Heredity*, 5th ed., Heath, 1957.)

carriers of the trait since the normal gene dominates the gene for hemophilia; this was the case with Queen Victoria $(X^N X^h)$. Since the Y chromosome lacks this locus a male shows the trait when only one gene is present $(X^h Y)$. In this kind of crisscross inheritance, sons can receive the defective gene only from their mother. The father passes the Y chromosome to his sons; his X chromosomes are transmitted to his daughters. Students can quickly solve such problems as: A girl is normal, but has a brother suffering from hemophilia. What are the possible genotypes of the parents? Can a man be a carrier of hemophilia?

Red-green color blindness About 8 percent of males have red-green color blindness, while only 0.4 percent of females show the trait. Have students diagram and explain why there is more color blindness among males. What is the genotype of a color-blind girl? What possible genotypes could her parents have?

Try the Ishihara tests for color blindness or the Holmgren test using skeins of different colored yarn. You may want to purchase color slides to test students (available from biological supply houses). Students may learn there are two types of red-green color blindness. Refer to "Visual

Pigments and Color Blindness," *Sci. Am.* (March 1975).

In total color blindness, individuals see only shades of gray and their vision improves in dim light. This rare condition is not sex-linked, but is an autosomal recessive.

Polygenes: additive effects

Among humans, skin color and height are good examples of inheritance in which several different genes for one trait produce an additive, or cumulative, effect. (Height in plants, and vitamin A content of some strains of corn are a few examples, among many.)

Skin color   There may be as many as five pairs of genes that affect the amount of melanin in the skin. However, for the sake of simplicity, let us assume that only two pairs of genes are involved. In this multiple factor inheritance let *AABB* represent Negro skin color and *aabb* represent white, with mulatto having the genotype *AaBb*. Any three of the color-producing genes would produce dark skin; two would produce medium coloring; and one gene would produce a light skin. What variations may occur in skin color of offspring of two mulattos: *AaBb* × *AaBb*?

Height   Assume there are four different genes for height: *A-B-C-D*. Then an individual with genotype *AaBBccDd* would be taller than one with gene pairs *AabbCcdd*, and so forth. As a result, height should show a normal distribution curve in random matings. However, environmental factors also affect height and cause variations in anticipated results.

Genetic diseases

The heredity of several disorders is well known so that individuals who have such genes in their family can be advised concerning the possible heredity of their offspring. Many defects are due to recessive lethal genes.

Galactosemia   This condition occurs in about 25 out of every million births. Individuals homozygous for the recessive gene lack the enzyme to metabolize galactose or lactose in milk. Fortunately, other sugars can be substituted in an infant's diet to avoid the effects of galactosemia, which are kidney and liver malfunctions, jaundice, and possible death. Infants that survive, however, may have mental deficiencies and stunted growth. (As in PKU and in diabetes, altering the environment reduces the deleterious effect of the genes.)

Tay-Sach's disease   In this disorder, fatty gangliosides accumulate causing a deterioration of the central nervous system so that muscle and nerve coordination is reduced; there is a wasting away of the body and the child soon dies. In many cases, dominance is not complete so that a carrier can be recognized. Homozygous individuals have two enzymes in their serum: hexosamidases A and B. In Tay-Sach's disease, homozygous recessive children lack the hexosamidase A. Heterozygous individuals are intermediate in their serum activity.

Using amniocentesis, it is possible to test whether a specific enzyme is lacking in a fetus homozygous for Tay-Sach's disease (amniocentesis techniques, p. 518).

Recessive genes are associated with other disorders: diabetes, muscular dystrophy, sickle-cell anemia (p. 509), and hemophilia (p. 511).

Alcaptonuria   The physician Garrod described this condition as an inborn error of metabolism (1909). In this recessive trait, homogentisic acid accumulates and is excreted in the urine, turning urine black. Normally the homogentisic acid would be broken down into acetoacetic acid by a gene that produces an enzyme in blood serum. There is also a hardening and darkening of cartilage in the body of the affected individuals. An increase in the consumption of tyrosine and phenylalanine by alcaptonurics leads to an increase in homogentisic acid since there are blocks in the normal sequence of enzyme-regulated breakdowns of the amino acids into acetoacetic acid and final conversion to carbon dioxide and water. Use Figure 7-8 to trace the complexity of this metabolic series that also includes tyrosine and albinism, and phenylalanine and phenylketonuria.

Phenylketonuria Another recessive gene causes a block in the metabolism of phenylpyruvic acid or of phenylalanine, resulting in extreme mental retardation. The lack of the specific enzyme which handles the increased consumption of these amino acids can be detected by the large amounts of phenylpyruvic acid excreted in urine. There is a dip test available that can be used to test urine of infants for PKU;[3] the diet of infants can be corrected to avoid the onset of the mental retardation.

Normal individuals possess the enzymes to oxidize phenylpyruvic acid into precursors of homogentisic acid (see Fig. 7-8). These errors in metabolism have revealed much of the picture of degradation of phenylalanine and of tyrosine. Notice that albinism is due to the lack of a gene that produces an enzyme effective in converting a precursor into melanin (skin color pigments).

Chromosomal abnormalities

Some individuals may have extra autosomal chromosomes, or extra (or fewer) sex chromosomes as a result of nondisjunction in meiosis or a translocation of a part of a chromosome.

Down's syndrome In this case an extra autosomal chromosome of pair #21 is present due to either nondisjunction or translocation of a part of chromosome 21 to another chromosome. Individuals with this disorder are mentally and physically retarded. The frequency of Down's syndrome births increases with the age of the mother. Among women under 30 years, the frequency is less than 1 per 2,000 births, increasing to 1 in 300 births for women 35 to 39 years, and in older women it may be as high as 1 in 35 births.

Many examples of disorders due to extra chromosomes are known; many may be lethal in the fetal stage.

Klinefelter's syndrome In meiosis, during gametogenesis, if nondisjunction occurs in the sex chromosomes an egg may end up having two X chromosomes or none; sperms also may have two sex chromosomes or none. In Klinefelter's syndrome, the zygote has two or more X chromosomes and a Y (for example, XXY). As a result of the presence of the Y chromosome, the individual has testes but he does not develop normally; the individual is sterile. There are also degrees of feminization, for example, breast enlargement and scant facial and pubic hair.

Turner's syndrome In this disorder, the genotype for the sex chromosomes is XO; there is only one X chromosome. Apparently a gamete with an X chromosome united with one lacking an X or a Y chromosome. As a result, the individual develops as a female with ovaries, but she is sterile and short of stature, and has the characteristic wide neck with fleshy folds of skin giving a webbed appearance.

Many other sex chromosome anomalies exist: for example, XYY, XXX, and XXXX.

Sexual dimorphism in cells

Drumsticks A structural difference found in undividing white blood cells is useful in distinguishing male and female genotypes. Davidson and Smith (1954) described a dimorphism in polymorphonuclear leukocytes of blood films from females and males. A bulblike projection attached by a thin fiber to the lobed nucleus is found in cells of females; these

---

[3] Phenistix is the trademark of a "dip and read" test for phenylketonuria in infants (Ames Co., Inc., Elkhart, IN).

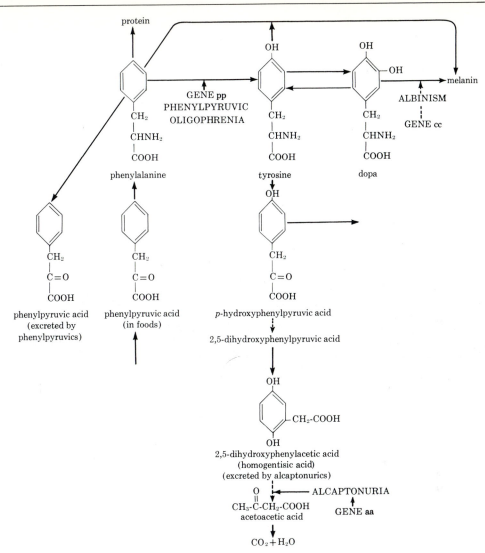

Fig. 7-8 Phenylalanine-tyrosine metabolism in humans. The metabolic steps assumed to be blocked in homozygotes for the three genes indicated are shown by arrows.

(After E. Sinnott, L.C. Dunn, and T. Dobzhansky, *Principles of Genetics*, 5th ed.; copyright, 1958, McGraw-Hill Book Co. Used by permission.)

structures, called drumsticks, are absent in blood cells of males (Fig. 7-9).

Barr bodies Sex differences can be determined by the presence of Barr bodies in cells (discovered by Barr in 1949). The Barr body is a dark-staining body that lies against the membrane of the resting nucleus. Only one X chromosome is active in the female. The distinctive pattern of heterochromatin matter of the inactive X chromosome is visible in epithelial cells of females. Therefore, it is used as a sex indicator (Fig. 7-10). Skin biopsies are made to secure deeper layers of epidermis, but more frequently oral smears are prepared of epithelial cells to show Barr bodies. Examine slides of epithelial cells of males and females.

In 1962, Mary Lyon proposed an ex-

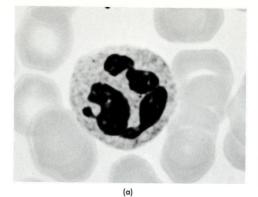

(a)

(b)

Fig. 7-9 White blood cells of normal humans: (a) cells from a female show drumstick on the nucleus; (b) cells from a male (both are magnified 2200 ×). (Photo courtesy Eastman Kodak Co.)

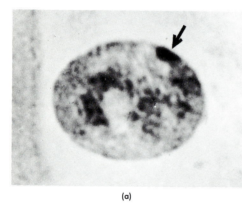

(a)

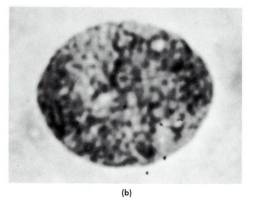

(b)

Fig. 7-10 Nucleus of normal squamous epithelium in humans: (a) cell of a female showing Barr body (arrow); (b) cell of a male (both are magnified 1750 ×). (Courtesy of Carolina Biological Supply Co.)

planation for the inactivation of one of the pair of XX chromosomes. Her hypothesis, often described as "lyonization" of an X chromosome, proposes that very early in embryological development, an X chromosome becomes inactive and results in a Barr body. In some cells, the inactive X may be from the father; in other cells, from the mother. There seems to be a random occurrence among X chromosomes. In subsequent mitotic divisions these cells continue with the same active X and inert Barr body. Normal women are then mosaics for sex-linked traits.

Compare the chromosome disorders and the number of Barr bodies found in cells in Table 7-7.

Can students explain why women with Turner's syndrome show a frequency of hemophilia and color blindness similar to that found in males?

Examining human chromosomes and karyotypes

In 1956 Tjio and Levan reported that the number of chromosomes of humans is 46, not 48;[4] this was soon confirmed by others.

Slides are available for the study of metaphase chromosomes in human leukocytes (magnification of at least 900× is needed for study of details.) A few millili-

[4] J. H. Tjio and A. Levan, "The Chromosome Number of Man," *Hereditas* (1956) 42:1-6.

TABLE 7-7   Barr bodies found in cells

	sex chromosomes	number of Barr bodies
normal man	XY	0
normal woman	XX	1
Turner's syndrome	XO	0
Klinefelter's syndrome	XXY	1

ters of blood are drawn from an individual and cultured and stimulated to divide by using an extract of red kidney beans called phytohemagglutinin. As white cells undergo mitosis, colchicine is added to arrest mitosis so that the chromatids do not separate; metaphase cells accumulate as a result. Cells are then placed in a hypotonic sodium citrate solution so that the cells become swollen as water enters; the chromosomes are then spread apart in the cells. Centrifuging results in scattering of the chromosomes in the cells. Finally, the cells are fixed, dropped on slides, and

rapidly dried in warm air. The procedure causes the swollen white blood cells to collapse, and (after staining) chromosomes can be counted in one plane. Each chromosome is actually two chromatids joined by a centromere (Figs. 7-11, 7-12).[5]

Photomicrographs are made of the chromosome spreads. In a karyotype, chromosomes are numbered according to a standardized system developed in 1959, which describes shape, length, and arrangement of arms and the distance of the chromosomes from the centromere. A karyotype of female somatic cells would show 23 pairs (22 autosomal pairs and one pair of XX); males would have 22 pairs of autosomes plus one unmatched pair of XY sex chromosomes (Fig. 7-12).

Notice that the chromosomes are numbered and that there are seven groups. You may want to enlarge a photograph of a chromosome spread, reproduce it, and have students cut out pairs of similar

[5] Summarized from H. Edgerton and R. Flagg, "Human Chromosomes, Nuclei, and Sex," *Carolina Tips* (January 1964) 27: 1-2.

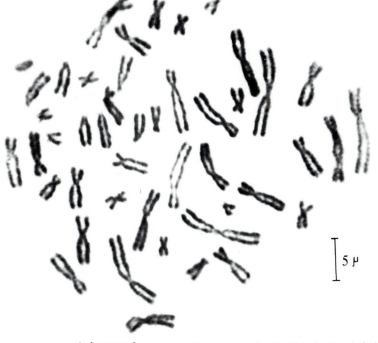

Fig. 7-11 Chromosome spread of a normal human male.

(Photo courtesy Carolina Biological Supply Co.)

5 μ

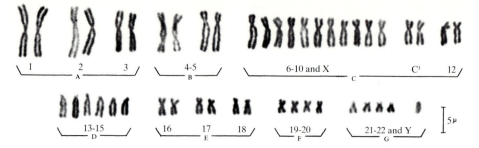

Fig. 7-12 Karyotype of a normal human male. Chromosome spread has been photographed, cut out, and arranged in groups according to size and shape. Notice there are seven groups. (Photo courtesy Carolina Biological Supply Co.)

chromosomes and develop a karyotype such as that shown in Figure 7-12. Have students determine whether the individual is male or female, or has Down's, Turner's, or Klinefelter's syndrome. This may be the time to begin a discussion of some aspects of genetic engineering (p. 542).

Amniocentesis

Many inborn errors of metabolism that often result in mental deficiencies can be diagnosed in the fetus in the amniotic sac. Cells from the fetus are found floating in amniotic fluid. The fluid and cells can be withdrawn by means of a syringe and cultured; subsequent cytological and biochemical studies of the cells reveal the possible presence of genotypes of deleterious conditions (for example, Down's syndrome and sex chromosome disorders).

You may want to have students trace the procedure used in amniocentesis, as shown in Figure 7-13. A small amount of amniotic fluid is drawn by puncturing the abdominal wall of the mother to reach the amnion and not harm the fetus. Cells may be examined immediately or cultured for later study. Immediate examination shows the sex of the fetus as determined by Barr bodies (p. 515) or the presence of the Y chromosome. More detailed studies are made from cultured cells, for example, biochemical studies of tissue cultures may reveal a fetus homozygous for Tay-Sach's disease if the cells lack the enzyme hexosamidase A. Radioactive tracers are also used to examine tissue cultures of fetal cells.

Genetic counseling plays an increasingly important role in assisting parents who are heterozygous for some condition in determining the possible frequency of the undesirable trait as homozygous in the offspring. In addition to giving frequencies of probability, amniocentesis can offer the evidence. You may want to assign reading of T. Friedmann's paper, "Prenatal Diagnosis of Genetic Disease," *Sci. Am.* (November 1971), and "Genetic Amniocentesis," F. Fuchs, *Sci. Am.* (June 1980). Many inborn metabolic errors that can be detected by amniocentesis are described in these papers.

Compare the percentage of concordance between monozygotic and dizygotic twins in traits depending on both genes and environmental factors (Table 7-8).

Students will find excellent reviews of older classic studies as well as current work in issues of *Scientific American*. For example, see "Visual Pigments and Color Blindness" (March 1975); "Sex Differences in Cells" (July 1963); "Mapping of Human Chromosomes" (April 1971); "The Isolation of Genes" (August 1973); "A Molecular Basis of Cancer" (November 1983); "Genetics of Antibody Diversity" (May 1982); "The Biological Effects of Low-Level Ionizing Radiation" (February 1982); "Oncogenesis" (March 1982); "Hereditary Fat-Metabolism Diseases" (August 1973); "Rabbit Hemoglobin from Frog Eggs" (August 1976); "Cytochrome C and the Evolution of Energy Metabo-

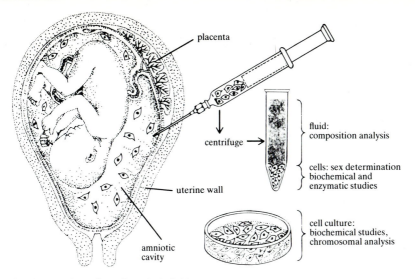

Fig. 7-13 Amniocentesis: cells and amniotic fluid removed from the amnion for chromosomal and biochemical studies. (From "Prenatal Diagnosis of Genetic Disease" by Theodore Friedmann. Copyright © 1971 by Scientific American, Inc. All rights reserved.)

TABLE 7-8 Comparison of monozygotic and dizygotic twins*

observed disease or behavior	% concordance	
	mz twins	*dz twins*
Tuberculosis	54	16
Cancer at the same site	7	3
Clubfoot	32	3
Measles	95	87
Scarlet fever	64	47
Rickets	88	22
Arterial hypertension	25	7
Manic-depressive syndrome	67	5
Death from infection	8	9
Rheumatoid arthritis	34	7
Schizophrenia (1930s)	68	11
Criminality (1930s)	72	34
Feeble-mindedness (1930s)	94	50

* From *Heredity, Evolution and Society*, 2nd ed., by I. Michael Lerner and William J. Libby. W. H. Freeman and Co. Copyright © 1976.

The percentage of concordance among twins for some traits depends on many genes and many environmental factors.

lism'' (March 1980). You may also want to refer to texts in genetics listed at the end of this chapter.

Heredity in some plants

Corn (maize)   Purchase hybrid kernels of corn containing recessive genes for albinism from biological supply houses; Texas Agricultural Experiment Station, College Station, TX; Meyers Hybrid Seed Corn Co., Hillsboro, OH; and Genetics Laboratory Supplies, Clinton, CT (among others). When germinated, these kernels should show 75 percent green shoots and 25 percent albino shoots, or a 3:1 ratio These seedlings show variations in the rate of formation of chlorophyll; further, they should be grown in sunlight so that the 75 percent show development of chlorophyll. Germinate kernels in the dark to show the shoots lack chlorophyll, then transfer to light to show the genetic and environmental bases for the appearance of chlorophyll in the seedlings.

GENETIC EARS OF CORN   Ears of corn offer excellent opportunities for counting segregating traits. Large numbers of kernels per ear give numbers that are close to theoretical Mendelian ratios. In many school situations, study of corn may be more practical than culturing and counting fruit flies.

Many types of traits in corn are available from supply houses. The ears can be coated with a thin layer of lacquer. Have students insert a pin at the starting point and tally rows around the corn ear.

Count the 3:1 ratio of purple aleurone color to yellow aleurone. These are the F_2 hybrids resulting from crossing colored × colorless (yellow) aleurone: $RR \times rr \rightarrow$ (F_1) $Rr \rightarrow$ (F_2) $1RR$: $2Rr$: $1rr$, or three purple to one yellow. Or show the results of a backcross of F_1 hybrid Rr to recessive parent—$Rr \times rr \rightarrow 1$ purple (Rr):1 yellow (rr).

Or count F_2 progeny for starchy-waxy endosperm (three starchy to one waxy) resulting from inbreeding hybrids in the F_1 progeny of a cross: $SS \times ss \rightarrow Ss$.

Independent assortment is shown in the 9:3:3:1 ratio which results from crossing dihybrids with traits for purple and yellow aleurone and the starchy-sweet endosperm.

WAXY-STARCHY POLLEN GRAINS Preserved tassels of corn plants that are heterozygous for nonwaxy ($Wxwx$) are available; slides of pollen can be made in class. When Gram's iodine solution is added to the pollen on a slide, some pollen grains (starchy) show a distinct dark blue color, and others do not stain with iodine (waxy). Students can count the proportion of both types on the slide. Since pollen grains of the heterozygous plants are gametophytic, a gamete ratio of 1:1 results.

STAINING KERNELS If kernels of similar heterozygous corn plants are cut and Gram's iodine applied to each, a ratio of three dark (starchy) to one waxy results— 1 $WXWX$: 2 $WXwx$: 1 $wxwx$.

Also refer to enzymatic differences in wrinkled and smooth peas, that is, ability to change sugar to starch (see below). Also refer to the role of gibberellic acid in altering dwarf peas, beans, and corn (p. 374). Beadle describes evidence that teosinte, a wild grass, may be a progenitor of modern corn in his paper, "The Ancestry of Corn," (*Sci. Am.*, January 1980).

Soybeans Obtain seeds that will produce heterozygous plants (whether pure or hybrid) that show a phenotypic difference:

Homozygous dominant shoots are green; heterozygous shoots are light green; and homozygous recessives are yellow. This gives a ratio of 1:2:1, not 3:1.[6]

Seeds germinate within ten days. The green and yellow variations are distinct soon after germination, but after about seven more days, the green color of the dominant heterozygous and dominant homozygous are differentiated.

Peas You may also obtain F_1 seeds from a cross between pure tall and dwarf peas. They will develop into hybrid tall pea plants. At the same time also purchase the seeds that are a result of a cross between two hybrid tall pea plants. Germinate the seeds in paper cups containing moist sand. As the seedlings grow, students will see the difference in height; a ratio of three tall to one short or dwarf will result.

GENETIC CHEMICAL VARIATIONS IN PEAS Wrinkled and smooth (or round) peas show variations in the type of starch grains they contain and in the rate of starch formation from a source of glucose-1-phosphate (dependent on the action of the enzyme starch phosphorylase).

Students can add scrapings from a cut surface of a soaked, wrinkled pea to a drop of Lugol's iodine solution on a microscope slide; prepare a slide of scrapings from a soaked smooth pea in a similar manner. Under the microscope, notice the oval starch grains of smooth peas and the grooved, round grains of wrinkled peas.

Also prepare fluid extracts of chopped or ground dried peas of the two types. Be sure to wash the food grinder, blender, or mortar and pestle when preparing crushed peas of both smooth and wrinkled types. Add about 50 ml of water for every 10 g of peas used. Centrifuge if necessary to separate the fluid containing the enzymes.

Test the extracts for potency (and/or quantity) of the starch-forming enzyme by examining how rapidly a glucose solution is converted to starch by each type of

---

[6] Such seeds are available from Carolina Biological Supply Co., Burlington, NC.

extract. Use a small quantity of glucose-1-phosphate (about 0.5 g) added to 100 ml of distilled water. Add Lugol's iodine to test sample drops of the glucose solution on a slide at 10-min intervals. The enzyme extract from wrinkled peas shows a darker bluish black color. (Use a white background for the slides.)

You may prefer to add 2 g of plain agar to the glucose-1-phosphate and distilled water. Bring this to a boil and plate Petri dishes. Pour the extracts on separate dishes containing a film of glucose and agar. Test with Lugol's solution for the appearance of starch.

Sorghum   Red-stemmed sorghum is dominant over green-stemmed (when grown in the presence of sunlight). Stress the role of the environment and the genetic composition. Seeds produced from inbred hybrid plants may be obtained from supply houses. These seedlings will germinate in about five days, but the differences in stem color are most obvious in about seven to ten days. When large numbers of seeds have been germinated, the 3:1 ratio of red-stemmed to green-stemmed plants is readily apparent.

Also notice the difference in pollen grains heterozygous for "waxy" (Fig. 7-14).

Tobacco   Germinate seeds of tobacco—normal and albino—on agar plates. To increase visibility, Carolina Biological recommends using a 3 percent agar to which carbon black is added in the proportion of 6 tsp of carbon black to 1 liter of water. You may want to add a mold inhibitor such as Tegosept M or Methyl Parasepts;

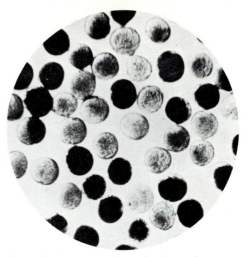

Fig. 7-14   Normal and "waxy" pollen grains in sorghum (magnified 270 ×, reduced by 39 percent for reproduction). When pollen grains heterozygous for the waxy gene are treated with iodine, half the grains (the normal grains) stain blue (dark in the photo) and half (the waxy grains) stain red. (Courtesy of R. E. Karper, Texas Agricultural Experiment Station, from *J. Heredity* 24:6, June 1933.)

there are many such inhibitors under different trade names.

Devise a demonstration to show that the appearance of chlorophyll is due both to heredity and to the proper environment, in this case, light.

Irradiated seeds   Supply houses, such as Carolina Biological, have seeds available that have been irradiated with doses ranging from 15,000 to 30,000 roentgens. Many of the young seedlings will show variations resulting from radiation (Fig. 7-15). To determine whether mutations have occurred, at least two subsequent

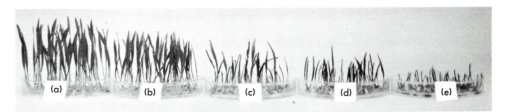

Fig. 7-15   Seedlings of seven-day-old Himalayan barley (*Hordeum vulgare*) germinated in water after the following doses of irradiation: (a) no irradiation; (b) 10,000 R; (c) 20,000 R; (d) 30,000 R;

(e) 50,000 R. Dosages b through e are sufficient to produce mutations. (Photo by H. Luippold, courtesy of Oak Ridge National Laboratory.)

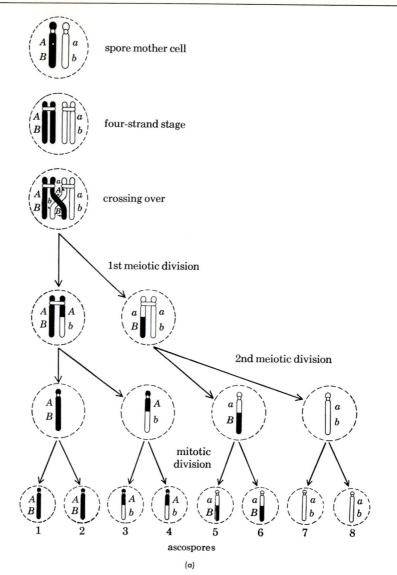

spore mother cell

four-strand stage

crossing over

1st meiotic division

2nd meiotic division

mitotic division

1   2   3   4   5   6   7   8

ascospores

(a)

Fig. 7-16 Heredity in *Neurospora*: (a) meiosis using one pair of chromosomes only (note the arrangement of the eight ascospores as they would lie in one ascus); (b) exchange between chromatids results in second-division segregation within an ascus. (a, from E. Sinnott, L. C. Dunn, and T. Dobzhansky, *Principles of Genetics,* 5th ed., copyright, 1958, McGraw-Hill Book Co., used by permission; b, after A. Srb and R. Owen, *General Genetics,* Freeman, 1952.)

generations of the plants would have to be studied.

Neurospora A series of laboratory activities may be planned to study heredity in this pink mold. Inoculate one side of an agar film (containing a minimal medium, described on p. 524) in a Petri dish with plus type *Neurospora crassa*. On the other half of the film add the opposite mating type, minus (these two types can be called *A* and *a*). A growth of asexual monoploid hyphae result with monoploid asexual spores, or conidia. However, conjugation, sexual fusion, or fertilization occurs when either monoploid filaments or

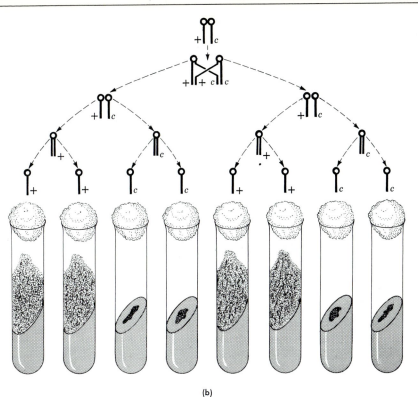

(b)

Fig. 7-16 *(continued)*

nuclei of conidia of opposite mating types fuse. As a result, a diploid fruiting body (a saclike perithecium) forms (see Fig. 6-57).

Mount a perithecium on a slide in a drop of alcohol. Use fine jeweler's forceps or very thin wire transfer needles. In alcohol, the perithecium softens and the long, tubular asci separate in the perithecium. The diploid fusion nucleus in each ascus undergoes two divisions in meiosis resulting in four monoploid nuclei; these undergo a mitotic division so that eight monoploid ascospores are lined up in each ascus in the perithecium.

Dodge showed that each ascospore is lined up so that four at one end are the product of one mating type, and the other four the product of the other mating type. The type of genes they contain will vary, depending on whether the alleles segregated at the first or second meiotic division. Lindegren showed that when two pairs of linked genes are crossed over, an exchange occurs in any one region between two of the four chromatids (between the one chromatid of one homologue and the chromatid of the other homologue) as shown in Figure 7-16.

Students interested in special crosses, such as normal wild *Neurospora* with albino, may practice opening a fruiting body or perithecium, and separating out a single ascus. Carefully separate each of the eight ascospores, keeping track of the order in which they lie in the ascus. Transfer each spore with a transfer loop of sterile water (to avoid loss of dry ascospores or contamination) to an agar slant or a Petri dish containing the special nutrient medium needed for *Neurospora* to grow. Heat in a water bath of 55° C (131° F) for half an hour; cool and then incubate. Complete details are provided in the fine laboratory manual of N. Horowitz, *Chemical Genetics* (Division of Biology, California Institute of Technology, Pasadena, 1960). Also refer

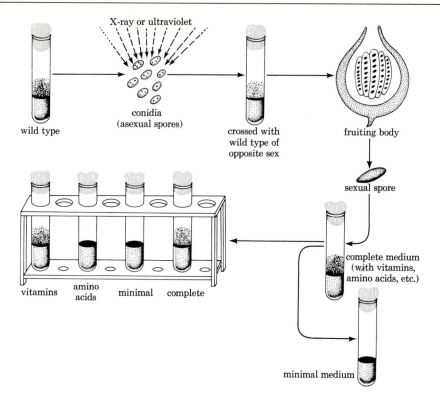

Fig. 7-17 Detection of biochemical mutants in *Neurospora*. The mutant lacks the ability to synthesize one of the vitamins. Notice that the mutant does not grow on minimal medium, nor on minimal medium to which amino acids are added. However, it does grow on minimal medium when vitamins are added. (After G. W. Beadle, *Science in Progress*, Yale Univ. Press.)

to the BSCS film *Neurospora Techniques* (Thorne).

DETECTION OF BIOCHEMICAL MUTANTS
The steps used by Beadle and Tatum to detect nutritional mutants are illustrated in Figures 7-17 and 7-18. *Neurospora* normally synthesizes its amino acids and other materials from a *minimal* diet. To prepare this diet, which may be purchased from Difco (see Appendix), combine the following:

ammonium nitrate	1.0 g
ammonium tartrate	5.0 g
monopotassium phosphate	1.0 g
magnesium sulfate	0.5 g
calcium chloride	0.1 g
sodium chloride	0.1 g
ferric chloride	10 drops of 1% solution
zinc sulfate	10 drops of 1% solution
sucrose	15.0 g
biotin	5.0 mg
distilled water	1.0 liter

The medium should be sterilized at 15 lb pressure for 10 min; since sucrose is present, avoid overheating. If solid medium is needed, plain agar can be added; the agar should be a select grade to avoid contamination of the medium with other minerals. It is preferable to grow strains on liquid media (see minimal diet above) to avoid unknown contaminants in the agar.

When a nutritional deficiency occurred, due to a mutation, Beadle and Tatum could locate it quickly because the spores did not grow on the minimal medium. If the spores could grow on a minimal medium to which one amino acid or one vitamin had been added as a supplement,

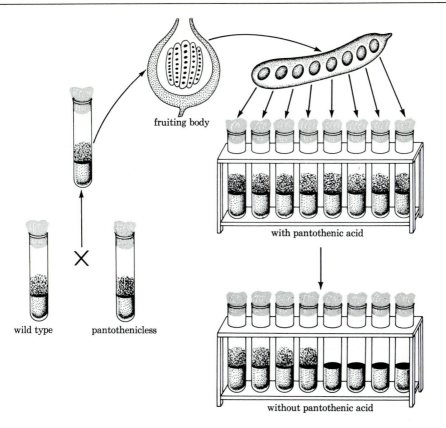

fruiting body

with pantothenic acid

without pantothenic acid

wild type

pantothenicless

Fig. 7-18 Determining inheritance of a biochemical mutant in *Neurospora*. Conidia are transferred from medium containing pantothenic acid to medium lacking the test substance. Notice the segregation of the mutant in tubes lacking pantothenic acid. (After G. W. Beadle, *Science in Progress*, Yale Univ. Press.)

this was the factor that the mutant could no longer synthesize; apparently it had lost the enzyme to synthesize this product. Ascospores from treated (by radiation or chemicals) and from untreated fungi were placed in individual test tubes containing a complete medium where they were allowed to produce mycelia so that the plant became established stock. Transfers were then made to tubes with a variety of minimal media.

This tedious, yet elegantly simple technique was the basis for the "one gene, one enzyme" theory that led to the discoveries of biochemical processes that gained for Beadle and Tatum, along with Lederberg, the 1958 Nobel Prize.

Seven linkage groups have been found in *Neurospora crassa*. Laboratory work may lead to fairly complex studies for small groups of highly motivated students. Extreme caution must be practiced in the laboratory for *Neurospora* can easily contaminate the whole laboratory (the spores are highly resistant).

Also read a description of a cross in *Neurospora* that shows independent assortment (genes on different chromosomes) of two sets of traits, a pattern of spreading growth, and its allele, a pattern of colonial growth; these alleles are on chromosome 1. Another pair of alleles found on chromosome 2 is for color—pink or orange, and albino conidia. Students will want to read further in current texts in genetics.

Chlamydomonas The biochemical genetics of sexuality has been widely studied in this flagellated green alga (Fig. 6-54). Monoploid cells reproduce by fission. Diploid zygotes are formed in some species under specific conditions and then undergo meiotic divisions resulting in four monoploid cells.

Moewus and his group (1950), working with *Chlamydomonas reinhardi*, showed that, under gene control, chemicals were produced that influenced potencies or valences of sexuality. They described five valences of maleness and five of femaleness. Furthermore, they postulated that certain hormonelike substances initiated growth of flagella, control of conjugation, or determination of valence of sexuality, and so on.

There are two physiologically different mating types among *Chlamydomonas*. Members of a clone arising from one cell are all "plus" (mt$^+$) type; members of another clone are "minus" (mt$^-$) mating type. Pairing occurs when algae of opposite mating types are brought together in a test tube of medium, and the fusions result in diploid zygotes. After meiotic divisions monoploid cells (or gametes, if you will) are formed, the typical *Chlamydomonas* cells. Two cells of the tetrad are mt$^+$ and two are mt$^-$. These cells show no structural differences, as all cells of *Chlamydomonas* look alike, but there is a physiological difference, so they may also be considered isogametes.

One trait that has been studied is the wild, or normal, flagellated form as compared with paralyzed (nonmotile) flagella. When a mating type of the strain producing normal flagella is crossed with the opposite mating type of a strain having paralyzed flagella, zygotes that are heterozygous are formed. As a result of meiosis, four cells are formed—two that have flagella and two that lack motile flagella. In a liquid medium in a test tube, the paralyzed cells sink to the bottom of the test tube so that the two types can be separated, thereby obtaining individual cells from which pure clones can be developed (barring mutations). Like most unicellular forms and fungi, the predominating stage in the life cycle of *Chlamydomo-*

nas is monoploid. Further data is available in Levine's *Genetics* (2ed. New York: Holt, Rinehart and Winston, 1968).

Acetabularia   There may be the opportunity to show that the nucleus of a cell is the source of genetic control and that the cytoplasm is affected by the genetic composition of the nucleus. *Acetabularia* (Fig. 7-19) is a one-celled marine alga differentiated into an umbrellalike cap on a stalk, which has a holdfast, or rhizome. (Remember this is a unicellular organism.) The nucleus of this alga is in the rhizome. Regeneration experiments by Hämmerling (see p. 527) showed that when the stalk and cap were removed, the rhizome containing the nucleus regenerated a new cap and stalk. Further, notice in Figure 7-19 that grafting experiments exchanging reciprocal parts of two species of *Acetabularia* resulted in producing the shape of cap controlled by the genetic contents of the nucleus of the holdfast. Substances emanate from the nucleus, pass through the intervening cytoplasm of a different species and affect the shape of the cap. (Also refer to page 436.)

Yeasts and bacteria   Plus and minus mating strains are found in some yeasts. (Fig. 6-58.) You may want to obtain yeast cultures that ferment galactose (producing carbon dioxide that can be measured) and cross these with cells that are unable to ferment this sugar. Students will want to refer to the classic work of Lindegren on yeasts (as well as on *Neurospora*).[7]

The literature on sexuality in bacteria is extensive, as is the study of mutations among bacteria. Also refer to genetics texts describing studies with *Pneumococcus* and transformation, studies of transduction, and work in phage and *Escherichia coli*. *Pathogenic bacteria should not be used in introductory laboratory work.*

Mutations   We have mentioned the possibility of studying the effects of radiation

[7] C. Lindegren, *The Yeast Cell: Its Genetics and Cytology* (St Louis, MO: Educational Publishers, 1949).

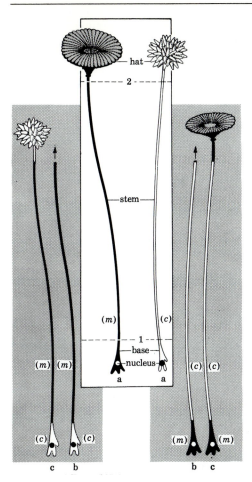

Fig. 7-19 Grafting experiments in *Acetabularia*. Grafting *Acetabularia mediterranea* stem to the nucleated base of *A. crenulata* (dark, left-hand panel) results in the generation of a deeply indented hat characteristic of *A. crenulata* (center panel, left). The reciprocal graft (dark, right-hand panel) produces an umbrella-shaped hat consistent with the character of *Acetabularia mediterranea* (center panel, right). (After A. Srb, R. Owen, and R. Edgar, *General Genetics*, 2nd ed., Freeman, 1965.)

and of chemical treatment on the growth of seedlings, of bacteria, of *Neurospora*, and others. Some students may wish to describe the results of treating plants with colchicine (*caution: toxic and especially dangerous to the eyes*). Squash smears may be examined of root tips of onions to show the doubling of chromosomes due to inhibition of spindle formation and the production of new cell walls. Refer to the preparation of

squash slides to show mitosis and effects of colchicine (p. 398). (*Note special cautions*) Also study prepared slides of mutant forms of fruit flies, or prepare flies by mounting them in clear Karo syrup (see chapter 2).

Law of probability

It is important to illustrate the fact that a 3:1 ratio among the offspring of hybrids is the result of random combinations of eggs and sperms. There are several ways to show this symbolically. Among these are the following:

1. Have students toss two similar coins simultaneously on the desk and keep tallies of the number of combinations of two heads, two tails, and a head-and-tail toss. Have students toss coins 10 times, then 50 times, and finally 100 times. Compile all the results and show that only when large numbers are used will the 1:2:1 ratio result with some accuracy. The two sides of a coin represent two possibilities similar to the segregation of a pair of genes in a hybrid, such as $TD \times TD$. Each pair TD separates into T and D. Thus, in recombination at fertilization this ratio appears: 25 percent TT, 50 percent TD, and 25 percent DD.

2. Mix together in a large bowl 100 green and 100 yellow dried peas. Or use black and white marbles, or beads. Have a student pick out two at a time, without looking. A partner may keep a tally of the number of times each of the three possible combinations appears. Each pea represents an egg or sperm (carrying one gene of a pair); two together represent a fertilization. For instance, approximately 25 percent of the times a student will pick up two green peas; 25 percent of the times, two yellow ones; and 50 percent of the times, a combination of green and yellow peas. Of course, a large number of selections is needed to approximate this 1:2:1 ratio.

You may want to have some students use the following chi-square method for determining goodness of fit. Also refer to the Hardy-Weinberg law, and other work on population genetics as it is referred to in tracing the migration of alleles for blood groups among humans (p. 509).

Chi-square (χ^2) method Chi-square is a criterion for testing the discrepancy between a set of observed values and the corresponding theoretical values expected on the basis of some hypothesis concerning a population.

In a random "picking" of yellow and green peas, or in the tossing of coins (as described above), results do not exactly follow the theoretical expected numbers. How great is the difference between the theoretical results (based on a proposed hypothesis) and the actual data obtained? Can the difference in results still be explained on the basis of chance, or is there a larger difference, significant enough, so that one may suspect that some factor other than chance is involved in the deviations observed? (*Note:* These are qualitative differences; quantitative differences in a population are not tested with this chi-square method.)

To test how often chance alone can be expected to account for the deviation, the formula

$$\chi^2 = \sum \frac{d^2}{c}$$

is used (d = deviation, c = the theoretical expectation, and Σ = the summation, or sum of). This simplified formula is derived from

$$\chi^2 = \sum \left[\frac{(o - c)^2}{c} \right]$$

where o = observed data. The deviation squared is

$$d^2 = (o - c)^2$$

Suppose that a student picks up at random several handfuls of yellow and green peas mixed thoroughly in a contain-

er. Upon counting his sample of 104 peas he does not find 52 yellow and 52 green as would be expected on the hypothesis of a 1:1 ratio. Instead he finds 56 yellow and 48 green peas. The deviation of the observed results from the anticipated is +4 and −4. Substitute in the formula:

$$\chi^2 = \sum \frac{d^2}{c} = \frac{(4)^2}{52} + \frac{(-4)^2}{52}$$

$$= \frac{16}{52} + \frac{16}{52} = 0.31 + 0.31 = 0.62$$

Now that the value of χ^2 in a given experiment is known, the next step is to calculate how often, on the basis of chance alone, this value χ^2 is likely to occur; that is, what is the probability that the deviation obtained will occur by chance alone. The number of degrees of freedom referred to in Table 7-9 relating to χ^2 values is one less than the number of classes in the ratio. For example, in the example given above either green or yellow seeds will occur; if green is counted, there is only one freedom, the chance of yellow. Hence, in this case the degree of freedom is 1. Notice that χ^2 value of 0.62 falls between probability, or P, values of 0.50 and 0.80. Therefore it is not unusual for a sampling to have 56:48 instead of the theoretical 52:52 of the total of 104 in the sample.

When two coins are tossed the expected results are 1 (tail-tail) : 2 (head-tail) : 1 (head-head). In an actual tally of 200 flips of the two coins, the expected results should be 50:100:50. Can the deviation of the actual data be accounted for on the basis of chance? A test of a 1:2:1 ratio would have two degrees of freedom. (A 3:1 ratio would have one degree of freedom.) Calculate the χ^2 value and use Table 7-9, knowing the number of degrees of freedom at which to enter the table. Notice that the probability decreases as the value of χ^2 increases. (A probability of 0.05 is 1 in 20; a probability of 0.01 is 1 in 100.)[8]

---

[8] There are excellent chapters on statistics useful in genetics in A. Srb, R. Owen, and R. Edgar, *General Genetics*, 2nd ed. (San Francisco: Freeman, 1965).

TABLE 7-9 Table of Chi-square*

degrees of freedom	P = 0.99	0.95	0.80	0.50	0.20	0.05 (1 in 20)	0.01 (1 in 100)
1	0.000157	0.00393	0.0642	0.455	1.642	3.841	6.635
2	0.0201	0.103	0.446	1.386	3.219	5.991	9.210
3	0.115	0.352	1.005	2.366	4.642	7.815	11.341
4	0.297	0.711	1.649	3.357	5.989	9.488	13.277
5	0.554	1.145	2.343	4.351	7.289	11.070	15.086
6	0.872	1.635	3.070	5.348	8.558	12.592	16.812
7	1.239	2.167	3.822	6.346	9.803	14.067	18.475
8	1.646	2.733	4.594	7.344	11.030	15.507	20.090
9	2.088	3.325	5.380	8.343	12.242	16.919	21.666
10	2.558	3.940	6.179	9.342	13.442	18.307	23.209
15	5.229	7.261	10.307	14.339	19.311	24.996	30.578
20	8.260	10.851	14.578	19.337	25.038	31.410	37.566
25	11.524	14.611	18.940	24.337	30.675	37.652	44.314
30	14.953	18.493	23.364	29.336	36.250	43.773	50.892

* From A. Srb and R. Owen, *General Genetics*, Freeman, San Francisco, 1952 (abridged from Table III of *Statistical Methods for Research Workers* by R. Fisher, Oliver & Boyd, Edinburgh).

A reading from Mendel's work

There may be times when a reading from an original source is appropriate. Such a translation as the one given below of Mendel's paper might be reproduced and developed in class as part of a case history in heredity. This example describes several methods used by scientists. You may want to add questions based on the reading and test the ability of students to interpret what they read.

THE GENERATION BRED FROM THE HYBRIDS

In this generation there reappear, together with the dominant characters, also the recessive ones with their peculiarities fully developed, and this occurs in the definitely expressed average proportion of three to one, so that among each four plants of this generation three display the dominant character and one the recessive. This relates without exception to all the characters which were investigated in the experiments. The angular wrinkled form of the seed, the green colour of the albumen, the white colour of the seed-coats and the flowers, the constrictions of the pods, the yellow colour of the unripe pod, of the stalk, of the calyx, and of the leaf venation, the umbel-like form of the inflorescence, and the dwarfed stem, all reappear in the numerical proportion given, with-

out any essential alteration. *Transitional forms were not observed in any experiment.*

Expt. 1. Form of seed—From 253 hybrids 7,342 seeds were obtained in the second trial year. Among them were 5,474 round or roundish ones and 1,850 angular wrinkled ones. Therefrom the ratio 2.96 to 1 is deduced.

Expt. 2. Colour of albumen—258 plants yielded 8,023 seeds, 6,022 yellow, and 2,001 green; their ratio, therefore, is as 3.01 to 1.

Expt. 3. Colour of the seed-coats—Among 929 plants 705 bore violet-red flowers and grey-brown seed-coats; 224 had white flowers and white seed-coats, giving the proportion 3.15 to 1.[9]

The entire translation is a splendid example of reporting the results of experiments and presenting a working hypothesis to explain inheritance in garden peas.

Demonstration Punnett square

Students may want to construct this device by screwing cup hooks into a large square

[9] From Gregor Mendel, *Experiments in Plant-Hybridisation*, trans. by Royal Horticultural Soc. of London (Cambridge, MA: Harvard Univ. Press, 1933) p. 321. Reprinted by permission of the publisher.

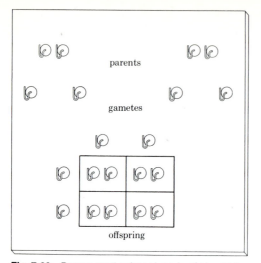

Fig. 7-20 Demonstration board made of plywood and cup screws for simple genetic crosses.

of plywood as in Figure 7-20. Two small hooks close together represent the diploid condition of the parent organisms; single hooks represent the gametes, or the monoploid possibilities. Below this, show a Punnett square with four boxes. In each box show two hooks, representing a diploid organism. Along the top and left side of the square, insert single hooks, representing the gametes.

Then students might make models of flowers out of plywood or cardboard and color them with crayon. Also model small animals such as guinea pigs or rats, black and white ones. Use small circles of color to indicate the genes in the gametes. Make holes in these models and hang them on the hooks. Have students make a variety of crosses in front of the class. Errors are easily detected as they occur.

Heredity in rats, mice, and hamsters

You may be able to purchase pedigreed hooded rats from an experimental station or college. Cross hooded with albino rats. Hooded seems to be dominant (with a few white, or albino, offspring appearing in the F_2, or second, generation after inbreeding

the hybrid hooded F_1 generation). This assumes the original cross was pure hooded × albino.

Pedigreed mice may be purchased from supply houses (see Appendix). Directions for possible crossings will be supplied by the biological supply house. (For ways to maintain mice or rats in the laboratory, see chapter 9.) Make crosses between pure black and albino mice; the F_1 offspring will be all black. These are hybrid black, each animal carrying a recessive gene for albinism.

When these F_1 offspring are inbred, the second filial generation, F_2, should show the ratio of three black to one white.

Students may make test crosses (backcrosses) by mating the black offspring in the second generation with an albino parent to determine whether the black animal is pure black or hybrid black.

Any of these ratios, of course, can be expected to hold only when large numbers of offspring are produced. In practice, a male would be mated to a large number of genetically similar females in order to get many offspring.

R. Whitney recommends hamsters as subjects for genetic study.[10] Some mutations that have been studied are partial albino (white coat, dark ears, red or black eyes), cream (yellow coat, black ears and eyes), microphthalmia (minute eyes), tawny (tan coat with black markings, black eyes and ears), brown or cinnamon (cinnamon coat, peach-gray ears, and eyes a rich red color), and white band (coat with a broad white band circling the body just behind the shoulders).

Co-dominance

Many examples of incomplete dominance, or co-dominance, exist among plants and animals. Recall that certain white cattle crossed with red-furred cattle have roan colored offspring.

[10] R. Whitney, "Hamsters as Genetic Subjects." *Turtox News* (1962) 40:4.

Tailless cats crossed with cats that have a normal tail have offspring that are short-tailed or bob-tailed. Among Andalusian fowl, slate or "blue" offspring result from a cross of black and white parents ($BB \times WW$).

Among snapdragons and in four-o'clocks (among many other flowers) there is an intermediate color in the offspring. What would the offspring of two pink-flowered four-o'clocks be ($RW \times RW$)? Recall, too, that the genes for antigens A and B are both expressed in people with AB blood type—another example of co-dominance.

Genetics of fish

There is an extensive literature on heredity in the guppy, *Lebistes reticulatus*. (For ways to maintain tropical fish, see chapter 9.) Blond and golden are alleles of the wild type and illustrate Mendel's laws.

For detailed studies see H. B. Goodrich et al., "The Cellular Expression and Genetics of Two New Genes in *Lebistes reticulatus*," *Genetics* (1944) 29:584–92. Also look into O. Winge and E. Ditlevsen, "Color Inheritance and Sex Determination in *Lebistes*," *Heredity* (1947) 1:65–83.

Some students may have pure strains of platys—unspotted and the dominant, spotted. Swordtails show readily observable traits such as body color and color pattern of the wagtail.

Useful pamphlets, prepared by M. Gordon, may be available.[11]

Heredity in Drosophila

Pure stocks of *Drosophila* may be obtained from a supply house or college laboratory. Some suggested stocks for studies in heredity are (see map, Fig. 7-21):

---

[11] M. Gordon, *Platies, Siamese Fighting Fish, and Swordtails*, Am. Museum of Natural History, New York. Also refer to M. Gordon, "Distribution in Time and Space of Seven Dominant Multiple Alleles in *Platypoecilus maculatus*," *Advances in Genetics* (1947) 1:95–132.

1. *Wild:* These are the typical *Drosophila* found in nature, with red eyes, gray body, and normal wings.

2. *Vestigial wings:* This trait is recessive to normal wings. Mate winged males to virgin vestigial-winged females (since the vestigial forms are unable to fly).

3. *Curly wings:* This is dominant over wild-type normal wings. The wings curl up at the tips, making it difficult for the insects to fly.

4. *White eyes:* This is a sex-linked gene found in the X chromosome. Make reciprocal crosses; that is, cross red-eyed females with white-eyed males, and white-eyed females with red-eyed males (Figs. 7-26 and 7-27). Flies must be pure for these traits.

5. *Yellow body:* This is a sex-linked gene, found in the X chromosome.

6. *Black body:* Flies with this characteristic can be used to show complete linkage and crossing over in one region since the black (or the wild-type) body gene is located in the same chromosome as vestigial (and/or long, normal wild-type) wings. See map, Figure 7-21. Cross black-bodied, long-winged flies with gray-bodied, vestigial-winged flies. Also cross black-bodied, vestigial-winged flies with normal gray-bodied, wild-type flies having normal long wings (Table 7-10). (See linkage, pp. 538–39.)

Mating fruit flies Students should refer to the map of the chromosomes of *Drosophila* for an indication of the position of the genes on the chromosomes of the organism (Fig. 7-21). In this way students will discover which genes are linked on the same chromosome.

Virgin female flies are needed when mating flies of two different types, for example, red eyes × white eyes. Since females retain sperms for a considerable time, females must be isolated as they emerge from pupae.

Many workers isolate virgin female flies in separate vials for three days. Flies that

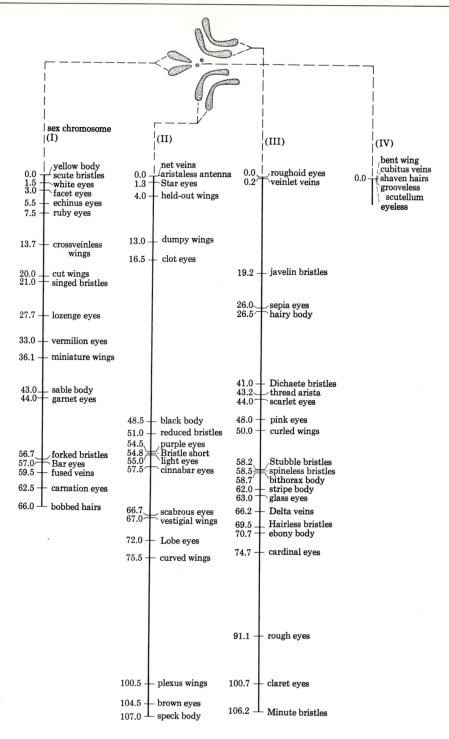

Fig. 7-21 Map of the four linkage groups or chromosomes of *Drosophila* showing loci of genes.

(After E. Sinnott, L.C. Dunn, and T. Dobzhansky, *Principles of Genetics*, 5th ed., McGraw-Hill, 1958.)

TABLE 7-10   Autosomal linkage in *Drosophila

Key: B = gray body; b = black body
V = long wings; v = vestigial wings
♂ = male, ♀ = female

(a)

Parents:	$(bV)(bV)$	×	$(Bv)(Bv)$
Gametes:	bV		Bv
Offspring:		$(bV)(Bv)$	

(b)

Parents:	F_1 ♂ $(bV)(Bv)$	×	$(bv)(bv)$ ♀
Gametes:	bV Bv		bv
Offspring:	$(bV)(bv)$		$(Bv)(bv)$

(c)

Parents:	F_1 ♀ $(bV)(Bv)$	×	$(bv)(bv)$ ♂
Gametes:	bV Bv BV bv		
Offspring:	$(bv)(bV)$ $(bv)(Bv)$ $(bv)(BV)$ $(bv)(bv)$		

* (a) Pure black-bodied, long-winged × pure gray-bodied, vestigial-winged; (b) F_1 male (gray-bodied, long-winged) × double recessive female (black-bodied, vestigial-winged); (c) reciprocal cross to b: F_1 female × double recessive male. Note that, while the usual way of symbolizing the parents would be $(bbVV)$, since the genes are linked we use the symbols $(bV)(bV)$ instead.

have been so "aged" are ready to lay eggs after mating. Larvae will hatch within a day and consume any mold that may have been carried into the new culture bottles.

When pupae appear, remove the parents by etherizing them and then placing them in a "morgue," that is, a stoppered bottle containing some mineral oil.

Usually females do not mate for some 12 hr after hatching. A more casual method is to remove the original parents as pupae begin to form in the culture bottles. Then the emerging flies will be virgin for the first 12 hr.

Parent flies should be removed in any case, as they may create errors in determining the results (when counting offspring). When the flies are kept at 20° C (68° F), counts of offspring may be continued up to the twentieth day. Thereafter, the second generation will emerge and errors in count will occur.

Examining and crossing fruit flies
Students will quickly learn to identify male and female flies (Fig. 7-22). The abdomen of the male is bluntly rounded with a wide band of dark pigment; the ventral posterior end of the abdomen shows a dark spot which is easily seen under low magnification, or with the naked eye. The female's abdomen is elongated, with thinner pigment bands. Females distended with eggs are easily recognized. With a hand lens it is possible to distinguish seven abdominal segments in the female and five in the male. The females have an ovipositor; the males possess sex combs (ten short black bristles) on the end of the tarsal joint of the first pair of legs.

It is difficult to look for banding in newly hatched flies, for pigment is not well developed. Also, dark-colored flies need to be identified principally by possession of sex combs and by shape of the body.

Etherizing flies   The flies must be etherized so that students may examine them

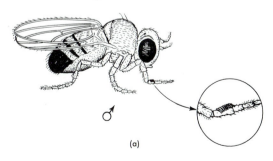

♂

(a)

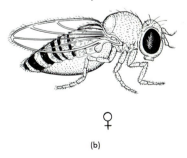

♀

(b)

Fig. 7-22   *Drosophila:* (a) note position of sex combs on first pair of legs of male (inset); (b) note the thin bands along the abdomen and larger-size body of female. (After W.G. Whaley, O.P. Breland, *et al.*, *Principles of Biology*, Harper & Row, 1954.)

with ease. To make an etherizing container, obtain a bottle having the same circumference as the mouth of the stock bottles. Insert a nail about 3-cm long into the cork stopper of the vial. Then cover the nail with layers of cotton and tie the cotton to the nail with cord (Fig. 7-23b). Put a few drops of ether on the cotton when you plan to etherize the flies (*caution*).

To transfer the flies, first tap the culture bottle on the table. When the flies fall to the bottom (this will be but for a moment), quickly remove the cover and hold the mouth of the etherizer closely against the mouth of the stock bottle (Fig. 7-23a). Invert the stock bottle and tap it slightly, so that the flies move into the etherizing bottle. Since the flies show a positive phototaxis, and a negative geotaxis, an electric light may be placed near the etherizer to encourage the flies to move into the empty bottle. Then quickly separate the bottles, and stopper both (Fig. 7-23b). The flies will be anesthetized within a few seconds. Guard against an excess of ether or too rapid etherization to avoid killing the flies. (When the wings stand out at an angle, the flies are dead.) A better device to use is a plastic anesthetizer available from supply houses.

Spill the anesthetized flies out onto white paper for examination. If the flies recover before the examination is completed, they must be re-etherized. Have on hand a Petri dish cover in which a strip of filter paper is held fast with adhesive tape. Add a drop of ether to the filter paper and place the dish over the flies for a few seconds.

Since the flies are fragile, use a camel hair brush or a metal strip to handle them during examinations and to transfer them from place to place. When you plan to cross two different genetic types, select several pairs of flies of the stock with which you are dealing and place them in a small cone of paper. Then introduce them into new bottles of culture medium for stock. Otherwise, anesthetized flies dropped into the medium will adhere to the moist medium and drown.

The stock cultures should be kept at a temperature range of 20° to 25° C (68° to 77° F). At 25° C the life cycle takes ten days; at 20° C the cycle is lengthened to fifteen days. Third molt, that is, third instar larvae slowly crawl up to the dry sides of the bottle or onto the paper strip inserted into the bottle. Here the larvae pupate. At 25° C the pupa stage lasts about four days. Lower temperatures prolong the life cycle. High temperatures increase sterility and reduce viability. The temperature within the bottles will be a bit higher than the room temperature because of fermentation of the yeast-sugar medium.

Subculture or transfer flies to new stock bottles with fresh medium each month, or more frequently, depending upon the temperature and the amount of evaporation of fluid from the medium. Prepare

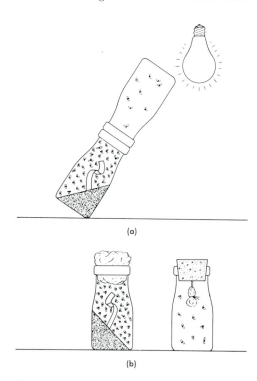

(a)

(b)

Fig. 7-23 Technique for transferring and etherizing fruit flies: (a) flies show phototaxis and leave culture bottle for new one; (b) old stoppered bottle with cotton plug and new one with etherizing stopper.

several duplicate stock cultures. Label the bottles and record the date of transfer.

Establishing stock cultures   Fruit flies are attracted to soft grapes, plums, bananas—in fact, to any fermenting fruits. Larvae feed freely on yeast and other microorganisms in the fermenting fruit juices; clearly, a fermenting medium must be prepared. For rapid, temporary cultures, where little handling will occur, the simplest medium is prepared by dipping a piece of ripe banana into a suspension of yeast (made from a quarter of a package of yeast dissolved in 100 ml of water). Insert this piece of banana along with a strip of paper toweling into a clean glass vial or bottle. In season, this may be left open out-of-doors to attract fruit flies, or you may introduce flies into the bottles. Then plug with cotton wrapped in cheesecloth, or with plastic caps. However, this medium is not recommended for careful work; its main use is to collect wild-type flies. In the two-week life span of fruit flies the medium described above will become a mash with the flies embedded in the soft material. Also, molds will grow in profusion.

Several media for Drosophila   Demerec and Kaufmann describe several media to which agar is added so that bottles may be inverted and transfers made more easily.[12] To prepare one or more of the following media, add a mold inhibitor available from supply houses (avoid using an excess since it will reduce the growth of yeast). Instant medium is also available from supply houses. Mix equal amounts of powder and water to solidify within minutes.

In the following recipes, consider about 50 to 60 ml of medium for a half-pint culture bottle.

BANANA MEDIUM   Dissolve 15 g of agar in 480 ml of water by bringing it to a boil; stir well. To this add 500 g of banana pulp

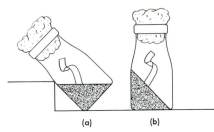

Fig. 7-24   Culture bottles for fruit flies: (a) bottle containing soft medium and strip of paper toweling is slanted to cool; (b) bottle containing solidified medium with increased surface area, ready for flies.

made by mashing a banana with a fork or by putting it through a strainer.

CREAM OF WHEAT MEDIUM   This preparation eliminates the agar. Measure out 775 ml of water, 115 ml of molasses or Karo, and 100 g of Cream of Wheat. Add the molasses to two-thirds of the water; bring to a boil. Mix the Cream of Wheat with the remaining third of cold water and add this to the boiling mixture. Continue to stir, and cook for 5 min after boiling begins. Add mold inhibitor and pour the medium into sterilized bottles, add strips of toweling, stopper the bottles, and tilt them, as in Figure 7-24.

CORNMEAL MEDIUM   This medium uses agar. Dissolve 15 g of agar in 500 ml of water and heat. After this comes to a boil, add 135 ml of corn syrup (Karo) or molasses. Combine 100 g of cornmeal in 250 ml of cold water and add to the heated mixture. Boil this *slowly* for about 5 min. Add mold inhibitor. Then pour this medium into sterilized bottles or vials, insert toweling as before, and plug the bottles with cotton or cover with caps. This quantity will fill 25 culture bottles.

CORNMEAL—MOLASSES—ROLLED-OATS MEDIUM   This recipe requires no agar. Measure out 730 ml of water, 110 ml of molasses or Karo, 150 g of cornmeal, and 16 g of rolled oats (not quick-cooking). Add rolled oats and molasses to two-thirds of the water and bring this to a boil. Mix the cornmeal with the remaining cold water and introduce this into the boiling mixture until it thickens but still can be

[12] M. Demerec and B. R. Kaufmann, *Drosophila Guide*, 7th ed. (Washington, DC: Carnegie Institution of Washington, 1961).

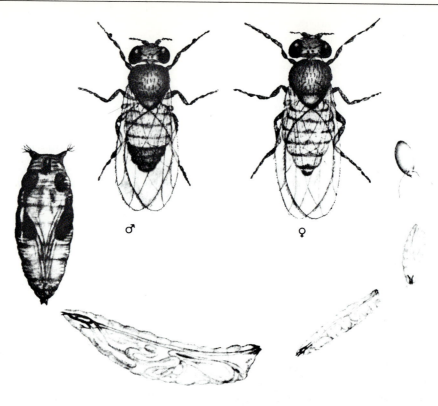

Fig. 7-25 Life cycle of *Drosophila melanogaster*.   (Courtesy Carolina Biological Supply Co.)

poured. Then add the mold inhibitor. You may want to add agar to the medium to prevent it from softening with a rise in room temperature.

In fact, for work during the summer months, agar should be included in any medium you use. (Should molds appear in the cultures of *Drosophila*, paint the surface of the culture medium with 95 percent alcohol or with a solution made from 1 part of carbolic acid to 8 parts of water. *Caution:* Remove the flies from the bottle before this treatment.)

Preparation of bottles Pour the medium into half-pint cream bottles or glass vials to a depth of 2 cm. It is safer to sterilize the bottles before introducing the medium. Then insert a strip of paper toweling into the medium while it is still soft to provide additional surface for egg-laying and pupation. Cover the bottles with cotton wrapped in cheesecloth and fastened with string, or use plastic covers. Tilt the bottles against a ledge to make slants to increase the surface; allow the medium to solidify (see Fig. 7-24a). You may want to pour medium from the cooking pot through a funnel into the bottles so that the medium does not spill along the sides of the bottles. Store the bottles in a refrigerator until the flies are to be introduced. Just before using the bottles to accept the flies, add 2 or 3 drops of a rich yeast suspension to the surface of the solid medium. Or add crumbs of dried yeast; this will dissolve in the surface moisture.

Life cycle Some teachers have students keep their own vials of flies to study the entire life cycle of *Drosophila* (Fig. 7-25). To accommodate large number of students, prepare the medium and use a plastic spoon to hold a small quantity of medium. The small spoon is then inserted into a vial. If several different cultures are given to the students, they will have materials to make crosses of newly emerging flies. They may transfer several pairs of flies into vials with fresh medium on a new spoon.

Since the spoons have handles, the handling of transfers is greatly facilitated; further, less cooking is needed.

Have students record the time for development at 10° C (50° F) and at 25° C (77° F)—optimum for development in ten days. Observe larval molts and the third slow-moving instars that climb the sides of the container to pupate. Examine pupae on a slide under the microscope to observe eyes, limbs, and wing formation. From the ventral surface, males can be distinguished from females by inspecting the first tarsal segment of each forelimb. Only males have sex combs on the forelimbs. Crosses may be made using pupal stages rather than waiting for adults to emerge.

Some demonstration crosses Many kinds of *Drosophila* are available for crosses; examine the chromosome maps to decide which types to use (Fig. 7-21).

LAW OF DOMINANCE   Cross a wild-type fruit fly (that is, gray-bodied, long-winged) with a black-bodied, long-winged fly. The first generation should be 100 percent hybrid gray-bodied flies if both parents were pure for these traits.

Cross pure long-winged with vestigial-winged flies to get all long-winged offspring (hybrid).

Or cross pure long-winged flies with curly-winged ones to get 100 percent curly-winged flies. In all these crosses students may observe the appearance in the offspring of one of the pair of contrasting traits.

LAW OF SEGREGATION   Cross the offspring (that is, the F_1 hybrids of the above crosses). When these hybrids are crossed, students should find that the F_2, or second generation, reveals a 3:1 ratio. Seventy-five percent will show the dominant trait and 25 percent will show the recessive trait. Students should see that the recessive trait which was not apparent in the F_1 hybrids has reappeared in the second generation due to the recombination of genes at fertilization.

TEST CROSS OR BACKCROSS   In order to determine whether an organism showing the dominant gene is pure or hybrid, the "unknown" should be crossed with the recessive type. The answer lies among the offspring of this cross, provided there are enough of them. For example, if we wanted to find out whether some long-winged flies are hybrid and carry a recessive gene for vestigial, we would cross them with vestigial-winged flies. If they were pure long-winged or homozygous for long wings, then there could be no vestigial-winged flies among the offspring. However, if the long-winged flies were hybrid for the trait and they were crossed with vestigial, 50 percent of the offspring should show the recessive trait, vestigial wings. In a two-factor backcross, the unknown type would be crossed with the double recessive. In an actual cross, apply the chi-square method (page 528).

LAW OF INDEPENDENT ASSORTMENT OF FACTORS   When two pairs of genes are located in different pairs of chromosomes, they are inherited independently of each other. Each gene behaves as a unit since the chromosomes segregate in maturation of eggs and sperms (meiosis), and are recombined in fertilization. At least two factors must be studied. For example, cross fruit flies pure for wild or gray body and pink eyes with flies pure for black body and red eyes (Note in Figure 7-21 that these genes are on different chromosomes.)

Upon inspection, students should be able to locate the $9/16$ of the flies that have gray body with red eyes; $3/16$, gray body with pink eyes; $3/16$, black body with red eyes; and $1/16$, black body with pink eyes. A dihybrid cross such as this one yields a 9:3:3:1 ratio. (In a trihybrid cross, multiply the 9:3:3:1 ratio by a 3:1 ratio, which represents the added hybrid, so that the offspring would yield a ratio of

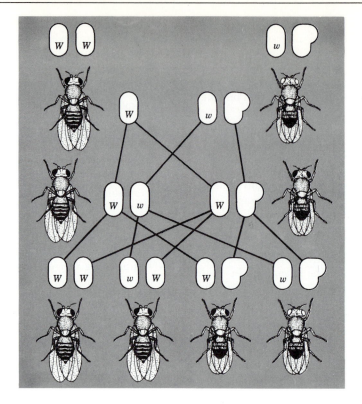

Fig. 7-26 Sex linkage in fruit flies: red-eyed female crossed with white-eyed male.

(After E. Sinnott and L.C. Dunn, *Principles of Genetics*, 2nd ed., McGraw-Hill, 1932.)

27:9:9:9:3:3:3:1. In this case, $^{27}/_{64}$ of the offspring should show the combination of three dominant traits and $^{1}/_{64}$ should show the three recessive traits.)

Apply the chi-square test; remember that with four classes of offspring you enter the table (Table 7-9) at three degrees of freedom.

SEX-LINKAGE After the students have learned how maleness or femaleness is inherited on the basis of XY or XX chromosome pairs, some students may want to cross red-eyed females with white-eyed males (Fig. 7-26). The gene for red or white eye color is located on the X chromosome. In this cross the offspring will all be red-eyed. But in the reciprocal cross (Fig. 7-27), white-eyed females crossed with red-eyed males, the females in the first generation are all red-eyed and the males are white-eyed. Trace the results as

shown in both cases in the second generation. You will recall this is similar to the inheritance of color blindness and hemophilia among human beings (pp. 511–12).

AUTOSOMAL LINKAGE Since there are thousands of genes in *Drosophila* and only four linkage groups, or chromosomes, many genes are found together or linked on the same chromosome. We focus now on autosomal linkage, not linkage on the X chromosome.

To show the linkage of two genes found on the same chromosome students might try this cross. Cross a pure black-bodied, normal long-winged fly with a gray-bodied, vestigial-winged fly. (See chromosome map Fig. 7-21, to find on which chromosomes the genes are located.) The offspring of pure parents should all be gray-bodied and long-winged (Table 7-10a, p. 533).

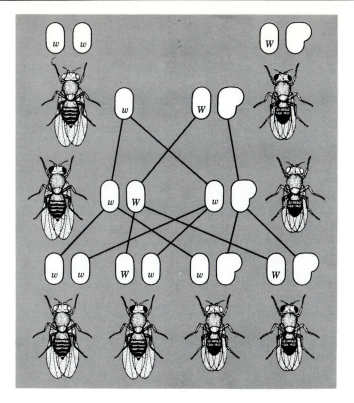

Fig. 7-27 Sex linkage in fruit flies: reciprocal cross showing white-eyed female mated with red-eyed male. (After E. Sinnott and L. C. Dunn, *Principles of Genetics*, 2nd ed., McGraw-Hill, 1932.)

Students who are interested may want to go further. They may show that there usually is not a complete linkage in the female since there is a crossing over of chromosomes. In the *male* fruit fly, however, there is *no crossover* of chromosomes. As a result, different ratios will be produced, depending upon whether a male or a female of the above F_1 generation is used. If the male F_1 is used (Table 7-10b), the offspring are 50 percent gray-bodied, vestigial-winged, and 50 percent black-bodied, long-winged. The expected types of crossover, gray-long and black-vestigial, do not appear. However, when the reciprocal cross is made (F_1 female with black, vestigial males, Table 7-10c), the four expected types appear.

In 1919 Morgan showed that the percentages from the cross shown in (c) were as follows:

	noncrossovers
gray, vestigial	41.5%
black, long	41.5%
total	83.0%
	crossovers
black, vestigial	8.5%
gray, long	8.5%
total	17.0%

Chromatograms of Drosophila eye pigments Pteridines are substances found in eye color in *Drosophila* as well as in some pigments of many other invertebrates, especially butterflies and moths, as well as plants, fish, and amphibians. The wild-type eye color of *Drosophila* contains several pteridines. Under ultraviolet light, they show different fluorescent patterns so that red eye color can be compared with several mutant eye colors such as sepia, lozenge, brown, rosy, vermillion, and

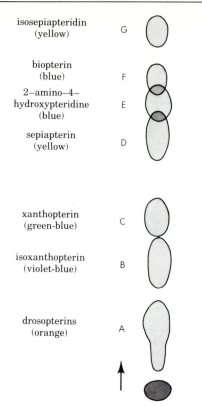

isosepiapteridin (yellow)	G
biopterin (blue)	F
2–amino–4–hydroxypteridine (blue)	E
sepiapterin (yellow)	D
xanthopterin (green-blue)	C
isoxanthopterin (violet-blue)	B
drosopterins (orange)	A

Fig. 7-28 Chromatogram of seven pteridines for eye color in wild type *Drosophila* as found by Hadorn. (From "Fractioning the Fruit Fly" by Ernst Hadorn. Copyright © 1962 by Scientific American, Inc. All rights reserved.)

white eyes. In addition, different amounts of the pteridines are found in larvae and pupae stages leading to many interesting observations.

Students may want to read Ernst Hadorn's paper in *Scientific American* describing the variety of pteridines, and also showing color reproductions of the chromatogram of the wild-type eye color.[13] Refer to Figure 7-28 for the sequence of pteridines, as described by Hadorn when propanol and ammonia are used as the solvent.

[13] E. Hadorn, "Fractionating the Fruit Fly," *Sci. Am.* (April 1962) 206 also refer to "Properties of Mutants of *Drosophila melanogaster* and Changes During Development as Revealed by Paper Chromatography," *Proc. Nat. Acad. Science* (1951) 37:650–55.

Abramoff and Thomson also describe the technique for fractionating eye pigments.[14]

White- and brown-eyed flies lack pteridines. However, in the brown-eyed mutant, the pigments are present in the larvae and pupae but disappear in the adult. Different amounts of pteridines are found when the chromatogram for wild-type red eye color is compared with vermillion, rosy, and lozenge eye colors.[15]

Summary: genetics problems

Have students solve a variety of problems—a review in a new form:

1. In summer squash, white fruit (W) is dominant over colored fruit (w). Disk-shaped fruit (D) is dominant over sphere shaped fruit (d). What colors and fruit shapes should appear in F_1 of a cross between white disk and colored sphere? What phenotypes should appear in F_2? What genotypes?

2. Among garden peas, what are the possible offspring of a cross of red flowered, yellow seeded plants ($RRYY$) and white flowered, green seeded plants ($rryy$)? What are the possible phenotypes when the F_1 are inbred?

3. Among sheep, an offspring of a white ewe and a white ram was black furred. Explain.

4. When solid-colored rabbits were crossed with spotted ones, the offspring were 100 percent spotted. When these were inbred, the F_2 showed 35 spotted

[14] P. Abramoff, and R. Thomson, "Biochemical Genetics," offprint 863, of the Freeman Library of Laboratory Separates, Freeman & Co., 660 Market Street, San Francisco, CA 94104.

[15] Sepia and lozenge lack drosopterins; rosy lacks isoxanthopterin, yet sepia has a larger amount of xanthopterin and sepiapterin. In addition, sepia has an increase in 2-amino-4-hydroxypteridine and of biopterin (as does rosy) in comparison with wild type.

and 11 solid colored. List the possible genotypes of the spotted rabbits.

5. Red hornless cattle crossed to white horned cattle produced offspring that are roan hornless. Explain the heredity of these F_1 offspring. What are the possible genotypes in the F_2 generation? And the possible phenotypes?

6. A girl with normal color vision whose father was colorblind marries a man with normal color vision. Among four children in the family, one son was colorblind, and another son had normal color vision. What is the genotype of the two sons? And of the parents? Why weren't either of their two daughters colorblind?

7. An infant with O Rh+ blood could not be the child of parents who are AB Rh+ and O Rh−. Explain.

8. $Hb^s Hb^s$ is the genotype of a person with sickle cell anemia. How can two people with sickle cell trait have a child who develops sickle cell anemia?

9. What are the possible genotypes of parents who have a daughter who is colorblind?

10. Five children in a family have a range of blood types: O, AB, A, and B. What possible genotypes in parents can produce this diversity?

Heredity in wasps

Habrobracon   Whiting has used the small parasitic wasp *Habrobracon* for studies in genetics. The wasp is about the size of the fruit fly. It is also recommended by Whiting for studies of complete metamorphosis, and to show parthenogenesis of eggs of unmated females.[16] The effect of environment on the development of body color can also be observed readily by students; at higher temperates (30°C

[86°F]) the body color is light honey-yellow, and black at lower temperatures (20° C [68° F]).

Since *Habrobracon* is parasitic on the larval or caterpillar stage of its host, the flour moth, *Ephestia*, must also be cultured.

The life cycle of *Habrobracon* is 10 days at 30° C (86° F), about 14 days at room temperature. Males of *Habrobracon* feed on honey, as do the females, but females do not lay eggs until they have fed on the host larvae. When the host larvae are inserted into the vials, observe the host–parasite relationship.

Both host and parasite can be refrigerated and maintained for months without the preparation of media customary in the maintenance of fruit flies.

Mormoniella   This wasp, smaller than fruit flies, is parasitic on pupae of a fly. Males are monoploid so gametic ratios are shown directly in the offspring of unmated females.[17]

The female has five pairs of chromosomes. Since males arise parthenogenetically, they receive all their genes from their mother. You will want to have students read the extensive coverage of *Mormoniella* genetics in the pamphlet available from Carolina Biological and in a fine paper by Richard Best, "*Mormoniella* Genetics," *Carolina Tips* (September 1977).

Carolina Biological maintains stocks of the parasitic wasp *Mormoniella* and one genus of the host, *Sarcophaga*, which is parasitized in the pupa stage.

The life cycle covers 10 days at 27° C (81° F); 14 days at 25° C (77° F). There is a minimum of preparation of culture media, for, as with *Habrobracon* and its host, the larvae of the host fly can also be refrigerated and maintained for months. Cultures are kept in small shell vials stoppered with cotton.

---

[16] P. W. Whiting, "The Parasitic Wasp, *Habrobracon*, as Class Material," *Ward's Natural Sci. Bull.* (February 1943).

[17] P. W. Whiting, "A Parasitic Wasp and Its Host for Genetics Instruction and for General Biology Courses," *Carolina Tips* (April 1955).

There are many stocks available, a good number with easily observed mutants of eye color.

Heredity in protozoa

Those students who become deeply interested in heredity of microorganisms may investigate the elegant experiments of Jennings, Sonneborn, Kimball, Chen, Kidder, and others on the identification of the mating types in some protozoa. There are some 30 mating types in *Paramecium aurelia*; in *P. bursaria* (the green one), Jennings, Chen, and others have described 6 varieties containing 23 mating types. In *Tetrahymena* some 40 mating types have been found; there may be more. In fact, one variety has at least 9 mating types. Kimball has described at least 6 mating types in *Euplotes patella*. The algalike flagel-

TABLE 7-11 **Number of chromosomes in some protozoa***

protozoa	number of chromosomes
Chlamydomonas	10 (monoploid)
Euglena viridis	30 or more
Phacus pyrum	30–40
Trichonympha campanula	52, or 26 doubles
Amoeba proteus	500–600
Endamoeba histolytica	6
Actinophrys sol	44 (diploid)
Didinium nasutum	16 (diploid)
Paramecium aurelia	30–40
Paramecium caudatum	about 36
Stentor coeruleus	28 (diploid)
Uroleptus halseyi	24 (diploid)
Stylonychia pustulata	6
Euplotes patella	6 (diploid)
Vorticella microstoma	4
Carchesium polypinum	16 (diploid)

* Adapted from R. Kudo, *Protozoology*, 4th ed., 1954; courtesy of Charles C. Thomas, Publisher, Springfield, Ill.

late *Chlamydomonas* has already been described in this text (p. 404, Fig. 6-20). Also see *Tetrahymena*, p. 404. Also refer to current texts in genetics and protozoology.

Recombinant DNA

Genetic engineering, or manipulation of genes, is mainly underway using *Echerichia coli*, a bacterium whose heredity is well established. There are some four thousand genes in the circular DNA in a cell of *E. coli*.[18] Other small, independent double layered loops, called plasmids, are found in the cytoplasm. These plasmids replicate independently of the main chromosomal DNA. In genetic engineering, a segment of DNA can be "snipped" from the DNA of a virus, bacterium, plant, animal, or human and spliced into the small plasmids. As a result, a recombinant DNA in a "transformed" cell of *E. coli* replicates itself at each fission producing clones of cells with the "foreign gene segment." When the foreign DNA is responsible for producing an enzyme, growth hormone, insulin, interferon, etc., its presence is immediately known.

There is good history behind the efforts of many scientists to manipulate genes, from attempts to transfer genes from one virus to another, to the use of *E. coli*, and the knowledge of a technique to split open a plasmid with enzymes and "cement on," or splice, a bit of foreign DNA. Have students read some of the history, such as the paper by Stanley Cohen, who named the transformed plasmids "chimeras" in "The Manipulation of Genes," *Sci. Am.* (July 1975). Also refer to S. Cohen and J. Shapiro, "Transposable Genetic Elements," *Sci. Am.* (February 1980); R. Novick, "Plasmids," *Sci. Am.* (December 1980); and a *Time* magazine story of the "technology" of gene splicing, "Shaping Life in the Lab" (March 9, 1981).

[18] Since bacteria possess only one chromosome, there is no reshuffling of genes.

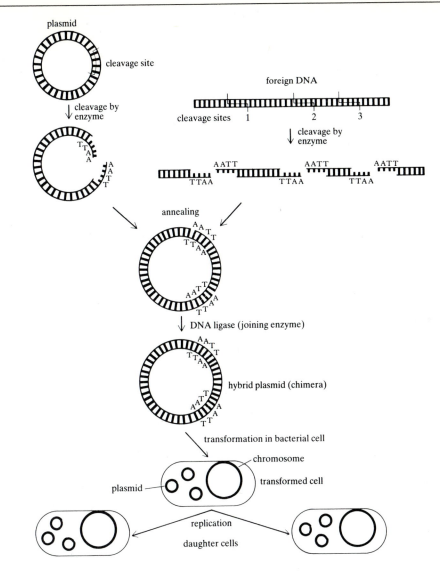

Fig. 7-29 Cleavage of a plasmid to introduce foreign DNA before plasmid is introduced into a bacterium. The foreign DNA is replicated along with the bacterial chromosome shortly before cell division. (From "Manipulation of Genes" by Stanley N. Cohen. Copyright © 1975 by Scientific American, Inc. All rights reserved.)

In a very general description of the technique, bacteria (*E. coli*) are centrifuged with special chemicals so that plasmids are separated from the cells. Then the plasmids are treated with enzymes that break open the loop at specific known regions. The ends of the loop are "sticky." Now mix with DNA of another organism, then add another enzyme to cement, or splice, to make new or "transformed" loops. These chimeras (mixes of genes from different species) are introduced into normal *E. coli* at a specific time, and molecular cloning occurs whenever *E. coli* undergo fission (Fig. 7-29).

With this technique, it is possible slowly to identify functions of segments of DNA or genes on chromosomes, including chromosomes of humans. Plasmid chimeras are now producing growth hormone, insulin, interferon, and other substances. It also may be possible to transfer nitrogen-fixing ability to a higher organism—such as a corn plant—that requires a high quantity of nitrogen for growth. The production of proteins such as insulin and interferon is described by Gilbert and Villa-Komaroff, "Proteins from Recombinant Bacteria," *Sci. Am.* (April 1980).

Evolving systems: organic molecules and heterotrophs

You may want to begin, perhaps, with chemical evolution and readings concerning theories of the first forms of life. Then develop this sequence: theories concerning the slow evolution of organic molecules, the first heterotrophs, and the difference between prokaryotic and eukaryotic cells. Finally, establish the divisions of our current organisms into five kingdoms as suggested by Whittaker.[19]

Chemical evolution

Review Gamow's "Big Bang" theory describing the expansion and explosion of the solar system some 16 billion years ago. Take a giant step to the time the earth's crust cooled to describe the gases in the atmosphere some 4 billion years ago: hydrogen (H_2), methane (CH_4), ammonia (NH_3), and H_2O. Students may report on the experiments of Stanley Miller in the 1950s in which Miller sent electric sparks through a mix of these gases and over a week or so obtained amino acids. These experiments support the 1930 hypothesis of Oparin. There may be time to offer students a reading, such as that of Oparin describing the first kinds of organisms as heterotrophs that consumed ready-made organic molecules in a primeval "soup"

---

[19] R. H. Whittaker, "New Concepts of Kingdoms of Organisms," *Science* (January 10, 1969). Also refer to pages 2 and 3.

(see below). Develop questions concerning the readings to test understanding of scientific writings.

Have students hypothesize about the requirements of a cell: What is the recipe for a cell? A mass of organic molecules needs a cell membrane to confine its organized contents; a means for utilizing the organic molecules for energy and growth (enzymes and ATP); a means for replicating itself (DNA-RNA systems). Might this type of evolution be underway on other planets?

A reading: concerning the first living things

A passage from A. Oparin's *Life: Its Nature, Origin, and Development*, such as the one that follows, or from a different source may be useful. Through reading such passages, students become aware of evolving hypotheses concerning early forms of life.

A committee of students may report on the complete little book, much of which they will understand. Or you may want to reproduce the passage and compose a series of questions, as shown for the selection by van Helmont on page 233.

THE FIRST LIVING THINGS

WERE ANAEROBIC HETEROTROPHS

By trying to detect in the tremendous variety of systems of metabolism in different organisms those similarities, those features of organisation which are most widespread among all living things, and therefore most ancient, we can, in the first place, establish two cardinal principles.

First, the metabolism of all contemporary living things is based on systems which are designed for the utilisation of ready-made organic substances as their primary building material for biosynthesis and as the source of that energy which is necessary for life although one might, theoretically, postulate many other satisfactory metabolic pathways.

In the second place, all contemporary living organisms have a biochemical system for obtaining energy from organic substances which is based on the anaerobic degradation of these substances although, with free oxygen in the atmosphere, as it is now, it would be perfectly rational for them to be oxidised directly.

It is self-evident and generally accepted that the overwhelming majority of the biological species which now inhabit our planet can only exist if they are constantly supplied with ready-made organic substances. This applies to all animals, both higher and lower, including most of the protozoa, the great majority of bacteria and all fungi. This fact alone is extremely suggestive. One can hardly imagine that all these evolved simply as Batesonian simplifications, as a complete loss of the autotrophic abilities which they once had. This is also contradicted by the intensive biochemical studies of the metabolic systems of these organisms. We do not find in them the least trace or vestige of those specific enzymic complexes or groups of reactions which are required by autotrophic forms of life while, on the other hand, the metabolism of autotrophs is always based on the same internal chemical mechanisms as that of all those other organisms which can only exist by consuming organic substances. The specific autotrophic mechanisms are merely superstructures on this foundation. It is just this sort of organisation of their metabolisms which allows autotrophs, under certain conditions, to revert entirely to the consumption of ready-made organic substances.

This can be demonstrated particularly clearly in the case of the least highly organised photoautotrophs—the algae—both in natural conditions and in the laboratory. By means of such experiments it has long ago been shown that if organic substances are introduced into sterile cultures of algae they will assimilate these substances directly. This may go on at the same time as photosynthesis but, in some cases, photosynthesis may stop altogether and the algae go over to a completely saprophytic way of life.

Under these conditions one gets a very luxuriant growth of blue-green algae such as *Nostoc* and the diatoms as well as such green algae as, for example, *Spirogyra*. Many forms of blue-green and other algae must obviously assimilate organic compounds directly even under natural conditions, when they live in dirty ponds. This is suggested by the fact that they grow especially luxuriantly in stagnant waters and such places where organic substances abound.

A heterotrophic basis for nutrition may be found, not only in the algae, but also in higher plants, although their photosynthesis apparatus has here reached the acme of its development. It is, however, only present in the chlorophyll-bearing cells of higher plants; the metabolism of all the rest of the tissues, which are colorless, is based, like that of all other living things, on the use of the organic substances supplied to them, in this case, by the photosynthesizing organs. Furthermore, even leaves revert to this form of metabolism when they are without light.

Thus the metabolism of all higher plants is entirely based on the heterotrophic mechanism of assimilation of organic substances, but in the green tissues this mechanism is accompanied by a supplementary, specific superstructure which has the task of supplying the whole organism with ready-made organic substances. If these substances reach the plant from outside in one way or another, then it can exist even without its photosynthetic superstructure, as may be observed under normal, natural conditions, especially in the germination of seeds. This may also be demonstrated artificially in cultures of vegetables tissues or in the growing of a complete adult plant of sugar beet in the dark from roots grown the previous year. In these cases either a complete higher plant or its tissues live in the total absence of any activity by their photosynthetic apparatus by the assimilation of exogenous organic substances. If, however, one breaks even one link in the chain of heterotrophic metabolism (by introducing a specific inhibitor, for example) then all the vital activities of the plant are brought to a standstill and it is destroyed.

Hence it is perfectly clear that the vital processes of the photoautotrophs are founded on the original and ancient form of metabolism based on the use of ready-made organic substances and that they developed the power of photosynthesis considerably later, as an accessory to their earlier heterotrophic metabolic mechanism.[20]

Evolving organisms through time past

Develop the parade of living things through time using the geologists' system of classifying strata in Eras, Periods, and

---

[20] From A. Oparin, *Life: Its Nature, Origin, and Development*, trans. by A. Synge (New York: Academic, 1962), 100–102. Reprinted by permission of Oliver & Boyd Ltd. and the author.

Epochs. The lowest layers of rock are the oldest, the fossils within are dated. Have students report on current methods of dating rocks and fossils using radioactive substances such as uranium-lead, potassium-argon, carbon-nitrogen, rubidium-stontium, and helium. For example, the half-life of uranium^{238} is 4.5 billion years; C^{14} is changed to n^{14} over a half-life of 5,700 years. The C^{14} method is used to calculate the age of objects less than 50,000 years.

Where did the first forms of life appear? When did back-boned organisms appear in the waters? And when did organisms step onto, or more correctly, crawl onto land? What kinds of organs did animals need to live on land? Why was animal life on land dependent upon the flourishing of green plants?

Establish the concept of evolutionary time. Dinosaurs thrived for a span of 140 million years; humans about 3.5 million years. Consider the enormous size of the dinosaurs of the Jurassic period (195 to 140 million years ago) that lived in the low plains of the United States in a climate that was almost subtropical. The dinosaurs that followed in the Cretaceous were faster moving; much of the western United States was then under water (140 to 45 million years ago).

When did mammals appear? What are some possible causes for the mass extinction of animals and plants some 65 million years ago? Have students read the paper by D. Russell, "The Mass Extinctions of the Late Mesozoic" in *Sci. Am.* (January 1982). At what times were there drastic changes in climate as caused by drifting of continents (p. 562)?

Fossils as evidence of past ecosystems

Why are fossil remains found in sedimentary rocks rather than other types of rock formation? Examine specimens of shale, sandstone, and other types of sedimentary rock. How were they formed? Why are the fossil evidences found as

teeth, bones, or spines? What are the requirements for fossil formation? How were fossils in amber or in tar pits formed? Some simulations may be made by students, and/or fossil specimens may be available in some areas. Perhaps it is possible to plan field trips to fruitful areas: limestone, coal, shale, or sandstone deposits. Have students start a collection of fossil specimens—some students may already have the beginnings of collections. Also obtain metal casts and ceramic or plastic models of prehistoric animals for display cabinets.

Plan to show films and/or filmstrips on the formation of fossils, as well as some fine 2×2 slides available from supply houses that trace the evolution of some kinds of horses, furnish reconstructions of prehistoric life in the seas, or illustrate the emergence of plant phyla, insects, and land vertebrates.

Also simulate the formation of fossils and of sedimentary rock. Several procedures follow. Organisms embedded in clear plastic may serve as models of how fossils have been formed in amber.

Simulating formation of fossils

"FOSSILS" IN PLASTER   Make imprints of leaves or shells to show how some fossils were formed in wet mud or sand. Embed leaves or shells in plaster of paris. Compare the plaster to sediments deposited by slow-moving streams or to a muddy embankment or swamp.

Prepare a mixture of plaster of paris and water, the consistency of pancake dough; stir until smooth, and pour a layer 3 cm deep into a cake pan. Then coat a variety of leaves or shells with petroleum jelly and lay them on the plaster. Cover these with another layer of plaster mix. After it has hardened, remove the plaster from the pan and crack it open to find the casts or imprints made by the shells or leaves. The actual molds, especially of shells, can be compared to what might be found when cracking open a fragment of sedimentary rock.

Make an imprint of the human hand. Coat the hand lightly with petroleum jelly and make an impression in soft plaster. Let the impression harden without adding another layer of plaster. Why are footprints or tracks of slithering animals considered fossil evidence?

"FOSSILS" IN ICE   Students often have small figures of animals as toys or charms. Place these in water in the compartments of an ice tray in a refrigerator; each ice cube shows the "animal fossil" embedded in ice. Or better, use fresh pieces of fruit in the same way to show that no decay takes place when bacteria of decay do not have access to the material or when conditions are unfavorable for their growth. Recall the well preserved mammoths in ice.

"FOSSILS" IN "AMBER"   Melt a few pieces of resin over a low flame; avoid boiling. Then make small paper boxes about 3-cm square, stapled or clipped together. Now pour the molten resin into several paper boxes. Completely submerge a hard-bodied insect, such as a beetle, in the resin. If air bubbles appear in the resin, heat a dissecting needle and insert it into the air bubbles.

Let the resin harden at room temperature (rapid cooling causes cracking). After the resin has cooled, slough off the paper covering by soaking the box in water. The product is an insect "fossil" which simulates the manner in which ancient insects, mainly ants and small flies, became embedded in drops of resin that trickled down the bark of evergreens; the fossil resin is amber. Or embed small insects in a drop of clear Karo syrup or in balsam on a glass slide to simulate the way insects were preserved.

"FOSSILS" IN GELATIN   Gelatin may be used as an embedding material instead of resin. Add ¼ cup of cold water to 1 tsp of gelatin. When the water has been absorbed, dissolve the mass in ¾ cup of hot water. Then pour this into paper boxes or Syracuse dishes. Embed small objects

such as fruits, in the gelatin before it hardens.

Simulating formation of sedimentary rock

Show how layers of sedimentary rock are formed. Partially fill large jars with fine sand, coarse gravel, small rocks, humus, and clay. If possible, try to include several different textures and colors. Include some shells or toy dinosaurs in this muddy mixture to simulate remains of organisms. Fill the container with water so that all this material represents a muddy stream. Shake the jar and then let it stand; note the size and kinds of materials that settle to the bottom first. Have students explain how shells or imprints of fish skeletons or leaf prints may be found in sedimentary rock.

Students may discuss their reading of J. Schopf's paper, "Evolution of the Earliest Cells," *Sci. Am.* (September 1978), and G. Vidal's "The Oldest Eukaryotic Cells," *Sci. Am.* (February 1984), in which he traces back 1.4 billion years to plankton.

Similarity and variation

Similarity and variation in related plants

Observe firsthand the early embryology of some plants. In what ways are all monocots similar?

Monocots   Soak the seeds of monocotyledonous plants such as oat, rye, barley, bluegrass, and corn. Plant them in flats of moist sand or light soil and label the types of seeds. As the seedlings begin to grow, observe the first leaves. Since there is a similar genetic pattern based on common ancestry, these first leaves closely resemble each other. How do they compare with dicots?

Dicots   At the same time, show the similarity among the first leaves of dicotyledonous seedlings. For example, soak seeds of mustard, lettuce, pumpkin, marigold, melon, radish, tomato, pansy, and sweet peas. Also plant these seeds in flats of moistened sand and compare their first leaves. (Be sure to label the seeds or students will fail to recognize the plants.)

Gymnosperms Scar or file seeds of pine, hemlock, spruce, arbor vitae, and fir; soak overnight and plant in moist, light soil. While they germinate slowly, the results are rewarding. Compare the first leaves of the seedlings; note their similarity. Then continue to observe the specific differences which develop later, showing that seeds contain different as well as earlier similar genetic patterns.

Mustard family Compare seedlings of members of the same *family* of plants. Plant soaked seeds of cabbage, brussel sprouts, kale, cauliflower, kohl-rabi, and collards; all of these forms are descendants of the wild cabbage. Do other families of plants show similarities—leguminous plants, for example?

Variations in plants of one species

Measuring seeds and leaves Examine variations among seeds of *one kind*. For example, use several pounds of dried lima beans or other seeds, such as mimosa seeds or locust. Have each student measure the length of five seeds in millimeters, and then sort seeds according to their length. Arrange 8 to 10 large jars or test tubes of uniform size; then add seeds and label the jars in millimeter intervals corresponding to the range of distribution in size you might find among the seeds. When all the seeds are placed in the jars or test tubes (as in Figure 7-30), a continuous normal curve of distribution results—a three-dimensional histogram. If these seeds are kept dry, they can be saved from year to year for this activity. At other times, distribute a few pounds of fresh pea pods. Measure

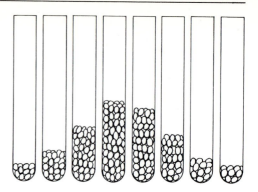

Fig. 7-30 Test tubes containing seeds sorted according to length. Note that the seed levels approximate the normal distribution curve.

the length of each pod, and perhaps the weight of each pod. Find the average length and compare this with the length of individual pods; have students plot their data. Make another quantitative study based on the weight of pea pods.

Sunflower seeds You may find sunflower seeds useful in studying other variations, such as striping. Also consider measuring seed clusters of *Tilia* (linden), *Ailanthus*, *Liriodendron* (tulip tree), and maples as well as the cones of evergreens, or the acorns of one oak tree.

Needle length in gymnosperms Examine variations in the length of needles from one evergreen (or the diameter of leaves of a broad-leaved tree). Collect a large number of needles from one kind of pine, or a large number of leaves from a single maple, oak, elm, or sassafras tree. With a centimeter ruler, measure the width of the leaves from the one tree. Plot the frequency distribution of the sizes.

Variations in seed germination Also show some qualitative differences among seeds of the *same species*. Soak seeds overnight, and place them between pieces of wet blotting paper ruled off in squares. Place an equal number of seeds in each square, the number depending upon the size of the seeds. Or use flats for germinating seeds. (Prepare several dishes or flats containing different kinds of seeds.) Then keep a daily record of the rate of germination. Students may estimate the percentage of seeds which germinated in 24 hr, 36 hr, 48 hr, and so on. Also count the number of nonviable seeds. You may want to test viability of embryos using the tetrazolium test (p. 472).

Further variations within a species

Variations among mollusks On a field trip, students may be able to collect large numbers of mollusk shells of *one species*: perhaps shells of scallop (*Pecten*), oyster (*Ostrea vulgaris*), periwinkle (*Littorina*), slipper shells (*Crepidula*), moonsnails (*Polinices*), or freshwater mussels (*Anodonta mutabilis*). Have students measure the width in centimeters of the shells of one species. Then plot a graph of the frequency distribution, using the measurement as the abscissa, and the frequency as the ordinate. Note that while the basic pattern of DNA is similar within a species, there are also variations due to recombinations of genes. Also examine variations in color shadings in scallops, periwinkles, or slipper shells.

Variations among human beings A classroom of students may want to study some differences among themselves. For instance, all the males of a certain age might stand. The rest of the class could record the height and the weight of these individuals, then plot the data on a graph. Perhaps a small committee of students might gather the same facts—height and weight—for large numbers of males of the same age in school. Then add these to the graph. Does a curve of normal distribution hold for human beings as well as for beans and shells? If we could plot a graph of the I.Q., or the grades in school, would they fall into a similar pattern when the assortment was as random as in the study of the height?

Variations among frogs Moore describes variations within a species; for example, he compares the characteristics of *Rana pipiens* from northern states with those of *Rana pipiens* from southern states.[21] Students may also study variations within the population of one locality.

If possible, order *Rana pipiens* from different supply houses. Compare patterns of spots and coloring and anatomical details. Moore also compares adaptations of the globular shape of the egg masses of early spring egg-layers (in cold water), such as *Rana pipiens*, *R. palustris*, and *R. sylvatica*, with species such as *R. clamitans* and *R. catesbiana* that breed later (in warmer water) and lay eggs that float on the surface as a thin film. Attempt to duplicate and study the effects on eggs of a variation in oxygen supply by allowing some large egg masses to develop in battery jars; compare these with smaller masses of eggs left to develop in other containers.

An interesting sidelight to this study would be to compare the parasites of frogs from different regions. (See pp. 623–29 and the many figures.)

Variations due to differences in the environment

Members of the *same* species of animal or plant may show variations due to environmental factors. Recall that development of chlorophyll in seedlings depends on light. Seedlings grown in the dark are yellow and spindly (etiolated). Variegated coleus plants develop patterns of brilliant color in sunlight. Observe the reddish pigment in barberry bushes compared to the dark green of leaves in the shade. On field trips, students may observe the differences in the shape of leaves of many plants growing along a stream or lake. Many plants show variations depending on whether the leaves are growing in water or in air. Collect specimens of crowfoot (*Ranunculus*), several of which have filamentous leaves when growing in water. Also collect some specimens which grow farther back

---

[21] J. Moore, "*Rana pipiens* and Problems of Evolution," *Ward's Natural Sci. Bull.* (Fall 1957) 31:3–7.

from the water's edge, and examine their broad leaves. In beggar's-ticks (*Bidens beckii*) the lower submerged leaves are deeply cleft, while farther up on the stem, well above the water level, the leaves are almost entire; whereas in the water parsnip the submerged leaves are feathery while the aerial leaves are almost entire. The submerged leaves of arrowhead (*Sagittaria*) are thin and linear, while the leaves above water are arrow-shaped. In addition many transitional stages can be found.

From these observations, students will learn that, in many cases, environment affects the expression of the genetic pattern, thereby altering the appearance of an organism. In other words, the phenotype of an organism depends on the combined effect of a specific DNA acting in a specific environment.

In effect, an organism does not "become adapted" to an environment. Instead, it already *has* the special adaptation, pre-adaptation, or genetic variation. If this variation is not a hindrance, the organism survives and reproduces more of its own kind. Thus the organism is said to be "adapted" to its environment. In fact, the variation may be beneficial and thereby have special survival value. For example, what "preadaptation" did the first fish have that enabled survival on land? Students might hypothesize as to how such adaptations as spines on cactus developed; how *Poinsettia* plants have leaves modified as red "petals." How have cases of mimicry or camouflage developed and helped organisms survive?

Himalayan rabbit   Discuss the possible internal environment or physiological mechanisms that may alter the organism. For example, when white fur is shaved from the back of a Himalayan rabbit, the hair grows back black if an ice pack is applied to the area. At what temperature are enzymes for pigment active: at high or low body temperature?

Peppered moth   There are films and filmstrips available that recount the well-documented case of the peppered moths. When Kettlewell studied museum specimens, he found that in Manchester, black moths comprised only about 1 percent of the population of peppered moths in 1848, while 50 years later, 99 percent of the moths were black. Have students develop hypotheses to explain possible reasons for this change. Recall that when Kettlewell tested his hypothesis, he released "marked" moths of both light and dark shades in equal numbers in Dorset and in Birmingham. Why did the birds eat the dark ones in Dorset? And the light ones in Birmingham? Predict the distribution in color of these moths when cleaner fuels are used.

Effect of a moisture-laden atmosphere on the form of growth of herbaceous plants   Dandelion plants, *Sempervivum*, and the like may be grown in a moist terrarium or under a bell jar in which the air has been saturated with moisture by inserting a wet sponge inside. In controls, develop dry conditions by placing the plants under a bell jar with an open glass container of calcium chloride. Other plants may be maintained under conditions of usual humidity. The long internodes and broad leaf blades which develop under conditions of high moisture contents should be apparent. In dry conditions students should find that the plants have short internodes and small leaf blades.

Among dandelion plants grown under normal conditions, leaves usually grow some 15 cm in length; in a moist atmosphere they may grow to 60 cm. Under moist conditions, *Sempervivum* loses its low growth form and becomes spindly with smaller leaves and a thinner cuticle.

Effect of darkness on potato plants   A group of students may try to reproduce the classic experiment performed by Vöchting in 1887, described by Palladin.[22] Terminal buds of potato plants were enclosed in a

---

[22] V. Palladin, *Plant Physiology*, Blakiston (New York: McGraw-Hill, 1926).

Fig. 7-31 Effect of darkness on growth of potato plants. Comparison of etiolated plant at left grown in the dark with normal plant of same age. (Photo by Dr. Carl L. Wilson, Dartmouth College.)

darkened box supported by a ring stand; aerial tubers developed. Other plants were placed in darkened boxes with the upper branches exposed to light; here, potato tubers formed above the soil on the darkened stem portions (Fig. 7-31).

Effect of mineral deficiencies on plant growth To study the effect of mineral deficiencies on the growth and form of plants, students may grow the young shoots of one kind of plant in different solutions of chemicals (nutrient solutions pp. 479–82).

Some adaptations of sun/shade leaves Leaves of many plants grown in the sun are thicker and better differentiated with shorter petioles, more pronounced venation, a hairier surface, and may have deeper lobes than leaves grown in the shade. Students may observe some of these adaptations in cross sections of leaves under the microscope.

Twining plants often lose this ability when grown in the dark. Stem elongation and leaf growth also vary when plants are grown in the dark. Etiolation is different in red and blue wavelengths of light. In blue light, the leaf blade area is often increased and petioles are shorter. In longer wavelengths (the red region) cells are elongated. (Blue wavelengths and white light prevent cell elongation.)

Wavelength and cell size Similarly, in blue and white light, fern prothallia grow in compact groups of cells; in red wavelengths, the cells are more elongated. Many fungi, too, are elongated in red light, while more compact cells are found in those grown in blue or white light.

Temperature and wing length in Drosophila Genes have their effect within a specific environment. In our desire to make the study of genetics simple we sometimes put stress on genes for a specific trait. We speak, for example, of the gene for vestigial wings in fruit flies. Yet if the eggs and the hatching larvae of flies homozygous for vestigial (*vgvg*) develop at a higher temperature (32° C instead of the usual 25° C), longer wings develop, not the vestigial condition (Fig. 7-32).

Effect of temperature on Serratia marcescens Prepare nutrient agar slants and inoculate them with a loopful of *Serratia marcescens*, bacteria that form red colonies. Incubate some slants at 37° C (99° F), others at room temperature. Examine the colonies after 48 hr; observe that colonies grown at the higher temperature are colorless.

Light and chlorophyll Albino plants lack the genes to produce chlorophyll. Yet, when plants have the genotype for chlorophyll, sunlight is needed for chlorophyll to develop. Obtain albino and green corn grains; grow kernels in the dark. When they germinate they lack chlorophyll. Now

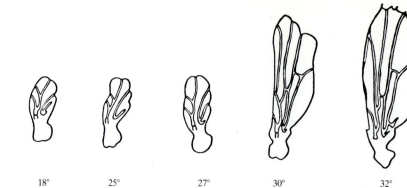

18° 25° 27° 30° 32°

Fig. 7-32 Effect of temperature on wing size in *Drosophila* males recessive for vestigial (*vg*) and dimorphos, a sex-linked specific modifier. Note the differences in wing size when flies are raised at different temperatures, ranging from 18° to 32° C.

(From *An Introduction to Genetics* by A. H. Sturtevant and G. W. Beadle. Copyright © 1939 by W. B. Saunders Company. Reprinted by permission of CBS College Publishing.)

shift the pots of labeled plants to light. Why do some turn green and some remain white? What is the difference in genotype? What environmental factor is needed for the DNA for chlorophyll to be expressed? The plastids are present, but light is needed for pigment production.

Genetically dwarf corn Treat genetically dwarf corn seedlings with gibberellic acid and observe the effects (p. 374).

Body color of Habrobracon Raise the pupae of these wasps at 20° C (68° F) and at 30° C (86° F). Note the effect of temperature on body color (p. 541).

Grand Rapids lettuce Light is a factor in the germination of some seeds. Germinate some lettuce seeds in the dark and some in red light, far red light, and daylight (pp. 377–79).

Tobacco seeds are also light sensitive. Soak seeds on filter paper in Petri dishes; expose the seeds to light for 3 to 6 hr daily until radicles are evident, then transfer to the dark for germination at 24° to 28° C (75° to 82° F).

Further studies In the laboratory or classroom it is difficult to demonstrate that

a phenotype is the result of the expression of many genes acting together. In some advanced seminars, a discussion of complementary genes, epistasis, cumulative effect, and sex-limited effects may be effective.

Speciation

Two major concepts need to be stressed: First, an organism is the product of its heredity and environment, and second, constant variability among offspring provides the raw materials for emerging new species. A species is composed of organisms that share a similar gene pool and interbreed. Today there are some 1.5 million species of living things.

Effect of isolation on species formation Explain how a species *A* may be separated by a body of water, or a mountain, and gradually develop into species *A* and *B*. In addition, how may different times of flowering, of seed germination, and of light for germination (as in Grand Rapids lettuce, p. 377) act as barriers that may lead to new species?

What is the role of mutations and DNA recombinations in the evolution of new species?

Some environmental pressures

Overproduction among organisms Collect the cones of pine trees, the pods of milkweeds, the head of a dandelion gone to seed, or the heads of sunflowers. On a visit around the school grounds, or to a park, have students count the number of seeds produced by several kinds of plants. Count the number of seeds in one pine cone, for example, and estimate the number of cones on one tree. What would be the total number of seeds produced by one tree? What are the chances of survival for each seed? If students have estimated the number of seeds on a maple tree, have them return later to count the number of seedlings nearby. Why are there so few? Begin to relate these observations to the steps developed in the theories of Darwin-Wallace and of De Vries.

Count the number of seeds in one pod of a milkweed and calculate the total number of seeds one plant could produce. Similarly, how many seeds might one sunflower produce?

Examine the roe of one fish and the eggs of one frog; look up the number of eggs laid by one shrimp, one oyster, one insect, and so forth. What are the chances for each egg to be fertilized? To grow into an embryo? To reach adulthood?

What conditions operate *against* the development of a seedling or spore? The fertilization of an egg and its development into an embryo?

Overcrowding in seedlings Is space a possible limiting factor in the growth of seedlings? Soak a large number of seeds overnight—use mustard, radish, oats, beans, or corn grains. Plant 20 seeds of *one* species in a small flowerpot, and another 20 in a pot twice as large. Keep the pots in a dark place and water them regularly. Over the next three weeks, count the number of seedlings that grow and survive in each flowerpot.

Intraspecies competition among seedlings Prepare several flats of garden soil. Mark off grids with string. In two flats have grid distances 10 cm apart, in two other flats make grid distances 5 cm apart, in a third set of flats 1 cm apart. Plant one species of plants, either young seedlings of beans or sunflowers. Place all flats in the same environmental conditions and observe growth of the young plants over several weeks until the results are obvious.

Or use the same size flowerpots, 5 cm wide, and plant fast growing, soaked seeds such as sunflower or pinto bean. Plant 2 seeds in pot *A*, 8 in pot *B*, 16 in pot *C*. Maintain all at the same environmental conditions and water daily. Cover the pots with plastic bags until the seeds germinate. After a month, compare the effects of competition for space on growth of seedlings.

Intraspecies competition among Drosophila Does overcrowding of organisms in one area affect their survival? Examine how the food supply can be a limiting factor in the survival of *Drosophila*.

Prepare fruit fly medium and pour into culture bottles (stock cultures, p. 535). This time pour the medium into several quart containers and also into several small vials. Then introduce four pairs of fruit flies into each container. (Students should be reminded to sprinkle the surface with yeast before introducing the flies.) Over a three to four week period, have students count the number of offspring surviving in each quart bottle and in each small vial. Plot the growth curves of the populations. What is the effect of increasing the carrying capacity of a given environment, that is, the maximum number of organisms that can be supported in a special environment?

Perhaps students may try this variation as individual study. Prepare several half pint bottles, each containing food and four pairs of flies. In this way there will be an increase in the total number of flies within a uniform environment. In about two months, when the population is at an ebb, count the total number of flies in all these bottles, distinguishing the dead ones from the living. Determine the total number of dead and living flies in all bottles. Now transfer all of the living flies in each bottle into bottles of fresh culture medium. The total population expands each time a new food supply is added at the end of every reproductive cycle. Then make counts of the initial increase in population in each bottle with an increase in food. Again, after two months, when the food supply has reached a low point, count the number of living flies. Is there a point at which the food supply is not large enough to sustain a further increase in population? (Also see a population study, Fig. 6-80, and p. 462.)

Overcrowding among praying mantids
If egg masses of the praying mantis are available, they may be kept in a small terrarium. Provide a source of water (or sugar water), such as a small sponge inserted halfway into a vial of water. Students may examine and estimate the tremendous number of nymphs that hatch out of each egg mass. Each day watch the decrease in population as cannibalism increases. You may later provide fruit flies as food for the nymphs or release them in a garden or park, since they are beneficial insects.

Intraspecies competition in Tribolium
What is the effect of a limited food supply and crowding on a population of *Tribolium* beetles? Begin with two pairs of *Tribolium confusum* in a jar of 50 g of enriched whole wheat flour. Make counts of all the stages in the population after two weeks. Use a sifter to separate beetles from the flour. Record the numbers; then return the beetles and all stages to a fresh container of 50 g of whole wheat flour. After two more weeks, again count and record all stages. What are the numbers of larvae, pupae, and adults after a month? Over a two-month period? Does a limited food supply and crowding affect any particular stage in the life cycle? Have students graph their data. Superimpose growth rate curves of larvae, pupae, and adults on one graph.

Or using a simpler technique, observe the effect of a *limited* food supply. Maintain

two pairs of beetles in 20 g of flour in one container, and another two pairs in 60 g flour in another container. Examine at two-week intervals and count the number of larvae, pupae, and adults.

At a temperature of 32° C (90° F) and some 70 to 75 percent relative humidity, *Tribolium confusum* develop from egg to adult in 27 days as follows: egg stage 4 days; larval stage 5 days; pupal stage 18 days. The life span of adult beetles is from 1 to 3 years.

Maintain the beetles at 25° to 30° C (77° to 86° F) in gallon containers of enriched whole wheat flour in the laboratory. Cover securely to avoid escape of beetles; use fine mesh. From time to time, add powdered brewers' yeast which seems to increase the rate of reproduction (also see culturing, p. 676).

Interspecies competition in Tribolium
Compare the competition between two species, *Tribolium confusum* and *T. castaneum* as reported by Park.[23] After four years of many experiments rearing these two species together under varying conditions of temperature and relative humidity, Park found that *T. castaneum* survives in a hot-wet climate, while *T. confusum* survives in a cool-dry environment (see Table 7-12).

Both species could survive under either climate alone but not in competition with the other species. At some time, you may want to try to repeat this experiment using temperature and humidity controlled chambers.

Growth in yeast populations　　Prepare a very dilute yeast culture in a medium such as dilute molasses to which a pea has been added to supply a protein requirement. Maintain the culture in a warm place.[24] Make serial dilutions[25] until a 1 ml drop on a scanned slide shows a small starting

[23] T. Park, "Experimental Studies of Interspecific Competition II. Temperature, Humidity and Competition in Two Species of *Tribolium*," *Physiol. Zool.* (1954) 27:177–238.
[24] Also refer to pages 311, 313 and 461 for culturing yeast in specific media.
[25] Serial dilutions, see technique, page 458.

TABLE 7-12   Competition between *Tribolium* species*   Park.

climate	temperature (° C)	relative humidity	Resul Results of competition (%)	
			T. castaneum wins	T. confusum wins
hot-wet	34	70	100	0
hot-dry	34	30	10	90
warm-wet	29	70	86	14
warm-dry	29	30	13	87
cool-wet	24	70	31	69
cool-dry	24	30	0	100

* From T. Park, "Experimental Studies of Interspecific Competition II Temperature, Humidity, and Competition in Two Species of *Tribolium*." *Physiol. Zol.* (1954) 27:177–238.

population of about 10 yeast cells. Maintain the culture in a warm place. Use a 1 ml pipette to make counts of the number of yeast cells at 2-hr intervals over an 8-hr period. Use a mechanical counter to tally the number of cells in a drop of the well mixed culture tube.

Have groups of students record their data, average the results, and plot the growth curve. Plot time on the X-axis and the total number of cells along the Y-axis; a sigmoid growth curve results (see p. 456).

Also plot a *growth rate* curve, that represents the number of cells added to the population at given time intervals; it gives a bell-shaped curve.

Other examples of competition

Roots of walnut trees   Andrews describes an example in which the roots of walnut trees make soil unfavorable for other plants' growth.[26] The roots secrete a substance that inhibits the growth of other kinds of plants; the hulls of walnuts contain the same substance. Grind hulls of a dozen walnuts into a fine powder and mix well into the soil of one flat, leave soil in a second flat untreated. Plant seedlings of a variety of plants in both flats marked off in grids of 10-cm squares for easy counting. Compare the growth of seedlings in each flat. Uproot the seedlings in the flats of

[26] W. A. Andrews, *A Guide to the Study of Terrestrial Ecology* (Englewood Cliffs, NJ: Prentice-Hall, 1974).

treated soil. Wash the seedlings, and replant them in untreated soil. What happens to the seedlings?

Photoperiods in plants   In the competition of seedlings for space and light, is there value in different photoperiods? Select young plants about five weeks old that have different photoperiods. For example, use long-day plants such as oats, red clover, radish, or others listed in Table 5-2; short-day plants such as aster, soybean, chrysanthemum; and neutral day plants such as tomato or corn. Label the pots and place one set of the three types of plants in a tall carton so that light from a 100-watt bulb enters only from the top. Cover with a black cloth for 16 hr of darkness followed by 8 hr of light. Arrange a second carton containing a second set of pots of the same three types; expose these plants to light for 16 hr then cover for 8 hr. Measure and record the initial height of each plant and the average width of a given number of leaves. Examine the plants as they grow over several weeks until they show signs of flowering or the failure to flower. Compare the same plants grown under short-day and long-day conditions as indicated.

Variations in structure and adaptation

What are possible explanations for diversified adaptations in beaks and in limbs of birds? For adaptive radiation (Fig. 7-33)? Students should also offer explanations for

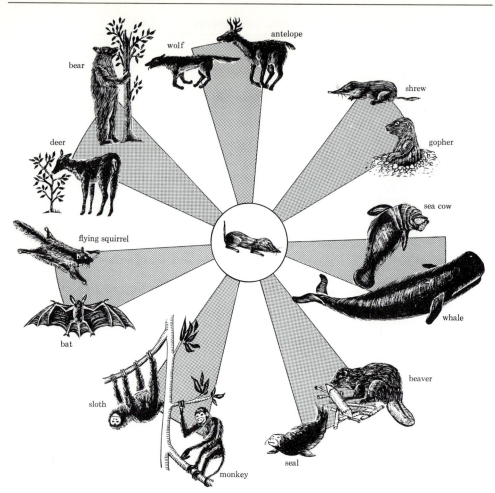

Fig. 7-33 One example of adaptive radiation among mammals.

(After C. Villee, *Biology*, 3rd ed., Saunders, 1957.)

convergent evolution, as shown in Figures 7-34 and 7-35.

Darwin's finches and adaptive radiation Perhaps offprints of D. Lack's classic paper "Darwin's Finches," *Sci. Am.* (April 1953) can be distributed to the class so that students may read and discuss this example of adaptive radiation (Fig. 7-36). Explain how finches varied. What are some possible explanations?

Migration and isolating mechanisms What factors affect migration of plants and animals? How may physical barriers, such as oceans, mountains, or a clearing, serve as isolating mechanisms that may lead to new species of plants or animals?

What is the effect of introducing a species into a new area where it has no natural enemies?

Use data on the distribution of blood groups among humans of different racial or national origins as the basis for a discussion of human migration and isolation. For additional material also refer to page 509. What is a gene pool? What is genetic drift? When are the frequencies of genes stable in a population? (See Hardy-Weinberg principle, p. 562.)

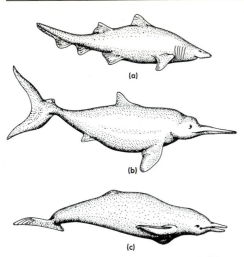

Fig. 7-34 Convergent evolution: (a) shark; (b) a fossil aquatic ichthyosaur (reptile); (c) dolphin. (After C. Villee, *Biology*, 3rd ed., Saunders, 1957.)

Evidences of relationships

Comparative anatomy   In the laboratory, dissect several vertebrates to compare their organ systems. Use charts or models to trace the evolution of the vertebrate heart (Fig. 7-37), the brain, respiratory, excretory, and/or the reproductive systems.

Similarly, make comparative studies of the evolution of conductive and supporting tissues in plants tracing the support systems of mosses, ferns, and seed plants. Trace the evolution of sexuality among the five kingdoms.

Vestigial structures   Additional evidence of relationships through common ancestry is the possession of vestiges. Have students list the structures in their body for which there is no use. How many vestigial structures can they name? What is the genetic basis for vestigial organs (Fig. 7-38)? Name some in the horse, in the male frog, in birds, and in the whale.

Comparative embryology   Refer to studies of developing eggs and embryos in chapter 6. How can we explain, in terms of DNA, the fact that eggs of animals go through similar cleavage stages, gastrula-

tion, and differentiation? Further, students may examine preserved specimens of vertebrate embryos in jars that have the labels concealed. How many vertebrates can students recognize in the early stages (as in Figure 7-39)? Why do closely related forms remain similar for the longest time?

Comparative study of skeletons   When possible, exhibit a variety of vertebrate skeletons. Elicit possible explanations for

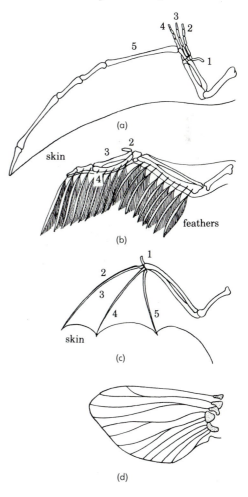

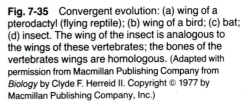

Fig. 7-35 Convergent evolution: (a) wing of a pterodactyl (flying reptile); (b) wing of a bird; (c) bat; (d) insect. The wing of the insect is analogous to the wings of these vertebrates; the bones of the vertebrates wings are homologous. (Adapted with permission from Macmillan Publishing Company from *Biology* by Clyde F. Herreid II. Copyright © 1977 by Macmillan Publishing Company, Inc.)

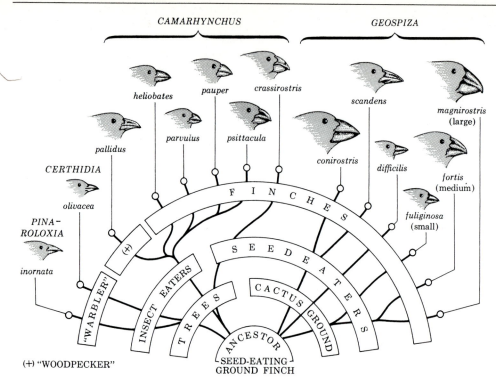

Fig. 7-36 Adaptive radiation among Darwin's finches on Galápagos Islands. (After G. G. Simpson and W. S. Beck, *Life: An Introduction to Biology*, 2nd ed. Harcourt Brace Jovanovich, 1965.)

the similarities in bone structure. What are the possible causes of the variations as special adaptations? Consider the likenesses in the forelimbs of vertebrates (Fig. 7-40). What are possible explanations for the origin and survival value of these specializations? Recognize homologous structures that have a similar embryonic history, yet which have specialized functions—as, among mammals, for flying, swimming, running, digging, and grasping.

Also compare parallel or convergent evolution in which similar environments "act" on unrelated organisms to select similar adaptations. Consider flight organs in bats, insects, birds, and ancient pterodactyls (Fig. 7-35), and in fishlike body shape among some aquatic vertebrates (Fig. 7-34).

Mimicry Have students prepare an exhibit case of other kinds of adaptations, such as mimicry as shown in the *Kallima*

butterfly (Fig. 7-41). Or show monarch and viceroy butterflies or walking sticks, a few of many cases of mimicry and camouflage. Recall the peppered moth, p. 550.

How might a special variation like the "eyes" on the underwings of the eyed hawk moth have developed? Offprints are available of papers in *Sci. Am.*: see "Mimicry in Parasitic Birds," by J. Nicolai (October 1974); "Coevolution of a Butterfly and a Vine," (August 1982), and "Moths, Melanin and Clean Air," by J. Bishop and L. Cook (January 1975).

A species' niche Are there two species of birds living in the same area? Are there several species of frogs in a given pond? Or several species of insects in the same 5 sq ft of ground somewhere near the school? Or two species of *Paramecium* in the same culture? If both species can survive in the same setting, they must have different ecological niches. (Also refer to Darwin's finches, Fig. 7-36.)

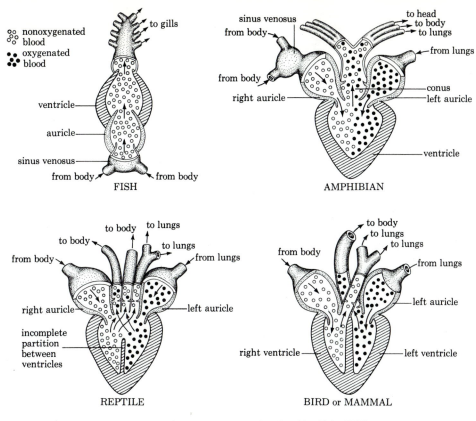

Fig. 7-37 Evolution of the vertebrate heart. (After R. W. Hegner and K. A. Stiles, *College Zoology*, 7th ed., Macmillan, New York, 1959.)

How might out-of-doors investigations be planned for the study of frogs at a pond, of insects in a vacant lot, or of two species of *Paramecium*? Under what hypothesis are the students operating? What testable predictions can be made? What is the design of the experiment? (Be sure to refer to the studies of succession (p. 587), especially the use of microaquaria on a slide for studies of protists; also see "Preludes to Inquiry" on page 230, and on page 615.)

Many studies of ecological relationships may be found among marine, freshwater, and land organisms. Refer to texts in ecology, and to activities in chapter 8.

Theories to explain new species

How did Lamarck, Darwin, Wallace, and De Vries explain the origin of species? Compare natural selection—that is, the environmental stresses that affect survival of genotypes—with artificial selection, the breeding methods people use to produce new kinds of roses, or other flowers, fruits, or dwarf trees. Catalogs, such as W. Atlee Burpee & Co., (300 Park Avenue, Warminster, PA 18974) have magnificent arrays of colorful breeds. Refer to breeding methods—hybridizing, inbreeding, and selection of plants and animals. Also refer to cloning of desirable types of plants. Recall the work some 20 years ago of F. C. Steward in propagating carrot plants from single cells (p. 488). Students may want to read J. Shepard, "Regeneration of Potato Plants from Leaf-cell Protoplasts," *Sci. Am.* (May 1982). In it the author describes enzymes such as pectinase and cellulase that remove cell walls so

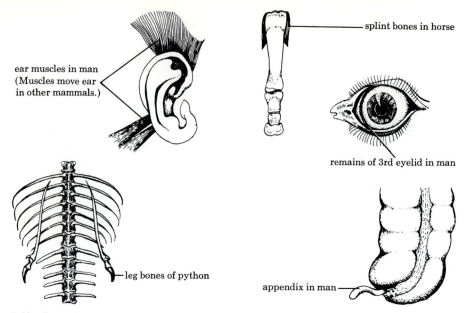

ear muscles in man
(Muscles move ear
in other mammals.)

splint bones in horse

remains of 3rd eyelid in man

leg bones of python

appendix in man

Fig. 7-38 Examples of vestigial structures in animals. (Drawings by Marion A. Cox. From *Experiences in Biology*, new ed., by Evelyn Morholt, copyright 1954, © 1960, by Harcourt Brace & World, Inc., and reproduced by permission of the artist.)

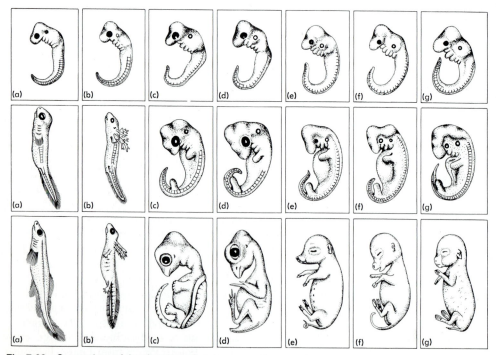

(a) (b) (c) (d) (e) (f) (g)

(a) (b) (c) (d) (e) (f) (g)

(a) (b) (c) (d) (e) (f) (g)

Fig. 7-39 Comparison of development of vertebrate embryos: (a) fish; (b) amphibia; (c) turtle; (d) bird; (e) pig; (f) sheep; (g) rabbit. (After G. J. Romanes, *Darwin and After Darwin*, Open Court Publishing Co.)

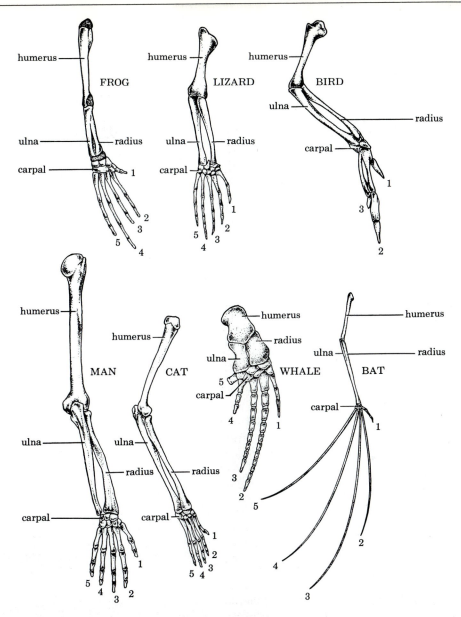

Fig. 7-40 Homologous structures: comparison of skeletal structure of forelimbs of some vertebrates.

(From C.A. Villee, *Biology*, 4th ed., W.B. Saunders Co., Philadelphia, 1962.)

that spherical cells called protoplasts are available for cloning new plants (using tissue culture techniques and nutrient solutions that contain necessary ratios of growth hormones, see Figs. 6-107, 6-108).

At some time, offer a stat of Darwin's writings on the long neck of giraffes and compare with a hypothesis Lamarck would have offered (Fig. 7-42). How would De Vries have explained this using his term "mutations"? How would each have "explained" the changes in the horse from *Hyracotherium* to *Equus* (Figs. 7-43 and 7-44)?

(a)   (b)

Fig. 7-41 *Kallima,* the dead-leaf butterfly, is a classic example of adaptation of form and coloration: (a) upper surface of the wings is brightly colored; (b) underside of the wings has the shape and coloration of a dead leaf. (Courtesy of the American Museum of Natural History, New York.)

Continental drifts

Develop the concept of evolving continents and the climatic changes that resulted as continents drifted away from, or toward, the equator. What effects resulted in prehistoric ecosystems? Recall the theory advanced by Alfred Wegener in the 1920s describing a single land mass (Pangaea) that split apart. Compare the east coast of South America and the west coast of Africa. Examine the movements of continents drifting on rigid subterranean plates (plate tectonics) as developed in Figure 7-45. Note that Pangaea, a large land mass, began drifting apart some 200 million years ago, in Permian times.

Have students draw the shapes of continents on cardboard and then cut out the continents as jigsaw pieces. Glue small magnets to the underside of the shapes. Have students fasten the continent jigsaw pieces to a magnetic board. Begin with the land mass Pangaea and draw the equator as a reference point. Move the continents as shown in Figure 7-45. Notice how the land mass that is now India probably pushed up the Himalayas. Computer constructions today show the evidences of similar fossils on originally connected land masses.

Relate the findings reported in *The New York Times,* March 21, 1982, of a land mammal fossil, a small marsupial found in the Antarctic. The small marsupial dates back to Eocene times, 40 million years ago, and supports the theory that marsupials reached Australia from South America and came across Antarctica which was a land bridge, before the continents drifted apart some 200 million years ago. In other words, marsupials are believed to have originated in the Americas and migrated to Australia. (Why the higher mammals, the placentates, did not migrate is still unanswered.) Earlier discoveries have been made in Antarctica of fossils of amphibians, birds, and reptiles. Some 65 million years ago, the Antarctic was a rain forest, still connected to South America.

Hardy–Weinberg principle

If dark hair is dominant over blonde, why doesn't blonde hair eventually disappear

Fig. 7-42 Lamarck's explanation for the increased length of neck of the giraffe. (From *Experiences in Biology,* revised edition, by Evelyn

Morholt. Copyright © 1967 Harcourt Brace & World. Reprinted with permission of the publisher.)

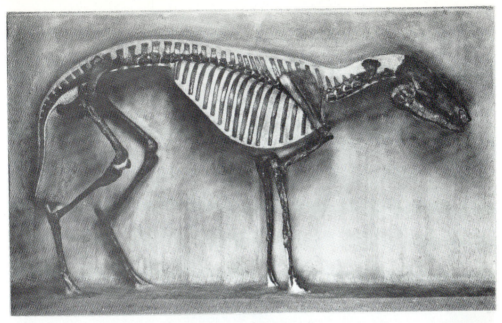

a

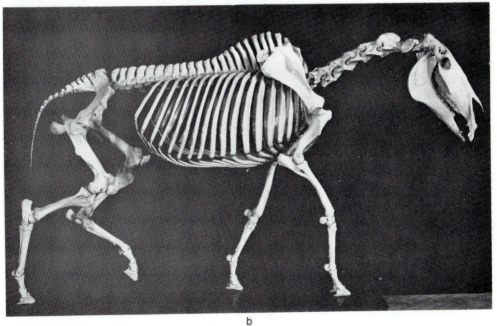

b

Fig. 7-43 Evolution of the horse: (a) skeleton of *Hyracotherium* (or *Eohippus*), about the size of a dog; (b) modern horse (*Equus*). (Photos courtesy American Museum of Natural History.)

in a population? In random mating in a large population, the frequency of a gene remains constant; that is, the proportion of homozygous to heterozygous remains constant. However, in a small population that interbreeds, chance, rather than selection,

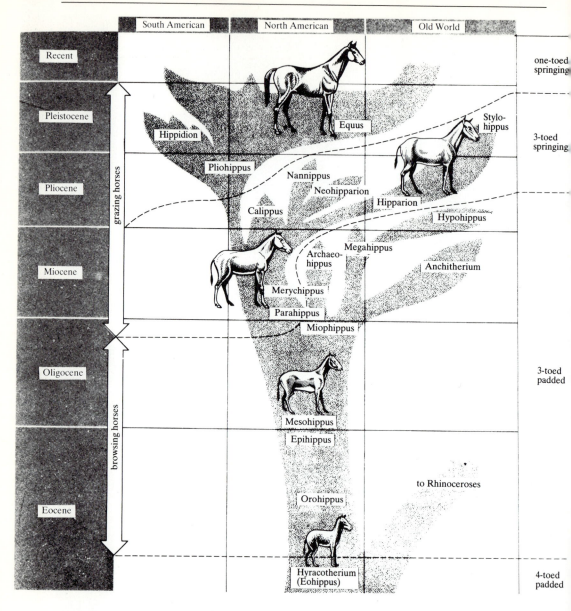

Fig. 7-44 Evolutionary lines among horses: note that some branches became extinct. (Adapted from *Biology* by Grover C. Stephens and Barbara Best North.

may shift heterozygous gene pairs so that they become homozygous. If the homozygous gene pair is harmful, it may disappear from the small interbreeding population. This example of *genetic drift* is an exception to the principle developed by Hardy and Weinberg that holds for ran-

dom mating. This principle was independently formulated by the Englishman G. H. Hardy and the German physician W. Weinberg.

In a large population then, there is no evolution; the gene frequency is constant after the first generation as long as there is

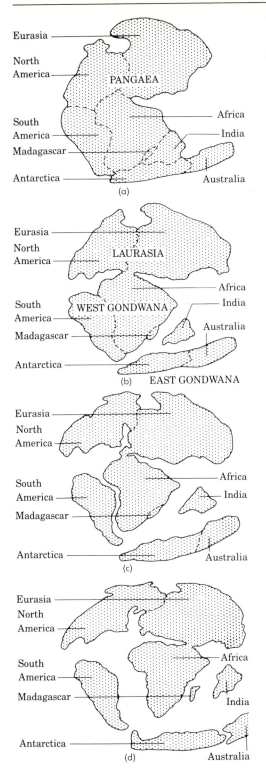

(a)

(b)   EAST GONDWANA

(c)

(d)

TABLE 7-13 Random mating in a large population according to the Hardy–Weinberg principle

		sperms	
		$p(D)$	$q(d)$
eggs	$p(D)$	$p^2(DD)$	$pq(Dd)$
	$q(d)$	$pq(Dd)$	$q^2(dd)$

random breeding, no natural selection due to environmental pressures, and no mutations.

If D and d made up a pair of alleles, the distribution of the genotypes in a large population is DD, Dd, dd. Hardy and Weinberg recognized this frequency as a quadratic equation:

p = frequency of D gene

q = frequency of d gene

$p + q = 1$ (since the allele of the pair will be either D or d)

Random mating in a large population (Table 7-13) results in the following algebraic sum:

$$p^2(DD) + 2pq(Dd) + q^2(dd)$$

This algebraic sum represents the frequency of the gene considered in any generation showing the stability of a pair of alleles in a population. (The assumption is that selection is not at work; see genetic drift, p. 564.)

Consider an example. Some people show T, the PTC taster gene, and some show t in the gene pool of a given population. In a sample it is found that 36 percent of the people have the dominant gene and 64 percent show the recessive trait. Which is pure or homozygous? Of course, 64 percent of the people have tt genes. This is the frequency of the recessive gene q in the binomial equation. Now

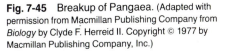

Fig. 7-45 Breakup of Pangaea. (Adapted with permission from Macmillan Publishing Company from *Biology* by Clyde F. Herreid II. Copyright © 1977 by Macmillan Publishing Company, Inc.)

p and q are gametes; $(p + q)^2$ are the kinds of possible zygotes. Expand the binomial:

$$(p + q)^2 = p^2 + 2pq + q^2$$
$$= 1 \text{ (100 percent possibilities)}$$

Since 64 percent of the people are tt, this is q^2:

$$q^2 = 64, \text{ and } q = \sqrt{64},$$

or 8 out of 10 gametes carry t gene.

Now $p = 2$ or 2 out of 10 gametes carry T gene. If $q = 8$ and $p = 2$, how many in the population are Tt, or heterozygous, for this trait?

$$p^2 + 2pq + q^2 = 1$$

$$(2)^2 + 2(2 \times 8) + (8)^2 = 1$$

$$4\% + 32\% + 64\% = 1$$

In summary, 32 percent of the population will be heterozygous (Tt) for the given trait under study.

What percent of the next generation will be homozygous for the dominant trait, the T trait? A population in equilibrium has:

genotypes	frequency
$TT =$	p^2
$Tt =$	$2pq$
$tt =$	q^2

The frequencies remain constant in a large population with no mutations or natural selection.

Classification of plants and animals

You may decide to bring to class samples of animals or plants. Better still, take your class outdoors to examine a biome, such as a pond, beach, desert, or wooded area. How do you help students develop an idea of order among the many kinds of plants and animals so that they can identify or classify them for reference? Some students may suggest grouping according to habitats, or habits of getting food. Certainly organisms differing as sharply as the animal inhabitants of a pond (fish, frogs, water

striders, *Daphnia*, protozoa, snails, planarians, water snakes, perhaps a beaver) cannot be placed or classified together. Students can readily name one outstanding trait that makes a fish different from a frog, a snake, a planarian, or other inhabitants of the pond.

Compare the categories (kingdom, phylum, order, family, genus, species) with the categories used by a librarian to classify books. This is familiar ground.

Practice putting similar animals or plants into broad categories and then breaking these down into smaller ones. For example, are all frogs in a pond the same species? What is a species?

Using a key for identification and classification

When students' observation of fine details has been sharpened, take them outdoors to learn the nature of a key as a means for identifying or classifying a plant or an animal. Suppose we begin with plants, since they are available year round.

An elementary approach might be used for a study of trees in winter. Have students examine trees and shrubs and select one *single*, obvious difference among them. Obviously, some trees have leaves (most evergreens) and some are deciduous. Begin to build a key by pairing off these evident single characteristics. Look more closely at the trees that do have leaves in the winter. Find one single characteristic that sets one tree or shrub apart from another. Perhaps one tree has clusters of needles, another has needles growing out in all directions from the stem, some have needles growing in one plane so that they look 2-ranked, another kind has especially small, fine needles (Fig. 7-46).

Then have students examine one group more carefully, perhaps those with needles in clusters (the pines); what differences can be observed? Do some clusters have two needles? Others three or five needles in a bundle, or fascicle? Distinguish these differences within the genus: the species differences (Fig. 7-47). Both Figures 7-46 and 7-47 are highly simplified keys and

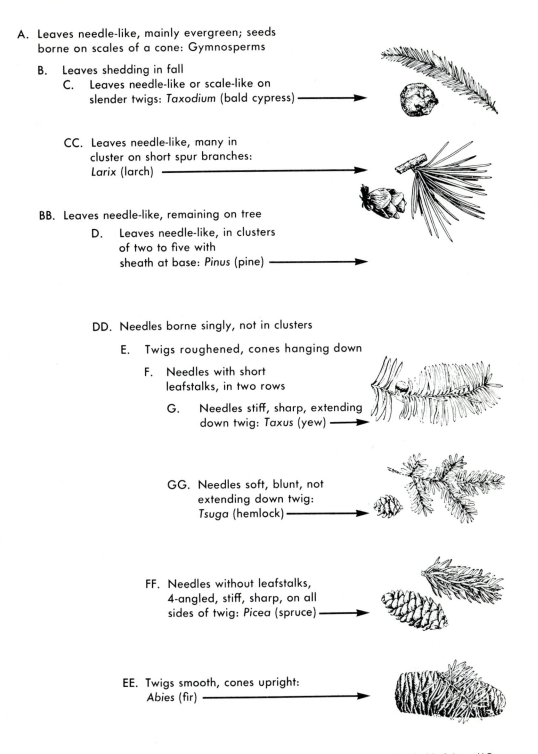

A. Leaves needle-like, mainly evergreen; seeds borne on scales of a cone: Gymnosperms

  B.   Leaves shedding in fall
      C.   Leaves needle-like or scale-like on slender twigs: *Taxodium* (bald cypress)

      CC. Leaves needle-like, many in cluster on short spur branches: *Larix* (larch)

  BB. Leaves needle-like, remaining on tree
      D.   Leaves needle-like, in clusters of two to five with sheath at base: *Pinus* (pine)

      DD. Needles borne singly, not in clusters
         E.   Twigs roughened, cones hanging down
            F.   Needles with short leafstalks, in two rows
               G.   Needles stiff, sharp, extending down twig: *Taxus* (yew)

               GG. Needles soft, blunt, not extending down twig: *Tsuga* (hemlock)

            FF. Needles without leafstalks, 4-angled, stiff, sharp, on all sides of twig: *Picea* (spruce)

         EE. Twigs smooth, cones upright: *Abies* (fir)

Fig. 7-46 Key to begin identification of gymnosperms. (Drawings from *Trees, The Yearbook of Agriculture*, U.S. Dept. of Agriculture, 1949.)

D. Leaves needle-like, in clusters
of two to five with
sheath at base: *Pinus* (pine)

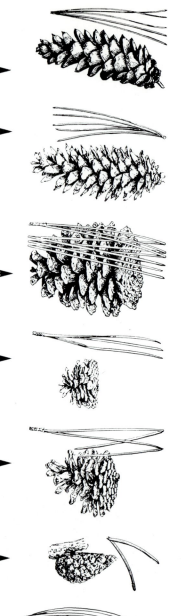

    a.   Five needles in cluster:
        Pinus strobus L. (eastern white pine) ⟶

        Pinus monticola Dougl. (western white pine) ⟶

  aa. Two to three needles in cluster
      (yellow, hard, or pitch pines)

    b.   Three needles in cluster;
       cone scales armed with spine

       c.   Needles greater than 8 in. long:
           Pinus palustris
           (long leaf, or southern pine) ⟶

       cc. Needles less than 8 in. long:
           Pinus rigida (pitch pine) ⟶

   bb. Two needles in cluster; cone scales unarmed

       d.   Needles greater than 3 in. long:
           Pinus resinosa
           (red pine) ⟶

       dd. Needles less than 3 in. long:
           Pinus banksiana
           (northern scrub pine, or jack pine) ⟶

  bbb. Three or two and three needles in cluster,
      4 to 7 in. long, cone scales armed:
      Pinus ponderosa Laws (ponderosa pine) ⟶

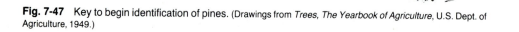

Fig. 7-47 Key to begin identification of pines. (Drawings from *Trees, The Yearbook of Agriculture*, U.S. Dept. of Agriculture, 1949.)

only one way that students may approach the problem of variations among living things and glean a basis for classification of organisms. Students may also use a standard key from plant and animal guides.

A simplified key to Protozoa is given in Figure 9-4 and to insects in Figure 8-59; from these, students may recognize basic differences and begin to build their own simplified keys, which they may compare with standardized keys in textbooks.

The five kingdoms

Consider the five kingdoms of organisms: Monera, Protista, Fungi, Plantae, and Animalia (see page 3). In this classification of kingdoms, the Monera (prokaryotes) are believed to have given rise to Protista. The Protista in turn evolved into three kingdoms of many-celled organisms: Fungi, Plantae (photosynthesizing organisms), and Animalia. (The origin of viruses and their placement are not included in this schema.)

Today the binomial system for classifying plants and animals developed by Carolus Linnaeus in the eighteenth century needs clarification; with our knowledge of molecular biology, we can now trace the evolution of, and similarities in amino acids, in enzyme systems, and DNA.

Besides the familiar taxonomy that relies on evolution from a common ancestor and adaptations or fitness to changing environments, there are new taxonomic schemes. Have students report on a controversial system called cladistics, which stresses the branching within a family of species on an evolutionary or taxonomic branch. Students should recognize that classification is not a "finished" business.

CAPSULE LESSONS: WAYS TO GET STARTED IN CLASS

7-1. How does the DNA code in the nucleus of a cell control the synthesis of specific proteins? Recall the formation of a single strand of mRNA as it is coded from a DNA strand template. Develop the role of tRNA and ribosomes in the synthesis of polypeptides for structural and functional proteins (enzymes).

7-2. Begin by showing a short film illustrating fertilization. Watch how a living sperm cell seems to be drawn into an egg. Notice that only one sperm enters, while the others are rejected after a fertilization membrane lifts off the surface of the egg. Elicit that a sperm carries half the DNA code, as does an egg. From this fusion the next generation arises. The class is now into a study of where egg and sperms "originate." What is the structure of sperms and eggs? How does a fertilized egg develop? How do specialized tissues and organs develop?

7-3. Construct a magnetic board from a sheet of iron and make cutouts of nucleotides. Use colored cardboard for the nucleotides; glue small magnets to one side so they can be moved and positioned on the upright magnetic board to show replication of strands of DNA. (The models must be lightweight so they do not slide off the tilted board.) Also have students use these materials to describe meiosis and the formation of gametes.

7-4. Show a film or filmstrip on some aspect of genetics. Refer to current A-V catalogs; many new films, loops, filmstrips, and transparencies appear yearly.

7-5. Organize a committee of students with some creativity and some degree of dexterity to make models of synthesis of proteins from the DNA code in the nucleus of a fertilized egg. Hall cabinets might well be used to exhibit "flow charts" and models of "egg to organism."

Fine assembly programs have been organized around such materials; furthermore, students may report to the student audience about their work. This gives students some sense of audience reaction, and a chance to stimulate the thinking of younger students.

7-6. On a spring field trip around the school grounds, examine the parts of flowers on a tree or shrub. Notice some plants have flowers that lack petals; some are staminate only. Students may notice that flower parts come in multiples of three or five. Many students enjoy keying an unknown plant. You may need to bring specimens to class for a more detailed study.

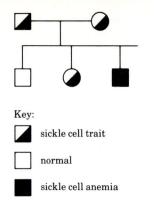

Key:

◨ sickle cell trait

□ normal

■ sickle cell anemia

Fig. 7-48 A pedigree with a key, for students to explain.

7-7. Without a preliminary explanation, draw a pedigree of sickle-cell anemia on the blackboard (Fig. 7-48). Have students describe the type of inheritance.

7-8. Some students may plan to try to cross different strains of guppies or platys. Or it may be possible to get mating types of *Paramecium bursaria* or of *Chlamydomonas* and study, first-hand, the behavior and heredity of living organisms.

7-9. Distribute squares of PTC paper and tally the number of students who taste the chemical and those who do not. Have some students test their parents too. How is the trait inherited? Develop the concept of the pairing of alleles and their segregation. You may want to follow this lesson with a committee report on Mendel's work. Once the need for reduction division of chromosomes has been developed, students may predict the phenotypes, the ratios to show dominance and segregation.

7-10. After students learn to use symbols for pedigree charts, draw several pedigrees such as the one in Figure 7-49; or copy others from the problems at the ends of chapters in genetics textbooks (see end-of-chapter listing). Draw these on the board, or reproduce some as part of an assignment. Have students explain the modes of heredity; include examples of sex-linked heredity.

7-11. Show culture bottles of fruit flies. Ask students to predict the approximate number of males and females in the bottle. Develop the concept of XX and XY chromosomes and how males and females are produced.

7-12. When students know the pattern of inheritance of X and Y chromosomes, develop how color blindness is inherited in humans. Do students understand that the gene is on the X chromosome? That a colorblind female has two genes for color blindness? Why is colorblindness (or hemophilia) more frequent in males? What evidence is there that the gene is *not* on the Y chromosome? Have students also diagram how hemophilia is inherited.

7-13. Have students type their own blood (as a laboratory lesson, p. 280). Before you begin, be certain to get notes of permission from parents. Develop the basic laws of heredity from the data secured. For example, if a student has type A blood what possible blood types could the parents have? What blood types could the parents not have? How might a child of type B parents have type O blood?

7-14. What is the difference between identical and fraternal twins? When student know that identical twins develop from one fertilized egg and therefore have the same DNA, they should be able to describe in what ways these twins resemble each other. What kinds of traits may be different in a set of identical twins? Students should recognize that learned behavior patterns result in differences between identical twins. Different environments may favor, or "select" different genetic factors.

7-15. Have students report on the inheritance of the Rh factor; while there are many genes involved in this complex inheritance, consider it due to the effect of one pair of genes as a simplistic introduction. Since 85 percent of

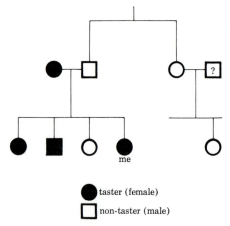

me

● taster (female)

□ non-taster (male)

Fig. 7-49 Pedigree of a student who is a taster of PTC. Have students explain the inheritance of the trait in the family.

the population tested in the United States have the Rh factor in their red corpuscles and about 15 percent of the people lack this factor and are called Rh negative, the Rh-positive factor is considered dominant. What percentage of the offspring might be Rh negative, if both the parents are hybrid for Rh positive?

7-16. Students might make a "flip book" to show how the two processes mitosis and meiosis differ. What effect might irregularities in movement of chromosomes have on the heredity of the offspring? Also have students explain possible causes of Down's syndrome, and of Klinefelter's and Turner's syndromes.

7-17. Time may be well spent in current readings on the effect of radiation, of drugs, herbicides, and so forth, on chromosomes.

7-18. Have students report to class on the value of amniocentesis (p. 518).

7-19. Committees of students might breed hooded and white rats, fruit flies, or *Mormoniella*. Or they might grow seeds resulting from a cross of two hybrid sorghum plants or corn plants; students gain skill in reasoning on the basis of the evidence. Kits of materials for making crosses are available from biological supply houses.

7-20. When possible, have students report on a visit to a county or state fair. How have new plants and animals been developed (studies of selection, hybridization and inbreeding)?

7-21. Try to obtain color slides from supply houses on variations in cattle (beef and dairy types), and in dogs. How do these forms differ from the ancestral wild type? How have these new types been developed? Perhaps some student in class breeds dogs, or a parent breeds horses, cattle, cats. Have the student describe some pedigrees.

7-22. Show a new plant or animal. Flip through pages in a seed catalog to show new types of flowers and fruits. How were they developed? Elicit the value of mutations or a change in a nucleotide sequence (deletion, substitution, and so on). What may cause a change in the translation of a DNA code in the synthesis of a protein? Perhaps use sickle-cell hemoglobin as an example.

How does a breeder select points for breeding? What does the consumer want in chickens, beef, milk, eggs? Develop the values of inbreeding, hybridization, and selective breeding.

7-23. Have students grow seedlings in the dark. Notice chlorophyll lacking? What evidence is there that chlorophyll-regulating genes need specific environmental conditions before an effect is produced in the plant? In light, how long does it take for seedlings to turn green? Develop examples of the expression of genes in a specific environment.

Or use several cloned coleus plants (same genetic makeup). Place some in moderate light, others in bright light. How does the environment affect the development of color in coleus plants? Look for similar examples around the school grounds.

7-24. Draw a Himalayan rabbit on the board to show its black pigmented nose, paws, and tail. Suppose a patch of fur on its back is shaved. What color fur will develop? Students usually answer white fur. Now draw an ice cube on the shaved skin of your diagram. What kind of fur color will grow back? Can students recognize that the extremities of the rabbit have a lower temperature? Develop the interaction of genes in a specific environment.

7-25. You may want to pose the following problem for solution. A group of science students has purchased a black male guinea pig for a study of heredity. How may the students (limited to one generation of guinea pigs) discover whether this animal is pure black or hybrid black? Elicit that the male might be mated to several white females (recessive). Should a white-furred offspring appear, the unknown must have been a hybrid; for if the unknown were pure black, then all the offspring should be black-furred (hybrid black).

7-26. A student might report to the class on exceptions to dominance—codominance. For example, when red and white four-o'clock flowers are crossed, the hybrid offspring have pink flowers. Many examples may be found in books on heredity. What is the heredity of roan cattle? Of Andalusian fowl?

7-27. At some time, have students demonstrate the way some fossils were formed. Use the techniques described in this chapter. Also make a sedimentation jar. Students may use plaster of Paris or ice as embedding materials. Establish the need for a quick burial, that is, removal from bacteria of decay to prevent decay of the plant or animal. Discuss other materials in which fossils have been found: ice, amber, tar, coal, and so forth. Plan a trip to a museum, if possible, or encourage students to visit on weekends.

7-28. Some students may be able to make clay models of dinosaurs (model-making, chapter 10) or to develop a diorama. Use such devices as these to explore this question: How much of the reconstruction of prehistoric animals or plants is fact and how much is the imagination of the artist? Plan an exhibit of "Life One Million Years Ago." Or have students prepare a phylogenetic collection of plants and animals.

7-29. Take your class through time by means of a series of slides describing dinosaurs, or ancient humans (available from museums of Natural History or supply houses). Also plan to read D. Mossman and W. Sarjeant, "The Footprints of Extinct Animals," *Sci. Am.* (January 1983).

How has the horse changed through time? What were dinosaurs like? How did Neanderthals differ from Cro-Magnon people?

7-30. In a laboratory lesson, have students study variations found in the length of seeds in a few pounds of lima beans. Measure the length of seeds and graph the findings. Or give each student an envelope containing a pine cone full of seeds. If possible, use cones from a single tree. Again, have students measure the length of the seeds. Elicit the possible causes of the variations. Estimate the total number of seeds produced by a single tree. What are the chances that all the seeds from one tree will develop into new pine trees?

7-31. Use a film as an introduction to the theories of evolution. Develop a discussion around these questions: How did the different kinds of mimicry originate? How is it that an animal matches its environment in pattern, form, and color? What are some adaptations of plants to their environment?

7-32. How might Darwin have modified his theory if he had known of Mendel's work? This might be an approach to a review of the points in Darwin's theory. You may want to provide a reading from Darwin's *Origin of Species* to show that he was not certain as to how variations occurred. Compare this with the concept that inheritable changes in plants and animals are based on changes in genes.

7-33. To one committee of students give a variety of seeds of one family of plants, such as the mustard family; to another committee give seeds of cereal plants; to a third committee give seeds of different members of the legume family. Do this at least a week in advance of the classroom discussion so that seeds begin to sprout. Then students may compare the shape, size, and time of appearance of the first leaves of the plants within each family. Have students notice the similarity. Develop the notion that these plants share a pool of genes from a common ancestor in that family group. Stu-

dents may notice that variations also exist in the rate of germination of seeds of the same kind. What is an advantage to the plant species of rapidly germinating seeds?

7-34. Begin a discussion of theories relating to evolution of organisms. Although every female frog lays several hundred eggs each spring, the population of frogs in a certain pond remains relatively constant. Establish overproduction, the role of enemies, unfavorable environment, and so forth. What is meant by the "fittest" organisms?

7-35. A committee of students might report to the class on evidence indicating relationships among plants or animals, for example, comparative embryology and vestigial remains. How have different *species* of oaks, maples, azaleas, and others developed?

7-36. Have students plan short debates or a panel discussion on some of these topics: (1) the role of mutations and recombinations in evolution; (2) the role of barriers in the formation of new species; (3) genetic drift and its effect on populations of organisms; (4) the differences between natural selection and artificial selection (that is, humans' use of selective breeding); and (5) the role of lethal mutations in a population.

7-37. After a study of the classic theories of Lamarck, Darwin, and De Vries, have students apply these theories to an explanation of this adaptation: Assume that all bears were originally brown; how might the origin of white polar bears have been explained by Lamarck? By Darwin? By De Vries?

How might each have accounted for the long neck of giraffes? How might they have explained camouflage in animals (walking sticks, moths the color of tree bark, green grasshoppers and brown ones, too, among many other examples)?

7-38. Plan a laboratory lesson with skeletons of several vertebrates (or study these in a visit to a museum). Compare the similarities and differences among the skeletons. Explain the similar pattern among the vertebrates and the differences. (See Figs. 7-35, 7-40.)

7-39. Show skulls of several mammals and have students try to identify the animals. How are they adapted to their specific food habits? Elicit a discussion of the relationship between structure and function; between homologous and analogous structures. Also show a series of models of brains of vertebrates. Have students place them in ascending order. What is their reasoning?

7-40. Using blood groups as evidence, have students trace human migration.

7-41. What kinds of fossils have been found in the Rancho La Brea tar pits? What is an explanation for the tremendous varieties of animals found here as fossils? Possibly describe the fossils in the La Brea fossils museum (part of the Los Angeles County Museum of Natural History).

7-42. Show a film on changes in the peppered moth. Develop the significance of the environment as the selecting factor in the appearance of favorable phenotypes (survival value).

7-43. Develop the theory of plate tectonics that accounts for drifting of continents. Trace the movement of continents after the breakup of Pangaea. As climates changed, so did ecological systems. Elicit possible changes in plant-animal food chains.

7-44. Have students report on theories of chemical evolution and the first heterotrophs.

7-45. Make available many offprints from *Scientific American* for enrichment of discussions on ancient people; for example, D. Pilbeam's paper, "The Descent of Hominoids and Hominids" (March 1984), and W. Rukang's and L. Shenglong's "Peking Man" (June 1983), and "The Archeology of Lascaux Cave," by A. Leroi-Gourhan (June 1982).

BOOK SHELF

Ambrose, E. *The Nature and Origin of the Biological World*. New York: Wiley, 1982.

Baer, Adela S. *A Genetic Perspective*. Philadelphia: Saunders, 1977.

Bentley, B. *The Biology of Nectaries*. New York: Columbia Univ. Press, 1982.

Bock, K. *Human Nature and History*. New York: Columbia Univ. Press, 1980.

Broda, P. *Plasmids*. San Francisco: Freeman, 1979.

Burns, G. *Science of Genetics*, 5th ed. New York: Macmillan, 1983.

Calow, P. *Invertebrate Biology*. New York: Wiley, 1981.

Colbert, E. *Evolution of the Vertebrates*, 3rd ed. New York: Wiley, 1980.

Cronquist, A. *An Integrated System of Classification of Flowering Plants*. New York: Columbia Univ. Press, 1981.

Dobzhansky, T., and E. Boesiger. *Human Culture*. New York: Columbia Univ. Press, 1983.

Ehrlich, P., and A. Ehrlich. *Extinction*. New York: Random House, 1981.

Ehrlich, P., and P. Raven. "Butterflies and Plants: A Study in Coevolution," *Evolution* 18: 586–608, 1964.

Eldredge, N., and I. Tattersall. *The Myths of Human Evolution*. New York: Columbia Univ. Press, 1982.

Forey, P., ed. *The Evolving Biosphere: Chance, Change and Challenge*. New York: Cambridge Univ. Press, 1981.

Frank, P. "A Laboratory Study of Intraspecies and Interspecies Competition in *Daphnia puliceria* and *Simocephalus vetulus*," *Physiol. Zool.* 25: 178–204, 1952.

Gardner, E. *Human Heredity*. New York: Wiley, 1983.

Gardner E., and D. Snustad. *Principles of Genetics*, 7th ed. New York: Wiley, 1984.

Grant, V. *Genetics of Flowering Plants*. New York: Columbia Univ. Press, 1975.

Harper, J. "A Darwinian Approach to Plant Ecology," *J. Ecology* 55: 247–270, 1967.

Harrison, G. et al. *Human Biology: An Introduction to Human Evolution, Variation, Growth and Ecology*, 2nd ed. New York: Oxford Univ. Press, 1977.

Herreid, C. *Biology*. New York: Macmillan, 1977.

Kornberg, A. *DNA Replication*. San Francisco: Freeman, 1980.

Lewin, B. *Genes*. New York: Wiley, 1983.

Lewis, W., and M. Elvin-Lewis. *Medical Botany*. New York: Wiley, 1982.

Mayr, Ernst. *Evolution and the Diversity of Life*. Cambridge: Harvard Univ. Press, 1976.

Margulis, L. *Symbiosis in Cell Evolution*. San Francisco: Freeman, 1981.

Mendel, G. *Experiments in Plant Hybridization*. Translated by the Royal Horticultural Society, London. Cambridge: Harvard Univ. Press, 1933.

Nitecki, M. H. ed. *Biochemical Aspects of Evolutionary Biology*. Chicago: Univ. of Chicago Press, 1982.

Novitski, E. *Human Genetics*. New York: Macmillan, 1977.

Pellegrino, C., and J. Stoff. *Darwin's Universe: Origins and Crises in the History of Life*. New York: Van Nostrand Reinhold, 1983.

Prance, G., ed. *Biological Diversification in the Tropics*. New York: Columbia Univ. Press, 1981.

Rothwell, N. *Understanding Genetics*, 3rd ed. New York: Oxford U. Press, 1983.

Scientific American, "Evolution," Sept. 1978.

Scientific American readings: *Industrial Microbiology and the Advent of Genetic Engineering*. San Francisco: Freeman, 1981.

Scientific American readings: *Recombinant DNA*. San Francisco: Freeman, 1978.

Sherman, I., and V. Sherman. *Biology: A Human Approach*. 3rd ed. New York: Oxford Univ. Press, 1983.

Singer, S. *Human Genetics*. San Francisco: Freeman, 1978.

Stebbins, G. L. *Darwin to DNA, Molecules to Humanity*. San Francisco: Freeman, 1982.

Stern, C. *Principles of Human Genetics*, 3rd ed. San Francisco: Freeman, 1973.

Stewart, A., and D. Hunt. *The Genetic Basis of Development*. New York: Wiley, 1981.

Thomas, B. *The Evolution of Plants and Flowers*. New York: St. Martin's Press, 1981.

Thompson, J. *Interaction and Coevolution*. New York: Wiley, 1982.

Watson, J., and J. Tooze. *The DNA Story* (documentary history of gene cloning). San Francisco: Freeman, 1981.

Welsh, J. *Fundamentals of Plant Genetics and Breeding*. New York: Wiley, 1981.

You may also want to send for booklets from the following sources:

Cystic Fibrosis Foundation, 6000 Executive Blvd., Rockville, MD 20852

National Foundation-March of Dimes, 1275 Momaroneck Ave., White Plains, NY 10685

National Heart, Lung, and Blood Institute (Sickle Cell Disease Branch), National Institutes of Health, Federal Bldg., Room 504, Bethesda, MD 20205

National Hemophilia Foundation, 25 W. 39th St., New York, NY 10018

National Institute of General Medical Sciences, National Institutes of Health, Bethesda, MD 20205; booklet "What are the Facts about Genetic Disease?" available from Supt. of Documents, U.S. Government Printing Office, Washington, D.C. 20402 (booklet DHEW #77-370).

8

The Biosphere:
Ecological Patterns and
the Limitations of Planet Earth

Several major conceptual schemes or patterns underlie a study of the biosphere. For example:

- The organism is the product of its heredity and environment
- An organism is interdependent with other organisms and its environment
- Constant slow changes, or evolutions, occur over time among populations as well as in their environments

In this chapter we consider the land areas, or biomes, named after the dominant plant types found there—regions characterized, in the main, by their climate. The major biomes of the biosphere are the tundra, taiga, temperate coniferous forest, temperate deciduous forest, savanna, prairie, desert, and tropical forest.

Some abiotic factors in the environment that characterize ecosystems are light, temperature, pH, O_2–CO_2 content, precipitation, available minerals, as well as pressures that may limit the range of populations.

These special habitats, and the role organisms play in specified niches in their habitats, offer many opportunities for study. Recall that E. Odum refers to a habitat as the "address" and the niche as the "profession" of the organism in an ecosystem.[1] There are opportunities to consider the functioning relationships, the interaction within food chains, food webs (Fig. 8-1); to build pyramids to trace the flow of energy from sunlight trapped by photosynthesizing producers through levels of consumers, with loss of heat at each level (Figs. 8-2, 8-3, and 8-4). Students may examine a variety of ecosystems, and their populations of organisms interacting within communities in a crack in a pavement, a bird bath, a rotting log, as well as communities in a quadrant of soil, a lake, pond, or ocean.

Methods for studying aquatic ecosystems (communities) in freshwaters as well as in marine regions–including beaches and mudflats—are described below. Students may observe food chains in a pond or minipond, in a depression slide or finger bowl: bacteria–*Paramecium*–*Didinium*. They may also tally populations of two species of *Paramecium* that occupy the same niche and share a common lifestyle: For example, in this interspecies competition, why does *P. caudatum* become extinct when *P. aurelia* is introduced into the same culture medium?

Field and laboratory studies

Where possible, examine the interaction of organisms with each other and their environment in field studies. A field trip need not be a weekend at a nature camp, nor a day's planned excursion to the woods or a beach. Consider brief field trips—a walk

[1] E. Odum, *Ecology*, (New York: Holt, Rinehart and Winston, 1963). Useful graphic models illustrate principles of ecology in the 1975, 2nd ed., and incorporates links between natural and social sciences.

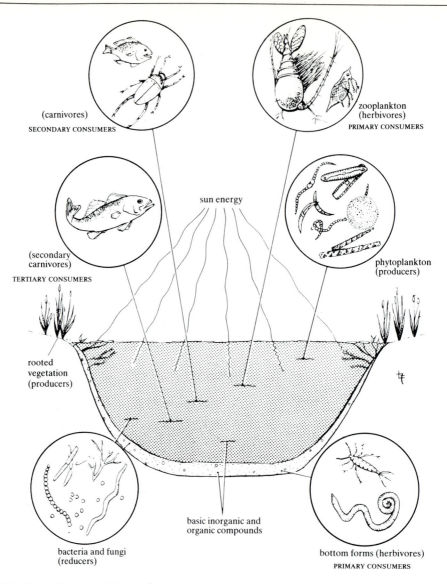

sun energy

(carnivores)
SECONDARY CONSUMERS

zooplankton
(herbivores)
PRIMARY CONSUMERS

(secondary
carnivores)
TERTIARY CONSUMERS

phytoplankton
(producers)

rooted
vegetation
(producers)

bacteria and fungi
(reducers)

basic inorganic and
organic compounds

bottom forms (herbivores)
PRIMARY CONSUMERS

Fig. 8-1 Food web in a pond. Trace the many food chains. (From *Biological Principles and Processes* by Claude A. Villee and Vincent G. Dethier. Copyright © 1971 by W. B. Saunders Company. Reprinted by permission of CBS College Publishing.)

around the school grounds, a vacant lot, the trees of the neighborhood.

Students need to recognize the living organisms in their environment. However brief or long the trip, it should be planned. The teacher surely needs to visit the area first. What is the purpose of the specific trip? What is the terrain? What is to be collected? Proper clothing and equipment are requisites along with hand lenses, containers, nets, and so forth. (Also refer to nature trails and school grounds, p. 638.)

Biotic communities in fresh waters

Aquatic forms in fresh waters are stratified

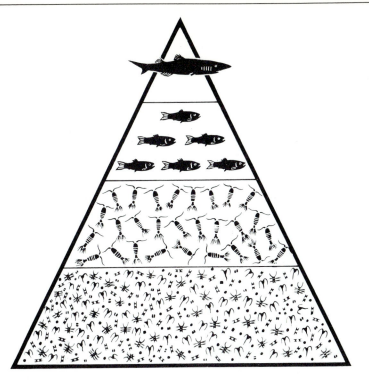

Fig. 8-2 A pyramid of numbers in the ocean.
(From *Basic Ecology* by Ralph Buchsbaum and Mildred
Buchsbaum. Copyright © 1957 by the Boxwood Press, Pacific Grove, CA.)

in regions. Marine forms also are found at different depths. Plankton are tiny free-floating forms with little or no method of locomotion; they include phytoplankton, zooplankton, and bacteria (Fig. 8-1). Use water sampling bottles to collect samples and keep well aerated. Slow-moving benthos forms live near the bottom. In fresh water these might include worms, mollusks, larvae and nymphs of insects, and aquatic insects. Nets and dredges (# 70 sieve) with mesh opening 0.21 mm are recommended for collecting benthos forms. Nekton forms are larger, free-swimming organisms that are collected in nets.[2] In introductory studies, improvise by using hoops made of wire hangers or plastic containers with their bottoms cut

away. Across these, stretch sheer dacron, nylon hose, curtain fabrics, or similar materials as sifters. Secure the fabric to the wire hoops with needle and thread or staples. Insert a bottle into the bottom of the net to collect water samples.

Collecting and identifying on field trips
Divide the class into teams before the field trip so that different groups will collect from several habitats, for example, mud along the shoreline or littoral region; submerged plants; organisms found under overturned rocks (be sure to replace the rocks for the inhabitants); surface of ponds and lakes, as well as bottom mud sediments where the CO_2 content is high and pH is more acidic. (Also refer to preparation of a Winogradsky column, p. 607, and to a laboratory study of succession of protozoa in mud sediment, p. 587.)

[2] Carolina Biological Supply Co. has sifters of varying sizes so that a rich sample of organisms may be sifted through progressively smaller mesh openings.

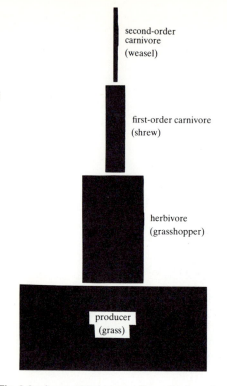

Fig. 8-3 A pyramid of energy. Trace the flow of energy along the trophic levels. (Adapted from *A Guide to Freshwater Ecology*, William A. Andrews, ed. Copyright © 1972 by Prentice-Hall, Publishers.)

Dip nets, kitchen sieves, and larger plankton nets can be used to collect zooplankton such as cladocerans, copepods, water-skimming insects, and mosquito larvae (Fig. 8-5). Use finer mesh nets to collect algae. Try to use fine mesh and coarse mesh nets that can be fastened on the same handle (a length no more than 3 ft). Dredges are used to collect samples from deeper lakes. (Some collectors use fishing lines to cast bottles far out in a lake to get samples of sediment from deeper lakes.)

Collect some of the mud from the bank, as well as specimens, in plastic pails or bags. Avoid overcrowding plant or animal specimens since they soon consume the available oxygen supply. Submerged plants or branches to which many microorganisms may be attached can be transported in plastic bags or wet newspaper. Within 5 hr, transfer the pond water and specimens into larger battery jars or aquaria.

After the mud settles, identify the swimming forms with a hand lens or dissecting microscope. In the spring you are likely to find *Daphnia* and larvae of mosquitoes, dragonflies, damsel flies, and mayflies (Fig. 8-5). Look for caddisfly cases as well as *Tubifex* worms in mud samples (Figs. 1-44, 1-45, 8-9, 8-10).

Microorganisms congregate at different levels in the battery jars—some are surface forms, others bottom dwellers. For a rapid inventory of the microscopic forms that have been collected, place some clean coverslips on the bottom of the containers and float others on the surface of the water. If these coverslips are left in these positions for several hours (or overnight), many microorganisms become attached to them. Also refer to submerged slides, for collecting freshwater (pp. 135, 586) or marine forms (p. 595). Carefully remove the coverslips with forceps, place them in a drop of water on clean slides, and examine under the microscope (see key to protozoa, chapter 1). Scrape the surfaces of submerged leaves and examine the scrapings under the microscope to find protozoa such as *Amoeba* (Fig. 1-13) and *Vorticella*

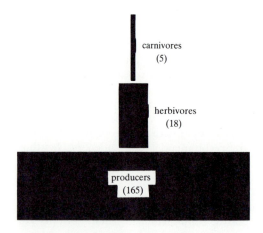

Fig. 8-4 A pyramid of biomass in a lake showing the number of grams of biomass per meter squared. (Adapted from *A Guide to Freshwater Ecology*, William A. Andrews, ed. Copyright © 1972 by Prentice-Hall, Publishers.)

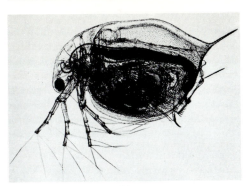

(a)

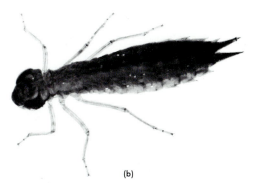

(b)

(c)

(d)

Fig. 8-5 Aquatic specimens that may be collected on a field trip: (a) *Daphnia*; (b) dragonfly nymph; (c) damselfly nymph; (d) mayfly nymph. (a, b, courtesy of Carolina Biological Supply Co.; c, d, photos by Hugh Spencer.)

(Figs. 1-11, 2-9, 2-10), and some flatworms such as planarians (Fig. 1-31), as well as insect eggs, rotifers, and some snails. Break apart swollen, rotting twigs which have been submerged. Look for snails, planarians, and hydras (Figs. 1-81, 1-31, 1-21) and, in clean water, specimens of the freshwater sponge *Spongilla* (p. 10).

Keep all these containers in moderate light. Maintain at a temperature below 24° C (75° F), and keep the containers covered to prevent evaporation. Plan to subculture certain organisms important to your work by isolating them in containers of aquarium water in which plants and animals have been growing (see culture methods, chapter 9). You may want to isolate some types of bacteria or algae by streaking agar medium in Petri dishes or minidishes (welled slides) are described on pages 459–62, 465.

Plant specimens may include diatoms, blue-green and green algae, elodea, duck-weed, *Cabomba*, and other types of water plants. Spores, eggs, and encysted forms develop quickly in the laboratory. Other forms become dormant for a time and reappear within a period of a month or so. This is excellent material for studies of food cycles and ecological patterns in different biomes. Population counts can be made (p. 465).

In some cases the collection of living material is not feasible, and the materials, such as soft-bodied specimens, might better be preserved in the field (pp. 446, 647).

A variety of algae   While some algae may be collected on land, especially in damp or boggy areas, most freshwater algae are abundant in ponds and lakes. (Marine algae may be studied immediately or kept for a short time in marine aquaria (pp. 589, 674),[3] but they cannot

---

[3] If students search for *Fucus* in the fall, they will find that fertilization is underway; antheridia are orange-red in color and the branches containing eggs are a dull green. Dry and preserve *Fucus* and other marine forms, such as *Ulva* and the delicate red algae, on cardboard or bristol board.

be maintained in class as readily as fresh-water forms.)

When collecting freshwater algae (Fig. 8-6), it is best to find specimens that thrive in slow-moving bodies of water; these forms have a better chance of surviving in an aquarium. Avoid overcrowding the specimens; transfer them as soon as possible into large containers in the laboratory. If you can refrigerate the specimens, you will increase their viability.

In shallow water you may find the slowmoving, blue-green filaments of *Oscillatoria* (Fig. 1-114). This form is also found in damp places (such as the outer black layer on damp flowerpots). The blue-black mats on the surface of damp soil in the pots are also likely to be *Oscillatoria*. *Euglena* (Fig. 1-2) may be found in shallow pools, and you may well find desmids of several kinds in the greenish mud of very shallow ponds and at the edge of a lake. The silky threads which float in sunny spots of ponds, lakes, and ditches, and have a slippery feel between the fingers, are likely to be those of *Spirogyra* (Fig. 1-116). And in the same locale you may find the water net *Hydrodictyon* (Fig. 1-119). The colonial green algae *Volvox* (Fig. 8-7), composed of hundreds to thousands of cells, are also inhabitants of lakes and ponds. *Volvox* are green spheres, about the size of the head of a pin, which can be more easily recognized with a hand lens.

You may also recognize *Vaucheria*, which is found as a green felt covering on rocks in ponds and ditches or on flowerpots. On the other hand, a green, fuzzy or hairy covering on rocks in ponds and slow-moving rivers may be the simple filamentous alga *Ulothrix*. Note the difference between *Vaucheria* (Fig. 6-53) and *Ulothrix* (Fig. 6-50). *Vaucheria* has branching filaments which lack cross-walls forming a multinucleated syncytium.

In damp places on land (especially in greenhouses) or on damp bark, fences, and rocks, look for a green covering of *Protococcus* (or *Pleurococcus*). This simple green alga (Fig. 8-8) seems to be more likely a re-

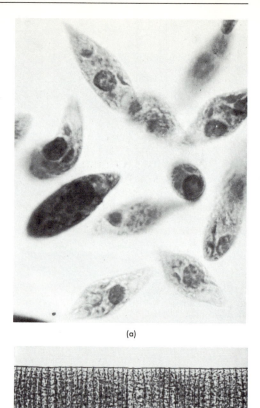

(a)

(b)

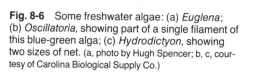

(c)

Fig. 8-6   Some freshwater algae: (a) *Euglena*; (b) *Oscillatoria*, showing part of a single filament of this blue-green alga; (c) *Hydrodictyon*, showing two sizes of net. (a, photo by Hugh Spencer; b, c, courtesy of Carolina Biological Supply Co.)

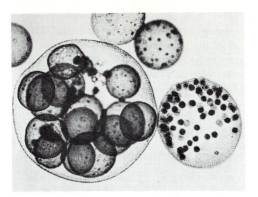

Fig. 8-7 *Volvox globator*, a colonial green alga with sexual stages and several daughter colonies. (Courtesy of Carolina Biological Supply Co.)

duced form, rather than a primitive alga. Yet it is useful for studies of a typical green plant cell.

In samples of pond water, along with desmids (which can be recognized by their symmetrical halves), you should find the primitive green alga *Chlamydomonas* (Fig. 8-26). Under high-power magnification look for the chloroplast, two contractile vacuoles, and two flagella in the anterior region. Some species also have a red eye spot.

Study of quiet ponds and lakes Have students try to identify both the macrofauna and flora as well as the microscopic organisms in a quiet pond or lake. Since ponds are shallower than lakes, light penetrates deeper so that ponds comprise inhabitants of the littoral zone. Examine the plankton; scrape the undersurface of duckweeds or other floating leaves. Look for *Daphnia*, tadpoles, small fish, and on the surface find water striders and whirligig beetles (Fig. 8-9). Brush through the rooted plants to catch small organisms that may be attached, or hide among the leaves as scuds do. Are there water lilies? Or water hyacinths? Are there dragonflies and damselflies, as well as mosquitoes? What birds may be present? What kinds of tadpoles and frogs?

Students might diagram a cross-section of the pond or lake to place the organisms at the level or zone in which they live. The littoral zone of lakes shows extensive types of plants ranging from plankton to floating plants and rooted plants; animal species include tadpoles and immature stages of insects like nymphs of damselflies and dragonflies that hide in the vegetation. In lower strata of lakes where light does not

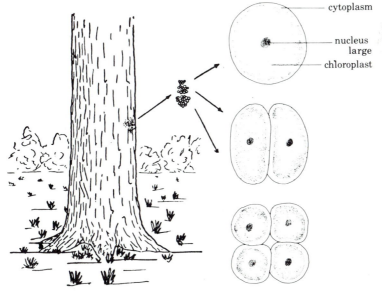

Fig. 8-8 *Protococcus* (*Pleurococcus*), a simple green alga found on damp tree trunks and stone walls. Note the large chloroplast in each cell. (From

W. Brown, *The Plant Kindgom*, New York: Ginn and Co., 1935.)

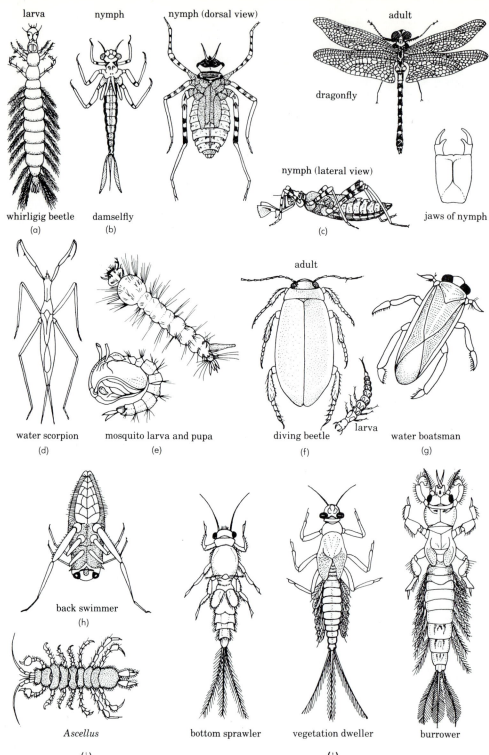

larva

nymph

nymph (dorsal view)

adult

dragonfly

nymph (lateral view)

jaws of nymph

whirligig beetle
(a)

damselfly
(b)

(c)

water scorpion
(d)

mosquito larva and pupa
(e)

adult

diving beetle
(f)

larva

water boatsman
(g)

back swimmer
(h)

Ascellus

(i)

bottom sprawler

vegetation dweller

burrower

(j)

Fig. 8-9 A variety of inhabitants of a pond: (a) larva whirligig beetle; (b) damselfly nymph; (c) dragonfly nymph and adult; (d) water scorpion; (e) mosquito larva and pupa stages; (f) larva and adult of a predaceous diving beetle; (g) water boatsman; (h) backswimmer hanging from the surface of the water; (i) an isopod (*Ascellus*); (j) nymphs of three kinds of mayflies. (Adapted from *A Guide to Freshwater Ecology*, William A. Andrews, ed. Copyright © 1972 by Prentice-Hall, Publishers.)

penetrate, bottom feeders may include many leeches, worms such as *Tubifex*, and small mollusks that can tolerate limited oxygen supplies. Many bacteria, fungi, and protozoa are also present.

What is the temperature of the water? Is it the same at the shoreline as further out, or deeper down? How deep is the pond or lake? Have students make contour maps to show its depth, its elevation in relation to land, and its area and shape.[4] What is the source of the water, the possibility of contamination? Is the water clear or turbid? Is the lake bed gravel, flat rock, hard clay, or mud? What is the pH of the water? Also consider a rough measurement of the oxygen-carbon dioxide exchange rate using dark and light bottles, page 586. Consider how equilibrium is established. Carbonic acid increases as CO_2, resulting from respiration of the many organisms, combines with water molecules. An equilibrium is reached, when cations, (mainly Ca^{++}) are present:

$$2CO_2 + 2H_2O \rightleftharpoons 2H^+ (HCO_3)^- + Ca^{++} \rightleftharpoons Ca^{++} (HCO_3)_2^- + Ca^{++} \rightleftharpoons 2Ca^{++} (CO_3)^=$$

Take photographs of the pond or lake at different seasons; over several years, they may show a succession of different plant and animal migrants. Perhaps plants invade the pond or lake so that gradually a swamp forms. What types of organisms survive in the new habitat? What kinds of pioneer plants first appear, to be followed by animal consumers?

[4] Excellent techniques are described in *A Guide to the Study of Fresh Water Ecology*, W. Andrews, ed., (Englewood Cliffs, N.J.: Prentice-Hall, 1972)

Have students determine the water quality using test kits to establish the biochemical oxygen demand (BOD) as well as the dissolved oxygen and chemical oxygen demand (COD). An introduction is given on page 597; the rapid field test kits are available from Hach as well as La Motte (see Appendix).

Study of fast-moving streams   It may also be possible to have students examine the inhabitants of a fast-moving stream characterized by well-aerated water.

Study the life in the stream. Which organisms live under submerged rocks? Which are free-swimming? Which attached? Are there living things found on the banks and shallow part of the stream? Which are predators? Which are competitors? Which are found in the deeper water? Do organisms move with the current or face into it? How does the oxygen supply of the stream compare with that of pond water? (Use La Motte or Hach test kits. Appendix.) Finally, trace the food webs that exist (Fig. 8-1). What factors may upset the balance in this web of life? Develop a food pyramid. Which are primary consumers? Secondary consumers (Figs. 8-2, 8-3, and 8-4)?

Find larvae or beetles, mayflies, stoneflies, flatworms, and snails, and then make a comparative study of the two aquatic communities: the well-aerated stream and the pond (Figs. 8-9, 8-10, and 8-11). Identify the plant and animal inhabitants of each and enter them in their proper location in a cross-sectional view. (See also Fig. 8-1; immersion slides, microorganisms, p. 35.)

Prepare a contour map of the stream to show the nature of the bed and banks. Also keep a record of the temperature, clarity, and relative velocity of the water in the stream, as well as the nature of the slope. Are the organisms active or slow-moving? A fast-moving stream often has a high oxygen content, which supports active organisms.

If students plan to transport living organisms from a stream to the classroom,

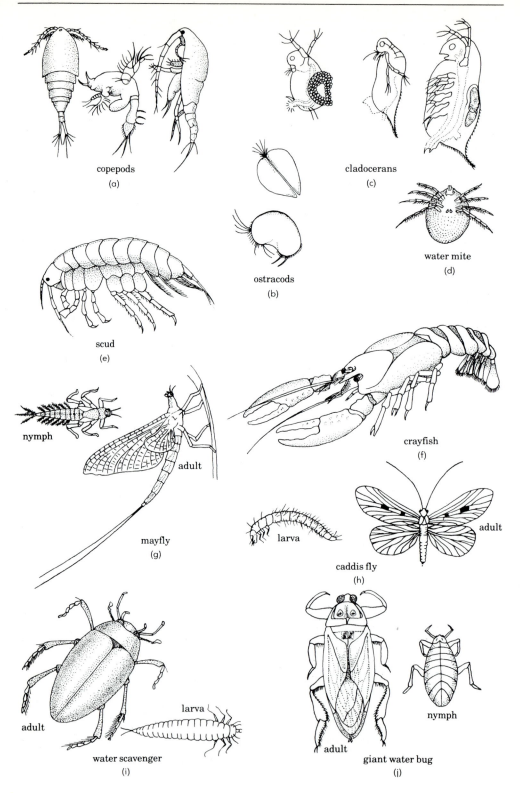

copepods
(a)

cladocerans
(c)

ostracods
(b)

water mite
(d)

scud
(e)

nymph

adult

mayfly
(g)

crayfish
(f)

larva

adult

caddis fly
(h)

adult

larva

water scavenger
(i)

adult

nymph

giant water bug
(j)

Fig. 8-10   (*left*) Other inhabitants of ponds and lakes: (a) copepods; (b) ostracods; (c) cladocerans; (d) water mite; (e) scud; (f) crayfish; (g) mayfly nymph and adult; (h) caddisfly larva and adult;

(i) water scavenger; (j) giant water bug nymph and adult. (Adapted from *A Guide to Freshwater Ecology*, William A. Andrews, ed. Copyright © 1972 by Prentice-Hall, Publishers.)

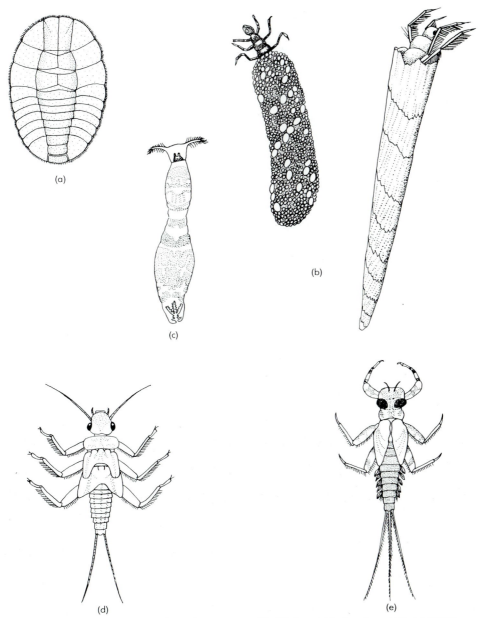

Fig. 8-11   Adaptations of larval and pupal stages of some insects found in fast-moving streams. Note the flattened shape or attachment regions. (a) water penny; (b) caddisfly larvae in cases; (c) blackfly larva; (d) stonefly nymph; (e) mayfly nymph. (Adapted from *A Guide to Freshwater Ecology*, William A. Andrews, ed. Copyright © 1972 by Prentice-Hall, Publishers.)

they will have greater success in keeping organisms alive if they select specimens from more quiet regions of a stream.

A bog, woods, beach at ebb tide

In some areas, it may be more appropriate to study a flood plain marsh, a bog, desert, burned-over area, or open meadow. The same procedures hold: Record the physical factors, the kinds of plants and animals, and collect mud, sand, or soil samples for microscopic examination in the laboratory. Refer to studies of soil organisms, such as nematodes, decomposers that may be yeasts and other fungi, and bacteria; prepare Berlese funnels, Fig. 8-22; Baermann funnels, Fig. 8-23; Winogradsky columns, p. 607, where applicable. Also consider microscopic studes, including streaking of organisms on growth enhancement media in sterile Petri dishes (p. 603).

Trace the food webs in the areas; are there natural enemies that hold some populations in check?

Submerged slides for long-range studies

Slides may be submerged in either freshwater or marine environments and inspected after a week, three weeks, a month, or two months to count the kinds of populations and the succession of populations over a long period.

Remove most of the bottom and top of slide boxes made of plastic or wood; insert slides into the remaining grooves. Hold fast with wire or string so that slides may be placed vertically and also horizontally so that water may flow between the slides. Attach the slide sets to submerged plants so they float undisturbed under water. Then carefully remove a box of slides after a week, another after a longer time, and so on, to examine succession of organisms. Try to identify the organisms, or diagram them, and make counts of the number of different populations that appear over a time sequence.

Refer to illustrations of plant and animal organisms in chapters 1 and 2 for identification, as well as within the context of other chapters. (Use the Index for

identification and staining macro – and microorganisms.) Also refer to the excellent guides for identification of specimens in the series edited by W. Andrews;[5] and consult texts such as those by Whittaker and Smith, and the handbooks by Lind and Darnell.[6]

Light and dark bottles

As part of a study of the *functioning* of an ecosystem, examine the productivity of a lake or pond as measured by its rate of photosynthesis in relation to respiration. Use equal volume samples of pond water algae, with no heterotrophs present, to fill 500-ml volume bottles. Seal the bottles; cover some with aluminum foil, or black paint. In the black bottles, respiration will be the ongoing process, while both respiration and photosynthesis occur in the light bottles.

Hang the bottles from a floating log, or submerge them by means of a strong cord attached to weights at *varying levels* of a lake or pond. Allow them to remain submerged for a given time, perhaps 1 to 24 hr or longer, and calculate the amount of dissolved oxygen in light and dark bottles in each of the separate levels of the water. Use a Hach or La Motte dissolved oxygen test kit (Appendix) or when available, use oxygen electrodes.

Calculate the change in oxygen content to determine the rate of photosynthesis and respiration in the algae per unit area. (Recall that one carbon atom is assimilated for each oxygen molecule released). What is the amount of oxygen produced in the light bottles? What is the amount of oxygen consumed in the black bottles, where respiration only occurs (assuming there are no heterotrophs present)? Add

---

[5] W. Andrews, ed. *A Guide to the Study of Fresh Water Ecology* (1972); *Terrestrial Ecology* (1974); *Soil Ecology* (1973); (Englewood Cliffs, N.J.: Prentice-Hall).

[6] R. H. Whittaker, *Communities and Ecosystems*, 2nd ed., (New York: Macmillan, 1975); R. L. Smith, *Ecology and Field Biology*, (New York: Harper & Row, 1966); O. Lind, *Handbook of Common Methods in Limnology*, 2nd ed., (St. Louis: Mosby, 1979); R. Darnell, *Organism and Environment, A Manual of Quantitative Ecology*, (San Francisco: Freeman, 1971).

the oxygen produced and the oxygen consumed to find the total oxygen produced over a given period of time per liter of water.

Studies of succession

The first method examines grass and attached soil in water exposed to light. Here students will find many protozoans that are often encysted, and resist drying, adverse temperature, or lack of food. The second method is a more rapid laboratory study of protozoan succession in the dark, using an enriched medium and seeding the medium.

Miniponds   Have students start their own miniponds by gathering dry grass and bits of attached soil from different locations in wide-mouthed baby food jars. To each jar, add two rice grains and fill half with water; leave jars open and set aside in moderate light at room temperature. The rice and grass provide sources of organic matter in these miniponds.

Since these cultures, or miniponds, are maintained in the light, algae as well as protozoa (and bacteria) will soon appear. Within two weeks, have students use capillary pipettes to examine samples from the bottom as well as the top scum of the water in the jars. After an initial proliferation of heterotrophic bacteria feeding on available organic molecules, there follows, in succession, small flagellates and herbivorous ciliates such as *Colpoda* and *Paramecium*, succeeded by carnivorous ciliates and amoebae. Early samples of the miniponds may be examined for bacteria using methylene blue stain, Loeffler's or Gram stain (p. 145).

Have students hypothesize about the chemical nature of the minipond environment: What factors affect the succession of organisms—light, pH, wastes of protozoa and bacteria, food supply, and ratio of O_2–CO_2? Explore cultures about two months old for *Vorticella* and rotifers. After about three months, the populations of

organisms decrease and die unless photosynthesizing algae are added. You may recall that J. Liebig, who studied chemical fertilizers for crops, found that the rate of growth of a crop was dependent on the nutrient (or other limiting factor) found in the smallest amount. (Studies are available on minerals and growth of plants, p. 478.)

Protozoa succession in an enrichment culture   At some time, you may want to determine the density of successive populations of microorganisms over a three– to four-week period. Examine 1-ml samples using a Sedgwick–Rafter counting chamber; make five random counts of each new type of organism found among the previously declining types. (Use low power only, so the slide isn't broken.) Make a count every other day during the first two weeks, and twice a week over the third and fourth weeks. As a guide to identification of the genera of protozoa, refer to figures in chapters 1, 2.

If the cultures are maintained *in the dark* and incubated at 30° C (86° F) green plants do not appear. The muddy bottom of a lake or pond is a rich source of bacteria, mainly motile *Bacillus subtilis*, fungi, encysted forms, eggs or larvae of invertebrates. Darnell suggests the following medium.[7] Boil 75 g of dry grass in 1 liter of distilled water for 15 min. Strain when cool, then dilute this grass broth with 50 ml of distilled water. (Refrigerate until needed.) Immediately before use, add 1 liter of distilled water to 100 ml of the broth and seed the medium with 3 g of lake or pond mud. If the cultures are maintained in the dark, the oxygen supply gradually decreases, and the pH becomes more acidic with a buildup of carbonic acid. In this laboratory succession (without green plants), the appearance of new populations occurs faster than in a natural setting with green producers in the water. It now seems that the type of succession is dependent on the increase of certain orga-

---

[7] Darnell, *Organism and Environment, op. cit.*

nic molecules released by one type of organism, that in turn provides the proper enviroment for a specific succeeding type of organism. Thus, the organic molecules in the medium that build up over time stimulate or inhibit the growth of a population of an oncoming species.

Also refer to the effects of other environmental pressures, such as overcrowding, temperature, and humidity on seedlings and on *Drosophila* and *Tribolium* (pp. 554, 551). What happens to the rate of reproduction of *Paramecium caudatum* during acclimatization to NaCl (pp. 139, 325)?

Interrelations in ecosystems

Organisms take substances from the non-living environment and, in turn, they return substances into the water, soil and/or air. As interactions among organisms and their environment shift, ecological successions occur. Clearly, as the forest floor becomes progressively more shaded due to a canopy of foliage, a succession of invading seedlings that require less light may take hold. Or changes in pH of the soil may affect the type of vegetation that can subsist and thereby affect the consumers that may survive.

Some shifts are periodical, such as temporary rain pools or pools formed by melting snow. Consider too the stratification from water to swamp, and then to land; or the dominance of one species, a climax species.

Trace the dynamics of interacting cycles of gases, minerals, and water, and the role of microorganisms in the nitrogen and mineral cycles from air to water to soil (refer to nitrogen-fixing bacteria, p.634), nitrification cycles (p. 608), and a Winogradsky column for study of bacterial decomposers (p. 607). Develop the fascinating modes of nutrition among living organisms within ecosystems (for example, symbiotic relationships, p. 631).

Consider using offprints of *Scientific American*: R. Revelle, "Carbon Dioxide and World Climate" (August 1982); M. Koehl, "Interaction of Moving Water and Sessile Organisms" (December 1982); "Dynamic Earth" (entire September 1983 issue).

Also refer to M. Link *Outdoor Education: A Manual for Teaching in Nature's Classroom*, (Englewood Cliffs, N J: Prentice-Hall, 1981).

A world in a sealed test tube   In a small way, focus attention on a green plant and a snail—a producer and a consumer—sealed together in a test tube of aquarium water. What interactions are under way between these living organisms and their environment? Prepare several sealed test tubes (soft glass, p. 300). You may want to add an indicator to the tubes before sealing them. Use dilute brom thymol blue which is just at its turning point to yellow due to an increase in CO_2. Exhale into a quantity of dilute brom thymol blue and add equal volumes to the aquarium water in the cooled pulled-out tubes. Add a healthy sprig of elodea and a snail, such as *Physa*; then return to the Bunsen burner and soften the pulled-out collar of the test tube to seal it (Fig. 4-33).

Students can follow the absorption of CO_2 in light by photosynthesizing elodea; as the medium becomes less acidic, the indicator turns to blue. When does the indicator turn to yellow again? How long can the snail live? The green plant? Which is a converter of radiant energy? Which converts chemical energy into movement? Do both organisms have ATP? What happens if the snail reproduces faster than the plant? How do the cells of the snail multiply? The same as in the plant? Are there similarities in metabolic processes in the snail and the green plant? A great deal of biology can be learned from a study of one small snail and an elodea sprig in a closed world.

Marine communities

There is something compelling about oceans that makes a study of their currents, nutrient upwellings, varying temperatures, chemistry, plankton, migrations, and intricate food webs fascinating for most students. Consider introducing students to Rachel Carson's beautiful books about life in the oceans.[8]

The world ocean, comprising the interconnecting large Pacific, Atlantic, Arctic, and Indian oceans, covers some 71 percent of the surface of this planet. Some 80 percent of the southern hemisphere is ocean.

Seawater is slightly alkaline (pH 7.5 to 8.4), maintained by the interactions of dissolved carbon dioxide and water which act as natural reversible buffering systems. The amount of hydrogen ions present shifts the reaction in either direction. An equilibrium exists among carbon dioxide, carbonic acid (H_2CO_3), and bicarbonate (HCO_3^-) and carbonate ($CO_3^=$).

$$CO_2 + H_2O \rightleftharpoons H_2CO_3 \text{ (carbonic acid)}$$

$$H_2CO_3 \rightleftharpoons HCO_3^- + H^+ \text{ (bicarbonate)}$$

$$HCO_3^- \rightleftharpoons CO_3^= + H^+ \text{ (carbonate)}$$

$$CO_2 + OH^- \rightleftharpoons HCO_3^-$$

$$H_2CO_3 + OH^- \rightleftharpoons HCO_3^- + H_2O$$

The formation of bicarbonate occurs mainly above pH 8. Hydroxyl ions com-

[8] Rachel Carson, *Under the Sea Wind* (1940); *The Sea Around Us* (1950); *The Edge of the Sea* (1955). All are available in paperback from New American Library Inc., 1633 Broadway, New York, NY 10019

bine with carbon dioxide at high alkalinity to form bicarbonate.

Life in the oceans Life in the oceans is dependent upon photosynthesizing phytoplankton found mainly in the upper photic zone through which light penetrates. This zone usually ranges from 50 to 125 m depending on the latitude, season, and clarity of the seawater. Wavelengths of light penetrating seawater are absorbed and scattered at different rates. For example, the orange reds (600 to 700 mμ) and violet (400 mμ) are readily absorbed in the intertidal zone by green algae. The blue and green wavelengths (450 to 550 mμ) penetrate deeper, especially in clear seawater; these are used by brown and red algae.

Golden brown diatoms make up the major part of the phytoplankton (Fig. 8-12). Many dinoflagellates are also part of the plankton (Fig. 8-13). Zooplankton is composed of larval forms of crustaceans, mainly copepods, that show a vertical migration, rising to the surface usually after sunset to graze on the mass of phytoplankton. Many temporary forms, whose numbers vary, are also found: larval stages of sponges, jellyfish, mollusks, annelids, echinoderms, hemichordates and fish. Describe some of their flotation adaptations (Fig. 8-14). Protozoa in zooplankton include flagellates, ciliates, radiolarians,

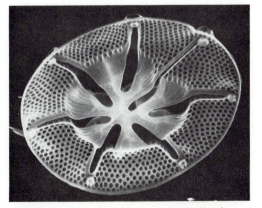

Fig. 8-12 Scanning electron micrographs of a diatom. (Photo courtesy Scripps Institute of Oceano-

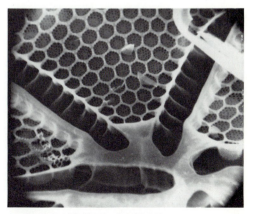

graphy, Univ. of California, San Diego.)

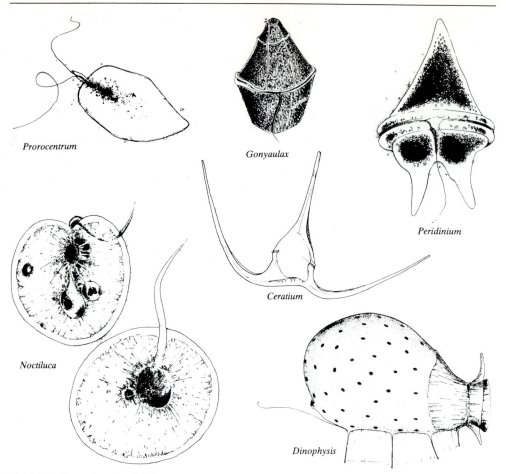

Prorocentrum

Gonyaulax

Peridinium

Ceratium

Noctiluca

Dinophysis

Fig. 8-13 Several marine dinoflagellates. (Adapted from *An Introduction to the Biology of Marine Life* by James L. Sumich. Reprinted with permission from W. C. Brown Co., Copyright © 1976.)

and foraminiferans (Fig. 8-15). *Globigerina* is an abundant foraminiferan found in plankton; its tests form the globigerina ooze on the seafloor (Fig. 8-16).

The average salinity of seawater is 35°/00 (i.e., 35 parts per thousand). The most abundant ions are Cl^- and Na^+. A cycle of dissolved nutrients, mainly nitrates and phosphates, are used by photosynthesizing plankton. The substances are returned in detritus to the floor of the ocean. Cyclic upwellings of these nutrients bring rich organic supplies to the surface.

Trace a generalized food pyramid (Fig. 8-17). Have students trace the many food chains that make up a generalized food web as shown in Figure 8-18. In coastal regions, plan field trips to beaches at low tide. How do some organisms survive at intertidal zones (Fig. 8-19)? Examine rocky areas and mudflats. Try to identify the variety of communities—the ecosystems of plants and animals interacting with each other and their environments. Collect and examine plankton. Refer to Figures 8-14 and 8-18, and illustrations in chapter 1 as guides for identification and dissection.

Bioluminescence Watch the waves break on a beach at night. Observe the bright glow of the rippling waters in many regions from luminescent species of dinoflagellates, mainly *Noctiluca* (Fig. 8-13).

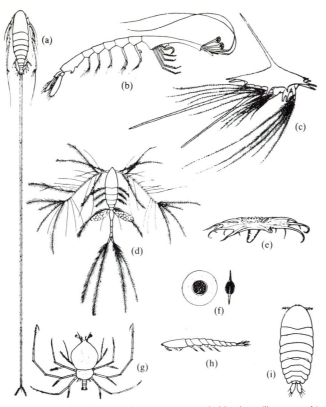

Fig. 8-14 Some adaptations for flotation found among zooplankton: (a) copepod; (b) decapod; (c) barnacle nauplius; (d) copepod; (e) holothurian; (f) pelagic egg of a copepod; (g) larva of lobster; (h) copepod side view; (i) copepod top view. (Adapted from *An Introduction to the Biology of Marine Life* by James L. Sumich. Reprinted with permission from W. C. Brown Co., Copyright © 1976.)

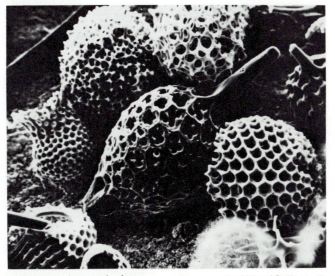

Fig. 8-15 Scanning electron micrograph of several radiolarians. (Photo courtesy Scripps Institute of Oceanography, Univ. of California, San Diego.)

Fig. 8-16 *Globigerina*, a common foraminiferan in plankton, with pseudopods extended. (Adapted from *An Introduction to the Biology of Marine Life* by James. L. Sumich. Reprinted with permission from W. C. Brown Co., Copyright © 1976.)

Some bioluminescence also results from symbiotic bacteria on fish; other types are due to special glandular cells, photophores, of some marine forms.

Extensive coverages of oceanography and marine life are available in texts such as J. Sumich, *Biology of Marine Life* (Dubuque, IA: Brown 1976); R. Zottoli, *Introduction to Marine Environments*, 2nd edition (St. Louis: Mosby 1978); H. Sverdrup et al., *The Oceans: Their Physics, Chemistry and General Biology* (Englewood Cliffs, NJ: Prentice-Hall 1942, revised 1970); and W. Russell-Hunter, *Aquatic Productivity* (New York: Macmillan 1970).

Also consider offprints of *Scientific American*; dominant marine food webs are pictured in "The Nature of Oceanic Life," by J. Isaacs in the September 1969 issue; "Marine Farming," by G. Pinchot (December 1970) provides background for more current papers. "Learning in a Marine Snail," by D. Alkon (July 1983) describes experiments in conditioning; "Harvesting of Interacting Species in

a Natural Environment," by J. Beddington and R. May (November 1982) describes the ecosystem of southern ocean waters, the Antarctic phytoplankton, and the role of shrimplike krill in affecting the abundance of penguins, seals, and whales. The role of certain bacteria that utilize sulfide, and the clams for which they serve as food is described in "Hot Springs on the Ocean Floor," by J. Edmond and K. Von Damm (April 1983). Consider how some spider crabs decorate themselves as camouflage in "Decorator Crabs," by M. Wicksten (February 1980). Examine behavior in schools of fish as in "The Structure and Function of Fish Schools," by B. Partridge (June 1982).

Marine decomposers

Only about 12 percent of bacterial species, and less than 1 percent of fungi are marine. *Pseudomonas* is the most common

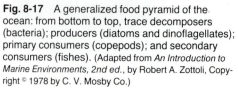

Fig. 8-17 A generalized food pyramid of the ocean: from bottom to top, trace decomposers (bacteria); producers (diatoms and dinoflagellates); primary consumers (copepods); and secondary consumers (fishes). (Adapted from *An Introduction to Marine Environments, 2nd ed.*, by Robert A. Zottoli, Copyright © 1978 by C. V. Mosby Co.)

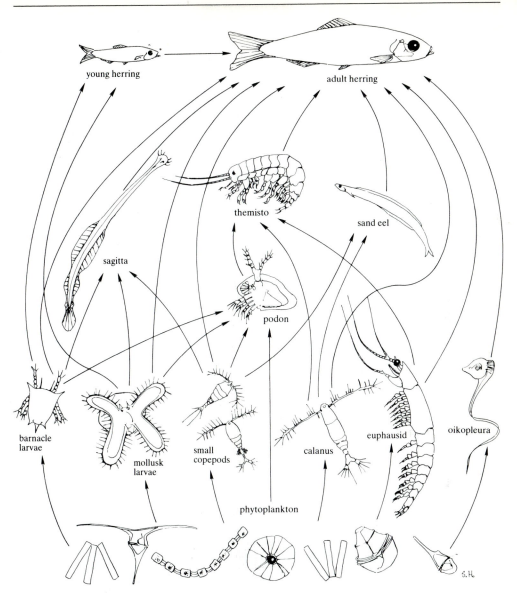

Fig. 8-18 A generalized food web in the ocean. Trace the interweaving food chains from phytoplankton to herring. (Adapted from *An Introduction to* the *Biology of Marine Life* by James L. Sumich. Reprinted with permission from W. C. Brown Co., Copyright © 1976.)

genus among the bacteria (flagellated, Gram-negative bacilli). The luminescent *Photobacterium* may be found on fish (Gram-negative bacilli). Use hanging drop techniques (p. 153), Gram stain (p. 145), and streak techniques on marine growth media to isolate different types (p. 594), and prepare a Winogradsky column for a study of anaerobic and aerobic microorganisms (p. 607).

For introductory studies, collect samples of beach sand or bay mud in separate sterile vials. Allow these to stand and use the water extract as an inoculum. (See techniques used for the study of soil microorganisms, p. 603.) With a sterile transfer loop or needle, streak the surface of sterile Petri dishes containing a marine growth medium (preparation follows). After a week's incubation, examine the plates for

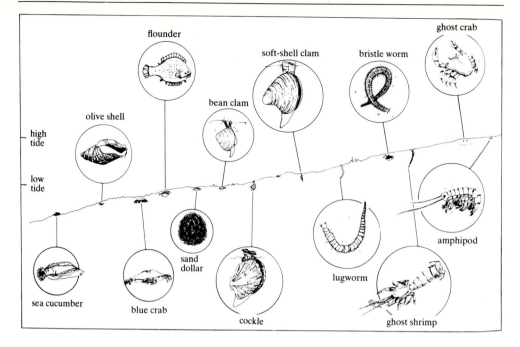

tide

low
tide

flounder

ghost crab

soft-shell clam

bristle worm

olive shell

bean clam

sea cucumber

blue crab

sand
dollar

cockle

lugworm

amphipod

ghost shrimp

Fig. 8-19 Intertidal zonation along a coast. Compare the organisms under water at low tide, with those at right on the dry part of the beach. (Adapted from *An Introduction to the Biology of Marine Life* by James L. Sumich. Reprinted with permission from W. C. Brown Co., Copyright © 1976.)

colonies. Along with typical bacterial colonies, expect to find some red and pink colonies, which may be marine yeasts; fuzzy growths are marine fungi (some of which are commonly parasitic on crustaceans).

Subculture bits from the different colonies; streak into fresh, sterile medium in Petri dishes and then incubate. Prepare smear slides and stain with Gram stain (p. 145). Also make hanging drop preparations (p. 153) of the Gram-negative rods of *Pseudomonas* to check for motility. In some mud samples the spirochaete *Cristispira* that lives in the gut of clams may be found.

Marine growth media   Sumich and Dudley suggest the following growth media for bacteria, and for culturing yeasts and molds.[9]

[9] J. Sumich and G. Dudley, *Laboratory and Field Investigations in Marine Biology*, Dubuque, IA; Kendall/Hunt, 1974.

Marine bacteria growth medium   Dissolve 10 g peptone and 15 g agar in 1 liter of sterile seawater (reduce quantities if desired). Plate in sterile Petri dishes. Use sterile techniques in handling bacterial colonies and transfer loops (p. 701).

Yeast and mold growth medium   Prepare two solutions and sterilize each separately; mix the solutions before the agar solidifies.

SOLUTION A: Dissolve 20 g of potato starch and 15 g agar in 750 ml of seawater.
SOLUTION B: Dissolve 10 g dextrose in 250 ml of distilled water. To this solution add three bacterial growth inhibitors: 10 mg streptomycin, 1 g sulfadiazine, and 25 mg tetracycline.

Have students observe and record the kinds of populations they find in the inoculums from bay mud with those from beach sand. Also distinguish colonies of bacteria, yeasts, or fungi by microscopic examination using the Gram stain, and by preparing hanging drop slides. Also refer to marine bacterial decomposers and the Winogradsky column (p. 607).

Artificial substrate sampling

If possible, examine communities of microorganisms that are attached to submerged surfaces (referred to as periphyton). Use plexiglass strips attached to submerged objects or a series of slides secured with snap clothes pins. Both phytoplankton and zooplankton can be examined on slides. A Sedgwick Rafter counting chamber ($50 \times 20 \times 1$ mm) may be used. Begin with a known volume, 50 or 100 ml of the periphyton suspension. Lay a coverslip diagonally across the Sedgwick Rafter counting cell; transfer 1 ml of suspension to the counting chamber and let it stand for 15 min to allow organisms to settle. Count the total of cells in each of 20 random fields. Millipore kits can also be used for counting colonies of bacteria (see Appendix).

Techniques used in water resources investigations of the United States Geological Survey are available with clear illustrations and descriptions in *Methods for Collection and Analysis of Aquatic Biological and Microbiological Samples*, K. V. Slack, et al., 1973, available from Supt. of Documents, U.S. Government Printing Office, Washington, D.C. 20402.

Chromatograms of marine algae

Marine plant pigments

While chlorophyll absorbs most light at the red region of the spectrum, other pigments extend the ability of marine plants to photosynthesize by absorbing and passing on energy from other wavelengths to chlorophyll. Sumich and Dudley describe one method for separating the chlorophyll pigments of marine algae from the accessory pigments which include phycobilins (blue phycocyanin in blue greens, and the reddish phycoerythrin of red algae).[10]

Carotenoids include carotenes and xanthophylls.

Place a quantity of *Ulva* (sea lettuce) or *Enteromorpha* in a blender and add 3 to 4 volumes of a solution of 20 percent ethyl alcohol and 80 percent acetone (*caution*). Blend for some 2 to 3 min until the plant tissues are homogenized, then centrifuge to obtain the green extract. Use the same technique for preparing strips of filter paper in a test tube as described for spinach or land plants, page 208 (Fig. 3-10). Apply spots of pigment repeatedly, allowing each drop to dry before applying the next. Some 50 or more drops may be needed depending on the concentration of the pigment extract. (*Caution*: work in a well-ventilated room.)

As a solvent in the test tube, use a solution (1:9) of 10 percent acetone and 90 percent petroleum ether (*caution*: good ventilation). In about 5 min, as the solvent moves up the strip, look for a narrow layer of carotene, the fastest moving pigment molecules, then the greenish yellow xanthophylls, the blue-green chlorophyll *a* and yellow-green chlorophyll *b*. Estimate the R_f of the different molecules.

Further readings

Techniques for collecting, preserving, and illustrating plants and animals are described in *Biological Techniques* by Jens Knudsen (New York: Harper and Row, 1966). For those who are seriously considering establishing marine aquaria, refer to the pamphlet, Carolina Marine Aquaria (available from Carolina Biological Supply Company).

Theoretical aspects of the forces governing ecological communities, mathematical models, as well as highly readable descriptions of organisms and the evolution of ecological systems are described by R. L. Smith in *Ecology and Field Biology*, (New York: Harper and Row, 1966). Students may be offered the *Scientific American* offprint, "Biological Clocks of the Tidal Zone," by J. Palmer (February 1975) for supplementary reading.

Preservative for flagellates

Prepare a modification of Lugol's solution by dissolving 10 g iodine crystals and 20 g potas-

---

[10] Sumich and Dudley, *Laboratory and Field Investigations in Marine Biology*. The authors also describe the separation of pigments of red algae in a chromatography column.

sium iodide in 200 ml of distilled water. A few days before use, add 20 ml of glacial acetic acid (*caution*); when stored in dark bottles the solution may keep up to a year. As a preservative, add this Lugol's solution to water samples in a ratio of 1 volume of solution to 100 volumes of seawater.

Water quality

A variety of kits comprising rapid tests are available from Hach[11] and LaMotte[12] to test water samples for hardness, pH, oxygen and carbon dioxide content, and for the presence of a number of minerals.

Hardness   The hardness of water is due mainly to the presence of calcium and magnesium ions. Drinking water averages about 250 mg/1 $CaCO_3$. When the alkalinity equals or exceeds the hardness, the hardness is caused by alkaline earth carbonate and bicarbonate salts.

Minerals and gases   While seawater contains traces of almost all elements, the most common ions are chloride (55 percent), sodium (30 percent), sulfate (7.7 percent), magnesium (3.7 percent) and potassium (1 percent). The Mohr method is used to determine chloride content; it involves titrating sodium chloride with silver nitrate. When potassium chromate is used as an indicator, an excess of silver precipitates as red silver chromate.

Carbon dioxide is present mainly as bicarbonates and carbonates. When test kits are not available, refer to *Methods for Collection and Analyses of Water Samples for Dissolved Minerals and Gases*, Book 5, Laboratory Analysis, by E. Brown et al., U.S. Geological Survey, 1970, available

from Supt. Documents, U.S. Government Printing Office, Washington, D.C. 20402.

You may recall that Sachs developed freshwater culture media for flowering plants. He found the main anions to be nitrates, sulfates, phosphates, and chlorides; the major cations are potassium, calcium, magnesium and sodium.

The amount of nitrogen and phosphorus influence the growth of phytoplankton in both marine and fresh waters. Collect a sample of water from a pond, lake, or ocean, then add a soil extract of nitrates and phosphates for enrichment. In a week or so there should be a good phytoplankton growth if the cultures are maintained in light and at an optimum temperature. Try to isolate the types of microorganisms by streaking the surface of sterile agar medium in Petri dishes to separate the algae accompanying the bacteria (p. 702). Also refer to water cultures, page 690, and to soil extract culture techniques, page 689.

pH of water   The pH of a substance indicates the ability of the substance to donate or to accept H^+ ions. Most waters have a pH range between 4 and 9. Many natural waters are slightly basic (pH 8), which is due mainly to the presence of carbonates and bicarbonates. The optimum pH for fish is between 7.8 and 8.5. (Also see buffering action, p. 159.)

Light penetration   Approximate readings of light penetration can be made with a Secchi disk. Lower the black and white disk into the water and record the depth at which the disk is no longer visible. Raise the disk and record the depth at which it is again visible. Average the two readings to determine the depth to which light penetrates.

Dip sticks   Use rapid test dip sticks for a semiquantitative determination of cations and anions in solution in drinking water, milk products, aquaria, meats (nitrate, nitrite) and in wine and food (sulfite); Information on EM Quants is available

[11] Hach Chemical Co., P.O. Box 389, Loveland, CO 80537, offers a good booklet describing 33 tests available for chemicals in water as well as pH, hardness, and dissolved oxygen, nitrites, and nitrates.

[12] La Motte Chemical Co., P.O. Box 329, Chestertown, MD 21620, offers booklets describing their tests for soil and water contents as well as test paper pH kits in 0.2 pH units and buffers.

from Manufacturing Chemists, Inc., 480 Democrat Rd., Gibbstown, NJ 08027.

Membrane filtration   Literature and listings of equipment for laboratory activities involving Millipore membrane filtration studies of air, water, cigarette smoke, and auto exhaust are available from Millipore Corp., New Bedford, MA 01730.

Dissolved oxygen (DO)   Oxygen content varies with the rate of photosynthesis by marine and freshwater plants and its availability from air. The solubility of oxygen varies with both temperature and mineral content. (Refer to light and dark bottle analysis, p. 586.)

The oxygen content of water decreases with a rise in temperature and density; the rate of flow of water also affects DO content. Test kits with modifications of the Winkler method for rapid field tests are available. However, the method is not recommended for heavily polluted waters containing a high content of chlorine or sulfites. Basically, the tests depend on the formation of a precipitate of manganous hydroxide. Manganous ions react with dissolved oxygen in an alkaline solution. Sodium azide is used to suppress nitrite interference with iodide. An acidified manganese floc is reduced by iodine to give free iodine (brownish color). Phenylarsine oxide is used to titrate iodine to its colorless endpoint.[13] A starch indicator is often added to increase the visibility of the change of light blue to colorless. Finally, the amount of dissolved oxygen is estimated from the amount of phenylarsine used in the titration.

The U.S. Geology Survey Booklet describes the Alsterberg azide method.[14] Many analytical tests are available in the literature. A fine authoritative source is *Methods for Analysis of Organic Substances in Water* by D. Goerlitz and E. Brown, U.S.

Geological Survey, 1972.[15] Other sources already mentioned are Sumich and Dudley,[16] and the fine Andrews series.[17]

Also refer to *Standard Methods for Examination of Water and Wastewater*, 15th ed., available from American Public Health Association; American Water Works Association, Water Pollution Control Federation. Further information concerning the publications is available from American Public Health Association, 1015 15th St. N.W., Washington, D.C. 20005.

Algae: pollution indicators   You may want to refer to C. M. Palmer, "A composite rating of algae tolerating organic pollution" (*J. Phycology* 1969 5:78) which contains a pollution index listing those algae that tolerate high quantities of organic matter and those found in clean water. Also refer to "Toxic Algae" in *Carolina Tips* (April 1974) for good background material on growth inhibitors produced by green algae, that affect their own growth as well as that of other algae (*Chlorella vulgaris* and members of the Volvocaceae).

In addition, consider pollution studies describing PCB's, sewage runoff, oil spills, detergents, pesticides, radioactive wastes, lead, mercury, and many more.

Air

Quality of air   Make both macroscopic and microscopic examinations of the discrete particulate substances in air such as pollen grains, spores, dust, encysted algae and protozoa. Line white paper plates with a thin film of petroleum jelly; coat microscope slides with a thin film; expose sterile Petri dishes of nutrient agar. Place some plates, slides and exposed dishes out

---

[13] Phenylarsine oxide, as a stable replacement for sodium thiosulfate, is available from Hach Co., P.O. Box 389, Loveland, CO 80537.

[14] E. Brown, *Methods for Gas Collection and Analysis*, Supt. Documents, Washington, D.C., 1970.

[15] Available from Supt. of Documents, U.S. Government Printing Office, Washington, D.C. 20402.

[16] Sumich and Dudley, *Laboratory and Field Investigations in Marine Biology*.

[17] Andrews, ed. *Contours: Studies of the Environment*, Four Guides.

of doors, others in the classroom, still others in streets of heavy traffic for a half hour. Incubate the Petri dishes along with an unexposed control plate for 48 hr. Examine the paper plate with a hand lens, and the slides under low and high power. When colonies appear in the incubated plates, distinguish colonies of bacteria from fuzzy colonies of molds. (There may also be airborne algae.) With a sterile loop, transfer samples to a drop of sterile water on a slide and spread into a thin film about the size of a dime. Fix by flaming the slides three or four times, then stain with Loeffler's methylene blue; also stain other slides with Gram stain. (When forceps are not available, use snap clothes pins to hold the slides.)

Materials for studies of pollutants in air as well as in water are available from the Millipore Corp.[18]

Biotic communities in soil

Rich organic soil teems with communities of both macrofauna and microorganisms. On a field trip, assign groups of students to collect soil from different habitats—surface leaf litter; surface soil to a depth of 15 cm; soil 30-cm deep; dry soil; boggy soil; moist soil; soil along a lake or pond shore; or other areas. Students will need to record the characteristics of the habitat, such as light intensity; compare soil exposed to constant light with soil of plants growing in shade. Obtain a small column or core of soil of known volume by inverting the open end of a can into loose soil.[19] Test pH and moisture content of the soil in the field. Use a kitchen sieve to shake out organisms from the leaf litter or use a box with a bottom of screening. Place each sample of soil in plastic bags; label the location and temperature. One group of students might collect samples of vegetation at different habitats and label them

for future reference. Spread leaves, stems, buds, flowers, seeds, and fruits in folds of newspaper and sheets of blotting paper and tie together in a plant press to dry the specimens for identification. When spreading leaves, fold one over so that the underside will be visible when dried. Use small nets to brush through vegetation to collect insects and larvae, spiders, sow bugs, termites, slugs, garden snails, and ticks (Fig. 8-20). Identify annelids—both earthworms and small white enchytraeids— often found in leaf litter. In class, use a Baermann funnel (p. 602) to collect enchytraeids.

Distinguish garden snails from land slugs (*Milax*); look for spiders, centipedes, millipedes, pill bugs, wood lice, and wingless forms, such as springtails and silverfish. There may well be ants and termites, as well as nymphs of cicadas. Many beetles and their larvae may be found— tiger, dung, and ground beetles. Also look for earwigs. Place soft-bodied forms in vials of alcohol (70 percent). Try to collect organisms of different types in separate vials so that a census may be taken in class after organisms are identified. Use Figures 8-20, 8-21, and 8-25 as aids in identification. Also refer to the survey of plant and animal kingdoms (pp. 2, 3). Consult such excellent guides in ecology as the series edited by Andrews[20], a laboratory manual for quantitative ecology by Darnell[21], a reference for collecting and preserving plants and animals by Knudsen[22], as well as the comprehensive reference by Smith[23], and Odum.[24] (Also refer to end of chapter listings.)

Macrofauna in soil and leaf litter   Spread out samples of soil in a large white tray to identify and count the

---

[18] Millipore Corp., New Bedford, MA 01730.
[19] M. Tanton, "A Corer for Sampling Soil and Litter Arthropods," *Ecology* (1969) 50:1.

[20] Andrews, ed. *Contours: Studies of the Environment*, A Series of Four Guides.
[21] Darnell, *Organism and Environment*.
[22] J. Knudsen, *Biological Techniques*, (New York: Harper and Row, 1966).
[23] Smith, *Ecology and Field Biology*.
[24] E. Odum, *Fundamentals of Ecology* 3rd ed., (Philadelphia: Saunders, 1971).

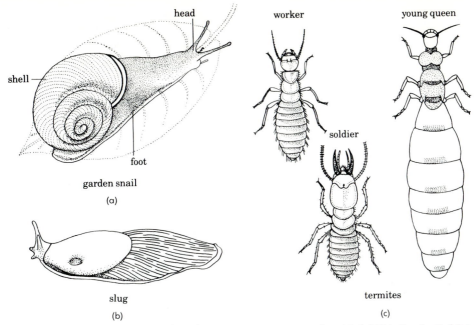

A. Andrews, ed. Copyright © 1973 by Prentice-Hall, Publishers.)

Fig. 8-20 Organisms found in soil: (a) garden snail; (b) common slug; (c) members of a termite colony. (Adapted from *The Study of Soil Ecology*, William

kinds of organisms in a 15-cm square of soil from two depths in the field (15 cm and 30 cm). With forceps, place actively moving organisms in a closed container for further study of behavior, or preserve in 70 percent alcohol (or killing jars for winged insects, p. 641). Label each jar as to habitat and time of day.

Extract small arthropods from leaf litter using a Berlese funnel (Fig. 8-22) or a modification, a Tullgren funnel. Prepare the Berlese funnel by rolling a file folder into a cone and stapling the edges as shown. Support the cone in a ring stand and support ring, and set a kitchen sieve or other bed of coarse mesh on top of the cone. Crumble leaf litter into the sieve. Shine a light with a 40-watt bulb about 8 cm above the leaf litter. Half fill a beaker with water and methanol (1 part methanol to 9 parts water) to catch the soil arthropods that move away from the heat source and fall through the funnel. Leave the funnel at room temperature; examine daily over a period of three to four days. Try to identify the specimens using Figures 8-20 and 8-21 as a guide.

Microorganisms in soil

Nematodes   Some of these small round worms in soil may be parasitic, but some are saprophytic. Isolate nematodes using a Baermann funnel apparatus as shown in Figure 8-23. Attach a length of rubber tubing to the funnel and apply a pinch clamp. Fill the glass funnel with water and place a 200 mesh sieve or other screen layer over the water. Larvae of nematodes are attracted to warm water, about 45°C (113°F). Line the sieve with wet-strength paper tissue or cloth and sprinkle a uniform layer of soil in the paper lining. Fold the tissue over the soil and pour enough water into the funnel so that the soil is just partially submerged. Over a period of 24 to 48 hrs nematodes should collect in the stem end of the funnel if they are present in the soil sample. Open the pinch clamp to get a quick flow of water into a tube or finger bowl. Examine the water for nematodes with a dissecting microscope; transfer some to a slide and apply a coverslip. Examine for details of anatomy under high power; use Figure 8-24 as a guide.

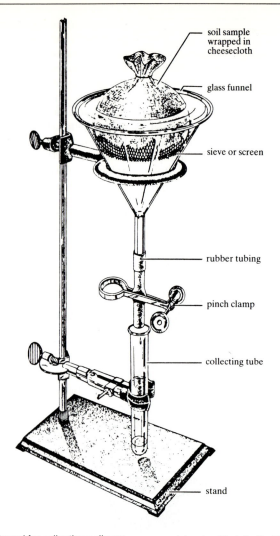

soil sample
wrapped in
cheesecloth

glass funnel

sieve or screen

rubber tubing

pinch clamp

collecting tube

stand

Fig. 8-23 Baermann funnel for collecting soil nematodes and other small worms such as enchytraeids. (Adapted from *Life in the Soil: High School Biology*, a laboratory block, by David Pramer. Copyright © 1965 by Heath Co.)

bacteria, add a loopful of soil water to a slide containing a dry drop of methylene blue. Apply a coverslip and watch the dye diffuse and stain the bacteria.

Population counts   Refer to dilution techniques (p. 458) and for plating samples in special medium (p. 603).

Soil water cultures may be maintained in small jars such as baby food jars to which a bit of hard-boiled egg yolk is added for enrichment. The enriched medium will support many microorganisms that are more limited in the soil community. Make counts of the types of microorganisms; use a scanning technique to make counts of 5 to 10 fields (p. 462).

Make population counts every third day and observe possible succession of microorganisms in this tiny ecosystem. As the bacteria population increases, small soil rhizopods and flagellates that live in films of water attached to soil particles increase in number. Few ciliates are found in

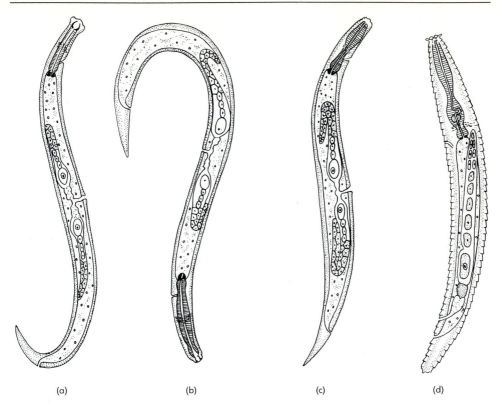

(a)   (b)   (c)   (d)

Fig. 8-24 Some types of soil nematodes:
(a) *Mononchus*; (b) *Plectus*; (c) *Rhabditis*;

(d) *Criconema*. (From *Plant Nematology* by Taylor
Jenkins. Copyright © 1967 by Reinhold Publishers.)

soil, but some may appear in this enriched
medium.

Some flagellates and amoebae that may
be found in soil are shown in Figure 8-25.
Some unusual soil amoeba, such as *Naegleria*, move by means of one pseudopodium; they encyst if the soil becomes dry.
In very wet soil, however, two posterior
flagella are extruded and away swim the
amoebae—a nice example of adaptation
to the environment.

A quantitative study of
microorganisms in different soils

Collect samples of soil from a field, a rich
garden soil, and a sandy loam soil. Weigh
out 50 g of each sample and transfer to
three separate 500-ml flasks of sterile
water. Shake well and let stand to allow
large particles to precipitate. Now dilute

each of the three labeled flasks of soil.
An acceptable limit for accurate counts
is 30 to 300 colonies per plate; serial
dilution is necessary before plating suspensions of high density. Use the dilution
technique as described on p. 458, and in
Figure 6-79. In the initial sample of 1 ml
suspension, use a dilution factor of 1: 10 or
1: 100 (that is, add 1 ml soil suspension to
9 ml sterile water, or 1 ml to 99 ml water).
Complete the dilutions of each sample one
at a time; or have teams of students work
together, each diluting one type of soil.

Plate all the Petri dishes with melted
sodium caseinate-glucose agar. (Use a water bath at 45 to 50° C (113 to 122° F).
Rock the seeded plates to gain a uniform
distribution of microorganisms in the agar.
Allow to solidify; invert the dishes and
incubate at some 25 to 30° C (77 to 86° F).
Use a hand lens to count colonies after

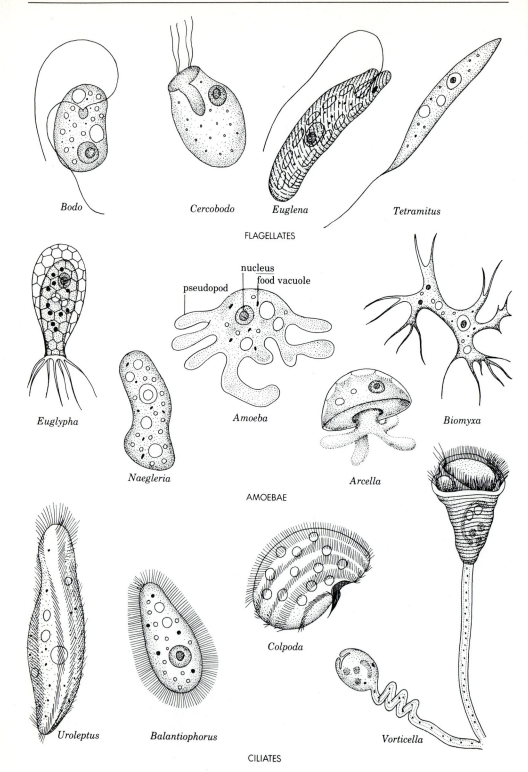

Bodo

Cercobodo

Euglena

Tetramitus

FLAGELLATES

Euglypha

Naegleria

pseudopod

nucleus

food vacuole

Amoeba

Biomyxa

Arcella

AMOEBAE

Uroleptus

Balantiophorus

Colpoda

Vorticella

CILIATES

Fig. 8-25   Some soil protozoa: (a) flagellates; (b) rhizopods; (c) ciliates. (Adapted from *Life in the Soil: High School Biology*, a laboratory block, by David Pramer. Copyright © 1965 by Heath Co.)

three and seven days. Place the dishes on a background of graph paper if a colony counter is not available. Try to use a mechanical tally counter.

Although the medium is not optimum for their growth, some colonies of molds and actinomycetes may appear. Molds will be fuzzy or cottony and actinomycetes often are dull white or colored, with small raised colonies. Tally the number of colonies of each kind of microorganism in the most diluted plates. Are certain soils of those tested more favorable for the growth of microorganisms? Have students tabulate their data and interpret the results.

PREPARATION OF SODIUM CASEINATE-GLUCOSE AGAR   Dehydrated medium is available from Difco, or combine the following:

sodium caseinate	2.0 g
glucose	1.0 g
dipotassium phosphate	0.2 g
magnesium sulfate	0.2 g
ferrous sulfate	trace
agar	15.0 g
tap water	1 liter

Also refer to the cultivation of soil algae and soil bacteria using Bold's or Pringsheim's soil water cultures (p. 689).

Soil algae   Sprinkle a little soil in sterile Petri dishes of Pringsheim's or Bristol's medium (p. 691); use either as liquid medium or solidify with agar. Maintain in light. Examine dishes for colonies after about a week. To isolate the algae, touch a sterile transfer loop to a green part of a colony (there are bacterial colonies too) and streak a sterile agar medium plate.

You may want students to make counts of cells in a given volume of suspension (1 ml pipette) using a slide scanning technique (p. 462), a hemacytometer (p. 276), or a Sedgwick Rafter chamber as used for counting phytoplankton. (Fig. 8-26.)

Slides submerged in soil   Pramer suggests the following simple technique to gain a qualitative examination of soil populations that interact in *close proximity* in their natural environment.[25] Prepare several beakers, each with 200 g of soil samples. Enrich the soil of some samples

[25] D. Pramer, *Life in the Soil*, BSCS. Laboratory Block, (Boston: Heath, 1968).

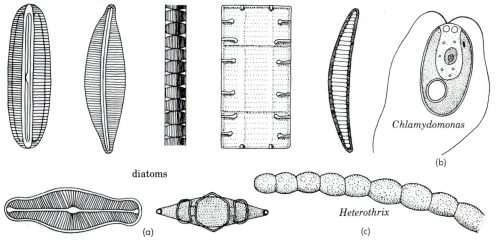

diatoms

Chlamydomonas

(b)

Heterothrix

(a)

(c)

Fig. 8-26   Some types of soil algae. (Adapted from *Life in the Soil: High School Biology*, a laboratory block, by David Pramer. Copyright © 1965 by Heath Co.)

by adding organic matter to some beakers, in this case, powdered cornflakes. Add 1 g of finely ground cornflakes per beaker and mix well with soil. To all the beakers add enough water to wet, but not to soak the soil. With a spatula make a slit deep into the soil in each beaker and insert clean slides in these slits. Close the beakers with plastic or aluminum foil and set aside at 28° C (82° F) for seven days. Tilt the slides when removing them from the soil so the attached populations of microorganisms on one side of each slide are not disturbed. Wipe the bottom of each slide and remove large soil particles by tapping the slides held at a slant. First scan the slides for living organisms that may be found in close proximity in their natural habitat. You may want to isolate populations for further examination. Divide a slide into four quadrants to partition off those microorganisms found in close association. With a sterile transfer loop, streak a sample from each quadrant across the surface of separate Petri dishes of sterile medium of soil extract agar enriched with an extract of corn meal. Label the dishes and incubate at 20 to 25° C (68 to 77° F) for two to four days.

Let other slides air-dry, then fix by passing them 3 or 4 times through a Bunsen flame. Stain with Loeffler's methylene blue, or use the Gram stain (page 145). Make a qualitative examination of each slide (preferably under oil) to observe the variety of populations that interact in close associations. Look for different shapes among bacteria as well as filaments of fungus plants.

Also prepare and stain slides of transfers from colonies isolated and incubated in separate Petri dishes (described above). Check for motility of microorganisms using a hanging drop preparation (p. 153). Compare all slides with controls (lacking enriched organic molecules). What is the effect of the organic enrichment on the populations of bacteria? Of fungi?

You may find nematode-trapping fungi among the microorganisms in some of the culture dishes. A bit of moist rotting wood in contact with soil can ensure a source of nematode-trapping fungi. Add a small bit of the wood to sterile dishes of corn meal extract agar medium. Scan the incubated plates; look especially for dead nematodes to locate the fungi that trap them.

Cellulose, soil bacteria, and fungi

Among the many soil microorganisms are bacteria, and mainly fungi, that secrete cellulase that decomposes the cellulose of dead plants in soil. Pramer suggests a method using filter paper to show the breakdown of cellulose residues by soil microorganisms as part of the carbon cycle.[26] Into each of 4 test tubes, place strips of filter paper 1 cm wide and 8 cm long. To each tube add 10 ml of the following salt solution:

K_2HPO_4	1.0 g
$NaNO_3$	0.5 g
$MgSO_4 \cdot 7H_2O$	0.5 g
KCl	0.5 g
$FeSO_4 \cdot 7H_2O$	0.01 g
distilled water	1 liter

Adjust to pH 7.5 with 0.1 M NaOH before autoclaving. Keep a part of the filter paper above the liquid as shown in Figure 8-27. Apply cotton plugs and sterilize at 15 lbs for 15 min.

In the meantime, prepare soil dilutions of 1/10, 1/100, 1/1000 and 1/10,000 of a number of collected soil samples. Use 1 ml pipettes to inoculate the sterile medium of each tube with the different soil dilutions. Use different pipettes for inoculating different dilutions in each tube. Incubate at 28° C (82° F). Examine every third day for two weeks for yellow or orange stains on the paper in the region of the liquid-air interface. Some soils may have a higher concentration of cellulose decomposers than others.

Tap the tubes that show cellulose decomposition against the palm of the hand; observe the filter paper begins to fall apart. For a microscopic examination of these decomposers, use forceps to transfer a bit

---

[26] Pramer, *Life in the Soil.*

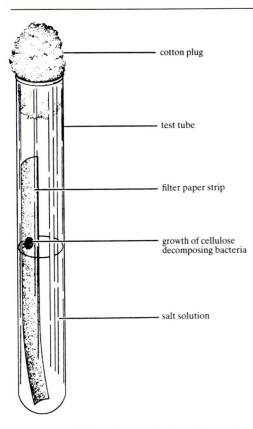

cotton plug

test tube

filter paper strip

growth of cellulose
decomposing bacteria

salt solution

Fig. 8-27 Cellulose-decomposing bacteria grow-ing on filter paper. (Adapted from *Life in the Soil: High School Biology*, a laboratory block, by David Pramer. Copyright © 1965 by Heath Co.)

of the fragmenting paper to a slide and tease apart the fibers. Apply ethanol to fix the slides. Stain with a solution of 1 per-cent erythrosin and 5 percent phenol in water. After 2 to 3-min wash and then counterstain with 0.1 percent aqueous solution of crystal violet for 5 to 10 min. Air-dry and examine under high power or oil.

Or use a chemical test. Add a few drops of zinc chloroiodide solution to the surface of a decomposing strip of filter paper. Note the light color where cellulose has been decomposed; also test the solution on a fresh piece of filter paper to show the usual purple color reaction given by cellulose.

You may prefer to use commercial cellu-lase in these investigations; or use a cell-free fungus culture filtrate as described by E. Cowling and A. Kelman in "Measure-ment of Cellulase Activity of Plant Patho-gens using the Cellulase Soil Technique" (*Sourcebook of Laboratory Exercises in Plant Pathology*, (San Francisco: Freeman, 1967). In the same text, refer to "An Alkali-Swelling Method with Cotton Fibers for Detecting Cellulase" by P. Marsh.

Starch and soil bacteria   Many soil bacteria break down starches. To demon-strate this, prepare a nutrient agar medium to which cornstarch, potato starch, or arrowroot starch has been added before it is sterilized (p. 746). Plate sterile Petri dishes and after cooling, streak the sur-faces of several plates with transfer needles that have been dipped into different sam-ples of soil. Incubate the plates; after 48 hr, spread a film of Lugol's iodine solu-tion over the surface of each dish. Why do some areas surrounding bacterial colonies show a clear zone? What class of enzymes are found in these soil bacteria?

Bacterial   decomposers—Winogradsky column   Bacterial decomposers are a vital part of the marine ecology of mudflats or coastal wetlands. They live either on the surface or as anaerobes in deep mud. You may want to culture a variety of these microorganisms found in the soils of fresh waters or from seashore samples. Among many other references, clear descriptions for the preparation of the Winogradsky column are given by Sumich and Dudley,[27] and also by Stechmeyer[28] to examine autotrophic bacteria.

Into 100 ml of mud sediment, mix 5 g of calcium carbonate and 5 g of calcium sulfate. To this add organic carbon in the form of 10 g of shredded newspaper. Pack this enriched sediment so that no air bub-bles are present in a 500-ml graduate cylinder or other glass or plastic tube until it is two-thirds full. The cylinder should be

[27] Sumich and Dudley, *Laboratory and Field Inves-tigations in Marine Biology.*
[28] Stechmeyer, "The Winogradsky Column," *Carolina Tips*, September, 1978.

some 5 cm in diameter and at least 17 cm high. To this add seawater if the sediment is from a marine source, or distilled water, to form a 3-cm layer; leave about 5 cm as a layer of air. Seal with plastic wrap or aluminum foil. Keep the cylinder in the dark for 10 to 14 days to kill off any photosynthesizing algae that may be in the sediment. Then transfer the column to sunlight (or fluorescent and infrared lighting) for at least a month.

A well-prepared column may contain chemoautotrophic bacteria that synthesize their nutrients from inorganic sources without light; those found in marine environments are anaerobic. There also may be anaerobic purple and green photosynthesizing bacteria that use hydrogen sulfide to reduce carbon dioxide and release sulfur. These bacteria lack chlorophyll; instead they contain the pigments grouped as bacteriochlorophylls and carotenoids. Locate the purple and green sulfur bacteria at the bottom layers where they use electron donors (hydrogen sulfide) to synthesize their carbohydrates. (In anaerobic respiration, sulfate-reducing bacteria produce sulfide, the electron donor.) Stechmeyer also describes the appearance of bacteria that decompose cellulose and other bacterial decomposition products, as well as the appearance of nonsulfur bacteria at the aerobic-anaerobic interface in the column.[29] Unlike Sumich and Dudley, she does not suggest leaving the column in the dark first to kill off algae. In her preparation, the watery top layer contains an aerobic layer of interesting green algae, blue-greens, fungi, protozoa, and aerobic bacteria—a variety of typical soil inhabitants.

After a month, examine the living components of the column. Use a sterile wire loop to transfer a sample of the green or purple photosynthesizing bacteria to a slide and spread as a thin film in a drop of sterile water; examine unstained specimens. Then fix the slide by flaming 3 to 4

times and apply Gram's stain (p 145). Also examine fresh samples for motility using the hanging drop technique (p. 153). These autotrophic bacteria are all Gram negative. However, the green bacteria in this group are bacilli which lack sulfur crystals within the cells, while the purple photosynthesizers, which may be bacilli, cocci, or spirilla, mainly store sulfur within their cells.

Most chemoautotrophic bacteria living on inorganic medium in the dark are Gram negative.

In addition, microscopic examination of other chemoautotrophic bacteria in marine mud reveal motile, short, curved bacilli or spirilla; these are the desulfovibrios. These are the organisms that produce hydrogen sulfide in respiration, used by the photosynthetic bacteria described earlier—which also give organic muds the odor of rotten eggs.

Soil bacteria and the nitrogen cycle

Nitrification by soil bacteria   Compare the nitrification processes in several types of soil, such as rich garden soil and a loam or field soil. Test for nitrite formation from ammonium salts, and then, through further reactions, test nitrate formation from nitrites.

$$2NH_3 + 3O_2 \longrightarrow 2HNO_2 + 2H_2O$$

$$2HNO_2 + O_2 \qquad 2HNO_3$$

Add 5 loopfuls of garden soil to a flask containing 25 ml of the ammonium solution (preparation below). Repeat with field soil. Incubate the two flasks at room temperature for six to eight weeks. Use Figure 8-28 as a guide for those studies.

AMMONIUM SULFATE SOLUTION   Weigh out the following (*except* for the magnesium carbonate), and add to 1 liter of water. Sterilize the magnesium carbonate separately and add to the rest of the solution using aseptic techniques.

[29] Stechmeyer, "The Winogradsky Column."

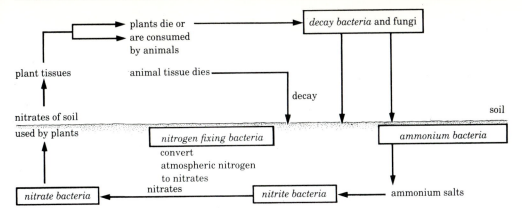

Fig. 8-28 Generalized nitrogen cycle in soil, freshwater, and oceans.

ammonium sulfate	1.0 g
dipotassium phosphate	1.0 g
sodium chloride	2.0 g
magnesium sulfate	0.5 g
ferrous sulfate	trace
magnesium carbonate	excess

Starting on the fourth week, test the flasks for the disappearance of ammonia and the *appearance* of nitrites. Use spot plates and Trommsdorf's reagent (preparation below) or a test kit as a test for nitrites. (Nessler reagent used as a test for the presence of ammonia is not recommended here since the reagent contains a saturated solution of toxic mercuric chloride.) On a spot plate, mix 2 drops of the reagent and 2 drops of H_2SO_4 (1:3). Now add a drop of the soil solution to be tested. The presence of nitrites gives a deep blue color.

PREPARATION OF TROMMSDORF'S RE-AGENT[30] Add 20 g of zinc chloride to 100 ml of water, heat and stir (avoid inhaling); slowly add 4 g of starch in water. Continue heating until the starch dissolves and the solution clears. Dilute with a little water, then add 2 g of zinc iodide. Now dilute to 1 liter and filter. Store in dark well-stoppered bottles.

After the fourth week, nitrites begin to disappear as nitrates are produced. When

[30] G. Peltier, C. Georgi, and L. Lindgren. *Laboratory Manual for General Bacteriology*, 4th ed., (New York: Wiley, 1952).

there is a *negative* Trommsdorf test for nitrites, test for nitrates. To test for the presence of nitrates add one drop of diphenylamine reagent and one drop of concentrated H_2SO_4 to a spot plate. To this add a drop of the soil suspension; a deep blue color results when nitrates are present. Since the presence of nitrites gives the same color, be sure to obtain a negative nitrite test before you test for nitrates.

It is advisable to purchase the diphenylamine reagent already prepared and follow the instructions for its use in testing for nitrates because the ingredients are toxic. You may want to demonstrate the test for nitrates and have students record the time needed for NH_3 to become NO_2 in both kinds of soils, and the time for NO_2 to be converted to NO_3.

Also consider having students prepare slides to examine the free-living nitrogen-fixing bacteria that are found in nodules on roots of legumes (see p. 634). Some students may want to inoculate soil with *Rhizobium* and sow seeds into this soil.

Antagonism between bacteria

While many bacteria are constantly added to soil, there is evidence that some bacteria produce substances that inhibit the growth of other bacteria. To demonstrate this, Carolina Biological Supply Co. offers a kit containing cultures of *Sarcina subflava* and a strain of *Bacillus subtilis* to be grown in one Petri dish. Carolina suggests the following procedure:

First, melt a sterile tube of prepared nutrient agar medium and let it cool to 45° C (113° F). Inoculate with a suspension of *S. subflava* and rotate it between your hands to distribute the bacteria evenly in the culture medium. Pour this uniformly seeded suspension into a sterile Petri dish and set it aside to solidify.

Next, inoculate this seeded plate, either by streaking or by touching a transfer needle at several places on the agar surface, with a strain of *B. subtilis*. Then incubate the plate or maintain it at room temperature. After 24 hr, inspect the plate; *S. subflava* cultures should be prominent over the agar except at the places where *B. subtilis* was inoculated. Look for the zones of inhibition (clear areas) around the areas where *B. subtilis* grows; apparently *subtilis* produces an antibiotic (a bacteriostatic action) that diffuses through the agar and inhibits the growth of *S. subflava*.

Effect of temperature on growth of bacteria

Prepare a nutrient agar medium to plate several Petri dishes (p. 700). Inoculate the cooling nutrient agar with a *pure* culture of harmless bacteria such as *Bacillus subtilis*, or the bacteria in sour milk. Then plate the Petri dishes so that each dish contains about the same quantity of bacteria and medium.

Place two or more dishes in a refrigerator at 4° to 8° C (39° to 46° F), two at room temperature (20° to 25° C [68° to 77° F]), and two in an incubator at 37° C (99° F). If you have a second incubator, keep two more Petri dishes at approximately 55° C (131° F). Check the temperature each day; observe and record the rate of development of visible colonies.

Effect of penicillin on growth of bacteria

Prepare several test tubes of nutrient agar medium (p. 700). While the medium is

cooling (45° C or 113° F) but still liquid, inoculate it with a pure culture of harmless bacteria and agitate by rolling between the palms. Then plate the seeded medium into several sterile Petri dishes. Prepared this way, the agar medium will have a uniform distribution of bacteria. As an inoculum, use a pure culture of nonpathogenic, Gram-positive bacteria (*Bacillus lactis* found in milk; or *B. subtilis* found in hay and soil), others with nonpathogenic, Gram-negative bacteria (*Pseudomonas hydrophilus* found on frogs suffering from red leg; *Achromobacter nitrificans* found in soil.

Now set two seeded Petri dishes aside as controls. To the other inoculated dishes add a few disks of filter paper about 5 to 6 mm in diameter that have been soaked in a solution of penicillin, prepared by dissolving a penicillin lozenge in cooled, sterile water; or soak the filter paper in a culture of *Penicillium* mold. Better still, purchase disks that have been soaked in specific antibiotics. After incubation of all dishes, students should find abundant growth of colonies only in the control cultures seeded with bacteria. The Petri dishes containing disks of the antibiotic, however, should show a clear zone, a "zone of inhibition," around the disks in some cases. Penicillin shows greater activity against Gram-positive bacteria, mainly cocci.

You may want to compare the action of several antibiotics at one time, or to find the quantitative differences in the spectrum of a single antibiotic. Bacto-Unidisks are sterile preparations containing specific amounts of commonly used antibiotics.[31] Many other commercial preparations are also available.

Effect of antiseptics on bacterial growth

You may also wish to demonstrate the bacteriocidal effect of antiseptics. Prepare nutrient agar medium (p. 700) and inoculate the cooling medium with a pure cul-

---

[31] Available from Difco Laboratories and other supply houses (see Appendix).

ture of harmless bacteria. (Refer to activity on penicillin, above.) Then plate the seeded agar in Petri dishes; in this way bacteria are distributed throughout the medium.

Now set aside several seeded dishes as controls. To the other dishes add selected antiseptics. To some dishes add a film of 2 percent tincture of iodine, or use filter paper disks that have been soaked in the antiseptic. To others add a film of 70 percent alcohol, to others a film (or disks) of full-strength Lysol, and to others diluted Lysol. You may want to include household mercurochrome (1 percent), as well as witch hazel and 3 percent hydrogen peroxide.

Incubate all the Petri dishes, including the control dishes. Look for the appearance of colonies. Compare the effectiveness of the various antiseptics used by counting the number of colonies and the size of each colony. Which chemicals were effective in a 24- to 36-hr period? Which chemicals were still effective after 72 hr?

Irradiation of bacteria

Wald et al. suggest an activity in which a dilute suspension of *Serratia marcescens* is spread over sterile agar.[32] The dish is divided into four quadrants by marking the bottom of the dish with a waxed pencil. Each quadrant is successively exposed to ultraviolet light (*caution*) of wavelengths near 260 mμ for 90, 120, and 150 sec (and one quadrant is left unexposed as a control). *Serratia marcescens* is especially effective for this study of possible mutations since several mutants occur that lack the normal brilliant red pigmentation. However all the Petri dishes must be incubated at the same room temperature (25° C or 77° F) since colorless colonies normally appear at higher temperatures (p. 551).

Atomic radiation In general, microorganisms have been found to be unusually

resistant to atomic radiation. A lethal dose for humans is less than 1,000 roentgens (r), whereas representative microorganisms have the following lethal doses:

Escherichia coli	10,000 r
yeast cells	30,000 r
Amoeba	100,000 r
Paramecium	300,000 r

Some characteristics of soil

The pH of soil Have students test the pH of small samples of soil from a swamp, burned-over area, woods, weedy region, soil along a fence row, a whitewashed wall, and rich garden soil. Cut a section of soil with a spade or tin can corer so that slices or cores are some 15-cm deep, and 30-cm deep. (A tin can corer provides dimensions for establishing the area of the sample.) Transport to class in a plastic bag to prevent drying out. Use soil testing kits such as La Motte's or Hach's to test for pH. Or, in a simple introduction, use litmus paper or brom cresol purple or other indicators to test pH and ways to modify the pH of soils.

Test several samples of soil from top and deeper layers. Lay moistened strips of indicator paper (if litmus, lay red and blue side by side) in Syracuse dishes or on slides or spot plates. Add a half teaspoon or less of the soil sample to the indicator strips. Add a drop of water to moisten the soil. Turn over the glass dish or slide to inspect the color of the indicator paper. Do deeper layers have the same pH as top layers of soil? Does pH have an effect on the absorption of mineral ions by root hairs?

Desert soils are usually alkaline; grasslands of the central United States are approximately neutral; and the forest regions of the eastern section of the country tend to be acidic. Soils that hold much rainwater tend to have sodium, calcium, and magnesium leached away and become acidic.

Demonstrate the effect of adding a bit of lime to a sample of acid soil. Repeat the

---

[32] G. Wald et al. *Twenty-six Afternoons of Biology,* (New York: Addison-Wesley, 1962).

pH test. You may want to "doctor" soil samples for laboratory studies in introductory work. Add a bit of vinegar, sulfur, tannic acid to a soil sample to show a test for an acid soil. Add lime (CaO) when you want alkaline soil samples.

Other indicators   Committees of students may use several different indicators to check soil pH. When using liquid indicators, better results are obtained if you use a soil water extract rather than soil samples.

Note the different pH ranges of indicators: use brom cresol green for a pH range of 3.6 to 5.2; brom cresol purple for a range of 5 to 7.5; brom thymol blue for a range of 6.0 to 7.6; phenol red for a range of 6.8 to 8.4. For other indicators, refer to page 156.

In testing, select 5 or 6 thin layers of soil some 15 to 30 cm from the surface. Sprinkle the sample with distilled water and let stand. Add 1 drop of the indicator from a dropper bottle to a white dish and 5 drops of the clear soil extract. Note that brom cresol purple has sharp color shifts within the pH range of most soils that support common plants: dark purple at pH 7.5 or higher, lavender at pH 7, reddish brown at 6, orange at pH 5.5 and yellow at pH 5 or lower.

To prepare the indicator, dissolve 0.2 g of brom cresol purple indicator powder in 50 ml of methyl or ethyl alcohol. Dilute with water to a volume of 500 ml. Add a few drops of limewater until a red color appears. Add a drop of indicator to 5 drops of the clear soil extract. (Add a few thymol crystals to the indicator if it is to be stored for some time.)

Students may recognize that the pH of soil—one factor among many—affects the type of plant community existing in different soils. In fact, an experienced observer can estimate approximately the pH of the soil from the kinds of plants growing in a given region. Table 9-2 shows some of the plants that thrive in soil of three different pH ranges.

In what way do the plant communities change when the minerals are constantly leached out of the soil? A student looking for an interesting investigation will want to read further in a text on soilless growth of plants to learn how the ability of plants to absorb minerals is affected by the pH of the soil (hydroponics, p. 478).

Soil profile   How does the top layer of soil differ from subsoil and bedrock? If possible, have students examine an excavation under way, or a hillside cut through so that more than 5 ft of soil in profile is exposed. If you live in a desert area, compare the difference between the top layer of sand and the more granular, rocky layers several feet below the surface. What factors contribute to the formation of fine-textured sand?

In a study of so-called average soil, students may recognize the comparatively thin layer of dark topsoil. Under this layer, about 1½ to 2 ft down, is a layer of subsoil. This rests on a deep (below 5 ft) layer of bedrock. An examination of the content of the soil will reveal that living organisms are found only in topsoil (p. 605).

How does the texture of the usual topsoil differ from subsoil? What kind of soil holds water best? You might want a committee of students to report on the way that bedrock, like granite and basalt, was formed. What factors change rock into soil? For example, what effect does each of the following conditions have on the breaking down of rock into soil: weathering (such as alternating heat and cold), winds, water in motion, plant roots, and ground water?

When they study fragmentation and exfoliation of rock, students can demonstrate some factors responsible for the breaking up of rock masses. Pour a dilute acid such as vinegar on limestone. This reaction might simulate the action of carbonic acid dissolving in ground water. Have students bring to class bits of rock on which mosses or lichens are growing. Notice how readily the rock crumbles? Can students find examples to show how the outer layers of rock "peel off" after they have been subjected to alternating contraction and expansion due to cold and heat? If a rock were cracked, what effect would freezing water that is trapped in the cracks have on the rock?

Conversely, how does soil form sedimentary rock?

How leaves add to the soil

Have students ever noticed the sponginess of soil, particularly in a wooded region? Perhaps such a region is near enough for a short field trip. Examine the cushionlike layers of leaves on the ground. Notice that the top layers of these leaves can be identified. Lift up layers of leaves. Observe that only veins of the underlying leaves are apparent. In layers below this, only fragments of leaves can be found. Now squeeze a mass of this material in your hands. Do you feel the moisture? Students may have had similar experiences in observing the changes in a compost heap (made of grass cuttings and leaves) around their own homes.

Which organisms in the soil aid in the rapid change of dead leaves into humus, giving this absorbent quality to the topsoil? How would the nitrogen cycle (as well as cycles of other minerals) be affected if bacteria of decay and soil fungi disappeared? For studies of the soil bacteria involved in the nitrogen cycle, refer to pages 634, 635. Also refer to minerals and mineral deficiencies in plant growth, page 479. Macronutrients for plants are commonly salts of calcium, magnesium, and potassium (cation); phosphorus, nitrogen and sulfur are used by plants in the form of phosphates, nitrates (also as ammonium ions), and sulfate anions.

CO_2 production by roots

Germinate fast-growing seeds such as sunflower, pinto bean, or pea. Into several test tubes, add an equal volume of water and 3 drops of brom thymol blue so that the water turns pale blue. Insert a thin layer of absorbent cotton into each tube to support seedlings; immerse the roots of the seedlings in water. Why does brom thymol blue turn yellow? Do different kinds of seeds give off the same volume of CO_2 over a given period? What effect would the increased CO_2 have on the pH of the soil? What is the effect on a bare rock as pioneer plants survive and establish themselves?

Soil temperature

How much of a difference is there in the temperature of soils in shaded and sunny regions? How might this affect the kind of community of living things that can survive?

Students may devise ways to use a thermometer to take readings of soil samples. For instance, check the readings of several thermometers beforehand by placing them in ice water and in boiling water. Keep a record of any degree of error in the readings among the different thermometers so that fractions of a degree (or degrees) may be added or subtracted from the readings students take. Now lay one thermometer so that the bulb touches soil in a shaded area, and a second on soil in a sunny region. Compare the temperatures of light-colored and dark-colored soil. Which surface warms up faster? Similarly, is there a difference between the temperature of soil on the surface layer and the soil several centimeters down? Tie a string on a thermometer and lower it into a hole from which a core of soil has been taken— some 15– to 30-cm deep; compare with the temperature of soil at the surface.

This kind of demonstration may be done at any time of the year. For example, students may take readings of soil temperatures under a blanket of snow—both at the surface and in lower regions.

Water-holding ability of soils

Prepare three funnels or glass chimneys with gauze taped or tied to the bottom. Pack one funnel with sand, another with clay, and a third with soil rich in humus. Then pour equal amounts of water into each funnel and watch the water run through the soil. If graduated cylinders are placed beneath the funnels, students can readily measure the amount of "runoff" water. Which soil permits water to run through most quickly? Which soil retains water?

Capillarity of soils   The texture of soil is a factor in determining how much water clings to particles of soil and is thereby readily accessible to the roots of plants.

Prepare the same materials as above but instead of pouring water into the funnels, pour an equal amount into each of the graduated cylinders. Be sure that the stem of the funnel reaches the water level in each case.

Watch the rate of water absorption in each preparation. Through which kind of soil does water rise readily?

Soil minerals   In his early studies of water cultures for flowering plants, Sachs found that the main anions for plant growth are nitrates, sulfates, phosphates, and chlorides. The major cations are potassium, calcium, magnesium, and sodium. The amounts of nitrogen and phosphorus influence the growth of all green plants, from algae to seed plants (as well as the phytoplankton in fresh and marine waters). Compare the growth of algae cultures in mineral deficient medium and mineral enriched medium. Also observe the growth of seedlings in deficient as well as mineral enriched soil or water cultures. Use seedlings of fast growing types such as mustard, pinto beans, sunflower, oats, or radish in small flower pots or in water cultures. Try growing seedlings in three kinds of soil: clay soil, sand, and rich loam containing humus. Use test kits to determine the amounts of minerals in each medium.

Compare the rate of growth of algae by cell counts. Measure in centimeters the height of seedlings receiving different amounts of minerals. Compare the greenness of the algal culture with controls and the color of leaves of seedlings. (Also refer to hydroponics, p. 478.)

Effect of nitrates on growing plants   To show the effect of mineral deficiencies on growing plants, have students place seedlings in jars of nutrient solutions (pp. 480–82), each of which lacks one mineral substance: nitrate, phosphate, or sulfate.

Compare the growth of seedlings in these deficient solutions with seedlings grown in complete nutrient solutions.

If a patch of sandy soil is available, an outdoor field activity may be devised. Locate an area where the grass seems pale (due perhaps to a deficiency of nitrates in the soil). Mark off a section several square feet in area, and evenly scatter over it about 10 to 30 g of sodium nitrate. (Emerson and Shields recommend about 30 g for a 10-ft^2 area.[33]) Then water this area immediately so that the nitrates will be absorbed into the soil. During the next few weeks look for differences in color of the grass in the treated area and the surrounding untreated area. Students should find that plants grow faster and that grass is greener than in the untreated control area.

You may want to relate this activity to crop rotation and the role of nitrogen-fixing bacteria in the soil and in the nodules of leguminous plants (p. 634).

Binding force of roots   Show that roots hold soil and thereby reduce erosion. Germinate many fast-growing seedlings in peat or plastic cups of moist sand or soil. Allow seedlings two weeks to grow, but water them sparingly, so that root systems become more extensive. Then, with a firm tug on the shoots, try to remove the plants from their containers. Observe how the sand (or soil) particles are held together by the root system so that the sand or soil takes the shape of the container. Why are ground cover plants grown on a slope?

USING   THE   BLACKBOARD   TO   SHOW TERRACING   Try this simple but effective way to show how soil absorbs water and helps to prevent floods. Dash some water on the blackboard and, as the small rivulets of water flow "downstream" along the board, dam these streams by pasting strips of wet toweling across them. Notice how the flow of water is stopped or slowed down. This demonstration may represent

---

[33] F. Emerson and L. Shields, *Laboratory and Field Exercises in Botany*, Blakiston (New York: McGraw-Hill, 1949).

terracing or strip cropping or the effect of a dam—whatever you wish.

RAINDROP EROSION What kind of soil is eroded easily by the force of raindrops? Students may set out disks of metal or wood on bare soil. After a heavy rainstorm they will probably find that the soil under the disks is at a slightly higher level than in the region of exposed soil (where some soil was carried away). Or check the amount of soil splashed up on stakes covered with white paper. In soil covered with vegetation there should be less splashed or transported soil than on uncovered soil. During what seasons of the year is it likely that more soil is transported from cultivated regions?

HOW MUDDY IS THE RUNOFF WATER? Another way to gain an understanding of how much soil is transported, or eroded, from any one place is to measure the amount of sediment in the runoff water. Water carries many minerals that have been leached from the soil in solution, but much of the soil is transported as a suspension.

After a rainstorm it may be possible for students to collect, in individual large bottles, samples of water running off from different regions. Allow the suspended particles to settle, and note the amount of sediment and the size of the particles carried by runoff water from soil where corn is cultivated, where wheat is growing, where there is no cover crop, where there is a gully, where land is plowed following its contour, and along a roadbank.

What is meant by sheet erosion? Elicit how silt is deposited at the foot of a slope. What is the value of cover crops in conserving soil? Explain how the continued existence of wells (and springs) is dependent upon cover crops.

STUDYING THE TREES IN YOUR COMMUNITY Students have learned that trees and cover crops help prevent floods and soil erosion. Does your community plant trees or conserve them in other ways?

On individual field trips, or on a group trip, students can make a count of trees, a tree census. What kinds of trees are there and how tall are they? Identify the trees and shrubs in the neighborhood. Which kinds of trees are used for special purposes, for example, for attracting birds to the gardens or fields, as windbreaks for cultivated land, as pioneers in occupying a new area?

List the many ways in which trees aid in conserving the resources of the region in which they live.

Preludes to inquiry: a study of a microecosystem

In addition to field work, or when outdoor field work is not feasible, small groups of students may work on several aspects of problems raised in the following series of investigations to support the concept that organisms are interrelated and interact with their environment.

Many major concepts in biology can be developed using a population of *one kind* of organism. Students may propose hypotheses, test their hypotheses by designing experiments, and develop skills in gathering valid data. In these investigations we use *Chlorella* because there is a wide research literature on this organism. Also, results of investigative work with *Chlorella* may be observed within a class period, or within a few days compared to changes in organisms such as seeds, frogs, and geraniums.

Chlorella pyrenoidosa is the species recommended for these investigations; at 30° C (86° F) it reproduces itself in some 20 hrs. There is a high-temperature strain that doubles in 7 hrs at 39° C (102° F). In good light, one milliliter of *Chlorella* cells of this strain can evolve close to 100 ml of oxygen/hr.

Somehow students are seldom really taught *how to inquire*; they are usually taught *how to confirm* what is known. They engage in "problem-doing," not "problem-solving."[34] For instance, consider the

[34] P. F. Brandwein, "Elements in a Strategy for Teaching Science in Elementary School," The Burton Lecture, (New York: Harcourt Brace Jovanovich, 1962).

study of photosynthesis and the use of the geranium plant in the classroom. Students "find out" that starch is produced (problem-doing); the solution to the problem is known and the student is, indeed, guided irrevocably to the known solution. Hence the activity may be considered "problem-doing" not "problem-solving." On the other hand, problem-solving should involve a confrontation of a "new" situation without a readily available solution in the text.

In the following block of investigations, a series of questions is given, and possible procedures are suggested, including some described in greater detail in other chapters. (See pp. 347, 349, 458, 462, 463, 465; and culturing, chapter 9.)

1. What is the average rate of reproduction of *Chlorella*?

    Isolate a few cells in a microaquarium (p. 460); make daily counts of the cells (p. 462) and clones (p. 465).

2. What are the effects of adding additional quantities of minerals such as nitrates, magnesium, or phosphates on growth and physiology?

    Prepare culture medium with *additional* amounts of these minerals in each stock preparation (see Knop's or others, p. 690). Conversely, what is the effect of nitrate *deficiency* or a lack of other substances on formation of chlorophyll? On rate of growth of *Chlorella*? Students may devise their own methods of investigation.

3. What is the effect of additional amounts of carbon dioxide on *Chlorella*?

    Introduce a small bubble of carbon dioxide into a microaquarium in a welled slide by means of a micropipette attached to a source of CO_2, such as a bottle of soda pop or dry ice. Be sure to establish controls. (See techniques for measuring growth, p. 455; preparation of welled slide, pp. 222, 460; indicators, p. 156.)

4. Do yeast cells have any effect on *Chlorella*?

Prepare a flask or bottle of yeast cells in dilute molasses to which two dried peas (as a protein source) have been added. Attach a delivery tube from this flask to a flask of *Chlorella*. Or on a sealed slide—a microaquarium—add a measured quantity of yeast cells to a drop of very dilute *Chlorella* suspension. Also prepare controls without the yeast cells. Is the difference between the controls and the *Chlorella* with yeast cells added measurable? For example, is there a greater intensity in the green color?)

5. For what organisms does *Chlorella* serve as a food niche?

    When students introduce *Paramecium* into a microaquarium of *Chlorella* and *Euglena*, they find that *Paramecium* feeds on *Chlorella* in preference to *Euglena*. What factors affect this preference? What is the preference of other heterotrophs such as *Vorticella*, rotifers, *Daphnia*, and *Aeolosoma*? What happens when the supply of *Chlorella* is consumed and *Euglena* is present in great numbers? What happens to the heterotrophs when the primary producers disappear?

6. How many *Chlorella* cells are needed to support a given number of heterotrophs? For example, to support 6 paramecia? Or 12 paramecia? Which multiplies faster, the food crop or the heterotroph? What factors affect balance in a microecosystem?

    Students may plan several designs for investigation using sealed welled slides: slide 1: 1 drop of rich *Chlorella* suspension added to 1 drop of a known number of *Paramecium*; slide 2: 1 drop of rich *Euglena*, plus 1 drop of a known number of *Paramecium*; slide 3: same as slide 1, but in a dark room; slide 4: same as slide 2, but in a dark room; slide 5: 1 drop of *Euglena*, plus 1 drop of *Chlorella*, plus three to six paramecia; slide 6: 1 drop of *Chlorella*, plus known quantities of *Paramecium*, plus *Didinium* (protozoan that preys upon *Paramecium*); slide 7: 1 drop of

Chlorella, plus *Aeolosoma* (segmented worm, p. 669) and *Daphnia*, or other mixed culture (p. 670); slide 8: same as slide 7, but without *Chlorella* (or controls) and in darkness.

7. Does *Chlorella* thrive better in a pure culture? Or are there biological or physical advantages in the combination of *Chlorella* with certain heterotrophs? Or with other autotrophs?

Here the design of the experimental techniques is left for students. What possible hypotheses may be testable?

8. Does *Chlorella* affect the pH of its environment? If so, does this affect the succession of other microorganisms in the habitat?

9. What gas is given off by *Chlorella* in sunlight? What factors affect this process?

Count the number of bubbles accumulating in a microaquarium (sealed welled slide) in a period of 15 min when it is exposed to light; also when it is kept in the dark. Other investigations concerning photosynthesis and wavelengths of light may be proposed; see chapter 3.

10. Is the usual biological succession of organisms affected (in a microaquarium or in a flask culture) by increasing the intensity of light? Or by decreasing the intensity of light? Could such alterations in the environment affect the survival of *Chlorella*? For example, growth of cover plants such as shrubs or trees might shade a small pond, or lightning might fell the shade trees.[35]

Students may design their own experiments. (See suggestions for measuring growth, chapter 6; photosynthesis, chapter 3; culturing solutions, p. 480.)

11. Can *Chlorella* synthesize any food materials in the dark?

[35] See J. Bonner, "The Upper Limit of a Crop Yield," *Science* (1962) 137:3523.

12. Are there photoperiods or rhythms of cell division in *Chlorella*? Does light intensity affect reproduction of *Chlorella*?

13. What is the optimum temperature for growth of *Chlorella*? What is the effect of culturing the cells at 40° C (104° F) for 6 hr? Can they thrive at 10° C (50° F)? Is it possible to adapt *Chlorella* to different temperature ranges? Is photosynthesis affected by the different temperature ranges? (See counting the bubbles of gas, p. 222.)

Students might carry on these investigations at home. Use a double boiler to maintain a raised temperature (use a thermometer), or add ice cubes to lower the temperature to 10° C (50° F).

14. Does washing a mass of *Chlorella* cells through fresh culture medium influence its rate of growth? What controls are needed in planning an investigation? (*Hint: Chlorella vulgaris* produces a toxic substance that inhibits its own growth rate (see soil bacteria, p. 609). Refer to R. Pratt "Some of the properties of growth inhibitors formed by *Chlorella* cells," *Amer. J. Bot.* (1942) 29 (142).

Students and teachers will surely think of many other problems that may be investigated using the many methods of intelligence. (Also refer to other Preludes, p. 229).

Modes of nutrition

Autotrophs or producers require a source of energy, water, carbon dioxide, and inorganic salts. Green plants and purple bacteria are photosynthetic autotrophs. However, some bacteria are chemosynthetic autotrophs obtaining energy from the oxidation of inorganic substances. For example, nitrite bacteria, *Nitrosomonas*, oxidize ammonium salts to nitrites; *Nitrobacter* oxidizes nitrites to nitrates. (Refer to flow chart, Fig. 8-28.) Some sulfur bacteria oxidize hydrogen sulfide to sulfates; iron bacteria oxidize ferrous iron to the ferric state. (Refer to Winogradsky column, p. 607.)

All animals, fungi, and most bacteria are heterotrophs requiring ready-made organic molecules as a food or energy source. Among those that are saprophytes, such as fungi and bacteria, soluble matter is absorbed through cell membranes. (Refer to extracellular digestion in molds, as well as in bacteria, p. 607.) Holozoic nutrition refers to the ingestion of solid food that needs to be digested.

Consider a variety of free-living heterotrophic modes of nutrition and the interrelationships of predator-prey, intraspecies competition (p. 553) and interspecies competition (p. 554). Also consider the special adaptations of organisms living together in symbiotic relations. Symbiosis is a general term including parasitism, commensalism, and mutualism. How have these special adaptations evolved in a coevolution so that certain DNA has survival value?

Intelligent development of the limited resources on earth requires a broad understanding on the part of students of the delicate strands that bind plants, animals, and their environment—air, soil, water.

Fig. 8-29   Indian pipe, a seed plant lacking chlorophyll, lives as a saprophyte.

Saprophytism

What is the role of saprophytes in food webs? Which ones are beneficial in that they return minerals to soil? Describe fermentation by yeast cells (pp. 310–13).

Expose bread, jams, boiled potatoes, or other foods such as citrus fruits to the air for several hours and then incubate. Or streak soil water extract across sterile nutrient medium in Petri dishes and observe the growth of soil fungi and bacteria (p. 601). Collect puffballs, mushrooms, shelf fungi. Examine cultures in sealed Petri dishes with a hand lens; make slides using transfer loops.

INDIAN PIPE   Find examples of the white seed plant Indian pipe (Fig. 8-29) growing in shaded regions, sometimes pushing up under masses of fallen leaves. It exists as a saprophyte, breaking down decaying vegetation and absorbing nutrient materials as mushrooms do.

Carefully transfer a whole cluster, with the soil in which it is growing, to a terrarium. When bruised, the plants turn blackish.

Symbiotic relationships

Plant parasites

Many bacteria and fungi that are parasitic on plants and animals have probably been described in context of other relationships in class. (Refer to *Agrobacterium tumefaciens* and crown gall formation in tomato plants (p. 619) as well as rusts and smuts (p. 619). The *Sourcebook of Laboratory Exercises in Plant Pathology*, compiled by a committee of the American Phytopathological Society (San Francisco: Freeman, 1967), has 227 exercises and covers a wide range of activities for many regions of the country.

Dodder   It may be possible to collect this orange-colored seed plant, which grows like a vine around a host plant such as goldenrod (Fig. 8-30). Dodder produces small clusters of white flowers and a large number of seeds. The seeds germinate and, as they grow, tightly entwine around a host plant. Small haustoria grow into the phloem tubes of the host and then the ground roots of the young parasite wilt. Examine a specimen and try to separate the parasite from its host plant. How is this relationship much like that of a tapeworm and its host? Why is dodder considered a parasite, while Indian pipe, also a seed plant lacking chlorophyll, is a saprophyte?

Some parasitic fungi   Blackberry rust is one of the easiest of the parasitic fungi to keep in the laboratory. Plant blackberries infected by rust in a window box; the rust, since it is autoecious, maintains itself on the plant.

Students also may collect several kinds of fungi, such as the powdery mildews which are parasitic on leaves of poplar, Virginia creeper, lilac, and willow. Others are found on cherry and apple leaves. Keep the leaves dried in envelopes until they are to be used; then soak the leaves in tap water for several hours and mount the scrapings carefully on clean slides in a drop of the fluid.

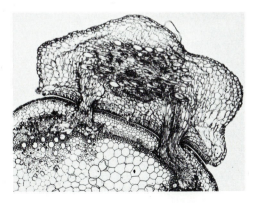

Fig. 8-30 Relationship of parasite to host: microscope slide showing haustoria of dodder growing into stem of host. (Photo courtesy of Carolina Biological Supply Co.)

Wheat rust is another important parasitic fungus. Trace its life history and economic importance.

Powdery mildew   Collect leaves of lilac or rose that show powdery mildew. Fold the edges of the leaves, which contain the mycelium of the powdery mildew on their surface, around a finger and use a razor blade to peel off a thin layer of epidermal cells. Transfer to a small drop of water for examination under high power. Also add a stain to the slide, such as cotton blue in lactophenol, to observe that haustoria of the mildew penetrate the epidermal cells.[36]

Many techniques for the study of plant pathogens are described in a pictorial *Atlas and Manual of Plant Pathology* by E. Barnes, (New York: Appleton–Century–Crofts, 1968). Current journals constantly offer new information and techniques.

Pathogenic bacteria and Koch's postulates

Show the action of a plant pathogen resulting in a plant tumor, or crown gall. Below is an activity illustrating Koch's postulates in identifying a specific organism responsible for a specific disease.

Use a 48-hr culture of *Agrobacterium tumefaciens* to inoculate tomato plants that are at least six to eight weeks old. However, at the start, make bacterial smears on slides about the size of a dime, a bit off center, so there is space for handling the slide. Air-dry the smears, then brush the slides through a flame some 5 to 6 times; stain with Gram's stain (p. 145). Label these slides and set them aside as reference slides of the *original* inoculum (Gram negative). Also examine hanging drop slides of these motile forms. Focus under high power (and oil if possible) on the edge of the drop and reduce the light.

Now to inoculate the young tomato plants, dip a sterile transfer needle in a

[36] C. Boothroyd, "Haustoria of the Powdery Mildew," in *Sourcebook of Laboratory Exercises in Plant Pathology*, (San Francisco: Freeman, 1967).

suspension of the bacteria and puncture the tissue near the stem nodes of a tomato plant. Flame the needle. Also puncture tissue of control plants with a plain sterile needle. Water the plants and maintain in sufficient light. After two to three weeks, look for swellings on the inoculated plants; these are the tumors or galls.

When galls have formed, it is necessary to compare their bacterial contents with the reference slides made from the original inoculum. To do so, use a sterile scalpel or razor blade to remove a gall from an infected plant. Dip the gall in household bleach[37] for a few seconds, then wash in water. Break open or crush a bit of the gall on a clean slide in a small loopful of sterile water. (It may be necessary to use a sterile mortar and pestle.) Again make thin smears, fix in a flame, and stain with Gram's. Examine these slides under oil and compare with the original reference slides. Also prepare hanging drop slides to compare with observations of the original inoculum. The bacteria in the galls of these infected plants should be the same as those in the original reference slides of the inoculum.

Maintain cultures of *Agrobacterium* by streaking several Petri dishes of sterile medium; invert and incubate at 30° C (86° F). Observe the plates every 24 hr for signs of colonies. After four to five days small colonies should be visible.

Animal parasites

Monocystis in Lumbricus   This gregarine (class Sporozoa) is a protozoan parasite found in the seminal vesicles of mature earthworms. There are many species and they differ in their life history. In one species, the adult trophozoite lives mainly in the coelom of the host; sporozoites move to the seminal vesicles, and there develop into mature trophozoites.

Narcotize an earthworm by placing it in 5 percent alcohol; then increase to 10

percent to kill the worms. Split open the body wall, between the 10th and 15th segments and pin back the skin to expose the white seminal vesicles (Fig. 1-47). Preserved earthworms may also be used. Mount a seminal vesicle in 0.7 percent sodium chloride solution and crush the contents on a slide which has a dry drop of methylene blue stain. Examine under low, and then high power. Look for elongated trophozoites. When sporocysts containing sporozoites found in soil are ingested by earthworms, these sporozoites penetrate the wall of the gut of each worm. They travel through the blood to the testes and become the trophozoite stage that grows in the seminal vesicles. Here the cigar-shaped trophozoites destroy sperm mother cells. Dormant sporozoites in cysts may also be found. Trophozoites later pair off and form round gametocytes. After fertilization, new sporozoites are formed.

Different species of large gregarines also inhabit the gut of grasshoppers and cockroaches. Slides may be prepared of the contents of the gut and mounted in saline at 22° C (72° F).

Rhabditis in earthworms   Look for this common nematode parasite in the nephridia of earthworms (Fig. 8-31) or use prepared slides.

Gregarina in Tenebrio   Gregarines are common parasites in the digestive system of many arthropods. A supply of mealworms (*Tenebrio*) provides readily available material. Place a larva of the beetle on a slide and with scissors cut off the last segment of the body. Use tweezers to tear off the head; the digestive system will be pulled out attached. Examine a length of the intestine on a slide without a coverslip under low power; and look for dark bodies of gregarines. If present, cut this section of the intestine and transfer to a drop of 0.5 percent saline solution on a clean slide; add a coverslip. It may be possible to observe the gliding trophozoite adult stage.

Nematodes parasitic in plants   Some larval nematodes penetrate stomates of

---

[37] 5 percent sodium hypochlorite

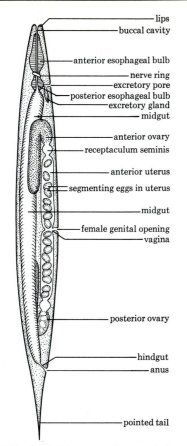

lips
buccal cavity

anterior esophageal bulb

nerve ring
excretory pore
posterior esophageal bulb
excretory gland
midgut

anterior ovary
receptaculum seminis

anterior uterus

segmenting eggs in uterus

midgut

female genital opening
vagina

posterior ovary

hindgut
anus

pointed tail

Fig. 8-31 *Rhabditis*, a parasitic nematode commonly found in nephridia of earthworms. (After W. S Bullough, *Practical Invertebrate Anatomy*, 2nd ed., Macmillan, London, 1958.)

leaves and feed on cell sap. However, most soil nematodes parasitize roots of plants or infect fruits.

Parasites in frogs

Frogs usually are heavily infected with many kinds of parasites—flukes, round worms, and protozoa. Dissect a freshly killed frog and separate the lungs, intestine and rectum, and urinary bladder into different Syracuse dishes or Petri dishes containing frog Ringer's solution (p. 155). Look for yellow cysts in the lining of the mouth; remove these and tease apart for examination under the microscope. Look

for encysted cercariae of the flatworm, *Clinostomum*; these become sexually mature in water birds that feed on frogs.

Flukes   Examine the lungs which may have black bulges containing flukes, or the flukes may be attached to the outside of the lungs. Tease apart these black regions from the elastic lungs to release the worms into saline solution. You may find the trematode, *Haematoloechus*, almost 1 cm long, a mottled fluke that may show red blood in the intestine consumed from its host (Fig. 8-32b). Some 40 species of this genus are found in frogs and toads. Dark areas in the fluke contain vast numbers of eggs. Why do parasites produce such large numbers of eggs?

Transfer one trematode to a dish of saline, another to a dish of spring water. After 5 to 10 min look for the extrusion of

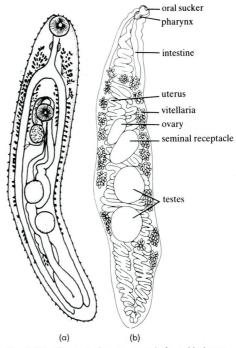

oral sucker
pharynx

intestine

uterus
vitellaria
ovary
seminal receptacle

testes

(a)   (b)

Fig. 8-32 Trematodes commonly found in lungs of frogs: (a) *Haplometra*; (b) *Haematoloechus*. (a, from *General Parasitology* by Thomas Clement Cheng. Reprinted with permission from Academic Press, Copyright© 1973. b, from *Laboratory Outlines in Biology—II* by Peter Abramoff and Robert G. Thomson. W.H. Freeman and Company. Copyright© 1972.)

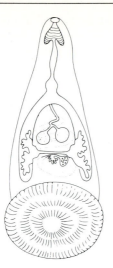

Fig. 8-33 *Diplodiscus*, a trematode found in rectum of amphibians. Note the large posterior sucker. (From *Laboratory Outlines in Biology—II* by Peter Abramoff and Robert G. Thomson. W. H. Freeman and Co. Copyright © 1972.)

eggs from the genital pore near the anterior sucker. In which solution are eggs extruded? The encysted cercariae of the fluke are found in dragonflies consumed by frogs. Under low power observe the muscular movement of the trematode. How many suckers are there? How does it respond to the stimulus of a probe? Also look for *Haplometra*, another trematode in the lungs of the frog (Fig. 8-32a). *Diplodiscus subclavatus*, another common trematode, is found in the rectum of amphibians (Fig. 8-33). Note the posterior sucker in this cone-shaped trematode. Cheng refers to *Megalodiscus temperatus* as one of the most frequently found trematode parasites of frogs in American laboratories;[38] it may be 6 mm long and 2 to 2.2 mm in thickness.

For an examination of intestinal and rectal flukes, cut open the intestine and rectum of the frog to dislodge flukes attached to the lining of the intestinal wall. Transfer a drop of the fluid to a slide; apply a coverslip and examine under high power.

[38] T. Cheng, *General Parasitology*, New York: Academic Press, 1973.

Some flukes have one anterior sucker, others also have a ventral sucker that is about one third of the way down from the anterior end of the body. Use Figures 8-34 and 8-35 to aid in identification of the parts of the flukes. Note especially the two cecae that comprise the intestine. (How do flukes differ from tapeworms in structure?)

Some six species of trematodes may be found in the urinary bladder of the frog. A common inhabitant is *Gorgodera* (Fig 8-34), a small trematode with a large ventral sucker. How does its movement, especially its anterior end, remind one of the character in Greek mythology?

When possible, also examine a prepared slide of a liver fluke, *Clonorchis* that inhabits the human liver (Fig. 8-35) note the bifurcated intestine. Compare with flukes of the frog.

Many flukes have an adult stage in a vertebrate host such as a frog, and other stages in their life cycle often in a mollusk. You may be able to tease apart a fluke on a slide in a drop of frog Ringer's solution to look for larval flukes or miracidia, the form that enters the mollusk host.

When time permits an extensive study, look for flukes in the kidneys, digestive

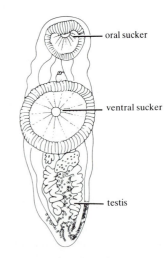

Fig. 8-34 *Gorgodera*, a parasitic trematode found in the urinary bladder of the frog. Note the large ventral sucker. (From *Laboratory Outlines in Biology—II* by Peter Abramoff and Robert G. Thomson. W. H. Freeman and Co. Copyright © 1972.)

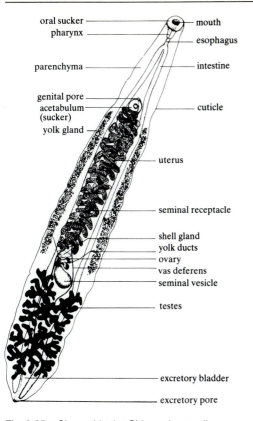

Fig. 8-35   *Clonorchis*, the Chinese human liver fluke. (Reprinted from *General Parasitology* by Thomas Clement Cheng. Reprint with permission from Academic Press, Copyright © 1973.)

glands, and pericardial cavity of marine snails and clams. *Cotylaspis* is but one example (Fig 8-36). If tree toads are available, look for the bladder fluke, *Polystoma* (Fig. 8-37).

If the West Coast sea urchin, *Strongylocentrotus*, is available, look for the parasitic turbellarian flatworm *Syndisyrinx* found in the intestine (Fig. 8-38).

Fascinating investigations may be undertaken into the protozoa parasites and the trematode parasites of many invertebrates and vertebrates, including sea urchins, clams, squids, and octopuses, as well as frogs, toads, and salamanders. Refer to the splendid work by Cheng, in *General Parasitology*, and to the laboratory manuals listed at the end of this chapter.

Nematodes in frogs   Look for many small wiggling nematodes in the lungs of the frog; *Rhabdias ranae* is one type present (Fig. 8-39). Other roundworms are found in the body cavity of the frog.

Protozoa in frogs

On the whole, there are fewer parasitic ciliates than parasitic amoebae and flagellates; the ciliates evolved later. Yet some ciliates are widespread in frogs.

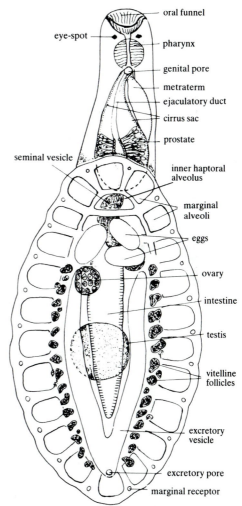

Fig. 8-36   *Cotylaspis insignis*, a fluke found in clams and snails. (Adapted and reprinted from *An Illustrated Laboratory Manual of Parasitology*, 5th ed., by Raymond M. Cable. Copyright © 1977 by Burgess Publishing Co., Minneapolis, MN.)

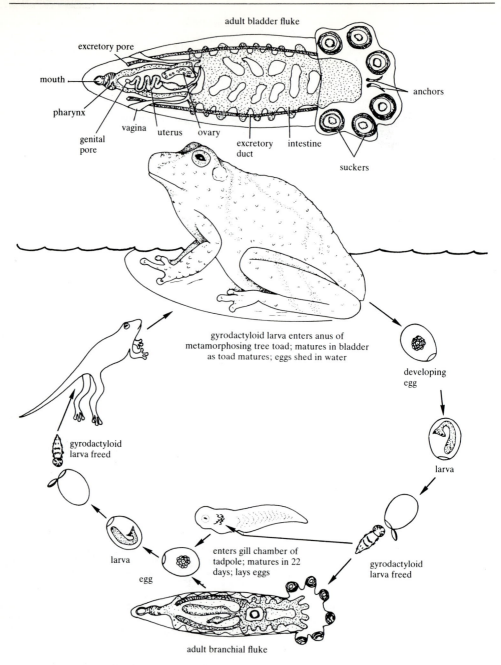

adult bladder fluke

excretory pore

mouth

pharynx

genital pore

vagina   uterus   ovary

excretory duct

intestine

suckers

anchors

gyrodactyloid larva enters anus of metamorphosing tree toad; matures in bladder as toad matures; eggs shed in water

developing egg

larva

gyrodactyloid larva freed

larva

egg

enters gill chamber of tadpole; matures in 22 days; lays eggs

gyrodactyloid larva freed

adult branchial fluke

Fig. 8-37 *Polystoma*: life cycle of the bladder fluke found in tree toads. (Adapted from *Biology of the* *Invertebrates* by Cleveland P. Hickman, 2nd ed. Copyright © 1973 by C. V. Mosby Company.)

Trichodina This ciliate may be found in the skin and gills of tadpoles, adult frogs, and toads (Fig. 8-40). These parasites also may be found attached to the gills of many freshwater fish. Some species also are found in the urinary bladder of frogs and toads.

Trichodina can be identified by the ring of hooked, toothlike structures on the disk or sucker found at the posterior end of the

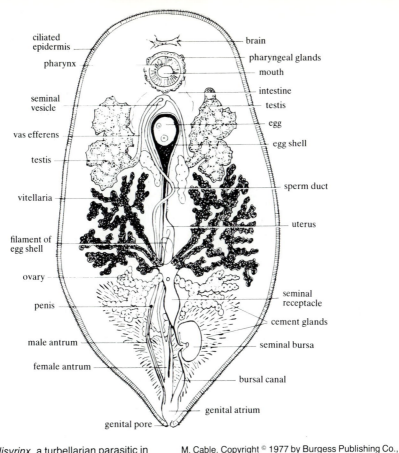

ciliated epidermis

pharynx

seminal vesicle

vas efferens

testis

vitellaria

filament of egg shell

ovary

penis

male antrum

female antrum

genital pore

brain

pharyngeal glands

mouth

intestine

testis

egg

egg shell

sperm duct

uterus

seminal receptacle

cement glands

seminal bursa

bursal canal

genital atrium

Fig. 8-38 *Syndisyrinx*, a turbellarian parasitic in sea urchins. (Adapted and reprinted from *An Illustrated Laboratory Manual of Parasitology*, 5th ed., by Raymond M. Cable. Copyright © 1977 by Burgess Publishing Co., Minneapolis, MN.)

cell. A large horseshoe-shaped macronucleus is clearly visible.

Opalina Examine a drop of fluid from the rectum of a frog under high power. Look for the large leaflike *Opalina ranarum*, which is flattened dorsoventrally and contains many small scattered nuclei (Fig. 8-41). It may be a commensal; it lacks an oral groove and a contractile vacuole. Some species may be 1-mm long and visible without a microscope. Opalina is covered with cilia arranged in diagonal rows so that it moves with a twisting motion. Another species, *Opalina hylaxena*, is found in the intestine of the tree frog, *Hyla versicolor*.

Opalinids are regarded as more primitive than true ciliates, more related to flagellates such as *Trichonympha*. They are often placed in a superclass of the Sarcomastigophora, organisms generally found in the colon of fish, amphibians, and reptiles.

Over evolutionary history, a finely tuned hormonal mechanism has been developed between parasite and host. During the nonbreeding stages of the host frog the active feeding trophozoite stage is found, which divides by fission (Fig. 8-42). However, in the spring, before egg-laying by the host frogs, these large opalinids multiply rapidly to produce many small forms that encyst, and are excreted in the wastes of the frog. These encysted forms, which sink to the bottom of the ponds, are the infectious stage consumed by tadpoles. Within the intestine of a tadpole, the

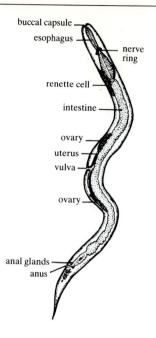

buccal capsule
esophagus
nerve ring
renette cell
intestine
ovary
uterus
vulva
ovary
anal glands
anus

Fig. 8-39 *Rhabdias ranae*, small parasitic nematode found in lungs of frogs. (From *General Parasitology* by Thomas Clement Cheng. Reprinted with permission from Academic Press, Copyright © 1973.)

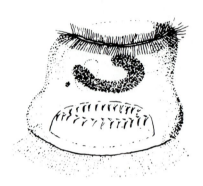

Fig. 8-40 *Trichodina*, a ciliate found in the skin and gills of tadpoles, of frogs, and toads. (From *General Parasitology* by Thomas Clement Cheng. Reprinted with permission from Academic Press, Copyright © 1973.)

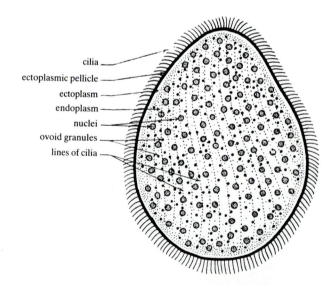

cilia
ectoplasmic pellicle
ectoplasm
endoplasm
nuclei
ovoid granules
lines of cilia

Fig. 8-41 *Opalina*, a parasitic ciliate found in the rectum of frogs. (Reprinted and adapted with permission of Macmillan Publishing Co. from *Practical Invertebrate Anatomy*, 2nd ed., by W. S. Bullough. Copyright© 1960 by W. S. Bullough.)

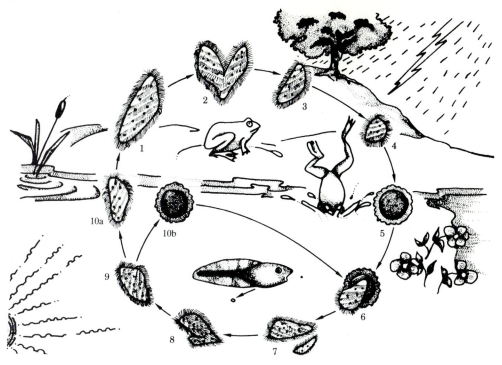

Fig. 8-42   Life cycle of *Opalina ranarum*: (1) to (3) active feeding trophozoite stages found in gut of frog; (4), (5) cyst forms, and is consumed by tadpole; (6), (7) cyst hatches with formation of gametes in tadpole's intestine; (8), (9) fertilization and zygote; zygote either continues as young asexual stage dividing by fission in gut of tadpole (10a); or encysts (10b) and is discharged into water and swallowed by another tadpole. (Reprinted and adapted with permission of Macmillan Publishing Co. from *College Zoology*, 10th ed., by Richard A. Boolootian and Karl A. Stiles. Copyright © 1981.)

LEFT | RIGHT

Fig. 8-43   Reproduction in *Opalina*. Left: fission; right: fusion of gametes (syngamy). (Reprinted and adapted with permission of Macmillan Publishing Co. from *College Zoology*, 10th ed., by Richard A. Boolootian and Karl A. Stiles. Copyright © 1981 by Macmillan Publishing Co., Inc.)

encysted opalinids emerge, go through a series of divisions that results in the formation of gametes, smaller microgametes that fertilize the larger macrogametes (Fig. 8-43). The resulting zygotes may encyst or develop into large populations of opalinids depending on the available food supply in the host's gut.

Cheng describes how the reproductive cycles of *Opalina* are influenced by secre-

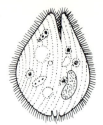

Fig. 8-44 *Balantidium*, a ciliate found in the colon of a frog. Note the anterior peristome. (From Laboratory *Outlines in Biology—II* by Peter Abramoff and Robert G. Thomson. W. H. Freeman and Co. Copyright © 1972.)

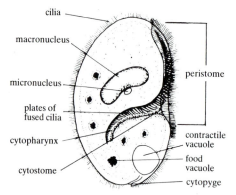

cilia

macronucleus

micronucleus

plates of
fused cilia

cytopharynx

cytostome

peristome

contractile
vacuole

food
vacuole

cytopyge

Fig. 8-45 *Nyctotherus cordiformis*, a ciliate found in the colon of a frog. (Adapted and reprinted from *An Illustrated Laboratory Manual of Parasitology*, 5th ed., by Ramond M. Cable. Copyright © 1977 by Burgess Publishing Co., Minneapolis, MN.)

tions of the gonadotropic hormones of the host frog[39]. Light and temperature trigger the hypothalamus, which stimulates the pituitary to produce FSH, LH (luteinizing hormone), and LTH (luteotropic hormone). The encystment of *Opalina* trophozoites is influenced by the secretion of these hormones. About two weeks before frogs enter the water to mate, the rate of fission of *Opalina* increases and small forms encyst.

The active opalinids, or trophozoites remain in the rectum of tadpoles through their metamorphosis into adult frogs.

Balantidium and Nyctotherus   Along with *Opalina*, two other protozoa may be found in the colon of the frog. One is *Balantidium duodeni*, an ovoid form with an anterior peristome circled with cilia

[39] Cheng, *General Parasitology*.

(Fig. 8-44). Other species are intestinal parasites of many invertebrates and vertebrates; *B. coli* for example, is found in humans.

Nyctotherus cordiformis, smaller than *Opalina*, is recognized by its lateral peristome opening into a cytostome and gullet. The region is lined with long cilia and a membranelle that follows the peristome to the cytostome (Fig. 8-45). Introduction of pregnancy hormones into host frogs results in the encystment of these ciliates. *Nyctotherus* is also found in cockroaches, centipedes, and millipedes.

Trichomonas, a flagellated protozoan, also may be found in the large intestine of the frog (Fig. 8-46).

As an in depth investigation, explore the parasites in frogs from different regions—northern and southern states—to compare both genus and species of flukes, and protozoa parasites.

Trypanosomes   From freshly killed frogs, use a pipette to draw up some blood from a large blood vessel near the heart and prepare a thin blood film on a clean slide. Examine the slide for flagellated trypanosomes (Fig. 8-47). Amoeboid parasites may be visible within red blood cells as in Figure 8-48.

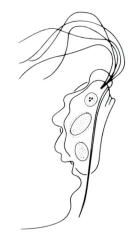

Fig. 8-46 *Trichomonas*, a flagellate found in the colon of frogs. (From *Laboratory Outlines in Biology—II* by Peter Abramoff and Robert G. Thomson. W. H. Freeman and Co. Copyright © 1972.)

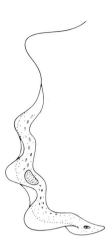

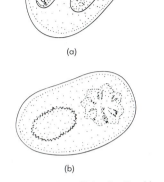

(a)

(b)

Fig. 8-47 Flagellated trypanosome. (From *Laboratory Outlines in Biology—II* by Peter Abramoff and Robert G. Thomson. W. H. Freeman and Co. Copyright © 1972.)

Fig. 8-48 Amoebae in red blood cells of frogs: (a) *Lankesterella*; (b) *Cytamoeba*. (From *Laboratory Outlines in Biology—II* by Peter Abramoff and Robert G. Thomson. W. H. Freeman and Co. Copyright © 1972.)

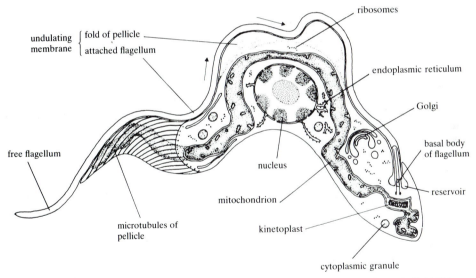

ribosomes

undulating { fold of pellicle
membrane { attached flagellum

endoplasmic reticulum

Golgi

basal body
of flagellum

free flagellum

reservoir

nucleus

mitochondrion

kinetoplast

microtubules of
pellicle

cytoplasmic granule

Fig. 8-49 Structure of a *Trypanosome* based on electron micrograph. (Adapted from *the Protozoa* by Keith Vickerman and Francis E. G. Cox, published by Houghton Mifflin Co. Copyright © 1967 by Keith Vickerman and Francis E. G. Cox.)

Parasites in humans

Plasmodium Use prepared slides to trace the complex life history of the specialized malarial plasmodium, *Plasmodium vivax*. One host is a cold-blooded, female mosquito, and the other a warm-blooded vertebrate in which the sexual cycle occurs. The mosquito is probably an evolutionary development.

Trypanosomes Examine prepared slides of blood infected with trypanosomes (Fig. 8-49). In cases of African sleeping

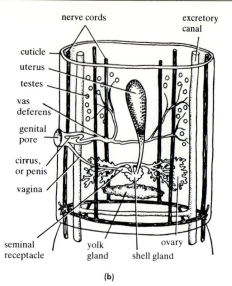

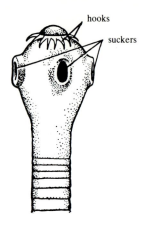

(a)

(b)

Fig. 8-50 *Taenia solium*, the pork tapeworm; (a) head or scolex showing hooks and three of the four suckers; (b) one segment or proglottid. (From *Life: An Introduction to Biology*, 2nd ed., by G.G. Simpson, C.S. Pittendrigh, and L.H. Tiffany. Used with permission of Harcourt Brace Jovanovich, Publishers. Copyright © 1957.)

sickness, the vector is the tsetse fly. In Chagas disease, trypanosomes reproduce in the heart muscle of humans after they are transmitted by blood-sucking bugs infected with the trypanosomes.

The undulating membrane and end flagellum of trypanosomes may be visible in some prepared slides of blood. In general, trypanosomes migrate from the gut to the salivary glands where they undergo several changes in development. In humans, the trypanosomes of West African sleeping sickness enter the cerebrospinal fluid and cause the coma of sleeping sickness. The same species is found in birds, reptiles, and domestic cattle.

Parasitic worms Use prepared slides to examine the proglottids and head of a tapeworm shown in Figure 8-50. Also identify the organelles of the Chinese human liver fluke, *Clonorchis sinensis*, shown in Figure 8-35. Compare its anatomy with observations of prepared slides of the sheep liver fluke, *Fasciola hepatica* (Fig. 8-51). Also compare with flatworm parasites of the frog.

Consider a study of human schistosomiasis which is prevalent in many parts of Asia, Africa and South America. The trematode *Schistosoma* is found in the blood of humans as well as other organisms (Fig. 8-52). Three species that infect humans are *Schistosoma mansoni*, prevalent in Puerto Rico, Dominican Republic, parts of Brazil and Africa; *S. japonicum*, found in the Philippines, China, and Japan; *S haematobium*, found mainly in North Africa in the valley of the Nile, Central and West Africa, the Near East, and southern Portugal. For extensive coverage of *Schistosoma* consult *General Parasitology*, by T. Cheng.

Develop a study of the life history of some of the roundworm parasites of humans. Examine prepared slides of *Trichinella* (Fig. 1-33) and the microfilarian worm *Wuchereria bancrofti* (Fig. 1-34). Compare these with *Ascaris* (Fig. 1-36), and with *Necator* (Fig. 1-32), the hookworm. Have students trace the life histories of these worm parasites that live in two hosts. For complete descriptions, refer to texts in protozoology, parasitology, and bacteriology. (See end of chapter listings for suggested authors and titles.)

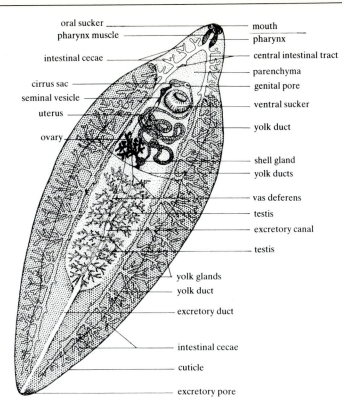

oral sucker
pharynx muscle
intestinal cecae
cirrus sac
seminal vesicle
uterus
ovary

mouth
pharynx
central intestinal tract
parenchyma
genital pore
ventral sucker
yolk duct
shell gland
yolk ducts
vas deferens
testis
excretory canal
testis
yolk glands
yolk duct
excretory duct
intestinal cecae
cuticle
excretory pore

Fig. 8-51 *Fasciola hepatica,* the sheep liver fluke. (Reprinted from *General Parasitology* by Thomas Clement Cheng. Reprint with permission from Academic Press, Copyright © 1973.)

Mutualism: symbiotic relationship

Symbiotic protozoans Some symbiotic protozoans may be found in the gut of earthworms, such as *Anoplophyra marylandensis*; others may be found in the intestine of *Stronglocentrotus*. The protozoan *Conchoptherus* is found in the mantle cavity of some clams. Crane describes the bacteriolytic effects of large numbers of the spirochaete *Cristispira* found in the crystalline style of certain pelecypods such as *Mercenaria mytilus* (cherrystone clam) and in *Donax*.[40] Transfer some of the spirochaetes into a young culture of *Sarcina luteum* in Petri dishes to study the bacteriolytic effect.

Trichonympha and termites The wood-digesting flagellate *Trichonympha* is found in the gut of many wood-eating insects that lack the enzymes to digest wood (Fig. 1-4). Many genera and species are found in termites as well as in woodroaches. It has been estimated that more than a quarter of the weight of worker termites is due to its flagellate contents. Here is a case of extreme mutualism in which neither organism can survive without the other.

It may be possible to locate a termite colony in a tree stump while on a field trip; such a colony can be maintained in the laboratory (p. 678).

Prepare a wet mount of these slow-moving, graceful flagellates in a drop of

[40] J. Crane, Jr. *Introduction to Marine Biology,* Columbus, OH: Charles Merrill, 1973.

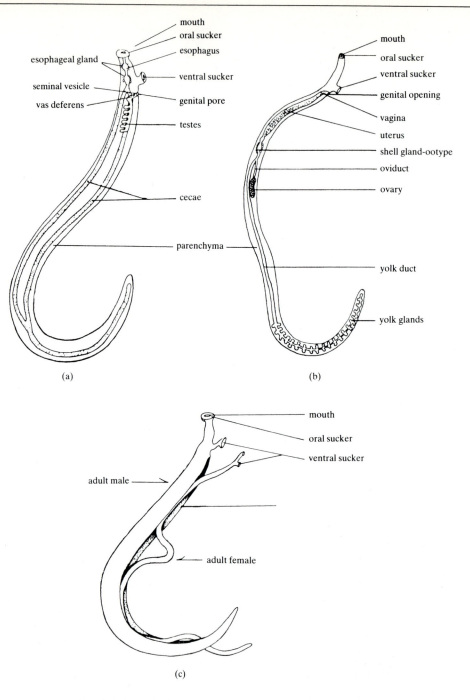

Fig. 8-52 *Schistosoma japonicum*, a trematode found in blood and other sites in humans and other organisms. (a) male; (b) female; (c) location of the female in the gynecophoric canal of male. (Reprinted from *General Parasitology* by Thomas Clement Cheng. Reprint with permission from Academic Press, Copyright © 1973.)

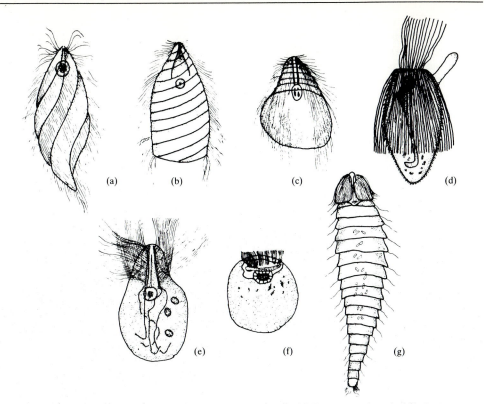

Fig. 8-53 Some flagellates in termites. (a) *Holomastigotes elongatum* (b) *Spirotrichonympha leidyi* (c) *Microspirotrichonympha porteri* (d) *Joenia annectens* (e) *Staurojoenina assimilis* (f) *Kofoidia loriculata* (g) *Teratonympha mirabilis* (Reprinted from *General Parasitology* by Thomas Clement Cheng. Reprint with permission from Academic Press, Copyright© 1973.)

saline solution (0.4 to 0.7 percent) or Ringer's solution (pp. 155–56). Using forceps, hold down the termite abdomen on the slide; with another forceps gently press down on the abdomen so that the gut contents flow into the salt solution. Or grasp the head of a termite with forceps and make an incision into the body to expose the small intestine and the many, many flagellates in its contents. Under high power, distinguish several species as well as genera among the flagellates. (Reduce the light to observe flagella; a drop of very dilute Lugol's iodine solution, or Giemsa stain may increase visibility.) Figure 8-53 illustrates some of the types that may be found in wood-eating insects.

Many of the flagellates belong to the Kofoidea (Order Hypermastigina) or to the Pyrsonympha (Order Polymastigina). They reproduce asexually by longitudinal fission and encysted forms can be found.

Back in 1925, Cleveland showed that when termites were defaunated they died within a few days.[41] However, when dying termites were reinfected with flagellates through oral feeding, digestion resumed. When termites molt, the chitonous lining of the gut is lost along with the

[41] L. R. Cleveland, "The Effects of Oxygenation and Starvation on Symbiosis Between the Termite *Termopsis* and Its Intestinal Flagellates," *Biol. Bull.* (1925) 48:455.

flagellates; after molting, reinfection occurs. *Trichonympha* is common in the woodroach, *Cryptocercus.* Encystment has been observed once a year when the woodroach molts. Small wood-eating beetles, sow bugs, and cockroaches have similar flagellates as well as amoebae or bacteria that serve the same function of digesting wood.

Adult flagellates found in the gut are monoploid; when these cells encyst, the chromosome number is replicated in immobile gametocytes—a very complex sexual cycle follows. Cheng[42] summarizes the experiments of Cleveland who found that when 100 units of ecdysone (the molting hormone that also stimulates development and differentiation of adult organs) were injected into 4th and 5th instar nymphs of *Cryptocercus*, the flagellate population in their gut completed its sexual cycle in seven to eight days instead of the usual forty or more days. Ecdysone, then, appears to initiate gametogenesis in the flagellates. The production of ecdysone is stimulated by ecdysiotropin, the brain or activation hormone of the insect.

Lichens   Collect the crusty gray patches of lichens from rocks or tree bark. (The green patches that may be found on tree bark are probably *Protococcus*, Figure 8-8.) Tease apart a bit of lichen in a drop of water in a Syracuse dish and mount on a slide for examination under low and high power. Locate the green algae enmeshed in filaments of fungi, a case of mutualism in which both organisms benefit: the algae carry on photosynthesis and the fungi absorb and hold water and minerals (Fig. 8-54). Together these two plants thrive on rocks, tree bark, or fence posts. Alone, each would have a limited range of survival (culturing, p. 713). Do different species of lichens have the same species of fungi and algae?

Rhizobium: Nitrogen-fixing   bacteria
These free-living bacteria exist in soil and

in nodules on the roots of legumes (Fig. 8-55). When decomposers in the soil break down the organic proteins of dead plants and animals, nitrogen is released in the form of positively charged ammonium ions; these cling to soil particles that are negatively charged. Roots of plants yield hydrogen ions that replace the positively charged ammonium ions (or others) that are absorbed by roots.

Thousands of nitrogen-fixing bacteria invade root tissue but do little harm. Examine *Rhizobium* in a study of mutualism by crushing a washed nodule between two glass slides. Add a drop of water to each slide and examine under high power. For more careful study, place a drop of methylene blue stain (p. 145) on each of several slides. Let the stain dry; store slides in a slide box so that they are ready for use. Add a drop of water containing nitrogen-fixing bacteria to the slide, apply a coverslip, and examine under high power. Notice how the stain gradually diffuses and stains the bacteria.

Or you may prefer to make a smear of the bacteria and stain the smear to prepare a permanent slide, as follows. Make a thin film of the bacteria by spreading the fluid of a wet mount along the slide with a dissecting needle or with another slide. Fix the bacteria to the slide by passing the slide through a Bunsen flame two or three times. Lay the slide across a Syracuse dish, and flood it with Loeffler's methylene blue for 1 min. Then rinse off the stain with water, let it dry, and examine under high power.

Inoculating soil with nitrogen-fixing bacteria (Rhizobium)   Add *Rhizobium* cultures to sterilized garden soil in several small flower pots and plant soaked peas, alfalfa, and/or clover seeds. Also plant similar seeds in pots of untreated, but sterilized soil.

Where is there a more luxuriant growth? When the plants are six to eight weeks old, uproot some plants and wash off the soil. Remove a nodule from the root with a sterile scalpel and forceps and crush in a

[42] Cheng, *General Parasitology.*

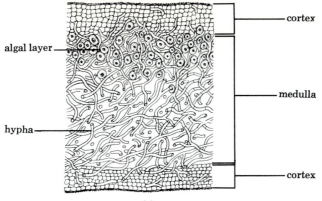

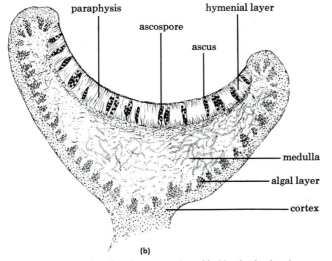

Fig. 8-54 *Physcia*, a lichen: (a) section showing association of algal and fungal elements; (b) sec- tion of fruiting body showing ascospores.

drop of water on a slide. Use a sterile transfer loop to mix in water; dilute with more water if the mix is cloudy. Make a thin smear over a clean slide about the size of a dime, air-dry, and fix in a flame. Stain with Loeffler's methylene blue (p. 145) for 2 to 4 min. Also prepare slides using Gram's stain (p. 145) and examine under oil.

Growth of colonies may be studied by brushing or streaking the surface of sterile agar medium (refer to an enrichment medium, p. 603). Also refer to studies in hydroponics, page 480.

Insectivorous plants   Several insectivorous plants may be found in bogs or purchased. Sundew, pitcher plant or Venus's flytrap (Fig. 9-31) grow well in acid soil.

In freshwater you may find *Utricularia*, a floating plant that has small bladder traps into which microorganisms swim and become captive (Fig. 8-56).

In all cases insectivorous plants are green and carry on photosynthesis. Yet there may be a deficiency in nitrates. The plants secrete enzymes that digest soft parts of the insects; this protein material

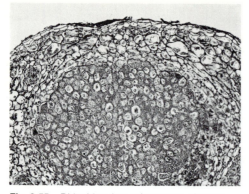

Fig. 8-55 *Rhizobium leguminosarum*, nitrogen-fixing bacteria. Slide shows cross section of nodule attached to root of leguminous plant.

apparently supplies the sources of nitrogen for plant growth.

Students may try to establish these plants in a terrarium that duplicates the acidic environment of a bog. Add some bits of charcoal to the tank to absorb odors. Introduce fruit flies, preferably vestigial ones, into the tank (p. 721).

When insectivorous plants are not available, you may select one of the many films that show the plants in action.

Other types of symbiosis

Part-time parasites Several organisms exist as partial parasites in the sense that they are not permanently attached to or do not live continuously within a host. For example, mistletoe carries on photosynthesis and absorbs water and minerals from the host plant; it sends roots that grow into tissues of the stem of the host tree. Compare the mode of nutrition in mistletoe and dodder (p. 619).

Many other examples come to mind; mosquitoes and leeches are also temporary parasites.

Parasites and several hosts Unraveling the complete food chain of some parasites and their hosts has taken years of intensive study. Often one stage of a para-

(a)

(b)

Fig. 8-56 *Utricularia*, the bladderwort: (a) view of aquatic insectivorous plant; (b) enlarged to show bladder traps. (Photo by Hugh Spencer.)

sitic worm might be found in a bird, another in a fish, and a third in a snail. In each host the parasitic worm was in a different stage of its life cycle, and these were not recognized as developmental stages of one organism. There are many examples of parasites that have two or three hosts. Have students report on the life history of the broad tapeworm of man (*Diphyllobothrium*), the pork tapeworm (*Taenia solium*), and the Oriental liver fluke (*Clonorchis*).[43]

What would be the effect on the food chain if a parasite destroyed its host? Explain why parasitic forms produce such large numbers of eggs.

Further examples   Students may come upon many cases of symbiosis, or commensalism, on field trips near school.

If there is a beach or bay in the vicinity, look for hermit crabs, which inhabit deserted gastropod shells—possibly whelk, periwinkle, or moonsnail. To the shells are attached sea anemones, sea lettuce, or other marine forms as a type of camouflage. The usually sessile anemones are carried about by the hermit crab to sources of food; and possibly the hermit crab has better chances of survival with these devices for camouflage. When the crabs are extricated from the shells, observe the degenerated abdomen of each crab. Notice how the abdomen is twisted to fit into the spirals of the mollusk shell. Students may also know of the harvestfish that swims among the tentacles of the Portuguese man-of-war.

In collecting specimens from ponds you may come upon green hydras or green paramecia (or they may be purchased from supply houses). In these animals live green algae in a symbiotic relationship. (See also *Paramecium bursaria*, p. 660 and green hydra, pp. 496, 666.)

Students may find ants and their "cows" or plant lice. These aphids or lice are tended by some species of ants that "milk" the aphids of a sugar solution they excrete.

[43] Refer to T. Cheng's *General Parasitology* for extensive in-depth studies.

A course in biomedical techniques

The following course outline describes briefly techniques in bacteriology and parasitology that have been successful with many students over several years. It is a companion course to the one described in chapter 4 in blood, blood cells, hormones and homeostasis, and kidney functions (p. 322).

Students discover for themselves whether they have the manipulative skills, patience, and ability to forego frustration in the introductory techniques offered. These two courses may be given over a school year as a form of practical career orientation.

I. *Bacteriology*

Identification of several bacteria from prepared slides, 2 × 2 slides, and film loops; preparation of slides from cultures of bacteria of decay, milk, sauerkraut juice, yoghurt, and also pigmented bacteria; Loeffler's methylene blue stain and Gram stain; preparation of sterile nutrient agar media; pouring slants and plating Petri dishes; identification of colonies; separation of mixed cultures by isolation techniques—streaking and pour plates (dilution); using a bacterium to repeat Koch's technique; transfer techniques; films and loops and readings on techniques and content.

II. *Control of Diseases*

Kinds of organisms causing diseases; classification of diseases; immunology (mechanisms of the body); cancer; cardio-vascular diseases; films, strips, loops, readings on individual diseases, and reports to the class on some caused by bacteria, viruses, protozoa, fungi, worms, genetic, and degenerative diseases.

III. *Current Science—Literacy in Science*

Readings and reports from current relevant papers in *Scientific American* as well as pamphlets from associations—respiratory diseases, cancer, environmental disorders, genetic disorders, and food additives.

IV. *Genetic Counseling*

Basic human genetics (human karyotypes); genetic disorders due to gene malfunction and extra chromosome disorders (or deficient chromosome conditions); role of amniocentesis; genetic counseling; recombinant DNA.

V. *Parasitology*

Life cycles—parasite and host among protozoa and worm diseases mainly in animals, and fungal and nematode diseases in plants; dissection of frogs for identification of trematodes, protozoa, and round worms; prepared slides of parasitic forms—trichina worm, hookworm, tapeworm; guest speakers from animal hospitals (often former students); films and loops; readings and sharing of information through oral reports.

VI. *Career Information in Health-Related Fields*

Where and how to get information about continuing schooling, cost; listings of associations that furnish information; job opportunities in civil service as well as the private sector.

VII. *Environmental Pollution*

Effect of pollutants in air, water, soil on the body; microscope examinations of effect of pollutants on ciliates, lichens, algae; effects of food additives, smoking, drugs, alcohol on the body with microscopic examination of their effects on ciliates; role of drugs in internal pollution of the body and social effects; population density and global view of sources of food and energy in the world today.

Going further

Nature trails around school

It may be possible to plan nature trails as a class or club activity. Some schools plant trees and shrubs that are especially useful in illustrating patterns of coexistence of organisms. Questions can be raised: Why do flowers reproduce at different times? Discuss photoperiods. Is there an advantage in having seeds germinate at different times? What kinds of variations exist among leaves on a tree? Is there a difference in the number of chloroplasts in cells of sun and shade plants? What kinds of plants can survive under the canopy of taller trees and shrubs? On a field trip, consider the evolution of flower patterns in magnolia, forsythia, cherry trees, maples of several varieties, dogwood, elm, oak, and many more. Try to include witch hazel which blossoms in the fall. Compare dandelions so readily available in many regions as complex composites or clusters of flowers, with the flower of the more primitive magnolia.

Students might build birdhouses in shop classes, or make attractive markers for a nature trail and erect a bulletin board showing a map of the trail or announcing interesting nature happenings for each month. And why not make a guide sheet for students to use on the nature trail?

Later, seedlings may be observed and tallied; investigate some aspects of the struggle for survival and favorable growth. Where there are no grounds or open lots to be found, surely students can find a crack in a pavement or rock that supports life. What are pioneer organisms? How is soil accumulated?

Try to transplant mosses, ferns, young trees collected from other regions to their comparable habitats on the school grounds. Perhaps there is a place for a make-shift lean-to greenhouse. A small pond may be constructed near school and stocked with plants and animals from a nearby lake.

A museum of plants and animals

It is useful to have a museum of living organisms in school. This may be a room or a portion of a laboratory, the size dependent upon the space available.

Students willingly bring living things to class; also, some living organisms may be purchased. Often some students know how to care for an aquarium. Student curators may be trained to maintain both freshwater and marine aquaria. They may be trained to prepare terraria of various kinds—one duplicating an arid environment with sand, cactus plants, horned "toads," and another containing a miniature woodland with humus in which ferns, mosses, lichens, and small wild flowers grow. Place a pan of water at one end of a terrarium and include frogs' eggs, tadpoles, or adult frogs or salamanders. (Toads mess a tank since they burrow.) A bog terrarium can be made from sphagnum moss and insectivorous plants. Add charcoal to absorb odors and reduce acidity (preparation, pp. 682, 685, 686).

Turtles, snakes, small lizards such as chameleons, and small mammals such as rats and hamsters may be cared for by students trained to handle them (student curators and squads, chapter 10; caring for animals, chapter 9).

Students can be taught to culture protozoa, fruit flies, many worms, and other organisms useful in the year's work in biology. A living-material center may grow out of these activities. Then one school may be able to supply nearby schools with living materials (also some exhibits of life cycles of plants or insects which may be preserved).

You may have facilities for arranging a small greenhouse or, more simply, a sandbox in which seeds or cuttings of plants and other kinds of vegetative propagation may grow. Bell jars or plastic bags can be used to cover young plants to reduce loss of moisture.

Many lessons may be taught in a museum of living organisms. In fact, you may find it useful to distribute a sheet of questions to help guide the observations and thinking of students. In this learning activity, students may study: among different ecosystems food chains, reproduction, behavior, interdependence of living things, camouflage, heredity, evolutionary relationships, and energy cycles (food-making and food-getting).

If it is at all possible, students might prepare for a school museum examples of the major plant and animal phyla. A listing of examples that might be collected is given below; of course, microscopic forms may be represented by models. Where possible, maintain living organisms or well-prepared preserved specimens in museum jars. Also, guides for identification can be prepared, such as those for insects (p. 643), protozoa (p. 665), and conifers (p. 556). See also suggestions for the preservation of representative organisms (p. 646) and the work of museum curators (p. 759).

Student-made documentaries

Individual students interested in photography may develop a fine documentary of many ecological relationships. They might select nearby areas that are in need of repair—barren slopes, places that have been affected by raindrop erosion, and the like—and then show the effects on these areas of planting cover crops, terracing, or contour plowing. Or they might show in film all the intricacies of a few selected examples of plant and animal relations in a biome—for example, a woods, an open field, a fence post, or a pond.

Another committee of students might write the script for this film, which could be shown to a large group of students, parents, or faculty members.

Some camera enthusiasts may combine their interest in ecology and talents in photography by collecting their specimens as photographs. They may get excellent results using close-up photography. Such a collection of portraits of plants and animals might well become part of an identification key that students could devise.

Long-range investigations   Students who are deeply committed to biology may want to explore areas of biology that are currently under investigation. Teachers will want to look into the BSCS investigations for secondary school students—*Research Problems in Biology: Investigations for Students*, Series One-Four (Anchor Books, Doubleday, 1963, 1965). A guide accom-

panies the investigations—P. F. Brand-
wein et al., *Teaching High School Biology: A
Guide to Working with Potential Biologists*
(BSCS Bull. 2, AIBS, 1962).

These investigations contributed by
biologists from over the country may be
used as "models." Each one includes a
description of the problem, the hypothesis
under study, and some approaches. A brief
guide to the literature is also included. The
investigations are as broad as life, and the
methods of approach or inquiry are as
varied as the scientists who prepared these
investigations for students.

Collecting land forms

Collecting fungi, mosses, ferns, and seed plants

These plants may be transported in plastic
bags along with some of the soil in which
they grow. Refer to pp. 706, 713, 721 for
suggestions on where to locate specific
plants and how to maintain them indoors.

Collecting insects

Collecting nets for capturing insects differ
in size from the water nets previously
described. One type, a light-weight collecting net, is used primarily for the collection of fragile insects such as butterflies,
moths, and dragonflies. The net is fine
mesh, possibly nylon, sewed to a rim about
30 cm in diameter. The net should be at
least twice as deep as the diameter of the
rim, so that a twist of the net can be made
to enclose a specimen in the bottom. This
makes it possible to transfer the specimen
to a collecting bottle. The handle should
be light-weight and about 1-m long.

Another type of net, a sweeping net,
is used to brush or sweep through tall
vegetation; different kinds of insects may
be captured, including many beetles. This
net should have a stout wire rim and
handle. The length of the handle depends
somewhat upon the individual, since increased length will need additional
strength. In fact, some nets on the market
have handles only about 15-cm long. The

sweep net may be made of white muslin,
nylon, or duck; the diameter of the wire
rim should be about 30 cm. (Again, this
depends upon the strength of the individual.) The depth of the net should be at
least one and one-half to two times the
diameter of the net.

When hunting insect specimens with a
net, twist the net to enclose the insects in
the bottom. Then transfer them into a
killing jar (described below). When flying
specimens or delicate-winged lepidoptera
have been captured, their wings may be
damaged as they struggle within the net;
this can be avoided by placing a drop of
ether or chloroform on the net. Then
transfer to a killing jar and later to an
insect case.

The culturing of certain insects useful in
classroom demonstrations and investigations is described in chapter 9.

Experience in field work demands the
development of some ingenious devices for
carrying the variety of bottles and jars that
should be on hand. A sturdy knapsack
holds jars of many sizes and leaves both
hands free for work. Plastic vials and
bottles are preferable to glass ones; these
make the backpack lighter and are practically unbreakable. If a section of the plastic containers is rubbed with sandpaper
or emery cloth to roughen it a bit, it is
possible to write on the bottles with pencil.
These markings may be washed off later in
the laboratory or classroom. (Some plastic
bottles, available from supply houses, are
coated with a special lining so that strong
chemicals may be stored in them.)

Small vials are useful in collecting
arachnids, larvae, and other soft-bodied
forms that must be preserved immediately
in alcohol. Two or three uniformly sized
vials should be prepared as killing jars,
others should contain alcohol (70 percent),
and the remainder might be left as transfers, to carry specimens from the net to a
large jar.

In the field you will also need medicine
droppers (carry them in one envelope), as
well as forceps, hand lenses, and prepared
envelopes in which to pack lepidopterans,
as described below.

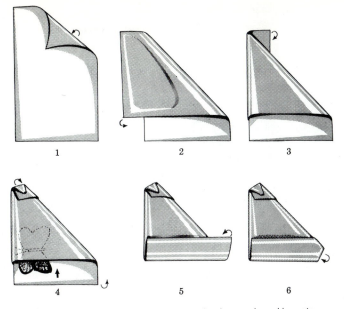

Fig. 8-57   Steps in preparing paper envelopes for transporting large-winged insects.

To repeat, in the field, larval stages and soft-bodied insects should be dropped into vials of 70 percent alcohol. Methods for preserving these forms permanently are described later in this chapter. However, lepidoptera, green insects, most beetles, flies, and bees (and other Hymenoptera) should not generally be put into alcohol.

Killing jars   Prepare both large and small killing jars for collecting insects in the field. Since the crystals and fumes of the potassium cyanide contents of the standard killing jar are deadly and poisonous to handle, it is advisable to purchase the jars ready-made.

As the bottles are used, moisture accumulates in the bottom from the secretions of captured insects. To avoid this, cover the plaster with thin blotting paper or cork sheeting.

Since the fumes act slowly, large fragile-winged specimens placed in killing jars may struggle furiously. Insert a bit of cotton with ether or chloroform (*caution*) into the killing jar, or dab it on the insect net before you transfer the captured animal to the killing jar so that the wings will not be damaged.

Paper envelopes for large-winged insects
Dragonflies, damsel flies, butterflies, and Dobson flies should be put into special envelopes to prevent wing damage (Fig. 8-57). In fact, these insects may be stored in the laboratory in these envelopes until you are ready to mount them. At that time they may be put into a relaxing jar, then placed on a wing spreader, and finally mounted (all three techniques are described later in this chapter.) You can purchase these envelopes, made of cellophane, from supply houses, or take slips of paper of various sizes, the largest about the size of a postcard, and fold them as shown. Students should have a good supply of these envelopes before starting out on a field trip. As they are filled, label all collected materials, giving location, date, time of day, and habitat of the collecting area.

Collecting other land animals   Spiders and other arachnids may also be collected by net. Earthworms, even earthworm cocoons (a bit larger than a wheat grain), slugs, land snails, and insect eggs (as well as vertebrates such as salamanders), should be collected in cigar boxes or jars

(a)

(b)

(c)

(d)

Fig. 8-58 Representatives of some orders of insects: (a) praying mantis; (b) June beetle; (c) cabbage butterfly; (d) damsel fly. (a, b, c, courtesy of Carolina Biological Supply Co.; d, photo by Harold V. Green.)

containing soil. Decaying tree stumps may be transported to the laboratory in moist newspaper and plastic bags and placed in covered containers (since termites may be in the decaying wood). The flagellates in the intestine of termites, sow bugs, and related forms are excellent for microscopic study. (See p. 631 for preparing slides of symbiotic flagellates.)

Identifying insects

Prior to a field trip, the main orders of insects may be distinguished by studying illustrations in field guides. On field trips many students like to be able to distinguish a beetle from a wasp or a cicada. An introduction to the methods for classifying animals and plants is given on page 566, where the categorization of living things can be compared to a library system of classifying books.

Students may devise their own simplified classification system to "key" out unknown insects when they know the basic differences among the orders. Broadly speaking, what kinds of variations distinguish the orders of insects? How would students classify the specimens in Figure 8-58?

Students might collect many insects and then sort them into probable similar groups. How does a beetle differ from a butterfly, a dragonfly, a bee, a fly, or a grasshopper? What are the distinguishing features that set apart one order from others? While students will recognize that insects, in general, have three body parts, two pairs of wings, and three pairs of jointed legs, there are also some large

INSECTA

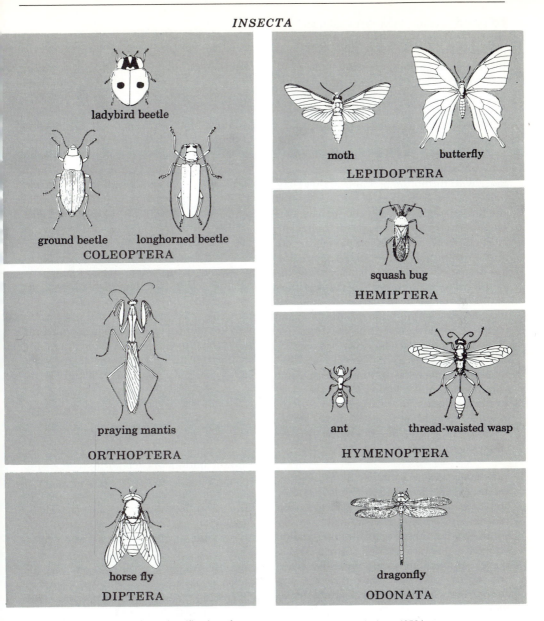

Fig. 8-59 Beginning guide to classification of orders of insects. (After *Insects, The Yearbook of Agriculture*, U.S. Dept. of Agriculture, 1952.)

differences. One order has front wings that are scalelike (beetles), another order has one pair of wings (flies), another has leathery front wings and color-patterned hind wings that pleat below the upper ones (grasshopper group), another has colored wings of overlapping scales (butterflies and moths), another has membranous wings and a constricted "waist" (wasp and bee group). Representatives of these orders and a few others are shown in Figure 8-59. There are many, many orders of insects that have not been included in this much abbreviated introduction.

Now, *within* an order of insects, such as the Coleoptera, students should be able to choose the traits that differentiate ground beetles from cucumber beetles, weevils, longhorned beetles, or ladybird beetles. Similarly, they should be able to distinguish families within the order Hymenoptera. Why are wasps in a different family from ants or bees?

When students have learned the traits of each order and have a nodding acquaintance with the notion of families, they may look again with sharpened observation for the distinguishing differences *within a family*. Students may turn to a professional key such as those found in books on insects and learn the genera that make up specific families of insects. Furthermore, *within one genus* (for example, among ladybird beetles or potato bugs) what variations may be counted?

At the same time you may want to use the studies of variations described among protozoa (chapter 9) or evergreens (chapter 7); or refer to other examples of variations and their causes developed in chapter 7.

Special techniques for mounting and displaying insects

Mounting insects   While it is best to mount insects soon after collecting, this step may be postponed. Keep the larger insects in the paper envelopes. However, in time insects become brittle, and antennae, legs, and wings break. Better specimens of lepidoptera are hatched from pupae rather than captured in nets.

Insects such as lepidoptera, dragonflies, and damsel flies are mounted with their wings fully extended. Grasshoppers are mounted with one wing outspread (Fig. 8-60), to show the color and pattern of the underlying wing, a point often required for identification. Frequently the wings of cicadas, lacewings, and Dobson flies are spread before they are mounted.

If insects have become brittle they may be softened in a relaxing jar, then placed on a spreading board if their wings are to be extended.

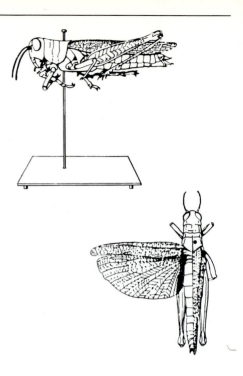

Fig. 8-60   Technique for mounting grasshoppers. Note that the wings on one side of the body are extended to show underwing coloration.

Relaxing jars   Prepare several jars of different sizes to accommodate the insects. Into the bottom of each jar insert a wad of wet cotton to which a few drops of carbolic acid have been added to inhibit the growth of molds. Cover this with a layer of blotting paper. (You may prefer to substitute moist sand at the bottom of the jar, in which case add a few drops of carbolic acid to the blotter.)

Place the dried insects on the blotter; cover the jar. After 24 hr the insects should be soft enough to mount or spread on a spreading board. Nevertheless, they must be handled with some care, as they are not as pliable as they were when freshly killed.

Spreading the wings of insects   The kind of spreader shown in Figure 8-61 may be purchased from biological supply houses, or constructed by students. The spreader consists of two side sections made of soft wood with a channel along the center just wide enough to accommo-

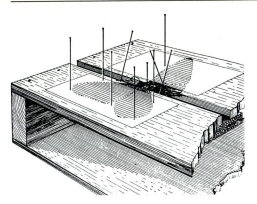

Fig. 8-61 Spreader for drying wings of such insects as butterflies, moths, and dragonflies.

date the body of an insect such as moth or grasshopper.

Students should learn to pin insects such as dragonflies, damsel flies, caddis flies, butterflies, and moths through the thorax and insert the mounting pin into the cork in the groove of the wing spreader. Use forceps to insert the pin into the cork so that the pin does not bend under the pressure of the fingers.

Arrange the legs of the insects as they are in life, and use forceps to spread the wings. Pin down the outspread wings by inserting pins along the front marginal part of the wings. Use strips of paper to hold the wings in position and fasten them with pins.

The wings will dry in a slightly tilted position, but their weight will cause them to droop a bit when the insects are finally mounted. You will find that the length of time needed for the insect to dry out depends upon the size of its body. It may vary from one day to two weeks.

Mounting insects for display   The usual method for mounting insects for observational study or exhibit is to set them on pins that can be inserted into a corklined box. Cigar boxes make fine display cases; even more attractive are clear plastic boxes with a bottom lining of pressed cork, corrugated paper, or balsa wood.

Use special black enamel insect-mounting pins about 3 to 4 cm long. The commonly used grades are Nos. 1 and 3.

There is a uniform method for pinning insects.

Most insects are pinned through the thorax. Moths and butterflies are pinned squarely in the center of the thorax, while grasshoppers, flies, bees, wasps, squash bugs, and the like are pinned to the right of center in the thorax. Beetles are not pinned through the thorax at all, but through the right wing cover about a quarter of the way back.

Very small beetles, such as weevils, are too small to be pinned. First, fasten these small insects to small triangles of lightweight stiff paper, such as library card stock, using thinned white shellac. Then insert the pin through the broad base of the paper triangle. Small fragile insects such as mosquitoes may be first pinned to a bit of cork (Fig. 8-62).

To label insects, prepare a small strip of paper (about $\frac{1}{4} \times \frac{1}{2}$ in) carrying the name, location, and date of collection. Place this under the insect by transfixing it on the same pin.

Care of insect collections   Some small insects attack preserved insects. To keep these museum pests under control, heat a number of pins and insert them through small chunks of paradichlorobenzene and pin these in two corners of each collection box.

Mounting specimens preserved in alcohol   At times it may be desirable to show the complete life history of a specific

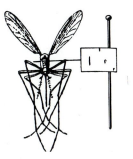

Fig. 8-62 Technique for pinning and labeling small insects. A mosquito is shown pinned with a "minuten nadeln" to a block of cork on a regular insect pin.

insect, especially the life history of insect pests or of special beneficial insects. Place each stage (eggs, larvae, and pupae) into separate small vials of 70 percent alcohol or 5 percent formalin. Seal the vials with paraffin. Keep the adult stage or nymphs dried.

Mount the small vials and the adult stage in Riker mounts or in homemade display boxes containing cotton. Students may make display cases by cutting away the paper cover of several boxes of the same size, leaving about 1 cm margin as a frame. Tape a sheet of glass or of plastic to the inside of the cut-away box cover, and line the box with cotton. Lay the specimens in the cotton and cover the box. Insert pins horizontally on each of the four sides of the box to fasten the cover to the box.

Additional projects Some students may prepare valuable collections of a single species of insect. For example, a study may be made of variations of striping among potato beetles, or the number of spots in one species of ladybird beetles. Some students have successfully embedded specimens in polyester resins.[44]

Students and teachers who want more advanced information on handling of insects are fortunate, for there is a tremendous literature on collecting, preserving, exhibiting, and identifying insects.

For further information on aquatic insects or on partially aquatic insects, refer to Ward and Whipple, *Fresh-Water Biology*, 2nd ed., edited by W. T. Edmondson (New York: Wiley, 1959) and R. Pennak, *Fresh-Water Invertebrates of the United States*, 2nd ed. (New York: Wiley, 1978). Other valuable books are listed in the *Book Shelf* at the end of this chapter.

Preserving animal specimens

You may want to refer to the descriptions that follow for some notes on the preserva-

---

[44] Most biological supply houses offer information and materials for this technique.

tion of animals in the laboratory or in the field. Preserved specimens are useful in demonstrating relationships among living things (classification and dissection).

Preserving soft-bodied specimens

In general, soft-bodied animals should be preserved first in 50 percent alcohol for a few days, then transferred to 70 percent, and finally to 90 percent alcohol. Formaldehyde may also be used as a preserving fluid, although it tends to make the specimens more brittle. Puncture the body of each animal at several widely separated places with a dissecting needle before you put it in the preserving fluid so that complete penetration of the preservative takes place. (See p. 167 for fixatives; or preserving fluids see below.)

Later, the preserved specimens can be mounted in jars. Each specimen can be supported against a glass slide and fastened to the glass with thread. Use uniformly sized, wide-mouthed jars and seal the jars with paraffin to prevent evaporation of the preserving fluid. Label the jars and cover the labels with plastic tape so that they will not become smudged in use.

In time, the specimens will bleach and fragment, a signal that new specimens and fresh fluid are needed.

In summary, the following soft-bodied forms may be preserved in alcohol or formaldehyde: (*caution*: avoid inhaling. See formalin, p. 167.)

Jellyfish and *hydras* should be anesthetized before preserving so that they will remain in an extended position. Techniques are described on pp. 167 and 168. The specimens will become opaque in preserving fluid.

Flatworms, roundworms, and *annelid worms* should be anesthetized before preserving. Flatworms can be flattened between two glass slides then inserted into alcohol.

Soft-bodied mollusks such as squids, become opaque and pinkish when preserved.

Crustaceans such as hermit crabs, fiddler crabs, and larger crabs, as well as barnacles and the shipworm (*Teredo*), may be preserved or mounted.

Soft-bodied insects such as grasshoppers, praying mantids, locusts, some beetles (and larvae and pupae stages), can be preserved in 70 percent alcohol.

Spiders, centipedes, millipedes, scorpions, mites, and *ticks* may be preserved in 70 percent alcohol.

CARL'S SOLUTION   is excellent for killing and preserving small, soft-bodied forms such as mites, centipedes, and millipedes. A small amount of glycerin may be added to the solution to prevent hard-bodied insects from becoming brittle in the preservative. Combine the first three of these substances; then add the glacial acetic acid to the solution just before using (*caution*).

alcohol (95%)	170 ml
formalin (40% formaldehyde)	60 ml
water	280 ml
glacial acetic acid	20 ml

Small fish, as well as their *eggs* and *fry,* may be preserved separately in 70 percent alcohol in small vials to make a life-history series.

Among *amphibia,* all the stages in the life history may be preserved, preferably in 70 percent alcohol.

Young bird embryos and *embryos of mammals* should be preserved in a fixative or transferred through a graded series of alcohol (70 to 80 to 90 percent) to prevent undue shrinkage.

Preserving hard-bodied specimens

Preserve the following, where possible, in the dry state.

Sponges may be mounted on bristol board after drying in the sun.

Starfish may be dried in the sun; weigh down in a flat position and then mount.

Sea urchins should be dried in the sun after removing the viscera; with a knife, cut away the mouth parts (Aristotle's lantern), and pull away the attached internal organs.

Sand dollars are quite flat and may be completely dried in the sun.

Crabs may be preserved whole in alcohol. Or the carapace may be dissected away carefully, and then the crab dried completely and mounted.

Mollusk shells must be dried out completely. Mount on bristol board or in boxes lined with cotton such as Riker mounts.

To preserve live mollusks, boil them in water, and let them remain in the cooling water. Bivalves are easy to clean by scraping out the organism. Use a bent wire to remove a snail's body by twisting the wire along the snail's spiral shell. Use a nail file to clean the outside of the shells. Immerse them in a household bleach solution (1 part bleach to 2 parts water) for 2 to 3 hr. Then dry the shells in the shade and apply a thin film of lubricating oil to preserve the shells' luster; repeat every few months. (Avoid using a lacquer or shellac.)

Small horseshoe crabs, in various stages of development, may be collected and allowed to dry thoroughly. Then the surfaces can be shellacked. (If the forms are not completely dry, the shellac will become opaque.) Larger forms of horseshoe crabs must be eviscerated. Then the cleaned shells can be dried.

Young crayfish may be dried in the sun.

Excellent chapters on preparing vertebrate skeletons, preserving and mounting birds, small mammals, reptiles, fish and amphibians, as well as invertebrates and plants are available in the reference by J. Knudsen, *Biological Techniques* (New York: Harper and Row, 1966).

Preserving plants

Land plants

Collect land plants, leaves, stems, flowers or fruits as they appear in season, in folds of newspaper in a vasculum or collecting case. Mosses and ferns can be pressed to make an attractive collection. Turn over a leaf specimen so that its undersurface is exposed. Include seeds, fruits and flowers. Some flowers should be split open to reveal the arrangement of stamens. Use newspaper or gray carpet paper as substitutes for specimen drying paper. Press the specimen papers together in a plant press. Bind tightly and keep in a warm place or in the sun. Corrugated sheets from cartons may be alternated with the specimen papers to permit the circulation of warm air through the plant press. Replace the wet specimen papers with fresh absorbent paper to speed drying every few days or as needed.

Mount the dry specimens in an attractive manner on herbarium paper and secure them with thin strips of masking tape or transparent tape. Some specimens may be glued to the paper. You may want to cover the entire arrangement with a plastic wrap after the specimens are thoroughly dried. Most specimens are worthless unless they are identified, with the location in which they were found, the abundance of the plants, the color of the petals, the odor of the flowers, and the date. A collector's good notes should help recall the natural lifestyle of the preserved specimens.

When collecting parasitic fungi, a specimen of the host plant should also be included.

When flowers or seeds are small, enclose them in an envelope, attached to the mounting paper. Fleshy petals may be pressed between wax paper to preserve their color before they are placed on absorbent paper in a plant press.

You may want to immerse some plant specimens in a preservative for display in museum jars.

Some forms, such as mushrooms and lichens may be preserved by drying; the fleshy fungi tend to shrink a bit. Lichens may best be stored in envelopes until ready for use. At that time, soak them in a finger bowl of water for a few hours to soften, then prepare slides for microscopic examination. To retain the color of fleshy fungi use a preservative recommended in a Turtox Leaflet (General Biological Supply House). Combine the following:

alcohol (50%)	100 ml
formalin (40%; caution: avoid inhaling)	6 ml

See page 167 for other preservatives described in the literature.

Aquatic plants

Floating seed plants can appear as a tangled mass; spread out on absorbent specimen paper in a natural position for drying. Bulbous regions may be split in half both crosswise and lengthwise.

Marine algae   As soon as possible after collecting marine algae, immerse the plants in FAA preservative (formalin-aceto-alcohol) made by combining:

formalin (*caution*: avoid inhaling)	100 ml
glacial acetic acid (*caution*)	100 ml
ethyl alcohol (50%)	1800 ml

To this add a teaspoon of cupric sulfate to preserve the color of green algae or Bismarck brown (0.5 g in 50 ml of tap water) for brown algae.[44] Immerse algae in the stain, rinse in water, and compare its color with freshly collected specimens. (A little experimentation may be needed to estimate timing.) Then store the algae in a 5 percent formalin solution (*caution*: avoid fumes) made with seawater to which borax is added to obtain an alkaline pH. (Check with an indicator.) Larger kelps may be better preserved in a solution comprising 3:3:3:1 of glycerin, ethyl alcohol (70 percent), seawater, and carbolic acid. Then transfer to pure glycerin for two to three days, wrap in absorbent toweling and store in wooden boxes in folds of toweling soaked in glycerin.

For examination, specimens may be floated in trays of water.

Red algae also lose color quickly. Float delicate specimens in water and slip absorbent paper under them; gently lift out the specimens to dry in their natural spread showing their feathery patterning. Other, thicker specimens, may well be preserved in glycerin. Or use formalin and seawater as described above for brown algae. This time, stain the red algae with safranin (0.5 g in 50 ml tap water). For examination, flatten the red algae in water in white trays.

[44] C. Hillson, "The Preservation of Marine Algae for Classroom Use," *Turtox News* January 1967.

Preserving specimens and preservatives

Green algae Another solution that will preserve the natural color of green algae and will not alter cellular structure for microscopic examination is described below. Combine the following:

potassium chrome alum	10 g
water	500 ml
formalin (40%)	
(*caution*: avoid fumes)	5 ml

Preservatives containing larger quantities of formalin will bleach the green color. Fleshy marine algae may be soaked in glycerin first, for several days, then dried.

Preservative for green plants This also prevents bleaching of chlorophyll in plant tissues. You can use the FAA fixative (p. 167) by adding enough copper sulfate to make a saturated solution, or you may prefer to make up this solution to have on hand as follows. Dissolve 0.2 g of copper sulfate in 35 ml of water. To this solution, add a solution composed of:

ethyl alcohol (95%)	50 ml
formalin (40% formaldehyde)	10 ml
(*caution*: avoid fumes)	
glacial acetic acid	5 ml

Specimens may be stored in this preservative indefinitely. Occasionally transfer specimens into fresh solution.

CAPSULE LESSONS: WAYS TO GET STARTED IN CLASS

8-1. Draw a diagram of the primary producers, primary consumers, secondary consumers, and reducers (decomposers) in a pond. Ask students to complete the picture and to explain the relationships. Trace food chains, cycles of oxygen and carbon dioxide, and the nitrogen cycle.

8-2. Make *Chlorella* central to a lesson. Use the lesson as a basis for introducing students to a block of preludes to investigations concerning *Chlorella* (pp. 229–32, also pp. 615–17).

8-3. Ask students to give an account of their experiences as fishermen. What did they catch? Was it a predator or prey? Upon what did it feed? Trace the food chain back to a primary producer. Be certain to include the decomposers in the cycles. What effect did the removal of the catch from the stream have on the plant and animal community in the stream?

8-4. Seal a snail in a test tube (Fig. 4-33 and ask how long it will live. Develop the idea of the dependence of living things on green plants as producers. Trace the energy flow in a food chain. Then develop different modes of nutrition among plants and animals (saprophytes, parasites, mutualism, and other special types as described in this chapter).

8-5. If possible, begin with a field trip to a specific community or habitat: pond, desert area, beach, woods, meadow, or bog. Have students describe the food chains in the habitat.

You will find useful the guides edited by Andrews (see Book Shelf).

8-6. Have students bring in samples of soil from different regions and examine soil and soil water under the microscope for small organisms. Streak sterile plates with soil water (p. 702); then transfer some soil water extract to enriched medium (p. 689). Develop the idea of interdependence of plants and animals. Which organisms are producers? Saprophytes? What substances do animal consumers feed upon? What might happen if one kind of microorganism disappeared?

8-7. There are many soil-testing kits available. Demonstrate some characteristics of soils around the school area (p. 611). How might living things be affected if the acid soil of swamps or bogs became more alkaline? Try to use compost formed from decaying leaves for tests of acid soil.

Develop the idea that plants require specific conditions for growth, such as pH, minerals, proper soil texture. Recall that the degree of acidity of soil determines, in some measure, the availability of minerals to the plants.

8-8. Develop the abiotic as well as biotic factors that affect populations of organisms in the soil, in freshwater and marine regions. How do these factors limit the range of populations, or possibly bring adaptive (genetic) variations into action?

8-9. Develop laboratory activities for students to observe the succession of microorganisms in a minipond firsthand (p. 587).

8-10. Plan to submerge slides in a quiet stream or in the ocean to examine populations of organisms as they exist in close proximity (methods, pp. 586, 595, 605).

8-11. Collect airborne microorganisms on open sterile Petri dishes or on slides coated with a thin film of petroleum jelly. Incubate the dishes and use sterile techniques to isolate different types: algae, molds, bacteria, and protozoa that may be present. Examine the slides coated with petroleum jelly for particulate particles including pollen grains.

8-12. Trace the nitrogen cycle and plan laboratory activities in studies of nitrogen-fixing bacteria and those in the nitrifying cycles (pp. 608, 634).

8-13. Develop the notion that in a biotic community each species has its own niche in the habitat. Dicuss food-getting in the daytime and nocturnal organisms; niches at different levels in a forest: floor, shrub level, nest-building at higher canopies of trees.

8-14. What roles do bacteria and fungi play as decomposers in recycling carbon compounds and minerals in soil and water?

8-15. Plan laboratory lessons on modes of nutrition: How do heterotrophs obtain their food (energy); saprophytes and parasites? Cite examples of mutualism, commensalism. Describe how insectivorous plants obtain their nitrogen supply. Also discuss the chemo-autotrophic bacteria.

8-16. Develop the concept of *populations* of organisms interacting in communities within an ecosystem. Describe the action of pioneer organisms; are they plant or animal organisms? What is a climax forest?

8-17. List the different plant and animal communities on earth. What adaptations are found among desert organisms? Those of a tundra? Of a tropical rain forest? A deciduous forest?

8-18. What is meant by biological magnification up a food chain (or pyramid)? Use DDT or mercury or other pollutant to trace the flow of the substance from producers to consumers. Debate the care of a small, fragile planet Earth.

8-19. Collect labels from food packages to list the additives. What are the purposes for including the different additives in foods? List the many categories.

8-20. Study intraspecies competition (p. 554) and interspecies competition (pp. 70, 554).

8-21. Plan special readings, especially from *Scientific American* offprints, concerning current ecological problems such as people vs insect pests; pollution of air, water, and soil. Discuss the use of integrated pest management methods (IPM), a combined use of biological, genetic, and chemical methods used to control insect pests without jeopardizing other living things in the ecosystem.

8-22. Begin a study of the relationship between water and soil by demonstrating the ability of different soils to hold water. Use funnels of different soils as described in this chapter. Have students explain how cover plants and forests prevent floods and the erosion of soil.

8-23. Plan to have a committee of students make a survey of good land practices in your community. They might also recommend conservation measures that could effectively be introduced. Perhaps they can write an article for a newspaper—either school or community—to report their findings and suggestions.

Students should be able to explain the values of crop rotation, strip cropping, ways to spread ground water, terracing, use of green manure, irrigation, contour plowing, and the use of fertilizers and the proportion of sulfates, phosphates, and nitrates these should contain for specific crops.

8-24. Take a field trip around the school grounds and list places that show erosion of soil by wind, water, or ice. What conservation practices could be used in each area? Have students explain the value of each of their suggestions.

In some cases a class might put some of these ideas into practice. They might terrace the lawn or a piece of land nearby to reduce a sharp angle of an existing slope. Or plant saplings or cover crops on a hillside or a burned-over area. If a photographic record is kept year by year, students in subsequent classes might continue the practices when they see how effective they have been.

8-25. Have students report on the destruction of soil from overgrazing. Have students report on the use of radioactive isotopes added to fertilizers to trace the amounts and the kinds of minerals that different crop plants take from the soil. Many reference materials citing new technologies are available from the U.S. Atomic Energy Commission (Appendix) and from current magazine articles.

8-26. As a group activity, investigate the value of birds in a community. For example, make a survey of the kinds of birds attracted to a given area. Then build birdhouses to attract birds to the school area. Although many may consume seeds, how effective in general are birds in holding down the insect population? Perhaps students can explain the practice of planting berry shrubs and trees along rows of cultivated crops.

While chicken hawks may consume some chickens, why is it a poor practice for a community to offer a bounty for dead hawks?

8-27. Perhaps your school is located in an area where you may solicit the help of foresters (or county agents). On a field trip, they may show how forests are handled as crops, the succession of trees in woods and climax forests, what kinds of seedlings are pioneers in a new region, and whether the prevailing winds carry light seeds or heavy seeds greater distances into new regions. Students might prepare questions in advance. What are some new practices in forest management?

You may ask a county agricultural agent to explain how to test crops and soil for mineral deficiencies in the laboratory.

8-28. Develop students' skills in reading charts, and interpreting data from graphs found in current literature. Have students write their interpretation or prepare questions based on the readings.

8-29. Take short field trips out of doors, around school grounds, around the block. Identify trees, observe variations, compare varieties of flowers. When a trip to the beach is possible, examine the variations among oyster shells, evidence of action of oyster drills, egg cases of whelks and beach insects. What types of vegetation exist in beach areas?

When possible, trace cycles of water, minerals, oxygen and carbon dioxide in soil and in the waters—both fresh and marine samples.

8-30. Can students recognize poisonous plants such as poison ivy or poison oak? Do they know the harmful plant and animal pests? Can they recognize poisonous spiders, scorpions, and snakes in the community?

8-31. Many fine films, color slides, video tapes, and filmstrips enable you to take your classes on a "field trip" when no other type of trip is possible. Compare organisms of a tundra, a desert, grasslands and other biomes.

8-32. Plan long-range laboratory studies of successions in a bird bath, a fallen log, or a crack in the pavement. Develop the concept of interrelations, and the checks and balances among living things. What are natural enemies? Look for examples of parasites, saprophytes, and symbionts.

8-33. What is the effect of adding 5 drops of a 3 percent solution of potassium phosphate or of potassium nitrate to 10 ml of a *Euglena* culture?

What effect does a 3 percent solution of a phosphate have on germinating pinto beans or radish seedlings? Students may hypothesize possible results and test their hypotheses.

8-34. You may want to introduce reproduction in plants by taking students on a short walk around the school grounds. Study the floral structure and compare the flowers of many different plants in bloom without picking the flowers from trees or shrubs (Fig. 1-132). Elicit through questions the essential organs needed for reproduction in flowering plants. How do pollen grains get to the stigma of flowers? How is a seed formed?

8-35. Study seed dispersal mechanisms outdoors. What are the chances for survival of seeds around the parent plants? What is the rate of migration for plants? What effect does the survival of seeds have on other plants and animals? Examine lawns for seedlings of maple, linden, and others.

8-36. Trace the path of a pesticide along the food webs—mercury, lead, or PCBs (polychlorinated biphenyls), used in paints, inks, and transformers that are soluble in fats. PCB's affect birds' feathers and water-proofing of feathers. What are its effects on humans? Consider using the *Scientific American* offprint (May 1971): "Mercury in the Environment" by L. Goldwater.

DDT lasts some 10 to 15 years or more in soil and food chains and is soluble in fatty tissues. Trace the soil residues that flow into streams, rivers, and oceans; freshwater fish are highly susceptible. How have penguins in the Antarctic come to have DDT in their tissues?

What is the effect of sewage run-off (which may have high amounts of phosphates and organic matter) on the growth of organisms in a lake or stream? What is eutrophication? Refer to I. G. Simmons, *The Ecology of Natural Resources*, 2nd ed. (New York: Wiley, 1981), and to L. Hodges *Environmental Pollution*, (New York: Holt, Rinehart and Winston, 1973), which stress physical and chemical principles. Also refer to the 1983 Yearbook of the Dept. of Agriculture *Using Our Natural Resources*.

BOOK SHELF

These are only a selected listing of references in the broad field of ecology. Also refer to footnotes in this chapter and to papers in *Scientific American* for current work.

Agricultural Research Service, *Selected Weeds of the United States*, Handbook 366, U.S. Dept. of Agriculture, Supt. of Documents, U.S. Government Printing Office, Washington, D.C. 20402.

Alexopoulos, C. J., and S. W. Mims. *Introductory Mycology*, 3rd ed. New York: Wiley, 1979.

American Public Health Assoc., A. Greenberg (chairman) *Standard Methods for Examination of Water and Waste Water*, 15th ed., American Public Health Assoc., 1015 15th St., N.W., Washington, D.C. 20005. 1980.

Anderson, J. M. *Ecology for Environmental Sciences*. New York: Wiley, 1981.

Andrews, W., ed. *Contours: Studies of the Environment*, Four Guides in a series: *Environmental Pollution* (1972); *Fresh Water Ecology* (1972); *Soil Ecology* (1973); *Terrestrial Ecology* (1974). Englewood Cliffs, NJ: Prentice-Hall.

Arnet, R. Jr., and R. Jacques Jr. *Insects*. New York: Simon and Schuster, 1981.

Bailey, J., and J. Mansfield, eds. *Phytoalexins* (antibiotics occurring naturally in plants). New York: Wiley, 1982.

Barnes, J. *Oceanography and Marine Biology*, a book of techniques. New York: Macmillan, 1959.

Benson, H. *Microbiological Applications: Laboratory Manual in General Microbiology*, 2nd ed. Dubuque, IA: William Brown, 1973.

Bohn, H., et al. *Soil Chemistry*, New York: Wiley, 1979.

Brenchley, D., and J. Towler. *Resource Handbook—the Environment: Activities and Explorations* (lists government agencies and resource organizations). Boston: Houghton Mifflin, 1975.

Briggs, J. *Marine Zoogeography*. New York: McGraw-Hill, 1974.

Brock, T. et al. *Biology of Microorganisms*, 4th ed. Englewood Cliffs, New Jersey: Prentice-Hall, 1984.

Cable, R. *An Illustrated Laboratory Manual of Parasitology*, 5th ed. Minneapolis: Burgess, 1977.

Callow, J. ed. *Biochemical Plant Pathology*, New York: Wiley, 1984.

Cheng, T. *General Parasitology*, New York: Academic, 1973.

Clapham, W. Jr. *Natural Ecosystems*, 2nd ed. New York: Macmillan, 1983.

Dawes, C. *Marine Botany*. New York: Wiley, 1981.

Depts. Air Force and Army Medical Service. *Clinical Laboratory Procedures—Parasitology*, AFM—160 48; TM 8–227 2, August 1974, Supt. of Documents, U.S. Government Printing Office, Washington, D.C. 20402.

Duffus, C., and J. Slaughter. *Seeds and Their Uses*. New York: Wiley, 1980.

Etherington, J. *Environment and Plant Ecology*, 2nd ed. New York: Wiley, 1982.

Fuerst, R., Frobisher and Fuerst's *Microbiology in Health and Disease*, 15th ed. Philadelphia: Saunders, 1983.

Goerlitz, D., and E. Brown. *Methods for Analysis of Organic Substances in Water*, U.S. Geological Survey, U.S. Dept. of Interior, Supt. of Documents, U.S. Government Printing Office, Washington, D.C. 20402, 1972.

Gribben, J. *Future Worlds* (food, energy, natural resources). New York: Plenum, 1980.

Griffin, D. *Fungal Physiology*. New York: Wiley, 1981.

Joklik, W., and H. Willett, eds. *Zinsser Microbiology*, 16th ed. New York: Appleton, Century, 1976.

Kapp, R. *How to Know Pollen and Spores*. Dubuque, IA: William Brown, 1969.

Kelman, A. (chairman) et al. *Sourcebook of Laboratory Exercises in Plant Pathology*, American Phytopathological Society. San Francisco: Freeman, 1967.

Ketchum, P. *Microbiology: Introduction for Health Professions*, New York: Wiley, 1984.

MacFarlane, B., and D. White. *Natural History of Infectious Disease*, 4th ed. New York: Cambridge Univ. Press, 1972.

McConnaughey, B., and R. Zottoli. *Introduction to Marine Biology*, 4th ed. St. Louis, MO: Mosby, 1983.

Nestor, W. *Into Winter: Discovering a Season*. Boston: Houghton Mifflin, 1982.

Noble, E., and G. Noble. *Parasitology: The Biology of Animal Parasites*, 5th ed. Philadelphia: Lea and Febiger, 1982.

Odum, E. *Basic Ecology*, New York: Saunders College Publishing, 1983.

Pielou, E. *Biogeography*. New York: Wiley, 1979.

Rheinheimer, G. *Aquatic Microbiology*, 2nd ed. New York: Wiley, 1981.

Schipper, L. *Lecture Outline of Preventive Veterinary Medicine for Animal Science Students*, 6th ed. Minneapolis: Burgess, 1982.

Simmons, I. G. *The Ecology of Natural Resources*, 2nd ed. New York: Wiley, 1981.

Singleton, P., and D. Sainsbury. *Introduction to Bacteria*. New York: Wiley, 1981.

Slack, K. et al. *Methods for Collection and Analysis of Aquatic Biological and Microbiological Samples*. U.S. Geological Survey, Supt. of Documents, U.S. Government Printing Office, Washington, D.C. 20402, 1973.

Smith, A. *Principles of Microbiology*, 9th ed. St. Louis, MO: Mosby, 1981.

Starr, C. and R. Taggart. *Biology: The Unity and Diversity of Life*, 3rd ed. Belmont, CA: Wadsworth, 1984.

Stephenson, T., and A. Stephenson. *Life Between the Tide Marks on Rocky Shores*. San Francisco: Freeman, 1971.

Sumich, J. *An Introduction to the Biology of Marine Life*. Dubuque, IA: William Brown, 1976.

Sumich, J., and G. Dudley. *Laboratory and Field Investigations in Marine Biology*. Dubuque, IA: Kendall/Hunt, 1974.

Vickery, M. *Ecology of Tropical Plants*. New York: Wiley, 1984.

Wallace, R. et al. *Biosphere, The Realm of Life*, Scott Foresman, 1984.

Wenger, K. ed. *Forestry Handbook*, 2nd ed. New York: Wiley, 1984.

Wilson, E. O. *Life: Cells, Organisms, Populations*. Sunderland, MA: Sinauer Associates, 1977.

Wratten, S. and G. Fry *Field and Laboratory Exercise in Ecology*, Baltimore, MD University Park Press, 1980.

Zottoli, R. *Introduction to Marine Environments*, 2nd ed. St Louis, MO: Mosby, 1978.

Many papers from *Scientific American* have been compiled in the Scientific American Book: *The Oceans*, (San Francisco: Freeman, 1969). The 10 chapters appeared in September 1969 issue and are also available as offprints. *Industrial Microbiology* is a September 1981 issue.

You may want to build a reference library that offers supplementary readings to students, or have students build their own science libraries.

For example, obtain a listing of offprints and/or books from the following:

Wm. C. Brown Group, 2460 Kerper Blvd., Dubuque, IA 52001 (Pictured Key Nature Series)

Doubleday/Natural History Press, 245 Park Ave., New York, NY 10167

Dover Publications, Inc., 180 Varick St., New York, NY 10014

Golden Guide Series. Western Publishing Co., Inc., 1220 Mound Ave., Racine, WI 53404

Houghton Mifflin Co., 2 Park St., Boston, MA 02108 (Peterson Field Guides and also their International Series of Field Guides on Europe, Asia, West Indies, Africa)

Alfred Knopf Inc., 201 E. 50th St., New York, NY 10022 (Audubon Society Field Guides)

Little, Brown and Co., 34 Beacon St., Boston, MA 02106 (Explorer's Notebook Series)

Macmillan Pocket Encyclopedias in Color. Macmillan Publishing Co., 866 Third Ave., New York, NY 10022

New American Library, Education Division, 1633 Broadway, New York, NY 10019

Oxford/Carolina Biology Readers. Carolina Biological Supply Co., 2700 York Rd., Burlington, NC 27215 (Some 100 titles; short 16 page monographs written by people active in the field for which they write)

Penguin Books Inc., 625 Madison Ave., New York, NY 10022

Scientific American. W. H. Freeman Co., 660 Market St., San Francisco, CA 94104

Simon and Schuster Guides (Nature). Simon & Schuster, 1230 Avenue of the Americas, New York, NY 10020

Time-Life Books Inc., Alexandria, VA 22314

9

Maintaining Organisms for Laboratory and Classroom Activities

In this chapter, we describe methods for collecting and cultivating invertebrates, vertebrates, and the divisions of plants. As each phylum is presented, the possible usefulness of these organisms in the laboratory and classroom is indicated. You may want to refer to chapter 8 for specific details on planning field trips, and for tips on collecting organisms that may be maintained in the laboratory.

The best methods for maintaining living things in a classroom or laboratory are those that reproduce the most favorable field conditions and eliminate natural enemies. This is, in essence, the principle upon which the construction of vivaria is based. Proper temperature and nutrients are the basic factors, and for green plants an adequate source of light.

Maintaining animals

Many invertebrates and some classes of vertebrates can be cultured in the laboratory or classroom for studies of behavior, ecology, classification, reproduction and variations, for comparative studies of organ systems and circulation of blood, and for observation of heartbeat. Many of these animals also serve as food for small vertebrates that often are reared in the laboratory.

Students usually bring in materials for identification when a resourceful teacher stimulates interest in living things. If a teacher places eggs of frogs in aquaria or bowls in the classroom, or brings in patches of different mosses, a fern or two, and a lichen for a terrarium, students will ask questions.

In many cases there will be a few knowledgeable students who can give short talks to classes or to clubs about the behavior of certain animals. There also will be students who would like to care for living materials (student squads, chapter 10). Often, students want to learn how to identify shells, trees, or flowers; how to start an insect collection; how to set up a terrarium or an aquarium; how to get more plants without using seeds; how to photograph animals and flowers; how to make prints and enlargements; and how to breed tropical fish.

Protozoa among the protists

Culturing protozoa is relatively simple when certain fundamental precautions are taken. The techniques described below have proved useful to the authors in the biology classroom and laboratory. First, consider the general conditions necessary for a culture center. Specific culture methods for the more common, easily obtained forms of protozoa follow.

Suggestions for maintaining a culture center

In general, the room or portion of the laboratory where cultures are to be kept should fulfill the following environmental conditions:

1. Maintain cultures at a constant temperature within an optimum range of 18° to 21° C (64° to 70° F). During warm weather stack finger bowls in a metal container that you can keep cool by putting it in a sink and circulating tap water to the level of the top dish. At temperatures above 25° C (77° F) or below 15° (59° F), cultures do not maintain themselves at maximum and may die off.

2. Keep cultures away from fumes of concentrated acids such as nitric acid, hydrochloric acid, and sulfuric acid and such alkalies as ammonium hydroxide.

3. Try to keep the cultures at a hydrogen-ion concentration approximately neutral (pH 7).

4. Keep the cultures in medium light; darkness is not detrimental. Direct sunlight is harmful since the temperature of the culture may be raised above the optimum. (Of course, this does not apply to the culturing of green forms.)

5. Aerate cultures to avoid surface scum that may appear as a result of bacterial growth since this may block oxygen exchange.

6. Avoid sudden drafts near the cultures, since they may carry spore contaminants, as well as some common chemicals in powdered form.

7. Keep glassware clean. Use low-suds detergent; rinse repeatedly. For a final rinsing, wash in distilled water if the tap water is not free of copper and chlorine.

The number of organisms in a beginning culture greatly affects the growth of the entire population. For this reason, when a limited number of organisms is available, begin cultures in Syracuse dishes; as the population density increases, transfer quantities into small finger bowls.

Pure strains of protists are available from Carolina Biological Supply Co., as well as others (see Appendix).

Specific culture methods

Rhizopods

Amoeba   The amoeba most commonly used in the classroom, *Amoeba proteus*, is not as common in nature as other amoebae. It may be found in ponds or pools that do not contain much organic matter, where the water is clear and not too alkaline, and where exceedingly swift currents are absent. Both *Amoeba proteus* and *A. dubia* (Fig. 9-1) are found among aquatic plants such as *Cabomba*, *Anacharis*, and *Myriophyllum*. Often scrapings from the base of a stalk of the cattail (*Typha*) or from the underside of a leaf of the water lily, *Nymphaea*, will yield many amoebae.

Place small amounts of these plants on which amoebae may be present into finger bowls, Petri dishes, or flat jars. Cover the material with pond water in which the plants had been growing, or add spring water. Be certain to maintain them at room temperature. To each container add two to four uncooked rice grains. Amoebae generally appear in one week to ten days in successful cultures. Examine with a dissecting microscope to locate amoebae in the cultures.

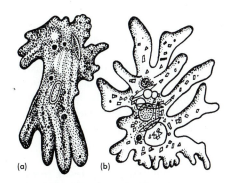

Fig. 9-1 Two species of *Amoeba* useful in the classroom: (a) *Amoeba proteus*; (b) *A. dubia* (not drawn to scale).

Amoebae congregate on the bottom or sides of a container. They may be removed in either of two ways. Carefully pour off the excess fluid into another dish. Then with a pipette pick up the amoebae attached to the bottom and place them in fresh spring water to which a few uncooked rice grains have been added. When the rice grains begin to decay and the bacteria population increases, small ciliates, such as *Chilomonas* begin to increase. *Chilomonas* furnishes a good food supply for *Amoeba*.

If the animals are not congregated on the bottom, swirl the finger bowl with a rotating motion, thereby causing the heavier particles to fall into the center of the dish. You may then pick up the amoebae with a pipette.

The method described above is useful for *temporary* cultures in the laboratory. When amoebae are to be cultured continuously, the following methods have been successful.

METHOD A: HAY INFUSION   Halsey[1] has described a typical method. Place eight, 3-cm lengths of timothy hay stalks in 100 ml of spring water. Boil this mixture for 10 min, and let it stand for 24 hr. Then add large quantities of *Colpidium* or *Chilomonas*. Let this medium stand for two to three days, then inoculate with amoebae. As the culture develops, the number of food organisms (that is, the ciliated forms) decreases. When this happens, remove half of the culture medium, and add an equal amount of fresh hay infusion to which *Colpidium* or *Chilomonas* have been added. Add two grains of uncooked rice or boiled wheat, or four 3-cm lengths of boiled timothy hay for every 50 ml of culture medium added. Cultures may last as long as six months. However, if large amounts of organic matter with accompanying large quantities of bacteria are present, they tend to cause the death of the

amoebae. A pH of 7 or slightly alkaline is optimum.

Many methods described in the literature are similar to that of Halsey (see papers by Jennings,[2] Kofoid,[3] Hyman,[4] Dawson,[5] and LeRoy and Ford[6]). In all these methods a medium that has no specific chemical composition is used. While these methods are successful in the hands of some workers, they offer pitfalls for the beginner. In fact, the very simplicity of the method is its undoing. The beginner often needs a method that takes care of all the variables that may cause failure, namely, medium, food, temperature, pH, and so forth. Such a preparation is described in method B.

METHOD B: SYNTHETIC POND WATER MEDIA   Chalkley,[7] Pace,[8] Brandwein,[9] and Hopkins and Pace[10] have described methods that make use of synthetic pond water of a specific chemical composition, instead of natural pond water or hay infusion. (In some cases a buffer may be needed.) In our experience, these methods are superior to those described in the preceding paragraphs. Although more time is spent in preparing this culture medium, it is fully repaid by the quantity of animals found in each culture.

We have found the methods of Chalkley and of Brandwein especially successful. The method described here is our mod-

---

[1] H. R. Halsey, "Culturing *Amoeba proteus* and *A. dubia*," in P. Galtsoff *et al.*, eds. (J. Needham, chairman), *Culture Methods for Invertebrate Animals*, Ithaca, NY Comstock Cornell Univ. Press, (1937). (Reprinted by Dover Press, NY)

[2] H. S. Jennings, "Methods of Cultivating Amoebae and Other Protozoa for Class Use," *J. Appl. Microbiol. and Lab. Methods* (1903) 6:2406.
[3] C. A. Kofoid, "A Reliable Method for Obtaining *Amoeba* for Class Use," *Trans. Am. Microscop. Soc.* (1915) 34:271.
[4] L. Hyman, "Methods of Securing and Cultivating Protozoa: General Statement and Methods," *Trans. Am. Microscop. Soc.* (1925) 44:216.
[5] J. A. Dawson, "The Culture of Large Free-living Amoebae," *Am. Naturalist* (1928) 62:453.
[6] W. LeRoy and N. Ford, "Amoeba," in Galtsoff et al., eds. *Culture Methods for Invertebrate Animals*.
[7] H. Chalkley, "Stock Cultures of *Amoeba proteus*," *Science* (1930) 71:442.
[8] D. Pace, "The Relation of Inorganic Salts to Growth and Reproduction in *A. proteus*," *Arch. Protistol.* (1933) 79:133.
[9] P. F. Brandwein, "Culture Methods for Protozoa," *Am. Naturalist* (1935) 69:628.
[10] D. L. Hopkins and D. Pace, "The Culture of *Amoeba proteus* Leidy Partim Schaefer," in Galtsoff et al., eds. *Culture Methods for Invertebrate Animals*.

ification of existing methods. This is selected because of our prolonged experience with it; it has also been successfully used by other teachers and students. The method has been used with many other protozoa and small invertebrates. Both methods depend upon a synthetic pond water prepared as follows.

BRANDWEIN'S SOLUTION A To prepare this synthetic pond water, weigh out the following salts, and dissolve them in distilled water to make 1 liter of solution:

NaCl	1.20 g
KCl	0.03 g
$CaCl_2$	0.04 g
$NaHCO_3$	0.02 g
phosphate buffer	50 ml
(pH 6.9–7.0)	

This is *stock* solution A. For use, it should be diluted 1:10 with distilled water. (For each milliliter of stock solution A, add 10 ml of distilled water.)

Rinse a number of finger bowls in hot water, then in cold. Next prepare a 1 percent aqueous solution of powdered nonnutrient agar in distilled water, or in solution A. Heat slowly until smooth, then pour while fluid a 1- to 2-mm layer into the bottom of the finger bowls. While the agar is still soft, embed five grains of rice in the agar.[11]

Introduce about 50 amoebae, together with 15 ml of the medium in which they have grown, into each bowl, and add about 30 ml of dilute solution A. During each of the next three days, add 15 ml of dilute solution A, until the total volume is about 90 ml. A few days after the cultures have been started, the layer of agar will separate from the bottom of the dish. Then amoebae may be found growing in layers on the upper and lower surfaces of the agar as well as on the glass surfaces.

After about two months of growth, the culture wanes and should be subcultured. This may be accomplished by dividing the contents of each finger bowl into four parts. Prepare fresh finger bowls contain-

[11] The agar, while not entirely necessary, helps to fix the rice.

ing a film of agar. Add one-fourth of the old culture to each freshly prepared finger bowl, and an equal volume of dilute solution A.

When the original source of amoebae is limited, as it may be when collected in the field, it may be necessary to start small cultures in Syracuse dishes rather than in the larger finger bowls. This apparently provides a better initial concentration of amoebae and makes the change of culture conditions less abrupt.

Prepare the Syracuse dishes with a thin layer of agar and embed two rice grains in each dish. Introduce organisms with about 4 ml of the water in which they were collected; then add 4 ml of dilute solution A. In a successful culture amoebae rapidly proliferate. When some 200 organisms are present, add the culture to a rice-agar finger bowl, with 20 ml of dilute solution A.

Caution: It is detrimental to have large ciliates such as *Stentor*, *Paramecium*, large hypotrichs (*Euplotes*), or *Spirostomun* in cultures of amoebae. Microscopic worms such as *Nais* or *Aeolosoma* (Fig. 1-41), or crustaceans such as *Cyclops* (Fig. 1-56, 1-59) or *Daphnia* (Fig. 1-60) also are harmful contaminants. Cultures containing such organisms may as well be discarded. Moderate populations of *Chilomonas* and *Colpidium* are beneficial as food organisms for amoebae, but these forms should not be present in such numbers that the medium is clouded by their presence. At times, a mold *Dictyuchus*, may grow about the rice, but this does not seem to be detrimental; in fact, amoebae may be found congregated in the mycelia.

Two other solutions for synthetic pond water that may be used as medium for amoebae follow:

CHALKLEY'S SOLUTION Combine the following with 1 liter of water:

NaCl	0.1 g
KCl	0.004 g
$CaCl_2$	0.006 g

Allies of Amoeba Many teachers have found that the large amoeba *Pelomyxa* (sometimes called *Chaos chaos*), which is 3 to 4 mm in length, is superior to *Amoeba proteus* (600 μ) or *A. dubia* (30 μ) (Fig. 9-2). The taxonomy of this giant form is still a matter of controversy.

PELOMYXA   Culture in Brandwein's solution A, using the rice-agar method described earlier. In addition, add a pipetteful of *Paramecium, Blepharisma,* or *Stentor* (see Figs. 1-5, 1-7, 1-8), or all three, as food. If a rich supply of *Paramecium* is available, there should be good growth of *Pelomyxa* within a month. Since *Pelomyxa* is omnivorous, its food vacuoles will be colored red by engulfed *Blepharisma* or green by trapped *Stentor*. Various stages of rotifers and small worms may also be found in the vacuoles of *Pelomyxa*. *Pelomyxa* is characterized by having many nuclei (as many as a thousand), and it may contain up to 12 contractile vacuoles. When it divides it often divides into three parts instead of two.

ARCELLA   This shelled amoeba may be cultured by method A or B.

DIFFLUGIA   Various species of this amoeba, such as *Difflugia oblonga, D. lobostoma,* and *D. constricta,* may be cultured by using a number of green algae as food organisms, using the following method described by Stump.[12] Place such algae as

---

[12] A. B. Stump, "Method of Culturing Testaceae," described in Galtsoff et al., eds. *Culture Methods for Invertebrate Animals.*

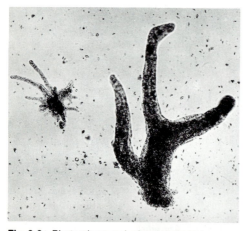

Fig. 9-2 Photomicrograph showing *Amoeba proteus* and *Pelomyxa* (*Chaos chaos*). (Photo by Walter Dawn.)

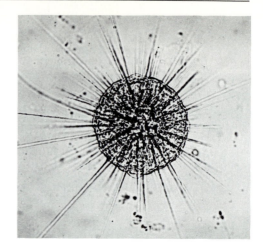

Fig. 9-3 *Actinosphaerium.* (Courtesy of Carolina Biological Supply Co.)

Spirogyra, Zygnema, or *Oedogonium* in Petri dishes or finger bowls. Cover the algae with spring or pond water. Thick cultures should develop in about ten days. Add small quantities of fine sand for these amoebae to build the intricate shells they carry about (see Fig. 1-15). In our experience, this method of cultivation is subject to many variables as described in method A, page 656. *Difflugia* can be readily cultured by method B, using synthetic pond water of a known chemical composition, provided small amounts of fine sand are also added to each culture.

ACTINOSPHAERIUM   To culture this form (Fig. 9-3) method B is preferable. For best results, assure the presence of a moderate amount of *Paramecium*. To 30 ml of solution A in a rice-agar finger bowl, add 20 ml of a culture of *Paramecium*, and inoculate with five to ten specimens of *Actinosphaerium*. Prolific cultures are usually obtained in about ten days to two weeks. At this time, it may be necessary to subculture.

KNOP'S SOLUTION   In addition to the synthetic pond water recipes offered in method B, both heliozoans, *Actinospaerium* and *Actinophrys,* may be cultured in a modification of Knop's solution prepared by dissolving the following salts in distilled water to make 1 liter.

MgSO$_4$	0.25 g
Ca(NO$_3$)$_2$	1.00 g
KH$_2$PO$_4$	0.25 g
KCl	0.12 g
FeCl$_3$	trace

When these cultures are examined from week to week, a succession of living forms is found. In fact, the amoebae are the last to appear; that is, they begin to increase after the ciliates have reached a peak and the culture is "declining." In such a culture a frequent succession may be: small flagellates, *Colpoda*, hypotrichs, *Paramecium*, *Vorticella*, then *Amoeba*. *Paramecium* often appear in a new culture during its second week, and amoebae after two to six weeks.

Slime molds

Many protozoologists consider slime molds to be colonial amoebae and classify them among the Sarcodina. However, most botany texts classify slime molds as a division in the Kingdom Fungi—subdivision Myxomycota.

However biologists wish to classify them, slime molds offer a fascinating study of a complex life history with a conspicuous multinucleated, amoebalike plasmodium. This moving plasmodium sweeps up bacteria, as well as organic debris found in the decaying wood in which it may be found. Refer to the life history of slime molds (p. 706); methods for cultivation are also given using a dried resistant sclerotia stage.

Ciliates

Methods similar to those described for the subclass Rhizopoda may be used in collecting ciliates. When only a small number of organisms has been collected, use Syracuse dishes as described in method B to obtain concentrations for inoculation into larger finger bowls.

General culture methods   Several general methods may be used with success to culture most ciliates. Methods A and B have been described above for amoebae (p. 656). However, when using method B (synthetic pond water media) for ciliates, subculture every two weeks by dividing the culture into two to four parts, then add fresh medium and rice grains. If a culture of *Chilomonas* is available, add this to the medium. After two days, inoculate with *Paramecium*, *Stentor*, *Blepharisma*, or other ciliates.

METHOD C: WHEAT GRAINS   In this method, add five grains of wheat to each 100 ml of boiled, then cooled, pond water. Adjust to pH 7 to 7.6. Let this mixture stand in exposed dishes for a day or so before inoculating with ciliates such as *Paramecium*. Add several pipettefuls of *Chilomonas* if available. Maintain at 21° C (70° F). A population peak should be attained within two weeks.

METHOD D: EGG YOLK   Prepare a thin, smooth paste by grinding 0.5 g of the yolk of a hardboiled egg with a small amount of tap water or boiled pond water. Then add this paste to 500 ml of Brandwein's solution A, or boiled pond water, or tap water. Let this mixture stand for two days before inoculating with the ciliated forms. Or inoculate the mixture immediately with *Chilomonas*, then add the ciliated forms to be cultivated, without the two days of delay.

METHOD E: RICE GRAINS   Synthetic pond water (p. 656) is an excellent medium for most of the ciliates used in the classroom. The only modification used here is a variation in the number of uncooked rice grains. In culturing ciliates use about eight rice grains per finger bowl.

METHOD F: YEAST   Add about ¼ package of dehydrated yeast (a package contains 7.5 g) to 250 ml of pond water or spring water. Mix well and allow this culture medium to stand exposed to the air for several hours. Inoculate with the culture of protozoa you plan to maintain. At room temperature, rich cultures develop within a week. Keep the cultures covered after they have been inoculated with protozoa to prevent evaporation and contamination.

Paramecium grown in this medium often develop a darkened gray shade and have blackish food vacuoles.

METHOD G: SKIM MILK In this method recommended by H. Frings, a pinch of skim-milk powder is added to 250 ml of spring water (or boiled, filtered pond water).[13] For careful work with protozoa, experimentation with volumes and weights may be necessary, but most protozoa tolerate a wide range of concentration of this milk powder.

Add a few pipettefuls of *Paramecium* to this culture medium, and maintain at about 22° C (72° F). In two to three days there should be a rapid proliferation with an abundance of forms undergoing fission. A population peak is reached in five days with this medium, and new cultures should be prepared within two weeks. At times, the addition of a pinch of milk powder to an old culture may be sufficient to renew the population peak for a short time.

Frings finds the protozoa reared in this medium to be large, with clear cytoplasm and clear food vacuoles; the macronucleus is usually visible. Other forms, such as *Tetrahymena* (Fig. 2-10), *Colpoda, Oxytricha, Lacrymaria, Halteria, Vorticella* (Fig. 1-11), *Colpidium, Euplotes* (Fig. 9-4), and *Stylonychia* (Fig. 1-12) also have been successfully cultured with this medium. Amoebae do not maintain themselves in it. On the other hand, plant forms such as *Monas, Chilomonas, Pandorina, Euglena* (Fig. 1-2), and *Peranema* thrive in this medium.

METHOD H: LETTUCE Rub the outer dry leaves of lettuce through a fine-mesh wire strainer. Boil pond water, and when it cools, add 1 tsp of the lettuce to 1 liter of pond water. Boil this mixture for 1 min; let the jar stand covered, overnight. Later, divide the medium into small finger bowls or baby-food jars, and inoculate with a culture of *Paramecium*. Keep the containers covered.

In this method, DuShane and Regnery suggest that a pinch of powdered milk be added to the culture after 12 days and then weekly.[14] After a month's growth, subculture by dividing the old culture into four parts and inoculate into fresh medium.

METHOD I: CEREAL GRAIN Another technique uses dehydrated cereal grain, cerophyll.[15] Combine the following:

cerophyll	10 g
calcium carbonate	5 pinches
sodium chloride	3.8 g
distilled water	5 l

Boil the medium and filter while still warm. When cool, inoculate with paramecia and/or other ciliates.

Culture methods for specific ciliates
Of the methods described above, these are especially recommended for specific ciliates.

PARAMECIUM AND COLPODA Methods E, H and I are superior.

PARAMECIUM BURSARIA This green form is readily cultured by method E. Keep the culture in medium light; *P. bursaria* congregate near the source of light. The green alga *Chlorella* lives in a symbiotic relationship with this species of *Paramecium*.

To demonstrate conjugation, two different pure-line cultures are needed. In sunlight, early in the afternoon, the most effective clumping occurs. Some 12 to 24 hr later, observe conjugating pairs.

BLEPHARISMA This pink ciliate (Fig. 1-7) grows well using methods C and E, although method I seems to be superior. In old cultures look for the giant carnivorous forms that show the presence of smaller *Blepharisma* in food vacuoles. Maintain in moderate light to retain the pink or rose color which fades in bright light.

VORTICELLA A modification of method D is the most satisfactory for this stalked form (Figs. 1-11, 9-4). Prepare the egg yolk-tap water medium as described (method D) and allow it to stand for two

---

[13] H. Frings, "Dried Skim Milk Powder for Rearing Protozoa," *Turtox News* (January 1948) 26:1.

[14] G. DuShane and D. Regnery, *Experiments in General Biology*, (San Francisco: Freeman, 1950).
[15] Available from Cerophyll Co., Kansas City, MO.

PROTOZOA

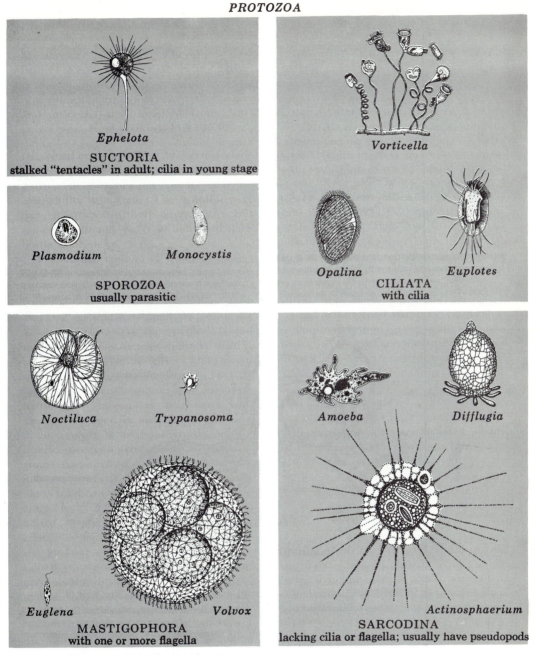

Ephelota
SUCTORIA
stalked "tentacles" in adult; cilia in young stage

Plasmodium **Monocystis**
SPOROZOA
usually parasitic

Vorticella

Opalina **Euplotes**
CILIATA
with cilia

Noctiluca **Trypanosoma**

Euglena **Volvox**
MASTIGOPHORA
with one or more flagella

Amoeba **Difflugia**

Actinosphaerium
SARCODINA
lacking cilia or flagella; usually have pseudopods

Fig. 9-4 A beginning classification of protozoa.

days. Do not add *Chilomonas*. Pour 40 ml of the supernatant fluid into a finger bowl and add *Vorticella*. The mature stalked forms adhere to the bottom of the finger bowls; young free-swimming forms often form a thick layer on the *surface* of the culture.

Subculture every two weeks. Scrape the bottom of the finger bowls to free the organisms and divide the culture into four

(a)

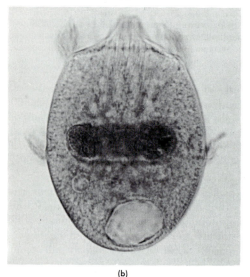

(b)

Fig. 9-5 Ciliates: (a) *Spirostomum*; (b) *Didinium* ingesting *Paramecium*.

parts. To each bowl add 30 ml of fresh medium. When a very heavy population is desired, pour off about 30 ml of the original liquid and add 30 ml of fresh liquid for replacement. In the same manner, remove contaminants by rinsing the bowls several times with Brandwein's solution A (since the *Vorticella* adhere to the bottom). Add fresh medium.

STENTOR Both *Stentor polymorphus* and *S. coeruleus* may be cultured by methods C, D

and E; method E is superior. Excellent cultures of *Stentor* may be obtained when pipettefuls of *Paramecium* or *Blepharisma* (or both), as well as more *Chilomonas*, are introduced into the culture periodically.

SPIROSTOMUM Culture this elongated ciliate (Fig. 9-5a) using methods C, D, and E. When the cultures become putrid, conditions seem to be favorable for this form.

DIDINIUM Introduce several of these carnivorous ciliates (Fig. 9-5b) into a rich culture of *Paramecium* (prepared by method C or E). *Didinium* feed upon *Paramecium* and are found in a similar habitat. Within a week the *Paramecium* in the culture will probably have been consumed. Then inoculate the *Didinium* into fresh cultures of *Paramecium*.

You may preserve *Didinium* for later use by filtering an old culture through filter paper to retain the cysts. Dry the filter paper in air and store in envelopes. When you wish to start a culture of *Didinium*, add a dried sheet of this paper to a thriving culture of *Paramecium*. You may prefer to allow a culture of *Didinium* to dry out in its container. When you wish to revive the culture, add a fresh *Paramecium* culture to the dish.

COLPIDIUM This small ciliate (Fig. 2-10) may be cultured in a medium made by boiling 100 ml of spring water or pond water to which 1.5 g of whole rye grains have been added. After 10 min of boiling, filter the fluid, then cool. Expose to air for a day, and then inoculate with about 10 ml of an culture of *Colpidium*. At a temperature of about 22° C (72° F) these cultures reach a peak in about a week. Also use method G or E.

Some interesting work has been done with bacteria-free cultures. Kudo gives Kidder's formula for cultivating bacteria-free cultures of *Colpidium* (also axenic cultures, p. 664).[16] In this method, add 10 g of brewer's yeast to 1 liter of distilled water. Then boil and filter through cotton. Again filter, this time through filter paper. To this solution, add 20 g of Difco dehy-

[16] R. Kudo, *Protozoology*, 5th ed., Springfield, IL: Thomas, 1973.

drated proteose-peptone medium. Sterilize the whole solution for 20 min at 15 lb pressure in an autoclave or pressure cooker. Inoculate with individual *Colpidium*.

STYLONYCHIA AND OXYTRICHA Method E may be used for culturing *Stylonychia* (Fig. 1-12) and *Oxytricha*; previous inoculation with *Chilomonas* is necessary.

EUPLOTES For culturing this form, use method C or D; inoculate with *Chilomonas* as a preliminary step.

TETRAHYMENA Method G (skim milk) is superior in culturing *Tetrahymena*, although method B is also effective. Or, prepare 1 percent solution of Difco dehydrated proteose-peptone medium (1 g/ 100 ml distilled water). Sterilize the solution in an autoclave or pressure cooker at 15 lb pressure for 15 min; when cool, inoculate with *Tetrahymena* (Fig. 2-10). Then maintain the cultures in the dark for a few days to obtain a rich culture. Cultures may be refrigerated.

One fact bears repetition; in raising certain carnivorous organisms, the best results are obtained when the food organism is raised separately and then added periodically to the culture medium. For example, *Stentor* grows best in media to which *Chilomonas* or *Colpidium* have been added. Similarly, *Didinium* cultures require *Paramecium* as a food organism, *Lionotus* feeds on *Colpidium*, and *Actinobolus* consumes *Halteria*.

Flagellates

This class includes protozoa with one or more flagella. Among the Mastigophora are some forms that may be classified as protozoa or as plant forms belonging to the Phytomastigina. An interesting feature of this group is its lengthwise fission in asexual reproduction.

The procedures described for collecting Rhizopoda and Ciliata should also be used for gathering flagellates. In addition, the green-surface "blooms" which may be found in ditches or ponds may often contain large numbers of *Euglena, Chlamydomonas*, and similar flagellates (Figs. 1-3, 1-116).

General methods for flagellates In general, three of the methods described previously are also recommended for cultivating flagellates. Of course, the forms that contain chlorophyll require moderate light.

Method B (described for rhizopods) and methods C and D (described for ciliates) are also successful for flagellates.

METHOD J: WHEAT GRAINS A modification of method C is also recommended for many flagellates; boil 4 wheat grains in 80 ml of pond water. When the medium cools, add a few milliliters of pond water containing the flagellates to be cultured.

METHOD K: MODIFIED KLEBS' SOLUTION This method is recommended for *Euglena* as well as other flagellates.[17] To 100 ml of this solution in a glass battery jar, add 20 rice grains (which have been boiled for 5 min) and 900 ml of distilled water. Let this mixture stand for two days. Inoculate the mixture with an old culture of *Euglena*, and keep the jar in indirect sunlight. Direct rays of the sun should not strike this culture for more than 1 hr a day. Inoculate the culture with *Euglena* three times at 3-day intervals. If an old culture of *Euglena* with encysted forms is available (these may be found on the sides of the jar) inoculate the cysts along with the motile forms. After two to three weeks, add an additional 10 mg of tryptophane powder that has been dissolved in 25 ml of the modified Klebs' solution.

To 1 liter of distilled water, add the following:

KNO_3	0.25 g
$MgSO_4$	0.25 g
KH_2PO_4	0.25 g
$Ca(NO_3)_2$	1.00 g
bacto-tryptophane broth powder (1-form)	0.01 g

METHOD L: STARR'S SOLUTION Starr recommends adding the following to 1 liter of distilled water to culture *Euglena*[18].

[17] Brandwein, "Culture Methods for Protozoa."
[18] R. C. Starr, *Amer. J. Bot.* (1964) 51:1013.

sodium acetate	1.0	g
beef extract	1.0	g
tryptone	2.0	g
yeast extract	2.0	g
calcium chloride	0.01	g

The medium may be solidified by adding 15 g agar per liter of culture medium. Prepare and refrigerate before use.

Culture methods for specific flagellates

EUGLENA Method K among others is useful; however, Jahn[19] and Hall[20] offer additional recommendations for culturing this autotroph. While these require sterile conditions[21] and are tedious to prepare, Difco offers a prepared dehydrated medium. Once they become established, *Euglena* may be maintained indefinitely in pure, that is, axenic cultures. Also refer to some other methods for green algae (pp. 688–96).

Euglena is classified in Family Euglenidae (Class Mastigophora) among the protists. When grown in light these flagellates photosynthesize food materials, as do typical autotrophs. However, since they sustain themselves and reproduce when grown in the dark when supplied with an organic source, they are often studied as examples of heterotrophs. Chloroplasts bleach out or fragment, and colorless forms persist as consumers of organic materials. *Euglena* can be bleached with streptomycin.

CHLAMYDOMONAS These small green flagellates, resembling algae, grow well under the conditions of methods K and B

and the other methods described for growing green algae.

PERANEMA Method B yields excellent results; method J is also suitable.

CHILOMONAS Use methods C, D, or E, all of which yield excellent results.

ENTOSIPHON Use method J or B, both of which are equally successful.

In this discussion we have omitted the Sporozoa and Suctoria since they are not commonly maintained in introductory biology classes. The sporozoan, *Monocystis*, may be found in the intestine of earthworms (p. 620).

Axenic cultures

Axenic cultures Methods for establishing axenic cultures of protozoa (pure cultures) were probably first devised by Parpart;[22] additional methods were developed by Claff,[23] Kidder,[24] and others.

An axenic culture of *Paramecium*, for example, can be established only after the organisms are washed free of bacteria, other protozoa, fungi, and multicellular microorganisms. After the desired forms are separated (by washing or use of antibiotics), the organisms must be grown in a suitable medium. Some of these synthetic media contain an extensive list of ingredients. Manwell[25] lists a minimal, chemically defined medium for the ciliate *Tetrahymena* (as used by Elliott[26]) comprising over 11 amino acids, 2 carbon sources, 4 nucleic acids, 7 growth factors, and 5 inorganic salts (quantities in milligrams per liter). (See also p. 663 for culturing *Tetrahymena* using a Difco proteose-peptone medium.)

Laboratory strains are available from Carolina Biological Supply Co.

[19] T. L. Jahn, "Studies on the Physiology of the Euglenoid Flagellates," Part III, "The Effect of Hydrogen Ion Concentration on the Growth of *Euglena gracilis* Klebs," (1931) *Biol. Bull.* 61:387.
[20] R. P. Hall, "On the Relation of Hydrogen Ion Concentration to the Growth of *Euglena anabaena* var. *minor* and *E. desos*," *Arch. Protistol.* (1933) 79:239.
[21] A. K. Parpart, "The Bacteriological Sterilization of *Paramecium*," *Biol. Bull.* (1928) 55: 113–20, and G. W. Kidder, "The Technique and Significance of Control in Protozoan Culture," in G. N. Calkins and F. M. Summers, eds., *Protozoa in Biological Research*, (New York: Columbia Univ. Press, 1941).

[22] A. K. Parpart, "The Bacteriological Sterilization of *Paramecium*."
[23] C. L. Claff, "A Migration-Dilution Apparatus for the Sterilization of Protozoa," *Physiol. Zool.* (1940) 13:334–40.
[24] G. W. Kidder, "The Technique and Significance of Control in Protozoan Culture." In Calkins and Summers, eds., *Protozoa in Biological Research*.
[25] R. Manwell, *Introduction to Protozoology*, (New York: St. Martin's (1964).
[26] A. M. Elliott, "Biology of *Tetrahymena*," *Ann. Rev. Microbiol.* (1959) 13:76–96.

Uses of protozoa in the classroom and laboratory   The organisms that can be cultured by the methods described above have many uses during the year. When cultures are maintained routinely, living materials are available at any time for such studies as

1. Microscopic examination of organelles (chapter 2)
2. Behavior (chapter 5)
3. Food-getting (chapter 3)
4. Reproduction and heredity (chapters 6 and 7)
5. Ecosystems (chapter 8)

Classification of protozoa   In some circumstances, students may learn to use a key to identify organisms or as an introduction to a study of variations among living things (see chapter 7). We shall not present detailed keys to classification of protozoa, but simply attempt, mainly with drawings, to focus observations on basic differences among the more common kinds of protozoa that might be found in pools of water, a ditch, or a container of rainwater.[27] For example, if students examined with a microscope a drop of pond water rich in protozoa they might find some of the forms shown in Fig. 9-4. Sketches of these representative forms could be drawn on the blackboard so differences may be observed. What is one distinguishing characteristic of each type? Observe that some forms are ciliated, some have one or more flagella, possibly some under view are amorphous masses of protoplasm with bulging false feet (pseudopodia), or some are highly organized cells with fine, spinelike projections.

Suppose we examine the flagellated forms closely. There may be single-celled types, such as *Euglena*, or colonial types, such as *Volvox*. Furthermore, some may have chlorophyll or other pigment (Sub-

class Phytomastigina); others may lack color bodies (Subclass Zoomastigina) and be free living or, like trypanosomes, parasitic.

A more complex problem of variations exists among the members of Class Ciliata. The location of the "mouth," or peristome, and the arrangement of cilia or long, fused cilia (membranelles) in relation to the peristome are the basic distinguishing features of the four subclasses of Ciliata; see *Paramecium*, *Euplotes*, and *Vorticella*.

Microscopic examination of the contents of the intestine of an earthworm may reveal the sporozoan *Monocystis* (p. 620). Prepared slides of plasmodia that cause malaria may also be studied at this time. You may also want to have students examine the flagellates (Zoomastigina) in the intestine of termites (p. 631).

Also locate the protozoa commonly found in the rectum of the frog: *Opalina*, *Nyctotherus*, and *Balantidium* (Figs. 8-41, 8-44, 8-45).

Students who want to go further in identifying genera of protozoa may consult the fine keys that are developed in textbooks of protozoology (see Book Shelf).

Freshwater invertebrates

Sponges

Only one family of sponges is found in freshwater, the Spongillidae. While marine sponges may be collected from considerable depths and dried for classroom use, they are difficult to keep alive without saltwater aquaria. In fact, freshwater forms are also very difficult to keep in the laboratory. Specimens of freshwater sponges may be found as crusty, brownish growths attached to submerged plants and rocks. Many of these sponges have the texture of raw liver. Some forms exposed to sunlight may be green due to *Chlorella* living symbiotically with the sponge.

Spongilla (Figs. 1-18, 1-20) is difficult to keep alive more than a few weeks in an aquarium, and this is possible only when

[27] We pass lightly over two major types of protozoa: Sporozoa (many of which are parasitic) and Suctoria (which are not widely abundant).

large specimens have been collected (as at the end of the summer and into fall). However, resistant gemmules of these sponges may be cultivated in depression slides. Some will grow and attach themselves to the glass surfaces. Keep them in darkness at 20° C (68° F).

Because sponges as a group have limited usefulness in introductory biology, they are given only brief mention here.

Recall that sponges have remarkable power to regenerate. When disorganized by pressing a sponge through cheesecloth, the cells reassemble within days to form newly organized sponges.

Hydra

While many jellyfish forms may be collected and preserved for examination, *Hydra* is usually the one that is maintained under fairly simple laboratory conditions. *Hydra* is most often studied for its feeding habits, methods of reproduction and regeneration.

Look for *Hydra* in lakes or ponds, attached to submerged stems of water plants or on the underside of floating leaves of water lilies or water hyacinths. In autumn, many budding hydras may be found (see Figs. 1-21, 1-22, and 1-24).

In the laboratory, transfer the plants upon which hydras are found into finger bowls or small aquaria. If available, use the water in which the hydras have been living or water from a thriving aquarium. Add the aquarium water slowly, a glassful per day, to the original water in which the hydras were found, so that the organisms will become acclimated. Brandwein's solution A (p. 657) may be used in the absence of satisfactory aquarium water. (Also refer to page 674 for marine coelenterate culture methods.)

Keep the containers in medium light or semidarkness, at a temperature below 20° C (68° F). Within a day or so the hydras will be found on the surface of the water. They can be picked up with a pipette and transferred into new containers to start fresh cultures. (If the hy-

dras remain contracted and fail to expand, the water that has been added is not suitable. Do not add more hydras to such a tank.)

Loomis and Lenhoff recommend the following method for culturing hydras[28]. Since they found that calcium ions are necessary for growth of hydras, they developed a chemically defined medium utilizing $CaCl_2$. To remove the toxic copper ions from tap water, they suggest using a chelating agent. In this solution the chelating agent is Versene (disodium ethylenediamine tetraacetate).

Prepare a stock solution of modified tap water:

$NaHCO_3$	20 g
Versene	10 g
$CaCl_2$	50 mg
tap water	1 liter

The sodium bicarbonate is used as a buffer to maintain the solution at pH 7.5 to 8.

Loomis and Lenhoff also found that as population density increased, sexual differentiation occurred (see also Fig. 1-21).

Keep the cultures in moderate light. Green hydras, *Chlorohydra viridissima*, in which *Chlorella* live symbiotically in the gastrodermis or endoderm, require more light. (Incidentally, these algae are passed to the next generation in the cytoplasm of the eggs of hydras.)

About twice a week feed the hydras a rich culture of *Dero*, *Tubifex*, *Daphnia*, or *Artemia* (culturing, pp. 669, 670, 672). Well-fed hydras grow rapidly; excess feeding often results in budding.

On the whole, the most successful hydra for cultivation in the laboratory aquarium is *Hydra oligactis*. Adequate green water plants are needed for a rich supply of oxygen.

[28] W. F. Loomis and H. M. Lenhoff, "Growth and Sexual Differentiation of *Hydra* in Mass Culture," *J. Exptl. Zool.* (1956), 132:555–74. Other culturing methods are given in Galtsoff et al., eds. *Culture Methods for Invertebrate Animals.* Also refer to W. S. Bullough, *Practical Invertebrate Anatomy*, 2nd ed., (New York: Macmillan, 1958).

At times periods of depression beset hydras; tentacles are contracted and the body becomes shortened. Depression may often be avoided by frequent changing of water.[29]

Uses in the classroom and laboratory
A student might design an investigation into ways to prevent depression of hydras. Also use hydras to investigate the following behaviors:

1. *Food-getting.* Examine how the hydra uses nematocysts (p. 15) to capture brine shrimp larvae or *Daphnia* (Figs. 1-22, 1-23). Use starved hydras that have not been fed for 24 hr to study ingestion (p. 238).

2. *Regeneration* (p. 496). Hydras might be used to study the nature of polarity. For example, can tentacles grow on the "wrong" end?

3. *Sexual differentiation* (p.18).

4. *Taxes* (p. 336). Demonstrate the responses to touch, to food, to weak acids, and to light.

5. *Glutathione feeding response* (p. 239).

6. *Rate of growth* (p. 238) and *grafting* (p. 496).

7. *Growth and differentiation* at the tissue level. Fulton suggests using a colonial hydroid, *Cordylophora lacustris*[30]. The organisms may be cultured on slides slanted in a beaker of culture medium. Doubling time of hydranths is three days.

Preserve some hydras to be used only when living hydras are not available. First, narcotize the forms by adding crystals of menthol or of Epsom salts (magnesium sulfate) to the finger bowl of water containing the hydras. When they are elongated and quiet, place them in 70 percent alcohol to fix and preserve them. (Refer also to p. 646.

[29] Refer to W. F. Loomis, "Sex Gas in Hydra," *Sci. Am.*, (April 1959).
[30] C. Fulton, "Culture of a Colonial Hydroid Under Controlled Conditions," *Science* (1960) 132:473–74.

Planaria

Look for these small flatworms on the underside of submerged logs and under stones in ponds and lakes. Several varieties may be found in clear running water, but the usual forms are the small, blackish *Planaria maculata* and the more frequent laboratory form, the brown *Dugesia tigrina*. When you find some in a submerged log, wrap the whole log in wet newspaper and bring to the laboratory. Submerge the log in a white enamel pan of water and peel off sections of wood; usually the planarians float to the top. Planarians may also be baited by submerging a piece of raw beef liver or hard-boiled egg yolk tied in cheesecloth attached to a string in a cold stream or lake. This method often attracts the larger form *Planaria dorotocephalia*. Brush off the gathered forms into collecting jars and submerge your bait in another part of the lake or stream.

Transfer the collected plant materials into larger glass jars and keep in moderate light. Soon planarians may be found clustering on the surface of the water or adhering to the sides of the jars. Then pick them off with a pipette or camel's hair brush and isolate them in separate culture containers.

Because planarians are photonegative, they should be maintained in black or opaque containers; enameled pans are excellent. Change the water frequently, with fresh additions of aquarium water, spring water, or Brandwein's solution A (p. 657). Keep them at a temperature about 18° C (64° F). Once a week feed the planarians finely chopped raw beef liver; better still, since live food does not foul the water, feed bits of worms (*Tubifex* or *Enchytraeus*; culturing, p. 668). At other times, feed them bits of hard-boiled egg yolk. Remove the excess food with a pipette after several hours to avoid fouling the water.

Uses in the classroom and laboratory
Planarians are studied as a representative of the platyhelminthes (chapter 1). These flatworms are classic material for studies in regeneration (p. 497). At times you

may find planarians reproducing by fragmentation. They rarely reproduce sexually in the laboratory; orange cocoons about the size of radish seeds, may be found that occasionally hatch out some 4 to 6 small planarians in about two weeks.

Use planarians to show taxes (pp. 22, 239). What happens when one is put on a slide and the glass is tapped? How do they respond to light? See also RNA and learning (p. 23) and ingestion (p. 239). Study locomotion, especially the rapid gliding movement due to ciliated epidermis.

For gross examination under the microscope, it may be necessary to narcotize the worms before putting them on a slide. To a small watch glass or Syracuse dish of pond water containing a few planarians, add a few crystals of Epsom salts or menthol (see p. 646). When the forms are quiet, lift them with a toothpick and arrange the animals on slides so that the proboscis is uppermost. (See also a technique for sandwiching planaria, p. 239.)

Vinegar eels

These nonparasitic roundworms (Fig. 1–35) feed upon the fungus "mother-of-vinegar." Because bottled vinegar has been pasteurized to inhibit the growth of these roundworms, bulk vinegar must be used as a source.

Add small quantities of bulk vinegar containing these worms to quart containers of pure, unadulterated cider vinegar. Then cover the cultures to prevent evaporation. Or add vinegar eels to small finger bowls containing two small cubes of raw apple in 150 ml of cider vinegar.

Wide fluctuations in temperature are tolerated by vinegar eels. Subculture the stock about four times during the year, adding a bit of the old culture to fresh cider vinegar. You may also purchase vinegar eels from supply houses (Appendix).

Uses in the classroom and laboratory

Roundworms may be used in studies of relationships among organisms in an ecosystem (chapter 8). Where parasitic roundworms are not available for study, the vinegar eel may be substituted (chapter 1).

Earthworms

These annelids are readily collected at night or after a good rain when they come to the surface.

Place several worms in wooden containers such as cigar boxes. Place 5 to 15 cm of moist, rich soil or peat moss (sphagnum moss) in the boxes and lightly dampen the soil. Keep the worms covered and in a cool place (at a temperature about 15° C (59° F). About twice a week feed them lettuce and bread soaked in milk; bury the food.

Uses in the classroom and laboratory

Earthworms are excellent organisms for studies of taxes. When the cover is removed so that the worms are exposed to light, watch the rapid burrowing movements. Earthworms may also be used to study chemotaxes (see chapter 5).

Earthworms are also classic "types" for dissection. The reproductive, digestive, and circulatory systems, as well as the ventral nerve cord, may be studied as the worm is dissected (Fig. 1-47). Of course, earthworms may also be cultivated as food for frogs and some reptiles maintained in the laboratory. (See also *Monocystis*, p. 620). Mature worms are also a source of living gametes (p. 48). The nematode *Rhabditis* may be found in the nephridia (Fig. 8-31).

Enchytraeus

These white, semiaquatic segmented worms (Fig. 1-46) may be purchased from an aquarium store, a supply house, or collected from samples of damp, rich garden soil. (Also refer to Baermann funnel, Fig. 8-23).

Cultivate these tiny annelids in the same way as earthworms. Keep in covered boxes containing 5 to 10 cm of rich garden soil. Feed them lettuce and potatoes boiled in their skins; alternate this with cooked oatmeal. On occasion, feed them bread soaked in milk. Keep several small cultures going rather than one large culture; maintain at temperatures about 20° C (68° F). Under these conditions, they multiply rapidly and many cocoons should be found among the masses of food.

Uses in the classroom and laboratory
Enchytraeid worms are used mainly as food for other laboratory animals such as fish, amphibia, and small reptiles. They may also be used to demonstrate taxes. Prepare wet mounts of *Enchytraeus*; examine the contraction of muscle, such as peristalsis along the length of the intestine.

Tubifex and other aquatic worms

Any of the oligochaete annelids described in the following paragraphs may be purchased from biological supply houses that distribute living materials (see Appendix).

Tubifex (Fig. 1-45) may also be obtained readily from aquarium shops. In the field, *Tubifex* may be collected from the muddy bottom and decaying leaves of streams and ponds. They form tubes of mud held together by a secretion from epidermal cells. These forms usually have a reddish color due to dissolved erythrocruorin in the blood.

Members of the family Naididae, such as *Nais* (Fig. 3-24) and *Dero* (Fig. 1-42), carry on respiration through ciliated gills in the anal region. Naididae are abundant in old cultures of protozoa. Under the microscope, *Aulophorus* can be distinguished from *Dero* by its two microscopic, fingerlike terminal processes, as well as gills (Fig. 1-43), which are not found in *Dero*. The Naididae are larger than members of the family Aeolosomatidae, which are found in similar places; they lack colored oil globules. *Nais* is about 3 mm long, comprising 15 to 37 segments. *Aeolosoma* (Fig. 1-41) consists of 8 to 10 segments and is only about 1 mm long. *Aeolosoma* reproduces asexually by transverse fission. The Naididae also lack dorsal bun-

dles of setae in the anterior segments. *Aeolosoma* is readily identified by minute yellow, greenish, or red globules in the epithelium.

Introduce *Tubifex* and *Nais* into well-established aquarium tanks that contain a few centimeters of muddy soil.

Culture *Nais*, *Dero*, *Aeolosoma*, and *Aulophorus* by the methods described earlier for maintaining protozoa, particularly method B or D.

Uses in the classroom and laboratory
These aquatic worms serve as a food supply for other laboratory animals, such as fish and hydras. They may be used in laboratory and classroom investigations in the following areas:

1. *Regeneration* (pp. 497–98).

2. *Circulation of blood*, which is visible in some forms, and peristaltic contractions, which are especially clear in wet mounts of these worms, under low and high power (p. 268).

3. *Asexual reproduction* (fragmentation and transverse fission); they are examples of hermaphroditic oligochaetes.

Rotifers

Half fill several jars with submerged plants from a pond, and then add pond water to fill the jars. In the laboratory, remove the covers and place the jars in moderate light. After a day or so rotifers (Fig. 1-37) will be found congregated on the surface, where there is an abundant supply of oxygen.

Use a pipette to pick the rotifers out of the jar, and introduce them into finger bowls of pond water. Change the culture water frequently. Feed the rotifers *Chlorella*, *Euglena*, or *Chlamydomonas*. Keep the finger bowls stacked to prevent evaporation of the medium.

In one simple procedure, dissolve 1 g of nonfat dried milk in 1 liter of spring water.[31] Use a phosphate buffer (p. 159)

[31] L. Lindner, H. Goldman, and P. Ruzicka, "Simple Methods for Rotifer Culture," *Turtox News* (1961) 39:74.

to maintain the culture at pH 7 to 7.5. The condition of the local tap water may make it necessary to use a chelating agent, such as Versene (p. 666) to remove the copper ions; or use spring water or Brandwein's solution A (p. 657).

Snails and clams

Among mollusks there are wide variations in shape. Such egg-laying forms as *Physa*, *Planorbis*, and *Lymnaea* (Fig. 9-6) may be found attached to water plants in ponds and lakes, or they may be purchased from aquarium stores. *Planorbis* has a shell coiled in one plane like a watch spring. It lays eggs in clusters of jelly. A popular form for aquaria is the imported red variety, which lays pinkish masses of eggs. When both kinds are bred together, the common brown variety seems to be dominant. *Physa*, recognized by its sinistral spiral shell, lays eggs in long ribbons of jelly. *Lymnaea* is brownish-black with a dextrally coiled shell. It lays egg masses in jelly, usually found attached to stems of aquatic plants. A larger form, *Campeloma* (Fig. 9-6d), is a live-bearer and may be found in lakes or rivers attached to rocks or plants. Keep the aquatic snails such as *Planorbis* in an established aquarium. They normally feed on algae, but when they increase in number they feed upon the aquatic plants (dissection, chapter 1).

Finally, *Helix* (Fig. 1-80), a land snail, may be found in moist but not too acid soil such as that in gardens or some wooded areas. Raise *Helix* in a cool place in a moist terrarium with occasional feeding of lettuce. At times, also add whole oats rolled in calcium carbonate powder.

Uses in the classroom and laboratory

When the snails reproduce, separate the developing eggs into covered Syracuse dishes. With a hand lens or dissecting microscope, examine cleavage stages and ciliated veliger larvae. In the young embryos notice the beating heart. Snails furnish a living source of material for studies of early embryology (pp. 407, 267).

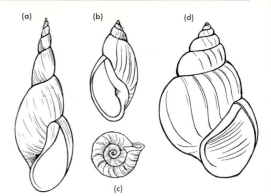

Fig. 9-6 Four freshwater snails that thrive in an aquarium: (a) *Lymnaea*; (b) *Physa*; (c) *Planorbis*; (d) *Campeloma*.

Snails should be kept in aquaria to hold down the abundant growth of algae and to remove decaying materials in tanks.

Freshwater mussels and clams as well as marine oysters and clams are also good material for dissection (chapter 1).

Land snails, which are available from some fish markets, can be distributed to students for many interesting studies of taxes (see chapter 5).

Some small crustaceans

Small freshwater crustaceans can be collected from lakes and ponds with fine mesh nets. Culture methods for most of these forms are similar to those described for *Daphnia*.

Daphnia

These small, laterally compressed water fleas (Order Cladocera) are characterized by a body enclosed in a transparent bivalve shell (Fig. 1-60). A cleft marks off the head from the rest of the body. Large second antennae are modified as swimming appendages to assist the four to six pairs of swimming legs. During the spring and summer, females are usually found; eggs generally develop parthenogenetically and may be seen in the brood pouch. In the autumn, males appear, and the "winter eggs" are fertilized in the brood pouch. At the next molt, the eggs are shed in a thickened brood pouch, called an ephippium. In the spring, these fertilized winter eggs of *Daphnia* hatch into parthenogenetic females.

Female *Daphnia* may be recognized by the curved shape of the end of the intestine. In the male, the intestine is a straight tube.

A great many successful methods have been described for maintaining *Daphnia*, which feed on bacteria and nonfilamentous algae such as *Chlorella*. Two methods that have proved successful are described here.

USING "GREEN WATER" Fill large battery jars with tap water and let them stand overnight to permit evaporation of gases that may be harmful. Then put the battery jars in strong sunlight and inoculate them with nonfilamentous algae from a "soupy green" aquarium. After this "green water" has been standing for about two to three days, add *Daphnia* and a few milliliters of hard-boiled egg yolk mashed into a paste with a bit of culture medium. You may also add a suspension of yeast to stimulate growth. This method produces a luxuriant growth of *Daphnia*. The temperature range may vary between 24° and 26° C (75° and 79° F). The sediment often contains viable eggs.

USING A MODIFICATION OF KNOP'S SOLUTION In this method, a 6 percent *stock* solution of a known chemical composition is prepared. Combine the following with 1 liter of distilled water:

KNO_3	1 g
$MgSO_4$	1 g
K_2HPO_4	1 g

Then add 3 g of calcium nitrate, $Ca(NO_3)_2$. As a result, a precipitate of calcium phosphate $Ca_3(PO_4)_2$ is formed.

For immediate use dilute by adding 5 liters of distilled water to 1 liter of the stock solution. This will yield a dilute 0.1 percent solution. Pour into several battery jars. When necessary, this may be further diluted with an additional 4 liters of distilled water. Even this weak solution will maintain *Daphnia* adequately when the culture medium has been inoculated with nonfilamentous algae and maintained in light until the water becomes tinged with a green color. About once a week, add a bit of hard-boiled egg yolk paste and a bit of yeast suspension.

Uses in the classroom and laboratory
Daphnia serve as excellent food for small fish, tadpoles, and hydras. Introduce a drop of a culture of these water fleas on a slide containing one or two hydras. Under the microscope watch ingestion by the hydra. What is the role of the nematocysts (p. 17)? Use *Daphnia* to clear an aquarium that has become soupy green; examine *Daphnia* under the microscope to find the green food tube.

Demonstrate the rapidly beating heart of *Daphnia* under low power of the microscope (some 120 times per minute, or faster, depending on the species, p. 267). Also demonstrate the effect of epinephrine and/or drugs on the heartbeat. Use a hanging drop preparation (Fig. 2-22), or put bits of broken coverslips near the *Daphnia* as you prepare a wet mount to avoid crushing them; or immobilize *Daphnia* in a dab of petroleum jelly in a depression slide.

Small amounts of epinephrine (adrenalin), and pituitary hormones such as commercial pituitrin, cause a spontaneous shedding of the eggs from the dorsal brood sac.

Use wet mounts of *Daphnia* for examination of circulation and heartbeat, method of ingestion, peristalsis, and behavior (pp. 239, 240). In some lake-dwelling *Daphnia*, there may be changes in head shape, from round to helmet-shaped, between spring and summer. What conditions cause this cyclomorphosis? Also refer to changes in color depending on oxygen content of the water. *Daphnia* reared in well-aerated media have light colored hemoglobin; orange *Daphnia* are found in poorly aerated media when more hemoglobin forms.

Fig. 9-7 *Cyclops*: female with egg sacs. (Photo by Hugh Spencer.)

Cypris

At first glance *Cypris* (Fig. 1-61) is often mistaken for *Daphnia*. However, it has an opaque shell which makes the study of its internal anatomy difficult.

Cypris is laterally compressed and completely enclosed in a bivalve shell. It usually has seven pairs of appendages, and its antennae protrude from the shells and are used in swimming. *Cypris* may be collected from ponds and streams.

Cyclops

This small elongated crustacean (Subclass Copepoda) lacks a shell and has no abdominal appendages (Figs. 1-59, 9-7). It is characterized by the single compound eye located in the center of the head; it uses antennae for locomotion. During the summer months, females may be found carrying two brood pouches posterior to the body, as shown in Fig. 9-7. *Cyclops* may be found in brackish water as well as in freshwater streams and lakes.

Culture methods are similar to those described for *Daphnia*. *Cyclops* is also a good organism for introductory work with a microscope since they are easy to find on a slide (p. 51). Discover the effect of epinephrine on this graceful organism. They can also be used as food for small invertebrates, fish, and amphibia.

Eubranchipus

These small crustaceans, called fairy

shrimp (Figs. 1-57, 9-8), Order Anostraca, swim with their ventral side uppermost by means of gracefully coordinated thoracic appendages. The head bears stalked eyes, and the body is transparent. Fairy shrimps are found in shallow, stagnant ponds that may dry up during the summer months. The resistant eggs settle in the mud until the next spring. Culture them like *Daphnia* and other small crustaceans.

Gammarus

These shrimps (Fig. 1-64) are found abundantly along freshwater streams and along the seashore. They are members of Order Amphipoda, distinguished by laterally compressed bodies with gills borne on the legs. The first three pairs of legs are used as swimming legs, with the last three pairs modified as stiff processes used in jumping. They swim in an upright motion, or crawl on decaying vegetation. *Gammarus* is easy to maintain in the laboratory when placed in an established tank of much vegetation and pond water; they reappear year after year and need no attention, except for a pea-sized bit of hard-boiled egg yolk added every month or so.

Reproduction in these freshwater organisms occurs in the spring and summer. Use *Gammarus* in studies of heredity—the inheritance of eye color has been well studied.

Artemia

Brine shrimps and their larvae (Fig. 1-58) are found in saline lakes, or they may be purchased as dried, resting eggs from aquarium shops or supply houses. They belong to Order Eubranchiopoda and are characterized by 10 to 30 pairs of leaflike swimming limbs. In development they have a nauplius stage like *Cypris* and *Cyclops*. (*Daphnia* has direct development.)

Artemia are sensitive to light and orient themselves so that the ventral surface is placed toward light. Thus they often swim with the ventral surface uppermost. In the female, lateral egg pouches are conspic-

Fig. 9-8 *Eubranchipus*, the fairy shrimp, a small crustacean that swims ventral side up.

(Photo by L. C. Peltier, courtesy of *Nature Magazine*, Washington, D.C.)

uous. Rapid beating of the limbs is characteristic of this form (150 to 200 beats per minute). After each molting, the females are ready for mating. Batches of eggs may be laid as often as every four or five days when ample food is available. Eggs laid with abundant secretion usually remain dormant for some time, often for several months. A period of drying out seems to shorten the time of hatching. Dried eggs retain their viability for several years, provided they are kept in a cool place. On the other hand, eggs that have a scant secretion when laid hatch out in one or two days as nauplius larvae.

HATCHING ARTEMIA EGGS   Dried brine shrimp eggs are available throughout the year and can readily be hatched to show a nauplius larva stage of crustaceans. Much effort has been given to develop standard conditions for hatching these eggs, since the larvae are used commercially as live food for *Hydra*. You may wish to feed live larvae to *Hydra* or to planarian worms in studies of ingestion (pp. 238, 239). After considerable experimentation, Loomis and Lenhoff have recommended the following procedure for hatching large quantities of *Artemia* eggs (reduce quantities to fit your laboratory needs).[32]

[32] Loomis and Lenhoff, "Growth and Sexual Differentiation of *Hydra* in Mass Culture."

Prepare a stock solution of saturated sodium chloride (360 g per liter) by dissolving 5 lb of commercial table salt in 2 gal of hot tap water. Cool this solution, and dilute it with tap water (1:100). Seed wide, shallow hatching dishes of the dilute salt solution with *Artemia* eggs (½ tsp of eggs per 500 ml of solution). Incubate for 48 hr at a constant temperature of 21° C (70° F). (At a temperature of 30° C [86° F] eggs hatch in one day; at 15° C [59° F] it takes three days.)

When the larvae are to be fed to hydras the larvae must be rinsed in aquarium water or conditioned water (see p. 679), or the salinity will kill the hydras. To collect the larvae, shine a light at one side of the container; then the phototactic larvae can be siphoned off with thin rubber tubing into a fine net (125-mesh).

Avoid overcrowding the culture of *Artemia* larvae, and supply them with non-filamentous algae and a yeast suspension. Algae scrapings from the sides of a tank have been found to develop as well in weak salt solutions and may be added to the culture of *Artemia*. However, there must be an adequate oxygen supply as well as food for the nauplius larvae to mature rapidly. Keep the containers in moderate light at a temperature that remains below 25° C (77° F).

(a)

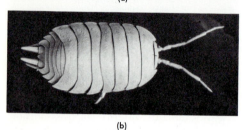

(b)

Fig. 9-9 (a) *Porcellio*, the sow bug; (b) *Armadillidium*, the pill bug. (a, photo by Hugh Spencer; b, courtesy of U.S. Dept. of Agriculture.)

Oniscus

These land isopods and the related genera *Porcellio*, the sow bug, and *Armadillidium*, the pill bug (Fig. 9-9), are all well adapted for their way of life and are widespread. They may be found under stones, boards, and logs and in other dark, moist, undisturbed places.

The pill bug shows a characteristic response by rolling itself into a ball when it is disturbed. Wood lice and pill bugs are found in the same ecological conditions and are cultivated in the same way.

Their usual habitats are best duplicated in the laboratory in a terrarium containing damp, rich humus with small rocks or logs under which the organisms can hide. Supply these isopods with bits of ripe fruit (apples), bits of lettuce, and at times pieces of raw potatoes. They will also accept small earthworms and insects.

Uses in the classroom and laboratory

Isopods are interesting to have on hand for studies of adaptation. Notice how their compressed bodies fit them for their environment. Use them in studies of taxes and general patterns of behavior (p. 330); also examine the contents of the intestine for flagellates (p. 631).

Marine aquaria

More and more, marine tanks are being maintained in biology laboratories; they make fascinating studies for students, especially those in inland cities. Instant synthetic seawater salt mixes can be purchased, along with small invertebrates and seaweeds from biological supply houses (Appendix). Many recipes exist for the preparation of your own synthetic seawater. Humason describes Hale's method for preparing seawater with a salinity of 34.33 0/00 and a chlorine content of 19 0/00.[33]

L. Brixius, curator of the Oregon State University Marine Science Center, suggests this recipe for preparing your own seawater.[34] Into 5 gallons of distilled water, dissolve:

> 2 level tsp potassium chloride
> 22.3 oz sodium chloride (Kosher salt)
> 3 ½ oz magnesium sulfate
> (Epsom salt)
> ¼ tsp potassium iodide
> ⅛ tsp sodium bromide
> 1 Tbl calcium chloride

Aerate the water for a week before adding to the tank.

If you live along the coast, collect small specimens and seaweed together with a gallon or so of seawater. Transport in styrofoam picnic coolers. Then maintain the specimens and seawater in a large, loosely covered battery jar. Within a week, crustaceans and diatoms should appear.

[33] G. Humason, *Animal Tissue Techniques* San Francisco: Freeman, 4th ed., 1979.
[34] L. Brixius, "Building and Maintaining a Cooled Salt Water Aquarium", Oregon State University, U.S. Dept. Agriculture, O.S.U. Marine Science Center.

Use all glass or plastic aquaria as marine tanks (no metal). Aeration is necessary. D. James describes the role of biological filtration;[35] bacteria found in the substrate of the filter bed convert ammonia and urea to nitrites and then to nitrates. To avoid the increased concentration of nitrates in the tank, replace ¼ of the tank water with fresh seawater once a month. Mechanical filtration removes particulate matter so the water remains clear, and also increases the amount of dissolved oxygen. (Many interesting microorganisms, including miracidia of parasitic forms, may be found in an examination of the fibrous matter in filters.) Chemical filtration involves ion exchange columns. A buffering system is needed to avoid drastic changes in pH; add calcium carbonate in the form of crushed limestone rock or crushed oyster shells.

Filter collected seawater before adding it to the tank. A hydrometer reading should be about 1.022. Since sea plants are not good oxygenators, attach an aerator and pump to the tank. Allow at least a week for the tank to become conditioned.

Since some salts are absorbed by the organisms, the salts must be replenished; each month add a level tsp of a 3:1 mixture of rock salt and Epsom salts to a 20-gal tank. Every five months or so, add a small piece of plaster of Paris. Watch the level of water in the tank. As evaporation occurs, add conditioned seawater. Also replace some water every few weeks to dilute the accumulated ammonia and urea. (Test kits are available from Hach and from La Motte to test mineral content; see Appendix.)

Collect small starfish, small crabs, small clams and/or oysters, sea anemones, and seaweeds. Use fluorescent lights to prevent a change in the color of the water. (Incandescent lights seem to increase the growth of microorganisms, turning the water yellow.) Many sea worms thrive in these tanks. To feed the starfish, clams,

oysters, and anemones, remove them from the tank and place them in small containers so that the aquarium does not become contaminated by uneaten food. Feed the starfish and small crabs bits of clams or oysters; brine shrimp larvae are suitable food for anemones.

Small barnacles, clams, and scallops which are filter feeders, consume brine shrimp larvae, algae, and protozoa. Small sea urchins consume seaweeds, lettuce and/or bits of chopped fish. Snails feed upon lettuce, fresh spinach, or seaweeds.

Observe the means of locomotion of small starfish along the glass of the aquarium tank. Place a small oyster in a separate container, add a bit of lampblack to the water, and study the action of the incurrent and excurrent siphons as water moves in and out of the animal (anatomy, chapter 1).

Marine algae may be cultured separately and added to the tank. Refer to L. Provasoli, J. McLaughlin, and M. Droop, "The Development of Artificial Media for Marine Algae," *Arch. Mikrobiol.* (1957) 25:392–428.

For study of marine invertebrates, refer to dissections of mollusks, echinoderms, and crustaceans as described and illustrated in chapter 1. Knuden's *Biological Techniques*[36] is especially recommended for methods useful in collecting and preserving marine forms (also refer here to p. 646). Also consult Bullough[37] for further enrichment. Boolootian and Stiles[38] and Sherman and Sherman[39] have splendid texts and laboratory guides for studies of internal anatomy, ecology, behavior, and embryology. Water quality studies, relating to ecosystems, are described in chapter 8.

---

[35] D. James, "Try a Marine Aquarium", *Carolina Tips*, January 1972.

[36] J. Knudsen, *Biological Techniques* (Collecting, Preserving and Illustrating Plants and Animals) (New York: Harper and Row, 1966).

[37] W. Bullough, *Practical Invertebrate Anatomy*, 2nd ed. (New York: Macmillan 1960).

[38] R. Boolootian and K. Stiles, *College Zoology*, 10th ed. (New York: Macmillan, 1981).

[39] I. Sherman and V. Sherman *The Invertebrates: Function and Form* A Laboratory Guide, 2nd ed. (New York: Macmillan, 1976).

Insects

Some insects may best be collected as pupae, others as adults or larvae. The planned classroom work will, no doubt, determine the kinds of insects that will be gathered and maintained. Following are some methods for keeping alive just a few kinds of insects; there are as many methods as there are kinds of insects.

Praying mantis

Collect egg masses in the fall or early spring. The egg cases are recognized as tan, foamlike masses attached to twigs. Or purchase eggs from biological supply houses.

Keep the egg cases in covered terraria. With a gradual increase in temperature, hundreds of nymphs emerge. Supply the nymphs with dilute sugar solution or honey served in low, flat containers.

Uses in the classroom and laboratory
In a study of reproduction the praying mantis is an illustration of a beneficial insect with incomplete metamorphosis: egg, nymph, adult. Some of the nymphs can be placed in corked vials and examined with a hand lens.

Moths and butterflies

When pupa cases are collected in the fall, they should be stored in a cool place throughout the winter months. Place them in a box that can be left outside a window. In the spring put them into a small screened box or terrarium. Include twigs as supports for the emerging adults. Live pupae are generally heavier than dead or parasitized forms are. (For instance, *Cecropia* moth pupae, a common form that students bring to class, is often parasitized by the Ichneumon fly.) Some teachers on the West Coast have their students observe metamorphosis of the mourning cloak butterfly (*Vanessa*). The changing of larvae into the chrysalis stage is almost completed within an hour; in

seven days the butterfly emerges. Compare this development with that of the praying mantis or the grasshopper.

Uses in the classroom and laboratory
In these examples of insects, students may compare types of metamorphosis. At times, students may be fortunate enough to see a butterfly or moth emerge from a pupal case.

Drosophila

Examine all stages in metamorphosis as well as genetic crosses and taxes.
Directions for raising fruit flies in class are given on pages 531–38.

Tribolium confusum

These beetles are often found in packaged flour and cereals; or cultures may be purchased. Keep both sexes in jars or finger bowls of slightly moistened whole-wheat flour, oatmeal, cornmeal, or bran. Where possible, adequate moisture may be supplied by attaching a moist sponge to the cover of the container.

Metamorphosis is complete within five to six weeks when cultures are kept at temperatures between 28° and 30° C (82° to 86° F). Start new colonies by placing pupae in fresh food medium.

Uses in the classroom and laboratory
Tribolium beetles show all the stages in complete metamorphosis: egg, larva, pupa, and adult. They are also fine organisms around which students may design studies in growth rate, genetics, behavior, and competition (p. 554).

Tenebrio beetles

Larvae of these beetles, called mealworms, (Fig. 9-10a) may be purchased from aquarium shops.

Culture the mealworms in battery jars half filled with moist bran or oatmeal, covered with fine mesh so that adult beetles cannot fly off. When adult beetles

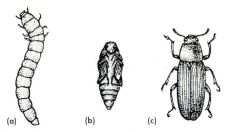

Fig. 9-10 Metamorphosis of *Tenebrio*: (a) larva; (b) pupa; (c) adult.

develop (Figs. 9-10b and c), they should be fed bits of raw carrots or potatoes. At a temperature around 30° C (86° F), the complete life cycle may take four to six months.

Uses in the classroom and laboratory
While these beetles show complete metamorphosis in a reproductive cycle, they are also bred as food for laboratory fish, amphibia, and some reptiles.

Aquatic insects

When collecting pond water, include submerged leaves and some bottom mud. Inspection of battery jars and finger bowls of pond water over a week or month will reveal larvae of many forms. Some of these forms were collected as immature aquatic larvae; others were collected as eggs laid on submerged leaves or in the mud. Possibly nymphs of stone flies, mayflies, or damsel flies or the predatory dragonfly nymphs will be found in the water (Figs. 8-5, 8-9, 9-11). Separate the predatory forms from the mayfly and damsel fly larvae, and maintain the larvae in gallon battery jars. Avoid overcrowding the larvae, since a good oxygen supply is required for their maintenance.

Adult insects can be maintained in large tanks covered with screening. Examine the adaptations of such forms as the water boatsman, upsidedown backswimmer, whirligig beetle, or water bug (Figs 9-12, 8-9).

Uses in the classroom and laboratory
You may want to observe the life cycle of several insects as part of a study of reproduction. Some insects also furnish an excellent example of predator-prey relations in ecological communities, particularly of a pond community (chapter 8).

Circulation may be observed in some small gill-bearing forms. When possible, add carmine powder and trace the path of water.

Activities of social insects

Directions for making a beehive may be found in several books on bees. Some teachers purchase wooden ant houses and observation boxes or an observation beehive from a biological supply house (see Appendix).

Making an ant colony   Students may prepare a temporary observation ant colony for class use. Partially fill a battery jar or large clear plastic box with slightly moistened sandy soil. Collect an ant hill from the field and place it in the container. Include in the colony some workers and a queen. Keep the container covered or in the dark. Galleries made by the ants may be visible through the glass sides of the container. When possible place the jar in a basin of water so that a moat is formed, preventing the escape of ants.

Feed the ants lettuce, carrots, and potatoes (to provide moisture as well as food),

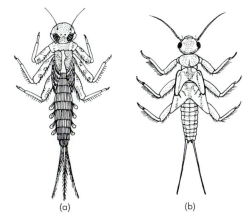

Fig. 9-11 Nymphs of two aquatic insects: (a) mayfly; (b) stonefly.

(a)

(b)

(c)

Fig. 9-12 Adult aquatic insects: (a) backswimmer; (b) whirligig beetle; (c) water bug (male carrying eggs). (a, courtesy of U.S. Dept. of Agriculture; b, c, photos by Lynwood M. Chace.)

bread crumbs and dilute molasses or honey. On occasion add some dead insects. Remove all excess food to prevent the growth of molds.

Keeping a termite colony Colonies of termites may be kept in the laboratory for a continuing study of social insects. A colony consists of a wingless large queen, winged males, wingless workers, and wingless soldiers. Inspect old tree stumps and wet logs for termite galleries. Gently strip off sections of the bark and wood to expose the insects and their eggs. Collect all stages with the wood in which they were thriving. In fact, move to the laboratory as much of the log as is practicable.

In the laboratory separate the termites from the wood with a camel's hair brush to avoid injuring their soft-bodies. Keep the insects, along with wood fragments, in covered finger bowls or Petri dishes. Add strips of moistened filter paper, then store in a dark place at room temperature. Keep moist by adding a few drops of water twice a week.

DeLong and Keagy describe several excellent methods for making observation termite colonies in the laboratory.[40] One simple method recommends the use of flat battery jars of the Delco type. Place a piece of balsa wood along the inside of each of the two wide sides of the jar. Then fill the jar about one-fourth full of soil. Place thin strips of balsa wood between the glass walls and the balsa wood sheets in order to leave a space for free movement of termites between the balsa wood layers and the glass walls. When termites are introduced into the jar, they establish themselves within a few hours. Tunneling may be observed in a short time.

Uses in the classroom and laboratory These insects are splendid for a study of social behavior; they also serve as a source of symbiotic flagellates that are found in

[40] D. DeLong and R. Keagy, "Termite Cultures in the Laboratory," *Turtox News* (May 1949) 27:5.

their intestines. Prepare wet mounts of these graceful flagellates as described on page 631.

Fish

There is scarcely a laboratory that does not have an aquarium; tanks are probably the best single device for maintaining algae, flagellates, water plants, snails, tadpoles, and small fish.

Types of freshwater aquaria

The healthiest fish are those that are placed in an aquarium of appropriate size. Overcrowding is usually detrimental. A pair of fish 3 cm long requires at least 1 gal of water. A 5-gal tank can house six pairs of fish, 3 cm or so long, together with the needed plants and other animals. In our experience, 4– to 5-gal tanks are the most suitable for aquaria in the laboratory or classroom. For study in the classroom the fish may be placed in smaller, more easily handled plastic tanks.

For most purposes, rectangular tanks with slate or glass bottoms and chromium or iron frames in which thick glass is fitted are desirable. Cheap tanks usually end up costing more than good ones (damaged tables, time spent in repair).

Certain tropical fish require special heating and other arrangements. Their care is well treated in many texts.

How to prepare a freshwater aquarium

Proper, careful preparation and planting are necessary for success in maintaining animals. Before the animals are added, the tank should be prepared and planted.

Wash the tank with coarse sand and warm water. Avoid using very hot water; in many tanks the preparations used to cement the sides to the frame may soften. After several rinsings with cold water the aquarium should be filled two-thirds with cold water and allowed to stand for a day or so. During this time any leaks may be detected and any soluble matter in the tank will be dissolved. Discard this water.

When the aquarium has been thoroughly cleaned, it is ready for plants and animals. Cover the bottom of the tank with a 2-cm layer of coarse sand (gravel) that has been washed in boiling water. Embed a clean piece of clam shell at each end of the 5-gal aquarium to help neutralize acidity and to furnish calcium salts for the shells of the snails. Over this, put another 2-cm layer of clean sand. Excessive growth of certain algae may be avoided by embedding a 5-cm^2 strip of copper (or several copper coins) in the sand. Next, lay a large sheet of paper on top of the sand before pouring water into the tank; the paper will prevent the sand from becoming stirred. Add water to a level of 3 to 5 cm from the top, and remove the paper. Let the tank stand for one or two days to bring the water to room temperature, to help dissolve air in the water, and to rid the water of chlorine.

It is good practice to add ½ gal of established aquarium water to a freshly prepared tank. A tank to which this "conditioned" water has been added develops more quickly than a completely new one. Better still, add 1 gal of a thriving *Daphnia* culture to the newly prepared aquarium. When neither of these is available, the water should stand for two or three days. Then plants may be added along with the water in which they have been purchased or collected. Select rooted plants as well as floating plants for display and protection for young fish. After a few days, add snails and fish.

Plants for an aquarium There is a variety of plants from which to choose. The common plants that grow well in a tank are *Lemna, Vallisneria,* corkscrew *Vallisneria, Sagittaria, Anacharis* (elodea), *Cabomba, Myriophyllum, Ludwigia, Potamogeton, Chrysosplenium, Herpestis, Utricularia, Lysimachia nummularia,* and *Cryptocoryne.* Plant *Anacharis, Cabomba,* or *Vallisneria* first, since these, in our experience, are the hardiest of the plants that may be purchased or collected. *Vallisneria* roots readily; it is pleasing to display and does not tangle (so that a fish net may be used easily). *Potamogeton*, *Chrysosplenium*, and *Lemna* are still hardier, but the first two must be collected and have a tendency to grow rapidly and crowd the tank. This may be beneficial to the fish but may not be satisfactory for display purposes.

The water ferns *Salvinia* and *Marsilia*, and algae such as *Nitella*, will do well after the water has been conditioned.

In any event, it is desirable to use a variety of plants of different leaf texture for display. Aquaria in decorative bottles are also popular. Place plants in the back of the tank where they will not interfere with examination and handling of the animals. Then add a rock or two for scenic effect and, more important, to afford a hiding place for the fish, especially for gravid females.

Snails   Place two snails in the aquarium for each gallon of water. The snails tend to keep the glass clean by removing encrusting algae. In addition, very young snails serve as food for some of the fish. *Lymnaea*, *Physa*, or *Planorbis* (Fig. 9-6) are suitable for this purpose. *Campeloma*, a very large snail, is excellent but requires more space than the others; three of them in a 5-gal tank are sufficient.

Light   Keep the aquarium in medium light; strong light favors the growth of algae which turn the water green. In general, northern or western exposure is most suitable. However, when a southern exposure is the only available location the portion of the aquarium facing the direct rays of the sun may be covered with paper, aluminum foil, or glazed glass, or painted green. Should algae turn the water green, add quantities of *Daphnia*. After the water is cleared of the algae, the fish will feed upon the *Daphnia*. However, when there are many fish in the tank remove the fish before adding *Daphnia*; otherwise the fish will eat the water fleas before they have a chance at the algae. Add more

snails if there is an excessive growth of filamentous algae.

Feeding   There are many fish foods on the market; most of them consist of dried, chopped shrimp, brine shrimp, ant pupae, or dried *Daphnia*. Others have dried vegetables added. Any one of these is satisfactory for tropical and native fish, provided live food is added now and then (once a week is fine). Some fish, like the Bettas, do very well on dried food, while others, like the Japanese medaka, reproduce regularly only when fed live food daily. (Also refer to reproduction in medaka, pp. 411, 681.)

Native fish prefer live food but will accept dried food. *Enchytraeus* (white worms)—either chopped or whole—*Tubifex* worms, *Daphnia*, and bits of fresh liver are also acceptable.

Avoid overfeeding fish, for the excess food will foul the water, killing the fish. A pinch of dry food daily is enough for tropical fish about 7– to 10-cm long. During the first few days watch how much food the fish consume. Reduce the quantity the following day if the food is not consumed. Remove the excess by siphoning. It is better to underfeed than to feed more than the fish consume in a day.

Day-to-day care   Besides feeding the fish, keep the water level constant and remove dead plants or animals. You may want to add a catfish to serve as a scavenger. When plants grow rapidly some should be removed to prevent overcrowding. This is especially true of *Lemna* and *Salvinia*, two forms that grow profusely. A floating thick mass of *Lemna* (duckweed) may cut off the surface oxygen supply.

Also remove excess snails as they increase in number, for an excess will destroy plants. It is not necessary, especially for tropical fish, to change the water in the tank except when fouling or special care requires it. There is no cause for alarm if the water becomes yellowish or greenish; this is good "conditioned" aquarium water. A suitable pH range may be as wide as 6.8 to 7.2 for the average tank.

Tropical fish   Tropical fish add beauty and variety to a tank. Many hybrid forms have been developed as knowledge of the genetics of fish has accumulated. Only a few kinds of fish, typical and useful in the classroom, are described here. Although the life histories of the "fighting fish" *Betta* and other bubblenest builders are of unusual interest, they are not of general value for classroom use.

The guppy (*Lebistes reticulatus*) is probably the most common tropical fish, and perhaps the most successful for the beginner to maintain. It is hardy and can withstand low room temperatures. It is a live-bearer (ovoviviparous) and reproduces readily.

Platys (*Platypoecilus*) are not as hardy as the guppy but are interesting for display and study. They are also live-bearers. A temperature range from 20° to 25° C (68° to 77° F) is optimum for these fish.

Japanese medaka (*Oryzias latipis*) do not need as careful temperature control as do other tropical fish, but they do need live food and specific hours of light for regular reproduction. When they are fed *Tubifex*, *Daphnia*, or *Enchytraeus*, the female produces about 8 to 20 eggs each morning. These eggs hang from the cloaca and can easily be removed if the fish is caught and transferred to a finger bowl of tank water. The eggs are clear and excellent for the study of embryological development. Students may examine these eggs under a dissecting microscope or magnifying glass. (Also see embryology, p. 411.)

Native fish   Killifish (*Fundulus heteroclitus*), the redbellied dace (*Chrosomus erythrogaster*), the stickleback, the banded sunfish (*Mesogonistius chaetodon*), and the blue-gill sunfish are desirable fish for the laboratory. These species should be kept separately in tanks well supplied with plants. In one sense they are easier to keep since they can withstand a wide range in temperature.

In general, these fish will not accept prepared foods although they may do so after some time. All live food and bits of raw meat, raw fish, or raw liver are readily accepted and eaten voraciously.

Goldfish (*Carassius auratus*) will thrive under the same conditions as tropical fish. Small goldfish are preferable, since the larger ones require larger tanks. Goldfish do not need careful temperature control. Optimum temperature conditions range from 10° to 25° C (50° to 77° F). They are omnivorous and feed on plants in the tank and bits of boiled spinach in addition to other food.

Goldfish and killifish, as well as some other small forms, are useful in demonstrations of circulation (p. 270) and the effect of drugs.

Diseases of aquarium fish   Fungi such as the water mold (*Saprolegnia*) and protozoa called "water itch" (*Ichthyophthyrius*) are two of the many parasites that attack fish. When fish show gray patches on the fins or scales they should be isolated quickly, for these are symptoms of disease.

Fish that show these patches should be immersed in a 10 percent solution of uniodized table salt. After 1 hr remove the fish and wash in water. Usually the patches disappear after this heroic treatment. Nevertheless, put the fish into a 0.5 percent solution of potassium permanganate for 15 min. Then quarantine the affected fish in separate tanks and watch for the reappearance of the symptoms. (Avoid sharp changes in the temperature of the water during this treatment.)

Actually, fish diseases are difficult to handle; it seems most desirable to discard fish that have become diseased. This has proved to be the best and the cheapest procedure (even after experts have prescribed the "cure").

Amphibia

Salamanders and frogs are especially desirable to have on hand for studies in circulation, behavior, adaptation, life history, and the development of eggs. (For studies of induction of ovulation and parthenogenesis, see page 421). Students may prepare vivaria so that the tanks duplicate the natural habitat of these amphibia. Stages in the development of frogs' eggs are shown in Figure 6-35; the effects of feeding thyroxin or iodine on tadpoles, page 421.

Salamanders

Preparing vivaria   Long, low vivaria covered with glass are most desirable for salamanders. However, whatever type of container is used, it should provide a pool with a "beach" for most kinds of salamanders (Fig. 9-13).

The vivarium must be cleaned as thoroughly as if it were to house fish. At one end of the tank put a small glass, plastic or aluminum foil pan to serve as a pool. Cover the rest of the tank with coarse pebbles together with a few pieces of charcoal to ensure good drainage. Then cover the pebbles with a loam soil rich in humus, slanting the layers of soil away from the small pool to a height of about 8 cm and keeping about 5 cm of water in the pool. Students can prepare an effective natural habitat by planting a "beach" around the pool, using layers of moss, such as *Sphagnum*, *Mnium*, *Dicranum*, or similar types. Place at least one rock in the water. Try planting the rest of the vivarium with small ferns, partridge berries, and a variety of other mosses that have different textures and shades of green.

Feeding   Almost all salamanders require live food such as *Daphnia*, *Tubifex*, *Lumbricus* (chopped), *Enchytraeus*, *Tenebrio* larvae, and *Drosophila*. On occasion, some forms such as *Triturus* (the red eft) may take fresh liver if it is dangled in front of them on a string. But dead animals and food that has not been eaten must be removed within an hour or so to avoid fouling the tank.

Temperature   The best temperature range seems to lie between 15° to 18° C

(a)

(b)

Fig. 9-13   *Diemictylus viridescens* (formerly called *Triturus*), the red-spotted newt: (a) aquatic phase; (b) terrestrial phase. (Photos by Hugh Spencer.)

(59° and 64° F). However, many salamanders can survive at temperatures as high as 25° C (77° F).

Red eft, or water newt (Triturus viridescens)   This salamander (Fig. 9-13), now often called *Diemictylus viridescens*, is the one most easily reared. It can be easily handled, and its slow movements make it desirable for observation and study. Students may readily observe the two phases in the life cycle of the water newt: the water and land phases. In the water phase, the animal is olive-green with carmine spots and yellow-speckled undersides. In this stage it may be kept with fish in an aquarium, although it is more desirable to keep it in a vivarium where it will remain in the pool of water. (Water plants should be included in the pool.) When *Triturus* becomes sexually mature, and, if both sexes are present, fertilized eggs may be deposited on the water plants. Larvae hatch in water and undergo metamorphosis into the red phase—the land stage. The land stage is a beautiful creature that moves slowly over land and feeds on small insects and worms. Provide a raft or flat rock and moss in the tank for this stage. After two to three years the red eft returns to the water and changes into the green phase.

In water, the newts feed on *Daphnia*, *Tubifex*, bits of *Enchytraeus*, and earthworms. On land, they consume larvae of *Drosophila* and *Tenebrio*.

The ambystomas (tiger, spotted, and Jefferson's salamanders) These animals are readily kept in vivaria. The young stage (axolotl) remains in water and feeds on *Daphnia*, *Tubifex*, earthworms, and *Enchytraeus*. Adults feed voraciously on earthworms, insects and *Tenebrio* larvae.

A related form, the Mexican axolotl, may also be maintained in a laboratory. It may provide eggs for study. Mexican axolotls are of interest because they do not undergo metamorphosis but retain larval external gills, reproduce, and remain aquatic permanently. This characteristic seems a special adaptation existing among forms that live in arid regions. When fed thyroid glands of beef, these axolotls undergo metamorphosis.

Other salamanders that may be kept in a similar tank are the red-backed *Plethodon cinereus*, the slimy *Plethodon glutinosios*, the dusky *Desmognathus fuscus*, the two-lined *Eurycea bislineata*, the red *Pseudotriton ruber*, and the Pacific salamander. (Over the country, these have different common names.) These woodland salamanders feed upon small pieces of liver and lean beef, which should be offered to them on the end of a toothpick. There are a few forms such as *Amphiuma* and the red-bellied *Triturus pyrrhogaster* which thrive better in an aquarium. They may be fed the same diet as suggested previously.

Diseases In the laboratory many salamanders may be affected by such fungi as *Saprolegnia* (p. 106). The infection is symptomized by patches of fuzzy white thread on the tail or over the entire body. While most animals infected by fungi do not recover, some first-aid measures may be attempted. Isolate the infected animal and place it in 5 percent potassium permanganate for 10 min. After washing with cold water, place the animal in a jar of water to which a sheet of copper, copper wire, or copper filings have been added. Or add a drop of 1 percent copper sulfate solution to 200 ml of water. Keep the animal in this solution for about two days. It may be necessary to disinfect the vivarium; then clean it thoroughly and replant it.

Frogs

Egg-laying time among frogs Frogs' eggs and tadpoles, as well as adult frogs, may be collected by students and cultivated in the laboratory or classroom for long-range enriching experiences. The time of egg-laying is different for each species, especially dependent on the temperature of the biomes. Therefore there are deviations from the time schedule given here; egg-laying occurs progressively later in the spring as one travels from the southern to the northern states.

As soon as the ice disappears, very early in March, *Rana sylvatica*, the wood frog, breeds laying about 600 eggs. At this time the water temperature may be as low as 5° C (41° F). The eggs should be kept in water maintained at 5° to 15° C (41° to 59° F). The common leopard frog, *Rana pipiens*, lays eggs in March and April; about 2000 eggs are deposited at a time. The pond temperature at this time is about 15° C (59° F). Egg masses of *Rana palustris* are brownish in color and larger than those of *R. pipiens*. This species breeds in April.

The green frog *Rana clamitans* and *R. catesbiana* the bullfrog, are both summer breeders. The males of both species can be distinguished from the females by the presence of a yellow throat and tympanic membranes larger than the eyes. In the females these membranes are the same size as the eyes. The eggs of *Rana clamitans* do not survive in temperatures below 12° C (54° F).

The spring peeper, *Hyla crucifer*, breeds in April, and its eggs are laid singly. Pairs found in amplexus will deposit eggs in the laboratory. This is the best way to collect these eggs.

Toads lay eggs in long strings of jelly. Those of *Bufo fowleri* and *B. americana* are found in early June, of *B. californica* in May. The young develop rapidly.

Vivaria The eggs should be kept in finger bowls or in shallow aquaria containing about 15 cm of water. Such a tank can be well stocked with plants to provide sufficient oxygen.

In our experience, it is best to hatch the eggs in finger bowls. Jelly masses can be cut with scissors so that about 50 eggs are put in each bowl. Remove the unfertilized eggs. These can be spotted readily, for the fertilized eggs orient themselves so that the pigmented portion is uppermost. Thus, within the jelly masses, when the black surfaces are uppermost the eggs have been fertilized and are probably undergoing cleavage.

Cleavage stages may be studied with a hand lens, or a dissecting microscope. The rate of development of frogs' eggs usually varies directly with the temperature. Viable temperatures vary from 15° to 24° C (59° to 75° F). As tadpoles hatch out they still have considerable yolk sac, and no feeding is needed (see stages in Fig. 6-35). However, as they grow older, add scraps of boiled lettuce or spinach. At times, raw lettuce and spinach are accepted, as well as aquatic plants. Change the water twice weekly. After several weeks, pieces of hard-boiled egg yolk or small bits of raw liver may be added to the water. But quickly remove the excess food to prevent fouling of the water. When the hind legs have appeared and the forelimbs are just breaking through the operculum and skin, place the tadpoles in a combination water-woodland vivarium (as described for sala-manders); they are ready to undergo metamorphosis into land forms.

As soon as tadpoles of the larger species, *Rana clamitans* and *R. catesbiana* attain a length of about 3 cm, transfer them to a larger aquarium having a 15 cm water level. Tadpoles of *Rana clamitans* do not complete metamorphosis until the following year. Bullfrog tadpoles take two to four years for complete metamorphosis. Tadpoles may be kept in a "balanced" aquarium. When they are kept separately, they can be fed scraps of lettuce, liver, or hard-boiled egg yolk.

When a few frogs are kept for display purposes, a woodland vivarium is desirable. Such forms as bullfrogs are an exception; they fare better in 5 cm of water in a clean aquarium. Change the water daily.

The problem of handling large quantities of frogs in the laboratory is a difficult one. The survival rate is high when they are kept at 10° C (50° F) in a granite sink containing about 3 cm of water. Add a small amount of uniodized salt to the water. Cover them with wire mesh; change the water daily. Better still, when facilities are available, keep the water running slowly in the sink. An unused sink in which the drain can be stoppered with a closed wire mesh tube about 15 cm high is ideal. A slanted board should be provided so the frogs may leave the water. Keep frogs away from zinc.

When possible, keep frogs in a drawer of a refrigerator in about 3 cm of water or in large aquaria with water at a level of 3 cm. Flush the frogs with a stream of water when the water in the tanks is changed each day; the water becomes fouled quickly. Dead frogs and frogs suffering from "red leg" should be isolated immediately since the infection is highly contagious. To reduce infection add a dilute solution of uniodized table salt (0.2 percent) to the water in which the frogs are kept.

Frogs that are most easily kept in beach vivaria as adults are *Rana pipiens*, *R. palustris*, and the green frog *R. clamitans*. They accept *Tenebrio* larvae, small earthworms, flies, and similar living matter as food.

Rana catesbiana should be kept in water that is about 5 to 7 cm deep. Provide a rock with a surface slightly above the water. These forms readily eat live animals, such as smaller frogs and earthworms. The mature male of *Rana pipiens* has thick dark thumb pads, while males of *Rana catesbiana* and *R. clamitans* have large tympanic membranes, almost twice the diameter of an eye.

The members of Hylidae, *Hyla crucifer* (spring peeper) and *H. versicolor* (tree frog, Fig. 9-14), may be kept in a beach vivarium. *Hyla crucifer* feeds on *Drosophila* and small mealworms. Provide a few stout twigs as supports for the tree frogs in the terrarium.

Among the toads (Bufonidae), *Bufo fowleri* and *B. americana* both require terraria similar to that of the Hylidae. However, it must be remembered that toads are active burrowers, and they disrupt a well-managed terrarium.

Reptiles

Turtles

Vivaria   Most turtles, with the exception of those described below, should be kept in aquaria containing about 5 to

Fig. 9-14 *Hyla versicolor*, the common tree frog. (Photo by Hugh Spencer.)

10 cm of water. Cork floats can be added or a flat rock placed in one corner of the vivarium as a useful resting place. Change the water twice weekly to keep it clear.

Painted turtles, wood turtles, and box turtles may be kept in water (as described) or in a beach vivarium. However, box turtles seem to prefer a moist terrarium or vivarium rather than water. Segregate adult snapping turtles from other turtles (small ones may be kept with other species).

Feeding   Most of the aquatic forms will accept bits of fish, ground raw meat, liver, earthworms, or dead frogs. In addition, most turtles will accept hard-boiled egg cut into slices, as well as lettuce and slices of apples. Box and wood turtles also take snails, slugs, and *Tenebrio* larvae.

When turtles become sluggish and show a tendency to hibernate they should be placed in a cool place. Forced feeding, especially at this time, is often detrimental.

Sex differences   The most uniform guide for distinguishing sexes in turtles is the shape of the plastron, the under shell. The plastron in the female of many species is slightly convex, while in the male, it is slightly concave. During the breeding season, there is a swelling of the anal region in the male. Eye color is a distinguishing characteristic among box turtles; males usually have red eyes and females have yellow eyes. Males usually have longer claws than females.

Lizards and alligators

Small specimens of the American alligator and several kinds of lizards can be reared in the laboratory, such as the horned "toad" (*Phrynosoma*, see Fig. 9-15), the skink (*Eumeces*), and the chameleon (*Anolis*).

Vivaria   Chameleons and the larger skinks should be housed in a large terrarium. Include some twigs so the animals have room to climb. Spray the plants in the terrarium daily to supply water for

Fig. 9-15 Lizard: *Phrynosoma*, the horned "toad." (Courtesy of the American Museum of Natural History.)

these lizards, since they seldom drink from a dish. They subsist mainly on live insects. It may be necessary to raise *Drosophila* for this purpose, especially during the winter months.

Young alligators survive in a vivarium of sand and rocks with a water trough embedded in the sand. Horned "toads" are maintained best in a similar desert vivarium containing about 12 cm of sand for burrowing, along with several rock piles for hiding. Embed a bowl of water up to the level of the sand. Students who care for the animals should provide several hours of direct sunlight but must also ventilate the tank so that the temperature does not exceed 26° C (79° F).

Feeding   All these reptiles feed upon live insects, *Tenebrio* larvae, bits of earthworms, *Enchytraeus*, or similar live food. In addition, young alligators may take small frogs and fish. Chameleons and skinks may learn to accept small bits of raw liver or meat which are dangled before them. They do not feed regularly at low temperatures; 18° to 26° C (64° to 79° F) is the most suitable range.

Snakes (nonpoisonous)

One way to break down the inordinate fear of snakes many students have is to maintain one or two snakes in the laboratory or project room. Those forms that are easiest to keep and to handle are the garter, ribbon, hog-nosed, black, DeKay, and ring-necked snakes. Many others such as the bull, milk, water and green snakes may also be maintained but in our experience the first-mentioned are the easiest to keep.

Vivaria   Mesh cages much like those used for mammals are best for housing snakes, although the mesh should be of smaller gauge so that the smallest snakes cannot escape. In addition, bottom pans of zinc are needed. Door openings at the top of the cage are the most convenient for handling the animals. Into such a vivarium place a pan or bowl of water and a few rocks. Keep the snakes at a temperature between 21° and 26° C (70° and 79° F).

Where all-mesh cages are not available, use ordinary aquarium tanks with tight-fitting zinc mesh tops. You may need a weight to hold down the cover. An aquarium completely enclosed by glass is undesirable since there is no provision for ventilation. When cages of mesh are used for DeKay and ring-necked snakes, make sure that the size of the mesh is small enough to prevent their escape. Snakes may be kept in terraria like those described for *Hyla* or *Bufo* (p. 685). All snakes should be washed weekly by flushing with water. Their cages should be cleaned at the same time.

Feeding   Most snakes described here feed upon readily available food. DeKay and ring-necked snakes feed on insects such as *Tenebrio* and on small earthworms; hog-nosed, garter, and ribbon snakes accept whole, large earthworms as well as insects, frogs or other amphibia, and lizards. Black snakes need live mammals; they may be fed a small rat every two weeks. On occasion, black snakes will accept a dead animal if it is waved in front of them.

General care   Some students are experts at handling snakes; they can do much to help other students overcome their fears. When snakes are handled gently each day, they will in turn become gentle. Large snakes should be handled with thick gloves. Grasp a snake behind the head with one hand while the other is used to support its body.

Birds

Parakeets and canaries require little space and more or less routine care. Directions for maintaining these birds are given by the dealer at the time of purchase. In our experience, birds other than parakeets and canaries need more space and care then can be provided in the average laboratory. However, where there are students who are expert in the handling of birds, this activity may become highly profitable.

On occasion, a crow may be housed in a large cage. Young ones seem to find the surroundings agreeable and can be conditioned by students studying behavior.

Mammals

White rats and hamsters are the easiest mammals to maintain in the laboratory. Students become quite adept at handling them. Guinea pigs and rabbits do not, ordinarily, demonstrate anything for which mice and rats are not suitable, and they require much more space and care. The gestation periods and breeding ages for all these mammals are given in Table 9-1.

The care of the rat will be described in some detail; care of the mouse is similar. However, good ventilation is needed when mice are reared since they have an offensive odor.

Cages Most of the cages available commercially are satisfactory. The larger cages are best since they allow for exercise.

A door at the top of a cage is most convenient since it permits easier handling. However, such cages cannot be stacked. Cages should be cleaned daily and fresh cedar shavings placed in the bottom pan. Cleaning a cage the first or second day

TABLE 9-1 Gestation periods of some mammals

	gestation period (in days)	Breeding age
White rats	21–22	Females may be bred when 4 months old. Wean the young and separate the sexes after 21 days.
Mice	20–22	Breed females when 60 days old. Wean the young after 21 days.
Guinea pigs	63	Breed when 9 months old. Wean the young and separate the sexes at 4 to 5 weeks.
Golden hamsters	16–19	Breed the females when 60 days old.
Rabbits	30–32	Females are ready for mating when 10 months old. Wean the young and separate the sexes after 8 weeks.

after delivery of a litter is not advisable. In fact, care must be taken for several weeks in order not to disturb unnecessarily the mother and young. Disposable plastic cages are also useful.

Male and female rats should be segregated when 50 days old and should be kept segregated until mating is desirable. Remove the male as soon as the female is pregnant.

General care Rats should be fed only once a day. They should be treated as pets (although many students will tend to overdo this). If they are, they will respond satisfactorily and reproduce readily. Rough treatment may result in viciousness and cause the mother to destroy her litter. From birth the rats should be handled gently and fondled. When this is done they do not bite, and, in fact, they are so conditioned that they will run forward to be handled. Such animals are desirable but students must be trained to handle them.

Feeding Rats may be fed a diet of bread, sometimes soaked in milk, in addition to lettuce, carrots, other vegetables, sunflower seeds, and similar foods. The bread should be broken and the carrots cut into portions equal to the number of rats in the cage. They will also accept hard-boiled eggs. Two or 3 drops of cod-liver oil on pieces of bread should be given twice a week; provide a bowl of milk weekly.

Water must be supplied at all times. Water fountains, blown of one solid piece of glass or plastic are available from biological supply houses (Appendix).

Synthetic diets for rats are available from supply houses. When the young are to be weaned, feed them milk, bread soaked in milk, and lettuce. After they are 30 days old they can be fed the same diet as the adults. Pellet foods may then be added to the diet.

Care during the pregnancy and nursing periods As soon as the female is pregnant, the male should be removed. A pregnant rat should be given strips of newspaper or paper toweling for nest building. Once a nest is built it is not necessary to change the paper, although this can be done after the young have hair and their eyes have opened (after 16 to 18 days).

The period of gestation is 21 to 22 days. The sex of the young rats may be distinguished by the fact that the distance between the anus and the genital papilla is greater in the male than in the female.

The young should be permitted to remain with the mother for 21 to 24 days, after which time they should be weaned. When the young are kept with the mother for a longer period there is a severe drain on the female.

A gentle female rat will respond favorably to handling during the nursing period and will not resent handling of the young, provided it is done by someone who has been responsible for her daily care.

However, as mentioned earlier, special care should be taken during this period to avoid unnecessary disturbance of the mother and young, and they should be handled gently; otherwise the mother may turn vicious and destroy her litter. The mother should be fed whenever she is removed from her young. The female, especially during this nursing period, may react unfavorably to strangers. A bit of chocolate or carrot given to her as she is returned to her young will help ease her distress.

Maintaining plants

Methods for culturing plants from algae to seed plants are described below. A wide selection of techniques offers opportunities for some experimentation in choosing the most useful procedures for a specific situation.

Algae

It might seem that algae should respond readily to culture. In fact, certain algae respond only to specific nutritional substances in cultures; but some species are not so selective. For example, *Oscillatoria*, *Chlorella*, and *Cladophora* respond to a variety of methods.

Most of the algae useful in the biology laboratory may be cultivated by one of the methods described in this chapter. A summary of the most suitable methods for each alga is also given under its genus name (p. 692).

Special considerations A successful culture medium for algae must provide the major nutrient salts, a usable source of nitrogen, a supply of carbon, some trace elements (micronutrients), and a suitable pH.

In addition, temperature and light conditions must be suitable for the successful culturing of algae. Most algae and plants grow well at 21° C (70° F). Care should be taken to avoid temperatures higher than

27° C (81° F). Light from a north window is preferable, but, especially during winter months, artificial light can be used to supplement daylight. A standard cool, white fluorescent light placed a few feet from the plants or cultures provides about 50 to 75 foot-candles of light intensity. For most successful growth of algae, a 16-hr light period should be alternated with an 8-hr period of darkness. Light intensity of some 200 foot-candles stimulates rapid growth, which is especially suitable for young cultures.

In special culture rooms or greenhouses, where artificial light is used exclusively, a balance should be maintained between incandescent and fluorescent lamps to provide light in both the red and blue wavelengths. Fluorescent light—rich in blue wavelengths and deficient in red wavelengths—encourages the growth of low vegetation. Incandescent light—rich in red wavelengths and poor in blue wavelengths—stimulates the growth of tall, spindly plants with little supporting tissue. Hence, a balanced combination of the two produces normal vegetative growth. (Also refer to greenhouses, p. 763).

The carbon dioxide content of a culture medium may be increased by adding several drops of 0.5 percent bicarbonate solution to each 10 ml culture. Or small amounts of CO_2 may be pumped into a medium. You may want to demonstrate, on a small scale, the effectiveness of increasing CO_2 content on the rate of photosynthesis and growth of *Chlorella* by preparing an actively growing yeast culture in a fermenting medium, such as molasses. Extend a delivery tube from the yeast culture to deliver the CO_2 output from yeast cells into the *Chlorella* medium. Note how much greener the *Chlorella* culture becomes in a few days compared to controls. (Also refer to p. 310.)

General methods

A great number of green algae can be successfully cultured by introducing them into an established aquarium containing freshwater plants and fish that have been thriving for about two months. However, avoid overcrowding.

A thriving culture of *Daphnia* is a good culture medium for coarse-filamented *Spirogyra* such as *Spirogyra nitida*[41] (see also p. 671). The method does not seem to work with fine-filamented species, however; the algae rapidly choke the container in which *Daphnia* is cultivated.

Keep the cultures in medium light, or in a north window. Each week add about 0.1 g of fresh, hard-boiled egg yolk smoothed between the fingers into a paste. Species of *Spirogyra* containing two or more spiral chloroplasts seem to respond more vigorously than species with single chloroplasts. (See also soil water culture medium for *Spirogyra* below; however, omit the $CaCO_3$.)

In our experience, almost all the algae commonly used show vigorous growth when cultivated by this method. An attempt to discover improved methods of culturing algae is a very good project for an interested student.

A temperature range of 18° to 27° C (65° to 80° F) is optimum in all the methods described.

Soil culture media

Pringsheim was one of the pioneers in introducing the use of soil-water cultures that closely duplicate a miniature artificial pond.[42] He added either garden soil with seasonal compost or soil from an arable field (especially clay soils) to water to make a mud phase. This mixture seemed to supply organic matter, growth factors, and trace elements in amounts that apparently were optimum for the growth of algae.

To the bottom of test tubes or other containers, add dry soil that has been

[41] P. F. Brandwein, "Preliminary Observations on the Culture of *Spirogyra*," *Am. J. Botany* (1940) 27:195–98.
[42] E. G. Pringsheim. *Pure Cultures of Algae: Their Preparation and Maintenance*, (New York: Macmillan (Cambridge Univ. Press, 1946).

treated in a steam sterilizer for an hour or more on two consecutive days (to remove contaminants). Carefully add sterile, distilled water. If necessary, alter the pH of the culture: add calcium carbonate to raise the pH or add some peat to lower the pH.

Pringsheim also describes a technique in which the soil and water are sterilized as described above, but the mixture is allowed to settle so that the soil component remains in the bottom of the tubes (avoid shaking tubes).

Soil cultures are recommended for pure cultures of algae. If students study soil flora, use this method, but omit the sterilization of the soil. After a rich growth of algae results, algae can be isolated from the mixed cultures and subcultured as unialgal cultures (see p. 695).

Bold describes a modified soil culture that uses a *clear filtrate* of soil, rather than soil.[43] Dissolve 500 g of good field soil in 1 liter of distilled water; sterilize for 2 hr. When cool, filter the solution, and decant it several times until a clear filtrate is obtained. This is called a *stock* solution of soil extract, and it must be diluted for use. In actual testing situations, start with a dilute solution, and gradually add more extract until good growth results. We have found the following dilution useful. Combine 5 ml of stock soil solution with 94 ml of sterile distilled water. To this, add 1 ml of a 5 percent solution of KNO_3. Pour this solution into small finger bowls, baby-food jars, or similar containers. Inoculate with the pure type of alga desired, and cover with aluminum foil. Maintain in moderate light at 20° to 24° C (68° to 75° F). Use a phosphate buffer if necessary (p. 159).

Solutions of known chemical composition

Benecke's solution is similar to Klebs' solution described earlier (p. 663); notice some salts differ in the recipes. We have found Benecke's solution successful especially with flagellates such as *Euglena*.

---

[43] H. Bold, "The Cultivation of Algae," *Botan. Rev.* (1942) 8:69–138.

BENECKE'S SOLUTION   Dissolve the following in 1 liter of distilled water:

$Ca(NO_3)_2$	0.5 g
$MgSO_4$	0.1 g
K_2HPO_4	0.2 g
$FeCl_3$ (1% solution)	trace (1 drop)

KNOP'S SOLUTION   Prepare a 0.6 percent Knop's solution for green algae by dissolving the following salts in 1 liter of distilled water, in this order:

KNO_3	1 g
$MgSO_4$	1 g
K_2HPO_4	1 g
$Ca(NO_3)_2$	3 g

In this solution, the calcium nitrate precipitates, forming a layer at the bottom of the bottle. Be sure to shake the bottle when using the solution. A 1 to 4 percent solution of sucrose is sometimes added to this dilution of Knop's solution to stimulate the formation of zoospores in some algae.

Use distilled water in preparing these solutions since tap water may contain chlorine or copper, both of which are highly toxic to algae.

The ability of these solutions to support the growth of various algae is greatly enhanced by adding to the culture jar about 3 cm of soil taken from the pond where the algae have been growing. Good garden soil is a useful substitute. First, boil the soil in distilled water to destroy contaminating algae. Then add this together with distilled water to the prepared container. Let the medium stand for a day or so before adding algae. Such soil-solution cultures have been very satisfactory since they contain microelements necessary for growth. You may need to subculture when the algae reach maximum growth. (See specific descriptions of soil cultures, p. 689.)

BEIJERINCK'S SOLUTION   Among the classic solutions for culturing algae is one prepared by Beijerinck. He isolated pure cultures of algae in the 1890s in 1.5 percent agar cultures. Add the following salts to 1 liter of distilled water:

NH_4NO_3	0.5 g
K_2PO_4	0.2 g
$MgSO_4 \cdot 7H_2O$	0.2 g
$CaCl_2 \cdot 2H_2O$	0.1 g

To this solution, Fogg suggests that the following minerals be added.[44]

Fe (as $FeCl_3$)	0.4	mg
Mn (as $MnSO_4$)	0.1	mg
Cu (as $CuSO_4$)	0.01	mg
Zn (as $ZnSO_4$)	0.01	mg

BRISTOL'S SOLUTION   This is a popular solution for growing algae. Bold recommends the following modification.[45]

Prepare *six stock solutions* by dissolving each of the following salts in 400 ml of distilled water. Refrigerate each solution.

1.	$NaNO_3$	10.0 g
2.	$CaCl_2$	1.0 g
3.	K_2HPO_4	3.0 g
4.	KH_2PO_4	7.0 g
5.	$MgSO_4$	3.0 g
6.	NaCl	1.0 g

To prepare a *dilute* solution for culturing algae, add 10 ml of each of all six stock solutions to 940 ml of distilled water. To this, add 1 drop of a 1 percent solution of $FeCl_3$ and 2 ml of Arnon's solution of microelements, described below.

In addition, Bold suggests that organic matter be added in the form of Difco yeast extract (1 g of yeast extract per liter of modified Bristol's solution); then sterilize.

Any of these solutions may be solidified for use in Petri dishes by adding 1.0 to 1.5 percent of agar (15 g to 1 liter of boiling water).

For culturing microorganisms in soil, use an enriched soil water extract and agar. Combine 960 ml Bristol's solution with 15 g agar; heat to dissolve the agar; add 40 ml of soil water extract. (Also refer to p. 690.)

Micronutrient solutions

Meyers suggests another successful method used in large-scale cultivation of *Chlorella* and in sustaining vigorous growth of other algae.[46] He recommends a freshly prepared Knop's solution, at pH 6.8, provided with necessary microelements by the addition of iron and Arnon's solution. A chelating agent is used to prevent the microelements from precipitating out of solution.

MEYERS' MODIFICATION OF KNOP'S SOLUTION[47]   Dissolve the salts listed in A in 1 liter of distilled water. To this, add the chelating agent in B. Dissolve the microelements in C in a small amount of dilute sulfuric acid and add the mixture to the rest of the solution. Finally, adjust the pH with buffers.

A. *Macroelements*

$MgSO_4 \cdot 7H_2O$	2.50 g
KNO_3	1.25 g
KH_2PO_4	1.25 g

B. *Chelating agent (ethylenediamine tetra-acetic acid)*

EDTA (or Versene)	0.50 g

C. *Microelements (Arnon's solution)*

		g/liter	ppm
Ca	(as $CaCl_2$)	0.084	30
B	(as H_3BO_3)	0.114	20
Fe	(as $FeSO_4 \cdot 7H_2O$)	0.050	10
Zn	(as $ZnSO_4 \cdot 7H_2O$)	0.088	20
Mn	(as $MnCl_2 \cdot H_2O$)	0.014	4
Mo	(as MoO_3)	0.007	4
Cu	(as $CuSO_4 \cdot 5H_2O$)	0.016	4
Co	(as $Co(NO_3)_2 \cdot 6H_2O$)	0.005	1

Cobalt, the last of the microelements listed above, is a specific requirement for the growth of some algae, such as *Euglena gracilis*. (Cobalt is also a constituent of vitamin B_{12}.)

An excessively high initial nitrogen content is toxic to algae; as nitrogen is

---

[44] G. E. Fogg, "Famous Plants: 4, *Chlorella*," in M. L. Johnson, M. Abercrombie, and G. E. Fogg, eds., *New Biology*, Vol. XV, Harmondsworth, Eng.: Penguin, 1953.
[45] Bold, "The Cultivation of Algae."

[46] J. Meyers, "Modified Knop's Solution for *Chlorella*," mimeo, personal communication.
[47] J. Meyers, "Modified Knop's Solution for *Chlorella*."

absorbed, renew it in the solution. Further, the uptake of nitrate by algae also causes a rise in pH of the medium; adjust pH to 6.3 to 6.8.

Solid agar media

Algae may also be cultivated on solid medium under sterile conditions. Two useful methods are described below. Agar may also be added to several other media described.

To prepare solid medium, mix the agar in the nutrient salt solution and gently bring to a boil. Agar solidifies at 4° C (39° F) and melts at 96 to 98° C (205 to 208° F). The usual concentration of agar added to a culture medium is 1.5 to 2 percent. (Use plain agar: 15 g to 1 liter of culture medium or water.) Heat in a double boiler and stir until the agar is dissolved. (See instructions for plating finger bowls or Petri dishes, page 700.)

MODIFIED PFEFFER'S SOLUTION Combine the following salts with 1 liter of distilled water. Add the $Ca(NO_3)_2$ last.

$Ca(NO_3)_2$	1.00 g
K_2HPO_4	0.25 g
$Fe_3(PO_4)_2$	0.05 g
$(NH_4)_2SO_4$	0.50 g

When this solution has been prepared, add 15 g of agar and 15 g of fructose and gently heat in a double boiler.

MODIFIED KNOP'S SOLUTION This is a slight modification of the recipe given on page 658: Combine the following salts with 1 liter of distilled water. Add the $Ca(NO_3)_2$ last.

$Ca(NO_3)_2$	1.00 g
KCl	0.25 g
$MgSO_4$	0.25 g
K_2HPO_4	0.25 g
$FeCl_3$	trace

To this solution add 20 g of agar and 20 g of glucose and gently heat to dissolve the agar.

Sources of algae

You may want to cultivate algae of a given

genetic homogeneity for special investigations. Some 150 kinds of unialgal cultures are available. Consult R. C. Starr's paper "The Culture Collection of Algae at Indiana University" (*Am. J. Botany* 1960 17:67–86). In it he lists cultures available for research and describes the culturing and induced sexual reproduction of specific algae. Starr's cultures are available now from Carolina Biological Supply Co.

When clonal cultures are not necessary for laboratory study, cultures may be purchased from many biological supply houses (see Appendix).

Methods for culturing marine algae are described on page 694.

Summary of methods for specific algae

The following methods are listed in order of their effectiveness for each type of alga.

Oscillatoria (Fig. 1-114) Use culture media of known composition such as Benecke's or Knop's or the general methods (p. 689) or soil culture methods (p. 689). This blue-green alga, a moneran, is found as a thin scum or sheet on the surface of stagnant water or damp soil, or also on flowerpots; it is dark green or even blackish in color. Culture *Oscillatoria* separately from green algae since the green algae are often denied an adequate supply of O_2 by the surface growth of *Oscillatoria*.

Nostoc (Fig. 1-113) This blue-green grows well in a soil water culture medium. Add 960 ml of Bristol's solution (p. 691) to 40 ml of soil water extract. To solidify, add 15 g of agar to the liter of culture medium.

Hydrodictyon (Fig. 1-119) Use general methods (p. 689), Beijerinck's or Bristol's solution. Collected from the surface of lakes, this green net *Hydrodictyon* grows well in bright light.

Vaucheria (Figs. 1-119, 6-53) Use general methods (p. 689). This green alga is most commonly found in the vegetative stage.

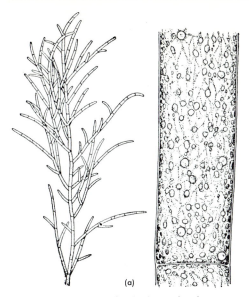

(a)

(b)

Fig. 9-16 Algae: (a) *Cladophora*, showing a portion of the plant (magnified 6x) and part of a mature cell with five large nuclei and many smaller oval pyrenoids; (b) *Nitella*. (a, adapted from *The Plant King-*

dom by William H. Brown, Ginn and Co., Publishers. Copyright © 1935 by William H. Brown; b, courtesy General Biological Supply House, Inc., Chicago.)

Klebs induced formation of antheridia and oogonia within four to five days by placing *Vaucheria* in a 2 to 4 percent solution of sucrose in *bright* light. The common form *Vaucheria sessilis* is found as green "felt" on flowerpots in green houses. In the spring, the common form in ponds and lakes is *V. geminata*.

Chara Use soil culture medium (p. 689), Bristol's solution or general methods for algae (p. 689). *Chara* grows well when about 3 cm of pond soil is added in the bottom of the culture jar.

Cladophora and Oedogonium (Figs. 6-55, 9-16) Both of these algae grow well in general culture methods media (p. 689), or in solutions such as Knop's or Benecke's.

Spirogyra Use the general methods described on p. 689 with *Daphnia*. Be sure *Daphnia* are thriving in the culture before adding *Spirogyra*.

Nitella (Fig. 9-16) Culture this green alga using general methods described on page 689 with *Daphnia*.

Desmids Use the culture media described under general methods, or Knop's with *Daphnia*.

Diatoms (Figs. 1-118, 8-12) Collect diatoms from submerged rocks and stems of plants, or from the surface of mud in ponds and ditches in the spring. Diatoms abound in the brownish coating on rocks and on stems that are slippery to the touch.

Scenedesmus (Fig. 1-116) Use soil water cultures (p. 689), Bristol's, or Knop's.

Volvox (Fig. 8-7) Culture by the general methods (p. 689), or use Knop's, Benecke's or Meyers' modification of Knop's with microelements (p. 691). When a soil water culture is used, add a pinch of $CaCO_3$ to a container of about 2 cm of garden soil and water (p. 689). Use cotton stoppers and steam without pressure for 1 hr.

Chlamydomonas (Figs. 1-3, 1-116) Use culture methods similar to those for flagellates (p. 663); or use soil culture media. Culture in flasks or finger bowls.

Eudorina, Pandorina Use soil culture medium or a general culture technique (p. 689), or one that contains micronutrients (p. 691). Add 1 percent agar to solidify the medium in Petri dishes.

Euglena (Fig. 1-2) Culture methods are described on page 664; also use a soil culture medium or follow the suggested medium offered by Starr (below). Culture in finger bowls or larger jars. To grow in the dark, add a carbon source; when *Euglena* is raised in the dark, its chloroplasts fragment, and the nongreen *Euglena* grows as a heterotroph.

Starr recommends the following culture medium for careful investigations.[48] Dissolve the following in distilled water up to 1 liter:

sodium acetate	1.00 g
beef extract (Difco)	1.00 g
yeast extract (Difco)	2.00 g
tryptone (Difco)	2.00 g
calcium chloride	0.01 g

Add 15 g of agar to solidify. Autoclave for 15 min or use a pressure cooker.

Chlorella (Fig. 1-116) Use Knop's with agar, a soil culture (p. 689), Meyers' modification of Knop's with microelements, Beijerinck's, or Bristol's. This tiny spherical alga is a ubiquitous contaminant of laboratory cultures. A supply bottle of Knop's solution standing on a shelf may turn green in two weeks if light is available. Green water in an aquarium often consists of *Chlorella*, as well as *Euglena*, *Scenedesmus*, and *Chlamydomonas*.

Chlorella also flourishes in the dark when an organic carbon source, such as acetate, or 10 percent glucose, or other organic acid is provided. It also maintains its bright green chloroplast. (Recall that *Euglena* loses its chloroplasts in darkness.)

Carolina Biological Supply Co. grows *Chlorella* at 20° C (68° F) in a medium containing beef extract, dextrose, and agar.

Difco offers a prepared medium that needs only to be rehydrated, poured into tubes or plates, and autoclaved.[49] Buffers are included to maintain a pH 6.8.

Marine algae

Methods for preparing and maintaining a marine aquarium are described on page 674. Store ocean water in gallon glass containers in moderate light. Within a week to ten days, marine algae become apparent, especially green, brown, and red algae. Collect *Fucus* from their attachment to rocks at low tide and *Ulva* (sea lettuce), an example of marine green algae that is two cells in thickness. Male and female *Fucus* (Fig. 6-52) together with instructions for a study of fertilization may be purchased from most supply houses (see Appendix). Further references for collecting and maintaining marine forms are described on page 592.

SCHREIBER'S SOLUTION This is one of several enriched seawater culture media suggested in the literature. (See seawater recipes, pp. 744, 674.) To 1 liter of seawater add:

$NaNO_3$	0.10 g
Na_2HPO_4	0.02 g
Soil extract (commercial fertilizer 4-10-4)	50 ml

Also see L. Provasoli, J. McLaughlin and M. Droop, "The Development of Artificial Media for Marine Algae," *Arch. Mikrobiol.* (1957) 25: 392-428.

Use *Fucus* as an example of a marine brown alga; the pigment fucoxanthanin may hide the chlorophyll. As a photosynthesizer *Fucus* produces polysaccharides, not starches, as green algae do. There is no asexual reproduction in *Fucus*; examine the fertile branching tips that contain cuplike structures with sperm cells or egg cells. (Refer to life cycle, p. 439.)

Collect feathery red algae containing the

---

[48] R. C. Starr, *Am. Jour. Bot.* (1964) 51:1013.

[49] Available from Difco Laboratories, Detroit, Michigan; 48201. Many specialized media are available for algae as well as fungi and bacteria.

pigment phycaerythrin, which masks the green of these photosynthesizers.

The golden brown algae include the diatoms of fresh and salt waters, and the marine dinoflagellates. These algae store oil rather than starch. They are studied for the beauty of their radial or bilateral patterns. Observe how the two halves of a diatom fit together like a Petri dish. Diatomaceous earth is composed of the silicon walls and oil of diatoms that have accumulated over millions of years.

Isolating algae

To isolate a pure culture from a mixed algal culture, use Beijerinck's method of streaking a transfer loop or needle on an agar film, similar to the method used in isolating bacteria (p. 702). Or use dilutions as pour plates (p. 703). For a more extensive discussion, refer to R. A. Lewin, "The Isolation of Algae," *Rev. Algolog.* (1959) 5:181–97.

Uses of green algae in the classroom and laboratory

Algae of many kinds are used to show the structure of green plant cells (wet preparations, chapter 2). *Spirogyra* (Fig. 1-116) is the most commonly studied form in introductory work. This is unfortunate because *Spirogyra*—with its web-like strands of cytoplasm, spiral chloroplasts, and pyrenoids—is one of the most complex forms to examine, and it is difficult to keep in the laboratory.

Protococcus (Fig. 8-8) shows a conspicuous chloroplast that you may want students to observe in cell studies. A drop of Lugol's solution (p. 176) added to a slide of *Protococcus* will also stain the nucleus. *Euglena* is a motile flagellate with several chloroplasts (Fig. 1-2). At times you may prefer to substitute *Peranema* for *Euglena*, because its long flagellum is distinctly visible under low power. Although it has no chloroplasts (it is holozoic or saprozoic), it does undergo many changes in form and exhibits "euglenoid motion." A

common form is *Peranema trichophorum* (Fig. 1-3), which is some 60 to 70 μ long. *Euglena* may be used to show positive phototropic responses as well (p. 330).

Euglena may also be studied as an example of an organism that reproduces by longitudinal fission. This form is sometimes used in assays of vitamins and other growth-promoting factors (p.482). *Euglena* is a good choice for a test organism in studies of photosynthesis, since both green and nongreen varieties of the same species are available (bleach with streptomycin); refer also to *Chlorella* (p. 694).

Study diatoms under the microscope for the beauty of the sculpturing in their silicaceous shells. *Spirogyra* may be studied to show conjugation as well as plant cell structure, while such forms as *Vaucheria* and *Chlamydomonas* may be used to observe the evolution of sexuality in plants. *Chlorella* may be used as a focus for developing key concepts in biology, as described on pages 229-32 and 615-17.

Other studies of algae will, of course, depend upon the nature of the course. In most introductory courses in biology few algae are studied: mainly *Spirogyra* and some other forms casually studied to illustrate the phylum. Some teachers, however, use a variety of algae to illustrate many aspects of biology; for example, *Volvox* may be used to illustrate heterosexuality and the prologue to metazoan structure, *Spirogyra* to illustrate pyrenoid activity as well as conjugation, *Ulothrix* to show zoospores, *Oedogonium* to show heterospory, and *Chlorella* and *Euglena* to introduce modes of nutrition and behavior.

Recall that in *Spirogyra* the entire cell of a monoploid filament acts as a gamete and moves across a bridge into a "passive" cell of an opposite filament. The fusion of cells results in a diploid zygospore.

In *Ulothrix*, biflagellated gametes are formed by the monoploid filaments, and a zygospore results from the fusion of these gametes, or zoospores. The evolution of different sized gametes—eggs and ciliated sperms—is found in *Oedogonium*. The alternation of an asexual sporophyte (diploid) generation with a monoploid gametophyte generation is a forerunner of the type of alternating generations found among mosses, ferns, and the seed plants. (Refer to life cycles in chapter 6.)

Viruses: noncellular organic particles

You may wish to demonstrate how the tobacco mosaic virus (TMV), a single strand of RNA covered by a protein sheath, can exist as both a crystalline dried substance, yet when introduced into living cells of a host can take control of the cell's metabolism and reproduce more virus particles. Use young tobacco and/or tomato plants grown from seed (about eight weeks old) as the host plants. Maintain plants in separate small pots for a variety of treatments.

Barnes describes the following procedure.[50] Soak 1 g of dried tobacco leaves infected with TMV in 10 ml of water in a Petri dish for 15 to 20 mins.[51] Grind the leaves in a mortar with a pestle; dilute the inoculum 1:100 for use. First injure or bruise the leaves of healthy young plants by sprinkling the surface of leaves with carborundum (400 to 600 mesh). Place the carborundum in pepper shakers and cover with a layer or two of cheesecloth. Use very small amounts to dust the leaves—so little it is not visible (avoid inhaling the carborundum).

Dip a cotton swab into the inoculum of TMV and gently brush the entire surface of several leaves. Support the leaves with toweling under each leaf so you can hold the leaf as you apply the inoculum. Brush only once; wash the leaves under a gentle flow of tap water. Keep control plants separate; apply the carborundum and tap water, but not the TMV. Use the same

technique to prepare an inoculum of soaked cigarette, cigar, and pipe tobacco in water. Apply this to a third group of young plants. Label each group of plants. Wash hands in alcohol to avoid contamination when using different materials.

Maintain plants in adequate light at a temperature of 26° C (79° F). Local infections should be apparent after 4 to 14 days. Systemic symptoms showing mottled, puckered leaves in a mosaic pattern should appear after three to four weeks. Every three weeks, add 10 ml of 2 percent sodium nitrate solution to each potted plant.

The leaves may be dried and stored in envelopes for additional investigations.

A microscopic study of viruses. Locate epidermal hairs on the petioles of infected leaves. Scrape some into a drop of water on a slide, apply a cover slip and examine under high power. Look for virus crystals of different shapes within the epidermal cells.

Kingdom Monera

The two groups of procaryotic cells in this kingdom are the blue-green algae (p. 99), and the bacteria (Schizomycetes). Among the 1500 or so species of bacteria, some are autotrophs, but most are heterotrophs.

Collecting bacteria

Yoghurt, sauerkraut, or pickle juice (from a barrel) are good sources of harmless bacteria available immediately for use. (See chapter 2 for preparation of wet mounts, staining, and hanging-drop slides to observe motility in bacteria.)

Decaying substances are another source of bacteria. Soak beans or peas in water for several days. As they decay, pipette off the cloudy liquid which will be teeming with large bacteria, probably motile *Bacillus subtilis*, among others. Or, put some timothy hay, meat, or boiled potato into distilled water and leave exposed to the air. After a few days stopper the test tubes or small flasks lightly with nonab-

[50] E. Barnes, *Atlas and Manual of Plant Pathology* New York: Appleton-Century-Crofts, 1968.
[51] Purchase dried tobacco mosaic virus (general strain 135) for high school students from American Type Culture Collection, 12301 Parklawn Drive, Rockville, MD 20852.

sorbent cotton, and keep them in a comparatively warm place for the next few days. Also examine the scum formed in the water of a vase of cut flowers.

Special kinds of bacteria, such as chromogenic or luminescent genera of microorganisms may be obtained from biological supply houses (Appendix). Obviously there is no need for pathogenic forms in a school laboratory. In fact, their use in high school laboratories is prohibited by law in many states. Furthermore, the literature in bacteriology indicates that under certain conditions some of the chromogenic forms may mutate and become pathogenic.

Also refer to nitrogen-fixing bacteria, *Rhizobium* (p. 634); nitrification in soil by bacteria tracing nitrite formation from ammonium salts, and nitrate formation (p. 608); also the kinds of bacteria, as well as abundance, in soil using dilution procedures (p. 602).

The preparation of bacterial smears and stains including the Gram-stain technique are described in chapter 2.

The preparation of a Winogradsky column is described on page 607 for studies of bacterial decomposers in soil water and marine waters. Further, submerged slide techniques are presented on page 595 and page 605 for studies of populations of microorganisms interacting in close proximity.

The effect of penicillin and/or other antibiotics is described on page 610; the effects of antiseptics on page 610; the role of temperature on page 610; replica plating, as developed by Lederberg is described on page 705.

Culture media for bacteria

When pure cultures are to be maintained and subcultured, a suitable medium must be used. Students may observe growth patterns of colonies—their color, shape, and texture—as well as a microscopic examination by staining bacterial smears with Loeffler's methylene blue as well as Gram's stain. (See staining, chapter 2; counting colonies, p. 462; turbidity,

p. 464; products of bacterial metabolism p. 706.

A word of caution Every precaution should be taken to avoid contamination of cultures with pathogenic forms. Coughing into sterile dishes and incubating thousands of bacteria from the respiratory tract is dangerous. In fact, each culture, even those obtained from soil, or from decaying matter *should be treated as though it were pathogenic.* Sterile techniques are required. Even when nonpathogenic colonies predominate, a small number of pathogens may be present; they may have failed to grow luxuriantly into conspicuous colonies because conditions were not optimum. They may require other salts, microelements, blood medium, special proteins, or a different pH. Yet they may be present in the culture. An old agar slant or Petri dish may also contain bacterial mutants.

The main concepts and techniques that students will need to know in introductory bacteriology may be gleaned from studies of *Bacillus subtilis,* a motile rod in decaying bean or pea scum and from pigmented bacteria such as *Sarcina lutea* (yellow colonies) and *Serratia marcescens* (red colonies), when cultured below 37° C (99° F); also see p. 551).

Bacteria everywhere To develop the notion that bacteria are found just about everywhere, open sterile Petri dishes of nutrient agar to the air for 15 min while a control is kept covered. In addition, students may press fingertips on the surface of the agar, imbed hair, or scrapings from the teeth. Wet cotton applicators in sterile isotonic salt solution and brush across a coin, a dollar bill, a door handle, a windowsill, and so forth. *Seal* the plates with masking tape; label and invert before incubating at 37° C (99° F) for a few days. (Invert the dishes so that condensed water droplets do not fall back on the surface and spoil the growth of colonies.)

Preparation of culture media for bacteria For limited uses in the class or laboratory, it probably is advantageous to purchase prepared sterilized Petri dishes and/or test tubes with medium from supply houses. Or purchase dehydrated agar medium or broth which can be readily prepared for plating. Use plain agar to solidify media of known chemical composition.[52] In the laboratory, have students measure accurately the required amount of dried medium, pour it in a dry flask, add distilled water, and autoclave. Then pour the medium into plates and test tubes (p. 700). These media contain buffers that approximate a pH 6.8 to 7. However, you may want students to know the contents of bacterial media and also to gain skills in techniques. To this end, many recipes are offered below in the event you want to prepare your own media.

A suitable culture medium should contain a source of nitrogen such as peptone, plus inorganic salts, carbon, and possibly some growth-promoting substances. Special substances such as dextrose, serum, or special salts may be required for successful growth of some genera of bacteria; these are also available commercially.

NUTRIENT BROTH   Weigh the following and combine in 1 liter of distilled water:

beef extract	3 g
peptone	5 g
NaCl	5 g

Heat this mixture slowly to 65° C (149° F), stirring until the substances are completely dissolved. Filter through paper or absorbent cotton and adjust to pH 6.8 to 7.0 by adding a bit of sodium bicarbonate.

Both nutrient broth and nutrient agar may be further enriched by adding the following for each liter of nutrient broth or agar:

peptone	15 g
yeast extract	5 g
glucose	2 g

---

[52] Available from Difco, as well as biological supply houses (Appendix). An instant medium, *sterigel*, is available from Carolina Biological; 100 ml prepares 10 Petri dishes (100 × 15 mm).

Pour the mixture through a funnel into test tubes, filling them one-third full; loosely stopper with nonabsorbent cotton and cover with aluminum foil to prevent the stoppers from popping out when autoclaved. Sterilize in an autoclave at 15 lb pressure for 15 min. This quantity should be sufficient to prepare 36 test tubes.

NUTRIENT AGAR   Use commercially available nutrient agar or add the following to 1 liter of water:

beef extract	3 g
peptone	5 g
agar	15 g

Heat in a double boiler or autoclave for 15 min.

A liquid broth may be solidified by adding agar. For example, add 15 to 20 g of agar to the recipe above for nutrient broth. Adjust the pH by adding a few drops of NaOH solution. Autoclave at 15 lb pressure for 15 min. The melting point of agar is about 99° C (210° F); it solidifies at about 39° C (102° F).

MALT EXTRACT AGAR   Dissolve the following in one liter of water:

malt extract	15 g
agar	15 g

Dissolve by heating and stirring in a double boiler. Autoclave for 15 min.

NUTRIENT GELATIN   While gelatin was one of the first substances added to solidify medium, it has limited usefulness since it melts at room temperature in warm weather. However, it can be used to identify bacteria that produce a protease.

To one liter of water add the following:

beef extract	3 g
peptone	5 g
gelatin (bacto)	120 g

Heat in a double boiler; restore the volume with water and adjust the pH to about 7.0 to 7.5. (Add a bit of sodium bicarbonate.)

The nutrient gelatin may be clarified by adding a raw egg mixed with a small amount of water to the gelatin medium. Heat very slowly until the egg becomes firmly solidified. Then filter the solution through cotton and pour into test tubes (about 10 ml per tube). Finally, sterilize the tubes in a double boiler (without steam pressure) for 20 min on three successive days. Cool the medium rapidly after each sterilization. This quantity should be sufficient for some 36 test tubes.

POTATO MEDIUM   Use a cork-borer to cut cylinders from large washed, peeled potatoes. Cut the cylinders obliquely into wedge-shaped portions; leave overnight in running water to reduce acidity. Then place a wedge of potato into each of several test tubes, or use slices in covered Petri dishes. Add 3 ml of distilled water to each test tube, and stopper the tubes with nonabsorbent cotton. Stand the tubes in a wire basket, but avoid packing them. Push down the cotton plugs so they won't pop out, or cover with caps of aluminum foil. Sterilize the tubes in an autoclave or pressure cooker for 20 min at 15 lb pressure. Be sure to allow air to escape from the pressure cooker before closing the valve. If a double boiler is used, heat to boiling for 1 hr.

You may want students to expose these tubes to air, or inoculate the surface of the potatoes with a platinum or Nichrome wire needle or loop laden with bacteria. Or brush cotton swabs dipped in sterile saline solution against a windowsill, skin, coin, or other source and inoculate a potato wedge. Place the inoculated dishes and/or tubes in a warm, dark place (or an incubator) for several days. Colonies of bacteria (and also molds probably) will appear as spots of different texture in white, cream, buff, yellow, orange, and other colors. Molds have a fuzzy texture that is quite different from bacterial colonies.

CZAPEK BROTH   The preparation for this medium of known chemical composition is given on p. 710.

Sterilizing culture media

Naturally, when specific bacteria are to be cultured, conditions must be sterile. There are several ways to sterilize the materials you will need.

DRY HEAT   Glassware, including Petri dishes, test tubes, and flasks, may be sterilized by placing them in an oven at a temperature of 160° to 190° C (320° to 374° F) for at least 1 hr. At temperatures over 175° C (347° F) cotton and paper begin to char. When glassware is tightly packed in containers, it should be sterilized in an oven for at least 2 hr at 160° to 190° C.

STEAM UNDER PRESSURE (AUTOCLAVE)   Sterilize culture media, cotton, rubber, and glassware in an autoclave or large pressure cooker. For small quantities of medium, use a smaller pressure cooker. When large quantities of media are sterilized in flasks or bottles, keep them in the autoclave for 30 min. Cover the cotton plugs on the flasks with aluminum foil to keep them dry. (Plastic dishes cannot be autoclaved.)

Follow the directions for using an autoclave or pressure cooker that accompany the apparatus. Open the escape or safety valve before turning on the steam. When steam escapes full force, close the valve so that only a small opening remains for steam to escape; regulate the valve to maintain the desired pressure.

STEAM WITHOUT PRESSURE   Some kinds of media cannot be sterilized under pressure. Milk, nutrient gelatin, and other materials containing carbohydrates may be hydrolyzed or otherwise altered by overheating. Sterilize these media in a double boiler or use a pressure cooker without applying the steam pressure. Heat for about 30 min, then cool the media to room temperature. Repeat this boiling process on two successive days—intermittent sterilization.

STERILIZATION WITH CHEMICALS   Discarded cultures should be treated with steam under pressure or the surface of the agar covered with a 2 to 5 percent solution of cresol or *Lysol*. Plastic dishes can be treated with *Lysol*, wrapped well and incinerated. Laboratory table tops should also be disinfected with cresol solution or *Lysol*. Wear rubber or plastic gloves when using strong disinfectants.

Boiling

Small pieces of glassware, syringes, and pipettes may be boiled for 30 min for effective sterilization.

Filling test tubes

Medium should be poured into test tubes so that it does not streak the sides of the tubes (Fig. 9-17a). Medium containing agar must be poured quickly so that the agar does not begin to solidify in the funnel or the delivering tube. (Keep flasks of medium in a hot water bath or saucepan of hot water on a hot plate.) When possible, set up a warmed glass funnel resting in a copper funnel warmer (Fig. 9-18) to permit circulation of hot water around the glass funnel. Place a cotton filter in the bottom of the glass funnel. Use a 10 cm square of absorbent cotton with the corners folded in toward the center to make a rounded filter. Wet the cotton, and press it

Fig. 9-18 Funnel warmer with a heavy copper double wall jacket. (Courtesy of Carolina Biological Supply Co.)

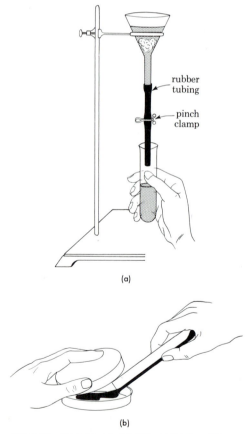

rubber tubing

pinch clamp

(a)

(b)

Fig. 9-17 Techniques: (a) filling test tube with warm culture medium (use pinch clamp to regulate flow); (b) plating a sterile Petri dish with warm medium.

into place. Attach a rubber tube and a clamp to the funnel. Line up the test tubes in racks. Hold one test tube at a time below the funnel, as shown in Fig. 9-17a, and insert the rubber tubing one-third down into the test tube. Pour the medium into the funnel; use the clamp to control the quantity of medium that flows into each test tube. Fill some tubes one-quarter full, others one-half full, and replace each cotton plug. Use nonabsorbent cotton in preparing rolled cotton plugs. Finally, sterilize all the tubes; stand them in wire baskets and place them in an autoclave for 20 min at 15 lb pressure. Remove the tubes and place those that are one quarter filled at an angle along a ledge so that the medium solidifies at a slant; this will provide a larger surface area on which to streak a bacteria inoculum. Use the test tubes that are half full to plate Petri dishes.

Plating Petri dishes

Sterilize Petri dishes by placing them in an autoclave for 15 min at 15 lb pressure, or wrap them individually in paper and place them in an oven. When the paper begins to char a bit, the temperature is high enough. Leave the dishes in the oven for several hours. These dishes may be stored in wrapped paper for several weeks.

Use the prepared test tubes that are half filled with medium for plating Petri dishes. If the agar has solidified, stand the tubes in a warm water bath on a hot plate to melt the medium. Wet down the table tops with a sponge and disinfectant; avoid air drafts while plating.

Light a Bunsen burner and work close to it. Remove one tube of the melted agar medium from the water bath and lift off the cotton plug with the fourth and fifth fingers so that the *outer* end of the plug faces the palm of the hand. Bring the mouth of the test tube across the Bunsen flame to sterilize the rim of the tube. Open the Petri dish only enough to pour the contents of the tube into the dish (Fig. 9-17b). Quickly cover the dish and tilt it so the medium is evenly distributed as a thin film in the bottom plate; set aside to solidify. When all these Petri dishes have solidified, invert the dishes to prevent condensing water vapor from falling back onto the agar surface. Wrap each dish in paper until you are ready to inoculate. Refrigerate the Petri dishes to prevent contamination; plates may be stored for a week or more before use.

Inoculating culture tubes and plates

Petri dishes or test tubes may be exposed to air to show that bacteria and mold spores are ever-present in the environment. But when you wish to subculture a pure culture, inoculate new sterile medium Petri dishes or tubes with a bit of the old culture. Transfer a bit of the bacterial culture using a transfer needle or loop that has been sterilized in a Bunsen flame.

Prepare an inoculating or transfer needle by holding the end of a piece of glass rod in a Bunsen flame until the glass is softened. With forceps, insert a 7 to 8 cm length of Nichrome wire (# 24) or a platinum wire into the melted glass rod. When solid material is to be transferred use a straight needle; for liquids, curve the wire into a small loop. Fire-polish the other end of the glass rod handle. You may prefer to purchase more durable transfer needles with metal handles.

To inoculate a sterile culture tube, heat the transfer wire until red hot in a flame to sterilize it. Let the needle cool before touching the culture from which a transfer is to be made. Almost simultaneously remove the cotton plug from the sterile culture tube and hold it between the fourth and fifth fingers as shown in Figure 9-19. Pass the mouth of the tube through a flame; dip the flamed sterile needle tip into the old culture. Remove the needle and transfer it into the sterile new tube, flask or Petri dish. Streak the needle lightly along the entire surface of the sterile agar— do not dig into the medium. Flame the mouth of the test tube again and the plugs (be careful that they not catch fire). Replace the plug on both old and new culture tubes. Flame the needle again before putting it aside in a needle holder. When sterilizing transfer loops or needles that have been used in inoculating a culture, insert the loop or needle into the *cool* part of the flame first, then bring it to the

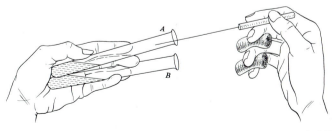

Fig. 9-19 Technique for subculturing: transfer from (A) into a second tube of sterile medium (B).

Note the cotton stoppers are held between the third, fourth, and fifth fingers.

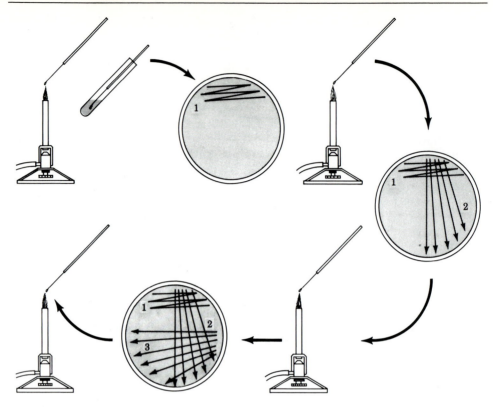

Fig. 9-20 Technique for streaking a sterile culture plate: flame the transfer loop or needle, zig-zag along (1) and flame the needle again; cut lines across (1) as lines (2) are made and flame needle; streaks (3)cut across lines (2) forming the most dilute streaks.

hot region to avoid splattering the microorganisms.

For patterns of streaking Petri plates to isolate bacteria see Figure 9-20.

Isolating bacteria

Recall that Robert Koch developed solid culture medium as a means for isolating the bacteria mixed in a suspension. You may want to use one or more techniques for streaking solid medium culture dishes. In addition, use easily identifiable bacteria so that colonies can be distinguished: bacteria can be recognized on stained slides as coccus, rod or spiral. For example, in a half test tube of sterile isotonic saline solution, mix a few loops of *Sarcina lutea* (a coccus that produces yellow colonies) and an equal quantity of *Bacillus subtilis* (a motile rod that produces whitish colonies).

Or use another color-producing form, *Serratia marcescens*, which produces red colonies at room temperature.[53]

Techniques for isolating bacteria

Separate the bacteria in a mixed culture by streaking the surface of sterile nutrient agar in Petri dishes or test tube slants. A dilution pour plate is another method; or you may want to roll a film of broth or other liquid across an agar surface using a glass rod bent at right angles.

Streak methods   Lift up the cover of a sterile Petri dish of nutrient agar just

---

[53] Other chromogenic bacteria include *Chromobacterium violaceum*, *Sarcina aurantiaca*, and *Micrococcus luteus*.

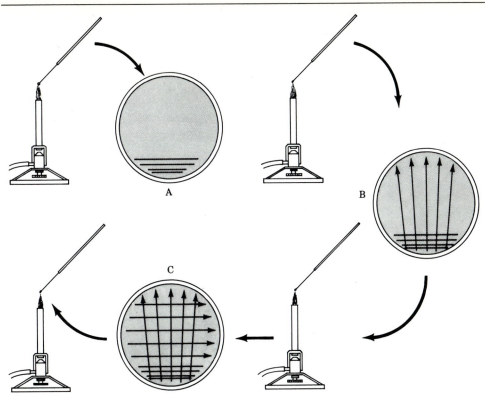

Fig. 9-21 A radiating streak technique: flame transfer needle or loop and make four parallel streaks (A); flame needle and make five radiating streaks (B); flame needle and cut across the radiating streaks so that most dilute inoculum results along lines (C).

enough to carry a loopful of the mixed inoculum to streak along the surface of the agar. Make several zigzag streaks lightly over the surface of quadrant 1; remove the loop and flame it. Then cut across the first streaks to dilute the inoculum in a zigzag fashion in quadrant 2. Refer to Figure 9-20. Then flame the loop and cut across the streaks of quadrant 2 so that the inoculum is further diluted in quadrant 3.

At some time use a *radiating* streak method (Fig. 9-21) to compare with the technique described above. With a transfer loop streak the mixed inoculum four times in *parallel* lines along one side of the surface of the agar in the Petri dish; this quadrant A has the first loopful of the inoculum. Flame the loop. Now cut through the parallel lines of quadrant A to make a series of streaks that fan out across the Petri dish surface. Again flame the

needle. Now turn the Petri dish around for step 3 and make a series of parallel streaks that cut across the radiating ones.

After several plates are prepared, invert the plates, tape the cover to the bottom, and label. Incubate at 37°C (99° F). (The dishes are inverted to avoid having condensed drops of water fall back on the surface of the growing colonies.) *Note*: If *Serratia marcescens* is incubated at 37° C (99° F) the colonies are white; they are red when incubated at room temperature about 25° C or 77° F. In fact, you may want students to recognize that DNA controls the heredity of the organisms. An interaction with the environment is a decisive factor in the phenotype that appears.

Loop dilution: pour plate method
Bacteria mixed in broth, milk, water, or

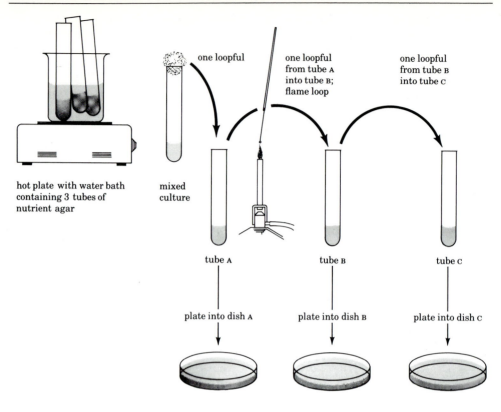

one loopful

one loopful
from tube A
into tube B;
flame loop

one loopful
from tube B
into tube C

hot plate with water bath
containing 3 tubes of
nutrient agar

mixed
culture

tube A

tube B

tube C

plate into dish A

plate into dish B

plate into dish C

Fig. 9-22   Summary of pour plate dilution method.

other fluids may also be isolated. Dilute one loopful of the culture and make a series of thin film plates. This is easier than streaking, yet a successful method. For quantitative plating to make bacterial population counts, see page 459.

Place three tubes one-third full of nutrient agar medium in a beaker of water on a hot plate; maintain at about 50° C (122° F). Label the tubes A, B and C. Now line up three sterile Petri dishes labeled *A*, *B* and *C*. Use a sterilized inoculating loop to carry a loopful of *mixed* culture from one test tube into melted nutrient agar medium in tube *A* (Fig. 9-22). Flame the needle. Roll the tube in the palms of the hands to mix the loopful of inoculum and return the tube to the hot water bath to avoid having the agar solidify. With a sterile loop transfer one loopful from tube *A* into a second tube of nutrient agar, tube

B. Roll the tube (*B*) in the hands after sterilizing the transfer loop. Now return tube *B* to the hot water bath. Again, with a sterile loop transfer a loopful of this more dilute culture medium into a third tube (*C*), flame the loop; rotate the tube in the hands to distribute the microorganisms. Now pour the diluted contents of tube *C* into Petri dish *C*. Plate Petri dish *A* with the contents of tube *A* (remember to flame the mouth of the test tube); plate Petri dish *B* with the contents of test tube *B*. When the three seeded plates have solidified, invert them and incubate at 37° C (99° F). After a day or two, many separate small colonies should be visible.

Bent rod: thin film technique   Glass or metal rods bent into a right angle may also be used to roll out a thin film of fluid inoculum on the surface of a sterile nutrient agar Petri dish. Without sterilizing the bent rod, roll it across a *second* sterile Petri dish medium for a more dilute inoculum that should show isolated colonies. (Also refer to counting colonies, pages 462–66, in chapter 6.)

Identifying isolated bacteria   Compare the degree of isolation of colonies in the Petri dishes using the several methods described. Barring contamination, there should be two kinds of colonies if a mixed culture of two genera of bacteria were originally used. Students can now make bacterial smears of each kind of colony, stain with Gram's stain and then compare their slides with smears originally prepared from the pure cultures of each type of bacteria. (See Gram stain technique, p. 145.)

You may also want students to subculture isolated colonies from Petri dishes to sterile nutrient agar slants (or broth). Use a sterile transfer needle rather than a loop and pick up a bit of the center of the colony to subculture on a slant. Streak once along the surface of the slant from the bottom to the top of the slant; avoid touching the walls of the test tube. Be sure to flame the mouth of the tube as well as the transfer needle.

Recall that in a microscopic examination (oil), students should be able to distinguish *Sarcina lutea*, a gram-positive coccus from *Serratia marcescens*, a gram-negative rod, and *Bacillus subtilis*, a gram-positive rod.

Replica plating

Lederberg developed a simplified technique, replica plating, for continuing colonies of bacteria of the same heredity without making transfers with needles or loops. A piece of velour was used to cover a cylinder of wood or other material. The diameter of the cylinder and velour covering was the same as that of a Petri dish, and the velour was held securely in place.

When colonies are grown in a Petri dish, each colony may be considered as having arisen from one organism. Hence all the cells are a clone, having the same heredity. Lederberg and his group inverted a growing Petri dish over the velour and wood cylinder, pressing lightly, so that a pattern of the colonial growths was formed on the velour. Then a fresh, sterile Petri dish of agar was inverted and touched to the "pattern" on the velour. When incubated, new colonies formed at the points of contact with the original colonies. In fact, if two dishes were placed on top of each other, the same pattern of distribution of clones could be matched. This, then, is also a way of marking off, or of identifying similar clones without the tedious task of making transfers. Further, it is one way to inspect for possible mutations.

What kinds of investigations might be undertaken with bacteria having the same heredity?

Products of bacteria

Change in pH   Changes in pH of a culture due to bacterial growth and metabolic products may be determined. Add a small quantity of an indicator, as suggested below, to the medium when it is originally prepared. Three indicators, among many, are offered.

BROM CRESOL PURPLE   Prepare a 0.2 percent solution by dissolving 0.2 g brom cresol purple in 50 ml ethyl alcohol (95 percent). Then add 50 ml of distilled water. The range of the indicator is yellow at pH 5.2 and purple at 6.8. Add 5 ml of the indicator solution to each liter of medium to be tested.

BROM THYMOL BLUE   Prepare a 0.04 percent solution by dissolving 0.04 g brom thymol blue in 50 ml of ethyl alcohol (95 percent); to this add 50 ml distilled water to make 100 ml. The indicator is yellow at pH 6 and blue at 7.6.

If the brom thymol blue indicator is to be added to a culture medium, make a stronger solution—use 0.5 g of brom blue indicator in preparing the solution. Add 5 ml of the 0.5 percent indicator solution to each liter of medium.

PHENOL RED Prepare this indicator solution (0.2 percent) for fermentation media. Dissolve 0.2 g phenol red in 50 ml of ethyl alcohol (95 percent); add 50 ml distilled water to prepare 100 ml. Add 10 ml of 0.2 percent solution to each liter of culture medium. The indicator is yellow at pH 6.8 and red at 8.4.

Metabolic product: indole Indole is a product of the degradation of proteins by bacteria. Prepare a tryptone broth (commercial preparation) and inoculate with *Bacillus subtilis* or other bacteria. Incubate for two to three days at 37° C (99° F) and then test a sample for indole with Kovac's reagent (preparation below).

Into a clean test tube pour 4 ml of the tryptone broth culture of *B. subtilis* and add 0.4 ml of Kovac's reagent. Shake the tube gently, then let it stand; observe that the reagent soon rises to the top of the contents. A deep red color in the reagent layer indicates the presence of indole.

KOVAC'S REAGENT (INDOLE TEST) Prepare the following:

n-amyl alcohol	75 ml
HCl (concentrated, 37%)	25 ml
p-dimethylamine benzaldehyde	5 g

Dissolve the aldehyde in alcohol and warm the mix gently using a water bath. When dissolved, carefully add HCl (*caution*) and stir. Then refrigerate.

Uses in the classroom and laboratory Bacteria, as beneficial flora in a soil and water ecosystem, are useful in basic studies in biology, and in studies of nitrification (p. 608) and nitrogen-fixation (p. 634). (For counts of microorganisms in soil, see pp. 463, 601.) Skills in sterile techniques are developed through culturing, plating, isolating pure types, examination of metabolic products of bacteria, studies of types of colonies, and microscopic examination of bacterial smears stained with Gram's stain (p. 145).

Students may study population growth rates of bacteria by quantitative plating (p. 459), or observe turbidity. The effectiveness of antibiotics may also be investigated using antibiotic disks applied to inoculated plates (p. 610). In addition, the effectiveness of preservatives such as salts, heavy sugar solutions, spices, vinegar and such chemical preservatives as calcium propionate may be readily investigated.

Attention should be given to plant diseases caused by bacteria. These pathogenic forms do not produce spores, are Gram negative, and cause such plant diseases as blights, galls (p. 619), wilts, leaf spots, and soft rots. The effects of the bacteria that cause crown gall in tomato plants can be observed over a month or two. Refer to plant parasites and other types of plant symbiosis under "Modes of Nutrition" in chapter 8.

Kingdom Fungi: subdivision Myxomycota

This subdivision includes plasmodial slime molds such as *Physarum*, which have a vegetative, multinucleated mass of protoplasm called a plasmodium, and a fruiting or sporulation stage. There are also cellular slime fungi such as *Dictyostelium*, which lack a flagellated cell stage, and for most of their life cycle retain the amoebalike plasmodium stage. Whether slime fungi should be considered plants or animals is a moot question. While slime molds have an amoeboid plasmodium, they also undergo sporulation.

Collecting slime fungi

Among the more than 400 species known, slime molds may be black, white, yellow, or red in color; find them under the bark of fallen logs or under boards in swamps or moist woods. They may be found balled up or in holes made by boring insects. A day or two after it rains is a good time to look for slime molds.

When conditions of moisture and food are adequate, the plasmodia increase in size and their nuclei divide repeatedly so that ultimately a myxomycete may be 5 to

8 cm in diameter and contain some thousand nuclei. Cell walls are lacking in this vegetative stage; the slow amoeboid movement can be observed under the microscope as it creeps over a surface scooping up food.

Under certain environmental conditions, the plasmodium becomes stationary and develops round fruiting bodies on short stalks. Light is one factor that induces sporulation, or fruiting bodies—a plantlike characteristic; removal from darkness for a 5-hr light interval initiates sporangia formation. Niacin is also necessary for this stage to occur. The spores are carried by wind and germinate in a moist environment to develop either flagellated swarm cells or small "amoebae"—myxamoebae. When enough water is present flagellated swarm cells form; in limited supplies of water the nonflagellated myxamoebae result. When two swarm cells unite, or two myxamoebae fuse, diploid zygotes are formed. A new plasmodium develops from the zygote and grows by many mitotic divisions without cell divisions so that a multinucleated plasmodium is again produced.

Small plasmodia may be obtained from the spores of the fruiting bodies of *Physarum* (Fig. 9-23), *Didymium, Fuligo,* or *Stemonitis.* These fruiting bodies may be crumbled on filter paper or on solid agar.

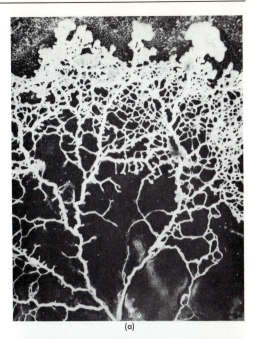

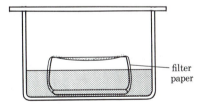

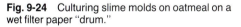

Fig. 9-23 Slime molds: (a) *Physarum polycephalum*, showing plasmodial stage (magnified 6 ×); (b) sporangial stage (magnified 9 ×). (Photos by Hugh Spencer.)

Filter paper culturing

Camp describes the following method for culturing slime molds on paper toweling and pulverized oatmeal.[54] Wrap a finger bowl or similar container in filter paper to make a "drum." Then place this prepared finger bowl flat within a larger battery jar; the exposed filter paper should have a flat and smooth surface. It is better to have the open or upper side of the finger bowl on top. Then add about 3 cm of water down the side of the battery jar.

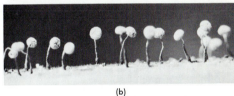

Fig. 9-24 Culturing slime molds on oatmeal on a wet filter paper "drum."

The filter paper acts as a wick to draw moisture to the top of the "drum" (Fig. 9-24).

Place the plasmodia on the filter paper; then sprinkle about 0.1 g of pulverized oatmeal (rolled oats which have been

[54] W. G. Camp, "A Method For Cultivating Myxomycete Plasmodia", *Bull. Torrey Botan. Club* (1936) 63:205–10.

ground in a mortar will suffice) onto the paper. Finally, cover the battery jar with a glass plate. Within 24 to 48 hr the oatmeal should be consumed by the spreading plasmodia. Avoid overfeeding, but as the plasmodia grow and spread over the paper, add larger quantities of oatmeal.

When the plasmodia have spread over the entire surface of the paper, place paper toweling on the inner surface of the battery jar so that, in effect, the inner surface is lined with paper. Then the toweling will also act as a wick to draw water. The plasmodia will move across the water moat between the filter paper and the toweling and begin to spread along the toweling. Place the oatmeal, however, on the central filter paper. Then after the plasmodia have spread over all the paper, remove the toweling and dry at room temperature. On the paper the plasmodia quickly form dormant sclerotia.

This dried paper may be cut into squares and stored in envelopes or sealed vials in a refrigerator. The plasmodia will remain viable for at least a year or two and may be used to prepare fresh cultures within this time.

To prepare a fresh culture, place a square of the dried culture paper on wet filter paper on the finger bowl arrangement already described. Allow 24 hr for the plasmodia to form, and place oatmeal powder in small quantities for food. Increase the amount of oatmeal as the plasmodia grow.

Students investigating the behavior of plasmodia may prefer to prepare a culture of slime molds by crumbling fruiting bodies on a medium containing agar, such as the one that follows.

Agar medium for plasmodia   Soak 6 g of uncooked oatmeal in 200 ml of water for 1 hr. Then strain the liquid. To 100 ml of the clear filtrate add 1.5 g of plain agar, heat gently until the agar is completely dissolved. Pour a thin layer of this mix into each of several Petri dishes. (It is not necessary to sterilize the agar dishes if they are to be used immediately.) Inoculate the

dishes with sporangia or sprinkle the agar with the dark spores; add 10 ml of sterile pond water to the surface of the agar. In two to three days, myxamoebae should appear. Subculture when necessary. Keep the cultures in a warm, dark place. Avoid high temperatures to control the growth of mold and bacteria contaminants.

Observe the formation of swarm cells from spores by placing spores in a depression slide with distilled water. Apply dabs of petroleum jelly at the corners of a coverslip to seal the slide. At 23° C (73° F) swarm cells should be observed within 24 hr.[55]

Uses in the classroom and laboratory In the laboratory, demonstrate myxamoebae and cytoplasmic streaming of slime molds under a dissecting microscope (30 ×). Factors that initiate differentiation among slime molds may provide the opportunity for long-range investigations.

Kingdom Fungi: subdivision Eumycota

The members of this subdivision are the true fungi, heterotrophs that require ready-made organic molecules and obtain their food as saprophytes, or as parasites that devastate food crops and cause diseases among animals as well as humans. Among the true fungi are the water molds, ascomycetes like yeast, *Aspergillus* and *Penicillium*, and the basidiomycetes—mushrooms, rusts, and smuts.

Collecting molds, yeasts, mushrooms
A variety of molds are readily available. Expose pieces of moist bread (preferably without a mold inhibitor) to air or dust against an open window. Maintain in covered containers lined with moist toweling (baby food jars take little space). Also soak some bread in a sugar solution and main-

[55] T. Register and W. West, "Plasmodial Slime Molds," *Carolina Tips* (March 1974) 37:3.

tain in the same way. Collect bruised fermenting plums, peaches, grapes; also citrus fruits such as lemons and grapefruits. Maintain all containers in an incubator at 30° C (86° F). Within a week, examine the bread to find *Rhizopus*, the common bread mold with a growth of mycelia and black sporangia (Figs. 2-17, 6-5). Yellow *Aspergillus* (Fig. 1-122) should be in the containers of bread soaked in sugar. *Penicillium* (Fig. 1-121) should be growing on the citrus fruit; *Saccharomyces* should be found on fermenting grapes and plums.

Several methods are described below for subculturing to obtain pure cultures. Other sources of fungi include *Penicillium* in Roquefort cheese. The species that flavor cheese are *P. roquefortii* and *P. camembertii*; the penicillin-producers are *P. notatum* and *P. chrysogenum*. Yeasts may also be found on the surface of sauerkraut or dill pickle juices, or may be purchased in dry form. Dung medium will yield many forms, such as *Mucor* (Fig. 9-25), which is a close relative of *Rhizopus*; also probably present will be *Pilobolus*, certain ascomycetes and basidiomycetes.

Collect or purchase edible mushrooms to study their structure and reproductive cycle.

Culturing molds

Most species of *Rhizopus* grow best at 30° C (86° F) in moderate light. *Rhizopus* and *Mucor* are often confused with each other, but they may be distinguished in this way. The sporangiophores in *Rhizopus* arise in a fascicle from a node on a stolon, while in *Mucor* there are single sporangiophores arising from mycelia (Fig. 9-25).

Asexual spores are abundant; you can also show sexual reproduction — conjugation in *Rhizopus* (Fig. 6-59). Two different mating strains are necessary for conjugation. (Purchase plus and minus strains from a biological supply house, listed in Appendix). When a plus strain is streaked across one side of a moistened slice of bread, or better, across a dish of potato dextrose agar or Sabouraud's dex-

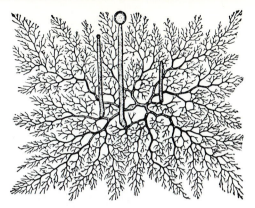

Fig. 9-25 *Mucor*, a common mold. (Adapted from *The Plant Kingdom* by William H. Brown, Ginn and Co., Publishers. Copyright© 1935 by William H. Brown.)

trose agar, and a minus strain streaked at the opposite end, sexual spores, zygospores, form along the center of the bread (Fig. 6-59). Zygospores are diploid and germinate in some six weeks. Meiosis occurs in the zygospores; each new sporangium produces asexual spores that germinate into plus or minus strains.

Culture media with agar   The following solid media are easily prepared. Agar may also be added to some other media described later. To subculture, inoculate sterile Petri dishes with spores or carry over small loopfuls of cultures from old plates. Streak the agar surface of the new plates using sterile techniques (p. 702).

OATMEAL AGAR   When pure cultures of one type of mold are to be prepared, sterile conditions must be observed. In this method, grind 5 g of oatmeal into 200 ml of water; boil this mixture for 20 min. Pour off the supernatant fluid through gauze to avoid including the oat residue. Add 4 g of agar to the fluid, and boil to dissolve. Finally, pour into sterile Petri dishes, cool, and later inoculate with spores of selected fungi.

POTATO-DEXTROSE AGAR   Peel and slice about 100 g of white potatoes and boil for 1 hr in 350 ml of distilled water. Strain through cheesecloth and restore the liquid to the original volume by adding more distilled water. To this 350 ml of potato filtrate add 1 g of dextrose and 10 g of agar; boil for 30 min in a double boiler. Strain and sterilize; then pour the liquid into sterile Petri dishes. Later, inoculate with desired mold.

Prepare potato-agar medium using the same procedure, omitting the dextrose.

Dehydrated medium is available from Difco and other biological supply houses (Appendix).

PRUNE AGAR   Boil about five prunes in water for 1 hr. Then pour off the supernatant liquid and make up to 200 ml. To this add 80 g of sucrose and 5 g of agar. This quantity (that is, 200 ml of fluid) will be enough for about seven Petri dishes. Multiply all amounts for larger quantities of the medium. Boil in a double boiler, strain and sterilize. Plate in sterile Petri dishes.

PEA AGAR   Boil about 80 dried peas for 1 hr. Then pour off the liquid and make up to 200 ml. To this add 5 g of agar. Boil in a double boiler then strain and sterilize. Plate the medium in sterile Petri dishes.

CORNMEAL DEXTROSE-PEPTONE AGAR To 1 liter of distilled water add:

cornmeal agar (dehydrated)	17 g
peptone	2 g
dextrose	10 g

Boil in a double boiler and sterilize.

DUNG AGAR   Soak some 200 g of horse, cow, or rabbit dung in water for three days. Then pour off the liquid and dilute with water until the liquid is straw colored. To this add about 2.5 g of agar for every 100 ml of diluted fluid. Boil the materials and plate into sterile culture dishes.

SABOURAUD'S MEDIUM   A useful medium for many kinds of fungi grown in the laboratory is Sabouraud's medium. Mix together and bring to a boil 10 g of peptone, 20 g of agar, and 1 liter of distilled water. To this add 40 g of maltose; filter if necessary. No adjustmet for pH is required. Finally, sterilize for 30 min at 8 lb pressure and pour into Petri dishes, test tubes or even Syracuse dishes. Pre-

pared media are also available from Difco and other supply houses (see Appendix).

Culture media without agar   Since it is often difficult to remove molds from solid media without damaging the rhizoids, several culture methods which omit agar may be useful.

KLEB'S SOLUTION   Kleb's solution (p. 663) can be considered one of these. Dilute the stock solution 1:10 with distilled water.

Another simple liquid culture medium is a 5 percent glucose solution.

JOHANSEN   recommends another medium.[56] Dissolve the following in 1 liter of distilled water:

sucrose (cane sugar)	30.0 g
NH_4Cl	6.0 g
$MgSO_4$	0.5 g
K_2HPO_4	0.5 g

BARNES' MEDIUM   Dissolve the following in 100 ml of distilled water.[57]

K_3PO_4	0.1 g
NH_4NO_3	0.1 g
KNO_3	0.1 g
glucose	0.1 g

(If desired, 2.5 g of agar may be added to this glucose-salts solution.)

CZAPEK'S MEDIUM: A SYNTHETIC MEDIUM Several genera of fungi, including many of the genus *Aspergillus* as well as soil bacteria grow well on this synthetic medium of known chemical composition. This medium contains inorganic nitrogen used by some bacteria and fungi as well as a carbon source, saccharose. Obtain from Difco in dehydrated form and rehydrate by dissolving 35 g of the broth in 1 liter of water (almost neutral, pH 7.3). Or you may want to prepare your own medium. To one liter of distilled water add:

saccharose (Difco)	30.0 g
sodium nitrate	3.0 g
dipotassium phosphate	1.0 g

[56] D. Johansen, *Plant Microtechnique*, New York: McGraw-Hill, 1940.
[57] H. C. Gwynne-Vaughan and B. Barnes, *The Structure and Development of the Fungi*, New York: Cambridge Univ. Press, 1927.

magnesium sulfate	0.5 g
potassium chloride	0.5 g
ferrous sulfate	0.01 g

A yeast extract (1 percent) may also be added to this medium. The medium may be solidified with 1.5 percent agar.

Culture methods for some marine yeasts and fungi are described in chapter 8.

Summary of methods for specific molds

Some molds are best cultured on specific media. For example, *Rhizopus* grows well on bread and potato-agar medium (with or without sugar). However, for careful work, especially in studies of zygospore formation use the potato agar medium. Observe conjugation in *Phycomyces blakesleeanus* across a potato-dextrose agar plate (Fig. 6-59).

The prune-agar medium is recommended as well as Czapek's medium for culturing *Aspergillus*. It's mold is also cultured readily on bread that has been exposed to the air for several hours and then soaked in a 10 percent solution of sucrose, grape juice, or prune juice. Keep in a covered jar at room temperature or in an incubator.

Penicillium is best kept on its source, Roquefort cheese or citrus fruit. For pure cultures use the prune-agar method. Some teachers prefer to culture this mold in flasks.

Penicillium and *Aspergillus*, both ascomycetes, do not seem to have sexual reproduction. Compare the cross-walls in these forms with *Rhizopus*. Contrast the dome-shaped sporangium of *Aspergillus* with the conidia at the tips of stalked conidiophores in *Penicillium* (Fig. 1-121).

Saprolegnia   This water mold grows abundantly on decaying insects, frogs' eggs, small fish, or decaying radish seeds in pond water. Observe how rapidly the infectious fuzzy mycelia spread over dead flies placed in pond water. Note that the many hyphae lack cross-walls so they are multinucleated. When the food supply becomes limited, mitosporangia develop at the ends of hyphae. In this vegetative stage, the mitosporangia produce biflagellated zoospores. In a sexual cycle, oogonia and antheridia differentiate on ends of wider hyphae. When antheridia penetrate the walls of oogonia male nuclei move into the egg cells and zygotes result. Flagellated meiospores develop from germinating zygotes.

Marasmius, Clitocybe, and Armillaria
Small hymenomycetes among the Basidiomycetes may be kept in flasks containing dung agar. Spores may be collected from the hymenium of the desired species.

Other forms, such as *Dictyuchus*, may be found and cultured in pond water to which wheat or rice grains have been added. The water must be changed when it becomes cloudy. *Dictyuchus* may also be found as a contaminant of protozoan cultures using rice.

Neurospora   This mold is easily cultured in a commercially prepared medium such as Difco's Bacto-Neurospora Culture Agar (pH 6.7). Add 1 liter of distilled water to 65 g of this powdered medium, which contains yeast extract, peptone, and maltose. Then heat it to boiling, pour tubes and sterilize in an autoclave for 15 min at 15 lb pressure.

Or prepare your own medium. To 1 liter of water add:

bacto-yeast extract	5 g
proteose peptone Difco	5 g
bacto-maltose	40 g
bacto agar	15 g

Inoculate with spores of *Neurospora* from a 48-hr culture. (Other fungi also grow well in this medium.)

Culturing molds on slides

Prepare individual slides for students using the following method recommended by

Vernon.[58] Boil together 6 g of oatmeal, 1 g of agar, and 100 ml of distilled water. Place a drop of the hot liquid on each of several glass slides. Cover with a coverslip until the medium has hardened. When cooled, remove the coverslip and cut the film of agar into several sections, separating each from the other to form little grooves. Inoculate with mold spores of similar or different kinds, and seal the coverslips with dabs of petroleum jelly applied at the four corners. Keep the slides in a moist chamber (preferably a Petri dish lined with damp filter paper). Aerial hyphae and sporangia or conidiophores will grow into the grooves, and they can be examined under the microscope.

Fenn recommends a technique that provides a less dense growth of mold plants so that students can transfer entire plants more easily to a microscope slide (Fig. 9-26).[59] Boil a sliced carrot in water for 20 min. After the liquid has cooled, soak strips of filter paper in the liquid. We recommend that you maintain strips in small stoppered vials or Petri dishes and use a sterile transfer loop to inoculate the wet strips with mold spores. Some students may be allergic to the spores. When they are distributed to students for observation with a hand lens or a dissecting microscope, the room will not become contaminated with mold spores. Maintain the dishes or vials for a few days at room temperature. Observe the pattern of growth and formation of spores. Transfer bits of mold to slides and mount in alcohol or glycerin; examine under high power.

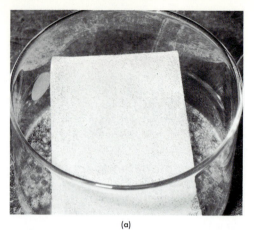

(a)

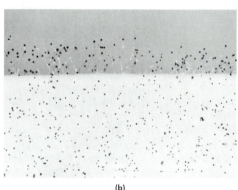

(b)

Fig. 9-26 Culturing *Rhizopus*: (a) a filter paper saturated with carrot water is laid flat against a glass square which is then slanted against the side of a crystallization dish; (b) magnification of the surface of the filter paper. (Courtesy of Robert H. Fenn, Manchester Community College, Manchester, Conn.)

Culturing yeast cells

To show budding as a means of asexual reproduction use *Saccharomyces ellipsoideus*, which is found on grapes, or *S. cerevisiae*, the commercial yeast of bread and beer. Prepare a 5 to 10 percent aqueous solution

of molasses or diluted grape juice. Add one-half of a package of dried yeast to 500 ml of the medium. To this add 1 g of commercial peptone or about 20 beans or peas for rapid fermentation and reproduction.

Insert a cotton plug into the container and set aside in a warm place—25° to 30° C (77° to 86° F). Within 6 to 24 hr rapidly budding cells should be found (Fig. 6-58). Subculture in a fermenting medium; for careful studies use Sabouraud's medium (p. 710).[60]

---

[58] T. R. Vernon, "An Improved Type of Moist Chamber for Studying Fungal Growth," *Ann. Botany* (1931), 45:733.

[59] R. H. Fenn, "A Simple Method for Growing *Rhizopus nigricans*," *Turtox News* (1962), 40:226–27.

[60] An acid medium (pH 5.6) is needed for the selective growth of fungi over bacteria in the culture plates.

Stain slides of yeast cells with methylene blue, described in chapter 2. There is a very slow diffusion of methylene blue into the cells so that living cells are stained selectively. Also refer to cell counts (p. 461) and anaerobic fermentation (p. 310).

Explore the complex reproductive cycles—asexual and sexual (conjugation), as found in some forms with the production of ascospores in an ascus (p. 443).

Lichens

Some lichens, such as the red-topped *Cladonia* or the gray, crusty *Parmelia* and *Physcia*, are found on the bark of trees, especially on fallen tree logs and tree stumps. Reindeer moss, *Cladonia rangiferina* forms the vegetation of tundra regions of the Arctic. Sections of bark covered with lichens may be put into a terrarium. Avoid excess moisture as molds grow readily; at times, keep the terrarium uncovered.

For examination, place dry, brittle specimens in water; after a few hours of soaking the lichens usually become quite flexible and green. Lichens are useful for study since they illustrate mutualism between an alga and a fungus (see p. 634). They are highly susceptible to pollutants in air. Tease apart a bit of lichen in a drop of glycerin or water on a glass slide and examine under the microscope. More resistant forms can be crushed in a mortar. The algae are photosynthesizers while the fungi are better able to hold water needed in the process. Also examine ascospores (Fig. 8-54a). Have students separate the symbionts and try to grow them alone. Can they be recombined?

Kingdom Plantae

Using the five kingdom classification of Whittaker,[61] the Kingdom Plantae comprises the algae (the green, red and brown,

already described) and the many-celled green photosynthesizing plants: the Bryophyta (including the mosses and liverworts), and the Tracheophytes (grouping together ferns, conifers, and flowering plants).

Division Bryophyta

Collecting and culturing mosses and liverworts When mosses and liverworts are collected, leave them attached to the small amount of soil on which they are found. Transport the specimens to the laboratory in plastic bags or newspaper in a vasculum; transfer to terraria in class or laboratory.

A woodland terrarium for mosses, and liverworts like *Marchantia* can be readily prepared. Place a 3-cm layer of coarse gravel or pebbles on the bottom of a tank. Add a few pieces of charcoal. Over this spread a 2-cm layer of sand. Then add a cover of garden loam about 3-cm deep. In this bed sod the mosses, and liverworts such as *Marchantia* (Fig. 9-27) which have been collected from the field. The water level within the terrarium should be halfway up the gravel layer, and the tank should be covered with glass. Keep the tank in medium light. When molds show a tendency to grow in the tank, reduce the amount of water and remove the cover until they disappear.[62] Maintain the terrarium with the least amount of water needed to keep the plants alive. During the winter months supply light from an electric light bulb.

Mosses such as *Mnium, Bryum, Fissidens, Dicranum,* and *Polytrichum* may be kept in this way. Liverworts such as *Pellia, Pallavacinia, Riccia, Marchantia* (Figs. 6-59, 6-60), *Conocephalum,* and *Lunularia* have grown abundantly under identical conditions. On the other hand, *Funaria* and *Polytrichum* can withstand a drier terrar-

---

[61] R. H. Whittaker, "New Concepts of Kingdoms of Organisms," *Science*, Jan. 10, 1969.

[62] Sometimes sprinkling the terrarium with a small amount of powdered sulfur will destroy the growth of molds.

(a)

(b)

(c)

Fig. 9-27 *Marchantia* life history: (a) thallus with gemmae; (b) thallus with antheridia; (c) thallus with sporophytes. (a, c, courtesy of Carolina Biological Supply Co.)

ium. Finally the aquatic liverworts, *Ricciocarpus* and *Riccia*, grow abundantly in an ordinary aquarium.

R. E. Anthony describes the careful development of a technique in "Greenhouse Culture of *Marchantia polymorpha* and Induction of Sexual Reproductive Structures" (*Turtox News* 1962 40:2–5). Anthony starts new cultures in flats either by using gemmae-cups which appear on the upper surface of the vegetative thallus, or by direct planting of the monoploid forked thallus. Flats of *Marchantia* show a preponderance of antheridial structures approximately 25 days after the start of light treatment. Both antheridial and archegonial structures are visible after some 35 days. Live sperms may be examined under the microscope. The resulting zygote grows into a diploid sporophyte that produces monoploid spores resulting from meiotic divisions.

Culturing moss protonemata

Several different nutrient solutions may be used for culturing moss protonemata. Five nutrient solutions are described below. Spores of mosses germinate well in a liquid solution; if desired, the medium may be solidified with the addition of agar.

Using Knop's solution   Crush a dry sporangium of a moss such as *Funaria* or *Catherinea* and liberate the spores onto the surface of Knop's solution that has been diluted to one-third of its original strength (p. 690). Maintain in Petri dishes or finger bowls in moderate light. Spores germinate in about two weeks, while branched protonemata are formed in four weeks. (Beijerinck's solutions, p. 690, can be substituted for Knop's solution.)

Using a solid medium   Some botanists prefer this method. Prepare 98 ml of dilute Knop's solution (one-third its original strength) or Beijerinck's solution. To this, add 2 g of agar. Heat until the agar is dissolved and then restore the volume to 100 ml with distilled water. Filter through absorbent cotton into Petri dishes so that a thin film covers the bottom of the dishes. Cover the Petri dishes and allow the agar to cool and solidify. Crush a clean sporangium over the medium, replace the cover, and place the Petri dishes in subdued light.

Several cultures should be prepared, since some may become contaminated. Germination usually takes place in about ten days. Buds and young gametophytes appear within two to three months.

Better cultures can be maintained using sterile medium and sterile dishes. And the unbroken sporangia may be surface-sterilized in sodium hypochlorite solution (Clorox solution, p. 470). Rinse the sporangia in sterile water, transfer to agar plates and crush with sterile forceps. The spores of *Funaria* grow especially well in the laboratory.

Using Shive's solution

Dissolve the following in 1 liter of distilled water:

$Ca(NO_3)_2 \cdot 4H_2O$	1.06 g
KH_2PO_4	0.31 g
$MgSO_4 \cdot 7H_2O$	0.55 g
$(NH_4)_2SO_4$	0.09 g
$FeSO_4 \cdot 7H_2O$	0.005 g

Prepare several Petri dishes with a thin layer of Shive's solution to which agar has been added. Blow spores from crushed moss capsules over the surface for growth of moss protonema. Then cover the Petri dishes and maintain in moderate light. (Algae, as well as fern prothallia, also grow well in Shive's solution.)

MODIFIED SHIVE'S SOLUTION Bonner and Galston recommend that micronutrients be added to Shive's solution.[63] For each liter of Shive's solution, 1 ml of the following solution should be added. Dissolve the following micronutrients in 1 liter of distilled water:

H_3BO_3	0.6	g
$MnCl_2 \cdot 4H_2O$	0.4	g

---

[63] J. Bonner and A. Galston, *Principles of Plant Physiology*, San Francisco: Freeman, 1952.

$CuSO_4 \cdot 5H_2O$	0.05 g
$ZnSO_4$	0.05 g
$H_2MoO_4 \cdot 4H_2O$	0.02 g

Also refer to Arnon's micronutrient solution (p. 481).

Flowerpot culture method

Protonemata of mosses may also be cultured using the method described for growing fern prothallia (p. 716). Gametophytes should appear within three weeks (life cycle, Fig. 6-61). If you also want to grow sporophytes, flood the gametophytes with water for about 1 hr so that fertilization may occur. Then let the excess water run off.

Examine the protonema stage under the microscope. Mount it directly in a 10 percent glycerin solution on a clean slide. The color of chlorophyll remains well preserved, but for a permanent slide the coverslip should be sealed.

Mount the gametophyte of the moss *Funaria hygrometrica* in water. Examine the "typical" plant cells with many chloroplasts and prominent cell walls. Students may notice that the gametophyte is just one layer of cells in thickness. Or mount leaves of the moss *Mnium*, which is also a single layer of cells in thickness.

Division Tracheophyta

Ferns

Collecting Terrestrial ferns may be kept in the laboratory if care is taken in transplanting them. Take plenty of soil along with the rhizomes and roots when ferns are uprooted. For large ferns, such as those of the Osmundaceae (for example, the cinnamon fern, *Osmunda cinnamomea*), dig a circle with a radius of about 15 cm around the fronds. (The rhizome extends about 15 cm into the soil.) For more delicate ferns, dig within a radius of 7 to 8 cm. Pack the roots and the rhizomes and the attached soil in moist newspaper, waxed paper, or plastic bags; if possible, transport the plants in a plant-carrying case (vasculum).

Culturing ferns   In the laboratory, transplant the ferns into sufficiently large clays pots to avoid crowding. Before transplanting, prepare the pots with 2 to 3 cm of coarse gravel, broken tile, or broken clay pot on the bottom, and then about an equal amount of garden loam. Put the fern on top of this, and continue to add soil composed of equal parts of sand, peat, and garden loam. If the ferns were removed with enough soil around the roots and rhizomes, the rhizomes probably have not been injured. Keep the plants in medium light and provide moisture by standing the flowerpots in water. Keep the soil moist but not wet. Such forms as the bracken fern (*Pteris aquilina*) and the hay-scented fern (*Dennstaedtia punctilobula*) can withstand fairly bright light but should not be put in direct or strong sunlight.

Water ferns such as *Salvinia, Azolla,* and *Marsilia* may be grown in aquaria which have been in use for a month or so. *Azolla* requires strong light, while *Salvinia* and *Marsilia* grow better in medium light.

Most of the small ferns need constant humidity, a condition that is best duplicated in the laboratory terrarium. In this way students may grow Maidenhair and Royal Osmunda with success. A suitable terrarium soil consists of 1 part of coarse sand, 1 part of fine peat moss, and 2 parts of good garden loam.

Several common ferns, with their fruiting fronds, are shown in Figure 9-28.

Culturing fern prothallia   Because it takes at least six weeks for fern gametophytes to grow, sow the spores well in advance of the time you will want gametophytes (life history, Figs. 6-64, 65, 66).

FLOWERPOT CULTURE OF FERNS   For this and the following method, use flowerpots and glass plates to maintain adequate moisture. Fill a large flowerpot with broken pieces of tile or porous clay flowerpots to within 5 cm of the top. Then cover this with 2 to 3 cm of rich loam, and then a thin layer of washed sand.

Sterilize the filled pot by placing it in boiling water for a few minutes or by pouring boiling water over the pot and its contents. However, keep the surface layer of fine sand intact. Next crush the sori found on the underside of ripe fern fronds (Fig. 9-28). Or spread ripe fern fronds on paper for a few days; the spores will fall on the paper. Sprinkle these spores over the sand surface of the flowerpot. Cover the pot with a glass plate to retain moisture and prevent contamination. Next stand the pot in a saucer of water. Add water to keep a level in the saucer and stand the pot in medium light. Should "damping off" by fungi (or mildewing) occur because the soil is too moist, water the pot with 0.01 percent potassium permanganate solution.

Prothallia develop from spores in three to four weeks, producing a five- to ten-cell stage. At this stage, the prothallia may be separated and subcultured in freshly prepared liquid or agar media (Fig. 9-29).

When gametophytes have been growing for about five months, they may be used to obtain sporophytes. Transfer them to the surface of sterile leaf mold packed into clay saucers. Water frequently and maintain a constant humidity. Within a few months sporophytes should begin to develop.

COSTELLO'S METHOD   In this method fern prothallia are grown over an inverted flowerpot. Clean a 10-cm clay flowerpot, and fill with sphagnum moss (peat moss) or paper toweling; moisten the moss (or toweling) and pack it tightly. Immerse all the materials for 10 min in boiling water to sterilize them. Allow the pot to cool, and invert it into a flowerpot saucer filled with water or Knop's solution diluted to half its original strength. The contents of the pot will act as a wick to draw up water.

Dust fern spores over the moist surface of the inverted flowerpot. Cover with a battery jar or bell jar, and place in a cool place in medium light. Within 10 to 20 days, small prothallia should be growing on the outer surface of the flowerpot.

AGAR MEDIUM   The sterile agar method successful for moss protonemata may be used here; use Knop's or Beijerinck's solutions (p. 690). Or use Murashige's solution, page 485. However, if small spor-

Fig. 9-28 Three common ferns, each with diagram of fruiting frond: (a) Christmas fern, *Polystichum acrosticoides*; (b) common polypody, *Polypodium vulgare*; (c) cinnamon fern, *Osmunda cinnamomea*. (Photographs: a, courtesy of U.S. Dept. of Agriculture; b, c, photos by Hugh Spencer; drawings: courtesy of General Biological Supply House, Inc., Chicago.)

ophytes are desired, use wide-mouthed low jars instead of Petri dishes. You may prefer to grow fern prothallia on agar slants in test tubes, flasks, or even in deep-welled slides.

PROTHALLIA ON SLIDES   At some time, you may want to sow spores on Knop's agar solution in welled slides so that the growth pattern of young prothallia may be observed under the microscope. Seal the coverslips with petroleum jelly and keep slides in a moist chamber. (Also refer to tissue culture techniques, page 483.)

Fern relatives   The club mosses include the two common genera *Lycopodium* (Fig. 9-30a) and *Selaginella*. Horsetails or scouring rushes, of which there is only one

Fig. 9-29 Fern prothallia growing in nutrient solution. (Courtesy of Carolina Biological Supply Co.)

genus, *Equisetum* (Fig. 9-30b) may be collected and kept in a terrarium for ferns and club mosses.

(a)

(b)

Fig. 9-30 Relatives of the ferns: (a) club moss, *Lycopodium obscurum*; (b) horsetail, *Equisetum arvense*. (a, photo by Hugh Spencer; b, Roche photo.)

Try to grow *Equisetum* in the laboratory. Spores germinate quickly when they are shed. Shake the spores onto soil and cover with a pane of plastic or glass. For best results, sterilize the soil either by heating in an oven or by watering it well with a dilute potassium permanganate solution (on the average, some 5 small crystals to 1 liter water). In three to five weeks, look for antheridia; archegonia appear a little later. (Life cycle of *Equisetum*, Fig. 6-62.)

Uses in the classroom and laboratory
Mosses are good organisms for study of alternation of generations. The leafy moss gametophytes produce monoploid eggs and sperms while the sporophyte is diploid. Meiosis occurs in the capsules producing monoploid spores. These, in turn, germinate into protonema as the cycle is repeated (Fig. 6-61).

Compare the life cycle of mosses with the separate gametophyte and sporophyte plants in ferns. Trace the increasing dominance of the sporophyte in alternation of generations (Figs. 6-64, 6-65, 6-66).

Gymnospermae and Angiospermae

Culturing seed plants   While most seed plants grow readily in porous flowerpots of suitable size, several precautions must be observed (also see hydroponics, pp. 478–82). There are many excellent books dealing with the care of specific house plants (see Book Shelf). However, let us consider briefly a few of the requirements for success in raising seed plants.

1. Avoid overwatering plants. The soil should not be wet (nor should it be very dry). One of the best ways of watering plants is to immerse the entire flowerpot in water for 10 min. In most cases this treatment is sufficient to ensure moisture for the roots for from four to five days.

2. Do not place plants over a radiator. In many cases, this is the reason for

poor growth of plants in the classroom and laboratory. Nor should a plant be kept too near the window, for in winter the side nearest the window is subjected to excessive cooling.

3. Examine the undersides of leaves for insect pests such as aphids, mealy bugs, and other scale insects. When these are found they should be washed off with soap and warm water. Spray with a nicotine-soap solution.

4. Every month or so, water the plants with about 100 ml of Knop's solution or add commercial mineral preparations to replenish the nutrient content of soil.

5. A good garden soil will result in the greatest growth. But it is equally important that each flowerpot has broken tile or clay at the bottom to ensure drainage. Compressed peat pellets are also available to start seedlings or young plants.

6. When a plant has outgrown the pot, it should be transferred to a larger one; tap the flowerpot to release the plant and remove the roots and soil to the new pot.

7. Plants may often be cloned by vegetative propagation; refer to page 492 for several methods. Also refer to tissue culture techniques, page 483.

8. Check the pH of the soil so that it is suitable for the specific plants. Table 9-2 has a selective listing.

Light requirements   Not all seed plants require equal amounts of light. In fact, teachers may be hard put to know the kinds of plants to grow in a northern exposure where there is little or no sunlight. A quick reference to the following lists[64] may help to establish a cheerful atmosphere in the classroom. Most of these plants may be used in many biological investigations throughout the year.

[64] Personal communication from M. Brooks.

TABLE 9-2   Plants and their pH requirements*

pH 5.0–7.0	pH 6.0–8.0	pH 4.0–5.0
azalea	ailanthus	alpine azalea
bayberry	alyssum	clubmoss
begonia	barberry	hydrangea
bittersweet	carnation	(blue)
bleeding	convolvulus	ladyslipper
heart	crocus	laurel, bog
calendula	dahlia	moss,
candytuft	daisy	sphagnum
chrysan-	forsythia	pepperbush,
themum	geranium	sweet
coleus	grape	pitcher plant
gladiolus	hyacinth	solomon's
goldenrod	hawthorn	seal, dwarf
iris	hibiscus	stagger bush
mountain	honeysuckle	stargrass
laurel	hydrangea	sundew
lily of the	(pink)	swamp pink
valley	lilac	trailing
lily	marigold	arbutus
privet	tulip	trillium
rose	violet	painted
broccoli	wistaria	venus flytrap
carrots	alfalfa	viburnum,
corn	barley	maple-leaf
lima beans	clover	blueberry
oats	sunflower	cranberry
potato	many grasses	huckleberry,
rye	cherry	black
soybeans	elm	many ferns
squash	locust	mountain ash
raspberry	sugar maple	white cedar
strawberry		oak (scrub)
tobacco		spruce (black
tomato		and red)
wheat		
watermelon		

* W. Lapp and E. Wherry "pH Preferences of Common Plants," *The Science Teacher*, April 1951.

PLANTS FOR FULL SUNLIGHT

Cactus	*Zygocactus truncatus*
Echeveria	*Echeveria*, sp.
Geranium	*Pelargonium*
Kalanchoë	*K. coccinea*
Oxalis	*O. rosea*
Patience plant	*Impatiens sultani* (or *Holstii*)
Primrose (fairy)	*Primula malacoides*
Spiraea	*Spiraea*

PLANTS FOR SUNLIGHT (2 TO 4 HR)

African violet	*Saintpaulia*
Asparagus "fern"	*Asparagus plumosus*
Begonias	Many varieties, *argenteo-guttata*
Coleus	*Coleus Blumei*
House iris (Apostle plant)	*Marica Northiana*
Nephthytis	*N. Afzelii*
Peperomia	*Peperomia maculosa* (and *arifolia*)
Pick-a-back	*Tolmiea Veitchii*
Spider plant	*Anthericum*, sp.
Strawberry begonia (saxifrage)	*Saxifraga*

PLANTS FOR WINDOWS WITHOUT SUN (NORTHERN EXPOSURE)

Aloe	*Aloe arborescens*
Aspidistra	*Aspidistra lurida*
Begonia	*Begonia rex*
Boston ferns	*Nephrolepis exaltata bostoniensis*
Chinese evergreen	*Aglaonema modestum*
Chinese rubber plant	*Crassula arborescens*
Date palm	*Phoenix dactylifera*
Dumbcane	*Dieffenbachia seguine*
Gold-dust Dracaena	*Dracaena Sanderiana*
Grape ivy	*Vitis*
Hen and chickens	*Sempervivum*, sp.
Holly fern	*Cyrtomium falcatum*
India rubber tree	*Ficus elastica*
Ivy, English	*Hedera Helix*
Philodendron	*Philodendron cordatum*
Snake plant	*Sansevieria*
Stonecrop (live-forever)	*Sedum*, sp.
Tradescantia	*Tradescantia zebrina*

Germinating seeds There are times when students may want to germinate seeds in the classroom either for class study or for individual studies. Seeds of gymnosperms germinate slowly, requiring two weeks to several months. Seeds of maples, oaks, or tulip trees grow more rapidly. When rapidly growing seeds are desired, use radish or mustard seeds for they germinate within 24 hr. The larger seeds of beans, peas, corn grains, squash, and castor beans germinate in one to two days. When seeds are soaked overnight at room temperature (before sowing) germination is hastened. (Some students may be interested in tissue cultures using embryos of germinating seeds, p. 483.)

All the stages in the germination of seeds can be observed by students. Line a small aluminum form with filter paper or sand. Then sow soaked seeds and dampen the paper or sand. Cover with a glass or plastic aquarium or bell jar and keep in moderate light. (Refer to pp. 471–78.)

Seed disinfectant When germinating seeds in large numbers disinfect the seed surfaces. Surface sterilize seeds by dipping them in 5 percent Clorox solution made of 50 ml of commercial Clorox and enough distilled water to make a liter. Or use dilute potassium permanganate solution (about 5 crystals to 1 liter of water). First dip seeds briefly in alcohol as a wetting agent, and then into the Clorox solution for 5 to 10 min (also see page 470).

Place the seeds in a cheesecloth bag and immerse in the disinfectant. For most seeds a 5-min immersion is sufficient; more hardy seeds can be immersed up to 10 min. Then wash the seeds several times in water for 10 to 15 min; spread them out to dry.

Seeds can be germinated on sphagnum moss to prevent "damping off." Grow young plants directly in shredded sphagnum moss (from a florist).

Plants which are ordinarily difficult to root have been grown successfully in vermiculite which is also sterile when fresh and has extensive adsorbent surfaces that hold good quantities of water as well as air.

Examine the development of germinating seeds by placing them between blotting paper and the glass walls of a tumbler. In this way students may study the developing root system as well as the shoots. For other demonstrations using germinating seeds refer to chapter 6; see also rate of

TABLE 9-3 Conditions for germination of some plants*

plant	medium	temperature (degrees C)	length (in days)
Abies (fir)	F, S	20	10 to 28
Achillea (yarrow)	F, P	20	4 to 10
Allium (onion)	F, S, P	10–20	5 to 14
Alnus (alder)	F, S, P	20–30	6 to 21
Asparagus	F, S, P	20	10 to 28
Avena (oats)	S, P	20	4 to 10
Beta (beet)	S	20–30	7 to 14
Betula (birch)	S, P	20–30	30
Brassica (mustard)	F, P	20–30	3 to 10
Daucus (wild carrot)	F, P	20–30	6 to 21
Fagus (beech)	S, P	20	6 to 28
Helianthus (sunflower)	F, S, P	20–30	4 to 10
Lathyrus (pea)	F, S	20	5 to 10
Lupinus (lupine)	S	20	4 to 10
Lycopersicum (tomato)	F, P	20–30	3 to 10
Medicago (alfalfa)	F, P	20	3 to 10
Nasturtium	F, S, P	20	4 to 10
Nicotiana (tobacco)	F, P	20–30	5 to 14
Pastinaca (parsnip)	F, S	20–30	5 to 14
Phaseolus (bean)	S	20	4 to 10
Phleum (timothy)	S	20	4 to 10
Pinus sylvestris (Scotch pine)	P	20	14 to 28
Pisum arvense (garden pea)	S	20	3 to 10
Quercus (oak)	S	20	28
Raphanus (radish)	F, P	20–30	3 to 10
Spinacia (spinach)	S	20–30	4 to 14
Trifolium, sp. (clover)	F, S, P	20	3 to 10
Triticum (wheat)	S, P	20	3 to 10
Zea (corn)	S, P	20–30	4 to 10

Key

F—Several pieces of wet filter or blotting paper placed below and above the seeds. Petri dishes are excellent.

S—Washed sand, thoroughly moistened but not wet. The seeds are embedded about 2 to 3 cm below the surface.

P—Loam soil in a pot. The seeds are planted about 2 to 3 cm below the surface. The soil should be loose.

* From H. C. Muller, "Methods for Establishing the Viability of Seeds of Various Plant Species," in E. Abderhalden, ed., *Handbuch der biologischen Arbeitsmethoden*, Urban & Schwarzenberg, Munich, 1924, Sec. XI, Vol. II, Part 121.

growth, page 471; root hairs studies, page 472.

Conditions for germination of some common seed plants are given in Table 9-3.

You may also be interested in establishing a greenhouse on the school lawn or as part of a laboratory. A U.S. Dept. of Agriculture booklet, *Home and Garden*, Bull. 163, describes minigardens for vegetables.

Seed plants: special methods

Insectivorous plants   These plants need special care. Plant them in a terrarium to maintain proper humidity. Feed *Drosera*, the sundew (Fig. 9-31a), vestigial-winged fruit flies, and for *Dionaea*, the Venus fly-trap, drop *Tenebrio* larvae (mealworms) or bits of meat into the traps (Fig. 5-8). *Sarracenia*, the pitcher plant, will survive in a humid terrarium without insect food (Fig. 9-31b). Water the plants once a month. Avoid more than 2 to 3 hr exposure to bright sunlight since the tank may become overheated. Some plants may be covered with plastic bags to retain moisture if they are grown in flowerpots. Culture in acidic garden soil with sphagnum moss over a bottom layer of broken clay flower pots. As the plants grow, and give rise to others, they may be repotted in a soil prepared by thoroughly mixing together 3 parts of garden loam, 1 part of

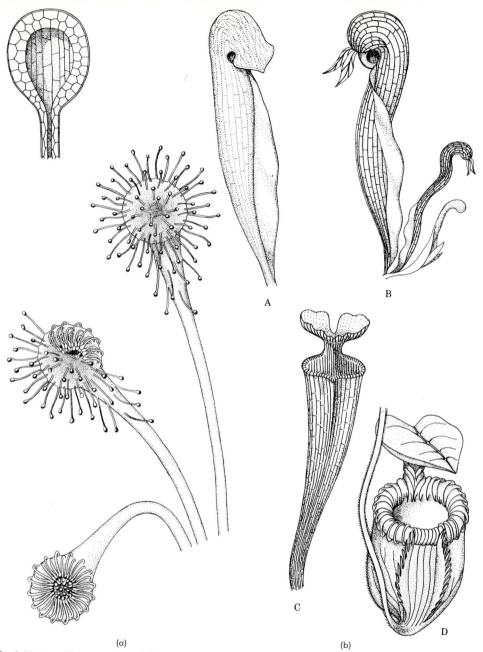

(a)   (b)

Fig. 9-31 Insectivorous plants: (a) Sundew, *Drosera*, showing lengthwise section of a gland located at the tip of a tentacle, stages in the curving in of tentacles over a trapped insect, and view of leaf with all tentacles extended: (b) variety in pitchers of pitcher plants: (A) *Sarracenia variolaris*, (B) *Darlingtonia californica*, (C) *Sarracenia drummondii*, (D) *Nepenthes villosa*. (Adapted from *The Plant Kingdom* by William H. Brown, Ginn and Co., Publishers. Copyright© 1935 by William H. Brown.)

peat moss, and 1 part of sand. The peat moss (sphagnum) will supply the needed acid condition.

The traps of insectivorous plants are strongly acid, approximately pH 3, and the fluid contains a protease. There is also bacterial activity within the plant traps. Young plants fed insects develop abundant flowers and seeds. Mature plants do not seem to require insects.

The floating, insectivorous filamentous water plant *Utricularia* produces yellow flowers which grow to the surface of the water. The plant lacks roots, but small bladders with trap-door devices, based on changes in pressure, catch the prey. Some small forms like *Euglena* may live and reproduce within the traps, apparently unharmed. Yet small worms, *Paramecium* and *Daphnia*, are digested.

Cactus plants   Cactus plants also require special treatment. *Opuntia*, the pear cactus, may be planted in ordinary garden soil but should be watered only once every two weeks. Other cactus plants, such as *Cereus*, *Echinocactus*, *Phyllocactus*, and related forms, should be potted in soil made up of 4 parts of sand and pebbles and 1 part of garden loam. A good watering every three weeks is sufficient. Seeds of cactus germinate slowly—two weeks to months—in flats of light, sandy loam.

Dodder (Cuscuta)   This bright orange-colored seed plant is often parasitic on clover and goldenrod, as well as on other plants (see Fig. 8-30). There is enough food in the seed for the young plant to begin to develop a slender stem and roots by which it becomes attached to a host. Leaves are reduced to scalelike structures. Then it sends wedgelike haustoria into the host's stem, forming a junction with the vascular system of its host. Finally, its connection with the ground is severed and it depends upon the host. Dodder produces abundant white flowers which hang in clusters in late summer.

Collected specimens wilt quickly but may be preserved in 70 percent alcohol. Pressed specimens may be useful for tem-porary display. Secure the specimens on mounting paper with transparent tape.

Mistletoe (Phoradendron)   This seed plant grows on branches of ash, elm, and hickory as well as many other trees. Mistletoe has green leaves and carries on photosynthesis, but it probably absorbs water and salts through its roots, which are embedded in the tree branches. Thus, it may be considered a partial parasite. Collect specimens and preserve them in the dry state on mounting paper.

Indian pipe (Monotropa)   This seed plant lacks chlorophyll and thus appears as a stark white stem with scalelike leaves (Fig. 8-29). It obtains its food from humus as a saprophyte, not a parasite.

When picked, the plant wilts quickly, developing a black-purple pigment. If the plants have been carefully transported undisturbed with their original soil and placed into a terrarium, they may remain white for several weeks. Specimens may be preserved in alcohol or formalin.

Wolffia   In this smallest of flowering plants, the floating plant is reduced to an oval or rounded green body (the stem). There are no leaves or roots present (Fig. 9-32a). Absorption occurs through the underside of the stem. New fronds, which are modified branches, grow out from depressions on the underside of one end of the oval frond.

Wolffia is found floating on the surface of lakes and ponds, at times very abundantly. The plants multiply rapidly and lend a pleasing appearance to aquaria in the laboratory and classroom. Students may devise independent investigations to study the growth of these plants under a variety of conditions, especially different wavelengths of light, as well as photoperiods. It is relatively easy to measure the growth of these simple plants. What is the effect of growth-promoting hormones?

Duckweed (Lemna)   *Lemna* (Fig. 9-32b) and a closely related form, *Spirodela*, are found as small floating plants in lakes and

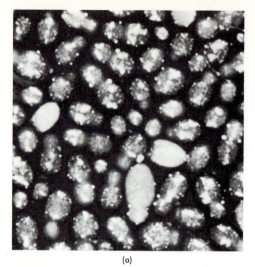

(a)

(b)

Fig. 9-32 (a) *Wolffia columbiana*; (b) *Lemna minor*. (Both are greatly magnified.) (Courtesy of General Biological Supply House, Inc., Chicago.)

ponds. The broad leaflike fronds are not leaves but stems. A single rootlet lacking root hairs extends from the underside of the frond. They grow well on the surface of an aquarium and add an attractive touch. However, they are not good oxygenators.

Study the reproductive cycle of *Lemna*, the effect of plant hormones and also the effect of light of different wavelengths and intensity. Refer also to a study of competition between species, page 469. Culture *Lemna* in nutrient solutions for luxuriant growth for these investigations.

Mimosa The sensitive plant *Mimosa pudica* grows readily from seeds planted in a light, sandy loam. Keep in sunlight, and invert a battery jar over the flowerpot in which the plants are contained. Young plants grow well at about 27° C (81° F). (Also refer to behavior, page 338.)

CAPSULE LESSONS: WAYS TO GET STARTED IN CLASS

9-1. There are times when you may want to develop a "parade" of the animal and plant kingdoms. This may perhaps be part of a survey of the many kinds of animals and plants available in your community. Try to use fresh materials, examine them under the microscope; maintain them in vivaria or aquaria for observation of behavior. Also try a series of vicarious field trips to a beach, a mudflat, or other area by means of films, 2 × 2 slides, or videotapes.

Examine the listings of living animals and plants available from a biological supply house

catalog; gather supplies for freshwater and marine tanks.

9-2. There may be space to maintain a center, a museum of living and preserved forms. In such cases, other teachers and their students may be invited for studies of behavior, classification, and/or aspects of ecological relationships; prepare a guide for visitors.

9-3. Encourage students to bring to class all kinds of living organisms you may be able to maintain. Student curators gain experience for a possible future career. Your school may be in a position to serve as a fresh materials culture center for algae, protozoa, and other small forms (Refer to student squads, pp. 756–62.)

9-4. Some students may want to plan their own investigations; many ideas for long-range investigations are to be found in context throughout the book.

9-5. Have each student bring to class a plant specimen for identification. Trace the evolution of plants from one-celled forms to seed plants. As a club activity or a small group project teach students to use an Identification key (Figs. 7-46, 7-47). Prepare a large bulletin board and mount dried specimens or drawings. In season, maintain flowering twigs in containers of water in a hall cabinet. Other plants may be kept in a terrarium.

9-6. Have students plan a field trip to explore the school grounds. Examine many kinds of flowering plants such as magnolia, *Forsythia*, violets, cherry, dandelion, and other plants that occupy special niches. (Suggestions for ways to develop a greenhouse and nature trails are given on p. 763).

Have the class develop a guide which outlines a specific trail and identifies plants along the way.

9-7. As a group project, have students put name plates on the trees and shrubs on the school grounds.

9-8. When planning a field trip, arrange to transplant ferns, mosses, possibly some spring flowers, small evergreens, and tree seedlings to a similar environment on the school grounds.

BOOK SHELF

The following are only a few of the broad references available. Refer also to the listings at the ends of chapters 1, 2, 5, and 8. A visit to a bookstore or perusal of a catalog from a biological supply house will reveal the wealth of field guides for identification of mollusks, insects, flowers, birds, fish, trees and shrubs, among other living things; some are listed at the end of chapter 8.

Armet, R. and R. Jacques, *Simon and Schuster's Guide to Insects*, New York: Simon and Schuster, 1981.

Barnes, R. *Invertebrate Zoology*, 4th ed., New York: Saunders College/Holt, Rinehart, and Winston, 1980.

Behler, J. and F. W. King, *The Audubon Society Field Guide to North American Reptiles and Amphibians.* New York: Knopf, 1979.

Brown, V. *The Amateur Naturalist's Handbook*, Englewood Cliffs, NJ, Prentice Hall, 1980.

Bullough, W. *Practical Invertebrate Anatomy*, 2nd ed., New York: Macmillan, 1960.

Collins, H. Jr. *Harper and Row's Complete Field Guide to North American Wildlife* (Eastern Edition) New York: Harper and Row, 1981.

Crum, H. *Mosses of Eastern North America*, 2 vols. New York: Columbia Univ. Press, 1981.

Dales, R. ed. *Practical Invertebrate Zoology*, 2nd ed., New York: Wiley, 1981.

Dasmann, R., *Environmental Conservation*, 5th ed., New York: Wiley, 1984.

Gadd, L. *Deadly Beautiful: The World's Most Poisonous Animals and Plants*, New York: Macmillan, 1980.

Hickman, C. Jr., L. Roberts, and F. Hickman. *Biology of Animals*, St. Louis, MO: Mosby, 1982.

Introduction to Basic Gardening, 3rd ed. by eds. of Sunset Books and Sunset Magazine. Menlo Park, CA: Lane Publishing Co., 1981.

Kinghorn, A. D., ed. *Toxic Plants.* New York: Columbia Univ. Press, 1980.

Kudo, R. *Protozoology*, 5th ed., Springfield, IL: C. C. Thomas, 1971.

Matthews, L. H. *The Natural History of the Whale.* New York: Columbia Univ. Press, 1978.

McCullagh, J., ed. *The Solar Greenhouse Book.* Emmaus, PA: Rodale Press, 1978.

Oliver, A. *The Larousse Guide to Shells of the World.* New York: Larousse, 1980.

Pennak, R. *Freshwater Invertebrates of United States*, 2nd ed., New York: Wiley, 1978.

Peterson, R. *A Field Guide to the Birds East of the Rockies*, 4th ed., Boston: Houghton Mifflin, 1980.

Peterson, R., *A Field Guide to Western Birds.* Boston: Houghton Mifflin, 1961.

Pirone, P. *Diseases and Pests of Ornamental Plants*, 5th ed., New York: Wiley, 1978.

Pyle, R. *The Audubon Society Field Guide to North American Butterflies.* New York: Knopf, 1981.

Ransom, J. *Harper and Row's Complete Field Guide to North American Wildlife.* (Western Edition). New York: Harper and Row, 1981.

Rehder, H. *The Audubon Society Field Guide to North American Sea Shells.* New York: Knopf, 1981.

Richardson, D. *The Biology of Mosses.* New York: Wiley, 1981.

Rickett, W. *Wild Flowers of America.* New York: Crown Publishers, 1978.

Singleton, P., and D. Sainsbury. *Introduction to Bacteria.* New York: Wiley, 1981.

Slack, A. *Carnivorous Plants.* Cambridge: MIT Press, 1980.

Stokes, D. *A Guide to Nature in Winter.* Boston: Little Brown, 1976.

Wilkinson, R. and H. Jacques, *How to Know the Weeds*, 3rd ed., Dubuque, IA: W. Brown, 1979.

Wistreich, G. and M. Lechtman *Microbiology*, 4th ed. New York: Macmillan, 1984.

The Brooklyn Botanic Garden (1000 Washington Avenue, Brooklyn, NY 11225) publishes a large number of illustrated handbooks in gardening and horticulture in general. For example, *Terrariums* 1975; *Gardening under artificial light* 1970; *Natural plant dyeing* 1973; *Home lawn Handbook* 1973. Send a postcard for the available Handbook brochures for Plants and Gardens.

The United States Dept. of Agriculture has available many well illustrated pamphlets: Lawn Diseases; House Plants; Flowering Perennials; Flowering Annuals; Indoor Gardens; Selecting House Plants and many more.

Lord & Burnham, manufacturers of glass greenhouses, have branches in many areas; write to Irvington-on-Hudson, N.Y. 10533 for regional branches.

10

Certain Useful Additional Techniques

Considering the great variety of activities available to the teacher of biology in this *Sourcebook*, and considering the fact that biology is increasingly finding its base in chemistry and certain probes involving skill in manipulation, it must be assumed that teachers or laboratory personnel will be familiar with the properties and behaviors of the substances and other materials they will utilize in their teaching. It is urged that before a substance or material is used, it should be standard practice to consult works such as *The Merck Index*, Kodak's *Laboratory Chemicals, Safety in the School Science Laboratory**, and others listed on page 737; that is, references concerned with the manufacturer's description of the properties of the substance and possible hazards in its use. For example,

we have utilized PTC (phenylthiocarbamide) in our classes with some 6,000 students or more without incident but we now find documented by Eastman Kodak certain hazards of which we were not aware earlier—that is, fatal if swallowed, harmful if absorbed through the skin or inhaled.

These strictures may seem unnecessarily binding but the counsels of safety remain paramount where teachers and their students are concerned.

* *The Merck Index*, 10th ed., Merck and Co., Inc., Rahway, NJ 1983.
Kodak Laboratory Chemicals, catalog 52, Eastman Kodak, Rochester, NY, 1985.
Safety in the School Science Laboratory, 250, Instructor's Resource Guide, National Institute for Occupational Safety and Health, Cincinnati, OH, 1979.

Safety in the classroom and laboratory

Science teachers are safety-conscious; they recognize that training students for activities and investigations is a responsibility that permits no exceptions, no shortcuts. In planning for many activities that involve the use of sharp dissection tools or using volatile substances, students need to be made aware of their responsibilities to themselves and to others in class.

Many boards of education over the country have prepared excellent handbooks describing safety procedures for student activities, teacher demonstrations, and the handling and storage of chemicals. Students on laboratory squads should also be trained in establishing responsible safe-

ty practices. However, since this *Sourcebook* is written for teachers, *not* for students, it is the teacher's responsibility to prepare solutions, stains, and equipment that may entail danger for students—inexperienced as they are.

Common accidents

Common accidents in the laboratory or classroom fall into four main categories: 1) cuts from broken glass; 2) burns from hot glass; 3) chemical burns; 4) errors in preparation of solutions, particularly the concentration of chemical components. Throughout the year students need to be reminded to observe safety procedures. A first-aid cabinet and a copy of the American Red Cross First Aid Textbook should be readily accessible.

Heating substances in test tubes, for

example, requires simple precautions: avoid looking into a test tube being heated; don't point the mouth of the tube at a person; apply heat first to the *top* part of the tube to heat the liquid, then gradually heat the lower part of the tube and liquid otherwise the bottom of the tube may blow out or hot material may splatter.

How can concentrated acid be handled safely? What precautions are advisable in handling bacteria? One among many of the curriculum bulletins available from school systems is called *Science Safety* (grades K-12).[1] Responsibilities of science teachers and students in maintaining safety in instruction are well described. Excerpts from this bulletin are detailed below.

FOR SCIENCE TEACHERS

1. Notify the science supervisor and laboratory specialist immediately of any hazard which comes to your attention.
2. Be familiar with first-aid treatment.
3. When an injury or accident occurs, give the permissible first aid and then:
   a. Report the accident immediately to the principal or his or her designated agent.
   b. Arrange for the completion and filing of the proper forms, signed by the injured party, witnesses, and the staff members concerned.
4. Perform classroom experiments and demonstrations only if you have previously tried them out or have been properly instructed by the supervisor or his or her representative.
5. Make sure that all rooms containing chemicals are properly ventilated and that there is no flame in the room when using flammable, volatile liquids such as alcohol, chloroform, acetone, or diethyl ether.
6. To prevent breakage, store bottles of dangerous chemicals, such as concentrated

acids and alkalis, on the floor in a protected area of the acid room or on the bottom of a steel cabinet. In no case should they be stored above eye level.

7. NEVER add water to concentrated acid. To prepare a diluted acid, wearing goggles:
   a. start with water in a heat-resistant container;
   b. place the container in a sink filled with cold water;
   c. slowly add small quantities of the concentrated acid to the water while stirring constantly.
8. Take extra precautions with concentrated sulfuric acid, nitric acid, and glacial acetic acid, and with concentrated solutions of caustic alkalis and other corrosive chemicals, such as bromine and iodine.
9. DO NOT use white phosphorus. If necessary, use red phosphorus instead.
10. Examine carefully and test for safety hazards all devices or equipment brought in by students before demonstrating them to the class.
11. Observe the following precautions when performing experiments where there is a possibility of splintering and flying glass, a flash, or an explosion:
    a. Keep students not less than eight feet away from the demonstration table.
    b. Place a safety shield between the students and yourself, and the demonstration.
12. Keep all the bottles and other containers tightly stoppered and labeled.
13. Do not return surplus chemicals to their containers. Do not pour excess reagent solutions back into their stock bottles.
14. Discard chemicals that are in unlabeled containers in a suitable manner. Wipe up spills immediately. Do not use paper towels to wipe up oxidizers or strong acids.
15. Use minimal quantities of chemicals in experiments and demonstrations.
16. Do not place any bottles, especially those containing acids or volatile organic liquids, near heating pipes nor allow them to stand in direct sunlight. Dangerous gas pressures may develop under such conditions.
17. Provide individual straws or mouthpieces for students who are asked to blow into tubes, balloons, or plastic bags.
18. Use tongs to handle dry ice. Do not place dry ice in a closed container, as the accumulated gas may lead to an explosion.

---

[1] Available from Bureau of Curriculum Development, Board of Education, New York City, Publications Sales Office, 110 Livingston Street, Brooklyn, NY 11201.

Excerpts from this bulletin adapted from *Science Safety* by permission of the Board of Education of the City of New York, 1983.

The container may be closed by placing a loose cotton plug in its mouth.

19. Do not use ether in a room where there is a flame or where flames may soon be lit. Use only small quantities at a time. The room should be well ventilated. If possible, use the ether near an open window. Allow containers in which ether has been used to air out thoroughly before being washed. Put ether-soaked cotton or rags in a fire-proof receptacle. Set the receptacle aside in a safe place to allow the ether to evaporate. These directions also apply to other volatile substances such as acetone, petroleum, and gasoline.[2]

20. Observe the following precautions when inserting glass tubing, thistle tubes, or glass rods into rubber stoppers:

   a. The ends of the tubing or rod being inserted should be fire-polished.
   b. The tubing should fit the stopper hole.
   c. Moisten the tubing and the hole with a lubricant such as water, soap solution, glycerine, or petrolatum.
   d. Use cloth or leather gloves to protect hands from injury in case the glass tubing breaks.
   e. Grasp the glass tubing very close to the rubber stopper.
   f. Insert the glass tubing through the hole in the stopper with a gentle twisting motion. NEVER force glass tubing.

21. When removing glass tubing, thistle tubes or glass rods from rubber stoppers, observe the following precautions:

   a. Remove glass tubing from rubber stopper, as soon as possible after use, to prevent the glass from freezing to the stopper.
   b. Remove glass tubing which has frozen in a rubber stopper by using a lubricated cork borer which is just large enough to slip over the glass. Protect hands with cloth or gloves. Slowly twist the cork borer through the stopper to bore the frozen glass tubing out of the stopper.
   c. Note that special care must be taken if a pocket knife or single-edge razor blade is used to split the stopper. Do not permit students to do this.

22. Try out a propane burner in the preparation room before using it in class. Under

[2] We should add the phrase "or for that matter, any inflammable substance."

no circumstances should students use a propane burner.

23. Demonstrate pipette techniques. Arrange for students to practice these techniques with water. DO NOT pipette anything by mouth.

24. Caution students who are about to heat material in test tubes not to look down into the tube and not to point the mouth of the tube toward themselves or others. Caution them to hold the test-tube holder in a way that will prevent material which boils over from hitting their hands. Caution them not to heat the holder. Test tube holders should hold test tubes near the lip.

25. Instruct students to slowly heat substances in test tubes, shaking them carefully at the same time. Otherwise, vapor meeting a mass of matter above it may cause the bottom of the tube to be blown out or the matter to be ejected violently.

26. Use only heat-resistant glassware to heat substances.

27. Instruct students in the proper use of Bunsen burners and alcohol lamps. Students with long hair should be given rubber bands to fasten their hair in pony-tail fashion to avoid igniting their hair. Ties and loose clothing should be tucked in. When a Bunsen burner "strikes back" (burns at the spud), shut off the gas immediately. Do not touch the hot barrel of the burner.

28. Instruct students to use appropriate protective devices, such as tongs, test-tube holders, and insulated gloves when heating objects.

29. Be sure that protective eyewear is worn by students, teachers, and visitors in the laboratory and during any demonstrations. Goggles or other protective eyewear must be available for all personnel. Protective eyewear used by one class should be properly sterilized by an approved Board of Health method before being distributed to the next class. It is recommended that each student be provided with a pair of goggles for his or her own personal use during the term through purchase or loan. Contact lenses and regular eyeglasses are not substitutes for safety goggles.

30. Use caution when heating any plastic item. It may be flammable or give off toxic or harmful vapors.

31. Familiarize yourself with the purposes and operation of the various kinds of fire extinguishers found in the science rooms. Do not direct a stream of water on oil or electrical fires; instead, use the carbon dioxide extinguisher or sand.

32. Cover a sterno can with the lid to extinguish it. If it is to be reignited, be careful when handling the hot lid and can.

33. IMMEDIATELY, properly dispose of all chipped, cracked, or starred glassware.

34. Prepare all dangerous or flammable gases under a fume hood.

35. Observe the following precautions when students are using concentrated acids and bases:

  a. Concentrated acids and bases should be dispensed by the teacher only, in quantities appropriate for a single use.

  b. HCl should be used in preference to H_2SO_4 wherever possible.

  c. Have available containers with saturated $NaHCO_3$ or slurry, and with 5% acetic acid.

  d. If acid or base is spilled on skin or clothing, wash immediately with large quantities of water, then neutralize.

36. Use only vinegar and lemon juice as acids in experiments conducted in grades K–5. Bicarbonate of soda is the only base to be used.

37. Students MUST NOT taste or smell chemicals unless specifically instructed to do so by the teacher.

38. Use alcohol lamp on a nonflammable surface, well away from the edge of the table. Make sure the cap is not so tight that students cannot readily remove it. Check to see that any spilled alcohol is wiped up before the lamp is ignited. Make certain that the lamp does not tilt or leak. In no case are alcohol lamps to be filled by the students. It is best that the lamps be filled the day before use, so that fumes are not present in the room. If filled the same day, wipe off all liquid from the outside of the lamp.

In no case should the supply of alcohol be present in the same room at the time the lamps are being used.

39. Be aware of special medical problems of students, such as asthma or high sensitivity to irritating fumes.

40. *In rendering first aid, the guiding principle is that the person is administering only immediate temporary care pending administration of competent medical care* ... First aid is "treatment which will protect the life and comfort of

the student until authorized treatment can be secured and is limited to *first treatment only,* following which, the child is placed under the care of his parents upon whom rests the responsibility for subsequent treatment." ... Procedures included in the American Red Cross official textbook should be followed.

Several especially relevant additional first-aid procedures follow:[3]

Burns from fires and chemicals. For chemical burns, wash and flush area with water and remove any clothing or jewelry that may have been in contact with the chemicals. For burns caused by fire or hot objects, apply cold water until the pain subsides.

Eye injuries from chemicals. Quickly flush eyes thoroughly with running water for at least 15 min. Be sure lids are kept open by holding them away from eyeball. Remove contact lenses if present. In first-aid treatment of eye, use *water only.* Summon medical attention, but do not interrupt the washing procedure.

Inhalation of gases. If a student inhales a toxic gas such as chlorine, hydrogen sulfide, sulfur dioxide, or nitrogen dioxide, remove the student to fresh air immediately.

Ingested poisonous chemicals. Read the label on the original container for specific first-aid. In general, if the student is conscious, give as much water as can be drunk.

Acids (acetic, hydrofluoric, hydrochloric, sulfuric, phosphoric, phenol): If the acid is swallowed, give large amounts of egg white obtained from the school kitchen. DO NOT INDUCE VOMITING. Do not give carbonates or sodium bicarbonate.

Bases, caustic alkalis (sodium hydroxide, potassium hydroxide, ammonium hydroxide, calcium oxide): DO NOT INDUCE VOMITING. Give large amount of 1% acetic acid or vinegar (1 part of vinegar to 4 parts of water), 1% citric acid, or lemon juice. Follow with milk.

---

[3] Contact the poison control center in your area. Keep the number posted conspicuously by the nearest telephones.

FOR CUSTODIANS

1. Provide proper ventilation and illumination of science storerooms, preparation rooms, laboratories and science classrooms.

2. See to it that a metal or earthenware waste crock, or fireproof plastic container is provided in every classroom where science experiments are performed. Such receptacles should be emptied daily.

3. Have fire extinguishers inspected each term and replenish as needed. Inspection date should be recorded on the attached tag.

BIOLOGY

Aquarium tank

Tanks filled with water are very heavy. Empty tanks by using a siphon or dipper before moving them to avoid cracking the glass or sustaining muscle strain or more serious injury.

Blood

Do not ask students in lower grades to volunteer blood samples. Use a student donor for a drop of blood only if specific permission has been granted by the parent. Cleanse skin with 70% alcohol and use only a sterile, disposable lancet. Do not use a lancet more than once. After drawing the sample, swab the site with alcohol and apply a band-aid or sterile gauze pad.

Centrifuge

Clamp the hand centrifuge securely to the table. In using the highspeed centrifuge, carefully balance the tube containing the materials with another tube containing water by using a platform balance to equalize the masses of the two tubes. Use only tubes designed for the centrifuge because these can withstand the large forces which are generated.

Clinitest tablets

These tablets are used to test for glucose. Do not permit students to hold them or swallow them.

Disinfectants

Creosote, cresols, and *Lysol* may cause burns of the skin and mucous membranes. Use rubber gloves.

Dissection

Issue clean, sharp instruments that are free of rust. Count the instruments issued and check to see that they are all returned before dismissing the class. Specimens preserved in formaldehyde should be thoroughly washed in running water before being used. Students should use gloves to avoid skin contact with the formaldehyde. Caution students about the possibility of formaldehyde squirting out of a specimen, such as a frog, on the first incision into the abdomen. Caution students to avoid rubbing their eyes with their fingers after handling formaldehyde. Make provision for pupils to wash their hands with soap and water after the dissection.

Epithelial cells

Direct students to secure epithelial cells by scraping the inside of the cheek carefully with the blunt edge of a flat toothpick. Do not use pointed instruments or any part of a scalpel. Toothpicks should be broken and discarded after use to avoid reuse by another student.

Ether

Order diethyl ether, which is used as a solvent and an anesthetic for laboratory animals, in one-pound tin cans or less. (This should not be confused with petroleum ether, a low-boiling hydrocarbon, which is also used as a solvent.) Prior to use, transfer the diethyl ether to a clean, dry, glass-stoppered, dark-glass bottle. Put some iron nails or iron wire in the bottle of ether to prevent the formation of explosive peroxides. Although this danger applies mainly to higher ethers, the precaution should also be taken with diethyl ether. The bottle label should show the date on which the ether was received and transferred. After one year, any remaining ether must be discarded and the bottle carefully washed and dried before fresh ether is added.

CAUTION
Keep diethylether away from open flames.
The same precautions apply to isopropyl ether and tetrahydrofuran. The formation of peroxides should be checked every six months with test strips.

Field trips

Plan field trips carefully, taking into consideration student safety as well as the scientific and educational objectives. Among the precautions to be observed are the following:

1. Obtain parent consent slips.

2. Keep the students under your direct supervision at all times.

3. Instruct students to wear clothing and shoes that are appropriate for the locale, the activity, and the season.

4. Carry a first-aid kit.

5. Provide special instruction for students using specialized tools such as the geologist's hammer. Students should wear goggles when chipping rocks, and hard hats if necessary.

6. Identify poison ivy, poison sumac, and other local poisonous plants and animals for the students.

7. Warn students against drinking water from lakes, streams, and ponds and against eating wild berries, nuts, and fruits.

8. Caution students against picking up snakes in the field and against reaching into holes. They should exercise caution in turning over rocks and logs.

9. Plan ahead to provide the proper number of collection bags or bottles for specimens.

10. The leader should visit and examine the trip route in advance.

Living things

1. Marine and freshwater animals
   a. Caution students against tasting or eating any organism collected in the field.
   b. Caution students concerning the dangers of collecting and handling specimens with claws, spines, or poisonous secretions, such as crayfish, Portuguese man-of-war, sea urchins, and jellyfish. Some students may be hypersensitive to stings by aquatic organisms. Boots or sneakers should be worn when collecting specimens.
   c. Discard decaying organisms. Preserve or refrigerate dead specimens that are to be used over an extended period of time.

2. Land animals
   a. Provide instruction on the collection, care, and handling of animals that are kept in the school or that might be encountered in the field. A laboratory animal is NOT A PET. If treated as a pet, its usefulness as a laboratory animal is seriously reduced.
   b. Clean and disinfect animals cages. Inspect them for hazards, such as frayed wires and sharp edges.
   c. Pick up rat cages by their handles, not by holding the wire mesh. A rat can inflict a severe bite through the spaces of the usual wire mesh.
   d. Wear gloves when handling animals that may bite or scratch, such as rabbits, rats, hamsters.

3. Plants
   a. Caution students against tasting or eating mushrooms, berries, seeds, and other parts of plants. Many common house and garden plants are TOXIC—azalea, crocus, daffodil, dieffenbachia, foxglove, mistletoe, poinsettia, etc.
   b. Before starting a field trip, provide advance identification of poison ivy, poison oak, and poison sumac if these are likely to be encountered. Caution students against contact with nettles, burrs and thorns. Students should not wear shorts on trips that will take them through fields and woods. Sturdy footwear should be worn.
   c. In lessons on flowers and bread mold, take care to prevent the excessive distribution of pollen or spores. Some students may be allergic to these materials. Culture dishes containing bread mold should be covered with saran or cellophane.

Microorganisms

Activities involving microorganisms constitute a valuable aspect of science instruction. However, the extent to which microorganisms can safely be used depends on the age and prior training of the students, the competence of the instructor, the availability of appropriate apparatus, and the kind of organism studied. Precautions for handling miscroorganisms include the following:

1. Do not culture pathogenic bacteria, algae, molds, protozoans, or viruses. Do not prepare a culture from a student's saliva or from his or her cough spray because this material may contain pathogens. Use only nonpathogenic stock cultures obtained from a reliable source. *Treat all bacterial cultures as if they were pathogenic* and use sterile techniques throughout.

2. Flame wire loops used for transferring bacteria before and after each use.

3. Strictly enforce rules against eating and drinking in the laboratory.

4. It is recommended that all exposed glass Petri dishes be sterilized by autoclaving. The use of sterile disposable plastic Petri dishes obviates this procedure.[4]

5. Pipettes used for transferring cultures or for making dilutions present a great potential hazard for untrained personnel. After use, pipettes should be placed into a cylinder containing disinfectant solution, washed in an automatic pipette washer, and then sterilized. To sterilize pipettes, place them in a sterilizing can and heat them for two hr in a hot-air oven at 160°-190° C. Keep pipettes in the sterilized can.

6. Use a bulb attachment rather than direct mouth pipetting for all fluids except water or saline. Do not blow a fluid containing microorganisms out of the pipette because spattering may result.

7. Sterilize all contaminated material that is to be discarded.

8. Always keep hands away from the face while working with microorganisms.

9. Seal with transparent tape any exposed Petri dishes that are passed around the class for inspection.

10. Do not permit a broth culture to wet the cotton plug or cap.

11. Do not permit fermentation to take place in a closed system or tightly sealed container.

Microtome

Keep fingers away from the blade while adjusting, cleaning, and using this instrument.

Photosynthesis demonstrations

Many of these demonstrations involve the heating of a leaf in hot alcohol or petroleum ether to extract chlorophyll. Observe the following precautions:

1. Use only heat-resistant glassware, such as pyrex or other borosilicate glass.

2. Use an electric hot plate. NEVER use an open flame where alcohol vapor or petroleum ether vapor is present.

[4] The authors of this *Sourcebook* recommend that exposed plastic Petri dishes be wrapped, sealed, then given to the custodian for incineration.

3. If an open flame is used to bring water to a boil in a large beaker, *shut off the flame* before pouring a small quantity of alcohol into a small beaker and submerging the leaf in alcohol. Then place the small beaker into the large beaker of hot water. The hot water is at a sufficiently high temperature to bring either alcohol or petroleum ether to a boil.

Pressure cooker, autoclave, sterilizer

In general, these should be operated only by teachers or laboratory specialists. Students should receive special instructions before being permitted to use these pieces of apparatus. Precautions for use of this equipment include the following:

1. Become familiar with the manufacturer's specific instructions before using the pressure cooker, autoclave, or sterilizer.

2. Examine the pressure release valve before using the apparatus and make sure the valve is in working order.

3. Keep the pressure under 20 lbs per square inch (psi).

4. Allow the apparatus to cool before removing the cover. Pressure should be down to normal and the stopcock should be open before the clamps are released.

5. Place a protective screen around hospital-type autoclaves that are in permanent locations.

6. Place a warning sign on the autoclave when it is in use.

Protein food test

Do not heat the materials in the xanthoproteic test for proteins after adding the nitric acid. Rinse the test tube and its contents with water before adding ammonium hydroxide (aqueous ammonia solution). Students should never handle a test tube or bottle containing concentrated nitric acid. The Biuret reagent used in testing for protein contains concentrated potassium hydroxide and should be handled with the usual precautions for strong alkalies.

Ultraviolet light

Make sure no one looks directly at the light source. Students may be momentarily exposed to this radiation as they observe its reflection from minerals, ores, teeth, and paints. Prolonged exposure, however, can cause serious burns of the retina. No one should look directly at the light source. For activities requiring more prolonged exposure, such as the observation of UV fluorescence from chromatograms of chloroplast pigments, use goggles which are appropriate for the wavelength of the UV light being used. Ordinary eyeglasses provide some protection against UV light.

Vacuum experiments

Use heavy-walled, round-bottom flasks or apparatus especially designed for vacuum work to prevent implosions. Use a safety shield.

X-rays

Do not operate or permit students to operate spectrum tubes, cathode-ray tubes, or other devices which may be potential X-ray emitters at higher voltage than is necessary for adequate operation.

Tools

Use only tools which are appropriate for the job to be done. For example, do not use a knife or scissors blade for prying, as it may snap.
Cut away from you when cutting with a knife or razor blade. Sheet metal should be cut only with sharp shears; file edges smooth or use crocus cloth (emery cloth). NEVER use a dull cutting tool. It may slip and cause serious injury.
Place hot soldering iron on stands to prevent fires. Use pliers to hold wires for soldering. Do not inhale fumes from soldering paste.

CHEMICALS DEEMED HAZARDOUS
(By Fire Department)

Explosives

An explosive is defined here as a compound or substance having properties of such a character that alone or in combination with other substances may decompose suddenly or generate heat or gas or pressure to produce flames or destructive blow to surrounding objects.

item	container and restrictions
Picric acid	g.s.b.[5]
Carbon disulphide	g.s.b.

[5] g.s.b. means glass stoppered bottle, or plastic screw cap.

item	container and restrictions
Collodions	g.s.b.
All gases under pressure	steel container

Flammable liquids

Flammable liquid is any liquid that will generate flammable vapors at a temperature below 100° F.

item	container and restrictions
Crude petroleum	tin can
Benzine, Benzol, or Naphtha of any kind	tin can
Coal tar	tin can
Coal tar oils (heavy)	tin can
Wood creosote	g.s.b.
Ether, Ethyl	tin can
Varnishes, Lacquers, etc.	tin can
Acetone	g.s.b.
Alcohol, Ethyl	tin can
Alcohol, Denatured	tin can
Alcohol, Methyl	tin can
Aldehyde, Ethyl	g.s.b.
Amyl acetate	g.s.b.
Amyl alcohol	g.s.b.
Aniline oil	g.s.b.
Kerosene	tin can or g.s.b.
Nitrobenzol	tin can
Turpentine	tin can
Toluol	tin can
Xylol	tin can
Essential oils	g.s.b.
Glycerine	g.s.b.

Combustible substances

Combustible substances are compounds or mixtures that emit flammable vapors at a temperature of 100° F to 300° F.

item	container and restrictions
Phosphorus, white	immersed in water in a glass bottle surrounded by sand
Phosphorus, red	g.s.b.
Sulphur	tin can
Metallic magnesium, including strip and powder	g.s.b.
Camphor	g.s.b.
Rosin	g.s.b.
Pitch (coal tar pitch)	tin can
Tar, refined (wood)	tin can
Venice turpentine	g.s.b.
Burgundy pitch	g.s.b.
Naphthalene	g.s.b.
Shellac	g.s.b. or tin can
Resin, Balsam, and other varnish gums	tin can

Pulverized charcoal	tin can
Lampblack	tin can
Cotton, absorbent	paper box
Cotton batting	paper box
Lycopodium	g.s.b.
Zinc dust	g.s.b.

Dangerously corrosive chemicals

item	container and restrictions
Anhydrous acetic acid	g.s.b.
Glacial acetic acid	g.s.b.
Hydrofluoric acid	wax bottle in an outside container of kaolin
Hydrochloric acid	g.s.b., 6 lbs each bottle
Sulphuric acid	g.s.b., 9 lbs each bottle
Phenol (carbolic acid)	g.s.b.
Sodium hydroxide	g.s.b.
Potassium hydroxide	g.s.b.
Acid, Chromic	g.s.b.
Acid, Nitric	g.s.b.
Acid, Nitric, Fuming	g.s.b.

Peroxides and other oxidizing agents

item	container and restrictions
Hydrogen peroxide, U.S.P.	g.s.b.
Sodium peroxide	tin box
Potassium peroxide	g.s.b.
Calcium peroxide	g.s.b.
Barium peroxide	g.s.b.
Other hydrogen peroxides over 3%; not to exceed 30%	g.s.b. (Keep sealed in original container with asbestos fiber packing, away from heat and sunlight.)
Potassium chlorate	g.s.b. or tin can
Sodium chlorate	g.s.b. or tin can
Barium chlorate	g.s.b.
Other metallic chlorates	g.s.b.
Potassium permanganate	g.s.b.
Sodium permanganate	g.s.b.
Other metallic permanganate	g.s.b.
Bismuth subnitrate	g.s.b.
Barium nitrate	g.s.b.
Strontium nitrate	g.s.b.
Cobalt nitrate	g.s.b.
Iron nitrate, Ferric	g.s.b.
Mercury nitrate (mercuric)	g.s.b.
Mercury nitrate (mercurous)	g.s.b.

Potassium nitrate	g.s.b. or tin can
Silver nitrate	g.s.b. (Use brown bottle.)
Sodium nitrate	g.s.b. or tin can
Other metallic nitrates	g.s.b

Substances made dangerous by contact with water

item	container and restrictions
Calcium carbide	g.s.b.
Metallic potassium	immersed in kerosene in bottle surrounded by sand
Metallic sodium	immersed in kerosene in bottle surrounded by sand
Quicklime	tin can
Sulfuric acid	g.s.b.

RULES AND REGULATIONS RELATING TO THE STORAGE AND USE OF LIMITED QUANTITIES OF CHEMICALS, ACIDS, FLAMMABLES, ETC., FOR INSTRUCTION PURPOSES IN PUBLIC HIGH SCHOOLS

(a)   That no liquefied chlorine may be stored in any school;

(b)   That no more than five gallons of volatile flammable oils derived from petroleum, shale oil or coal tar should be stored at any one time;

(c)   That no more than twenty-five pounds of potassium and/or sodium chlorate is permitted to be stored;

(d)   That no chemicals or substances as listed under C19-139.0 and C19-133.0 of the Code should be stored in a school;

(e)   That it shall be unlawful to store chemicals in close proximity to each other when they are of an explosive nature, or when one increases the energy of decomposition of the other, or when they are so constituted that they may react upon one another and become explosive or flammable;

(f)   That the storage of acids in containers should be confined to either the lowest shelves of soapstone cabinet, or within crockery or earthenware containers, so as to prevent spillage;

(g) That safety cans be provided for the storage of volatile flammable oils;

(h) That a bucket filled with sodium bicarbonate or soda ash be provided near storage of acids;

(i) That the storage of dangerous chemicals, volatile flammable oils and liquids be confined to metal cabinets vented at top and bottom. A card holder should be provided for a visible record of the contents and maximum amount stored therein; also, a caution sign, if applicable to read: "In case of fire do not use water";

(j) All preparation-, storage-, and class-rooms should be provided with portable fire extinguishers of a type suitable for chemical fires. Same should be examined frequently to make sure that they have not been tampered with or removed from their designated places, and at least once yearly all such devices must be examined for deterioration or injuries due to misuse, and recharged. The date of charging and signature of the person who performed it should appear on the tag attached to each extinguisher;

(k) Schools which use large quantities of dangerous chemicals, acids and/or flammable oils or liquids are not included in these regulations and applications for permits from such school should require special investigation by the Fire Department.

Note: The foregoing rules shall be the basis for the issuance of fire department permits to schools in New York City.

(1) The following is a list of the maximum quantities of combustibles and dangerous chemicals which may be stored in public high schools:

Maximum quantities of combustibles and dangerous chemicals that may be stored in schools

EXPLOSIVES

Picric acid	1 lb
Carbon disulphide	10 lbs
Carbon dioxide	1 lb
Anhydrous ammonia	1 lb
Sulphur dioxide	1 lb
Nitrous oxide	1 lb
Oxygen	1 lb

VOLATILE FLAMMABLE LIQUIDS (INSOLUBLE)

Crude petroleum	2 lbs
Benzine, Benzola, or Naphthas of any kind	2 lbs
Ether, Sulphuric	10 lbs
Varnishes, Lacquers, etc.	2 lbs

VOLATILE FLAMMABLE LIQUIDS (SOLUBLE)

Acetone	1 lb
Alcohol, Denatured	5 gals
Alcohol, Methyl	5 gals

NONVOLATILE FLAMMABLE LIQUIDS (INSOLUBLE)

Amyl acetate	2 lbs
Amyl alcohol	2 lbs
Aniline oil	1 lb
Kerosene	2 lbs
Turpentine	½ gal
Tolluol	1 gal
Xylol	1 gal
Essential oils	2 lbs

NONVOLATILE FLAMMABLE LIQUIDS (SOLUBLE)

Glycerine	5 lbs

COMBUSTIBLE SOLIDS

Phosphorus	¼ lb
Phosphorus, red	5 lbs
Sulphur	15 lbs
Metallic magnesium	1 lb

GUMS, RESINS, PITCH, ETC.

Camphor	1 lb
Resin	1 lb
Venice turpentine	1 lb
Naphthalene	1 lb
Shellac	1 lb

COMBUSTIBLE FIBERS AND POWDERS (VEGETABLE)

Pulverized charcoal	5 lbs
Lampblack	2 lbs
Cotton, Absorbent	5 lbs
Lycopodium	1 lb

DANGEROUSLY CORROSIVE ACIDS

Glacial acetic acid	5 gals
Hydrofluoric acid	1 lb
Hydrochloric acid	12 gals
Sulphuric acid	12 gals
Carbolic acid	1 lb

ACID

Acid, Chromic	1 lb
Acid, Nitric	12 gals

PEROXIDES

Hydrogen peroxide, U.S.P.	10 lbs
Sodium peroxide	2 lbs
Barium peroxide	2 lbs
Othe hydrogen peroxides over 3%, not to exceed 15%	5 lbs

PERMANGANATES

Potassium permanganate	1 lb

METALLIC OXIDES

Lead oxide (red)	5 lbs
Lead oxide (Litharge)	10 lbs
Oxide of mercury; red precipitate (mercuric)	10 lbs
Oxide of mercury; yellow precipitate (mercurous)	5 lbs

NITRATES

Barium nitrate	1 lb
Strontium nitrate	1 lb
Cobalt nitrate	1 lb
Copper nitrate	1 lb
Iron nitrate, Ferric mercury nitrate (mercuric)	1 lb
Mercury nitrate (mercurous)	1 lb
Potassium nitrate	10 lbs
Silver nitrate	5 lbs
Sodium nitrate	15 lbs
Other metallic nitrates	5 lbs

CHLORATES

Potassium chlorate	15 lbs

SUBSTANCES MADE DANGEROUS BY CONTACT WITH OTHER SUBSTANCES

Calcium carbide	5 lbs
Metallic potassium	½ lb
All other metals of the alkalies or alkaline earth	2 lbs
Metallic sodium	½ lb
Zinc dust	5 lbs
Slaked lime	25 lbs

Carcinogens*

aflatoxins
aminobiphenyl
arsenic compounds
asbestos
auramine
benzene
benzidine
bis (chloromethyl) ether

cadmium compounds (possibly cadmium oxide)
chlorampenicol
chloromethyl methyl ether
chromium and chromates (limited use)
cyclophosphamide

diesthylstilbestrol
hematite dust
isopropyl oils
melphalan
mustard gas
2-naphthylamine
nickel (limited use)
N,N-bis(2-chloroethyl)-2-naphthylamine

oxymetholone
phenacetin
phenytoin
soots, tars, and oils
vinyl chloride

* *United States Environmental Protection Agency list of toxic substances, January 1980, OPA 15/80.*

Be aware that many of these substances are known by a variety of names. If any of the substances listed above are on hand in the school, immediate steps must be taken for the removal of these chemicals. It is recommended that one of the area or regional offices of OSHA, NIOSH, or EPA be contacted to determine where the cancer causing substances may be taken for proper disposal.

Some biological supply houses, such as Fisher Scientific Co. (711 Forbes Ave., Pittsburgh, PA 15219) offer a safety manual. Consider too, the U.S. National Institute for Occupational Safety and Health (NIOSH) and the Occupational Safety and Health Administration (OSHA); both have regional offices in 10 cities. NIOSH headquarters is in the Department of Health, Education and Welfare (Cincinnati, OH) and the OSHA headquarters is in the Department of Labor (Washington, DC)

Other sources of information are: National Safety Council, 425 North Michigan Avenue, Chicago, IL 60611.
Manufacturing Chemists' Association, 1825 Connecticut Avenue, N.W., Washington, DC 20009.
Underwriters Laboratories, Inc., 207 East Ohio Street, Chicago, IL 60611.
American Public Health Association, 1015 18 Street, N.W., Washington, DC 20036.
American Chemical Society, 16 1155 Street, N.W., Washington, DC 20036.
American National Red Cross, Safety Services, 17 Street and D Street, N.W., Washington, DC 20006.

Films such as "Safe Handling of Laboratory Animals" (15 min sound, 16mm) are available from the National Medical

Audio-Visual Center, National Library of Medicine, Atlanta, GA 30324; the film "Laboratory Safety" (20 min sound), which deals with school laboratory safety, is also available from NIOSH (address above).

Solutions, stains, and general laboratory aids

The preparations of solutions, stains and other useful materials that have not been described in the context of appropriate chapters are listed here. (Use the index to locate a solution or stain in context.) *Refer to safety precautions before preparing solutions.*

Some general notes on the preparation of solution are described first.

Preparation of solutions

Concentrations of substances in a solution are given in terms of percent by weight, molality, molarity, or normality.

Percent by weight   Usually both solute and solvent of percentage solutions are measured by weight; their total weight is equal to 100 units, usually grams. This means grams of solute per 100 g solvent. Thus, when 10 g of NaCl are dissolved in 90 g (ml) water, the total volume is 100 ml. To reduce the concentration of a solution, recall that 1000 ml of the solution contains 1000 mg so that 1 ml contains 1 mg.

Percent by volume: making dilutions Percentage solutions of liquids are generally prepared by volume measurements. For example, 50 percent glycerol means 50 ml glycerol plus 50 ml water.

To make dilutions, measure in milliliters a volume of the higher percent solution that is equal in milliliters to the percentage needed for the new solution. For example, when you want to prepare 50 percent alcohol and you have 70 percent alcohol on hand, measure 50 ml of the 70 percent alcohol. To this, add enough distilled water to bring the volume in milliliters equal to the percentage of the original (in this case, 70 ml). In another example, dilute 95 percent alcohol to 70 percent.

Measure 70 ml of the 95 percent alcohol and add 25 ml of distilled water to make the total volume (ml) equal to the 95 percent alcohol:

95 percent alcohol	70 parts
distilled water	25 parts
	95 parts

Molality (m)   A molal solution is expressed in moles of solute per 1 Kg (1000 g) of solvent; this is, 1 gram-molecular weight in 1000 g of solute. One molal solution of NaCl (m) would be 1 gram formula weight, or 1 mole of NaCl plus 1000 g water.

To make mass comparisons among elements, the Avogadro number of atoms of each element is used. The standard comparison is carbon 12; the number of C 12 atoms needed to add up to 12 g is 6.02×10^{23}. The number 6.02×10^{23} represents the number of atoms of any element in one gram atomic weight of the element, or the number of formula units in one gram formula weight of a compound—the mole.

If the formula is known, the weight quantity of the substance can be indicated as moles. For example, the gram atomic weight of calcium is 40.08 grams; therefore 1 mole of calcium weighs 40.08 g.

In summary, then, a molal solution refers to the number of moles of solute *per kilogram* of solvent; a 1 m solution contains 1 mole of the solute in 1 Kg solvent. In aqueous solutions, recall that a *molar* solution has a final volume of 1 liter; however, a *molal* solution has a greater final volume since the solute has been added to 1 Kg (1000 g) solvent. Note that physical chemists use molal solutions since changes in temperature do not alter the molality or concentration of a solution.

Molarity (M)   A molar solution contains 1 mole (or 1 gram-molecular weight or 1 gram formula weight) of solute made up to 1 *liter* of solvent. A 1 *M* solution of H_2SO_4, for example, contains 98.08 g of H_2SO_4 (its gram formula weight) in 1 liter solution (Table 10-1). Use a volumetric flask for accurate measurements. To pre-

TABLE 10-1   Strengths of several concentrated acids and bases

compound	molecular weight	density g ml⁻¹	molarity
HCl	36.47	1.19	12.2
HNO_3	63.02	1.41	15.7
H_2SO_4	98.08	1.84	17.8
H_3PO_4	98.04	1.70	14.9
CH_3COOH	60.03	1.05	17.4
$NH_3(aq)$	17.03	0.90	14.8
NaOH	40.01	1.53	19.1
KOH	56.11	1.54	14.3
C_2H_5OH (absolute alcohol)	46.05	0.80	17.1

pare a liter of a 2 M solution, add 2 moles solute in a 1 liter volumetric flask, then add water up to the 1 liter mark. A 3 M solution of H_2SO_4 contains 3×98 (molecular weight), or 294 g H_2SO_4.

In another example, sodium chloride has a molecular weight of 58.45; thus a molar solution of NaCl (M) contains 58.45 g NaCl in 1 liter of solution. A 1 M HCl contains its molecular weight in grams (36.5) in 1 liter of solution. A 2 M solution then has twice the number of molecules of solute as a 1 M solution.

Dilutions of molar solutions are easily made. A 0.1 M solution of HCl contains $36.5 \times 0.1 = 3.65$ g of HCl per liter of solution. A 0.4 M solution of NaCl contains $58.45 \times 0.4 = 23.38$ g NaCl per liter solution.

A general equation may be used for determining the number of grams of solute needed to make a solution of a specific molarity:

$$\frac{\text{grams of}}{\text{solute}} = \frac{\text{weight of 1 mole}}{\text{(molecular weight)}}$$

$$\times \frac{\text{molarity}}{\text{needed}} \times \frac{\text{volume}}{\text{needed}}$$

For example: How is a 250 ml of a 0.2 M NaOH solution prepared? NaOH has a molecular weight of 40; 250 ml is 0.25 liter. Apply the equation:

$$\text{grams} = \frac{\text{molecular}}{\text{weight}} \times \text{molarity} \times \text{volume}$$

$$= 40 \times 0.2 \times 0.25$$
$$= 2$$

Weigh out 2 g NaOH and dissolve in some water; add enough water to bring the volume of the solution to 250 ml.

Normality (N)   A normal solution contains one gram-equivalent weight of solute in 1 liter of solution. To find the gram-equivalent weight of an element, divide its gram atomic weight by its valence:

$$\frac{\text{gram-equivalent weight}}{\text{valence}}$$

For example, sodium has an atomic weight 22.997 and a valence of 1; its gram-equivalent weight equals 22.997 g. Aluminum has an atomic weight of 26.97, and a valence of 3; its gram-equivalent weight is 8.99 g.

Normal solutions of *acids* (Table 10-2) contain 1 gram-formula weight of replaceable hydrogen (1.008 g) in 1 liter solution:

$$\frac{\text{normal solution}}{\text{acid}} = \frac{\text{gram formula weight}}{\substack{\text{number replaceable H} \\ \text{ions shown in formula}}}$$

Make up to 1 liter of solution.

For example, 1 N H_2SO_4 contains 2.016 g hydrogen, + 32 g sulfur + 64 g oxygen = 98.016 g H_2SO_4.

$$\frac{98.016}{2} = 49.008 \text{ g } H_2SO_4 \text{ in 1 liter solution.}$$

The solution contains 1.008 g of replaceable hydrogen ions.

In another example, 1 N H_3PO_4 contains 3.024 g hydrogen, 30.98 g phosphorus, and 64 g oxygen = 98.004 g H_3PO_4

$$\frac{98.004}{3} = 32.668 \text{ g } H_3PO_4 \text{ in 1 liter solution.}$$

The solution contains 1.008 g of replaceable hydrogen ions.

Most uses of normal solutions are in neutralization or acid-base reactions. In a neutralized reaction, one equivalent of an acid is the amount that liberates one mole of hydrogen ions in a water solution. (Half normal solutions (0.5 N) are half the amounts given in the examples.)

Similarly, a normal solution of a *base* contains 1 gram-formula weight of replaceable hydroxide ions (OH = 16.000 g + 1.008 g = 17.008 g). For example, a normal solution of NaOH contains 22.997 g sodium + 1 × (17.008) = 40.005 g NaOH in 1 liter solution. Of the 40.005 g of NaOH, 17.008 g make up 1 gram-formula weight of replaceable OH ions. If the solution contains half the amount it would be a 0.5 N. If it contains a tenth, it would be a 0.1 N solution. Equal volumes of acid and base solutions of the same normality will exactly neutralize each other.

$$V_A \times N_A = V_B \times N_B$$

At the equivalence point in a titration, the number of gram-equivalent weights of the reacting substances is the same.

In general, then, to prepare normal solutions, study the formula of the acid, base, or salt to be dissolved. When there is one hydrogen atom or one hydroxyl group or one of any ion that will combine with one hydrogen or hydroxyl, a normal solution is the same as a molar solution.

When two hydrogen atoms are present, as in H_2SO_4, a normal solution contains half as much H_2SO_4 as a molar solution, because there are 2 gram-equivalents in every gram-mole. This is also true for $Ca(OH)_2$, and a normal solution is prepared by making a 0.5 M solution of calcium hydroxide.

A *normal* solution of $FeCl_3$ (which has three chlorines, each of which could react with one hydrogen) would be a 0.33 M $FeCl_3$ solution.

In general, a normal solution is prepared by dissolving in 1 liter of solution a quantity of the acid, base, or salt determined in the following way:

$$\frac{\text{number of grams needed for 1 } M \text{ solution}}{\text{number of equivalents to 1 H in each molecule}} \quad or$$

$$\frac{\text{molecular weight of substance in grams}}{\text{valence}}$$

Refer to Table 2-4 for a listing of pH values of 0.1 N solutions of several acids and bases.

Parts per million (Ppm)   The designation parts per million refers to parts by weight of a given substance measured in milligrams of the substance in 1 liter of solution (1 mg/l = 1 ppm; 1000 mg/l = 1000 ppm).

Calibration of a test tube   There are times that approximations are handy. For example, an ordinary pipette can be substituted for a 1 ml pipette; 20 drops give a volume of approximately 1 ml. Students may measure the volume of a test tube and calculate how many milliliters fill half a test tube, or a third of a test tube. (However, not all test tubes are uniform.)

Rules of solubility   In preparing certain solutions, such as culture solutions for microorganisms or for studies in hydroponics, it is useful to know the relative solubilities of salts. The general rules of solubility are given in Table 10-3.

Among many references, the following offer excellent guides for the laboratory technician and teacher. S. Cherim, *Chemistry for Laboratory Technicians*, (Philadelphia: Saunders, 1971); J. Routh, *Mathematical Preparation for Laboratory Technicians*, (Philadelphia: Saunders, 1971); P. Perlman, *General Laboratory Techniques*, (Millston, NJ:

TABLE 10-2 Commonly used laboratory reagents—1 N solutions

reagent	formula mass	density g/cm³	percent	normality	volume reagent needed to prepare 1 liter with distilled water *
acetic acid	60.05	1.05	99.5	17.4	57.5
ammonium hydroxide	35.05	0.90	57.6	14.8	67.5
hydrochloric acid	36.46	1.19	37.0	12.1	83.0
nitric acid	63.02	1.42	69.5	15.7	64.0
sulfuric acid	98.08	1.84	96.0	36.0	28.0
phosphoric acid	97.99	1.69	85.0	44.0	23.0

* To make a final volume of 1 liter solution.

TABLE 10-3 General rules of solubility

soluble in water

Compounds of sodium, potassium, and ammonium.

Sulfates (except lead and barium sulfate; calcium, strontium, and silver sulfate are slightly soluble).

Chlorates, nitrates, and acetates.

Chlorides (except silver and mercury chloride; lead chloride is slightly soluble).

insoluble in water

Phosphates, carbonates, oxides, sulfides, sulfites, and silicates (except those of sodium, potassium, and ammonium).

Hydroxides (except those of sodium, ammonium, potassium; calcium, barium, and strontium hydroxides are slightly soluble).

Franklin, 1964), G. Humason, *Animal Tissue Techniques*, 3rd ed., (San Francisco: Freeman, 1972).

Some additional solutions for the stockroom

When possible, use distilled water in preparing solutions. Unless otherwise stated, temperature referred to is room temperature.

Acid-starch solution

Add 1 ml of dilute $NaNO_2$ solution and 1 ml of dilute H_2SO_4 to 10 ml of a starch solution just before it is to be used.

Adrenalin hydrochloride mix

Combine 100 mg adrenalin hydrochloride with 100 ml Ringer's solution.

Agar (starch)

Combine 10 g powdered starch, 10 g agar and 980 ml distilled water. Mix, heat slowly, and autoclave before use.

Buffer (standard)

Humason refers to the recipe of McIlvaine.[6] Prepare the following stock solutions of citric acid and disodium phosphate and refer to Table 10-4.

STOCK SOLUTIONS

0.1 M citric acid (anhydrous)

19.212 g made up to 1 liter with distilled water.

0.2 M disodium phosphate (anhydrous)

28.396 g ($\cdot 7H_2O$, 53.628 g) made up to 1 liter with distilled water.

For desired pH, mix correct amounts as indicated in Table 10-4.[7]

BENEDICT'S SOLUTION (QUALITATIVE)
This is used in a test for the presence of simple sugars in foods, blood, and urine. In the presence of simple sugars, a yellow or reddish precipitate of cuprous oxide forms when the reagent is heated with the

[6] G. Humason, *Animal Tissue Techniques* (San Francisco: Freeman, 1972)

[7] Also refer to tris buffer for pH in alkaline range (p. 649).

TABLE 10-4 Citric acid—Disodium phosphate buffer

pH	citric acid (ml)	disodium phosphate (ml)	pH	citric acid (ml)	disodium phosphate (ml)
2.2	19.6	0.4	5.2	9.28	10.72
2.4	17.76	1.24	5.4	8.85	11.15
2.6	17.82	2.18	5.6	8.4	11.6
2.8	16.83	3.17	5.8	7.91	12.09
3.0	15.89	4.11	6.0	7.37	12.63
3.2	15.06	4.94	6.2	6.78	13.22
3.4	14.3	5.7	6.4	6.15	13.85
3.6	13.56	6.44	6.6	5.45	14.55
3.8	12.9	7.1	6.8	4.55	15.45
4.0	12.29	7.71	7.0	3.53	16.47
4.2	11.72	8.28	7.2	2.61	17.39
4.4	11.18	8.82	7.4	1.83	18.17
4.6	10.65	9.35	7.6	1.27	18.73
4.8	10.14	9.86	7.8	0.85	19.15
5.0	9.7	10.3	8.0	0.55	19.45

"unknown." The test will detect 0.15 to 0.20 percent dextrose.

This solution can be purchased ready-made, or it can be prepared as follows.

Dissolve the carbonate and the citrate in 800 ml of water; warm slightly to speed solution. Then filter. Dissolve the copper sulfate in 100 ml of water and slowly pour into the first solution. Stir constantly; let this cool and add distilled water to make 1 liter.

sodium or potassium citrate	173.0 g
Na_2CO_3 (crystalline)	200.0 g
(or 100 g anhydrous)	
$CuSO_4$ (crystalline)	17.3 g

BIURET REAGENT AND PAPER Dissolve 2.5 g of copper sulfate in 1 liter of water to prepare a 0.01 M solution. Then prepare a second solution of 10 M sodium hydroxide by dissolving 440 g of sodium hydroxide (*caution*) in water and then making up to 1 liter. Before using, add about 25 ml of the copper sulfate solution to 1 liter of hydroxide solution. Or use Walker's modification of the Biuret reagent, prepared as follows.[8]

[8] B. Oser, *Hawk's Physiological Chemistry*, 14th ed., (New York: McGraw Hill, 1965).

Add 1 percent copper sulfate solution, a drop at a time, with stirring, to a 40 percent (approximate) solution of sodium hydroxide until the mixture becomes a deep blue color.

Then immerse filter paper in the reagent, dry it, and cut it into small strips for use in tests for proteins.

CATALASE DILUTIONS A number of dilutions can be prepared. For a 0.01 percent solution, add 10 mg catalase to 100 ml distilled water. Use 2.5 mg/100 ml water for 0.0025 percent dilution; 1 mg/100 ml water for a 0.001 percent dilution; 0.1 mg/100 ml for a catalase dilution of 0.0001 percent.

COBALT CHLORIDE PAPER Immerse sheets of filter paper in a 5 percent aqueous cobalt chloride solution (5 g cobalt chloride in 100 ml distilled water). Then remove and blot dry between other sheets of filter paper. Dry in an oven at 40° C (104° F). Cut the papers into strips. For immediate use, dry the strips quickly by placing them in a dry test tube and heating them over a flame until the paper turns from pink to blue.

Store cobalt chloride paper in wide-mouthed, tightly stoppered bottles containing a layer of anhydrous calcium chloride covered with cotton. Should the blue papers change to pink, indicating the presence of moisture, heat them again in a dry test tube.

COPPER ACETATE PRESERVATIVE This solution, which is especially effective for preserving the green coloring of plants, may be prepared as follows. Add enough copper acetate to 50 percent acetic acid to produce a saturated solution. To 250 ml of this saturated solution, add an equal volume of formalin (*caution*). Then make up to 4 liters with water. Refer to other preservatives, p. 649.

COPPER SULFATE (0.01 M) Dissolve 0.25 g copper sulfate in 100 ml water (total volume).

2,6-DICHLOROPHENOL-INDOPHENOL In 200 ml of water, dissolve 84 ml of sodium bicarbonate and 104 mg of the sodium salt of the indicator. Filter if necessary, and store in a refrigerator.

DIGITONIN SOLUTION   Dissolve 1 g of digitonin in 85 ml of 95 percent alcohol; add 15 ml of water, and mix thoroughly. This gives a 1 percent solution of digitonin in 80 percent alcohol.

EGG ALBUMEN (1 PERCENT)   Add 1 g powdered albumen to a small amount of water and stir carefully to *avoid foaming*; bring volume to 100 ml. For more dilute solutions, such as 0.1 percent, use one part of the 1 percent solution and 9 parts water.

FEHLING'S SOLUTION   Either purchase or prepare solutions 1 and 2 separately and store them separately in rubber-stoppered bottles. In testing for the presence of simple sugars add an equal amount of each solution to a test tube of the substance to be tested, and heat. A heavy yellow or reddish precipitate forms (cuprous oxide) if simple sugars are present.

solution 1

| CuSO$_4$ | 34.65 g |
| distilled water | 500 ml |

solution 2

KOH	125 g
potassium sodium tartrate	173 g
distilled water	500 ml

FORMALIN SOLUTION (5 PERCENT)   Add 1 part of commercial formaldehyde (37 to 40 percent) to 19 parts water. (Refer to safety precautions, p. 727.)

FORMALDEHYDE (0.5 PERCENT)   Dilute 10 ml of 40 percent formaldehyde solution with 800 ml of water; adjust to pH 7. (See also p. 167.) (*Caution:* avoid inhaling fumes.)

GLUTATHIONE[9]   (reduced) $10^{-5}M$

| reduced glutathione | 3 mg |
| distilled water | 10 ml |

Dilute 1 to 10 to give $10^{-5} M$.

GLYCEROL SOLUTION (15 PERCENT) Add 15 ml glycerol to 85 ml distilled water.

HYDROCHLORIC ACID (1 N)   Dilute 83 ml concentrated HCl (12 N) with water to bring the volume to 1 liter.

HYDROCHLORIC ACID (0.1 N)   Use 8.6 ml of concentrated HCl and make up to 1 liter with water.

HYDROCHLORIC ACID (0.2 N)   Bring 16.6 ml of concentrated HCl (about 12 N) to 1 liter with water.

HYDROCHLORIC ACID (1 M SOLUTION) Slowly add acid to distilled water: 36.5 g concentrated hydrochloric acid to make up to 1 liter of solution. (Recall that normal and molar solutions of HCl are the same; see 1 N solution above.)

HYDROCHLORIC ACID (0.1 M)   Slowly add 3.65 g (about 9.3 ml) concentrated hydrochloric acid to distilled water to make up to 1 liter solution.

HYDROGEN PEROXIDE SOLUTION (1 PERCENT)   Combine 33 ml hydrogen peroxide (3 percent) with 66 ml distilled water.

IODINE (TINCTURE OF)   Dissolve 70 g of iodine and 50 g of potassium iodide in 50 ml of distilled water. Then dilute to 1 liter with 95 percent alcohol.[10]

IODINE-POTASSIUM IODIDE SOLUTION (I$_2$KI)   Dissolve 3 g of potassium iodide in 25 ml of water. Then add 0.6 g of iodine, and stir until dissolved. Make up to 200 ml with distilled water. Store in a dark bottle. This dilute solution may be used in determining amylase activity (see Lugol's solution, p. 744).

LEAD ACETATE SOLUTION (0.2 M)   Dissolve 76 g hydrated lead acetate in water to make up to 1 liter of solution.

LIGNIN TEST   A saturated solution of phloroglucin[11] (1,3,5-trihydroxybenzene) in alcohol is recommended as a test for the presence of lignin. First, mount plant tissue sections in this solution for a few minutes, then transfer to a drop of water (containing a minute trace of hydrochloric acid). If lignin is present it stains a bright reddish-violet color.

LIMEWATER   To 1 liter of distilled water add an excess of calcium hydroxide. Cork

---

[9] I. Sherman, and V. Sherman *The Invertebrates: Function and Form,* A Laboratory Guide, 2nd ed., New York: Macmillan, 1976.

[10] Iodine stains may be removed from clothing by washing the stain with a 10 percent solution of sodium thiosulfate in water. Then rinse in water.
[11] F. Emerson and L. Shields, *Laboratory and Field Exercises in Botany,* New York: McGraw-Hill, 1949.

the bottle, shake well, and let it stand for 24 hr for the precipitate to settle. Then pour off the supernatant fluid (filter if necessary) and keep well stoppered.

The limewater should remain clear. When carbon dioxide is added, a milky precipitate of calcium carbonate is formed:

$$CO_2 + H_2O \longrightarrow H_2CO_3$$

$$Ca(OH)_2 + H_2CO_3 \longrightarrow CaCO_3 + 2H_2O$$

LITMUS SOLUTION To 25 g litmus powder add 250 ml boiling water; let stand 15 min, then decant liquid. Add 500 ml boiling water and let stand overnight, then filter.

LUGOL'S SOLUTION Humason describes several recipes from the literature.[12] For both solutions that follow, first dissolve the potassium iodide, the iodine will then go into solution.

Weigert variation. To 100 ml distilled water add 2 g potassium iodide, then 1 g iodine crystals.

Gram variation. To 300 ml distilled water add 2 g potassium iodide, then 1 g iodine.

METHYLENE BLUE STOCK SOLUTION Dissolve 1.5 g powdered methylene blue to 100 ml 95 percent ethyl alcohol. For use as a stain, dilute by adding 10 ml stock to 90 ml water.

METHYLENE BLUE (0.5 PERCENT) For a dilute aqueous solution, dissolve 0.5 g methylene blue in 100 ml distilled water.

METHYL CELLULOSE (10 PERCENT) Heat 100 ml water to 85°C (185°F; not boiling); shake 10 g methyl cellulose powder into the water; stir rapidly while cooling this mix in an ice bath to about 5°C (41°F). In this procedure, the solution is stable at room temperature and can be stored in tightly closed containers.

Other solutions in the literature suggest adding 10 g methyl cellulose to 90 ml water. (Some experimentation will be needed to find dilutions for slowing down ciliates; also refer to p. 142.)

NITRIC ACID (DILUTE) Add 1 part of concentrated nitric acid (*caution*) to 4 parts of water.

PHENYLTHIOCARBAMIDE (PTC) SOLUTION Dissolve 50 mg phenylthiocarbamide in 100 ml distilled water; dilute further to bring up to 1 liter solution.

PHLOROGLUCINOL: HCl SOLUTION (1 PERCENT) Prepare this 1 percent aqueous solution by mixing 1 g phloroglucinol in 100 ml water. Then add concentrated HCl (12 *N*) in a 1:1 ratio (*caution*). Use the solution in tests for pentose or galactose within a week.

POTASSIUM CHLORIDE (0.5 *M*) Dissolve 37 g potassium chloride in 1 liter distilled water.

POTASSIUM HYDROXIDE (0.1 *M*) Add 5.6 g KOH to water to make up to 1 liter.

POTASSIUM PHOSPHATE (0.1 *M*) Prepare this monopotassium buffer solution by dissolving 13.6 g KH_2PO_4 in water and bring volume to 1 liter.

POTASSIUM PHOSPHATE (0.1 *M*) Prepare this dipotassium phosphate buffer solution by dissolving 22.8 g K_2HPO_4 in water and bring volume to 1 liter.

PROTEOSE PEPTONE (1 PERCENT) Combine 1 g proteose peptone and 100 ml distilled water and adjust to pH 7.0 to 7.2.

PYROGALLOL SOLUTION One preparation, among many oxygen absorbents, is the following. Prepare a small quantity since it cannot be stored for any period of time. *Slowly* dissolve 160 g KOH pellets (*caution*) in 130 ml distilled water to make a 22 *N* hydroxide solution. Then carefully stir in 10 g pyrogallic acid. This gives about 200 ml pyrogallol.

QUININE SULFATE SOLUTION (0.1 PERCENT) Add 0.1 g quinine sulfate to 100 ml distilled water.

RENNIN SOLUTION Prepare a 0.1 percent solution by grinding 1 g rennin with 50 ml water to form a thin paste. Dilute with water to 1 liter.

SEAWATER (30 PERCENT) Combine 30 ml seawater with 70 ml distilled water.

SEAWATER (ARTIFICIAL) Humason describes Hale's procedure for preparing sea water with a salinity of 34.33 0/00 and a chlorinity of 19 0/00.[13] Dissolve the following salts in distilled water, and make

---

[12] G. Humason, *Animal Tissue Techniques.*

[13] G. Humason, *Animal Tissue Techniques.*

up to 1 liter. (This is not to be used for aquaria, but only for technical purposes.) Increase the quantities two- or four-fold as the need demands. Refer also to marine aquaria, p. 674.

NaCl	23.991 g
KCl	0.742 g
CaCl$_2$	1.135 g
or CaCl$_2 \cdot$ 6H$_2$O	2.240 g
MgCl$_2$	5.102 g
or MgCl$_2 \cdot$ 6H$_2$O	10.893 g
Na$_2$SO$_4$	4.012 g
or Na$_2$SO$_4 \cdot$ 10H$_2$O	9.1 g
NaHCO$_3$	0.197 g
NaBr	0.085 g
or NaBr $\cdot$ 2H$_2$O	0.115 g
SrCl$_2$	0.011 g
or SrCl$_2 \cdot$ 6H$_2$O	0.018 g
H$_3$BO$_3$	0.027 g

SEED DISINFECTANT   A description of seed germination (p. 470) suggests that seeds be dipped in household bleach (5 percent sodium hypochlorite; 1 part to 4 to 6 parts of water), or dilute potassium permanganate solution (about 5 crystals to 1 liter water). First dip seeds briefly into alcohol as a wetting agent, then into bleach for 5 to 10 min. Place the seeds in a cheesecloth bag and immerse in the liquid. For most seeds a 5-min immersion is sufficient. More hardy seeds can be immersed up to 10 min in this disinfectant. Then wash the seeds several times in water for 15 min; spread them out to dry.

SILVER NITRATE SOLUTION (1 *M*)   Dissolve 169.89 g silver nitrate in water and bring the volume to 1 liter.

SODIUM ACETATE-ACETIC ACID BUFFERS (pH 3.6 TO 5.8)   Arditti and Dunn suggest the following preparations.[14]

(a) Sodium acetate solutions: Prepare solutions of desired molarity as in Table 10-5.

(b) Acetic acid solutions: Prepare solutions of desired molarity as in Table 10-6.

Now to prepare the range of pH buffer desired, use *equal molarities* of both solutions. Refer to Table 10-7.

[14] J. Arditti and A. Dunn *Experimental Plant Physiology* New York: Holt, Rinehart & Winston, 1969.

TABLE 10-5   Sodium acetate solutions (*M*)

molarity (M)	NaAc g	final volume
1.0	82.0	1 liter
0.5	41.0	1 liter
0.2	16.4	1 liter
0.1	8.2	1 liter

TABLE 10-6   Acetate solutions (*M*)

molarity (M)	amount conc. HAC (ml)	final volume
1.0	60	1 liter
0.5	30	1 liter
0.2	12	1 liter
0.1	6	1 liter

TABLE 10-7   Preparation of pH ranges of Na acetate-acetate buffers

desired pH	Na acetate solution (ml)	acetic acid solution (ml)	final volume (ml)
3.6	75	925	1000
3.8	120	880	1000
4.0	180	820	1000
4.2	265	735	1000
4.4	370	630	1000
4.6	480	520	1000
4.8	590	410	1000
5.0	700	300	1000
5.2	790	210	1000
5.4	860	140	1000
5.6	910	90	1000
5.8	940	60	1000

SODIUM CARBONATE BUFFER (pH 11) Prepare two *stock* solutions:

1. sodium carbonate solution: dissolve 21.2 g sodium carbonate in 1 liter water

2. sodium bicarbonate solution: dissolve 16.8 g sodium bicarbonate in 1 liter water

Mix 40 ml sodium carbonate solution with 5 ml of the bicarbonate stock solution;

TABLE 10-8   Preparation of NaCl solutions*

desired molarity or normality	amount NaCl (g)	final volume (ml)
5.0	292.25	1000
2.5	146.13	1000
2.0	116.90	1000
1.0	58.45	1000
0.5	29.23	1000
0.2	11.69	1000
0.1	5.85	1000

* Adapted from J. Arditti and A. Dunn, *Experimental Plant Physiology* New York: Holt, Rinehart & Winston, 1969.

dilute with water to make 200 ml to get a pH 11.

SODIUM CHLORIDE SOLUTIONS (*M* AND *N*)   Prepare the desired molarity and normality as in Table 10-8.

SODIUM HYDROXIDE (1 *N*)   Dissolve 40 g NaOH in 1 liter water to make 1 *N* or 4 percent NaOH (*caution*).

SODIUM HYDROXIDE (10 *N*)   This 10 *N* or 40 percent sodium hydroxide solution is prepared by adding 40 g NaOH to 100 ml water. (Use a rubber-stoppered reagent bottle; *caution*.)

SODIUM HYDROXIDE (0.1 *N*)   Bring 4 g NaOH up to a liter with water. Or add 1 ml of 10 *N* NaOH to water to make 100 ml. Use a 1 ml pipette to transfer 1 ml of 10 *N* NaOH to 100 ml water.

SODIUM HYDROXIDE SOLUTION (1 *M*)   Dissolve 40 g of NaOH in 200 ml of water (*caution*); dilute with water to 1 liter of solution.

Prepare a 2 *M* solution by adding 80 g to enough water to make 1 liter of solution.

SODIUM PHOSPHATE (MONOBASIC 0.1 *M*)   Prepare this molar solution by adding 12 g NaH_2PO_4 to water and bring up to 1 liter.

SODIUM PHOSPHATE (DIBASIC 0.1 *M*)   Add 14.2 g Na_2HPO_4 to water and bring up to 1 liter.

STARCH PASTE (1 PERCENT)   Add 5 ml cold water to 1 g arrowroot starch and stir into a paste. Then add this to 95 ml of boiling water; stir constantly until smooth. Bring to a boil; let it cool. This is a satisfactory concentration for general use in demonstrations of salivary digestion.

SUCROSE SOLUTION (1 *M*)   Dissolve 342 g sucrose in water and bring volume to 1 liter.

SUCROSE SOLUTION (0.3 *M*)   Dissolve 103 g sucrose in distilled water and dilute to 1 liter. Add toluene as a preservative (see below).

SUCROSE SOLUTION (0.03 *M*)   Prepare by dissolving 10.3 g sucrose in water and dilute to 1 liter. This solution causes walls of guard cells to straighten, but does not cause plasmolysis of the cells.

SUCROSE SOLUTION (0.1 *M*)   Dissolve 34.2 g sucrose in water and make up to 1 liter. Add toluene as a preservative to sucrose solutions (see below).

SUCROSE SOLUTION (0.5 *M*)   Dissolve 171 g sucrose in water and dilute to 1 liter. Add toluene as a preservative (see below).

SUCROSE SOLUTIONS (DILUTIONS)   At times, you may want to prepare a 0.5 percent sucrose solution and use it for further dilutions. Prepare a 0.5 percent solution by adding 0.5 g sucrose to 99.5 ml water.

0.1 percent solution. Add 20 ml 0.5 percent solution and 80 ml water.

0.05 percent solution. Use 10 ml of 0.5 percent solution sucrose and 90 ml water.

0.01 percent solution. Add 20 ml 0.05 percent solution and 80 ml water.

0.005 percent solution. Add 10 ml 0.05 percent sucrose solution to 90 ml water.

0.001 percent solution. Add 20 ml 0.005 percent sucrose solution to 80 ml water.

SULFURIC ACID (0.1 *M* APPROXIMATELY)   *Slowly* pour 5.5 ml concentrated sulfuric acid to water and allow to cool (*caution*); mix well and dilute to 1 liter. Also refer to Table 10-9 for molarity and normality of H_2SO_4.

TOLUENE-THYMOL PRESERVATIVE   Dissolve 1 g thymol in 100 ml toluene (*caution*).

TRIS BUFFER (0.2 *M*)   Humason describes Hale's recipe for the following *stock* solutions of 0.2 *M* tris and the 0.1 *N* HCl.[15] Then prepare pH solutions in the alkaline range as in Table 10-10.

---

[15] G. Humason, *Animal Tissue Techniques*.

TABLE 10-9   Preparation of H_2SO_4 solutions (M and N)

desired molarity	desired normality	H_2SO_4 conc. ml	distilled water to ml
17.8	35.6	use as is	none
15.0	30.0	946.0	1000
10.0	20.0	564.0	1000
5.0	10.0	282.0	1000
2.5	5.0	141.0	1000
1.5	3.0	94.6	1000
1.0	2.0	56.4	1000
0.5	1.0	28.2	1000

TABLE 10-10   Tris-HCl buffer: alkaline range

pH	0.1 N HCl (ml)	pH	0.1 N HCl (ml)
7.19	45.0	8.14	25.0
7.36	42.5	8.23	22.5
7.54	40.0	8.32	20.0
7.66	37.5	8.41	17.5
7.77	35.0	8.51	15.0
7.87	32.5	8.62	12.5
7.96	30.0	8.74	10.0
8.05	27.5	8.92	7.5
		9.1	5.0

0.2 M tris (hydroxymethyl) aminomethane

24.228 g made up to 1 liter with distilled water.

0.1 N HCl (38% assay)

8.08 ml made up to 1 liter with distilled water.

To *25 ml 0.2 M tris* add 0.1 *N* HCl as indicated in Table 10-10 and dilute to 100 ml.

UREA **(0.1 M)**   Dissolve 6 g urea in 1 liter water.

UREA **(40 PERCENT)**   Dissolve 20 g urea in 50 ml water.

YEAST SUSPENSION   To 500 ml of a diluted molasses solution add 1 g peptone and ¼ to ½ package of yeast. Stir well; maintain the yeast suspension at 25° C to 30° C (77° F to 86° F) for 12 to 24 hr.

Laboratory aids

ADHESIVE FOR JOINING GLASS TO METAL
A solution of sodium silicate has been recommended as an adhesive for joining glass and metal.[16] A paper gasket may be soaked in a solution of sodium silicate and inserted between the edges of glass and metal cutout.

AQUARIUM CEMENT   This type of cement has been recommended for repair of aquaria. It will stick to glass, metal, stone, or wood. The first four ingredients may be

mixed together in the dry state. Then, just before using, add enough linseed oil to make a stiff putty. Allow three to four days for this cement to harden after it has been forced into crevices and smoothed over with a spatula. The following measurements are given as parts by weight.[17]

litharge	10 parts
plaster of Paris	10 parts
powdered rosin	1 part
dry white sand	10 parts
boiled linseed oil	

FERTILIZER FOR ACID-LOVING PLANTS[18]
Combine the following fertilizer ingredients. Then mix each pound of this with 5 cu ft of redwood or cypress sawdust.

$(NH_4)_2SO_4$	26 parts
superphosphate	31 parts
potash	190 parts

For a planting mixture, use half garden loam and half this mixture.

GRAFTING WAX (LIQUID)[19]   Reduce the quantities given to suit your specific needs.

rosin	4	lb
tallow	0.5	lb
isopropyl alcohol	0.75	qt
turpentine	0.25	pt

Melt together the rosin and tallow, then cool a bit and dilute with alcohol and turpentine (*caution*: no flames). Store in a tightly stoppered bottle.

[16] H. Bennett, ed., *Chemical Formulary,* (New York: Chemical Publishing Co, 1951).

[17] Bennett, ed., *Chemical Formulary.*
[18] Ibid.
[19] Ibid.

GRAFTING WAX (MALLEABLE)

rosin	4 parts
beeswax	2 parts
tallow	1 part

Heat the ingredients together, and pour the melted material into water. When cool, shape into balls with the fingers.

LUBRICANT (STOPCOCK)   Glycerin prevents sticking of ground-glass parts and is also useful in sealing ground-glass joints to prevent leaking of substances (such as ether) which are insoluble in it.

PHOTOGRAPHIC SOLUTION (NONCURLING) Immerse prints after washing in a solution made by combining 12 parts of glycerin, 5 parts of alcohol, and 83 parts of water.

SOIL ACIDIFIER[20]   Prepare the following and add it to the soil until the desired pH is obtained:

flowers of sulfur	1 part
$(NH_4)_2SO_4$	1 part
$Al_2(SO_4)_3$	1 part

Working with glass

Be sure to read the section on precautions in working with glass as described on page 729.

Cutting glass tubing

Capillary pipettes and glass bends can be made from glass tubing 6 mm in diameter. Cut glass by applying *one* firm scratch with a sharp triangular file on a section of glass tubing held securely on the top of a table. Wet the file with water if it is not sharp. Then hold the tubing as shown in Fig. 10-1, and as a precaution, wrap each end with a soft cloth or paper toweling. Hold the scratch away from the body; place the thumbs behind it. Pull on the tubing in opposite directions with the outside fingers and at the same time exert a forward push with the thumbs so that an even break in the tubing results. Fire-polish the cut ends in a flame. The bore of the opening of the tubing may be altered while the glass is

[20] Bennett, ed., *Chemical Formulary.*

Fig. 10-1   Correct way to hold glass tubing to break it after it has first been scratched.

heated. Overmelting of the glass will reduce the size; the opening can be increased by inserting the round end of the triangular file into the melted glass opening.

Use a glass cutting wheel to cut small sections of glass tubing, Pyrex glass tubings, or tubing of larger diameter.

Should the edges break unevenly, strike the tubing at a diagonal against a meshed wire square held in the hand. Then fire-polish in a flame.

Ordinary Bunsen or Tirrill burners are sufficient for working with soft glass; when a hotter flame is needed use a Meker burner with a metal grid that provides a wide flame.

The following precautions are in order when inserting glass tubing into rubber stoppers. First lubricate the glass tubing with water, petroleum jelly, or glycerin before inserting into the stopper. Hold the tubing in toweling or cloth as near to the part of the tubing to be inserted into the stopper as possible, then ease the tube into the stopper with a twisting motion using a little pressure at a time. (Be sure not to direct the tubing into the palm of the hand.)

Avoid freezing of tubing or thermometers in rubber stoppers by removing them after use. When a tube or thermometer is frozen in a rubber stopper, use a wet cork borer that is just large enough to slip over the thermometer or tubing. Slowly insert the borer through the stopper to release the glass. An easier method is to slit open the rubber stopper to release the glass.

Making a capillary pipette

Cut glass tubing into sections about 18 to 22 cm in length; fire-polish both ends.

Many sections of this tubing can be kept on hand until they are needed for making pipettes.

Between the thumb and first two fingers of both hands, rotate a center section of a length of glass in a flame until the glass is soft; remove the glass from the flame and pull it out. Hold it in position or place it on a metal mesh sheet until it cools. Too rapid a pull will produce capillary sections with too narrow a bore. Cut off the lengths desired and carefully fire-polish.

When a long tapered capillary tube is desired, rotate the glass tubing over a wing top or flame spreader.

The pulling of soft glass test tubes is described in chapter 4 in making a micro-aquarium (see Fig. 4-33).

Bending glass

A good bend retains the same size bore throughout the bend. Rotate tubing or glass rod holding both ends lightly, and turn the tubing evenly so that all sides are heated equally. Use a wing top when a broad U-shaped bend is desired so that a larger area of glass may be heated. Remove the tubing from the flame when it starts to sag, and bend it to the desired shape, applying equal pressure with each hand. Hold the tubing in one plane until it cools. Some workers remove the sagging tubing from the flame and take advantage of the natural sag by raising the ends upward so that the sag hangs down in a bend. Others lay the sagging tubing on metal mesh and make the bend from a flat surface, thus ensuring that the bend is in one plane. A good right-angle bend can easily be made in this way by using a square corner of the metal mesh as a pattern for the right angle. When bends of a specific angle are needed, cut a pattern or draw one on metal mesh. Set the tubing aside to cool. Students should learn to touch glassware cautiously before they grab it—hot glass looks exactly like cold glass!

Glass rods can be drawn out to varying thicknesses. Fine threads of glass are used for glass needles for careful work in embryology where bits of tissue are trans-

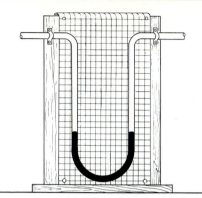

Fig. 10-2   A U-shaped tube mounted against graph paper to serve as a simple manometer. The tube may be filled with water colored with food dyes or red ink.

planted from a young embryo into an older one (see chapter 6, page 427).

Glass stirring rods may be easily made. Cut glass rods to desired lengths (depending on the size of the beakers used) and heat each end to melting; while the rod is soft, press it against a sheet of metal to flatten the ends.

A U-shaped tube: simple manometer

Use lengths of 1 mm glass tubing to make a manometer. Heat the tubing over a flame spreader (wing top) and shape into a symmetrical U-tube. At each end of the tube now make a right-angle bend, add water to which a food dye or red ink is added and mount the manometer on a board against a sheet of graph paper (Fig. 10-2). Attach rubber tubing to one end when ready to use.

Curved pipette

The curved pipette shown in Figure 10-3 is useful in transferring eggs of frogs or snails, protozoa, or small crustaceans from one finger bowl to another. It fits into the curve of the finger bowl, so that the tip does not break off so readily.

Rotate a long section of glass tubing over a fish tail or wing top. When the glass begins to sag, remove it from the flame and begin to make a U-shaped bend. However,

Fig. 10-3 A curved pipette convenient for transferring delicate specimens such as frogs' eggs out of small finger bowls.

quickly flick the wrists so that the palm of each hand is turned slightly inward forming a slight curve in another plane. After the glass cools, cut it in the center of the U-bend so that two pipettes result. Fire-polish the ends. The size of the bore should be made to fit the materials to be transferred; for example, it should be large if frogs' eggs are to be transferred.

Making a wash bottle

A wash bottle containing distilled water is a useful piece of equipment for careful work in the laboratory. While the plastic ones are available, you may want students to gain skill in handling glass.

Students will need a flat-bottomed Florence or Erlenmeyer flask, a two-hole stopper, several lengths of glass tubing, and a 4-cm section of rubber tubing to which a small glass nozzle is attached.

Prepare the glass bends and assemble as shown in the completed wash bottle (Fig. 10-4). Both the glass tubing and the stopper should be wet while the glass tubing is inserted into the stopper. Use toweling and hold the tubing next to the region inserted into the stopper. With a gentle twisting motion insert the glass tubing into the stopper. The glass tubing is safer to handle when the ends have been fire-polished beforehand.

Repairing damaged glassware

At times, some sharp, uneven edges of pipettes, flasks, beakers, and test tubes may be repaired for subsequent use by filing the chipped edges or rims and then fire-polishing. It may be necessary to reshape.

Making visual props

Clearly an effective teacher in the classroom evokes images for students—with words, sometimes with gestures, other times with useful props. While there are many kinds of props or aids, where possible, use living materials—the real props. These may be brought to class, or bring students to the living materials through field trips.

A three-dimensional model may be the next best experience. A slide, a photograph, a chart, a drawing on the board—all two-dimensional—also provide effective next-best experiences.

Only a brief introduction to the making of models, charts, and other props can be presented here.

A visual aid should be simple and used sparingly, to clarify confusion, to focus attention of all students to the same thing, to provide a new view of material already studied as reinforcements or as part of an evaluation of learning. Visual aids should not be "inserted" into a lesson because a good model of some sort is part of the equipment of the department. Students can become overwhelmed by seeing many

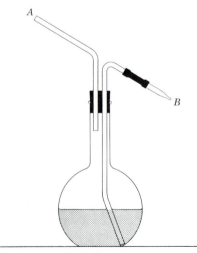

Fig. 10-4 A wash bottle can be used in two ways: water can be simply poured out of *A* or a fine stream of water can be obtained from *B* by flowing into *A*.

charts or models, complex in design, in one period. (All of us know of some favorite demonstration material that we think couldn't be simpler or more to the point. Yet we have found that all students do not have the same perception. They lack a common frame of reference. You may have found yourself explaining the model.)

Models

Real materials   A hen's egg is a simple model. (Boil it first as a safety measure!) Use it to focus questions: What kind of an animal arises from this egg? What does it contain that enables it to develop into a chicken, not an eagle? More than that— the egg develops into one kind of chicken —a Plymouth Rock, not a Leghorn.

Other times, students may bring the realia for the lessons: a beef heart, a sheep's brain or eye, frogs, insects, a beef or hog kidney, beef lungs with trachea (haslet), arteries and veins, seeds, flowers and fruits, leaves showing variations, cocoons, water from a pond, guppies, starfish, mollusk shells, worms, and many other materials mentioned throughout this book. On field trips to an overgrown lot, a park, a beach, students may observe the dynamic interrelationships among many living plants and animals in a natural setting—its ecosystem.

Sometimes professionally made models supplement the real things. Many times they simplify a concept by reducing the real thing to its essentials.

Models in clay   A ball of soft clay may be used to show cleavage stages. How does a fertilized egg begin to form a many-celled organism? With a spatula, cut the ball vertically into two, then each of these into two, to represent the four-celled stage. Recall the next cleavage is in a horizontal plane; each of the four cells divides and a ball of eight cells results.

Clay in different colors can be used in developing an early gastrula from a blastula. Use the clay ball to show the depression of the neural tube in the late gastrula stages, and how it is formed from the outer ectoderm of the gastrula.

Students have modeled many tissue cells, the heart, and the entire respiratory and digestive systems in clay. However, the models fall apart fairly easily so that clay seems to be most useful in manipulating models in class before the eyes of students. However, some students have successfully created dioramas showing farming practices that favor conservation of land, life among the dinosaurs, and many ideas concerning index fossils in layers or eras in the earth's history.

Models in glass   While you may draw an air sac of the lungs on the board using colored chalk, a round-bottomed Florence flask is a more effective three-dimensional model of an air sac, an alveolus. Draw capillaries on the glass with a red glass-marking pencil. How are gases exchanged between these capillaries and the alveolus? Or tie a meshwork of string around the bulb of the flask to represent a capillary network.

Make a villus model by inserting a small test tube (a lacteal) into a larger test tube (the outer membrane of the villus).

Pieces of string, wire, balloons, pipe cleaners   Tease apart some fibers of a square of gauze. Doesn't this resemble a capillary bed? Or use a length of jute cord; separate the fibers in the center region to represent capillaries. Dip one end of the cord in blue ink, the other end into red ink to gain an idea of an arteriole-capillary-venule.

We say that food materials pass along some 25 to 30 ft of digestive tract. Have you ever asked students to estimate a 30-foot length? Have students stretch a clothesline along the length of the room to show 30 ft. Now describe what happens to food as it passes along this long tube.

In chapter 6, Fig. 6-16, a description is given for showing "animated chromosomes" in mitosis, using short pieces of electrical wiring, rubber bands, and string.

How are alleles arranged on a pair of chromosomes? Are the same genes present in the different paired chromosomes? Wooden beads of different colors may be threaded on wires or pipe cleaners. Pairs of different colored pipe cleaners can represent paired chromosomes. Poppits are also excellent props if still available. These are plastic beads that fit into each other and can be pulled apart to be remade into strings of beads of different lengths. Many chromosome aberrations can be demonstrated with these beads: translocation of chromosome parts, fragmentation, crossover, and with the help of some modeling clay, nail polish or glue, nondisjunction of pairs of chromosomes.

Paper clips can hold together pipe cleaner "chromatids" as students trace mitosis and meiosis at their work stations. A limitless variety of materials may be used in modeling a molecule of DNA; refer to suggestions using cut-outs and small magnets, on page 389.

A cell model
Use agar to make a model of a cell. Prepare some agar in cold water, bring it to a boil, let it begin to cool. Pack the agar in a dialysis "cell membrane" bag. Embed small colored marbles or other particles in the cooling agar to represent nucleus, centrosome, and so forth. Or, if you wish to make a model of a plant cell, use green peas for chloroplasts.

Improvise in many ways. Suppose an indicator such as colorless phenolphthalein is added to the agar medium as it is being prepared. Later, if ammonium hydroxide is placed near the finished model of the "cell" wrapped in dialysis membrane, diffusion of the alkali into the "cell" will produce a pink color within the "cell," a useful demonstration of diffusion through a cell membrane.

An artificial cell
To a test tube containing a small amount of dilute albumin from a raw egg, add a few drops of chloroform (caution). Shake the materials together, mount a drop on a slide, and observe the artificial "cells" which form.

At times you may have reason to offer the following demonstration. All these materials are, of course, nonliving, but this demonstration may help students visualize how a cell grows. Place a crystal of copper sulfate in a solution of potassium ferrocyanide. As the copper sulfate passes into solution, a membrane forms around the crystal; this membrane of copper ferrocyanide forms on contact with the surrounding potassium ferrocyanide. Watch the membrane expand and stimulate the growth of a cell as water continues to pass through the membrane in one direction only. Note that the equal expansion in all directions is due to the entrance of water, not to the formation of new material (corresponding to new protoplasm which would occur in a living cell).

Models of plaster of paris
Plaster of Paris may be poured into a mold shaped of plasticine, cardboard, plastic or metal tins. Shape the plasticine to form a mold for a cell, such as an epithelial cell. If any parts are to protrude in plaster, they should be indentations now in the plasticine. For example, indent the region where a nucleus should be in the cell. Build up a cuff—the outer edges of the cell mold—about 3 cm high. Apply petroleum jelly to the clay if needed. Then prepare the plaster of paris. Add enough of the plaster (a good quality such as dental plaster) to water to form a consistency that is thick and creamy but thin enough to pour. Stir carefully to avoid introducing air bubbles in the plaster. Pour the plaster of paris into the mold slowly so that air bubbles are not formed. Allow this to set and harden. Later peel away the clay mold. If desired, these plaster models may be smoothed with fine sandpaper; you may want to apply shellac and paint in different colors.

Professional-looking models may often be made by students. These may become part of the stock of the science department. Latex models may be made; they have the advantages of being light in weight. Or you may want to use the rubber liquid now available at arts and crafts suppliers; this liquid may be used to make molds of various sorts. The advantage of these rubber molds is that they can be used repeatedly.

Models in papier-mâché In this method, strips of toweling or newspaper are dipped into a thin paste made of flour and water. Or strips of newspaper soaked in water may be worked over to make a malleable thick paste. Add 1 percent phenol to prevent spoilage. Then add this newspaper mash to flour and water to make an adhesive material. Construct the framework of the object to be modeled—such as a dinosaur, a cell, or cleavage stages—of cardboard and wire props. Then apply the mash of flour-soaked newspaper or apply overlapping strips of toweling soaked in flour paste. Apply successive layers after each preceding layer has dried thoroughly. About six or seven layers are needed. When the model is thoroughly dry, smooth it with fine sandpaper. These models may be painted in bright colors. An art teacher may offer alternate suggestions.

Models in wood Fine lightweight models may be made of balsa wood. Students have made three-dimensional models of microorganisms, cells, and the heart. These can have the added advantage of hinged halves so that the model can be opened, revealing brightly painted interiors—a section through an amoeba or paramecium, a generalized cell to show organelles, or the chambers of the heart. With a jigsaw, cut two-dimensional models—or, rather, plaques—of organs of the body or of tissue cells.

Models in clear plastic Attractive, convenient models are available of plant and animal specimens mounted in clear plastic. You may purchase models of representatives of the basic plant and animal phyla. Some students may want to try to duplicate these professionally made models. Biological supply houses sell the materials required for construction of these plastic models and provide complete instructions. Clear plastic is a tricky medium to work in, but some students have, after practice, made better-than-average models that could be used in teaching.

Preparation of skeletons

At times, a student may want to prepare a skeleton of a small bird or mammal. While a full description is beyond the scope of this book, students can get assistance from booklets available for purchase from biological supply houses, or from a text such as Knudsen's.[21]

Blueprints of leaves

Students may prepare a file to show diversity in types of leaves. Or, further, they may gather and dry specimens to serve for identification of trees and shrubs. This is one step in learning to identify plants, and it may result in an attractive display of materials. (Also refer to preserving plant specimens, p. 649.)

One way is to prepare leaf skeletons that can be taped between two glass slides and projected on a screen for class study. In another method, which young students also enjoy, blueprints are made of leaves, as follows. Tape one side of a 15-cm square of cardboard to one side of a sheet of glass cut to the same size. Insert a sheet of blueprint paper of the same size or smaller between the glass and cardboard backing. Lay a leaf on the blueprint paper and hold it flat by pressing the glass over it. Expose this to sunlight. A few trials will indicate the length of time needed. Remove the blueprint paper and rinse in water. Allow to dry, then flatten under a weight.

Outlines of leaves may be made on photographic paper, but developing the print is more time-consuming.

Skeletons of leaves

Students may wish to prepare skeletons of leaves that may be mounted between lantern-slide plates for projection. Such slides can be used for a series of discussions of plant identification.

It has been suggested[22] that leaves be immersed in a slow-boiling macerating

---

[21] J. Knudsen, *Biological Techniques*, (New York: Harper and Row, 1966).

[22] *Procedure for Preparing Leaf Skeletons*, General Biological Supply House, Chicago.

solution for about 2 min, until the leaves turn a dark brown. The solution can be prepared as follows. Boil together 16 oz of water, 4 oz of sodium carbonate, and 2 oz of calcium oxide. Cool and then filter. If you find it necessary to convert ounces to another unit of measure, see the tables in the Appendix.

Transfer the leaves from the slow-boiling solution into a wide, shallow tray containing a small amount of water. Rub the leaves with a soft brush to remove the tissue from the supporting tissue or veins. If the tissue does not separate easily from the veins, return the leaves to the macerating fluid.

Bleach the resulting skeletons of leaves by immersing them for a few minutes in the following solution. To 1 qt of water, add 1 tbsp of chloride of lime and a few drops of acetic acid. When the skeletons have turned white, they can be pressed and mounted.

Dried specimens

From time to time, students may add properly dried algae, fungi, evergreens, and flowering plant specimens to the botanical collection of a biology department. (Refer to preservation of plants, p. 649.) Also refer to Knudsen's reference.[23] Supply houses sell collecting cases, plant presses, driers, and mounting sheets.

Acrylic sprays form useful protective films to prevent fragmentation of delicate specimens of algae, leaves, and spore prints of mushrooms that may be in the school's herbarium collection. (*Caution:* Avoid inhaling the fumes of the spray.)

Freeze-dried specimens

Many biological specimens may be frozen solid so that ice crystals replace the tissue fluids. They then can be dried at sub-zero temperatures, resulting in permanently rigid, dried specimens that exhibit little shrinkage. Ward's Natural Science Establishment (see Appendix) offers a variety of freeze-dried specimens.

Taxidermy

There are many procedures for preserving skins of small animals. Knudsen's reference is also especially recommended for techniques.[24]

Slides and transparencies

Complete libraries of slides and transparencies are available from many biological supply houses (see Appendix). Habitat studies, field work, parasitology, hematology, genetic disorders, and all aspects of biology studied in high school and college are beautifully represented in either slides or transparencies.

At times, you may find it useful to flash daylight slides, perhaps of tissues, on the blackboard. A student may outline the projected cells in chalk on the blackboard. When the projector is turned off and the lights are turned on, the image will still be on the blackboard, where it may be examined in detail. This may help to develop some point in a lesson.

Students may also be taught to accompany their talks with demonstrations. Students will need to use a projector to show a short loop or a 2 × 2 slide carousel. Or it may be easier, at times, to prepare transparencies with overhead projectors. Because of their ease in preparation and dynamic flexibility in use, overlays made by students add a dimension to a lesson.

With a glass-making pencil or with Higgins inks, students may make simple drawings on dry, clear, glass slides (2 × 2 or 3¼ × 4). More successful slides may be made on ground glass using pencil, India ink, or crayons. A three-dimensional effect can be achieved with shading, or a blended color effect may be obtained by moistening crayons very slightly with water.

In addition, small thin specimens, such as skeletons of leaves, pressed flowers, or butterflies may be inserted between two slides and bound with gummed tape (see also p. 647). Kits are available for preparing transparencies.

---

[23] J. Knudsen, *Biological Techniques.*

[24] J. Knudsen, *Biological Techniques.*

Tissue mounted on 35-mm film

As an alternative to glass coverslips and microscope slides, film can be used for mounting tissue. Tissue can be mounted directly on flexible plastic strips of perforated 35-mm film, without emulsion, cut into 7.5 cm lengths or in lengths to roll on reels. The resulting "slidestrips" can then be observed under a microscope or projected with a filmstrip projector.

This "slidestrip" technique is described in *Carolina Tips* (27:4, April 1964); Carolina Biological also supplies a "slidestrip kit," a technique developed by V. Bush (1955) and improved by several other workers. In brief, paraffin sections of tissue are placed in a drop of water on a length of 35-mm film, without emulsion, to which albumin fixative has been applied. The sections of tissue and film are placed on a warming table to spread the tissue. After drying, the tissue and film are transferred to xylene to remove the paraffin. After being passed through alcohols, the tissue is stained, dehydrated, and finally cleared in xylene. Several layers of an acrylic spray are applied to the tissue while the material is still wet with xylene.

Prepared slides of serial sections of early embryos of the pig and the chick, as well as series in physiology, are available from biological supply houses (see Appendix).

Making charts

Drawings made on the blackboard with chalk of different colors are effective when they are developed *with* the class as part of a lesson.

Or permanent charts can be prepared by drawing on window shades with ink and crayons, or on muslin with wax crayons. When the wax-crayon drawing is finished, cover it with brown wrapping paper and iron the paper. The wax will melt and be absorbed into the muslin. (Protect the ironing board cover with paper too.) When they become soiled, these charts can be washed. The main advantage of these drawings on muslin is that they can be folded and kept in envelopes in a filing cabinet. Sew a 3 cm hem at the top of the chart and use metal eyelets so that the chart can be hung from hooks, or use thumbtacks. If you have a storage cabinet for other charts you may prefer to nail the muslin charts onto wooden rods like professional charts.

Supply houses provide professional charts of all kinds, and they also offer teachers leaflets and bulletins describing ways to make good-looking charts.

A picture file

With a little encouragement, students readily accept the responsibility for maintaining a current, interesting bulletin board. You may want to train a squad of students to establish a picture file. Give them file folders and space. Excellent pictures (and reference reading materials) may be found in *Scientific American, Natural History, National Geographic, Smithsonian, Discover* and similar magazines and bulletins.

Students will soon have complete sets of bulletin board pictures for any topic in the year's work: behavior, physiology, genetics and genetic engineering, disease prevention, reproduction, evolution, and ecology as well as the effects of pollutants.

Addenda

Making dissection pans   Rectangular metal cake pans of a uniform size make useful dissection pans. Melt together equal parts of paraffin and beeswax over a low flame. If a black wax is desired, add lampblack to the melted preparation. Then pour into metal pans to a depth of 3 cm and set aside to harden. Avoid stirring, for this creates air bubbles. Should air bubbles appear, they may be broken with a hot dissection needle while the wax is still warm.

Labeling rock specimens   Large rock specimens may be labeled by painting on a plaster of paris surface. Mix plaster of paris and water to a creamy consistency. Apply a thin, smooth layer on the side of the specimen which is to be labeled. When the plaster has "set," apply a thin solution of glue as sizing over the surface. Labels may be printed with a permanent ink on the sized dry surface of the plaster. Then protect the label by brushing over the dry ink with melted paraffin. Very small specimens may be partly embedded in plaster.

Permanent labels for specimen bottles Sometimes, to avoid smearing, printed labels can be placed inside a jar containing specimens preserved in alcohol.

Print labels with India ink and set aside to dry until the ink no longer glistens; use forceps to dip them into a jar containing 5 or 10 percent glacial acetic acid. Then drain the labels on blotting paper and insert into the specimen jars containing 70 percent ethyl alcohol. Rather strong paper should be used for labels because glacial acetic acid tends to soften the paper.

Restaining faded slides A procedure for removing coverslips from microscope slides and bleaching tissue for restaining is described in chapter 2.

Using student resources

Individual students have special talents that a teacher can put to use in many ways. In fact, almost every teacher has a student following of a sort. In the beginning, a teacher may plan a program of activities with this unorganized group of students who probably have many common interests. In this way a squad of students may develop.

How can a squad, that is, a small group of students, help an individual teacher or several teachers who make up a department? And how, in turn, does this activity help these students to grow in competence?

Co-curricular and extracurricular activities in school provide opportunities for young people to practice some of those so-called required learnings or developmental tasks that they all need to develop—skill in getting along with others, peers and adults; testing one's abilities;

getting practice in making judgments; self-evaluation of possible success in science as a vocation; manipulating tools of the scientist; gaining specific skills and a better understanding of what science is.

Many kinds of activities in which students may participate are listed here. They vary in their requirements for special skill or high ability. However, each type of job has a status-giving quality to help some student retain his or her individuality. Often through these informal relationships teachers establish a rapport that results in supportive guidance or mentoring.

Such activities benefit students in yet another way. Many teachers describe students' work in anecdotal accounts which are filed with the permanent cumulative record of students. These are useful as recommendations for college or for future employers. Several schools have student governments that sponsor school service programs and allot service points for all kinds of work done in the school and community. The accumulation of 50 points may lead to a certificate, 75 points to a service pin presented at commencement.

What are some ways that teachers have used students as resources to increase the effectiveness of activities in the stockroom, the laboratory, and the science classroom? A primary service is to assist in the laboratory and stockroom.

Laboratory squads

Probably the most pressing need for science teachers is assistance in the daily routine of getting materials from the stockroom, preparing solutions, setting up demonstrations or laboratory activities, purchasing perishable materials, dismantling equipment, washing glassware, and storing materials in some organized way. Students assistants, as *organized* laboratory squads, can be trained in these routine tasks. The work may be done before school, during school hours, or after school. Students can be helpful in the bookkeeping job of keeping track of equipment, mainly expendable materials. This is a necessity when it comes to ordering supplies and taking inventory.

In a small school science teachers probably prepare class demonstrations on the run, for they carry a heavy teaching load. Students must, of course, work under a teacher's supervision.

In larger schools many procedures can be used. Where several science teachers share a common stock of materials and equipment, one teacher may be responsible for ordering supplies and taking inventory. In large school systems laboratory assistants have this special responsibility. Any one plan we describe cannot be feasible for all schools, large and small; teachers will select the possibilities that best meet their own needs. Specifically, what goes on in a supply laboratory and stockroom in a school?

Filling daily orders   Students may fill "orders" that teachers have submitted in advance of their classes. However, this presumes that teachers in the department have organized themselves and have developed a routine for writing their orders for class teaching materials. This plan necessitates submitting orders a day in advance to a stockroom or laboratory center where the squad can have access to materials and can fill the orders: preparing solutions, obtaining models, and/or setting up demonstrations.

When teachers pool their own resources, this kind of procedure can be put into action. Under the guidance of a teacher (or a laboratory assistant), students can deliver this class order to the teacher's room when it is needed. When duplicate equipment is available, the teacher might get the materials in the morning and hold them for the day. Larger pieces of equipment, such as microscopes, a manikin, dissection pans and kits, or a microprojector, may be delivered to the teacher at the beginning of the period. Then the material can be picked up again toward the end of that same period and returned to the laboratory or supply center. Students trained as a laboratory squad can make the deliveries and pick-ups.

How do teachers write their daily orders for supplies to the supply center in the department? Consider initiating a standard system in which each teacher submits an order on uniform-sized paper or enters the "order" in a book a day in advance. Teachers can check their orders against those of others; or at the end of the day one squad of students can examine the orders. When two teachers order the same materials for the next day, they are informed of this and they agree to share the materials or one accepts a substitute or possibly changes a lesson for the next day. The advantage is that the teacher knows beforehand whether the materials needed will be available so that there is time to plan another lesson.

That afternoon (or the next morning), students locate the materials that teachers want for the next day and stack small items in a tote box or tray. Or they place large pieces or heavy equipment on a cart that can be wheeled through the halls to deliver to the teachers at the start of the day or the lesson. Bottles of strong acids and bases should be carried in wooden boxes as a safety precaution. (Refer to the section on safety in the laboratory, p. 727.) In addition, glassware, such as Petri dishes should be stacked in boxes.

For safety, avoid having the squad make deliveries when the student body is moving through the halls to their classes. For the same reason, when teachers are not going to keep their materials for the entire day, the pick-up of materials should occur a few minutes before the end of the class period. When this cannot be done, as occasionally happens, the teachers should assume the responsibility for getting the materials back to the common laboratory supply center. Otherwise the equipment will not be available for the next teacher who needs it.

When teachers need fresh materials such as elodea, frogs, protozoa cultures, a pig's uterus for dissection, blood-typing materials, germinating seedlings for studies in genetics, or a thriving culture of fruit flies, a week's advance notice must be given to the teacher and the squad serving the science teachers. Plan three weeks advance notice for materials from supply houses. In practice, this works out easily when teachers confer with each other from time to time each month and plan to share perishable materials at the same time. When teachers outline their work at the beginning of the school year, fresh materials can be purchased then, or living materials can be collected and cultured in the laboratory or classroom and made available when needed (see chapter 9).

Cleaning up   Two or three students in the laboratory should be responsible for dismantling the equipment returned by teachers and for washing glassware used in class. Then all the materials are returned to labeled shelves, cabinets, or drawers.

Laboratory housekeeping   Plants must be watered, and aquaria and terraria need care. Someone must dust open shelves of models and glassware. Microscopes must be checked and lenses cleaned from time to time. New labels need to be put on cabinets and museum jars, and charts may need mending. Some fish tanks need patching and some broken glassware could be put to new use if the jagged edges were cut off and filed or fire-polished smooth. The wax lining some of the dissection trays becomes cracked and should be melted slowly over a Bunsen flame.

Possibly one student could add water colors to some of the black and white charts to freshen them and enliven students' interest. Who would prepare cotton-lined display boxes (Riker mounts) for the unassorted fossil specimens on hand? Or partially imbed them in plaster of paris? Some student on your squad, perhaps.

Safety in the laboratory   Every month over the school year, student squads in the lab should undertake a fire drill of their own design. Can the students on the squad use the fire extinguishers? Do they know when to use the sand in the sand buckets in the lab? Do they know how to use the fire blanket? All of these materials should

be on hand. Also check the freshness of the supplies in the first-aid kit.

In the laboratory, students should not use a flame unless the teacher is present. Regulations should be designed concerning the storage of acids, bases, other dangerous chemicals such as chloroform or iodine, and so forth (refer to safety procedures, pp. 727–37).

Squads for special services

Students with special interests and skills may receive training of many kinds that cannot be given in "class time." Either squads or special club activities may be developed in the following areas. In many cases, there are opportunities for enrichment here and also situations for students to discover where their interests and abilities lie.

Making models, charts, and slides   Many teachers initiate a routine for listing materials or equipment they would like to have for class use. Many teachers regularly peruse the catalogs of supply houses. Often they list their ideas on library cards at the moment they think of them. Possibly these cards are filed in a small metal box. At a given time each teacher submits these cards to one person in the department who has assumed this responsibility. These cards serve as a guide at the time for ordering supplies. But the main use of these teachers' ideas is to provide the stimulus for students who like doing things with their hands and brains. These students may build a piece of equipment such as an incubator or a device to show a tropism, or they may make a series of models to show the cleavage stages of a developing egg. Other students may make charts to show comparative anatomy among vertebrates or variations in some flowers, fish, or other forms not easily obtainable for class use as firsthand experiences. (See p. 750 on making models and charts.)

Students may supplement the supply of prepared slides in the department. For example, a class or squad may be trained to prepare blood smears (p. 274), whole mounts of small forms (p. 171), stained slides of protozoa (p. 163), among many other kinds of slides.

Curators of living things   When space is available, a museum of living things, plants and animals, can be maintained over the school year so that teachers may bring their classes to this area for many lessons in biology.

Where space is at a premium a section of a classroom or a laboratory may be given over to three or four aquaria and at least one terrarium, and several shelves may be used to store cultures of protozoa and fruit flies. In another room (to avoid contamination of cultures of protozoa), different kinds of mold cultures may be grown. Perhaps there is also room for a cutting box—a box of sand or light soil— a mini greenhouse—in which cuttings of plants and other examples of vegetative propagation may be grown.

Students may be trained to culture protists, subculture them every month or so (technique, chapter 9), raise fruit flies (technique, chapter 7), prepare mold cultures, grow fern prothallia (p. 716), and so forth. Guppies and other small tropical fish can be kept in one tank with snails and water plants. In this way, life cycles of fish and of snails' eggs can be studied in class. Fresh materials for study of plant cells are available at any time. In separate aquaria students may maintain *Tubifex, Gammarus,* planaria, hydra, larvae of some insects, tadpoles, or other freshwater forms they may collect on field trips or from an aquarium shop (maintaining animals and plants, chapter 9).

In each classroom there should be at least one aquarium and one terrarium. If possible, establish a marine tank (p. 674). Clear plastic tanks or battery jars of different sizes may be used also to accommodate living things that students bring to class (for example, land snails or cocoons). Small cages may be made of screening and wood frames by some students; observation ant jars may be prepared (see page 677). Directions have been given for collecting living things, keeping them in class, and also for preserving biological materials (see chapters 8 and 9). When several classrooms house different living things, space has been "stretched" and the science department has living culture centers in each classroom.

Soil-testing squads   Many schools have established excellent relations with the community by serving as a center for testing soil samples. There are many soil-testing kits on the market (see supply house catalogs) and students may be trained to give suggestions for improving the soil; this is a useful conservation practice. It may be possible to get help from the agricultural county agent in the community.

Landscaping the school grounds   A squad—a small core of interested students—may be trained to plan a nature trail around school (see p. 638), build birdhouses to attract birds, and plant trees, shrubs, and flower beds useful to the teachers of biology. In this way, classes might be held on the school grounds for studies in reproduction, genetics, behavior, variations and evolution, interrelationships, and conservation. Students may label trees to help other students learn to identify some of the common trees and shrubs.

Book-room squad   A trained core of students can distribute books to classes at the beginning of the school year, or supplementary reference texts can be supplied to teachers whenever they are needed by a class. An up-to-the-minute inventory can be maintained if a tally sheet is posted on a bulletin board in the book room indicating books on shelves and books in use.

In fact, these students may take responsibility for an important enterprise in the department: encouraging students to build their own private science libraries. Some of the many excellent paperbound books might be sold (under a teacher's supervision) by a book-room squad before or after school one day a week as a special service.

Teachers will find ways to bring these readings into daily class use. Many teachers send for catalogs from publishers of paperbacks in science. Also refer to *Paperbound Books in Print* (R. R. Bowker Co., 1180 Avenue of Americas, New York, NY 10036) which has an author index, titles index, and an index by special categories of subjects.

Research groups   Some of these special services squads regard themselves as clubs and develop a pride in membership. They hold meetings and discuss problems of importance to their operation.

Students with a common interest may ask a teacher to sponsor their club, for they know the teacher shares their interest. In this way a microscopy club is formed under a teacher's supervision, or a tropical fish club, or a walking club which goes on field trips on Saturdays or to a weekend school camp.

Each teacher may sponsor at least one highly talented student and give him or her scope to work on some investigation that holds long-range interest. Gradually in such individual work deep interest and enthusiasm are born and spread so that soon a large group of students *select themselves* for a rigorous program of work. Students working on an original investigation gain skills in laboratory work, read more, and learn the methods of science by using them.[25] Here, too, gifted students can often discover whether they have the aptitudes, abilities, and other personality attributes needed for a career in science.

Some suggestions for student investigations offered by biologists over the country are detailed in *Research Problems in Biology: Investigations for Students*, Series One-Four BSCS, (New York: Anchor Books, Doubleday, 1963, 1965).

Students with some commitment to biology may participate in science fairs, seminars, and science talent search examinations for scholarships. They may also be guest speakers in classrooms, in school assembly programs, and at meetings of the PTA.[26]

Through these experiences students learn more about science and do more; they also learn that leadership involves responsibilities, and the ability to share their knowledge with a larger audience.

Science readings squad   This is a squad that can help to raise the interests of the entire student body if the members have a high level of initiative.

At the lowest level of operation, a small group of students might be assigned the task of writing a form letter to those industrial concerns (see Appendix) known to supply low-cost teaching aids such as vitamin booklets, charts of vitamin deficiencies, pamphlets that describe how to make something, and so forth. At a more advanced level, students can select magazine articles (for example, *National Geographic, Natural History, Scientific American* and *Smithsonian*) and the publications of nearby museums and state and federal governments.

Assign these students a bulletin board or hall case so they can post articles, pictures, and copies of new paperbacks. Offer suggestions about sources of information on specific topics such as computers in science careers, health services, air and water quality, safety and accident prevention, and identification of plants and animals in season (birds, insects, shells, and flowering plants). It may be possible for this squad of specialists to work with the librarians in preparing exhibits of books in science.

In many schools students have built attractive wall racks along which magazines may be inserted. This arrangement saves desk or table space in a classroom or laboratory.

Duplicating squad   Some students like to put things into print. When a small

[25] P. F. Brandwein, *The Gifted Student as Future Scientist*, (New York: Harcourt, Brace & World, 1955).

[26] P. F. Brandwein et al., *Teaching High School Biology: A Guide to Working with Potential Biologists*, BSCS Bull. 2, AIBS, Washington, D.C., 1962.

squad of reliable students is trained to use a copy machine they give the teacher time for more creative work.

Many teachers find it profitable to reproduce writings from original papers (as described on pp. 130, 232, 544). Statistical data for interpretation listed on transparencies make interesting tests. Devise a variety of tests that give some evidence of reading ability and skill in interpretation, hypothesis-making—all the skills needed in a world of technology and scientific research.

Publishing squad   In some classes, you may want to prepare a newsletter in biology or a small magazine for distribution in the department. Those students involved in some independent research or who enjoy writing for others might organize such a venture. This type of activity involves the creative efforts of many students: both writers and typists. In fact, several small schools could merge their efforts in a cooperative endeavor to widen their sales and influence. Here, mentors serve a special function.

Projectionist squads   In many schools a school aide or teacher is assigned to supervise the audio-visual program for the entire school. The A-V director is responsible for slides and film, and for keeping overhead projectors, cassette players, and microphones in good operating condition. Films must be ordered; a schedule of showings over the year must be planned so the films supplement the class activities; and a squad of projectionists must be trained.

Often one room is set apart, equipped with heavy drapes, screens, storage closets for films, filmstrips, loops, slides, projectors, computers and software. There should also be opaque projectors to supplement overhead projectors. Equipment for splicing and rewinding film is needed. Train students to maintain and store equipment.

Several students may be assigned to look through catalogs of free loan films, and record these on library cards to build a filing system of films available in many areas of biology. Evaluation sheets might be used after showings to indicate whether the film should be reordered.

Also keep files of instructional software for computers from companies such as Programs for Learning (P.O. Box 954, New Milford, CT 06776). Many packages such as "Fundamental Skills for General Chemistry," "Atomic Structure," and others have been developed. Genetics programs simulating patterns of inheritance in parakeets, in domestic cats, as well as others, are available from Cambridge Development Laboratory, 100 Fifth Avenue, Waltham, MA 02154. Also refer to *Books in Print* for new offerings in this expanding market.

Perpetuation of squads

We have discussed some ways students' skills can be put to use for a common purpose. How can squads already well trained be perpetuated? As they approach graduation, they should train younger students over at least a 3-month period to take over these tasks. In this way, although a teacher must supervise the students, the work is made routine. This saves time for the teacher, and students gain skill and self-confidence as they serve others in useful group work.

High grades in science should not be the decisive criterion for selecting students for these squads—each student brings his or her talent. Teachers agree that goodwill, fair play, and cooperation in sharing the load—many of the appreciations and attitudes we aim for in working with young people—often are more readily *caught* in this kind of work in groups than *taught* by teachers in classrooms.

Student squads in the classroom

Many students readily accept a share in improving the atmosphere of the classroom. They may turn an ordinary classroom into an attractive, distinctive place conducive to learning. These rooms can look like biology classrooms, places where interesting work is going on—different from the other classrooms students visit during the day. Volunteers may care for the plants in the classroom, the aquaria, the bulletin boards. Show cases may exhibit hobby collections of fossils, insects, shells.

Bring to class the flowering plants in season—especially the trees in flower—frogs' eggs, and cocoons. Place budding branches from trees and shrubs in water in containers of uniform size. On small library cards indicate the name of each plant, where it was found, and some interesting facts about the plant in the region.

Plan contests in which a number of plants, birds, or insects in the neighborhood must be identified. Prizes may be small paperbacks or field guides.

Tutoring squads   Students can offer a worthy service in tutoring others—there will be a student who is lacking experience in some area, or a student who takes a little longer to learn a concept, or students who have been absent for long periods due to accidents or illness, and who thereby fall behind in their work. Whenever possible, arrange to have other students who can communicate ideas tutor the students who need remedial help or temporary "briefing" on new work.

Laboratory aides   What kind of class management is efficient and effective for distributing slides, medicine droppers, cultures for examination, and microscopes? How can dissection pans, toweling, kits, magnifying lenses, and specimens for study be distributed without confusion, and collected without damage or loss? Some teachers organize laboratory activities for as many as 34 students they may meet in a classroom when a biology laboratory is not available. They assign some students as laboratory aides to distribute slides, coverslips, lens paper, dropper bottles of cultures and of stains on trays in different "spots" in the room so that the students do not crowd around one point of distribution. Microscopes are wheeled into class on a cart; when these are num-

bered it is possible to assign a specific microscope to each student or group of students. In this way students share responsibility for specific instruments.

Toward the end of the period, students return microscopes to the cart, and the laboratory aides collect the materials they distributed earlier in the period. Then the squad from the laboratory center picks up the materials (see laboratory squads, p. 756).

Space for work

In older schools, teachers have used their own good devices to find ways to make over old things. Gradually they have ordered new equipment. They have also improvised. When a room lacks running water, some teachers use a carboy of water with a rubber extension tube and clamp. Students willingly bring to class equipment to augment the facilities of the laboratory. In this way aluminum foil dishes, baby food jars, plastic boxes, and large plastic or glass jars can be gathered in place of more expensive labware.

Some schools have solved their individual problems concerning older classrooms not equipped for science teaching by purchasing a transportable demonstration table. These units give teachers an opportunity to provide more demonstrations and laboratory activities.

Newer schools are equipped with outlets in classroom and laboratory for water, gas, a vacuum jet, and one for compressed air.

Science teachers are often asked to suggest plans for new classrooms (or lists of ideas they would want to see in practice) which may be incorporated in blueprints for a new school.

What facilities would you request for a new school? What kind of tables and chairs, and for what age group? Do you favor separate adjoining closets or closets and cabinets in a central supply laboratory? Do you prefer a combination classroom-laboratory? Is there space for a darkroom? What provisions would you have for a greenhouse, a nature museum, or project room? (Refer to extensions of the classroom, below.)

The kinds of activities you plan will decide the work space needed. Below are suggestions for kinds of work space that teachers over the country find necessary in creative science teaching.

Classrooms and laboratories

If you are given the responsibility for ordering classroom furniture you will probably study supply house catalogs and those of companies which deal exclusively with school furniture.[27] Specific recommendations are given for space per cubic centimeter for each student; chairs and tables of different heights are needed to provide for the many individual differences in maturation among students.

Much has been written on kinds of furniture and floor layouts for science rooms which include cabinets and closet space, provisions for bulletin boards, and shelves for aquaria. The last word has not been said as to the "best" arrangement.

Extensions of the classroom

The greenhouse   In many schools around the country a greenhouse is associated with student investigations and field work activities either on the school grounds or at school camps. Many students find work in a school greenhouse an enriching experience. This work may be an extension of class activities or a means for supplying living materials for classroom study. The number of horticulture and indoor/outdoor gardening courses is increasing.[28] With some initiative, old

windows may be used to build slanted protective frames for growing plants.

Students may learn to grow algae, mosses, ferns, and seed plants. They can successfully raise plants by vegetative propagation. Some may go into soil-testing techniques while others begin studies in plant physiology including the effects of auxins and other plant hormones (pp. 364–74).

School grounds   Very often teachers of biology are asked to contribute ideas for landscaping the school grounds. When flowering trees and shrubs are specially selected for planting around school, a teacher can conduct a short field trip (within a class period) to introduce diverse types of flowers and fruits. At other seasons students may study seed formation, devices for seed dispersal, variations, and plant communities.

In fact, some teachers have laid out nature trails, have built artificial lakes that have later been stocked with plants and animals, and have, in general, stimulated the study of field biology. Gardening clubs are popular in many communities.

School camps   In some communities there is a cooperative bond between community and school to bring many firsthand outdoor nature experiences to students. Certainly students gain a greater appreciation of the biological world around them and the interacting factors in the conservation of living things. Often these are weekend excursions chaperoned by teachers and parents. When the school owns a camp, or rents one out of season, long-range activities can be planned.

School nature museum   Almost every teacher of biology grows some kinds of living things, and students have similar interests. When students and teachers pool their techniques, a small space can be found to exhibit living things for the whole school. Consider establishing a nature museum; train students as curators (p. 759). Invite teachers and students from other schools to visit these museums. Also

---

[27] Consult catalogs from Fisher Scientific, Toledo Metal Furniture Co., Ward's, Carolina Biological, among many others in your vicinity (see Appendix).

[28] Send for catalogs from manufacturers of greenhouses and supplies, such as Hobby Greenhouse Assoc. Inc., 45 Shady Drive, P.O. Box 695, Wallingford, CT 06492; or Barrington Industries, P.O. Box 133 FA, Barrington, IL. 60010; or Janco Greenhouses, Dept. HG-5, Davis Ave., Laurel, MD 20810; or Peter Reimuller, 980 17th Ave., Dept. 67C, Santa Cruz, CA 95062.

encourage students to give talks to elementary school students.

Living-culture center Many of the organisms for a school nature museum may be drawn from stocks that are maintained throughout the year in the biology classrooms or laboratories. In fact, a living-culture center which maintains protozoa, algae, *Drosophila* cultures, and other representative members of the invertebrates and vertebrates may serve schools in the community. Students who work on individual projects may also get their supplies of organisms from the school living-culture center.

Just as space and facilities are needed for a greenhouse, there is need to plan for an animal room—a place for tanks and cages to house fish, amphibia, reptiles, hamsters, white rats, and, as temporary guests, pigeons, or parakeets. (Directions for collecting and maintaining invertebrates and vertebrates are given in chapters 8 and 9.)

Project room In some schools, teachers have set aside space for individual students to pursue an interest, a research problem or a long-range "original" investigation in science, or to learn to use computers. Over a year or more these young people work under the supervision of a teacher; they use the methods of scientists. They need shelves for storage of materials, a work desk, and locker space for storage of their equipment. The facilities should include adequate electrical outlets and provisions for gas and water. A book shelf or cabinet for science reference books in the same room is also convenient.

Science library and committee room When space is not the limiting factor a room should be provided for a small science library and a separate place for students to meet for conferences or committee work associated with their classwork. This library does not compete with the school library but contains more advanced texts and magazines for students engaged in individual research work.

Film room Many schools set aside a room for showing films that can be used by all the classes in the school. This room should be equipped with dark shades, drapes, and a large screen. Specially trained student squads (described on p. 761) can run the projectors. These students also can keep the equipment in good condition in adequate storage facilities near the film room.

Catalogs of slides, films, filmstrips and loops, as well as computer software may be obtained from supply houses and from film distributors (see Appendix). You may also want to check on the availability of free loan films from industry.

Darkroom Many young people are camera enthusiasts and develop their own pictures. Thoughtful planning of space in a department or school may produce a spare closet which can be converted into a darkroom.

Teacher's corner Provisions for a small, snug, confined area for the teacher reaps high returns in teacher efficiency. While an "office" sounds formidable to some, a small nook to which a teacher may retreat to meet a student or colleague in semi-privacy, to work over records, or to read relieves the teacher of the strains that mount from the continued buzz and activity of many busy people in a school day.

Appendix

Reference Tables

Directory:
Equipment, Labware, Chemicals, Media

Directory:
Distributors of Audiovisual
Materials and/or Booklets

REFERENCE TABLES

Metric system

Length

The unit of length is the meter (m).

$$1 \text{ micron } (\mu) = 10^{-6} \text{ meter}$$
$$= 10^{-4} \text{ centimeter (cm)}$$
$$= 10^{-3} \text{ millimeter (mm)}$$
$$= 1,000 \text{ millimicrons (m}\mu)$$
$$= 10,000 \text{ Ångström units (Å)}$$
$$= \frac{1}{25,000} \text{ inch (in.)}$$

1 millimicron $= 10^{-6}$ millimeter
1 millimeter $= 1,000$ microns
1 Ångström unit $= 10^{-10}$ meter
$$= 10^{-8} \text{ centimeter}$$
$$= 10^{-7} \text{ millimeter}$$
1 centimeter $= 0.01$ meter
$$= 10 \text{ millimeters}$$
1 kilometer (km) $= 1,000$ meters

Weight

The unit of weight is the gram (g).

1 centigram (cg) $= 0.01$ gram
1 milligram (mg) $= 0.001$ gram
1 microgram (μg) $= 0.000001$ gram
1 kilogram (kg) $= 1,000$ grams

Volume

The unit of volume is the liter; this is the volume of 1 kilogram of water at 4°C at standard atmospheric pressure (equal to 1,000.028 cubic centimeters).

1 liter $= 1,000$ milliliters (ml)
1 milliliter $= 0.001$ liter (L)

United States system

Length

1 foot (ft) $= 12$ inches (in.)
1 yard (yd) $= 3$ feet $= 36$ inches
1 rod $= 5\frac{1}{2}$ yards $= 16\frac{1}{2}$ feet
1 mile (statute) $= 1,760$ yards $= 5,280$ feet
1 league $= 3$ miles
1 nautical mile
  (international) $= 6,076.1$ feet
  $= 7\frac{1}{3}$ cablelengths
1 fathom (mariners') $= 6$ feet
1 cablelength $= 120$ fathoms

Weight

TROY WEIGHT

1 pennyweight (dwt) $= 24$ grains (gr)
1 ounce (oz) $= 20$ pennyweights $= 480$ grains
1 pound (lb) $= 12$ ounces $= 5,760$ grains

APOTHECARIES' WEIGHT

1 scruple (sc) $= 20$ grains (gr)
1 dram (dr) $= 3$ scruples $= 60$ grains
1 ounce (oz) $= 8$ drams $= 480$ grains
1 pound (lb) $= 12$ ounces $= 5,760$ grains

AVOIRDUPOIS WEIGHT

1 dram (dr) $= 27^{11}/_{32}$ grains (gr)
1 ounce (oz) $= 16$ drams $= 437\frac{1}{2}$ grains
1 pound (lb) $= 16$ ounces $= 7,000$ grains

Volume

APOTHECARIES' FLUID MEASURE

1 fluid ounce (fl oz) $= 8$ fluid drams (fl dr)
  $= 480$ minims
1 pint (pt) $= 16$ fluid ounces
1 quart (qt) $= 2$ pints $= 32$ fluid ounces
1 gallon (gal) $= 4$ quarts
  $= 231$ cubic inches (cu in.)

DRY MEASURE

1 quart (qt) $= 2$ pints (pt)
  $= 67.2$ cubic inches (cu in.)
1 peck $= 8$ quarts $= 537.6$ cubic inches

Equivalents of metric and United States systems

Length

1 millimeter	=	0.03937	inch*
1 centimeter	=	0.3937	inch
1 meter	=	39.37	inches
1 meter	=	3.2808	feet
1 meter	=	1.09361	yards
1 meter	=	0.1988	rod
1 kilometer	=	0.6214	mile
1 inch	=	2.540	centimeters
1 foot	=	30.480	centimeters
1 foot	=	0.3048	meter
1 yard	=	91.440	centimeters
1 yard	=	0.9144	meter
1 rod	=	5.0292	meters
1 mile	=	1.6093	kilometers

* 1 mm $\cong \frac{1}{25}$ inch

Area

1 square centimeter	=	0.155	square inch
1 square meter	= 1,550.0		square inches
1 square meter	=	10.764	square feet
1 square meter	=	1.196	square yards
1 square inch	=	6.4516	square centimeters
1 square foot	=	929.0341	square centimeters
1 square foot	=	0.0929	square meter
1 square yard	= 8,361.31		square centimeters
1 square yard	=	0.8361	square meter

Capacity

1 milliliter	=	16.23	minims
1 milliliter	=	0.2705	fluid dram
1 milliliter	=	0.0338	fluid ounce
1 liter	=	33.8148	fluid ounces
1 liter	=	2.1134	pints
1 liter	=	1.0567	quarts
1 liter	=	0.2642	gallon
1 fluid dram	=	3.697	ml
1 fluid ounce	=	29.573	ml
1 pint	=	473.166	ml
1 quart	=	946.332	ml
1 gallon	=	3.785	liters
1 cubic inch	=	16.387	ml
1 cubic foot	=	28.316	liters

Volume

1 cubic centimeter	=	0.06102	cubic inch
1 cubic decimeter	=	61.0234	cubic inches
1 cubic meter	=	35.3145	cubic feet
1 cubic meter	=	1.3079	cubic yards
1 cubic inch	=	16.3872	cubic centimeters
1 cubic foot	=	28.317	cubic decimeters
1 cubic foot	=	0.02832	cubic meter
1 cubic yard	=	0.7646	cubic meter

Weight

1 milligram	=	0.015432	grain
1 gram	=	15.432	grains
1 gram	=	0.6430	pennyweight
1 gram	=	0.25720	apothecaries dram
1 gram	=	0.03527	avoirdupois ounce
1 gram	=	0.03215	apothecaries or troy ounce
1 kilogram	=	35.274	avoirdupois ounces
1 kilogram	=	32.151	apothecaries or troy ounces
1 kilogram	=	2.2046	avoirdupois pounds
1 kilogram	=	2.6792	apothecaries or troy pounds
1 grain	=	64.7989	milligrams
1 grain	=	0.0648	gram
1 pennyweight	=	1.5552	grams
1 apothecaries dram	=	3.8879	grams
1 avoirdupois ounce	=	28.3495	grams
1 apothecaries or troy ounce	=	31.1035	grams
1 avoirdupois pound	=	453.5924	grams
1 apothecaries or troy pound	=	373.2418	grams

Conversion tables (quick reference)

Weight

lb	kg	kg	lb
1	0.5	1	2.2
2	0.9	2	4.4
4	1.8	3	6.6
6	2.7	4	8.8
8	3.6	5	11.0
10	4.5	6	13.2
20	9.1	8	17.6
30	13.6	10	22
40	18.2	20	44
50	22.7	30	66
60	27.3	40	88
70	31.8	50	110
80	36.4	60	132
90	40.9	70	154
100	45.4	80	176
150	68.2	90	198
200	90.8	100	220

1 lb = 0.454 kg   1 kg = 2.204 lb

Length

in	cm	cm	in
1	2.5	1	0.4
2	5.1	2	0.8
4	10.2	3	1.2
6	15.2	4	1.6
8	20.3	5	2.0
12	30.5	6	2.4
18	46	8	3.1
24	61	10	3.9
30	76	20	7.9
36	91	30	11.8
42	107	40	15.7
48	122	50	19.7
54	137	60	23.6
60	152	70	27.6
66	168	80	31.5
72	183	90	35.4
78	198	100	39.4

1 inch = 2.54 cm   1 cm = 0.3937 inch

Roman and Arabic numerals

I	1	VII	7	XIII	13	XIX	19
II	2	VIII	8	XIV	14	XX	20
III	3	IX	9	XV	15	XXX	30
IV	4	X	10	XVI	16	XL	40
V	5	XI	11	XVII	17	L	50
VI	6	XII	12	XVIII	18	LX	60

LXX 70	D 500
LXXX 80	DC 600
XC 90	DCC 700
C 100	DCCC 800
CC 200	CM 900
CCC 300	M 1000
CCCC 400	MM 2000

Kitchen measures and other approximations

1 loopful ≅ 0.001 ml (depending on size of
    loop)
20 drops ≅ 1 ml (depending on bore of the
    pipette); 1 drop ≅ 0.05 ml
¼ small test tube ≅ 10 ml
   (20 × 150 mm)
1 small test tube ≅ 40 ml
   (20 × 150 mm)
1 teaspoon ≅ 4–5 ml
1 tablespoon ≅ 15 ml ≅ ½ fl. oz
   (3 teaspoons)
2 tablespoons ≅ 30 ml; 1 oz
1 fl. ounce ≅ 30 ml
1 quart ≅ 946 ml
1 gallon ≅ 3.8 L
1 teacup ≅ 120 ml ≅ 4 fl. oz
1 glassful ≅ 240 ml   ≅ 16 tabls ≅ 8 oz
       (or 0.24 L)
1 ounce ≅ 30 g
1 pound ≅ 500 g

Coins as weight substitutes

	approximate weight
dime	2.2 g
penny	3.1 g
nickel	5.0 g
quarter	5.7 g
half dollar	13.0 g
silver dollar	26.0 g

(a dime is about 1 mm thick)

Temperature conversion

It is simple to convert Fahrenheit into
Celsius temperatures (and vice versa)
without memorizing any formulas. All that
need be remembered is that $0°C = 32°F$
(the freezing point of water) and that each
Fahrenheit degree is only 5⁄9 as large as a
Celsius degree. This is because there are
100 degrees between the freezing point of
water on the Celsius scale, and 180 degrees
on the Fahrenheit scale: $100⁄180 = 5⁄9$. Thus
if you know the temperature in degrees
Fahrenheit, subtract 32 and take 5⁄9 of the
result to find the temperature in degrees
Celsius. If you know the Celsius temperature, multiply by 9⁄5 and add 32 to the
result to find the Fahrenheit temperature:

$$°F = °C × 9⁄5 + 32$$

$$°C = °F − 32 × 5⁄9$$

The illustration on page 770 is useful in
understanding the relationship between
these two temperature scales; or it can be
used for the conversion itself: Simply read
directly across from one thermometer scale
to the other.

Conversion tables

Temperature

°F	°C	°C	°F
0	−17.7	0	32.0
95	35.0	35.0	95.0
96	35.5	35.5	95.9
97	36.1	36.0	96.8
98	36.6	36.5	97.7
99	37.2	37.0	98.6
100	37.7	37.5	99.5
101	38.3	38.0	100.4
102	38.8	38.5	101.3
103	39.4	39.0	102.2
104	40.0	39.5	103.1
105	40.5	40.0	104.0
106	41.1	40.5	104.9
107	41.6	41.0	105.8
108	42.2	41.5	106.6
109	42.7	42.0	107.6
110	43.3	100.0	212.0

Greek alphabet

A	α	alpha	N	ν	nu
B	β	beta	Ξ	ξ	xi
Γ	γ	gamma	O	o	omicron
Δ	δ	delta	Π	π	pi
E	ε	epsilon	P	ρ	rho
Z	ζ	zeta	Σ	σ	sigma
H	η	eta	T	τ	tau
Θ	θ	theta	Y	υ	upsilon
I	ι	iota	Φ	φ	phi
K	κ	kappa	X	χ	chi
Λ	λ	lambda	Ψ	ψ	psi
M	μ	mu	Ω	ω	omega

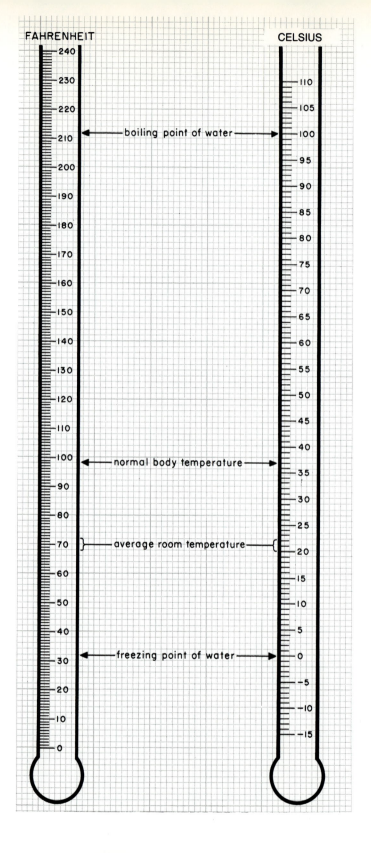

Powers of ten

$10^0 = 1$	$10^0 = 1$
$10^1 = 10$	$10^{-1} = 0.1$
$10^2 = 100$	$10^{-2} = 0.01$
$10^3 = 1,000$	$10^{-3} = 0.001$
$10^4 = 10,000$	$10^{-4} = 0.0001$
$10^6 = 1$ million	$10^{-6} =$ one millionth
$10^9 = 1$ billion	$10^{-9} =$ one billionth

Examples:

1. $(4.5 \times 10^8)(5 \times 10^4) = (4.5 \times 5) \times 10^{8+4}$
$$= 22.5 \times 10^{12}$$

2. $\dfrac{4.5 \times 10^{-8}}{5 \times 10^4} = \dfrac{4.5}{5} \times 10^{-8-4}$
$$= 0.9 \times 10^{-12}$$

Useful formulas

The following formulas are useful for calculating the volume or surface area of a cell:

$\pi = 3.1416$
Circumference of a circle $= \pi d = 2\pi r$
Area of a circle $= \pi r^2$
Surface area of a cylinder $= \pi dh = 2\pi rh$
Volume of a cylinder $= \pi r^2 h$
Surface area of a sphere $= 4\pi r^2$
Volume of a sphere $= \frac{1}{3}\pi r^3$

A typical cell such as the alga *Chlorella* is a sphere. Therefore, the volume of the cell may be determined as follows. Using an eyepiece micrometer (or methods described on pp. 125–26), measure the diameter of the cell. Divide the diameter by 2 to determine the radius r. Calculate the volume from the formula $\frac{1}{3}\pi r^3$. In a similar manner, the surface area of the nucleus of this cell may be computed from the formula $4\pi r^2$.

Wavelengths of various radiations*

radiation	Ångström units (Å)[†]
Cosmic rays	0.0005 and under
Gamma rays	0.005–1.40
X-rays	0.1–100
Ultraviolet	2920–4000
Visible spectrum	4000–7000
Violet	4000–4240
Blue	4240–4912
Green	4912–5750
Yellow	5750–5850
Orange	5850–6470
Red	6470–7000
Maximum visibility	5560
Infrared	over 7000

* From *Teaching High School Science: A Sourcebook for the Physical Sciences* by Joseph, Brandwein, Morholt, Pollack, and Castka, © 1961, by Harcourt, Brace & World, Inc., and reproduced with their permission.

[†] To convert to inches, multiply by 3.937×10^{-9}; to convert to centimeters, multiply by 1×10^{-8}.

Periodic table of the elements

Legend:
- ATOMIC NUMBER
- SYMBOL
- NAME
- ATOMIC WEIGHT (based on carbon–12)

Example:
1
H
Hydrogen
1.008

1	2	3	4	5	6	7	8	9	10	11	12	13	14	15	16	17	18
1 **H** Hydrogen 1.008																	2 **He** Helium 4.00
3 **Li** Lithium 6.94	4 **Be** Beryllium 9.01											5 **B** Boron 10.81	6 **C** Carbon 12.01	7 **N** Nitrogen 14.01	8 **O** Oxygen 16.00	9 **F** Fluorine 19.00	10 **Ne** Neon 20.18
11 **Na** Sodium 22.99	12 **Mg** Magnesium 24.31											13 **Al** Aluminum 26.98	14 **Si** Silicon 28.09	15 **P** Phosphorus 30.97	16 **S** Sulfur 32.06	17 **Cl** Chlorine 35.45	18 **Ar** Argon 39.95
19 **K** Potassium 39.10	20 **Ca** Calcium 40.08	21 **Sc** Scandium 44.96	22 **Ti** Titanium 47.90	23 **V** Vanadium 50.94	24 **Cr** Chromium 52.00	25 **Mn** Manganese 54.94	26 **Fe** Iron 55.85	27 **Co** Cobalt 58.93	28 **Ni** Nickel 58.71	29 **Cu** Copper 63.55	30 **Zn** Zinc 65.38	31 **Ga** Gallium 69.72	32 **Ge** Germanium 72.59	33 **As** Arsenic 74.92	34 **Se** Selenium 78.96	35 **Br** Bromine 79.90	36 **Kr** Krypton 83.80
37 **Rb** Rubidium 85.47	38 **Sr** Strontium 87.62	39 **Y** Yttrium 88.91	40 **Zr** Zirconium 91.22	41 **Nb** Niobium 92.91	42 **Mo** Molybdenum 95.94	43 **Tc** Technetium 98.91	44 **Ru** Ruthenium 101.07	45 **Rh** Rhodium 102.91	46 **Pd** Palladium 106.4	47 **Ag** Silver 107.87	48 **Cd** Cadmium 112.40	49 **In** Indium 114.82	50 **Sn** Tin 118.69	51 **Sb** Antimony 121.75	52 **Te** Tellurium 127.60	53 **I** Iodine 126.90	54 **Xe** Xenon 131.30
55 **Cs** Cesium 132.91	56 **Ba** Barium 137.34	57 **La** Lanthanum* 138.91	72 **Hf** Hafnium 178.49	73 **Ta** Tantalum 180.95	74 **W** Tungsten 183.85	75 **Re** Rhenium 186.2	76 **Os** Osmium 190.2	77 **Ir** Iridium 192.22	78 **Pt** Platinum 195.09	79 **Au** Gold 196.97	80 **Hg** Mercury 200.59	81 **Tl** Thallium 204.37	82 **Pb** Lead 207.2	83 **Bi** Bismuth 208.98	84 **Po** Polonium (210)	85 **At** Astatine (210)	86 **Rn** Radon (222)
87 **Fr** Francium (223)	88 **Ra** Radium 226.03	89 **Ac** Actinium† (227)	104 **Rf** Rutherfordium (257)	105 **Ha** Hahnium (260)													

LANTHANIDE* SERIES

58 **Ce** Cerium 140.12	59 **Pr** Praseodymium 140.91	60 **Nd** Neodymium 144.24	61 **Pm** Promethium (147)	62 **Sm** Samarium 150.4	63 **Eu** Europium 151.96	64 **Gd** Gadolinium 157.25	65 **Tb** Terbium 158.93	66 **Dy** Dysprosium 162.50	67 **Ho** Holmium 164.93	68 **Er** Erbium 167.26	69 **Tm** Thulium 168.93	70 **Yb** Ytterbium 173.04	71 **Lu** Lutetium 174.97

ACTINIDE† SERIES

90 **Th** Thorium 232.04	91 **Pa** Protactinium 231.04	92 **U** Uranium 238.03	93 **Np** Neptunium 237.05	94 **Pu** Plutonium (244)	95 **Am** Americium (243)	96 **Cm** Curium (247)	97 **Bk** Berkelium (247)	98 **Cf** Californium (251)	99 **Es** Einsteinium (254)	100 **Fm** Fermium (257)	101 **Md** Mendelevium (258)	102 **No** Nobelium (255)	103 **Lr** Lawrencium (256)

Some radioactive isotopes

isotope	chemical formula	half life	beta energy (Mev)	gamma energy (Mev)
Iodine-131	NaI	8.05 days	0.61	0.364
Phosphorus-32	NaH_2PO_4	14.3 days	1.71	none
Iron-59	$FeCl_3$	45 days	0.46	1.102, 1.290
Strontium-89	$SrCl_2$	50.4 days	1.46	none
Sulfur-35	H_2SO_4	86.7 days	0.167	none
Calcium-45	$CaCl_2$	165 days	0.25	none
Zinc-65	$ZnCl_2$	245 days	0.33	0.511, 1.114
Sodium-22	NaCl	2.58 years	0.54	1.277
Carbon-14	Na_2CO_3	5,770 years	0.156	none
Chlorine-36	KCl	300,000 years	0.71	none

DIRECTORY: EQUIPMENT, LABWARE, CHEMICALS, MEDIA

Although only one address is given in this directory for each company, many large companies have regional offices that are easier to reach for speedy deliveries. Send for catalogs from supply houses.

American Association for the Advancement of Science, *Access to Science* (periodical on opportunities for handicapped—wheelchairs, chemical, and biological laboratories), 1515 Massachusetts Ave., N.W., Washington, DC 20005.

Ace Scientific Supply Co., Inc., P.O. Box 1018, East Brunswick, NJ 08816.

Alconox Inc., 215 Park Ave. South, New York, NY 10003.

All Tech Associates, Inc. (Applied Science labs), 2051 Waukegan Road, Deerfield, IL 60015.

American Nuclear Products, Inc., 1232 East Commercial, Springfield, MO 65803.

American Optical, Scientific Instrument Division, Buffalo, NY 14215.

American Hospital Supply Corp., American Scientific Products, 1430 Waukegan Rd., McGaw Park, IL 60085.

Ames Laboratories Inc., 200 Rock Lane, Box 3024, Milford, CT 06460.

American Type Culture Collection, 12301 Parklawn Dr., Rockville, MD 20852.

Anderson Laboratories, Inc. (stains, reagents), P.O. Box 8429, 5901 Fitzhugh, Fort Worth, TX 76119.

Apple Computer, Inc., 20525 Mariani Ave., Cupertino, CA 95014.

W. Atlee Burpee Co., 300 Park Ave., Warminster, PA 18974.

J. T. Baker Chemical Co., 222 Red School Lane, Phillipsburg, NJ 08865.

Bausch & Lomb, Instruments and Systems Division, 42 East Ave., Rochester, NY 14603.

Becton-Dickinson, Labware, 1950 Williams Dr., Oxnard, CA 93030.

Becton, Dickinson, BBL Microbiology Systems, P.O. Box 243, Cockeysville, MD 21030.

Bel-Art Products, 6 Industrial Rd., Pequannock, NJ 07440.

Boekel Industries, Inc., 509 Vine St., Philadelphia, PA 19106.

Carolina Biological Supply Co., 2700 York Rd., Burlington, NC 27215; also Box 187, Gladstone, OR 97027.

Cole-Parmer Instrument Co., 7425 North Oak Park Ave., Chicago IL 60648.

College Biological Supply Co., 8857 Mount Israel Rd., Escondido, CA 92025.

Connecticut Valley Biological Supply Co., 82 Valley Rd., Southampton, MA 01073.

Corning Science Products, Corning, NY, 14831 (distributors over the country).

Difco Laboratories, P.O. Box 1058, Detroit, MI 48232 9931.

E. I. DuPont de Nemours & Co., Photo Products Dept., Concord Plaza, Wilmington, DE 19898.

Duralab Equipment Corp., 107–23 Farragut Rd., Brooklyn, NY 11236.

Eastern Scientific Products (metal lab furniture), 52 Skyline Dr., Ringwood, NJ 07456.

Eastman Kodak Co., 343 State St., Rochester, NY 14650.

Fisher Scientific Co. (unitized laboratory furniture), 653 Fisher Building, Pittsburgh, PA 15219.

Gratnell System Inc. (storage equipment, carts, and cupboards), 27777 Allen Parkway, Houston, TX 77019.

Gulf Specimen Co., Inc., P.O. Box 237, Panacea, FL 32346.

Hach Co., P.O. Box 389, Loveland, CO 80539.

Hellige Inc., Scientific Instruments, (labware, reagents) 877 Stewart Ave., Garden City, NY 11530.

Hoffmann-LaRoche Inc., Nutley, NJ 07110.

Jewel Industries, Inc. (aquaria, terraria), 5005 West Armitage Ave., Chicago, IL 60639.

Kewaunee Scientific Equipment Corp., South Center St., Adrian, MI 49221.

Kons Scientific Co. (living materials: protists, invertebrates, vertebrates), P.O. Box 3, Germantown, WI 53022–0003.

La Motte Chemical Products, P.O. Box 329, Chestertown, MD 21620.

Lane Science Equipment Corp., 225 West 34 Street, New York, NY 10022.

LaPine Scientific Co., 6001 South Knox Ave., Chicago, IL 60629.

E. Leitz, Inc., Instrument Division, 24 Link Dr., Rockleigh, NJ 07647.

Macmillan Science Co., Inc. (Turtox/Cambosco), 8200 South Hoyne St., Chicago, IL 60620.

Marine Biological Laboratory, Supply Dept., Woods Hole, MA 02543.

Markson Science, 7815 South 46th Street, Phoenix, AZ 85040.

Curtin Matheson Scientific, Inc., P.O. Box 1546, Houston, TX 77251.

MCB Manufacturing Chemists, Inc. (coast to coast distributors of reagents), 480 Democrat Rd., Gibbstown, NJ 08027.

Medical Plastics Laboratory Inc., P.O. Box 38, Gatesville, TX 76528.

Millipore Corp., Bedford, MA 01732.

Mogul-Ed, Laboratory Park, Chagrin Falls, OH 44022.

MSI (Micron Separations Inc.), 58 North Main St., Honeoye Falls, NY 14472 (membrane filters).

Nalge Co., Nalgene Labware Dept., 75 Panorama Creek Rd., P.O. Box 365, Rochester, NY 14602.

Ohaus Scale Corp. 29 Hanover Rd., Florham Park, NJ 07932.

Owens-Illinois Inc. (Kimble Division), P.O. Box 1035, Toledo, OH 43666.

Pacific Bio-Marine Supply Co., P.O. Box 285, Venice, CA 90293.

Perkin Elmer Data Systems Group, 2 Crescent Pl., Oceanport, NJ 07757.

Perkin Elmer Corp., Instrument Division, Main Avenue, Norwalk, CT 06856.

Pfaltz & Bauer, Inc., 375 Fairfield Ave., Stamford, CT 06902.

Sargent-Welch Scientific Co., 7300 North Linder Avenue, Skokie, IL 60077.

Arthur Thomas Co., Vine St. and Third, P.O. Box 779, Philadelphia, PA 19105.

Spectrum Medical Industries, Inc., 60916 Terminal Annex, Los Angeles, CA 90054.

Toledo Metal Furniture Co., 1100 Hastings St., Toledo, OH 43607.

Triarch Inc. (microscopes, prepared slides). Ripon, WI 54971.

Ward's Natural Science Establishment, Inc., P.O. Box 1712, Rochester, NY 14603; also P.O. Box 1749, Monterey, CA 93940.

Whatman Laboratory Products Inc., Whatman Paper Div., 9 Bridewell Place, Clifton, NJ 07014.

Carl Zeiss Inc., 1 Zeiss Drive, Thornwood, NY 10594.

DIRECTORY: DISTRIBUTORS OF AUDIOVISUAL MATERIALS AND/OR BOOKLETS

You will want to send for catalogs for specific materials such as films, loops, slides and filmstrips, software, or booklets that are available. However, in this directory, we have listed one address for each company; some of the larger ones maintain branch offices in many regions of the country, or are affiliated with film libraries. Therefore, in ordering materials try to locate a nearby branch office or film library to speed your order. Also check film libraries of colleges, government and state agencies such as wildlife, agriculture, careers, and so on, in your own area for available free films and/or filmstrips.

You will also want to refer to the *Educators' Guide to Free Science Materials* (films, filmstrips, video materials) edited by M. Saterstrom, Educators' Progress Service Inc., 214 Center St., Randolph, WI 53956. Also available from the same source is the *Educators' Guide to Free Tapes, Scripts, and Transcriptions.*

Another source, this time for paperbacks to help students build their own libraries, is *Paperbound Books in Print*, R.R. Bowker Co., 1180 Avenue of the Americas, New York, NY 10036.

An excellent source of reprints of many classic references, many on identification of plants and animals, is Dover Publications, Inc., 31 East Second St., Mineola, NY 10014. Send for their catalog.

Films, especially free films in popular demand, need to be ordered weeks, even months, in advance. When possible, plan large units of work for the school year early in September, and mark off several alternate days for each area of work for showing films. Then order early and submit alternate choices of days whenever feasible. In this way films reach the school a day or two before classes so that they supplement the day's lesson, rather than detract from it because they do not fit the sequence of work.

Films should be previewed for relevance to the day's work, suitability for the age level of students, and appropriateness in your community. Many teachers develop a file of their own approved films in this way, using library cards indicating the source of the film, rental or free loan, and a brief summary of the film or filmstrip (or attach the guide).

There are many techniques in showing films. A film or filmstrip need not be shown in its entirety; start the film at the place you want to use in your lesson to illustrate a point that cannot be demonstrated firsthand. In using a filmstrip, a teacher may ask questions, direct the observation of students, reverse and repeat a section for closer study and interpretation, or give students the time to ask questions while the filmstrip is still on hand. Some of the very short loops or films illustrating a technique, such as that of handling fruit flies, may be shown a second time to present the fine details after a broad overview.

At the end of the film or filmstrip, have students summarize the main ideas to develop in words the concepts that were progressively developed.

American Heart Assoc., 44 E. 23 St., New York, NY 10010.

American Medical Assoc., Motion Picture Library, 535 North Dearborn St., Chicago, IL 60610.

American Society for Microbiology (slide/tape programs, safety) 1913 I St. N.W., Washington, DC 20006.

Association Films, Inc., 2221 South Olive St., Los Angeles, CA 90007.

Association-Sterling Films, 866 Third Avenue, New York NY 10022.

Biology Media, (computer software), 918 Parker St., Berkeley, CA 94710.

Cambridge Development Laboratory, 100 Fifth Avenue, Waltham MA 02154.

Chevron Chemical Co., 200 Bush St., San Francisco, CA 94120.

Churchill Films, 662 North Robertson Blvd., Los Angeles, CA 90069.

Coronet Films, 65 East South Water Street, Chicago, IL 60601.

E. I. Du Pont de Nemours & Co., Audio-visual section, Wilmington, DE 19898.

Educational Computing Systems Inc., 136 Fairbanks, Oak Ridge, TN 37830.

Encyclopedia Britannica Educational Corp., 1150 Wilmette Ave., Wilmette, IL 60091.

Heinemann Educational Books, Inc., 4 Front St., Exeter, NH 03833.

Hubbard Scientific (anatomical models as well as film loops, slide sets, transparencies), 1946 Raymond Rd., Northbrook, IL 60062.

International Film Bureau Inc., 332 South Michigan Ave., Chicago, IL 60604.

E. Lilly & Co., P.O. Box 618, Indianapolis, IN 42606.

McGraw Hill Inc. Text Film Divis., 1221 Avenue of the Americas, New York, NY 10020.

Modern Talking Picture Service, 2323 New Hyde Park Rd., New Hyde Park, NY 11040; also 5000 Park St. North, St. Petersburg, FL 33709.

Moody Institute of Science, 12000 East Washington Blvd., Whittier, CA 90606.

Nasco (lab manual of dissection of *Xenopus*), 901 Janesville Ave., Fort Atkinson, WI 53538.

National Agricultural Chemicals Assoc., 1155 15 St. N.W., Washington, DC 20005; send for catalog of many companies that offer free loan of films, however, National Agricultural Chemicals does not book films.

National Audubon Society, 950 Third Ave., New York, NY 10022.

National Committee for Careers in Medical Laboratories, 9650 Rockville Pike, Bethesda, MD 20014.

National Education Television Film Service, Indiana University, Bloomington, IN 47405.

National Geographic Society Educational Services (film and video catalogs), 17th and M St., N.W., Washington, DC 20036.

New York University Film Library, 26 Washington Place, New York, NY 10003.

Science Books International, 51 Sleeper St., Boston, MA 02210.

Science and Mankind, Communications Park, Mount Kisco, NY 10549.

Science Software Systems Inc., 11899 West Pico Blvd., West Los Angeles, CA 90064.

Shell Film Library, 450 N. Meridian St., Indianapolis, IN 46204.

Society for Visual Education Inc., 1345 Diversey Parkway, Chicago, IL 60614.

Sutherland Educational Films, 8425 Third St., Los Angeles, CA 90048.

Thorne Films, 1229 University Ave., Boulder, CO 80302.

Time-Life Multimedia, Time/Life Building, Rockefeller Center, New York, NY 10020.

U.S. Atomic Energy Commission, Office Public Information, Germantown, MD 20767.

United Transparencies Inc., P.O. Box 688, Binghamton, NY 13902.

United World Films Inc., 221 Park Avenue South, New York, NY 10003.

Xerox Films, Xerox, Stamford, CT 06904.

Index

Page numbers in *italics* refer to illustrations.

D

Daffodil, *116. See also* Flowers
Damselfly, *63,* 581, *582, 642. See also* Insects, orders of
Dandelion, *118*
Daphnia, 51–53, *55*
 anesthetization of, 142–43
 for beginning studies, 128–29
 collecting of, 577–79, *579,* 581
 culturing of, 671
 digestion in, 239
 effect of epinephrine on heart
  of, 360–61
 heartbeat, 267–68
 ingestion in, 239
 in microaquarium, 301
 photoperiodism in, 379
 phototaxis in, 333
 reproductive behavior in,
  405–406
 use of in classroom and
  laboratory, 670–71
Dark field microscope, 123, *123*
Darkness
 and carbon dioxide production
  Anacharis, 305–306
  Chlorella, 305
  leafy plants, 306, *307*
 and leaves, 551
 for microprojectors, 126, *127*
 and potato plants, 550–51, *551*
Dark reaction in photosynthesis,
 197
Darkroom for students, 764
*Darlingtonia californica, 722. See
 also* Insectivorous plants
Darwin, C., experiments with
  plants. *See also* Evolution
 hormones of, 364, *366*
 sensitivity to light in, 334
Datura (jimson weed), *117*
Deaf-mutism, 511
Decapods, 57, *591*
Decaying substances as source of
  bacteria, 696–97
Decomposers
 bacterial, in soil, 607–608
 marine, 592–94, *592*
Dedifferentiation. *See also*
  Regeneration
 and regeneration, 491
Dehydration of tissue cells, 168
Dehydrogenase, role of, 188
Delafield's hematoxylin for tissue
  cells, 170
Dentalium, dissection of, 76
Dentate leaf, *110. See also* Leaves
Deoxygenated versus
  oxygenated blood,
  269–70, *269*
Depression slides, 136
Dermaptera, 56, *59, 63. See also*
  Insects, orders of

Dero for regeneration studies,
 498
Desert, biotic communities in,
 586
Desmids, culturing of, 693
De Vries, 559
Dialysis, 175–76
Dialyzing tubing, 255*n*
Diameter of field for microscope,
 124
Diaphragm and breathing action,
 319–20, *320*
Diastase
 extraction of from seedlings,
  245–46
 for starch hydrolysis on a slide,
  246–47
Diatoms, 101, 104, *104. See also*
  Algae
 collecting of, 693
 electron micrograph of, *591*
 in soil, *605*
2,-6-Dichlorophenol-indophenol,
 742
Dicots, 547–48
Dictyostelium, 706. See also Slime
  molds
Didinium, 14. See also Ciliates
 culturing for, 662
 food vacuoles in, 140
 ingesting *Paramecium,* 662
Diet, adequate, *259. See also*
  Nutrient content of food
Differential blood counts, human
  blood values in, 278, *278,*
  279, 280
Differentiation in embryos. *See*
  Embryos, of frog, sea
  urchin differentiation in
 of shoots and roots, *490*
Difflugia, 10, *16*
 culturing of, 658
Diffusion
 dialysis for, 175–76
 of gases, 241
 through membrane, 241–43,
  243
  collodion, 244–45, *245*
  different sized molecules,
   242–43
  frog intestine, 244, *244*
  ionized dyes, 244
  osmosis and osmotic
   pressure, 243–44, *243*
  solutions and colloids, 243
 versus osmosis, 293
 thistle tube model of, *244,* 296
Digestion
 in animals
  enzymes, 248
  proteins, test for, 251
  starch and saliva, 248–51,
   248

 temperature, effect of,
  249–50
 in *Daphnia,* 239
 diffusion in. *See* Diffusion
  through membrane
 food tube and, 240
 in *Paramecium,* 237–38, *237*
 in plants
  amylase activity, 245
  diastase extraction, 245–46
  invertase, 247–48
  proteins, 248
  starch agar film, 246
  starch hydrolysis on slide,
   246–47
  temperature, effect of, 247
 in small intestine
  amylase activity of
   pancreatic juice, 254
  emulsification of fats, 254
  hydrolysis of fats, 254–55
  invertase, 254
  trypsin, 253
Digitonin solution, 743. *See also*
  Ciliates
Dinoflagellates, 101, *103,* 589,
  590. See also Algae
 luminescent, 590, *590*
Dinophysis, 590
Diplodiscus subclavatus, 622, 622
Dip sticks
 to test urine, 324
 for water quality tests, 596–97
Diptera, *63, 643. See also* Insects,
  orders of
Disaccharides, test for, 177
Diseases
 in monozygotic and dizygotic
  twins, 519
 of salamanders in vivarium, 683
Disinfectant
 safe handling of, 731
 for seeds, 720–21
Display, insect, 645
Dissecting microscope, *122,* 123
Dissection. *See also* Anatomy
 of *Anodonta,* 68, 70, 73–74
 of *Arbacia,* 82
 of bee, 299
 of *Blatta,* 61–62
 of brain, 356
 of *Busycon,* 68
 of *Cambarus,* 57–58
 of clam, 267
 of *Cucumaria,* 86
 of earthworm, 37–38, 268
 of fish, 87–88
  ovaries and testes, 411
 of flowers, 448
 of fly, 299
 of frog, 89–90, 265–66, 272
  brain and spinal cord,
   346–47, *348*

Date Due